D0620965

# CRC Handbook
# of
# Laboratory
# Safety

## 3rd Edition

Editor

**A. Keith Furr, Ph.D.**
Professor of Nuclear Science and Engineering
Director
Department of Health and Safety
Virginia Polytechnic Institute
and State University
Blacksburg, Virginia

Boston

**Library of Congress Cataloging-in-Publication Data**

Handbook of laboratory safety / editor, A. Keith Furr. — 3rd ed.
     p.  cm.
     Rev. ed. of: CRC handbook of laboratory safety, 2nd ed. [1971].
     Includes bibliographical references.
     ISBN 0-8493-0353-2
     1. Chemical laboratories — Safety measures.    I. Furr, A. Keith,
    1932—              II. Chemical Rubber Company.       III. CRC handbook of
    laboratory safety.
QD51.H27       1989
542'.1'0289—dc20                                     89-22129
                                                              CIP

This book represents information obtained from authentic and highly regarded sources. Reprinted material is quoted with permission, and sources are indicated. A wide variety of references are listed. Every reasonable effort has been made to give reliable data and information, but the author and the publisher cannot assume responsibility for the validity of all materials or for the consequences of their use.

All rights reserved. This book, or an⬛⬛⬛⬛⬛⬛⬛⬛⬛⬛⬛⬛⬛⬛duced in any form without written consent from the publisher.

Direct all inquiries to CRC Press, Inc⬛⬛⬛⬛⬛⬛⬛⬛⬛⬛⬛⬛⬛⬛, Florida, 33431.

©

International ⬛

Library o⬛

# TABLE OF CONTENTS

# CONTRIBUTORS

**Daniel R. Conlon**
Instruments for Research & Industry
Cheltenham, Pennsylvania

**Richard F. Desjardins, M.D.**
University Occupational Physician, Ret.
Department of Health and Safety
Virginia Polytechnic and State
  University
Blacksburg, Virginia

**Lawrence G. Doucet, P.E.**
Doucet & Mainka, P.C.
Peekskill, New York

**Caldwell N. Dugan**
Division of Institutional Resources
National Science Foundation
Washington, D.C.

**Frank A. Graf, Jr.**
Research Scientist
Westinghouse Hanford Company
Richland, Washington

**Scott A. Heider**
Division of Institutional Resources
National Science Foundation
Washington, D.C.

**G. H. Herrmann, M.D.**
Medical Director, Pontchartrain Plant
E.I. du Pont de Nemours & Co.
La Place, Louisiana

**Harold Horowitz**
Division of Institutional Resources
National Science Foundation
Washington, D.C.

**Alvin B. Kaufman**
Litton Systems Division
Litton Industries
Woodland Hills, California

**Edwin N. Kaufman**
Senior Scientist
Douglas Aircraft Co.
Woodland Hills, California

**David M. Moore, D.V.M.**
Director, Laboratory Animal Resources
College of Veterinary Medicine
Virginia Polytechnic Institute and State
  University
Blacksburg, Virginia

**Gail P. Smith, Ph.D.**
Director (Ret.), Technical Staff
  Services
Corning Glass Co.
Corning, New York

**Eric W. Spencer**
Brown University
Providence, Rhode Island

**William L. Sprout, M.D.**
Medical Consultant
Haskell Laboratory
E.I. du Pont de Nemours & Co.
Newark, Delaware

**Norman V. Steere**
Laboratory Safety and Design
  Consultant
Minneapolis, Minnesota

**M. A. Trevino, M.D.**
Medical Director
Quimca Fluor, S.A. de C.V.
Matamoras, Tamaulipas, Mexico

**Barbara Tucker, M.T., ASCP Ret.**
Chief Medical Technologist
Northwestern Hospital
Minneapolis, Minnesota

**Paul Woodruff**
Environmental Resources Management,
  Inc.
West Chester, Pennsylvania

# THE EDITOR

**A. Keith Furr, Ph.D.,** is Head of the Department of Health and Safety at Virginia Polytechnic Institute and State University, Blacksburg, Virginia and Professor of Nuclear Science and Engineering. He received an A.B. degree, cum laude, from Catawba College in 1954, an M.S. Degree from Emory University in 1955, and the doctorate from Duke University in 1962. Since 1960 he has held successive appointments at VPI & SU as Assistant Professor of Physics, Associate Professor of Physics and Professor of Physics as well as his current appointment as Professor of Nuclear Science and Engineering. In addition to other assignments, he was Director of the University's Research Reactor and was Head of their Neutron Activation Analysis Laoratory. During the 1970s, Dr. Furr developed an undergraduate program in Radiation Health and Safety at VPI & SU. He belongs to the Health Physics Society, the Campus Safety Associaton, the National Safety Council, and the National Fire Protection Association. He has published over 60 papers in refereed journals in a wide range of areas, including physics, nuclear engineering and the role of trace elements in the environment, and has presented papers at numerous national and international meetings. His current interests are primarily concerned with expanding the scope and improving the effectiveness of safety and health programs in research facilities. Dr. Furr is also active in working with the public in defining the impact of research operations on the local environment.

term exposures to low levels of chemicals. The volume was amended on numerous occasions during its preparation as new information became available. Subjects such as AIDS, control of laboratory infections by many other organisms, recombinant DNA, carcinogenic effects, teratogenic effects, medical surveillance programs, etc., were not covered in previous editions because they either were not known or were not widely recognized as topics that should be addressed. These and other health concerns are covered in this edition.

As editor, author, or preparer of much of the new material, with the assistance, input, and influence of the members of my department and many of the research scientists and support personnel within our organization, I have had to develop a program for our university which is responsive to these laboratory and support personnel in all of the areas covered, and others as well. Funds have not always been plentiful and individuals have not always understood why practices which they have followed for years need to be changed. This personal experience is reflected in this text. Programs need to be flexible, and sometimes compromises need to be made which solve most of the problem, while resources are sought to solve the remainder. It is hoped that the information contained in this edition will answer many safety questions and will aid readers in developing responsive and effective programs within their own organizations.

**A. Keith Furr**
Blacksburg, Virginia

# FOREWORD

The third edition of the *Handbook of Laboratory Safety* builds upon the strengths of the earlier editions, but takes some new directions which I believe will make it an especially practical and useful work for laboratory personnel. It is not primarily intended for the safety and health specialist, but instead is meant to guide research personnel in working with safety and health professionals to implement effective health and safety programs in their facilities

Since the previous edition was prepared, there have been many new regulatory requirements which working scientists need to incorporate in their laboratory procedures. At some point, some of these regulations may be made to apply only indirectly, i.e., a performance standard approach may be adopted. However, the safety and health programs which might be designed for an individual laboratory would need to be comparably effective to a program designed to meet the regulatory standards, and it would be difficult for a laboratory manager to design a program without having a basis for comparison. In our current litigious society, it would be difficult for a person to offer as a defense a lack of awareness of what safety practices are acceptable.

Laboratory safety does not begin and end at the laboratory door, nor does it cover only the period during which work is actively in progress. This edition attempts to extend the scope of the areas covered so that research personnel can become aware of factors outside their immediate operations which impact on these activities. For example: laboratory operations must be performed in code complying facilities; hazardous waste disposal is now highly regulated, and, integrated over the entire operation of a university or corporate research facility, the relatively small contributions of individual laboratories can be comparable to a substantial chemical manufacturing operation; right-to-know legislation covers not only risks to laboratory personnel but the nearby community if a release or spill escapes off-site.

This edition is not limited to coverage of regulations but provides substantial new information on safe operations. It is not intended to teach chemistry to chemists or to make the individual scientist into an industrial hygienist, an architect, a ventilation engineer, a fire inspector, or a health physicist, but it does provide sufficient information so that the laboratory manager or worker can communicate effectively with these persons and comprehend the reasons behind many of the technical decisions which safety personnel are expected to make. In most large organizations, each of these functions will be assigned to internal groups outside of laboratories. However, in smaller facilities, enough information should be available to make it possible for conscientious individuals to design adequate programs, or to guide them to sources where this information is available. References in the text are designed to direct the readers to generalized source documents in most cases rather than to specialized technical articles on individual points within a field. Most readers are expected to be primarily interested in topics outside the area of safety but do wish and need information on safety and health issues. It is unreasonable to simultaneously expect a scientist to remain equally adept in the increasingly more sophisticated field of safety and health as in his own discipline.

A major portion of this edition is concerned with the health aspect of health and safety programs. This is still one of the areas where there is much uncertainty and a relatively scarce amount of data is available, especially concerning the effects of long

# DEDICATION

This book is dedicated primarily to my wife for her patience during the preparation of the manuscript. It is also dedicated to many of the faculty and staff of my University for their cooperation in establishing many of the programs on which much of the material is based, and to the staff of my department whose daily efforts have made many of these programs effective.

## Chapter 5
## Nonchemical Laboratories

Chapter 6
**Personal Protective Equipment**

# Assignment of Responsibilities 1

## 1.0. INTRODUCTION

Until relatively recently, if a formal safety program existed at all in a laboratory, the scope and the amount of effort placed on the program was largely at the discretion of the individual laboratory supervisor or laboratory director, especially in academic institutions. Some industrial organizations, recognizing the value to themselves and their employees of keeping the employees safe and well, have had excellent companywide safety programs for many years, including the laboratory environment. Many smaller firms did not have comparable programs. The same is true of academic institutions. Some of the more advanced universities have also had good safety programs of long standing. However, even some of the larger institutions, and a very high percentage of the moderate to smaller schools and research laboratories, have had minimal safety programs, usually limited to a fire safety group in the maintenance departments. If the school used radioactive materials, it had to have a radiation safety officer or committee because these were mandated by the Nuclear Regulatory Commission or its antecedents for any facility working with radioactive materials. Consequently, safety was usually not stressed and very few students received any formal safety training. Since attitudes, once established, are very difficult to change, the attitude that safety is of secondary importance has been carried over into professional careers for a significant fraction of current laboratory research personnel.

Part of the difficulty in establishing an effective laboratory safety program is the nature of laboratories. The activities conducted within them are extremely varied, and frequently change. The processes and materials in use may present unidentified problems. Research materials may be being synthesized for the first time or may be being used in novel ways. Probably the most common class of chemicals used are flammable solvents, and there are ample sources of ignition in most laboratories. Because of the changing needs in the laboratory and the scale of most reactions, over a relatively short period of time, even a well-managed facility tends to accumulate a large and varied inventory of chemicals. The equipment which is used is often fabricated or modified within the laboratory or at the instrument shop. If so, it obviously is not tested by any safety organization. Laboratory facilities which may have been well designed for their initial use may easily become wholly inadequate in terms of electrical services, ventilation, or special equipment such as hoods as programs change or as new occupants move into the space. Buildings and spaces originally built to house classrooms are often converted to serve expanding research programs, but they rarely serve well as laboratory structures without large, and difficult to obtain, sums being spent on renovation. Even with the best of intentions, physical solutions to safety problems which may work well in large-scale industrial processes are often extremely difficult to scale to the laboratory environment and alternatives may be expensive.

The individuals working within laboratories are usually very capable persons who tend to be strongly goal oriented, often to the unintended exclusion of other factors. Sometimes the person in charge, in solving a problem, inadvertently neglects some of the peripheral factors, such as safety, because they are not directly involved with attaining the research goal, or forgets that some of the technicians or students do not have the years of experience and training that the laboratory director does. It is true, too, that familiarity all too often makes one careless. Also, and unfortunately, like any other large group, the research community is not immune to the presence of a few persons who cut corners to serve their own interests. The growing competitiveness in finding funds for research, to publish, and to obtain tenure in academic environments cannot help but exacerbate this problem. It is probably unwise to depend wholly on the professional expertise of the research scientist or his voluntary implementation of safety programs. There are too many other priorities.

The organizational structure of most research institutions, especially at academic institutions, also contributes to the variability and general weakness in institutional safety programs, due to fragmentation of responsibility. The strong line organization found in businesses and factories is often very weak in universities. Individual research programs in an academic laboratory are often defined solely by the laboratory director, virtually independently of his department head. The typical academic department head may, or may not, establish the broad areas within the department which will be emphasized, but will intervene to a minimal degree in the conduct of the research or day to day operations. Intervention would likely be construed as an infringement of academic freedom. Similar gaps exist between the department head and his dean, and so on through successive administrative layers. Thus, implementation of a safety program is a difficult logistical problem in colleges and universities.

Two major factors have contributed to the changes which have occurred in the last few decades. None of the difficulties cited in the previous paragraphs have disappeared; indeed, some have become worse. However, new mandatory regulations and standards have been enacted, so "doing what has to be done" has become a much larger task. Secondly, and perhaps more importantly, the acceptance of the status quo has become increasingly unfashionable and unpalatable. The expectations of laboratory workers have increased. They are concerned about the effects of their environment on their health and safety, and they expect something to be done to eliminate perceived problems. If not, litigation is a possibility that management cannot ignore. A good case can be made that without these expectations and possible repercussions, many of the regulations governing health and safety in the workplace may not have been passed.

As a result of the new regulations and new attitudes, it has become necessary for the administration in research institutions to take an active role in safety management. They have had to define comprehensive universal safety and health standards for their organizations, and to take positive steps to encourage employees at all levels to actively support and apply standard safety policies for everyone. Individual laboratory directors still have the primary responsibility of managing their operations and, in fact, are more important than ever as key persons in implementing the institutional safety program, but they can no longer legally choose to have a safety program or not. This is not a role to which many of them are accustomed, nor are many of them comfortable in it. The regulations in the context of research facilities are too new; they are still evolving, and it is not always easy to know what is expected to be done. Old habits are difficult to shed, and there will be a lengthy transition period before safety is a matter of course in most laboratories, even under the new standards.

It is a serious fallacy to assume that to establish an effective safety program, one need only create a safety department and charge it with making things safe. Presumably, everyone will then immediately change their habits and accidents will no longer happen. The millennium will have arrived. Of course, creation of a central internal organization is an important first step, but it is only the first step. Much more important is the creation of the right atmosphere. Top management must support safety and let it be known that they do. Establishment of a safety department does indicate this posture to a degree, but management cannot take the position that their role is finished with this action. The safety department can provide the supportive structure, but it cannot make things safe. It takes the combined efforts of the entire organization to create an effective safety program.

One of the first things safety personnel must understand in defining their role is that they are not the first order of business for the organization of which they are a part. A widget manufacturer makes widgets as their first priority. A university has students to educate and scholarly research to perform. However, there is nothing to prevent these things from being done safely, and surprisingly, doing them safely usually means doing them better and more economically. It is important for a safety department to perform its function so that this is true and is perceived to be true. A safety department needs to establish itself as a service department, with a strong emphasis on service. This is not to say that enforcement and regulation are not appropriate for safety personnel, but even these functions can be done in a nonadversarial manner. Service as an operating premise also does not mean that the safety department is to take the safety burden entirely upon itself. The department must provide the appropriate structure for the remainder of the organization to naturally do things safely, although there are certainly support functions that are done best by safety professionals. Without the support of management, no safety department can succeed. Support entails many factors. Resource support is obviously very important, not only in terms of what can be done with the resources, but, if the safety department is not provided with a reasonable level of support, protestations of support by management will seem insincere and hollow. Conversely, if management "puts its money where its mouth is", the conclusion will be drawn by the employees that management does indeed see value in the programs it supports. There are other ways that management can illustrate support. To whom does the head of the department report? Is it a lower middle manager, or is it a senior person such as a vice president? If a confrontation does arise, and these unfortunately may be fairly frequent in an academic research laboratory where full professors may be accustomed to operating with few restraints, will the safety professional be supported when he is right? Does the safety department participate in making operational decisions or policy decisions? If the answers to these questions are positive, the status of the safety department will be enhanced. If the answers are negative, the effectiveness of the organization's safety program will be seriously diminished.

The clients, i.e., the employees, also must cooperate. They must see the actual value of the safety program to themselves. Of course, no one is going to seriously advocate not doing things safely, but many will take the position that, if left alone, they would have no safety problems. Many conscientiously feel that a formal safety program, with the attendant rules, regulations, and all the accompanying "red-tape", is not necessary. An astute safety professional, with the active support of management, can overcome this attitude. This can be done by education, so that people know what is expected of them and why; by actively involving the employees in setting the standards; by minimizing the administrative burden on individuals; by demonstrating that the safety program helps the organization's personnel solve their problems; and by making sure that the safety program clearly provides what it purports to do, a safer environment. Despite the difficulties which have been described, a good, sound laboratory safety program is possible, and under today's regulations, is mandatory. The intent of this handbook is to help define the requirements of a successful program. In most instances, the recommendations will be based on current and anticipated regulations or accepted practices, but in some cases will go beyond these where it appears a more conservative program might be prudent or where there appears to be an occasion for an innovative approach. In the latter case, the differences between the required procedures and the modified approach should be clear. The area of emphasis for much of the material will be the chemistry laboratory, but there will be separate sections for several other important areas. Several of these will be new to this edition, or greatly expanded or revised versions of material found in the second edition. In some cases, such as recombinant DNA, the subject was virtually unknown when the previous edition was prepared, and in other cases, there have been dramatic new developments.

## 1.1. LABORATORY SAFETY AS A COMPREHENSIVE RESPONSIBILITY

Laboratory safety is now the responsibility of many departments within an organization instead of being limited to the laboratory itself, management, and the safety department. Due to the advent of regulations, a number of other departments have specific responsibilities which directly affect laboratory safety programs or are affected by them. Among these other departments or offices are employee relations or personnel, frequently a separate office to guarantee equality of employment opportunities and nondiscrimination, the legal office, risk management and insurance, planning and engineering, possibly a separate office for architecture and design, purchasing, and the physical plant/maintenance department.

The personnel department is the first contact that most employees have with an organization and, in a well-planned orientation program, has the responsibility of informing the new employee of the basic policies of the organization. It is important that they include the rights and responsibilities of the employee regarding safety as well as working hours, pension program, etc. The employee should receive written statements of the overall safety policies of the organization, where additional information can be obtained, and to whom the employee can turn to express safety concerns. The initial impression of the importance placed on safety is made at this time, and every effort should be made to insure that it is positive. A brief description of the organization's Hazard Communication Program should be provided at the orientation for new employees. Detailed safety procedures should not be covered at this point where laboratory safety is concerned since laboratory activities vary widely from laboratory to laboratory. Information on safety procedures for a given laboratory should be provided by the person in charge of the facility. The personnel department is also the usual department to handle workman's compensation claims. Thus, it has a direct concern in the reduction of injuries in order to minimize the cost of such claims.

In many organizations, the group charged with assuring that no discrimination occurs and that everyone is assured of an equal opportunity is in the personnel department. In others, this group is autonomous or reports through a different chain of command in order to avoid any conflict of interest. Although there are obvious safety implications for handicapped persons, one of the biggest safety issues for the equal employment opportunity (EEO) group is that of fetal vulnerability to chemicals in the laboratory. In some areas, information is sufficiently developed to allow well-defined policies. In radiation safety, for example, the rights of fertile women have been clearly spelled out. On the other hand, information on the effect on the fetus or the reproductive process for the thousands of chemicals in use is very limited. In such a situation, women may have legitimate concerns about their exposure as well as possible unfair exclusion from some areas of employment. It is the responsibility of the EEO office to provide them with a means of expressing these concerns and to work with the employees and other parts of the organization to provide adequate safeguards to protect the rights of women employees.

There are many occasions for the legal office of an organization to concern itself with laboratory safety. Many of these are straightforward, concerned with contract terms and conditions for sponsored research. Contracts for public academic institutions, for example, usually cannot contain a "hold harmless and indemnification clause" since this, in effect, waives the protection provided by the immunity to suit claimed for many state agencies. The legal counsel must find alternative language or an alternative procedure for satisfying this contract condition. There are numerous restrictions built into laws or policies for various fund-granting agencies, and it is the responsibility of the legal office to insure that these terms are met. The question of personal liability often arises as well. Laboratory managers often ask what their personal exposure to litigation is by the manner in which they enforce the organization's safety program. It is obviously impossible to monitor every action by laboratory staff to assure compliance with safety and still have time to do research. However, what are the boundaries of adequate supervision? Finally, most organizations have grievance and disciplinary procedures for actions which involve alleged violations of safety procedures. The legal counsel will necessarily become involved in these.

Liability insurance, the ability to obtain it or not and the cost of the insurance when obtainable, has become a major issue. The insurance manager has often become a key person in processing contracts where safety is implicated. There have been occasions where the safety issues have been so significant that failure to obtain liability insurance has resulted in a decision not to perform certain research projects because the alternative would have been to expose or pledge a substantial portion of an organization's assets in the event of an incident. Because of the difficulty in obtaining insurance, many companies as well as state and federal agencies are initiating surveys and studies to determine their liability exposure and to determine the cost of abatement or preemptive actions. Frequently, these surveys and studies are coordinated or at least reviewed by the risk manager. The insurance manager also usually is eager to work with the safety group and personnel departments to stem the rising tide of workman's compensation claims and other safety related claims.

As previously stated, a comprehensive safety program is relatively new to many organizations, and nowhere is this more true than in large parts of academia. Except for adherence to a review of facilities by a building code official at the time of construction, academic facilities at the college and university level often have been relatively immune to further review, although many colleges and universities did maintain a position or two in the physical plant department for fire inspections. Most of the inspectors in these positions confined themselves to performing routine inspections of the contents of the buildings to ensure the halls did not become too cluttered, that exit signs and lights were all right, that fire extinguishers were charged and in place, and that the buildings were maintained in fairly good physical condition. Even during the construction review process, many building codes in the past were primitive at best and adherence to safety standards by the architect and contractor was not difficult. Subsequent to the original construction, renovation projects, whether carried out by the central department charged with this function or by laboratory or departmental workers, rarely received a safety review. Thus, unless it was extremely clear that the project was not safe, it was built and became part of the building. As a result, there is a huge backlog of existing improper and unsafe construction on many campuses.

It is the responsibility of a conscientious planning and engineering group to avoid creating any new problems and to correct existing conditions as resources permit. New renovation absolutely should conform to current safety standards. There should be no exceptions, except in some instances where, with the approval of building officials, it can be demonstrated that the exception is safer than the actual standard. In terms of laboratories, many architects and engineers are relatively unsophisticated in terms of design for safe operations. Many desirable laboratory safety features are not defined by code requirements. It is in this area of integrating safety considerations into project design that some of the best opportunities to enhance the working environment exist for a safety professional by working with the planning group and the program managers.

It would be highly desirable if safety, building function, and architectural elements were integrated from the very beginning in new construction. Where independent architecture and safety departments work together with the user groups from the beginning through the completed design and construction, the optimum situation can be achieved. This integrated approach often facilitates working with outside building officials such as the local planning and architectural review board or, for public agencies, with state engineering agencies and the state fire marshal to achieve a design which will satisfy everyone's needs. Such a cooperative approach will usually mean faster processing through the system, fewer problems during construction, and a more efficient and economical building. As in the case of renovation projects, relatively few architects and engineers have designed laboratories and are aware of all the safety features which need to be incorporated in the building. The majority of architectural firms are very good at designing offices, classroom buildings, storage facilities, etc. and they are very glib about the building making an architectural statement in harmony with the existing structures, but relatively few firms have built many laboratory facilities and surprisingly few employ code experts. Similarly, the users know what is needed in order to do the intended research, but they typically know virtually nothing of building codes and, unfortunately, few science professionals have had very much formal safety training. It is the role of the safety professional to be sufficiently knowledgeable about both areas to be able to see that the final design meets all needs safely.

The cooperation of purchasing will make the creation of safe laboratory conditions much easier. With their cooperation, it is possible to add safety specifications and in some cases to limit the eligibility of vendors of products to insure that the items ordered will function safely. Sometimes, this will substantially increase costs, but will significantly enhance safety. A good example of this involves the choice of chemical splash goggles. Chemical splash goggles which will minimally meet ANSI standards can be purchased very cheaply. However, the least expensive units are typically uncomfortable, or are hot, or fog quickly, or reflect a combination of these undesirable qualities. For short-term, very limited wear, they are reasonably satisfactory, but after a short interval the wearer usually removes them or slides them up on the forehead. In either case, they would not be offering the needed protection. At the author's institution, with the cooperation of the purchasing department, based on extensive evaluation and testing, several brands were determined to provide superior protection. Of these, three were substantially less expensive than the remainder, although still nearly five times more expensive than the cheapest brand which met ANSI standards. However, our institution now limits our purchases to these three brands until such time as subsequent testing can show that other models which might come on the market are comparable or better. This still allows for competitive bidding among the three brands and among distributors.

The physical plant department has the responsibility to provide heat, light, utilities, and custodial services, to repair broken equipment and make renovations, and to provide all the other services that are needed to keep a building functional. However, few laboratory workers really appreciate what they are asking of these persons. Many persons have had the experience of custodial workers being afraid to enter a laboratory with a radiation sign on the door and make allowances, but a mechanic is expected to work on a possibly chemically contaminated exhaust motor to a fume hood without question. These workers are becoming increasingly concerned about exposing themselves to toxic materials while performing maintenance in laboratory facilities. With the new hazard communication standard having gone into effect on May 25, 1986, maintenance personnel will become more aware of potential problems and may become even more reluctant to perform laboratory maintenance unless they can be assured of the lack of risk to themselves or are protected from exposure. In their own areas, maintenance personnel have many skills and considerable knowledge, but they are not scientists. Their lack of familiarity with the potential risks of the research being conducted in the facility where they work may lead to perhaps unreasonable fears. It will require a combination of education, tact, and accommodation on the part of laboratory personnel to assuage these fears in order that they may continue receiving needed laboratory maintenance services.

The preceding brief documentation of the involvement of numerous departments in laboratory safety illustrates that a good program is not limited to the establishment of good laboratory practices. Many other factors are involved and the role of the laboratory manager is becoming increasingly complex.

## 1.2. ORGANIZATION

When programing safety for an organization, it is important to consider the type of motivation of the personnel within the organization. There are, as noted earlier, likely to be important differences between the motivation of typical manufacturing employees and laboratory employees. If the laboratory is in an academic environment, the differences will be even more pronounced. The manufacturing employee is usually closely supervised, and will generally try to follow practices, including those involving safety, which have been set by management or agreed upon by management and the labor organization to which the employee belongs. The laboratory environment leads to much more self-supervision, especially for the laboratory director or manager, who is typically judged only by the results of his work. A laboratory represents a considerable investment for academic and other research institutions, so success is critical to the career of a research scientist. This difference may lead to the laboratory worker taking excessive risks to achieve desired results.

A second factor which is important to consider in programing safety is the structure of the organization. Laboratory supervisors in universities and in many research institutions do not have the same degree of communication with, or responsibilities to, higher level management as do typical industrial supervisors. They are much more independent and tend to resent "unwarranted" interference. They especially do not want outsiders to attempt to intervene in their program of research. Thus, a major role for safety personnel is to define the organizational constraints for safe practices for the laboratory supervisor, who must then adapt the research program to work within these boundaries.

At one time, the safety professional rarely went beyond an advisory or consulting role. With the advent of formal safety regulations, the safety professional has had to take a much more active role as a monitor of the safety performance of laboratory personnel and as an enforcer of the regulations. In addition, some of the support functions previously assigned to the maintenance department now require a level of expertise such that the safety department has assumed some of these support functions.

The proliferation of additional management positions over the last few decades has complicated the organizational structure so that laboratory personnel in most cases now report through completely different channels than do support personnel. In fact, the professional research chain of command may be differentiated into several autonomous chains of commands such as, for example, the different colleges in a university. The interactions of the safety group with laboratories thus cross several internal lines of responsibility. Both of these groups may, in turn, have to contend with a third independent entity represented by the financial managers, the result being that program decisions may, in fact, be determined by the senior financial officer. Therefore, where such a situation exists, programing for safety must take it into account.

There are any number of possible working arrangements for a safety program, but to be effective, they will have several things in common. Among these are

- Assignment of safety responsibility to a senior-level person such as a vice president, with the head of safety reporting directly to this person.
- The formation of a safety department, staffed by professionals, and with sufficient resources to perform their function. In an academic environment, there may be some advantages for the head of the safety department to have academic credentials. Since virtually all laboratory heads at a university have a doctorate and judge themselves and others by their scholarly activity, an individual with similar credentials may find it easier to gain and hold their attention. On the other hand, it is more important that the safety department perform professionally instead of simply having a given set of credentials.
- The creation of one or more safety committees at the institutional level. In an industrial situation, one safety committee will usually suffice, but in a research facility, especially one with a wide variety of programs, it would be preferable to have specialized committees such as a radiation safety committee, an institutional biosafety committee, a general laboratory safety committee, a human subjects review board, and an animal care committee. All of these are either mandated or recommended by various standards if the organization has research programs in corresponding areas. Some of these special interest committees will have more authority than others due to the underlying strength of the regulatory standard. For example, the use of radioactive materials is very strictly regulated by the Nuclear Regulatory Commission. Failure on the part of the institution to comply with the regulations and the terms of the institution's license can, and frequently does, result in substantial fines as well as national publicity. An overall committee, formed of the heads of the special interest committees may be a useful vehicle to set broad safety policies to insure consistency in the various areas of responsibilities. Each committee should have a definite, written charge as well as definite rules of procedures.
- Because of the independence of the various divisions or schools in larger organizations, the separate internal agencies may find it desirable to establish their own safety committees to take the overall institutional safety policies which must be followed and work out their own procedures to implement them most effectively within their own areas of responsibility.
- Each technical department should identify a single individual to act as liaison with the institutional safety department and the safety committees. In many organizations, serving in such a position is not especially beneficial to the career of the individual and, in fact, some organizations tend to shift such responsibilities onto less productive individuals. This is not appropriate. The assignment should be rotated among the most active and productive individuals.

Selection of suitable members of the committees at the various levels is extremely critical. In academic institutions, once tenure has been obtained, the drive of some faculty members appears to diminish and they become less productive. There is a tendency to give the less glamorous and more demanding committee assignments to these persons. Moreover, some of the very active and productive persons are reluctant to accept responsibilities not immediately germane to their own programs. Some of the same tendencies exist among nonacademic laboratory personnel as well. If safety committees are chosen based on these selection criteria, they will not be effective. Lackadaisical research persons cannot be expected to become energetic and concerned committee members, and the more productive personnel will not respect the committee's efforts. The members should be drawn from the most active and productive laboratory personnel, for limited terms if necessary, but the persons whose programs will be most affected should play a role in defining policies.

The heavy demands on the time of the latter group can be turned into an advantage in getting them to accept the assignment. First, let it be known that the selection is to be made from the "best" personnel since they have most to gain from an effective program and will be most adversely affected by a poor program. Candidates for a committee position should be encouraged to look at the assignment as an opportunity to positively influence a program that is important to them. If this is the approach taken, most persons will be flattered to be asked and will accept.

Some care must be taken to define "productive" personnel. The most productive individuals are not always the most visible or the most vocal. The actual achievers may be too busy to promote themselves and simply are quietly effective. The latter class should be sought out and the former passed over. Also, a safety committee member must be able to consider issues objectively and be willing to act even if, on some occasions, their own immediate self-interests might be adversely affected. The background, credentials, and references of candidates for membership on a safety committee should be checked as carefully as if the candidate were a job applicant.

Nothing will destroy a committee more rapidly than for the members to feel that the work of the committee serves no useful purpose. There must be meaningful work for the committee to do and each meeting must be structured, with a definite agenda and definite goals. Much of the responsibility for making a committee work depends upon an effective chairman and good staff work. The staff must prepare the working documents and distribute them in sufficient time for review by the members and to be prepared at the time of the meeting. The chairman has the responsibility to conduct the meeting fairly, with everyone having ample opportunity to contribute, but he must see that the business is conducted expeditiously. Meetings should be no longer than necessary.

The chairmen of the various safety committees should have administrative experience. They should have experience in managing a budget and managing people so that they can guide the committee should it take actions which might demand resources that would be hard to obtain or cause personnel problems. They should, of course, be knowledgeable in the area of the committee's responsibility. There are occasions when not all of these qualities can be found in a single individual, usually the specialized knowledge being the missing factor. In such cases, senior professional persons who have had to obtain and manage grants would be satisfactory alternatives to administrators. It normally would be less desirable for the safety

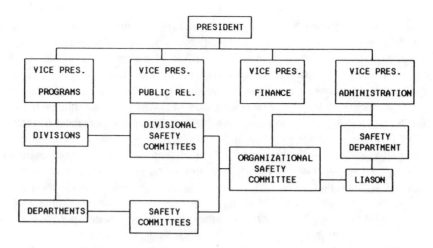

FIGURE 1.1.   Organizational chart featuring safety-related components.

professional to be committee chairman because of a possible perception of a conflict of interest, although there are circumstances, due to experience and training, when exceptions to this "rule" are appropriate. There also are logistic factors, when the chairman must take an active role in a safety area, which would make it desirable for a safety professional to chair a committee.

There has been considerable emphasis placed on the committees in the preceding paragraphs, but in a research institution it is unlikely that the busy professionals making up the committee would be willing to meet more than once a quarter, except in emergency situations. Therefore, the safety committees cannot be assigned the task of directly managing the safety program. They are, in effect, somewhat analogous to the legislative branch (and, on occasion, the judicial branch) of the safety "government" and the safety department is the administrative branch. Between meetings, the safety department must have the authority to act as necessary, with the knowledge that they will be held accountable to their superiors and, to an extent, the committees. It is up to the head of the safety department to prepare budgets, administer the programs and personnel assigned to the department, and to provide leadership in the area of safety.

Figure 1.1 presents a simplified organizational chart that embodies the concepts discussed in this section. In a small firm or academic institution, some of the duties and layers of responsibility might be combined, while in very large organizations, additional functions might be needed. Note that the safety committees are appointed by a vice president and have direct access to him.

## 1.3.  THE SAFETY DEPARTMENT

Since so many aspects of laboratory safety have been codified or affected by regulations, in order to provide the focus for the corporate or university safety program the safety department has had to assume a much stronger role. Uniform policies must be developed which cross internal territorial boundaries and which are administered evenly and impartially. This requires a central organization through which the institution can operate to implement the policies. Since it is the voice of the organization with respect to safety, the safety department has assumed the character of an administrative, rather than advisory, department.

Maintaining an awareness of the current status of all the regulations which affect operations demands more time and specialized knowledge than a busy laboratory director or most administrators have available. In fact, it has become necessary to develop specialists, even within safety, because of the growing complexity of the field. This again has led to a strengthening of the role of the safety department because both management and the client groups are often forced to defer to the expertise of the department.

The Safety Department is often forced to assume another unfamiliar role in directly providing support functions, rather than simply monitoring to see that operations are performed correctly, because of the skills, knowledge, and specialized equipment that are needed. A comprehensive safety department needs to have laboratory facilities of its own, with its own skilled personnel. Since a direct consequence of the provision of services is the need to fund these services, resources to meet safety needs are typically funneled through the safety department rather than being directly allocated to the research departments.

## Functions Appropriate to the Safety Department
- **Training**
  Safety orientation
  General work practices
  Supervisor training for hazard communication standard
  Training maintenance personnel for work involving laboratories
  Emergency procedures
  Fire
  First aid
  Cardiopulmonary resuscitation
  Serious injuries
  Chemical, radioactive material, and biological toxin spills
  Radiation safety
  Publishing safety manuals
  Chemical laboratory safety
  Biological laboratory safety
  Recombinant DNA laboratory safety
  Animal care facilities
  Chemical waste procedures
  Personal protective equipment usage and care
  Other Occupational Safety and Health Administration (OSHA)-required training
- **Waste disposal**
  Chemical wastes (including permissible processing and redistribution)
  Radioactive wastes
  Animal tissues
  Contingency planning
  Liaison with federal, state, and/or local waste officials

- **Building safety**
  Building inspections
  Fire system inspections, testing, calibration, and maintenance
  Monitoring building evacuation programs
  Interacting with local fire and emergency response groups
  Code review of all renovations and new construction
  Consulting on building design, participation on building committees
  Interacting with on-site construction contractors and architects
  Liaison with fire marshals and building code officials
- **Environmental protection and industrial hygiene**
  Air quality testing
  Testing fume hoods and biological safety cabinets
  Laboratory inspections
  Reviewing laboratory design proposals
  Reviewing research proposals for potential problems
  Reviewing chemical purchases for special regulated chemicals
  Maintaining controlled substance licenses
  Reviewing and specifying personal protective equipment
  Reviewing selected equipment purchases for safety features
  Maintaining laboratory census for safety features
  Providing consultation services
  Investigating laboratory accidents
  Asbestos testing (bulk and air sampling)
  Monitoring asbestos removal projects
  Maintaining records
- **Medical program**
  Setting criteria for employee participation in medical surveillance
  Assisting departments in selecting departmental participants
  Providing staff support for occupational physician
  Working with personnel department for preemployment examinations
  Analyzing accident records, preparation of reports
  Investigating serious accidents
  Liaison with local medical emergency groups
- **Radiation safety**
  Inspecting and surveying radiation-using facilities
  Reviewing all internal applications for use of ionizing radiation
  Reviewing use of nonionizing radiation
  Processing all orders for radioactive material
  Receiving and checking all radioactive material deliveries
  Maintaining radioactive material inventory
  Monitoring all personnel exposures
  Maintaining calibration of all survey instruments
  Maintaining all required bioassays
  Managing or overseeing all radioactive waste disposal
  Maintaining all required records
  Liaison with regulatory agencies

- **Institutional hazard awareness program**
  Maintaining material safety data sheet files
  Tracking chemical purchases/employee participation in training programs
  Serving as organization's emergency coordinator
  Representing organization on Superfund Amendments and Reauthorization Act
    (SARA) Title III local emergency planning committee
- **Miscellaneous responsibilities**
  Participating in all institutional-wide safety committees
  Provision of staff support to safety committees
  Investigation of employee complaints

This rather extensive list may seem too detailed and perhaps too comprehensive to many practicing safety professionals. It certainly reflects a strong, direct involvement of safety departments in the affairs of their home organizations. Because of the increasingly complex knowledge, skills, and resources needed to meet current safety standards, there is really no choice except to centralize the responsibility for dealing with many of these standards. The alternative of making everyone comparably knowledgeable and able to manage safety to avoid potential liability for themselves, as well as be technically proficient, is impractical. The goal is to preserve, in the area of safety, as much local autonomy and sense of personal responsibility as possible among the organization's other employees.

Despite the sense of the last paragraph, the safety department should not be too eager to accept additional responsibilities from other service groups. Frequently, responsibility will be shifted without a commensurate transfer of resources so that the departmental resources, including manpower as well as physical and financial resources, will be strained severely. Available resources may be difficult to obtain, especially in times of retrenchment, for most public and private organizations, and care must be taken to set achievable goals and priorities.

## 1.4. DEPARTMENTAL RESPONSIBILITIES

The local department is an entity within an organization with which individuals with more or less similar interests and goals identify themselves, and are recognized by the organization as a unit for administrative purposes. As such, they are the logical unit to establish and administer common safety-related procedures and practices suitable for the laboratory programs in the discipline identified with the department. The leadership of the department provides a natural channel for communication with the laboratory personnel and provides a vehicle for enforcing organizational policies or to direct concerns by the department about these policies to higher administrative levels. In the context of the present discussion, the major virtue of a department's leadership is that it is a physically present, recognized source of authority. No safety department is sufficiently large to be present everywhere. Safety personnel must rely on the help of local departments to implement safety practices and policies. The local department's leadership must accept the need for a strong safety program and see that a concern for safety is the standard expected of all departmental personnel.

At a minimum, the department should designate an individual to be a safety coordinator for the department, who may also act as liaison to the institutional or corporate safety committee. Preferably, however, the department should establish an internal safety committee with representatives from each major division within the department. Included on the committee should be individuals at various levels, such as laboratory supervisors and senior and junior technicians to ensure that all points of view are fairly represented. In an academic department, it might be desirable to include an experienced graduate student. One function of the committee would be to interpret the corporate or institutional policies in the context of the departmental operations and the operations of the individual laboratories. The committee could also advise individuals and make recommendations on improvements for safe operations in laboratories. The support of a committee can often facilitate obtaining needed resources, where an individual might not succeed. Finally, the committee should establish a frequent schedule of inspection of facilities, including support services, in addition to the actual laboratories. If conditions are found which need correction, the supervisor responsible for the problem area should be informed and a follow-up inspection should be performed after a reasonable interval to insure that the corrections had been made.

The department also can and should function as a resource center. For example, the Hazard Communication Standard, under OSHA, requires that certain information be readily available to the employees in the department, specifically material safety data sheets. A central hard-copy file of these in a large research organization or university could well encompass 5,000 to 10,000 records in a massive and difficult to manage file. A departmental file would be far more logical and manageable. Similarly, the department would be a logical unit to provide the training required for the same standard for the department's laboratory employees since, as noted above, the department is normally organized around a common discipline.

The department should also accept some responsibility for the safety of its personnel in terms of resources. There are legitimate questions concerning the limits of the institutional or organization's responsibility and those of the department or laboratory. It is clear, for example, that such things as utilities, custodial services, and building maintenance are the provenance of the corporation or university, while specialized equipment such as an electron microscope or a highly specialized laboratory environment is probably equally clearly the responsibility of the local department. However, who is responsible for eyewash stations and deluge showers, fume hoods, glove boxes, chemical waste disposal, respirators, safety glasses, flammable material storage cabinets and refrigerators, etc.? Some of these would not be needed were it not for specific research programs or grants. Should not the department or laboratory manager incorporate these costs, or at least a portion thereof, in his own budget? To some degree, the imposition of safety standards and regulations has inculcated in many persons a sense that safety is not their own responsibility and has created an attitude that someone besides themselves is responsible not only for the rules, but also for providing the means to obey the rules. An awareness that safety is its own reward seems to have been lost to an extent. As a minimum, it would appear that in soliciting a grant the investigator and the departmental leadership must bear the responsibility of assuring in advance that the resources are available or attainable so that the research can be performed safely and with appropriate regard for the protection of the environment.

## 1.5. LABORATORY RESPONSIBILITIES

Ultimately, the responsibility for being safe and working safely falls to the laboratory personnel. The laboratory supervisor, who may or may not be an active participant in the work of the laboratory, still must set the standards of performance expected of everyone within the laboratory. The laboratory supervisor must make it clear by example or by direction that carelessness in safety is no more acceptable than careless and sloppy science. The supervisor has a responsibility to those under his direction to establish safe work practices and to ensure that the employees are informed and fully understand any risks associated with the program of work. No one should be asked to perform an unsafe act. This is not to say that the work must be made totally free of risk because this desirable goal is unattainable, but it does mean that all reasonable and practical steps to minimize risks must be taken. The equipment must be maintained in good repair and adequately designed to work properly. Individuals must be fully trained in the procedures they are to do and the work must be carefully analyzed to foresee potential accidents or failures. Contingency plans must be developed to meet, at least, the most likely emergencies. To fail to do these things could expose the laboratory director and, in turn, the department and the university or corporation to charges of willful negligence.

The employees, on the other hand, have an equal share in the responsibility. They must adhere to the safety policies that have been established. They must make sure that they are knowledgeable about good laboratory safety practices, and they must not diverge from these practices because they are too time consuming, too much trouble, or not convenient at the moment. If they are uncertain of the proper procedure, they must not be reluctant to admit it and seek clarification. Often, especially if the laboratory supervisor is not an active participant in the actual work, a skilled and knowledgeable person may understand the potential risks better than his supervisor and has the duty to make suggestions to improve the safety of the program. To some extent, this situation is a bit unrealistic in that many subordinates are sometimes wary of "making waves" or contradicting their superior and, to be truthful, not all supervisors are appreciative of suggestions from subordinates. It is this situation for which the ability to make anonymous complaints, which are certain to be investigated, was incorporated in the federal OSHA Act and subsequently passed on to the state standards, where states have adopted their own OSHA statutes. Ideally, the team of a caring, conscientious laboratory supervisor working in cooperation with competent, intelligent, and imaginative laboratory workers should make it possible to perform laboratory work with minimal risk.

## REFERENCES

1. **Petersen, D.,** *Techniques of Safety Management,* 2nd ed., McGraw-Hill, New York, 1978.
2. *Prudent Practices for Handling Hazardous Chemicals in Laboratories,* National Academy Press, Washington, D.C., 1981.
3. *Safety in Academic Chemistry Laboratories,* 4th ed., American Chemical Society, Washington, D.C., 1985.
4. **Steere, N. V.,** Responsibility for laboratory safety, in *Handbook of Laboratory Safety,* 2nd ed., Steere, N.V., Ed., CRC Press, Boca Raton, FL, 1971, 3.

5.  **Becker, E. I. and Gatwood, G. T.,** Organization for safety in laboratories, in *Handbook of Laboratory Safety,* Steere, N. V., Ed., 2nd ed., CRC Press, Boca Raton, FL, 1971, 11.

6.  **Songer, J. R.,** Laboratory safety program organization, in *Laboratory Safety: Principles and Practices,* Miller, B. M., Gröschel, D. H. M., Richardson, J. H., Vesley, D., Songer, J. H., Housewright, R. D., and Barkley, W. E., Eds., American Society for Microbiology, Washington, D.C., 1986, 1.

7.  **Bilsom, R. E.,** *Torts Among the Ivy: Some Aspects of the Civil Liability of Universities,* University of Saskatchewan, Saskatoon, Canada, 1986.

8.  *Occupational Safety and Health Administration Proposed Standard for Laboratories Using Toxic Substances,* 51 FR 26660, OSHA, Washington, D.C., July 24, 1986.

9.  **Jefferson, E. G.,** Safety is good business, *Chemical and Engineering News,* p. 3, November 17, 1986.

10. **Koshland, D. E., Jr.,** The DNA Dragon 1, *Science,* 237, 1397(No. 4821), 1987.

11. *Occupational Safety and Health Administration Hazard Communication Standard,* 52 31852, OSHA, Washington, D.C., August 24, 1987.

# Emergency
# Programs  2

## 2.0. EMERGENCIES

Emergencies are, by definition, not planned. However, planning for emergencies cannot only be done, but is an essential component of laboratory safety. Where a building incorporates a hazardous material storage facility, an emergency contingency plan is required by the Resource Conservation and Recovery Act under the Environmental Protection Agency (EPA). This plan must cover emergency evacuation and response plans, emergency equipment to be kept on hand, security, and the involvement of outside organizations. Virtually every organization which generates more than 100 kg of hazardous chemical wastes each month comes under this act. Even lesser quantities of waste from many common commercial chemicals can invoke the provisions of the Act. The Act itself will be discussed in a later chapter, but the emergency planning provision simply codifies what every responsible laboratory organization should do anyway — prepare for any reasonably likely emergency.

A realistic appraisal of the circumstances which can lead to emergencies in a laboratory will reveal many foreseeable and controllable problems. Some of the problems which can be expected to occur might include:

- Fires
- Chemical spills
- Release of radioactive materials
- Release of compressed toxic and corrosive gases
- Release of pathogens and restricted biological materials
- Power failure
- Explosions
- Physical injuries to individuals
- Consequences of natural disasters

This list is not intended to be complete. In fact, the two most serious and dangerous emergencies in which this editor has been involved would not necessarily have been included in the above list and they were definitely not foreseen, even though a comprehensive and, it was thought, exhaustive safety analysis had been prepared, evaluated, and approved by both the University Safety Committee and by outside regulatory agencies. The inability to foresee all possible emergencies should not inhibit the development of plans to cope with those which can be anticipated.

The scope of this chapter will be to provide the general principles of emergency preparedness to serve as a guide for the preparation of individual, specific action plans and to provide some useful information to be used in various classes of emergencies. Planning and preparation are necessary to insure that the response to laboratory emergencies is prompt, correct, and effective. Injuries and property damage can be limited if emergency procedures are established and practiced regularly and if adequate resources are available. Plans which are developed and then filed away are worse than useless. They can provide a false sense of security.

Before developing the theme indicated in the preceding paragraph, a note of caution needs to be interjected which is applicable to the contents of this entire chapter. There are advantages in not overreacting. In one of the two emergencies mentioned above, a relatively minor incident was turned into a major and expensive incident because well-meaning and knowledgeable individuals acted too quickly, without full awareness of the total situation and without consulting other persons involved. No serious worsening of the situation would have resulted in doing absolutely nothing until the situation had been discussed and a plan of action developed. This is often the case and it is exactly the situation which can be avoided by anticipation, planning, and practice.

One other note of caution: no one is expected in the normal course of their work to go to extreme measures, risking their own lives, in order to cope with an emergency. Indeed, the more responsible action in many cases is to leave the scene as soon as it is clear that the situation is beyond an individual's capabilities in order to insure that emergency response groups will have a competent source of information about the situation when they arrive.

Despite the preceding cautionary paragraphs, there are steps that individuals and local groups can and should take, when appropriate, to confine and minimize the impact of emergencies. The first few moments of an emergency are frequently the most crucial. However, actions should be based on training, knowledge, and a due regard for priorities. Protection of life and health should come before protection of property or reputation. Unfortunately, the fear of being found responsible for a problem has often resulted in serious repercussions, often allowing the situation to worsen until out of control.

## 2.1. COMPONENTS OF EMERGENCY PREPAREDNESS

Emergency preparedness should not be considered the responsibility of any one individual or group. In an actual emergency, there are key persons or groups who will usually be the ones to respond to the situation. However, basic conditions should have been met in order to facilitate meeting their responsibilities. Some of these conditions stem from before the building involved was constructed. In the initial planning, the building should have been designed by architects, in cooperation with the persons responsible for the programs to be housed in the building, to incorporate safety codes and regulations. Appropriate fixed and movable equipment must be installed or provided, consistent with the concept of a facility which could be operated safely. Emergency equipment must be available, but decisions must be made as to what design features and equipment should be mandatory, what is desirable, and what would be a luxury. Decisions must be made as to who is responsible for providing this equipment and, obviously, with this decision comes the need to determine the source of funds. If all of the correct decisions were made so that the building, the furnishings, and the appropriate emergency equipment are all available, then it is necessary to designate which individuals and groups should have the responsibility for emergency planning and emergency response. The responsibilities of these key individuals and groups must be delineated and boundaries established between local responsibility, institutional responsibility, and outside emergency response agencies. Finally, an emergency response plan should be adopted with specified responses to potential emergencies.

### 2.1.1. FACILITIES, FIXED AND MOVABLE EQUIPMENT

In order to facilitate the design and construction of safe buildings, fire and building codes have been established in most localities which govern new construction and renovations to existing buildings. Generally, under these codes, research laboratories come under the classification of a business use occupancy or possibly as a hazardous use occupancy, each of which incorporate different safety features. OSHA also has standards in the area of fire safety, as well as ventilation, which must be met. The OSHA standards are consistent in every state, but building codes vary from locality to locality, often depending upon interpretations of a local code official. For a number of other types of risk, special regulations, such as the classification system for recombinant DNA research facilities, also have safety restrictions which must be included in the building design. The latter set of safety restrictions will be reserved to later chapters dealing with these special topics.

Concerns which should be addressed in the design of laboratory buildings to enhance emergency response depend upon the classification. For example, if the building is a hazardous use occupancy, most codes will require a sprinkler or other fire suppression systems. If a sprinkler or alarm system is required by a local fire code, then OSHA 1910.37(m & n) requires maintenance and testing. Also, OSHA will require under 1910.37(f)(2) that the doors swing in the direction of exit travel, yet most building codes have restrictions on doors swinging into corridors to avoid creating obstructions to corridor traffic. In order to satisfy both requirements, doors should be recessed into alcoves inside the laboratory. To meet some code standards, even existing facilities may have to be upgraded. In the state of Virginia, for example, a regulation became law in 1986 — that existing academic buildings with four or more classrooms or occupied by 50 or more persons must have a code-conforming fire alarm system— which had previously been a requirement only for new construction. (Note, this requirement unfortunately has been repealed.)

The size of a building, the number of floors, and the relationship to other structures all enter into code decisions affecting safety in emergency situations. Addition of equipment to a laboratory, such as a hood, can have serious fire safety implications. Is there adequate make-up air? If not, where can it be obtained? Halls should not be used as a plenum or as a supply of make-up air for more than a few hundred cubic feet per minute (cfm) for each laboratory space. Even a small, 4-ft fume hood discharges about 800 cfm, so one cannot draw the required make-up air in through louvers in the door. Usually, one must go outside, but what is the relation of this new inlet air intake to the exhaust system? Toxic fumes could be drawn back into a building. A fume exhaust duct penetrating a floor could allow a fire to spread from one floor to another. Therefore, most codes require fume hood ducts to be enclosed in a fire-rated chase. Because of the expense of constructing a chase, the cost of avoiding worsening the fire separation in a building could preclude installation of the hood, which, in turn, could preclude using the space for the intended research.

The interior arrangements of a laboratory are critical in permitting safe evacuation from the laboratory. In Section 2.0, six of the eight types of accidents listed could be much more serious should they occur between an individual and the exit from the room. A simple solution for these potential emergencies for larger laboratories is to have two well-separated exits. This is not always possible, so, alternatively, the solution is to evaluate what components of a laboratory are most likely to be involved in an incident and which would increase the hazard if it became involved in an ongoing emergency. These components should be located so that an escape route from the normal work area does not pass by them. Also, portable fire extinguishers, fire blankets, respirators, and other emergency equipment should be located on this same escape route. Eyewash stations and deluge showers should be located close to where injuries are likely to occur so an individual will not have to move substantial distances while in intense pain or blinded. Aisles should be wide (typically a minimum of 42 to 48 in.), straight, and uncluttered with excess equipment to facilitate movement in emergencies. A laboratory should have emergency lighting, but many do not. The considerable dangers posed to an individual stumbling around in a pitch dark laboratory should the power fail are obvious.

Many of the regulations found in OSHA standards include features which will minimize the scope and impact of an emergency such as a fire. For example, restrictions in 1910.106 on container sizes of flammable liquids and the amounts of these materials which are permitted to be stored outside flammable material storage cabinets are designed to limit the amount of fuel available to a fire and to extend the time before the material could become involved.

Every action should be considered in terms of what would ensue if the worst happened. In large projects, this is often part of a formal hazard analysis, but this concept should be extended to virtually every decision. For example, a common piece of equipment found in most laboratories is a refrigerator. A unit suitable for storing flammables, i.e., containing no internal sources of ignition, costs about twice as much as a similar unit designed for home use. It is tempting, especially if money is tight and the immediate need does not require storage of flammables, to save the difference. However, the average lifetime of a refrigeration unit is on the order of 15 to 20 years, and who can say what materials research programs will entail over such a long period? If an internal explosion should occur, the employees could be injured or killed and the laboratory, the building, and the product of years of research could be destroyed.

Many actions are influenced by the costs involved, as in the preceding example. A continuing question involves who should be responsible for paying for safety facilities and equipment. There are some straightforward guidelines which can be used:

1.  For new construction, safety should be integrated into the design and the choice of all fixed equipment, which should be incorporated in the building furniture and equipment package. This would include such major items as fume hoods since these are relatively expensive units to retrofit.

2.  Certain equipment and operational items common to the entire organization, e.g., fire extinguishers, emergency lighting, deluge showers, eyewash stations, fire alarm systems, and maintenance of these items should be just as much an institutional responsibility as provision of utilities.

3.  Items which are the result of operations unique to the individual laboratory or operations should be a local responsibility. This would include equipment such as flammable material refrigeration units, flammable material storage cabinets (if these are not built in), and specialized safety equipment such as radiation monitors, gas monitors, etc. This could include some major items which might be included under fixed equipment in new construction, if renovation of a space were to be involved. For example, it might be necessary to construct a shaft to enclose a fume hood duct and to provide a source of additional makeup air for the hood. Personal protective equipment such as goggles, face masks, respirators, gloves, and similar items should also be provided at either the laboratory or departmental level.

It is unlikely that any individual, whether it is the laboratory supervisor, safety professional, planner, or architect, will, alone, be sufficiently knowledgeable or have the requisite skills to make appropriate decisions for all of the factors discussed in this section. In addition, every one of these persons will have their own agenda. The inclusion of emergency preparedness features should be explicitly included as one of the charges to the building or project design committee so that these needs can be integrated with function, efficiency, esthetics, and cost.

It was not the intent at this point to elaborate on all the implications of codes as safety issues, but, rather, through the use of a few examples, to draw attention to the idea that the root cause of an emergency and the potential for successfully dealing with it could well lie with decisions made years earlier. The point that was intended to be made was that laboratory safety and the capability to respond to emergencies does not start with teaching good laboratory technique and the adoption of an emergency response plan after beginning operations.

## 2.2. INSTITUTIONAL OR CORPORATE EMERGENCY COMMITTEE

In most organizations, there are a number of support groups which have been assigned specific responsibilities for dealing with emergencies which extend beyond those associated only with laboratories. Among these are safety, police or security, maintenance, communications, and media or public relations. Unlike the laboratory supervisor, departmental chairman, or individual laboratory employee, who are primarily concerned with their research or administrative duties, these groups are directly concerned with one or more aspects of emergency response. In larger organizations, fire departments, physicians, medical services, or even more specialized groups may exist. Each of these groups have their own expertise, their own dedicated resources, and their own contacts with outside agencies. Representatives from these agencies will be the ones called immediately to the scene of an emergency and will usually be the ones expected to cope with the situation. This group should form the nucleus of the emergency planning committee. It should have input from the remainder of the organization and, in the current context, this input should include comprehensive coverage of the various areas of the corporate or institutional research programs. The committee should have direct access to upper levels of management and it should also interact closely with safety committees associated with each broad research area, e.g., chemical, radiation, biosafety, and animal care.

The emergency committee should meet periodically, at least once a year and preferably more often, to review the status of the organization's emergency preparedness, to plan for practice sessions, to review drills that have been conducted, and to investigate and review incidents that occur. Reports of these meetings, along with the findings, should be presented to management and to the individual safety committees.

## 2.3. EMERGENCY PLAN

The initial order of business for the emergency committee is to develop an emergency response plan (ERP). In developing the ERP, the committee should analyze the types of emergencies which could happen, their relative seriousness, and their relative probability of occurrence. The emergencies to be considered should specifically include releases of chemicals to the environment, as required under SARA, Title III. Once the classes of emergencies have been defined, each should be analyzed as to the resources, equipment, training, and manpower which would be needed for an adequate response. An integral part of this analysis would be provisional plans for using these resources to respond to potential emergencies. The analysis should include both internal and external resources. Finally, a critical evaluation should be made of the current status of the institutional resources and a recommendation made to correct deficiencies. Based on the preliminary studies, the final plan should be drafted, circulated for review, amended if required, and implemented. The support of management is critical or this exercise will be futile.

The actual plan should be operative at two levels. The first level which should be widely publicized, should be short, easy to grasp and to implement. No one has the time to search through an elaborate document when faced with a pressing emergency. Detailed plans have their uses for organized emergency groups but, for the use of the general public, a basic emergency plan is to evacuate the area or building, and call for emergency help. Often, evacuation will be more than is actually needed but it is a conservative and safe approach. The essential information to enable this can be placed on a single page for a facility, although if the evacuation needs to involve a larger area, this will involve major and comprehensive planning. Normally, planning for large-scale emergencies will be the responsibility of the corporate or institutional Emergency Coordinator, working with internal groups and the Local Emergency Planning Committee (required under SARA, Title III) and nearby support agencies.

At the second level, all of the groups likely to be involved in the emergency response should possess a copy and be familiar with the organization's emergency response manual, which spells out in detail, but still as simply and as flexibly as

FOR

FIRE

OR

OTHER

EMERGENCIES

CALL

911

FIGURE 2.1.   Emergency poster.

possible, the correct response to the classes of emergencies incorporated in the ERP. The ERP should include a staged response — a first-response protocol to ensure the maximum safety of the largest number of persons, with subsequent options to confine the emergency as rapidly as possible— to begin abating the situation and to terminate the problem. Finally, every plan should contain provisions for an evaluation after the conclusion of the emergency response to determine not only the cause of the problem and the measures needed to prevent similar incidents from reoccurring, but also to critique the way it was handled.

It is always the intent of every organization that no emergency will ever occur and for the more unusual situations considered in the ERP, long intervals may pass between incidents. However, it is essential to include in every emergency plan provisions for periodic review and practice.

### 2.3.1. EMERGENCY PLAN

**A.** The simplest possible plan would be to prominently post in every room an eye-catching sign similar to that shown in Figure 2.1. In serious emergencies, this is sufficient and entirely appropriate. However, it is too simplistic for most laboratory situations. In laboratory situations, where knowledgeable persons are usually present, it is often possible and even desirable to include a substantial amount of self-help and local actions to limit and correct the emergency instead of immediate evacuation.

However, guidelines need to be incorporated in the publicized plan to limit the local effort for the good of the individuals directly involved and the greater good of others.

Many situations can be handled by calling specific persons or groups for advice or aid. As a minimum, a list should be posted in every laboratory of some important telephone numbers, including:

- Emergency telephone number — 911 if available in the area
- University police or corporate security, if not available through 911 number
- Local government police, if not available through 911 number
- Fire department number, if not available through 911 number
- Emergency medical care (rescue squad), if not available through 911 number
- Nearest poison control center
- Nearest hospital
- Safety department
- Spill control group, if not available through 911 dispatcher or safety department
- Maintenance department number(s)
- Laboratory supervisor business and home telephone number
- Secondary laboratory authorities business and home telephone numbers
- Departmental or building authority numbers

These same numbers should be posted on the outside of each laboratory so that they would be available in the event of an incident requiring immediate evacuation of the laboratory area. The assistance available through these resources is likely to be needed in any emergency and does not significantly depend upon the activities in a given laboratory.

The last three sets of numbers should also be maintained in a master list maintained by the organization's security and safety groups and should be immediately available at their central offices.

In addition to the structured internal departments, a major resource available at any research-oriented institution are the scientists and technicians who work there. The ones most likely to be helpful for the types of emergencies anticipated in developing the emergency plan should be identified and a master list of their office and home telephone numbers maintained. A copy of the current list should be maintained by the key internal organizations involved in the emergency response plan. A copy of the list should also be personally maintained by the key individuals in the latter organizations, both in their offices and at home. Alternates should always be designated for these key persons so that backups are available at all times. A beeper system to allow these key persons to be reached when not at their usual locations would be highly desirable.

A library of reference materials should be maintained for the use of the emergency responders. The following is a short summary of some of the more useful references, many of which are revised frequently. Although these are mainly printed books, today, primarily as a result of information needs evoked by the OSHA Hazard Communication Standard, a number of other types of information sources are becoming widely available for chemical products. An example of this, included in the list, are material safety data sheets, available directly from the chemical product manufacturer. These are provided when the chemical is first purchased and when significant new information becomes available. Compilations of these are sold as hard-bound or loose-leaf volumes, on microfiche, or as computer software programs which are suitable for either mainframe computers or minicomputers. In addition, a number of firms provide newsletters and reference services.

- ACGIH, American Conference of Industrial Hygienists— Threshold Limit Values (TLV) for Chemical and Physical Substances
  P.O. Box 1937
  Cincinnati, OH 45201
- Chemical Hazards Response Information (CHRIS manuals, 4 books)
  U.S. Coast Guard
  400 7th St., S.W.
  Washington, D.C. 20590
- Dangerous Properties of Industrial Materials
  Sax, Irving
  Van Nostrand Reinhold
  New York, NY 10602
- Department of Transportation Emergency Response Guidebook, DOT Pub. No. P5800.3
  Materials Transportation Bureau
  Research and Special Programs Administration
  U.S. Department of Transportation
  Washington, D.C. 20590
- Effects of Exposure to Toxic Gases, First Aid & Medical Treatment
  Matheson Gas Products
  P.O. Box 85
  East Rutherford, NJ 07073
- Emergency Medical Treatment for Poisoning
  National Poison Center Network
  125 Desoto St.
  Pittsburgh, PA 15213
- Farm Chemicals Handbook
  Meister Publishing
  Willoughby, OH 44094
- Fire Prevention Guide on Hazardous Materials, National Fire Protection Association (NFPA)
  Batterymarch Park
  Quincy, MA 02269
- First Aid Manual for Chemical Accidents
  Lefevre, Marc J.
  Dowden, Hutchinson & Ross
  Stroudsburg, PA 18360
- Guide to Safe Handling of Compressed Gases
  Matheson Gas Products
  P.O. Box 85
  East Rutherford, NJ 07073

- Hazardous Materials
    Department of Transportation
    Office of Secretary Transportation
    Washington, D.C. 20590
- Hazardous Materials Emergency Planning Guide— NTR-1
    Hazmat Planning Guide WH-562A
    401 M St., S.W.
    Washington, D.C. 20460
- Material safety data sheets master file for chemicals in use at institution
    (Available from chemical manufacturer or generic data base)
- Merck Index
    Merck & Company
    Rahway, NJ 07065
- NIOSH/OSHA Pocket Guide to Chemical Hazards, DHHS(NIOSH) Pub. No
    78-210
    U.S. Government Printing Office
    Washington, DC 20402
- Physicians Desk Reference
    Medical Economics Company
    Oradell, NJ 07649
- Prudent Practices for Handling Hazardous Chemicals in Laboratories
    National Academy Press
    2101 Constitution Ave., N.W.
    Washington, D.C. 20418
- Handbook of Chemistry and Physics
    CRC Press
    2000 Corporate Blvd. NW
    Boca Raton, FL 33431
- Laboratory Safety, Principles and Practice
    American Society for Microbiology
    1913 I St., N.W.
    Washington, D.C. 20006

**B.** Internal resources will not always be sufficient to handle an emergency. Therefore, a list of external emergency organizations should be maintained by the organizational emergency groups as well. The following are among the most likely to be useful and available. Any others that might be useful to you and are available should be identified and added to the list. Currently available telephone numbers are given in some cases. These are subject to change and should be verified before incorporating them in a plan.

- Regional emergency group/coordinator
- Arson and/or bomb squad, if not otherwise identified
- Civil defense coordinator, if not otherwise identified
- Commercial analytical laboratories
- Commercial environmental emergency response firms

- Law enforcement organizations, e.g., city or county police chief or sheriff, state police, FBI
- Center for Disease Control, phone no. 404-633-5313
- CHEMTREC (for chemical and pesticide spills), phone no. 800-424-9300
- Compressed Gas Association, phone no. 212-354-1130
- Manufacturing Chemists Association, phone no. 202-483-6126
- National Fire Prevention Association, phone no. 617-328-9290
- National Response Center (USCG and EPA), phone no. 800-424-8802
- Nuclear Regulatory Commission, phone no. 301-492-7000
  (also state or regional federal office)
- Occupational Safety and Health Administration, phone no. 202-245-3045
  (also state or regional federal office)
- Poison Control Center, phone no. 502-432-9516

Many of these are sources of information only and will not provide actual assistance for the emergency response. The ones which are likely to have the capability to do so are the first six. However, the two commercial groups listed represent profit making organizations and the institution or corporation must be willing to pay for their services. Therefore, provision must be made for one or more members of the internal emergency organization to have the authority to commit at least some funds for the emergency response.

### 2.3.2. EMERGENCY EQUIPMENT

Another important step in preparing for an emergency is acquiring appropriate equipment, which is kept readily available for use. Some of this should be located in the laboratory area and every laboratory should be furnished with it. Other equipment, because of the cost and relatively rare occasions when it is likely to be needed, should be maintained at a central location. Even at the central location, the equipment maintained needs to be realistically selected. It is neither necessary nor desirable for every organization to maintain an expensive, fully equipped hazardous-material emergency response team. Some very large organizations may find them essential, but most institutions will not be able to justify the cost.

Some of the emergency equipment needs to be built in as part of the fixed equipment in the laboratory. Included in this group are the following items:

**Eyewash stations** — At least one of these should be in every laboratory and placed in an easily accessible location. It should be mounted on a plumbed water line rather than the small squeeze bottles that are sometimes used for the purpose. The bottles do not contain enough water to be effective. Since cold water can be uncomfortable to the eye, if possible, the eyewash supply should have a holding tank to insure that the water is at least near room temperature. During the academic year in many of the colder areas of the country, tap water is frequently well below room temperature.

**Deluge shower** — Eyewash stations and deluge showers ideally should be installed as a unit. Although the eyes are the most critical organs, chemicals splashed on the face may also splash on the body. A deluge shower should be capable of delivering about 1 gal/sec, with a water pressure of 20 to 50 psi. A common error is to plumb the unit into too small a line, which is incapable of delivering an adequate flow. The water should flow through at least a 1-in. line. Although a floor drain is desirable, it is not essential. One can always mop up afterwards.* However, care must be taken to insure that water from the shower cannot come into contact with electrical wiring. Again, the units should always be placed in an easily accessible location. Care is essential to maintain clear accessibility.

**Fire extinguishers** — OSHA requires that every flammable material storage area be equipped with a portable class B fire extinguisher. The standard does not specify the amount of flammable material which makes a room a storage facility, so, in effect, most laboratories face the need to comply with the standard. The unit should be at least a 12-lb unit and it should not be necessary to travel more than 25 ft to reach it from any point in the laboratory. Class B extinguishers are, of course, intended for flammable solvents. Other classes of fire extinguishers are class A, intended for combustible solid materials such as paper or wood, class C, where electrically live equipment is involved, and class D, where reactive metals such as sodium, are used. Combination units such as AB or ABC are available which, although not equally effective for all types of fires, can be used where mixed fuels are involved. More information on fire extinguishers will be found in a later section.

**Fire blanket** — A fire blanket is a desirable unit to have permanently mounted in a laboratory. These are usually installed in a vertical orientation so that users need only grasp the handle and roll themselves up in it in order to smother the fire. Some blankets include asbestos in their manufacture; these should not be installed and existing units replaced. The concern is that they could become a source of airborne asbestos fibers, which have known carcinogenic properties. It is likely that in the not too distant future, any blanket containing asbestos will be required to be removed. Unfortunately, heavy woolen blankets which are used as alternatives often are likely to be stolen.

**Emergency lights** — Emergency light must be provided by some mechanism. One alternative is to have two sources of power to the lighting circuits in a building. This can be a second source external to the building or secondary power sources within the building. There are several alternative types of internal power sources, including emergency generators, large uninteruptible power supplies to provide power for lights for a substantial area which depend on batteries to provide power for a fairly limited interval, and individual, trickle-charged, battery-powered lights in individual laboratories.

**First aid kit** — One of these needs to be in every laboratory and, again, because they can be taken, should be permanently mounted to a wall. They should be relatively small units. Packaged units are sold which are adequate for five or six persons. There is little value in having larger units since in the event of an emergency involving more persons, help will be needed from professional groups. Present in the kits should be a variety of bandages, adhesive tapes, alcohol swabs, gauze, and a few cold packs. Absent should be such items as iodine, merthiolate, and tourniquets. It is essential that a maintenance program be established to insure that the kit is always

---

* There should be a timed cutoff, however, at about 15 to 20 min, after which the unit would need to be reactivated. An instance was recently brought to the author's attention where, as an act of vandalism, a deluge shower was activated and rigged so that it would continue to run. Before it was discovered, over 100,000 gal of water flooded the facility. The unit was in the hall outside the laboratory.

adequately supplied. It is all too easy to use up the supplies without replenishing them.

**Fire alarm pull station** — The location of the nearest pull station should be familiar to everyone in the laboratory.

**Special safety equipment** — There are many specialized research areas which require special safety items. These may include, for example, explosion-proof wiring, combustible gas monitors, and explosion venting for laboratories working with highly explosive gases. The possibilities are too many to dwell on at this point.

Some emergency equipment need not be built in, but should be available. Among these items are the following:

**Absorptive material** — Probably the most common laboratory accident is a spill from a beaker or a chemical container. The volume is typically fairly small, rarely exceeding more than 4 or 5 l and usually much less. Of course, there are spills which would require immediate evacuation of the area or even the building, but more frequently the spilled material simply must be contained and cleaned up as quickly as possible. **THIS IS NOT THE RESPONSIBILITY OF THE CUSTODIAL STAFF.** They are not trained to do it properly or safely. Spill kit packages are available commercially to neutralize acids and bases and to absorb solvents or mercury. Although it is possible to put together similar packages, the commercial packages are convenient to obtain and store. After being used, the materials should be collected and disposed of as hazardous waste.

**Janitorial supplies** — Several miscellaneous items are needed to clean up an area. Among these are plastic and metal buckets, mops, brooms, dust pans, large heavy-duty polyethylene bags, kraft paper boxes (for broken glass), plastic coated coveralls, shoe covers, duct tape, and an assortment of gloves. If not kept in an individual laboratory, at least one set should be kept on each hall or floor of a building.

**Respirators** — Fumes and vapors from many irritating and dangerous materials can be protected against by the use of respirators with appropriate cartridges or filters. If operations are sufficiently standardized so that a standard respirator combination would be effective, they should be kept in an emergency kit. However, cartridge respirators are not intended for protection against materials which are immediately dangerous to life and health (IDLH). Respirators should be assigned to specific individuals.

**Supplied air escape units** — Supplied air units such as emergency squads might use are expensive and require a significant level of training to be able to put them on quickly and use them properly. However, small air-supplied units are available at very reasonable prices which only need to be pulled over one's head and activated to provide 5 min of air. This is usually sufficient time in which to escape the immediate area of an accident.

Virtually any small to moderate chemical emergency can be handled with the equipment described above.

A few major items of equipment should be readily available from the fire department, security force, or perhaps the emergency medical team. This is by no means certain and the institution or corporation should maintain a set of these major items. Many of these items require special training to be able to use safely.

**Oxygen meter** — A portable meter should be available to insure that the oxygen level is above the acceptable limit of 19.5%. It is important to be able to detect oxygen-deficient atmospheres where the levels are significantly less than the acceptable level.

**Combustible gas and toxic fume testing equipment** — A number of different types of equipment are sold to test for the presence of toxic fumes. A common type, frequently combined with an oxygen meter, is a device to detect combustible gases. Other specialized units are built to detect other gases such as carbon monoxide and hydrogen sulfide. Very elaborate and, consequently, expensive units, such as portable infrared spectrometers, gas chromatographs, and atomic absorption units, can detect a much greater variety of chemicals, often to very low concentrations. A less expensive alternative is a hand pump, used to pull known quantities of air through detector tubes containing chemicals selected to undergo a color change upon exposure to a specific chemical. All of these can be used to obtain an instantaneous or "grab" reading. Where a longer duration sample is desired, powered pumps can be used to collect samples, and for some chemicals, passive dosimeters can be worn which can be analyzed in a laboratory. Equipment selected to meet local needs should be selected.

**Supplied air breathing units** — These are not to be confused with the escape units previously described and usually will be available from the fire department. There are two types, one of which provides air from a compressed air tank, as does a SCUBA outfit. The most common-sized tank is rated at 30 min, which, under conditions of heavy exertion, may last only 20 min or less. The second type uses pure oxygen which is recirculated through a chemical scrubber to extend the life of the supply to 1 h, or longer for some units. The first of these two types have been available for a longer period and more emergency personnel have been trained to use them. The second offers a significantly longer working interval. This could be very important. Pairs of either type should be owned so that in the event an individual entering an emergency area is overcome, it would be possible to effect a rescue.

**Fire resistant suits** — Special fire-resistant suits are needed to enter burning areas. There are different grades of these which provide varying degrees of protection from fire. Some protect against steam or hot liquids as well. They normally require a self-contained air system to be worn during use.

**Chemical-resistant suits** — Protection is frequently needed in chemistry incidents for protection against corrosive liquids and vapors. In standardized situations, these can be custom selected for maximum protection against the specific chemicals of concern. Where a variety of chemicals such as acids, bases, and frequently used solvents are involved, a butyl rubber suit is often a reasonable choice. Combination units of these and fire-entry suits are available.

**Clean air supply system** — An alternative to self-contained air or oxygen tanks is a compressor system capable of delivering clean air through hoses from outside the area involved in the incident. Persons inside the work area would wear masks connected to the system.

**High-efficiency particulate and aerosol (HEPA)-filtered vacuum cleaner** — Ordinary vacuum cleaners, including wet-shop vacuums, do not remove very small particulates from the air. They remove larger particles, but the smaller ones pass through the internal container or filter and return to the room. In several instances, this has actually worsened the situation. For example, droplets from a mercury spill have been disbursed back into the air in the form of much smaller droplets and caused the mercury vapor pressure in the air to increase (a mercury vacuum has additional special requirements). In another actual case, in a carpeted room where large quantities of forms and computer paper were processed, vacuuming with an ordinary vacuum cleaner during normal working hours increased the number of respirable particles suspended in the air to such a level that several individuals who were allergic to dust had to be sent out of the area. HEPA filters will remove 0.9997 of all particles from the air which have a diameter of 0.3 μm or greater.

**Radios** — Radio communication between persons entering an accident area and those outside is highly desirable. Emergency groups will have portable radios with frequencies specifically assigned to them.

**Fire-suppressant materials** — In addition to water and the usual materials available in portable fire extinguishers, most fire departments now have available foam generators which can be used to saturate a fire area.

**Containment materials** — In order to prevent the spread of large amounts of liquid chemicals, a supply of diking materials needs to be maintained. Ready access to a supply of bales of straw is a great asset. Straw is cheap, easily handled, and easy to clean up afterwards. In the event of a spill reaching a stream, floating booms and skimmers are useful in containing and cleaning up the spill.

**Radiation emergency** — Many laboratories use radioisotopes. For emergencies involving these units, in addition to the other emergency equipment, radiation survey instruments should be maintained in an emergency kit separate from those used in normal daily activities. These should include instruments capable of detecting both gammas and low-energy betas to low levels as well as high-range instruments.

**Miscellaneous clothing** — Items needed include a variety of coveralls, including chemically resistant suits in a range of grades, and disposable Tyvek™ coveralls, gloves with different chemical resistances, regular work gloves, Kevlar™ or Zetex™ gloves for hot use, rubber and neoprene boots and shoe covers, head covers, hard hats, chemical splash goggles, safety glasses, and masks.

**Miscellaneous tools and paraphernalia** — A variety of small tools will be needed as well as shovels, pickaxes, axes, rope, flares, emergency lights, sawhorses, a bullhorn, a chainsaw, a metal cutting saw, a bolt cutter, and a "jaws of life" metal spreader.

**Victim protection** — In equipping an emergency kit, the emphasis is usually on protecting the emergency response personnel. In order to bring a victim out through a fire or chemically dangerous area, blankets, disposable, coated Tyvek™ overalls, loose-fitting chemically resistant gloves, and the 5-min escape air units should be available.

All of the equipment listed in this section must be maintained properly and a definite maintenance schedule must be established. For example, the integrity of the chemical protective suits must be verified on a 6-month schedule. A maintenance log must be kept in order to confirm that the maintenance program has been done on schedule.

Note that nowhere in this list of equipment is there mention of a fire hose. Although standards are provided in OSHA for fire brigades, in general, if a fire is sufficiently large to require a fire hose to control it, it is usually too large for anyone except professionals. Although building codes frequently require installation of 1.5-in emergency hose connections, in many instances building officials encourage the owners to request a variance to permit this requirement to be deleted. Many fire departments also question the value of such connections. Those institutions or corporations that choose to establish a fire brigade will need to provide training beyond the scope of this book.

### 2.3.3. BASIC EMERGENCY PROCEDURES

A list of several common types of emergencies that might occur in a laboratory was given in the introductory section to this chapter. There are some standard response procedures which will be common to all of these types of emergencies, as well as others, which are important to do **FIRST**.

1.  Make sure everyone in the immediate vicinity is aware of the problem. In a busy, active laboratory, an accident can occur in one part of the laboratory and others within the same laboratory could be temporarily unaware of the event. This is especially critical if the laboratory area is subdivided into multiple rooms.

2.  Confine the emergency, if reasonably achievable. Many emergencies can be readily confined if quick action is taken. Small quantities of a spilled chemical can be contained with absorbent materials or toweling by the persons directly involved if the chemical is not IDLH before it spreads too far or before it catches on fire. Small fires can and should be put out with portable fire extinguishers, but a very serious question of judgement is involved. What, precisely, is a small fire? Unless there is a reasonable certainty that the fire can be controlled, then evacuation of the building should be strongly considered and implemented as soon as the situation appears to be deteriorating. If more than one person is available, there may be more flexibility. One or more persons may attempt to contain the fire while others are taking initial steps to cause the building to be evacuated. Where it is necessary to evacuate an area larger than a single laboratory, definite procedures should be established to insure that all spaces are checked, including restrooms, janitors closets, etc.

    Evacuation is a conservative step and, whenever any doubt exists as to the severity of the situation at hand, should be implemented. It is inconvenient and disruptive to work activities, but the alternative is far worse if the fire cannot be controlled. The first few minutes of a fire are very important and any significant delay can make the job of the fire department much more difficult. Even if the fire is out before they arrive, there are things that the fire department needs to do. They need to check the area to insure that it is really out. They need to determine the cause of the fire in order to prepare an accurate report. The information they obtain will be needed to determine how to prevent subsequent fires due to the same cause and also usually will be needed by the insurer of the property.

3.  Evacuate the building. Whenever the situation is obviously serious, such as a major fire, a moderate to large spill of an IDLH material, or large spills of ordinarily dangerous materials, such as strong acids, then evacuation procedures for the area or the building must be initiated as soon as possible. Any measures taken in such a case to confine the emergency situation should provide extra time for the evacuation to be carried out safely.

Normally, primary evacuation routes from an area within a building should follow the shortest and most direct route, along corridors designed and constructed to meet standards for exitways. However, since in an emergency any given path may be blocked, alternate secondary routes should be designated. In no instance should an evacuation plan include elevators as part of the evacuation procedure, even for the handicapped. In the case of a fire, elevators should be designed to immediately go to the ground floor and be interlocked to stay there until the danger is over.

In any evacuation procedure, provisions for closing down operations should be included, presuming there is sufficient time to implement them. Gas should be turned off, along with electric and other types of heaters. Consideration should be given to closing valves on gas cylinders and turning off electrical apparatus. Closing sashes on fume hoods may be desirable. Certainly, any flammable material storage cabinets should be closed.

Even in the worst situation, there are some simple things which can be done by individuals evacuating the building to confine and minimize the emergency. The highest priority is to protect personnel, so the first thing is to actuate the building alarm, assuming that one exists. If not, air horns should be used or, failing that, then verbal warning. Doors to the laboratory should be closed on the way out. Doors between floors should be closed behind those evacuating. If the building has been built according to code, as briefly discussed earlier in this chapter, then these last two simple steps can significantly retard the spread of a fire or spread of fumes.

If a laboratory is under negative pressure, as most should be, then this will also tend to confine the emergency to a single room. In order to maintain a negative pressure, it may be desirable to leave the sash of a hood open or to leave the hood working, even though there might appear to be concern about the fire spreading through the hood duct. If a hood has been installed properly, the exhaust will be at a negative pressure with respect to the space surrounding the exhaust duct and, as noted earlier, is either going directly outside without passing through an intervening floor level or is enclosed in a fire-rated chase. Under these conditions, fire being drawn through a hood exhaust should not cause fire to spread to other floors. The door to the room being closed will further reduce the amount of fresh air available to support a fire. If an exhaust is turned off, any air intake should also be turned off to avoid creating a positive pressure in the room and thus possibly causing extension of the emergency by leakage into corridors.

Evacuation should be done as quickly as possible, but in such a way as to not engender a panic situation. This can best be achieved by having it be a frequently practiced procedure so that everyone is familiar with the routes. In a corporate situation with a stable personnel complement in the building, drills two or three times a year will quickly accomplish the purpose. In an academic environment, the problem is much more complicated. In most colleges and universities, as many as 12 classes per d may be held in the same classroom. Classrooms may be assigned by some central authority. This may result, for example, in a professor of economics being assigned a class in a chemistry building for one quarter during a year and no classes in that building for the

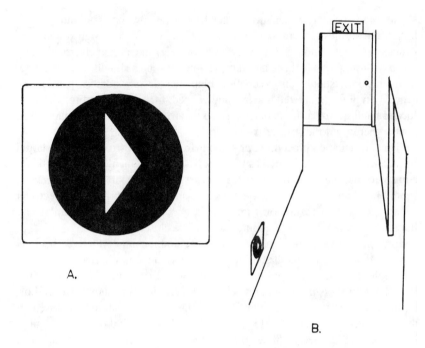

FIGURE 2.2.   (A) Emergency evacuation direction sign; (B) sign opposite each room door.

remainder of the year. During the course of an academic year, the population in the building may, in large part, change every quarter or semester. Because of all these complicating factors, a single drill per academic session could prepare as little as 10% of the population in a building for an actual emergency. Under these circumstances, unusual care should be taken to clearly mark evacuation routes from buildings and to train those individuals who form the "permanent" population in the building to take charge during an evacuation.

A simple but very effective evacuation system is illustrated in Figure 2.2A and B. A distinctive, high contrast, standardized symbol employed only for marking primary evacuation routes is placed on the corridor wall directly across from every door opening onto the corridor at appropriate intervals (about 50 ft) along the corridor and at every branching point along the path of egress. A person totally unfamiliar with a building need only follow the symbols to be conducted to the nearest exit. Since smoke tends to rise, these should be placed a short distance off the floor so that they will remain visible when signs placed above doors might be obscured. Power can fail, even in buildings equipped with separate emergency power for lights, so the directional symbol can be made with a phosphorescent paint which will remain visible for 2 or more hours, this being ample time to evacuate almost any building. Printing of the signs on a fragile substrate which cannot be removed intact will minimize theft of the signs for use as decorations in dormitories and residence halls. This system is used as a supplement to a code-conforming system of exit lights, rather than a substitute.

A standard part of any emergency evacuation includes a prechosen point of assembly for those evacuating. This should be a location which is generally upwind from the building being evacuated. Obviously, the wind does not always blow from the same direction, so alternative locations should be chosen. Those individuals most directly involved with the emergency and presumably the most aware of the circumstances should remain at the evacuation location and make themselves known to the emergency response groups when they arrive in order to assist these groups.

There should be a clearly defined line of communication among the persons responsible for the facility. Individuals in authority and with assigned responsibility for the space involved should also go to the assembly point and remain available to assist emergency personnel in managing the evacuees. If other senior corporate or institutional offficials involve themselves, they should inform the building authorities that they have arrived to assume responsibility. Procedures should be employed to account for the individuals who are known to have been in the building and reports of any persons still believed to be in the building should be made to the authority figures to pass on to the emergency responders. Any decision to allow reentrance to the building approved by the fire department or other emergency group should be disseminated by these persons.

The law currently requires access to most areas to be available to the handicapped. Often, this involves making spaces above the ground floor available by elevator, which should not be used in an emergency evacuation. Provisions should be made to assist the handicapped to evacuate a facility using the normal routes of egress. This could involve designating specific individuals or groups of individuals to provide this assistance.

4.  Summon aid. Unless the building alarm system is connected to a central station, pulling the alarm will not alert any external agencies. Moreover, the correct kind of assistance might not be obtained even if it does go to a central station. In most emergencies, the sooner the responding agency arrives at the scene, the more effective it can be at controlling the situation. Thus, it is imperative that requests for assistance be initiated quickly. However, in a serious, life-threatening emergency, evacuation should not be delayed to call for assistance. Calls can be made from a point outside the area affected by the emergency. Where personnel are available, one individual can be designated to make the appropriate telephone calls while others are engaged in other aspects of the emergency response.

A key group in responding to most institutional or corporate emergencies is the police department. If they are not the group contacted, it is probable that they monitor the emergency radio frequencies and will arrive either at nearly the same time as the emergency group summoned or even before. They should have training in a number of key areas, such as how to use fire extinguishers most effectively and how to give first aid and CPR, and should have the capability of independently causing a building to be evacuated. Once the fire department, rescue squad, or other emergency group arrives and assumes responsibility for their duties, the police are needed for crowd control and communications.

In any type of incident in which a spill of a hazardous material has occurred, the standard procedure is to establish a command center outside the periphery of the area affected by the accident and establish a controlled access point for the decontamination personnel entering the area. All materials and personnel entering and leaving the area should pass through the control point. Unless there are overriding considerations, decontamination should begin at the periphery and the work program designed to progressively constrict the affected area. Everything collected at the control point, including materials, contaminated clothing, and equipment which cannot be cleaned and reclaimed, should be immediately packaged for disposal according to standards applicable to the contaminant. Information and status reports should flow to the command point and overall direction of the response should come from the command center. To be effective, the emergency response needs central coordination and a clearly defined chain of command.

### 2.3.4. EMERGENCY PROCEDURES FOR SELECTED EMERGENCIES
### 1. Spills
A chemical spill is probably the most common accident in the laboratory and in most cases can be cleaned up by laboratory personnel with minimal effort or risk. According to the requirements of the OSHA Hazard Communication Standard, laboratory personnel should have been trained in the risks associated with the chemicals with which they are working and should know if it would be safe to clean up a minor spill. Workers should be especially sure to be familiar with the risks and the corrective actions to be taken in an emergency for chemicals labeled on the container "DANGER" or "WARNING".

Paper towels or absorbent and/or neutralizing materials can be used to clean up minor spills, with the residue being placed in an appropriate container for later disposal. In order to avoid decontamination of work surfaces, it is often convenient to protect them with plastic-backed absorbent paper. Relatively few materials associated with a cleanup should be placed in the trash. Most should be disposed of as hazardous waste. Chemical containers should not be placed in ordinary trash for disposal. In one actual instance, two different, partially empty containers placed in the ordinary trash in two separate buildings combined in the trash vehicle and caused the worker on the truck to be overcome with fumes, requiring emergency medical treatment. The nature of the fumes was unknown until later, when it was possible to retrieve the containers. It should be a general policy that no chemical container should be disposed of in the general trash.

Even small spills can often be dangerous if the spilled chemical comes into contact with a person's skin or gets into the eye. Strong acids and bases, as is well known, can cause serious chemical burns to tissue, but chemicals can cause serious injury by absorption through the skin as well. For example, phenol is readily absorbed through the skin and in relatively small quantities is quite toxic. Vapors from some spilled materials are IDLH by inhalation, even in small quantities. Obviously, work with such materials should be done in a hood where the sash can be lowered should a spill occur. However, if an accident occurs outside a hood, the area should be evacuated, the door to the laboratory closed, and help sought from persons trained and equipped to cope with such dangerous materials. In some instances, if a material is sufficiently volatile to give off enough vapors to be dangerous from a small spill, it often is sufficiently volatile to evaporate quickly and evacuation will allow time alone to effect a remedy.

In most cases, flushing the area of the body affected by a splash of liquid chemicals with copious amounts of low-pressure water for 15 to 30 min is the best treatment. The best source would be an eyewash station or deluge shower, but in an emergency, if these are not available, any other source of running water should be used. If the exposure is to the eyes, check for contact lenses and remove them if found, although contact lenses should never be worn in a chemical laboratory, and hold the eyes open while they are being flushed with water. Any clothing or jewelry in the affected area should be removed to ensure thorough cleansing. No neutralizing agents should be employed. If the original exposure was due to a dry chemical, loose material should be brushed off first and then follow the same course of action.

While washing is taking place, emergency medical help should be summoned. If a severe physical injury has occurred in addition to the chemical exposure, appropriate first aid measures should be taken while waiting for aid. In order of priority, restoration of breathing and restoration of blood circulation, stopping of severe bleeding, and treatment for shock should be done first. These injuries are life threatening.

Persons involved in the accident or the subsequent treatment of the injured person or persons should remain at the scene until emergency medical aid arrives. It is important that those treating the victim know what chemical was involved and, in addition, the persons providing assistance can provide emotional support to the victim. Generally, it is preferable that transport to a hospital be done by the emergency rescue personnel. They not only are trained and qualified to handle many types of medical emergencies, but will also have communication capability with an emergency medical treatment center. Through this radio contact, they can advise the emergency center physician of the situation and the physician can instruct the emergency team of actions they can initiate immediately. In addition, if special preparations are needed to treat the injured person upon arrival at the emergency center, these can be started during the transport interval.

Some materials such as mercury do not appear to pose much of an immediate hazard upon a spill and a cursory clean up may seem to be sufficient. However, mercury can divide into extremely small droplets and get into cracks and seams in the floor and laboratory furniture. Spilled mercury remains in the metallic form for a long time after a spill, capable of creating a significant concentration of mercury vapor pressure in a confined, poorly ventilated space. Exposure to these fumes over an extended period can lead to mercury poisoning. After gross visible quantities have been cleaned up mechanically or with an aspirator, absorbent material should be spread on the floor and left there for several hours. Afterwards, the area of the spill should be vacuumed with a HEPA-filtered vacuum cleaner adapted for mercury cleanup. A penknife can be used to check seams in floor tiles and cracks to check if the cleanup has been thoroughly done.

The preceding material on spills assumed that the incident only involved one chemical. Figure 2.3 shows what could have been, but miraculously was not, a major disaster which could have injured several persons and even have resulted in the loss of a major academic building. A set of wall shelves loaded with a large variety of chemicals collapsed while no one was in the laboratory working in the area. Unlike

FIGURE 2.3.    Accident due to poorly installed and insufficiently strong shelves.

the incident referred to earlier where two small quantities of chemicals, discarded in two different buildings, reacted and injured a nonlaboratory worker, here, although several bottles broke and chemicals became mixed, no reaction occurred and the damage was limited to the loss of the chemicals. If a vigorous reaction had occurred between the contents of any two of the broken bottles, the resulting heat might well have caused some of the remaining unbroken containers to have ruptured and a major disaster could have resulted. Where multiple chemicals are involved, the same techniques as with a simple incident should be applied, with the stipulation that unnecessary mixing of chemicals should be carefully avoided.

While all corrective measures are being done, the affected area should be secured to insure that no one is allowed in who is not needed. "Tourists" are not welcome. If necessary, help should be obtained from security to exclude unnecessary persons.

## 2. Fire

A second common laboratory emergency involves fire. Laboratory fires stem from many sources— the ubiquitous bunsen burner, runaway chemical reactions, electrical heating units, failure of temperature controls on equipment left unattended, such as heat baths, stills, etc., overloaded electrical circuits, and other equipment. With a fire, the possibility of the immediate laboratory personnel being qualified and able to cope with the emergency depends very strongly on the size of the fire. As indicated earlier, only if it is clear that the fire can be safely put out with portable extinguishers should a real attempt be made by laboratory personnel to do so. However, portable extinguishers can be used to gain time to initiate evacuation procedures.

In order to use an extinguisher effectively, laboratory personnel should receive training in their use. They should be familiar with the different types and for the type of fires for which they would be effective.

Class A extinguishers are intended to be used on fires involving solid fuels such as paper, wood, and plastics. Generally, a Class A extinguisher contains water under pressure. Water acts to cool the fuel during the extinguishing process, which has the advantage that the fuel has to be brought back up to kindling temperature once the fire has been put out. The large amount of energy required to convert liquid water into vapor places an added burden on the energy requirement to rekindle the fire in wet fuel. An extinguisher rated 1A is intended to be able to put out a fire of 64 ft$^2$ if used properly. A typical extinguisher will throw a stream of water up to 30 to 40 ft for approximately 1 min.

Class B extinguishers, intended for use on petroleum and solvent fires, usually contain carbon dioxide or a dry chemical such as potassium or sodium bicarbonate. The first of these puts out the fire by removing one of the essential components of a fire, oxygen, by displacing the air in the vicinity of the fire. The second uses a chemical in direct contact with the burning material. Some chemical extinguishers contain materials such as monoammonium phosphate or potassium carbamate which, even in small sizes, have very impressive ratings for putting out a solvent fire. The chemical extinguishers are messy and can damage electronic equipment. A third type of unit, which does not have this last negative characteristic, contains a halogenated compound. There are two types, Halon™ 1211 and Halon™ 1301, distinguished chiefly by the fact that the first of these operates at a lower pressure than the second and thus is more common as a portable extinguisher. Permanently installed systems tend to be Halon™ 1301. Both of these contain compounds involving chlorine and bromine and work by interrupting the chemistry of the fire. However, the Halon,™ being gaseous, can be easily dissipated and, once the air concentration falls below the level at which it is effective, no longer provides any residual fire protection. One way in which the Halon™ units can be used effectively is to install them in small storage rooms as ceiling-mounted units. Reasonably priced units can be installed which will go off automatically at temperatures set by fusible links in the heads of the units. Typical dry chemical or carbon dioxide units last on the order of 15 to 30 sec and, in the case of carbon dioxide units, it is necessary to be within 10 ft of the fire to use them effectively.

Class C extinguishers are intended for electrical fires, which preclude the use of water because of the shock hazard. Many Class B extinguishers are also rated for use on electrical fires.

Class D extinguishers are used primarily for reactive metal fires and a few other specialized applications. Due to the extra cost of these units, only those laboratories which actively use reactive metals need to be equipped with Class D units.

It requires training to use a portable extinguisher effectively since they last less than 1 min in most cases. To be most effective, the extinguishing material should be aimed at the base of the fire and worked from the point immediately in front of the extinguisher operator progressively toward the rear of the fire, away from the operator. If more than one person is present, additional extinguishers should be brought to the scene so that as one is used up, another can be quickly brought into use. About half of all fires which can be put out with portable extinguishers require only one, but conversely, the other half require more than one.

To be effective, an extinguisher must be full. Unfortunately, too many individuals with juvenile mentalities apparently feel that extinguishers are toys, provided for their amusement. This seems to be an attitude especially prevalent on college and university campuses. Therefore, the extinguishers should be checked frequently by laboratory personnel as well as by fire safety staff. If the unit has a gauge, it should be in the proper range. Empty and full weights are indicated on the extinguisher, so weighing will confirm if the unit is full or not. Breakable wire or plastic loops through the handles, which are broken when the unit is used, should be checked to see if they are intact. Any units which are found to be discharged should be replaced immediately.

Since a hood is where most hazardous laboratory operations are carried out, a substantial number of laboratory fires occur in them. In the event of a fire in a hood, the simplest procedure is to close the sash. This serves two purposes: it isolates the fire from the laboratory and reduces the amount of air available to support combustion. Since a properly installed hood exhausts either immediately to the outside or through a fire-rated chase, in most instances a fire in a hood can safely be left to burn itself out or at least can reasonably be counted upon not to spread while an extinguisher is obtained.

In the event a person's clothing catches on fire, it is important not to run because this provides air to support the flames. Many authorities recommend that a person with his clothes on fire should roll on the floor to attempt to smother the flames as the most effective method to smother the fire. A deluge shower is an effective way to put out the fire if it is in the immediate area or, if a fire blanket is close by, the fire can be smothered by the person quickly wrapping himself in it. If others are present, they can help smother the flames as the person rolls on the floor or they might employ a fire extinguisher to put the fire out. As with any other type of injury or burn, call for emergency medical assistance as quickly as possible. Perform whatever first aid is indicated while waiting for assistance.

### 3. Explosions

Among other possibilities, an explosion may result from a runaway chemical reaction, a ruptured high pressure vessel, or perhaps ignition of confined gases or fumes. Fortunately, these are less common in the laboratory than a fire, but they still occur frequently. Where the potential is known to be appreciable, protective shields and personal protective equipment should be mandatory. When an explosion does occur, in addition to the shock wave and the extreme air pressures which may occur, there is flying debris and possibly secondary fires and spilled chemicals which may exacerbate the situation and feed a fire or lead to further reactions. Often, there are toxic fumes released, which may be the most serious hazard involved, not only to the persons immediately involved, but also to others outside the area and emergency personnel. Initiation of procedures to handle fires and chemical spills are appropriate if the situation is manageable. The most likely physical complications are personal injuries, including injuries to the eye, lacerations, contusions, broken bones, and loss

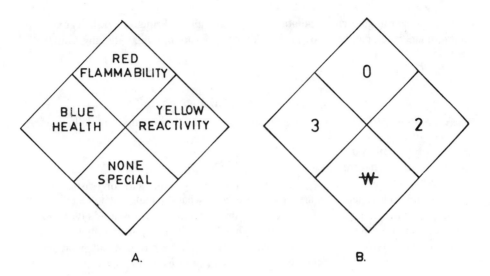

FIGURE 2.4.   (A) NFPA hazard diamond; (B) example for sulfuric acid.

of consciousness. Toxic fumes may cause respiratory injuries, possibly leading to long-lasting, permanent effects and even death. In addition, chemicals may be splashed over the body even more extensively than in a spill, so it may be even more imperative to wash them off. However, it is essential to establish priorities. If breathing is impaired, artificial respiration should be administered, and if heavy bleeding is occurring, pressure should be applied to the wound to stop it. These two problems are immediately life threatening. If there is time, and if it appears safe to do so, i.e., it does not appear that the spine has been injured or that other injuries will not be worsened by the movement, then injured persons should be removed from the immediate vicinity of the accident. This is partially to protect the rescuer as well as the victim from the effects of chemicals, fumes, and smoke. Basically, the same criteria apply as in a fire, unless it is possible to safely handle the situation with the personnel present; then at least the immediate area should be evacuated, if necessary, the building as well, and the fire department and other professional aid summoned.

For most fire departments, a fire or an explosion in a laboratory represents an uncommon occurrence. It would be highly desirable, in the absence of a knowledgeable person immediately on the scene, if information on the contents of the laboratory could be found posted either on the door or close by. Preferably, this information should be brief, legible from a distance, and in a format already familiar to fire personnel. Many localities have attempted to meet these needs by requiring the laboratory to be posted with the NFPA universal hazard diamond in which the degree of danger for reactivity, flammability, and health effects are indicated by a numerical rating, with the numerical rating referring to the contents of the laboratory instead of a specific chemical.

An example of an NFPA symbol is shown in Figure 2.4. There are four small diamonds, which together are assembled into a larger one. The four smaller diamonds are blue, for health or toxicity, red for flammability, yellow for reactivity, and white for special warnings such as radiation or carcinogenicity. Printed in each segment is a prominent black number showing the degree of hazard involved, ranging from 0 to 4.

The numerical ratings are

- 0 = according to present data, no known hazard
- 1 = slight hazard
- 2 = moderate hazard
- 3 = severe hazard
- 4 = extreme hazard

Although this system appears simple, it is somewhat difficult to implement. In a typical laboratory, there may literally be hundreds of chemicals on the shelves. How shall the rating for the laboratory be established? Shall it be determined by the rating of the worst material present for each category or shall the rating also depend upon the total amount of each of the chemicals present? For example, if the most flammable chemical present in a lab were ether, there would be a substantial difference in risk to firemen responding to a laboratory fire where the amount present was a single 500-ml container compared to one in which several 200-l containers were present. On a worst case criteria, both would have the same flammability rating. An alternative would be a subjective rating, combining both the worst case type of chemical with the amount present, to give a rating which in the judgment of the individual doing the rating properly takes into account both factors.

Another problem with the use of the NFPA symbol alone is that it may be too concise. Obviously, it does not inform fire personnel of what is present. In the new Hazardous Communication Standard, corporations and institutions will be required to provide information to the fire department on the location and quantities of their hazardous chemical holdings. This appears to alleviate the problem with the lack of detail. However, this may be too much information and it would be difficult to ensure its accuracy. In a major research institution, there may be literally hundreds of laboratories, each with potentially hundreds of different chemicals, with the inventory changing daily. Although a record of the contents would be helpful, even if not completely current, it would be too clumsy to use as a first-response tool.

A system which elaborates on the NFPA symbol by adding additional symbols is illustrated in Figure 2.5. The symbols can be added individually to the laboratory door and provide useful information on the major classes of chemicals being used in the laboratory and the type of activity involved. The symbols are readily visible from several feet away and are self explanatory.

### 4. Toxic Air Quality

An emergency situation that needs to be mentioned, primarily because it often leads to a fatality, is the danger of entering a space filled with a toxic gas or which is deficient in oxygen. The second of these circumstances is of real concern in maintenance operations when it is necessary to work in confined spaces, but it would be rare to encounter this situation in laboratory incidents. However, as a result of a fire, a spill of an IDLH substance, a leaking gas cylinder, or an improperly vented experiment releasing toxic fumes, it easily would be possible for a laboratory to be

| RADIATION OR RADIOACTIVE MATERIAL | INFECTIOUS AGENT | NO SMOKING | FLAMMABLE LIQUID |

| CARCINOGEN | SAFETY GOGGLES REQUIRED | EXPLOSIVE MATERIAL | TOXIC OR POISON MATERIAL |

FIGURE 2.5.   Representative hazard warning symbols.

full of fumes and gases which would be fatal. Not all gases which may be used fairly commonly in the laboratory have adequate warning properties. No one should enter a space where this could conceivably be the situation without using a self-contained air breathing apparatus, nor should an individual go in such a space without others being aware of it. There should always be a backup set of equipment, with a person available, trained, and able to use it to effect a rescue if necessary.

### 5. Radioactive and Contagious Biological Material Releases

Releases of radioactive material and active contagious biological materials represent two different types of emergencies which cause unusual concern because of the potential danger, perceived by the public, of the problem spreading beyond the immediate scene. In almost every instance, the levels of these two classes of materials used in ordinary laboratories are sufficiently small that the risk to the general public is minimal. However, in both cases, because of the public sensitivity to risks associated with these two classes of materials, unusual care must be taken in responding to the emergency.

*5A. Biological Accident*

In recent years, a system of classification of laboratories for biological safety has been established, defining biological safety levels 1 through 4. Research with organisms posing little or moderate risk requires only level 1 or 2 facilities, which are essentially open laboratories. Work with organisms which do pose considerable or substantial risks requires level 3 or 4 facilities. A characteristic of facilities at both level 3 and 4 laboratories is that they are essentially self-contained, with the entrance being through an anteroom or airlock and access restricted to authorized personnel.

This has greatly limited the possibility of an accident spreading beyond the confines of the facility. The major risks are accidents which cause direct exposure to individuals working in the laboratories. The facilities, especially those intended for higher risk use, are built to allow ease of decontamination to minimize the chances of a continuing source of infection in the event of a spill. Whenever a possibly infectious spill occurs, the immediate emergency procedure is to obtain medical care for the potentially exposed person as quickly as possible and to perform tests to determine if, in fact, the person involved has received the suspected exposure. A baseline medical examination (including a medical history) for each employee at the time of employment, with a serum sample taken for storage at that time, is of great value for comparison at the time of an accident. Because there may be delayed effects, records of any suspected incident need to be maintained indefinitely. As long as contaminated materials removed from the facility are autoclaved or double-bagged, followed by incineration, there is little risk to the general public from laboratory research involving biological materials.

Individuals not involved directly in the accident should evacuate the laboratory, and the area should be decontaminated by persons wearing proper protective clothing. It may be necessary to chemically decontaminate the entire exposed space. However, each incident needs to be treated on a case-by-case basis.

*5B. Radiation Incident*

Radioactive spills represent another class of accident that is of special concern. Again, there are circumstances that ameliorate the risk in actual accidents. Although laboratories in which radioactive materials are used are not classified as to the degree of risk, as are laboratories using pathogens, they do operate under unusually tight regulations which tend to minimize the amount of material involved in a single incident and to limit the number of persons involved to authorized and experienced personnel. As a result, an individual involved in a spill generally knows to restrict access to the area of the accident and to avoid spreading the material to uncontaminated areas. Every institution licensed to use radioactive materials is required to have a radiation safety program and a radiation safety officer who should be notified immediately in case of an accident. In obtaining the license to use radioactivity, the institution or corporation must demonstrate to the Nuclear Regulatory Commission (NRC) that it has the capability of managing accidents properly. In addition, there are requirements governing reports to the NRC or to the equivalent state agency in an "Agreement" state, spelled out in Title 10 of the Code of Federal Regulations, Part 20, when an accident occurs. Thus, the response to an emergency involving a release of radioactive material is relatively straightforward. The clothes and skin of persons in the area and those allowed to leave can be checked with survey meters, which should be present in laboratories using radioactive materials or brought to the scene by radiation safety personnel. Surface contamination on both the laboratory and personnel can be cleaned up with little risk, using proper personal protective equipment to protect those doing it. The protective equipment normally would consist of a cartridge respirator and filter, coveralls of Tyvek™ or a similar material, head and foot covers (these may need to be impregnated with an appropriate plastic material), and "impermeable" gloves. If the possibility exists that anyone ingested or inhaled radioactive material, then the individual should undergo further testing. This would include a bioassay for radioactive materials and, possibly, whole-body counting at a facility with this capability. A major advantage of radioactive materials is that instruments exist which can detect radiation from spilled materials to levels well below any known risk. In addition, many radiation workers wear personnel dosimeters which can be used to determine external doses.

A situation in which personal injury is accompanied by a spill of radioactive material onto that person introduces some complications in the emergency medical response. Radioactive material may have entered the body through a wound, and there is a possibility that both the emergency transport vehicle and the emergency room at the hospital could become contaminated. Again, due to the small quantities used in most laboratories, the contamination is unlikely to actually be a serious problem, but could be perceived as one by the emergency medical personnel. In order to reassure them, a radiation safety person should accompany the victim to the emergency center, if available, and be able to provide information on the nature of the radioactive material, the radiation levels to be expected, and advice on the risks posed by exposure to the patient and to others. The type of radiation and the chemical or material in which it is present can have a major impact on the actions of the emergency room personnel. Some materials are much worse than others if they have entered the body. As noted above, in order to ascertain that no internal contamination exists, a bioassay, other specialized tests, and a whole-body count of the victim may be needed.

A sheet of plastic placed between the injured person and the backboard or stretcher and brought up around the person will effectively keep the ambulance and the equipment being used from being contaminated and will serve the same purpose later at the emergency room. If it is felt to be necessary, emergency personnel can wear particulate masks or respirators to avoid any inhalation of any contaminants. The patient should be separated from any other occupants of the emergency reception area to avoid any unnecessary exposures, even if they are well within safe limits, again because of the public concern to exposures to radioactivity at any level. In the event that substantial levels of radiation might be involved, the victim should be placed in an isolated room and emergency equipment brought to the room, rather than using the normal emergency room. A possible location would be the morgue. In such an incident, it is important to document exposures for everyone involved in the emergency response. Even in low-activity situations, it is good practice to survey the interior of the ambulance, the parts of the emergency facility which might have been contaminated, equipment that may have been used, and the emergency personnel involved and make wipe tests for loose contamination. All radiation survey data should be carefully recorded and the records maintained for future use. The records should include estimated dose levels, based on the proximity to the radiation and the duration of the exposure.

### 6. Multiple Class Emergencies*

Emergency response procedures will need to incorporate sufficient flexibility to serve in many nonstandard situations. Unfortunately, one cannot depend upon an

---

* This section and the remainder of this chapter was coauthored by Dr. Richard F. Desjardins and Dr. A. Keith Furr.

accident being of a single type or even limited to one or two complicating factors. Consider the following hypothetical situation: a laboratory worker puts a beaker containing a volatile solvent, to which a radioactive compound has been added, into an ordinary refrigerator. Due to carelessness, it is not covered tightly. During the next several hours, the concentration of vapors builds up in the confined space and, at some point, the refrigerator goes through a defrost cycle. The vapor ignites explosively and the refrigerator door is blown off, striking a worker and knocking several bottles of chemicals off a shelf. Chemicals from the broken bottles spill onto the floor and the injured person. The solvent in the beaker, as well as that in several other containers, spills on the floor and ignites. The radioactive material in the beaker and in some of the other containers is spread throughout the laboratory and into adjacent rooms. Although this is posed as a hypothetical situation, it could easily happen.

In a complicated incident such as the one described above, the first priority is preservation of life, even ahead of possible future complications. In the presence of a fire which, in a laboratory containing solvents, always has at least the potential of spreading uncontrollably, evacuation of the injured party should be considered as the first priority, followed or paralleled by initiation of evacuation of the rest of the building. Treatment of the physical and chemical injuries to the victim and summoning the fire department and other emergency personnel can be done outside the building. Once everyone is safe, efforts to reduce property losses can take priority. Assuming that the fire will have been controllable, steps can be taken for cleanup and decontamination of the spilled chemicals and radioactive material. Unless there appears to be a risk that the contaminated area will spread, perhaps due to the runoff of water used in fighting the fire, it is not necessary for these last steps be done in any haste. However, the surrounding area must be cordoned off until measurements and surveys are completed by trained radiation safety and, perhaps, chemical safety personnel. This isolation must be maintained until a *formal* release of the area by the individual in charge, based on the information provided by the safety specialists.

After the incident is over, a review of the causes of the accident and the emergency response should be conducted by the appropriate safety committee or committees. In this case, the laboratory safety committee and the radiation safety committee would probably jointly conduct the review. Basically, there were two root causes of the incident. Solvents should not be stored in any container which cannot be tightly sealed, but this would not have caused the explosion if the refrigerator had been designed to be suitable for the storage of flammable materials. These are commercially available.

In the subsequent review, consideration should consider if anything could have made the incident worse. For example, in the hypothetical accident, the worker could have been alone, although this was not assumed to be the case. In academic research laboratories, research workers, and especially graduate students, tend to work unusual hours as they try to work around their class schedules and meet deadlines imposed by the framework of timetables, deadlines for submission of theses and dissertations, etc. If the injured person had been alone, the potential would have existed for a loss of life.

The situation described in the previous paragraphs illustrates not only that emergencies can be very complicated in the real world, but also that some emergency responses can wait, while others cannot. Components of the emergency that are immediately life threatening must be dealt with promptly, but others, such as cleaning up, can wait to be done carefully and properly after appropriate planning. Any incident should also be treated as a learning opportunity. There were basic operational errors leading to the postulated incident which could be repeated in other laboratories. There were aspects to the incident which would have permitted it to be worse. These should be factored into the emergency plan if they had not already been considered. If violations of policy had occurred, then the review should point these out and recommend courses of action to prevent future violations. An emergency plan should not only cover responses to classes of emergencies which have occurred, but should also have the capability of reducing the possibility that emergencies will occur.

### 2.3.5. FIRST AID, ARTIFICIAL RESPIRATION, AND CARDIOPULMONARY RESUSCITATION (CPR)

In several examples of responses to various emergencies, allusions were made to emergency medical procedures which should be performed. Most of these procedures require prior training. Because of the relatively high probability of accidents in laboratories, it would be desirable if at least a cadre of trained persons were available in every laboratory building.

Both first aid and CPR classes are taught by a number of organizations in almost every community. Among these are the Red Cross, American Heart Association, rescue squads, other volunteer organizations, and many hospitals. Usually, except for a small fee to cover the cost of materials, the classes are free. In addition, new labeling regulations and the OSHA Hazardous Communication Standard require that emergency information be made available on the labels of chemical containers and as part of the training programs. Since in most cases involving a chemical injury the chemical causing the injury will be known and, thus, information will be available, the following material on first aid for chemical injuries will be restricted to the case of basic first aid for an injury caused by an unknown chemical. Similarly, since formal class instruction in CPR, which will also cover artificial respiration, is almost always available, the material on CPR will be very basic. CPR should only be done by properly trained individuals, with the training including practice on manikins. Certification in CPR is easily and readily acquired. It is also important to periodically become recertified, as new concepts and procedures are frequently being evolved and presented in the training programs.

In all the following sections, it is assumed that emergency medical assistance will be called for immediately. Emergency medical personnel are trained to begin appropriate treatment upon their arrival. Depending upon the level of training and the availability of telemetry, they will have radio contact with a hospital emergency facility or a trauma center and can receive further instruction from a physician, while providing immediate care and during transit to the treatment center.

The following material is a composite of the information provided by a number of different sources. Where sources differed slightly, the more conservative approach was taken, i.e., that approach which appeared to offer the most protection to an injured person, with a second priority being the approach offering the least risk to the individuals providing the assistance. A third criterion was simplicity and the feasibility of performing the procedure with materials likely to be available. It was compiled explicitly in the context of injuries that are likely to occur as a result of laboratory accidents and is not intended to provide a comprehensive treatment of emergency medical care. It has been reviewed and revised by a physician.

Except where mandated by the nature of the problem, such as removal from a toxic atmosphere, no stress was placed on evacuation. Unless there are obvious fractures, there may be injuries to the spine or broken bones which may puncture vital organs that are not immediately apparent. If it is essential to move the victim, do so very carefully. Use a backboard or as close an equivalent as possible to keep the body straight, and support the head so that it does not shift. Any inappropriate movement of a fractured neck may sever the spinal cord and result in paralysis, death, or compromising of the patient's airway.

To repeat, before performing any of the more complicated first aid procedures, formal training classes should be taken from certified instructors. It is possible for an inexperienced person to cause additional injuries.

### 2.3.5.1. Artificial Respiration

Lack of oxygen is the most serious problem that might be encountered. If the victim is not breathing or the heart is not beating, then oxygen will not be delivered to the brain. If this condition persists for more than 4 to 6 min, it is likely that brain damage will occur. In this first section, it will be assumed that the heart is beating, but that the victim is not breathing.

### A. Artificial Respiration, Manual Method

Although mouth-to-mouth or mouth-to-nose artificial respiration is much more effective, an alternative method of artificial respiration will be discussed first. There are rare occasions when it is not safe to perform direct mouth-to-mouth resuscitation, such as when poisoning by an unknown or dangerous chemical substance is involved or when the victim has suffered major facial injuries which make it impossible to perform this resuscitation. Since the first of these conditions can be expected to occur in some laboratory accidents, it is well to know that there is an alternative procedure available. The method considered the best alternative is described below.

1.  Check the victim's mouth for foreign matter. To do this, insert the middle and forefinger into the mouth, inside one cheek, and then probe deeply into the mouth to the base of the tongue and the back of the throat, finally sliding your fingers out the opposite side of the mouth. Be aware that a semiconscious patient may bite down on your fingers. It would be wise to insert a folded towel or object that would not break teeth between the teeth while you are doing your examination.
2.  Place the victim on his back in a face-up position. Problems with aspirating vomitus can be reduced by having the head slightly lower than the trunk of the body. An open airway is essential and can be maintained by placing something, such as a rolled up jacket, under the victim's shoulders to raise them several inches. This will permit the head to drop backward and tilt the chin up. Turn the head to the side. Important! Do not do this if there is any suspicion of neck trauma.
3.  Kneel just behind the victim's head, take the victim's wrists, and fold the victim's arms across the lower chest.

4.  Lean forward, holding onto the wrists, and use the weight of your upper body to exert steady, even pressure on the victim's chest. Your arms should be approximately straight up when in the forward position. This will cause air to be forced out of the victim's chest. Perform this step in a smooth, flowing motion.

5.  As soon as step 4 is completed, take your weight off the victim's chest by straightening up and simultaneously pulling the victim's arms upward and backward over his head as far as possible. This will cause air to flow back into the lungs, thereby completing one equivalent breathing cycle.

6.  Steps 4 and 5 should be repeated 12 to 15 times per min. to pump air into and out of the victim's lungs. Stop and check frequently that no vomitus or foreign matter has been brought up into the airway. If a helper is available, let the helper do this while you continue the pumping process.

7.  Continue the procedure until normal breathing is established or until emergency medical personnel arrive. Be ready to recommence the process if breathing difficulties reoccur.

This less effective method was given first since it is human nature to question the need to learn a less effective method if a better exists and it has already been covered. However, it is important to limit the number of injured parties as well as to treat those already injured. Since, in the case of an unknown toxic or an especially dangerous substance, the person giving emergency treatment could be exposed to the same material, this procedure will serve in such cases.

### B. Artificial Respiration, Mouth-to-Mouth Method

There is consensus that this is the most effective method of artificial respiration. To be effective, it should be begun as quickly as possible.

1.  Check the victim's mouth for foreign matter. Clear out any that is found with your fingers. A cloth such as a towel or handkerchief covering the fingers helps in the removal of objects, or even some chemical, which would slide off wet fingers. The towel also will serve to protect the fingers should the patient bite down on them.

2.  The victim's air passage must be open. Put your hand under the person's neck and lift. Place your other hand on the victim's forehead and tilt the victims head back as far as it will reasonably go, essentially straightening the airway. A folded jacket or coat under the shoulders will aid in keeping the head back.

3.  Maintain this position and, with the fingers of the hand being used to tilt the head backward, pinch the nostrils closed.

4.  Open your mouth wide, take a deep breath, place your mouth firmly around the victim's mouth to get a good seal, and blow into the victim's mouth.

5.  Try to get a good volume of air into the victim's lungs with each breath.

6.  Watch to see the chest rise. When it has expanded, stop blowing and remove your mouth from the victim and check for exhalation by listening for air escaping from the mouth and watching to see the chest fall. This occurs naturally from the inherent elasticity of the chest wall.

7.  Steps 4 through 6 constitute one breathing cycle. Assuming it has been successful, i.e., there has been no significant resistance to the flow of air during the

inflow cycle and air has been exhaled properly, repeat steps 4 through 6 at a rate of 12 to 15 times per min.

8.  If you are not getting a proper exchange of air, check again to see if there is anything obstructing the air passage and make sure you are holding the head properly so as to prevent the tongue from blocking the flow of air.

9.  Continue the procedure until normal breathing is established or until emergency medical personnel arrive. Be ready to recommence the process if breathing difficulties recur.

### 2.3.5.2 Cardiopulmonary Resuscitation

Cardiopulmonary resuscitation combines artificial respiration with techniques to manually and artificially provide blood circulation and is to be used where the heart is not beating. It requires special training and practice and should not be done except by trained, qualified individuals. As noted earlier, receiving the training is not difficult since it is readily available from several organizations. The material presented here is intended to provide information as to the general procedures, but neither it nor the previous procedures are intended as an instructional guide.

One-person CPR will be the procedure described. In this procedure, the person performing it will provide both artificial breathing and blood circulation. When two persons are available, each person can take the responsibility for one of these functions and can switch from time to time to relieve fatigue.

### A. Initial Steps

1.  Check for breathing. Do this by placing one hand under the victim's neck and the other on the forehead. Lift with the hand under the neck and tilt the head back. While doing this, place your ear near the victim's mouth and look toward the chest. If the victim is breathing, you should be able to feel air on your skin as it is being exhaled, you should be able to hear the victim breathe and you should be able to see the chest rise and fall. Do this for at least 5 sec.

2.  If the victim is not breathing, while holding the victim as in step 1, pinch the nostrils closed, open your mouth wide, place it over the victim's mouth, and give two quick, full breaths into the victim's mouth.

3.  After step 2, repeat step 1 to see if breathing has started. If not, proceed to step 4.

4.  Check the victim's pulse with the hand that had been under the victim's neck. Keep the head tilted back with the other hand on the forehead. Check the pulse by sliding the tips of the fingers into the groove on the victim's neck to the side of the adam's apple nearest you. Again check for at least 5 sec.

### B. Formal CPR Procedures

The victim must be on a firm surface. Otherwise, when pressure is applied to the chest, the heart will not be compressed against the backbone as the backbone is pressed into the yielding soft surface. The head of the victim should not be higher than the heart, so that blood will flow to the brain as is needed to avoid brain damage. Although the brain averages about 2% of your body weight, it requires 20% of the oxygen you breathe, as well as at least 40 mg% of blood dextrose. Any concentration less than this will result in unconsciousness and progressive brain damage. Preferably, the feet and legs should be adjusted to be higher than the heart to facilitate blood flowing back to the heart, but this has a lower priority than commencing CPR procedures.

1. Kneel beside the victim at breast height. Rest on your knees, not your heels.
2. Locate the victim's breastbone. Place the heel of one of your hands on the breastbone so that the lower edge of the hand is about two finger widths up from the bottom tip of the breastbone. Put your other hand on top of the first hand. Lift your fingers or otherwise keep from pressing with them. Improper placing of the hands can cause damage during the compression cycle.
3. Place your shoulders directly over the breastbone. Keep your arms straight.
4. To initiate the compression cycle, push straight down, pivoting at the hips.
5. Push firmly and steadily down until the chest has been compressed about 4 to 5 cm. Then smoothly relax the pressure until the chest rebounds and is no longer compressed and start the compression cycle again. The compression-relaxation cycle should be a smooth, continuous process.
6. Continue the chest compression procedure for 15 cycles at the rate of 80 per min. This should take between 11 and 12 sec. Then quickly place your mouth over the mouth of the victim and give two quick, full breaths. Then return to compressing the chest for another 15 cycles. Be sure to locate your hands properly and compress the chest as in steps 4 and 5.
7. Continue step 6, alternating compressing the chest and providing artificial respiration.
8. Quickly check for a pulse after 1 min. and then every few minutes thereafter. Watch for any signs of recovery.
9. If a pulse is found, check for breathing. If necessary, give artificial respiration only, but check frequently to be sure that the heart is still beating.
10. Continue with whatever portion of the procedure is necessary until the victim is functioning on his own, emergency medical personnel arrive, or it is obvious that efforts will not succeed. A half hour is not unreasonable as a period of actively using CPR.

A person who has not been properly trained in the techniques of CPR can injure the victim, either by driving the lower end of the breastbone into the liver or by breaking ribs if the hands are not placed correctly or the pressure is exerted incorrectly. However, remember that ribs will heal, but a nonbreathing or heart-stopped patient dies. Thus, to repeat once more, the techniques for artificial respiration and CPR are not difficult, but do require training and practice. Familiarity with these techniques is likely to be of value at any time.

### 2.3.5.3 First Aid
#### A. Severe Bleeding

A person may bleed to death in a very short time from severe or heavy bleeding, so whenever this problem is involved in an accident, it is extremely important to stop it as soon as possible. Arterial bleeding may be frightening, but the muscular artery wall usually contracts to diminish or stop the flow. Venous bleeding is more insidi-

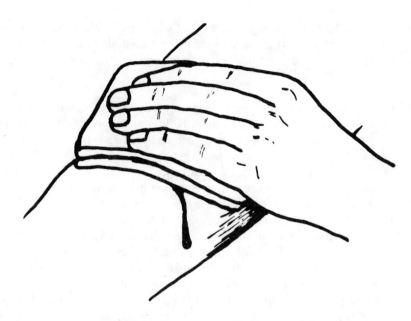

FIGURE 2.6.    Apply pressure directly to a wound.

ous, as it flows steadily. The relative absence of muscle in the vein wall does not help to stem the flow.

1. The most effective treatment is pressure applied directly to the wound, over which a sterile dressing has been placed (Figure 2.6).
2. Apply a sterile dressing if immediately available, a handkerchief if not, or some other cloth to the wound. Then place the palm of the hand directly over the wound and apply pressure. If nothing is available, use your bare hand, but try to find something to use as a dressing as soon as possible.
3. A dressing will help staunch the flow of blood by absorbing the blood and permitting it to clot. Do not remove a dressing if it becomes blood soaked, but leave it in place and apply an additional one on top of the first in order not to disturb any clotting that may have started. Keep pressure on with the hand until you have time to place a pressure bandage over the dressing to keep it in place.
4. Unless there are other injuries such as a fracture or the possibility of a spinal injury, in which case the victim should be disturbed as little as possible, the wound should be elevated so that the injured part of the body is higher than the heart. This will reduce the blood pressure to the area of the wound.
5. If bleeding persists and cannot be stopped by direct pressure, putting pressure on the arteries supplying the blood to the area may be needed. In this technique, pressure is applied to the arteries by compressing the artery between the wound and the heart against a bone at the points indicated in Figure 2.7. Since this stops all circulation to points beyond the point of compression, it can cause additional injuries if continued too long. For this reason, it should be discontinued as quickly as possible and one should return to using direct pressure and elevation to control the bleeding, unless this is the only effective technique.

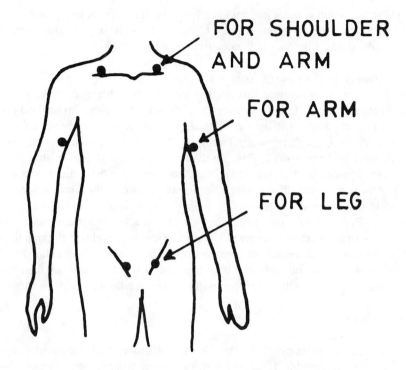

FIGURE 2.7.   Points at which to apply pressure to stop blood flow.

6.   As a last resort, since it stops the flow of blood to the limb beyond the point of application, a tourniquet, which should be at least 2 in wide, can be applied. An example where the use of a tourniquet might be indicated is to stop bleeding from a severed limb.

***B. Shock***

Shock may accompany almost any type of severe injury, exposure to toxic chemicals, a heart attack, loss of blood, burns, or any other severe trauma. It can be recognized from a number of characteristic symptoms: skin cold to the touch, possibly clammy and bluish or pale, weakness, a rapid, weak pulse, rapid irregular breathing, restlessness, and exhibition of unusual signs of thirst. As the condition worsens, the victim will become unresponsive and the eyes may become widely dilated.

The treatment for shock is

1.   The victim should be lying down, although the type of injury may determine what is the best position. If uncertain, allow the victim to lie flat on his back. Unless it is painful or makes it harder for the victim to breathe, it will help if the feet are raised 20 to 30 cm high.

2.   Use blankets to keep the victim from losing body heat, but do not try to add heat.

3.   If the victim is conscious and not vomiting, nor appears likely to do so, then about half a glass of liquid every 15 min or so will be helpful. However, do not give fluids if the victim is unconscious or nauseated.

### C. Poisoning by Unknown Chemicals

The rationale in limiting this section to unknown chemicals is that first aid information is readily found when the chemical is known, either immediately on the label of the container or in the material safety data sheet for the chemical (note that current law requires chemical producers to provide MSDSs to those who purchase chemicals from them). An MSDS file should be maintained in every laboratory.

Even if the chemical causing the injury is not known and the victim is unconscious so that no direct information is available, an examination of the circumstances of the accident and an examination of the victim's lips, skin, mouth, and tongue could provide helpful information on whether the victim swallowed a poisonous substance, inhaled a toxic vapor, or whether the injury was due to absorbtion through the skin. If the abdomen is distended, and pressing on it causes pain, the likelihood is that a corrosive or caustic substance has been ingested. Any information on the nature of the harmful material will be helpful to the physician who will treat the victim.

*Poisoning by Inhalation*

1.   When poisoning by inhalation is suspected, evacuate the victim to a safe area as soon as possible. If fumes are still suspected to be present, a rescuer should wear a self-contained respirator. Do not take a chance which might result in a second victim.
2.   Check for unusual breath odors if the victim is breathing.
3.   Loosen tight clothing around the victim's neck and waist.
4.   Maintain an open airway.
5.   If the victim is not breathing, perform artificial respiration using the manual method. It is dangerous to the person providing aid to give mouth-to-mouth artificial respiration if the toxic material is not known.

*Poisoning by Ingestion*

1.   Examine the lips and mouth to ascertain if the tissues are damaged— a possible indicator that the poison was ingested, although the absence of such signs is not conclusive.
2.   Check the mouth and remove any dentures.
3.   If the victim is not breathing, perform manual artificial respiration.
4.   If the victim becomes conscious, try to get the victim to vomit, unless it is possible that the poisoning is due to strong acids, caustics, petroleum products, or hydrogen peroxide, in which case additional injuries would be caused to the upper throat, esophagus, and larynx. Vomiting may be induced by tickling the back of the throat. Lower the head so that the vomit will not reenter the mouth and throat. Dilute the poison in the stomach with water or milk.
5.   If the victim has already vomited, collect a sample of the vomit, if possible, for analysis.

6. If convulsions occur, do not restrain the victim, but do remove objects with which he might injure himself or orient the victim to prevent his striking fixed heavy objects.
7. Watch for an obstruction in the victim's mouth. Remove if possible, but do not force fingers or a hard object between the victim's teeth. If a soft pad can be inserted between the victim's teeth, it will protect the tongue from being bitten. A badly bleeding tongue immensely complicates the patient's problems.
8. Loosen tight clothing such as a collar, tie, belt, or waistband.
9. If the convulsions cease, turn the victim on his side or face down so that any fluids in the mouth will drain.
10. Treat for shock if the symptoms for shock are noted.

*Poisoning by Contact*

1. If the chemical got into the victim's eyes, check for and remove any contact lenses. Immediately take the victim to an eyewash fountain (if one is not available, to a shower or even a sink, otherwise) and wash the eyes, making sure that the eyelids are held wide open. Wash for at least 15 min. If the chemical is a caustic rather than an acid, the victim may not feel as much pain (an acid causes pain due to precipitation of a protein complex) and may wish to quit earlier. An alkali or caustic chemical is more serious as it does not precipitate protein and continues to penetrate the globe, even to global rupture. It is imperative that the chemical be flushed out thoroughly.
2. Do not use an eye ointment or neutralizing agent.
3. If the chemical only got on the victim's exposed skin, such as the hands, wash thoroughly until the chemical is totally removed.
4. If the chemical got onto the clothed portion of the body, remove the contaminated clothes as quickly as possible, protecting your own hands and body, and place the victim under a deluge shower. If the eyes were not affected initially, protect them while washing the contaminated areas. Be careful not to damage the affected skin areas by rubbing too firmly. Let the flowing water rinse the chemical off. A detergent is sometimes used, but be careful not to carry the offending chemical to other parts of the body. Be particularly careful to clean folds, crevices, creases, and groin.
5. If both the eyes and portions of the body were exposed, a combination eyewash and deluge shower unit should be used, if available. If not, take the victim to the deluge shower and tilt his head back, holding the eyelids wide open, and wash the entire body.

### D. Burns (This Section Applies to Burns due to Heat)
**First degree (minor)**— Painful and red. No blisters. Skin elastic. Minimal swelling, involves epidermis only.

1. Apply cold water to relieve pain and facilitate healing. Avoid reexposure as the already injured skin can be more susceptible to further damage than normal skin.

**Second degree** — Severe and painful, but no immediate tissue damage. Pale to red, weeping blisters, vesicles. Marked swelling, involves epidermis and dermis.

1. Immerse the affected area in cold water to abate the pain.
2. Apply cold, clean cloths to the burned area.
3. Carefully blot dry.
4. Do not break blisters.
5. If legs or arms are involved, keep them elevated with respect to the trunk of the body.

**Third degree** — Deep, severe burns, likely tissue damage. White, red, or black, dry inelastic tissue. No pain, involves full thickness of skin and may involve subcutaneous tissue, muscle, and bone.

1. Do not remove burnt clothing from the burned area.
2. Cover the burned area with a thick sterile dressing or clean cloths.
3. Do not immerse an extensive burned area in cold water because this could exacerbate the potential for shock and introduce infection. A coldpack may be used on limited areas such as the face.
4. If the hands, feet, or legs are involved, keep them elevated with respect to the trunk of the body.
5. Third-degree burns must be treated by a physician and/or hospital. They may need reconstruction, skin grafting, and prolonged care. Control of infection is mandatory.

# REFERENCES

Note that many of the basic references were incorporated in the text as materials needed to plan or facilitate an effective emergency program. The following are additional references used in preparing the material.

1. **Laughlin, J. W., Ed.,** *Private Fire Protection and Detection,* ISFTA 210, International Fire Training Association, Fire Protection Publications, Oklahoma State University, Stillwater, 1979.
2. *Emergency Eyewash and Shower Equipment,* ANSI Z358.1-1981, American National Standards Institute, New York, 1981.
3. **Srachta, B. J.,** in *Safety and Health,* National Safety Council, Chicago, 1987, 50.
4. **Steere, N. V.,** Fire, emergency, and rescue procedures, in *Handbook of Laboratory Safety,* Steere, N. V., Ed., CRC Press, Boca Raton, FL, 1971, 15.
5. *Practice for Occupational and Educational Eye and Face Protection.* ANSI Z87.1-1979, American National Standards Institute, New York, 1979.
6. **Schwope, A. D., Costas, P. P., Jackson, J. O., Stull, J. O., and Weitzman, D. J., Eds.,** *Guidelines for the Selection of Chemical Protective Clothing,* 3rd ed., Arthur D. Little, Cambridge, 1987.
7. *Practices for Respiratory Protection,* ANSI Z88.2-1980, American National Standards Institute, New York, 1980.
8. **Kairys, C. J.,** Hazmat protection improves with equipment documentation, in *Occup. Health Saf.,* 56(12), 20, 1987.
9. **Still, S. and Still, J. M., Jr.,** *Burning Issues,* Humana Hospital, Augusta, GA (charts).
10. **Schmelzer. L. L.,** Emergency procedures and protocols, in *Cancer Research Safety Workshop Workbook,* Office of Research Safety, National Cancer Institute, Bethesda, MD, 1978, 106.

11. **Gröschel, D. H. M., Dwork, K. G., Wenzel, R. P., and Schiebel, L. W.**, Laboratory accidents with infectious agents, in *Laboratory Safety, Principles and Practices,* Miller, B. M., Gröschel, D. H. M., Richardson, J. H., Vesley, D., Songer, J. R., Housewright, R. D., and Barkley, W. E., Eds., American Society Of Microbiology, Washington, D.C., 1986, 261.

12. **Edlich, R. F., Levesque, E., Morgan, R. F., Kenney, J. G., Silloway, K. A., and Thacker, J. G.,** Laboratory personnel as first responders, in *Laboratory Safety, Principles and Practices,* Miller, B. M., Gröschel, D. H. M., Richardson, J. H., Vesley, D., Songer, J. R., Housewright, R. D., and Barkley, W. E., Eds., American Society Of Microbiology, Washington, D.C., 1986, 279.

13. **Miller, B. M., Gröschel, D. H. M., Richardson, J. H., Vesley, D., Songer, J. R., Housewright, R. D., and Barkley, W. E., Eds.,** Emergency first aid guide, Appendix 4, in *Laboratory Safety, Principles and Practices,* American Society Of Microbiology, Washington, D.C., 1986, 348.

14. *Safety in Academic Chemistry Laboratories,* 4th ed., American Chemical Society, Washington, D.C., 1985.

15. *Multimedia Standard First Aid, Student Workbook,* American Red Cross, 1981.

16. *Standard First Aid & Personal Safety,* 2nd ed., American Red Cross, Washington, D.C., 1979.

17. *Adult CPR, Workbook,* American Red Cross, Washington, D.C., 1987.

18. **Hafen, B. Q. and Karren, K. J.,** *First Aid and Emergency Care Workbook,* 3rd ed., Morton Publishing, Englewood, CO, 1984.

# Laboratory Facilities: Design and Equipment 3

## 3.1. LABORATORY DESIGN

The design of a laboratory facility depends upon both function and program needs. Although there are differences among engineering, life sciences, and chemistry laboratories — and within the field of chemistry, between laboratories intended for physical chemistry and polymer synthesis, to take two examples — the similarities outweigh the differences. Approximately the same amount of space is required. Certain utilities are invariably needed. Ventilation is needed to eliminate odors and vapors from the air which might have the potential to adversely affect the health of the employees as well as to provide tempered air for comfort. Provision is needed for stocking reasonable quantities of chemicals and supplies. As these are used over a period of time, chemical wastes are generated and provisions must be made for temporary storage and disposal of these according to regulatory standards. They must provide suitable work space for laboratory workers. Many of these items, as well as others, vary only in degree. Most differences are relatively superficial and are represented primarily by the equipment which each laboratory contains and the selection of reagents which are used.

Not only are laboratories basically similar, there is a growing need for "generic" laboratory spaces which are readily adaptable to different research programs. This is due in part to the manner in which most research is funded today. In industry, laboratory operations are generally goal oriented, i.e., they exist to develop a product or to perform basic research in a field relevant to the company's commercial interests. In the academic field, research is primarily funded by grants. These grants can be from any number of public and private sources, but with only a moderate number of exceptions, the research program is based on the submission of a proposal to perform research toward a specific end during a stipulated period of time. At the end of this period, the grant may or may not be renewed and, if not, other uses must be found for the space. Laboratory space is too limited and too expensive to be allowed to remain idle. The result has been to design relatively small laboratories, typically suitable for four persons working simultaneously, with connections to adjacent rooms to permit growth if needed. Under these circumstances, it will be appropriate in most of this chapter to base the discussion upon a standard module. One potential result of this flexibility may be an eventual breakdown of the idea of department-owned space for research buildings, i.e., the concept of a chemistry or biology building, as least as far as research space, and it may eventually be abandoned, at least in part, in favor of assignable spaces based on current needs.

Instructional laboratories are an exception since they are intended for relatively stable basic programs and normally would be somewhat larger than is needed for research programs. Also, except at advanced levels, instructional laboratories do not conduct experiments or use chemicals having the same degree of risk as research laboratories. However, even in the case of instructional laboratories, many of the design features needed for safety remain the same.

### 3.1.1. ENGINEERING AND ARCHITECTURAL PRINCIPLES

The increasing cost of sophisticated laboratory space dictates a number of design considerations. It is essential that space be used to maximum advantage. Due to the necessity for mechanical services, closets, columns, wall thicknesses, halls, stairs, elevators, and rest rooms the percentage of net assignable space in even a well-designed, efficient building is generally on the order of about 65%. Due to the large number of fume hoods in a typical laboratory building, as well as the increasingly stringent temperature and humidity constraints imposed by solid-state circuitry in laboratory apparatus and computers, heating and ventilation (HVAC) systems are becoming more difficult to design which accommodate these needs as well as the need to provide personnel comfort, conserve energy, and provide low life-cycle maintenance costs. Regulatory requirements to accommodate the handicapped in virtually every program impose additional constraints on accessibility and provisions for emergencies. The design needs to be sufficiently flexible not only to accommodate different uses based on current technology, but also to accommodate technological innovations. For example, provision for installation of additional data and voice lines beyond the number currently needed is almost certainly desirable. Interaction of the occupants of the building with each other, outside services, and other disciplines also mandate a number of design parameters. This last set of parameters is the most dependent upon the specific programs using the building and will require substantial input by the users. Typically, all of these design needs must be accommodated within a previously established construction budget, so the design process is a constant series of compromises.

Although the concepts briefly enumerated above are the most important from a pragmatic viewpoint, they are not likely to be the most important to an architect, although, of course, they must be factored into his concept. To the architect, the most important thing is that the building meet all needs in an attractive way. There is nothing wrong with this idea if compromises in the interest of the architectural statement do not compromise the basic needs of the users. Generally, the most efficient space is a cube, with no more than the minimally required penetrations of the walls and with no embellishments. No one would truly like to see this become the standard. Buildings should fit into their environment in an aesthetic and congenial manner, but function and use factors should be preeminent.

No mention has been made up to this point of health and safety design factors.

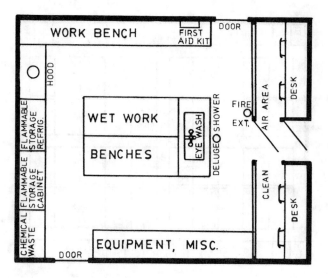

FIGURE 3.1.  Standard laboratory module.

They must be incorporated into virtually every other design feature. The location of a building, access to the building, the materials of construction and interior finish, size and quality of doors, width and length of corridors, number of floors, number of square feet per floor, selection of equipment, and utilities, to name a few, all are impacted by safety and health requirements. Although it would be anticipated that architects and engineers would be thoroughly familiar with all these considerations, experience has shown that this is not necessarily so, especially where they involve safety concepts other than those relating to fire or strength of materials. Even in these areas, the wide range of variability in interpretation of codes results in a tendency to liberally interpret the codes in favor of increasing the amount of usable space or enhancing the visual aspects of the design. It is surprising how few architectural firms maintain dedicated expertise on their permanent staff in the areas of building code compliance and health and safety.

Since relatively few laboratories are built, compared to the numbers of other types of buildings, comparatively few firms are really well prepared to design them for maximum safety, especially in terms of environmental air quality and laboratory hazards. For this reason, the eventual owners of a building being planned should be sure to include persons knowledgeable in health and safety requirements in the group representing them while working with architects and contractors. Where this expertise is not available in-house, they should not hesitate to hire appropriate consultants to review the plans and specifications prior to soliciting bids.

Shown in Figure 3.1 is a standard laboratory module which forms the basis for much of the material in this chapter. Although simply a representative example, this design provides a significant number of safety features. Some variations on this design omit the central workbenches down the center of the facility. The laboratories on either side are mirror images of this one and the alternating pattern can be repeated

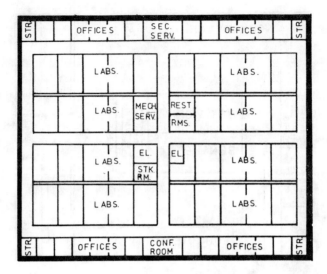

FIGURE 3.2.   Typical floor of research building using modular laboratory concept.

to fill the available space. The two side doors can be operational doors if the space allocated to the research is inadequate and access to additional lab modules is needed, or either or both can be constructed as breakaway emergency exits. As a result, from every point within the room, even at the end of a sequence of laboratories, there are always at least two exits available in an emergency, unless the entire room is directly involved.

In this basic design, the areas where the likelihood of a violent accident are greatest are at the far end of the laboratory, away from the corridor entrance, and are well separated from stored flammable materials and other reagents. The desk areas are separated from the work areas by a transparent barrier which, with the door to the laboratory properly closed, isolates the workers, when not actively engaged in their research, from the possible effects of an accident and a continuous exposure to laboratory atmospheric pollutants. The latter factor is enhanced by the normal negative atmospheric pressure between the laboratory and the corridor, so that the air in the desk areas should be as clean as the corridor air. It would thus represent a safe space for employees or students to socialize, study, or have a drink or snack.

Figure 3.2 provides a simplified illustration of how the modular approach can be integrated into an efficient and safe building design. This figure represents a typical upper floor of a research building. The ground floor would have a loading dock, receiving area, and probably some administrative offices. This plan represents a relatively large building, with provisions for 36 laboratories and 24 offices. For this size facility, the four stairs shown would be needed to accommodate the approximately 175 persons on this floor alone. The arrangement is such that every individual would have at least two alternate routes to follow in an emergency, and the simplicity of the arrangement is such that it would be unlikely to be confusing to even transient occupants of the building. Further, the arrangement lends itself readily to either up or downscaling in size. Even the smallest unit, i.e., one quadrant, of the floor plan could accommodate the same basic safety and other design features. In some instances, this concept results in space which is somewhat overdesigned for the application for which it is being used, but never underdesigned. This is a more costly approach initially, but long-range costs should be substantially less as the need for upgrading should arise only rarely. By varying only the laboratory furniture configuration, resulting in differing amounts of bench space and storage, the basic modular space could provide room for additional hoods or wall space for larger pieces of equipment.

The arrangement of laboratories back to back, with a space between them, as shown in Figure 3.2, provides a natural opportunity to provide utilities, fresh air and exhaust ventilation ducts, and space for exhaust ducts for hoods in a manner which will facilitate compliance with building codes. The location of offices on the outer walls provides access to the outdoors, which should be appealing to the individuals assigned to these rooms. There are other configurations which would also serve, such as having the laboratory module groups arranged transversely to the major axis of the building. The use of compactly arranged modules, which these naturally permit, allows the architect and building owner to achieve an efficient building in terms of both net assignable square feet and utilization.

An aspect of the above design which may not be immediately apparent is that it is especially appropriate for adding to an older facility which was originally designed to meet less demanding standards than those of today. The newer component, situated adjacent to the original structure and designed to meet current sophisticated research requirements, can be connected to the older one at appropriate places. By proper construction and separations, it would be possible to treat the old and new components as separate buildings, even though they are joined, so that it would not be necessary to renovate the older building to current construction standards. Less demanding operations, such as instructional laboratories and offices, could remain in the older component and activities requiring more and higher level services, higher construction standards, etc. would be located in the new area. All of a department's operations would be in the "same" building, which has important logistical and personnel implications, but construction of an entirely new building for a department would be unnecessary. This concept is called an "infill" approach and provides some important financial savings as it extends the lifetime of some older facilities. The methods of joining and maintaining separations between the two components also provide opportunities for architects to express themselves, such as making the spaces between the two sections into attractive atria to be used as social spaces.

## REFERENCES (SECTIONS 3.1 TO 3.1.1)

1. **Barker, J. H.,** Designing for Safer Laboratories, CDC Laboratory Facilities Planning Committee, Chamblee Facility, Atlanta, GA.
2. **Earl Wall and Associates,** Basic Program of Space Requirements, Department of Chemistry, VPI & SU, Laboratory Layout Studies, Blacksburg, VA, 1980.

### 3.1.2. BUILDING CODES AND REGULATORY REQUIREMENTS*

There are many codes and standards applicable to building construction. Many of these are incorporated by reference in the OSHA standards. A large number of the codes grew out of a concern for fire safety and, hence, this general area is relatively mature. Existing health codes generally address only acute exposures and immediate toxic effects. Since concern for long-term systemic effects have only recently been addressed in standards, there are fewer of them. OSHA proposed replacing the detailed OSHA standards, which were intended primarily for industrial situations, with a performance standard which would require laboratories to prepare and voluntarily comply with an industrial hygiene plan, applicable to the specific laboratory operations, providing adequate personnel protection. There have been a number of objections to this standard by a variety of organizations which feel that voluntary compliance with an internally developed plan does not provide adequate protection, while other groups feel that research laboratories, as small-scale operations and for a number of other reasons, should be exempt from the proposed regulations. Because this issue has not been resolved and since any final ruling will almost certainly be tested in court, the material to be presented here will assume the present state of OSHA standards. In any event, the intent of the proposed standard is to provide protection at least equivalent to the existing standards. Further, the Hazard Communication Standard which is now applicable to laboratories should serve to insure that laboratory personnel will receive training and be made aware of the risks to which they are exposed.

There are two specific sources from which a substantial portion of the material in this section is derived or to which it is compared. For a building code, the information will be drawn primarily from the BOCA (Building Officials and Code Administrators) code. This is not used universally, but is a widely accepted building code which, where applicable, provides mandatory standards. Other regional building codes are not conceptually different. Standard 45 of the National Fire Protection Association is specifically labeled as a laboratory safety standard. Generally, it has not been adopted as a formal legal requirement in most localities, but does provide valuable guidance in certain areas for goals against which both existing and proposed laboratories can be measured. These two sources are primarily concerned with fire and construction safety. The materials cited are those which most directly affect the safety of building occupants or which may be useful to persons discussing building design with architects and contractors.

### 3.1.2.1. Building Classification**

Standard 45 and the BOCA code do not always agree, or at least they sometimes lead to different interpretations. The classification of the structure or building in

---

* The material in this chapter concerned with building code requirements has been reviewed by Mr. Howard W. Summers, Chief Fire Marshal for the State of Virginia.

** Tables or definitions in this section or subsections are derived from The 1984 BOCA Code which has been formally adopted by the state in which the author lives, although they may be paraphrased or shortened where it is felt that, in the context of laboratory structures, clarity might be improved. In each instance that a code is directly quoted or closely paraphrased, the following copyright notice applies. In most cases, the notice will be repeated, but if not, it does apply. "BOCA Basic/National Building Code 1984 edition, copyright 1983, Building Officials and Code Administrators International, Inc. Published by arrangements with the author. All rights reserved."

which testing or research laboratories are operated is usually designated under BOCA as an educational or business use occupancy, although if the degree of hazard meets a number of specific criteria, it may be designated as a hazardous use facility. In NFPA Standard 45, buildings used for the purpose of instruction by six or more persons are classified as an educational occupancy. The classification is not a trivial question since it evokes a number of different design and construction constraints. A building used primarily for instruction, which might include instructional laboratories, and some testing and research laboratories might be considered primarily an educational occupancy if the research areas were separated properly. An educational occupancy is more restrictive than a business occupancy, but less so than a hazard use classification. Standard 45 classifies laboratories as class A, B, or C, according to the quantities of flammable and combustible liquids contained within them, with A being the most hazardous and C being the least. As will be discussed later, a system of ratings has been developed in the life sciences to designate laboratories according to four safety levels, with classes 1 and 2 meeting the needs of most laboratory operations, while 3 and 4 are restrictive and very restrictive, respectively. This concept for consistency, might eventually be adapted to laboratories of all types. In a later section of this handbook, a proposed classification scheme for chemical laboratories is presented.

For the purposes of this section, the BOCA code will be used as the reference to define the classification of a building. The basic classification, therefore, will be either as an educational or business use occupancy. However, since some laboratories and ancillary spaces such as store rooms may meet the BOCA definitions of high hazard use (Group H), the following material, extracted from section 306 of the 1984 BOCA code, can be used as a guide to whether a given building or facility should be considered as a Group H occupancy.

Section 306 begins with the general statement:*

> All buildings and structures, or parts thereof, shall be classified in Use Group H which are used for the manufacturing, processing, generation or storage of corrosive, highly toxic, highly combustible, flammable or explosive materials that constitute a high fire or explosion hazard, including loose combustible fibers, dust and unstable materials.

Following this general definition, the BOCA manual provides a table of indicators employed in Use Group H classifications. The portion of this table which would be applicable to laboratories is given below in Table 3.1

There are a number of exceptions in the BOCA code which would exempt a building from being classified in Use Group H, even though the materials listed in Table 3.1 might be found within it. The one most applicable to laboratories (modified slightly) is*

> Any building or portion of a building containing less than the exempt amount of those materials shown in Table 3.2 when maintained in accordance with the BOCA B/NFPC Fire Prevention Code.

---

* BOCA Basic/National Building Code 1984 edition, copyright 1983, Building Officials and Code Administrators International, Inc. Published by arrangements with the author. All rights reserved.

## TABLE 3.1
## Use Group H, High Hazard Uses

Combustible liquids having a flash point[a] at or above 100°F (38°C) shall be divided as follows:
    Class 2 liquids shall include those having flash points at or above 100°F (38°C), and below 140°F (60°C)
    Class 3 liquids shall include those having flash points at or above 140°F (60°C) and below 200°F (93°C)
Corrosive liquids which, when in contact with living tissue, will cause severe damage to such tissue by chemical action or are liable to cause fire when in contact with organic matter or with certain chemicals such as acids and alkaline caustic liquids
Explosive material and any chemical compound, mixture, or device, the primary and common purpose of which is to function by explosion with substantially simultaneous release of gas and heat, the resultant pressure being capable of destructive effects
Flammable liquids having a flash point below 100°F (38°C) and a vapor pressure not exceeding 40 psia (276 kPa) at 100°F (38°C). Class 1 liquids shall include those having flash points below 100°F (38°C) and may be subdivided as follows:
    Class 1A shall include those having flash points below 73°F (23°C) and having a boiling point below 100°F (38°C)
    Class 1B shall include those having flash points below 73°F (23°C) and having a boiling point at or above 100°F (38°C)
    Class 1C shall include those having flash points at or above 73°F (23°C) and below 100°F (38°C)
Flammable gas having a flammability range with air greater than 1% by volume which is a liquid while under pressure and having a vapor pressure in excess of 27 psia (186 kPa) at a temperature of 100°F (38°C)

[a]    Flash point is the minimum temperature at which a liquid gives off vapor in sufficient concentration to form an ignitable mixture with air near the surface of the liquid within the vessel, as specified by appropriate test procedures and apparatus.

It is possible to have different areas in a building classified differently. If this is so, then the requirements for each use area shall be met in those areas and where provisions differ, the requirements providing the greater degree of safety shall apply to the entire building or a complete fire separation shall be provided between the two sections. Generally, the most restrictive height and area restrictions will still apply to the entire building.

### 3.1.2.2. Types of Construction*

There are five classifications of types of construction. Basically, without providing complete definitions, the construction materials in Types 1 and 2 are considered noncombustible. Type 3 may include some combustible materials, not including, however, exterior, fire, and party walls. Type 4 requires the exterior walls to be of noncombustible materials, while the interior structural members may be of heavy solid or laminated timber without concealed spaces (the allowable dimensions of the wooden components are specified in the BOCA Standard). Type 5 constructions may be made of any materials permitted by the code that have appropriate fire ratings, which are dependent on a number of parameters. With the exception of Type 4, these

* BOCA Basic/National Building Code 1984 edition, copyright 1983, Building Officials and Code Administrators International, Inc. Published by arrangements with the author. All rights reserved.

**TABLE 3.2**
**Exempt Amounts of Hazardous Materials, Liquids,**
**and Chemicals**

| Materials | Maximum quantities[a] |
|---|---|
| Flammable liquids | |
| Class 1A | 30 gal[b] |
| Class 1B | 60 gal[b] |
| Class 1C | 90 gal[b] |
| Combustible liquids[c] | |
| Class 2 | 120 gal[b] |
| Class 3A | 250 gal[b] |
| Combination flammable liquids[d] | 120 gal[b] |
| Flammable gases | 3000 ft³ at 1 atm of pressure at 70°F |
| Liquified flammable gases | 60 gal |
| Combustible fibers — loose | 100 ft³ |
| Combustible fibers — baled | 1000 ft³ |
| Flammable solids | 500 lb |
| Unstable materials | No exemptions |
| Corrosive liquids | 55 gal |
| Oxidizing material — gases | 6000 ft³ |
| Oxidizing material — liquids | 50 gal |
| Oxidizing material — solids | 500 lb |
| Organic peroxides | 10 lb |
| Nitromethane (unstable materials) | No exemptions |
| Ammonium nitrate | 1000 lb |
| Ammonium nitrate mixtures containing more than 60% nitrate by weight | 1000 lb |
| Highly toxic material and poisonous gas | No exemptions |

[a] 1 gal = 0.00379 m³; 1 ft³ = 0.028 m³; 1 lb = 0.454 kg; 70°F = 21.1°C.
[b] Quantities may be increased by 100% in areas which are not accessible to the public. In buildings where automatic fire suppression systems are installed, the quantities may be increased 100% in the areas accessible to the public.
[c] Tank storage up to 660 gal for fuel burning equipment meeting the require ments of the BOCA mechanical and fire prevention codes shall be permitted.
[d] Containing not more than exempt quantities of Class 1A, 1B or 1C flammable liquids.

classifications are subdivided further into subclasses. As will be noted later in Table 3.4, for the three types of occupancy laboratories are likely to fall under, the height and area restrictions for Type 5 construction would make it unlikely that a typical chemical research laboratory would use this level of construction. Therefore, in Table 3.3, where the fire resistance ratings of the various structural components in each type of construction are given, the data for Type 5 construction are deleted.

In order to facilitate the use of the following tables, a number of definitions are in order (paraphrased slightly from those given in BOCA).

• Fire resistance rating: the time in hours or minutes that materials or their assemblies will resist fire exposure, as determined by the fire test specified in

ASTM E119 or, for concrete materials, the CRSI book, Reinforced Concrete Fire Resistance, or PCI MNL 124-77 or, for steel columns, the method given in AISI, Designing Fire Protection for Steel Materials.

- Fire separation (exterior): the distance in feet measured from the building face to the closest interior lot line, to the center line of a street or public way, or to an imaginary line between two buildings on the same property.
- Protected: construction in which all structural members are constructed or protected in such a manner that the individual unit or the combined assemblage of all such units has the requisite fire resistance rating for its specific use or application.
- Wall
  — Bearing wall: any wall supporting a vertical load in addition to its own weight.
  — Fire wall: a fire resistance-rated wall, which may have protected openings, which is intended to restrict the spread of a fire and which is continuous from the foundation to or through the roof of a building.
  — Fire separation wall: similar to a fire wall in that it is intended to restrict the spread of a fire, but does not include the requirement of extending from the foundation to the roof of a building.
  — Party wall: a fire wall on an interior lot line used for joint service between two buildings.

A number of points are immediately obvious upon examining Table 3.3. Qualification for Type 1 construction requires fire resistance ratings substantially higher than those for other classes. It will cost more to achieve these ratings.

Note that Table 3.3 above gives the basic data. There are a number of qualifications and exceptions in the full table. Ratings for bearing walls are much higher than those for walls which serve only to separate spaces. Walls which serve to prevent the spread of fire have ratings as high as or higher than bearing walls. Not given in the table is the additional requirement that fire walls must have structural stability sufficient to survive the collapse of the structure on either side. An important point to note is that shafts, such as those that might enclose a fume hood duct, penetrating a fire separation such as a floor must be of noncombustible material and provide a 2-h fire resistance rating. This is not to say that the duct itself must meet this requirement, but it must be enclosed in a shaft which does. Note that Type 2C construction has no fire resistance rating requirements for structural components, except for those items directly related to fire separation or components associated with paths of egress. Type 3B construction is comparably tolerant for most internal components, although it is similar to Type 2A and 2B for exterior walls.

### 3.1.2.3. Area and Height Limitations*

The following table, again a shorter version of the one given in BOCA, gives the maximum heights, number of stories above grade, and areas (for one- or two-story buildings facing on a street or public space 30 ft or more in width) for the classes of

---

* BOCA Basic/National Building Code 1984 edition, copyright 1983, Building Officials and Code Administrators International, Inc. Published by arrangements with the author. All rights reserved.

**TABLE 3.3**
**Fire Resistance Ratings of Structure Ratings in Hours**
**(abbreviated version of Table 401, BOCA Building Code)**

| | Type of construction | | | | | | | |
|---|---|---|---|---|---|---|---|---|
| | Noncombustible | | | | | Noncombustible/combustible | | |
| | Type 1 | | Type 2 | | | Type 3 | | Type 4 |
| | Protected | | Protected | Unprotected | | Protected | Unprotected | (Heavy timber) |
| Structure element | 1A | 1B | 2A | 2B | 2C | 3A | 3B | 4 |
| Exterior walls | | | | | | | | |
| Fire separation of 30 ft | | | | | | | | |
|   Bearing | 4 | 3 | 2 | 1 | 0 | 2 | 2 | 2 |
|   Nonbearing | 0 | 0 | 0 | 0 | 0 | 0 | 0 | 0 |
| Fire separation of 11 ft or more, but less than 30 ft | | | | | | | | |
|   Bearing | 4 | 3 | 2 | 1 | 0 | 2 | 2 | 2 |
|   Nonbearing | 1.5 | 1.5 | 1 | 1 | 0 | 1.5 | 1.5 | 2 |
| Fire and party walls | 4 | 3 | 2 | 2 | 2 | 2 | 2 | 2 |
| | ( —— Not less than fire grading of use group —— ) | | | | | | | |
| | 1.5 for educational, 2 for business, 4 for high hazard | | | | | | | |
| Fire separation assemblies | ( —— Fire resistance ratings corresponding to fire grading of use group —— ) | | | | | | | |
| Fire enclosures of exits, exit hallways, and stairways | 2 | 2 | 2 | 2 | 2 | 2 | 2 | 2 |
| Shafts (other than exits) and elevator hoistways | 2 | 2 | 2 | 2 | 2 | 2 | 2 | 2 |
| | ( —— Noncombustible —— ) | | | | | | | |
| Exit access corridors | 1 | 1 | 1 | 1 | 1 | 1 | 1 | 1 |
| Interior bearing walls, partitions, columns, trusses (other than roof trusses), and framing[a] | | | | | | | | |
|   Supporting more than one floor | 4 | 3 | 2 | 1 | 9 | 2 | 9 | |
|   Supporting only one floor | 3 | 2 | 1.5 | 1 | 0 | 1 | 0 | |
|   Supporting a roof only | 3 | 2 | 1.5 | 1 | 0 | 1 | 0 | |
| Structural members supporting wall | 3 | 2 | 1.5 | 1 | 0 | 1 | 0 | 1 |
| | ( —— Not less than fire resistance of wall supported —— ) | | | | | | | |
| Floor construction, including beams[b] | 3 | 2 | 1.5 | 1 | 0 | 1 | 0 | 1 |
| Roof construction including beams, trusses and framing, and arches and roof deck | | | | | | | | |
|   15 ft or less in height to lowest member | 2 | 1.5 | 1 | 1 | 0 | 1 | 0 | |

**TABLE 3.3 (continued)**
**Fire Resistance Ratings of Structure Ratings in Hours**
**(abbreviated version of Table 401, BOCA Building Code)**

| | Type of construction | | | | | | | |
|---|---|---|---|---|---|---|---|---|
| | Noncombustible | | | | | Noncombustible/combustible | | |
| | Type 1 | | Type 2 | | | Type 3 | | Type 4 |
| | Protected | | Protected | | Unprotected | Protected | Unprotected | (Heavy timber) |
| Structure element | 1A | 1B | 2A | 2B | 2C | 3A | 3B | 4 |
| Roof construction including beams, trusses and framing, and arches and roof deck | | | | | | | | |
| More than 15 ft, but less than 20 ft in height to lowest member | 1 | 1 | 1 | 0 | 0 | 0 | 0 | |
| 20 ft or more in height to lowest member | 0 | 0 | 0 | 0 | 0 | 0 | 0 | |

a    See full definition of Type 4 construction in BOCA Code Manual.
b    For materials other than heavy timbers.

occupancy which are the most probable for laboratory facilities — business, educational, high and moderate hazard storage. Note that the areas are for the maximum horizontal projection of the building. There are provisions which would allow these limits to be exceeded or would allow area to be traded for additional height. For example, except for the hazard use group, the areas in Table 3.4 can be increased by 200% for one- and two-story buildings and 100% for buildings with more than two stories if they are equipped with an automatic fire alarm system. Similarly, for each 1% of a building's perimeter in excess of 25% fronting on a street or a vacant space, 30 ft or more wide, and accessible by a legitimate (more than 18 ft wide) marked fire lane, the building's area can be increased by 2%. An automatic fire suppression system also allows the height of a building to be increased by one additional story or 20 ft, again with the exception of a hazard use group building. If land values are high or space is limited, this is an option. However, the difficulty in evacuating multistory buildings argues against buildings of excessive height, especially, as noted earlier for academic buildings, where the likelihood is that many of the occupants will not be familiar with evacuation routes, coupled with the possibility that a laboratory fire may involve toxic fumes. This concept is supported in the Code by the disparity between business and educational use occupancies in Table 3.4. For example, although the areas are the same for each, one or two fewer stories are permitted for each type of construction for a building classified as an educational occupancy structure.

**TABLE 3.4**
**Height and Area Limits**

| Type of construction | Use class | | | |
| --- | --- | --- | --- | --- |
| | Business | Educational | High hazard | Moderate storage |
| Noncombustible | | | | |
| 1A (protected) | Not limited | Not limited | 16,800 | Not limited |
| | | | 5 stories, 65 ft | |
| 1B (protected) | Not limited | Not limited | 14,400 | Not limited |
| | | | 3 stories, 40 ft | |
| 2A (protected) | 34,200 | 34,200 | 11,400 | 19,950 |
| | 7 stories, 85 ft | 5 stories, 65 ft | 3 stories, 40 ft | 5 stories, 65 ft |
| 2B (protected) | 22,500 | 22,500 | 7,500 | 13,125 |
| | 5 stories, 65 ft | 3 stories, 40 ft | 2 stories, 30 ft | 4 stories, 50 ft |
| 2C (unprotected) | 14,400 | 14,400 | 4,800 | 8,400 |
| | 3 stories, 40 ft | 2 stories, 30 ft | 1 stories, 20 ft | 2 stories, 30 ft |
| Noncombustible/ combustible | | | | |
| 3A (protected) | 19,800 | 19,800 | 6,600 | 11,550 |
| | 4 stories, 50 ft | 3 stories, 40 ft | 2 stories, 30 ft | 3 stories, 40 ft |
| 3B (unprotected) | 14,400 | 14,400 | 4,800 | 8,400 |
| | 3 stories, 40 ft | 2 stories, 30 ft | 1 stories, 20 ft | 2 stories, 30 ft |
| 4 (heavy timber) | 21,600 | 21,600 | 7,200 | 12,600 |
| | 5 stories, 65 ft | 3 stories, 40 ft | 2 stories, 30 ft | 4 stories, 50 ft |

In order to erect a building more than the one or two stories for which the area figures in Table 3.4 are relevant, a penalty must be paid in terms of allowable area per floor, except for 1A and 1B construction. For 2C construction, this penalty amounts to 5% for each story above two. For all other types of construction, there is a 20% penalty for three- and four-story buildings and an additional penalty of 10% more for each story above four.

Although laboratory towers have been built, it is clear that a heavy penalty in either area per floor or in construction costs, due to building to a higher resistance rating, must be paid for erecting a code-conforming, high-rise laboratory building. Since there are significant reductions in the fire resistance ratings required for structural components for 2B construction and still more for 2C, compared to 2A, and since there are, as noted earlier, problems associated with evacuating multistory laboratory buildings, a good choice for a reasonable size laboratory facility might well be a 2B business-use-occupancy structure.

There are other factors and conditions which may become involved in determining the allowable area, height, etc. in addition to the ones discussed in this section. However, the intent here is not to provide a course in code review, which often involves much more sophisticated details than would be possible to cover here, but, rather, to provide sufficient information as a foundation of basic information for laboratory personnel. The details must be negotiated among the architect, contractor, building official, and representatives of the owner. The participation of laboratory personnel is desirable to define their program needs in the context of what is permissible under the building code. Code issues are not always clear cut, with much of the actual language subject to interpretation. Also, there are often alternative ways to provide equivalent protection, so that requests for variances based on this concept are frequently acceptable.

Additional code issues will be addressed in many of the following sections, where specific design features will be discussed in detail.

### 3.1.3. LABORATORY CLASSIFICATION

There are no universal safety criteria to classify laboratories which take into account all types of risks. Standard 45 of the National Fire Prevention Association designates chemical laboratories of different degrees of risk, based essentially on fire safety factors, regulating the amount of solvents which each class may contain. The Centers for Disease Control has published a set of guidelines establishing a Biological Safety Level rating system for laboratories in the life sciences and those using animals, based on a number of parameters relating to the infectiousness to humans of the organisms used in the facility. This system parallels an earlier four-level classification scheme developed for those working in recombinant DNA research. The standards associated with biological organisms are concerned with the potential risk to the public at large as well as to the laboratory workers. The Department of Agriculture regulates the importation, possession, or use of a number of nonindigenous pathogens of domestic animals. The Drug Enforcement Agency licenses and sets standards for facilities in which controlled substances are employed to ensure that they are used safely and to guard against their loss or theft. The Nuclear Regulatory Commission licenses agencies or individuals using radioactive materials to ensure that neither the workers nor the general public are adversely affected by the use of radiation. To obtain a license, one must demonstrate the competence to use the material safely and to be able and willing to meet an extremely detailed set of performance standards. All of these standards have been developed essentially independently and, where a regulatory agency is involved, are administered separately. In many instances, laboratory operations will be affected by many of these standards. However, even if all of the standards were imposed simultaneously, there would still be many safety factors which would not be included. Thus it is at least partially the responsibility of the institution or corporation to establish additional criteria to properly evaluate the degree of risk in a research program and to assign it to a suitable classified space.

Materials with low risk potentials will obviously be much more tolerant of poor facilities or procedures than one involving a high risk but not totally so. Even a small quantity of an 1A flammable such as ether, used in an inadequate facility could lead to a serious accident while the same quantity, used in a hood by a careful worker following sound safety procedures could be used quite safely.

OSHA is currently proposing that performance standards be established which would require that every laboratory develop a safety program which would ensure that the employees would be as well protected as those working in industrial situations, for which detailed safety standards have been published under OSHA. This would bypass, at least as far as OSHA is concerned, the need for any sort of laboratory classification scheme, leaving the responsibility primarily to the local laboratory or organization. It would not replace the biological guidelines or standards since the proposed OSHA standard does not, at this time, include pathogens as a possible risk, nor would it supercede radiation safety standards. However, much opposition has developed to this proposed standard based on the lack of experience of most research personnel in such areas as toxicology and other specialties which would affect the safety of laboratory personnel. Although it is conceded by most of those opposing the proposed standard that experienced laboratory personnel are very knowledgeable about the chemistry of the material with which they work, it is felt by many that this knowledge usually does not cover health and safety factors sufficiently. As a result of the objections which have been raised, it appears that a laboratory safety standard from OSHA may be delayed.

Although it is unlikely that a formal system of classifying laboratories according to a comprehensive safety standard is imminent, it does appear incumbent upon an institution or corporation to ensure that research is assigned to space which is suitably designed and equipped so that research can be performed with a reasonable assurance of safety. If research programs are evaluated properly, it should be possible to assign them to laboratories classified into low, moderate, substantial, and high risk categories. This type of classification seems to be the simplest and most practicable to use and has the further advantage of already being employed in life science laboratories. Before examining the features which might be incorporated into each category, which will depend somewhat upon the area of research involved, it might be well to list at least some of the parameters that should be considered in evaluating research programs.

### 3.1.3.1. Program-Related Factors

Evaluation of programs to permit assignment to the appropriate class of facility should depend upon several factors:

I.    Materials
    A.    Recognized risks
        1.    Flammable
        2.    Reactive
        3.    Explosive
        4.    Acute toxicity
        5.    Known systemic or chronic health effects
            a.    Carcinogens, onconogens
            b.    Mutagens, terratogens
            c.    Affect reproduction/fertility
            d.    Radiation
            e.    Pathogens
            f.    Affect the respiratory system
            g.    Neurotoxic
            h.    Known strong allergens
            i.    Other known health effects
        6.    Physical risks
            a.    Electrical
            b.    High pressure
            c.    Heat and cold
            d.    Sound

          e.    Nonionizing radiation/light
          f.    Other physical risk factors
    B.   Quantities/scale of operations
    C.   Procedures

II.   Information/training
    A.   Health and safety training
        1.   Documentation of safety and health training for laboratory managers
        2.   Procedures to train new personnel
        3.   Procedures to train all personnel when new materials/new procedures are used
    B.   Material safety data sheets available for all chemicals
    C.   General health and safety program in effect

III.  Personnel protection
    A.   Exposure monitoring
    B.   Personal protective equipment available
    C.   Health assurance program available

The information in Part I above is, in effect, an evaluation of possible negative aspects of the program under consideration while positive information under each of the items in Parts II and III can be used to offset, to some degree, the needs which must be met by the facility. It is preferable, however, to design in safety rather than depending upon procedures and administrative rules.

### 3.1.3.2. Laboratory Class Characteristics

In the following four sections, oriented primarily toward chemistry laboratories, the reader already familiar with laboratory classification guidelines established by the Centers for Disease Control will note that, in many respects, the recommendations for low, moderate, substantial, and high risk categories closely parallel those for biosafety levels 1 to 4. It will be noted that this system will involve classifying laboratory facilities by much more than the configuration of bricks and mortar of which they are built, or their contents of a single type of hazardous materials, although these aspects will be important. Separate major sections in Chapter 5 will be devoted to laboratories in the life sciences, animal facilities, and radiation so reference to topics relevant to those areas will be deferred to those sections.

### 3.1.3.2.1. Low-Risk Facility

A low risk facility is used for work with materials, equipment, or classes of operations with no known or minimal risk to the workers, the general public, or to the environment. It is possible to work on open benches. No special protection or enclosures are needed for the equipment or operations. Laboratory workers have been properly trained in laboratory procedures and are supervised by a trained and knowledgeable person. If there are any potential risks, the employees have been informed of them, how to detect them if they are not immediately obvious, and emergency procedures. Although the laboratory design requirements are not stringent, features which would be difficult to change if the utilization should become one which would require a higher classification should be built to a higher level. Examples, marked with an asterisk (∗) are provisions for easily cleaned and decontaminated floors and laboratory furniture and good ventilation.

## A. Standard Practices

1.  Access to the laboratory is limited at the discretion of the laboratory supervisor, as needed.
2.  A program exists to ensure that reagents are stored according to compatibility.
3.  An annual chemical inventory will be performed and sent to a central data collection point. Outdated and obsolete chemicals will be disposed of through a chemical waste disposal program.
4.  The laboratory will be maintained in an orderly fashion.
5.  Although it is anticipated that the amount of hazardous chemicals used in a low risk facility will be very limited, all secondary containers, containing materials incorporating more than 1.0% of a hazardous component or combination of hazardous components, which will be used more than a single work day, shall be labeled with a label listing the hazardous components.
6.  Any chemical wastes are placed in appropriate and properly identified containers for disposal through a chemical waste disposal program. Broken glass is disposed of in heavy cardboard or kraftboard boxes labeled "broken glass". Only ordinary solid, nonhazardous waste may be placed in ordinary trash containers.
7.  Eating, drinking, and smoking are not permitted in the work area.
8.  No food or drink can be placed in refrigeration units used in the laboratory.
9.  The telephone numbers of the laboratory supervisor, any alternates, and the department head shall be posted on the outside of the laboratory door or the adjacent wall.

## B. Special Practices
There are no special practices associated with a low risk laboratory.

## C. Special Safety Equipment

1.  Any refrigerators or freezers shall be rated as acceptable for "Flammable Material Storage" except for ultra-low temperature units.∗
2.  No other special safety equipment is needed.

## D. Laboratory Facilities

1.  The floor of the laboratory is designed to be easily cleaned. Seamless floors and curved junctures to walls aid in accomplishing this.∗
2.  Bench tops should be resistant to the effects of acids, bases, solvents, moderate heat, and should not absorb water. The tops should have few seams or crevices to facilitate cleaning.∗
3.  Furniture should be designed to be sturdy and designed for convenient utilization. Storage spaces should be easily accessible.
4.  Aisle spaces should be 40 to 48 in. wide and shall not be constricted to less than 28 in. by any temporary obstacles.

5.   Electrical outlets shall be three-wire outlets with high quality, low resistance ground connections. Circuits should be clearly identified to correlate with labels in breaker panels.
6.   The laboratory should be supplied with a sink. The plumbing shall be sized to accommodate a deluge shower and eyewash station.* With average water pressure, this would normally be a one-inch line or larger.
7.   Normal building ventilation is sufficient. However, it is recommended that 6 air changes per hour of 100% fresh air be provided as standard.*

### 3.1.3.2.2. Moderate Risk Facility

A moderate risk facility involves material, practices, and use of equipment such that improper use could pose some danger to the employees, the general public, or the environment. Generally, the materials used would have health, reactivity, or flammability ratings, according to NFPA Standard 704 of 2 or less. Small quantities of materials with higher ratings might be involved in work being performed in chemical fume hoods or in closed systems. Work with special risks, such as with carcinogens, would not be performed in a moderate risk facility. Equipment which could pose a physical hazard should have adequate safeguards or interlocks. However, in general, most operations could be safely carried out on an open work bench or without unusual precautions. The amounts of flammables kept in the laboratory meet NFPA standard 45 for Class A laboratories (or less), and when not in use, are stored in either a suitable flammable material storage cabinet or other storage unit.

The person responsible for the work being performed in the laboratory is to be a competent scientist. This individual shall develop and implement a safety and health program for the facility. The individual workers are to be fully trained in the laboratory procedures being employed and to have received special training in the risks specifically associated with the materials or work being performed. The workers are to be informed about the means available to them to detect hazardous conditions and the emergency procedures which should be followed, should an incident occur.

### A. Standard Practices

1.   Access to the laboratory work area is limited during the periods work is actively in progress, at the discretion of the laboratory supervisor.
2.   A program exists to ensure that chemicals are stored properly, according to compatibility. Quantities of chemicals with hazard ratings of 3 or greater are limited to the amount needed for use in a 2-week interval, or in accordance with NFPA standard 45 for flammables, whichever is less.
3.   An annual chemical inventory will be performed and sent to a central data collection point. Outdated and obsolete chemicals will be disposed of through a chemical waste disposal program. Ethers and other materials which degrade to unstable compounds shall be shelf dated for disposal 6 months after being opened, but no more than 12 months after purchase, even if unopened, unless processed to remove the dangerous compounds.
4.   A material safety data sheet (MSDS) file will be maintained for all chemicals purchased in the laboratory, for which they are available. The file will be

accessible to the employees in the laboratory. All laboratory workers shall be trained in how to interpret the information in an MSDS.

5. All secondary containers, containing materials containing more than 1.0% of a hazardous component or combination of hazardous components, which will be used more than a single work day shall be labeled with a label listing the hazardous components.

6. Any chemical wastes are placed in appropriate and properly identified containers for disposal through a chemical waste disposal program. Broken glass is disposed of in heavy cardboard or kraftboard boxes labeled "broken glass". Only ordinary solid, nonhazardous waste may be placed in ordinary trash containers.

7. The laboratory will be maintained in an orderly fashion.

8. No food or drink can be placed in refrigeration units used in the laboratory.

9. A placard or other warning device shall be placed on the door or on the wall immediately adjacent to the door identifying the major classes of hazards in the laboratory (see Section 2.3.4).

10. The telephone numbers of the laboratory supervisor, any alternates, and the department head shall be posted on the outside of the laboratory door or the adjacent wall.

## B. Special Practices

1. Work with materials with safety and health ratings of 3 or greater in any category shall be performed in a functioning fume hood.

2. Work with substantial amounts of materials with hazard ratings of 1 or 2 shall be formed in a hood, or in an assembly designed to be safe in the event of a worst-case failure.

3. Appropriate personal protective equipment shall be worn in the work area. Because eyes are critical organs which are very susceptible to chemical injuries or minor explosions, it is recommended that chemical splash goggles be worn whenever the work involved offers any possibility of eye injury. Wearing of contact lenses should be avoided, but if an individual must wear them for medical reasons, then they must wear chemical splash goggles at all times they are in the laboratory.

## C. Special Safety Equipment

1. Any refrigerators or freezers shall be rated as acceptable for "Flammable Material Storage" except for ultra-low temperature units.

2. A flammable material storage cabinet, either built-in or free standing, shall be used for the storage of flammable materials.

3. The laboratory shall be equipped with a fume hood.

4. The laboratory shall be equipped with an eyewash station and a deluge shower.

5. The laboratory shall be provided with one or more Class 12 ABC fire extinguishers.

6. Any special equipment mandated by the research program shall be provided.

## D. Laboratory Facilities

1.    The floor of the laboratory is designed to be easily cleaned. Seamless floors and curved junctures to walls aid in accomplishing this.
2.    Bench tops should be resistant to the effects of acids, bases, solvents, moderate heat, and should not absorb water. To facilitate cleaning, the tops should have few seams or crevices.
3.    Furniture should be designed to be sturdy and designed for convenient utilization. Storage spaces should be easily accessible.
4.    Aisle spaces should be 40 to 48 in. wide and shall not be constricted to less than 28 in. by any temporary obstacles.
5.    Electrical outlets shall be three-wire outlets with high quality, low resistance ground connections. Circuits should be clearly identified to correlate with labels in breaker panels.
6.    The laboratory shall be supplied with a sink. The trap should be of corrosion-resistant material. The plumbing shall be sized to accommodate the deluge shower and eyewash station. With average water pressure, this would normally be a 1-in. line or larger.
7.    Six air changes per hour of 100% fresh air shall be supplied to the facility. No air shall be recirculated. The ventilation system shall be designed such that the room air balance is maintained at a small negative pressure with respect to the corridors whether the fume hood is on or off.
8.    It is recommended that the facility include a separation of work spaces and desk areas as well as a second exit, as shown in the standard laboratory module, Figure 3.1 (see Section 3.1.1).

### 3.1.3.2.3. Substantial-Risk Facility

For the two lower risk categories, it is possible to be almost completely general since they are specifically intended to be used for only limited risks. However, for both substantial risk and high risk facilities, the nature of the risk will dictate specific safety-related aspects of the facility. Most of these can be accommodated at the substantial risk level within the standard laboratory module, appropriately modified and equipped.

The use of highly toxic, highly reactive, or highly flammable chemicals or gases would mandate the work being conducted within at least a substantial risk facility. If explosives are involved, then the laboratory should be designed with this in mind. Explosion venting may be required. The location of the facility may be dictated by the need to contain or control the debris or fragments from an explosion. The level of construction may need to be enhanced to make the walls stronger to increase their explosion resistance. The use of toxic or explosive gases may require continuous air monitoring with alarms designed to alert the occupants of levels approaching an action level, which should be no higher than 50% of the level representing either a Permissible Exposure Limit (PEL) or the Lower Explosive Limit (LEL). The alarms must be connected to the building alarm system. Highly flammable materials may require special automatic extinguisher systems, using high speed fire detectors, such as ultraviolet light sensors coupled with dry chemical or halon fire suppression systems. There are, of course, other risks, as tabulated in Section 3.1.3.1, which would require other precautions.

Access to a substantial risk facility shall be restricted during operations, and at other times at the discretion of the laboratory supervisor, to authorized personnel only. The laboratory supervisor shall be a competent scientist, having specific knowledge and training relevant to the risks associated with the program of research in the laboratory. Each person authorized to enter the laboratory shall have received specific safety training appropriate to the work and to the materials employed. A formal, written emergency plan shall be developed, and practiced at least annually. A copy of the emergency plan shall be provided to all agencies, including those outside the immediate facility, who would be called upon to respond to an incident. The emergency plan shall include a list of all personnel in the facility with business and home telephone numbers.

## A. Standard Practices

1.  Access to the laboratory is limited to authorized personnel only during operations, and to others at times and under such conditions as designated by written rules or as established by the laboratory supervisor.
2.  All chemicals must be stored properly, according to compatibility. Any chemicals which pose a special hazard or risk shall be limited to the minimum quantities needed for the research program, and materials not in actual use shall be stored under appropriate safety conditions. For example, excess quantities of explosives should be stored in magazines, away from the immediate facility.
3.  An annual chemical inventory shall be performed and sent to a central data collection point. A continuous inventory log shall be maintained of chemicals posing special risks. Outdated and obsolete chemicals will be disposed of through a chemical waste disposal program. Ethers and other materials which degrade to unstable compounds shall be shelf dated for disposal 6 months after being opened, but no more than 12 months after purchase, even if unopened, unless processed to prevent dangerous levels of the unstable compounds from forming.
4.  An MSDS file will be maintained for all chemicals purchased in the laboratory for which they are available. The file will be accessible to the employees in the laboratory. All laboratory workers shall be trained in how to interpret the information in an MSDS.
5.  All secondary containers, holding materials with more than 1% of a hazardous component or combination of hazardous components (0.1% for carcinogens), which will be used more than a single work day shall be labeled with a listing of the hazardous components.
6.  Any chemical wastes are placed in appropriate and properly identified containers for disposal through a chemical waste disposal program. Any wastes which pose a special hazard, or fall under special regulations, and require special handling shall be isolated and a program developed to dispose of them safely and legally. Broken glass is disposed of in heavy cardboard or kraftboard boxes labeled "broken glass". Only ordinary solid, nonhazardous waste may be placed in ordinary trash containers.
7.  The laboratory will be maintained in an orderly fashion. Any spills or accidents will be promptly cleaned up and the affected area decontaminated or rendered safe.

8. No food or drink can be brought into the operational areas of the laboratory, nor can anyone smoke or apply cosmetics.

9. Any required signage or posting mandated by any regulatory agency shall be posted on the outside of the door to the entrance to the laboratory. In addition, a placard or other warning device shall be placed on the door or on the wall immediately adjacent to the door identifying any other major classes of hazards in the laboratory (see Section 2.3.4). A sign shall be placed on the door stating in prominent letters, meeting any regulatory standards, "AUTHORIZED ADMISSION ONLY".

10. The telephone numbers of the laboratory supervisor, any alternates, and the department head shall be posted on the outside of the laboratory door or the adjacent wall.

## B. Special Practices

1. Specific policies, depending upon the nature of the hazard, shall be adopted and scrupulously followed to minimize the risk to laboratory personnel, the general public, and the environment. Several examples of laboratory practices for various hazards are given below. This list is not intended to be comprehensive but instead represents some of the more likely special precautions needed for a variety of types of risks.

   a. All work with hazardous kinds or quantities of materials shall be performed in a fume hood, or in specially designed and totally enclosed systems. It may be desirable for the hood to be equipped with a permanent internal fire suppression system.

   b. Work with explosives shall be limited to the minimum quantities needed. For small quantities used in a hood, an explosion barrier in the hood with personnel wearing protective eye wear, face masks, and hand protection may be sufficient protection. For larger quantities, the facility must be specifically designed for the research program.

   c. Some gases, such as fluorine, burn with an invisible flame. Apparatus for work with such materials should be placed behind a barrier to protect against an inadvertent introduction of a hand or other part of the body, so as to prevent burns.

   d. Systems containing toxic gases or gases which allowed to escape could pose an explosive hazard, shall be leak tested prior to use and after any maintenance or modification which could affect the integrity of the system. Where feasible, the gas cylinders may be placed external to the facility and the gases piped into the laboratory to help minimize the quantity of gas available to an incident. Permanently installed gas sensors, capable of detecting levels of gas well below the danger limits may be needed in some cases.

   e. Vacuum systems, capable of imploding, resulting in substantial quantities of glass shrapnel or flying debris, shall be protected with cages or barriers, or for smaller systems, shall be wrapped in tape.

   f. Systems representing other physical hazards, such as high voltage, radiation, intense laser light beams, high pressure, etc. shall be interlocked so as to prevent inadvertent injuries. The interlocks shall be designed to be fail-safe such that no one failure of a component would render the safety interlock system inoperative.

2.   Activities in which the attention of the worker is not normally engaged with laboratory operations, such as record maintenance, calculations, discussions, study, relaxation, etc. shall not be performed in the laboratory proper but shall be performed in an area isolated from the active work area. The segregated desk area of the standard laboratory module is specifically intended to serve this purpose. Depending upon the nature of the hazard, it is usually feasible to make at least a portion of the barrier separating the two sections of the laboratory transparent so that operations can be viewed if necessary.

3.   Workers in the laboratory should participate in a medical surveillance program if they actively use materials for a significant portion of their work week, which would pose a significant short or long term risk to their health. Employees shall be provided medical examinations if they work with any material requiring participation in a medical program by OSHA or other regulatory agency under conditions which do not qualify for an exemption. Employees shall notify the laboratory supervisor as soon as possible of any illness that might be attributable to their work environment. Records shall be maintained of any such incident.

4.   No safety feature or interlock of any equipment in the facility shall be disabled without written approval of the laboratory supervisor. Any operations which depend upon the continuing function of a critical piece of safety equipment, such as a fume hood, shall be discontinued should the equipment need to be temporarily removed from service for maintenance. Any such item of equipment out of service shall be clearly tagged with a signed "Out of Service" tag. Only the person originally signing the tag, or a specific alternate, shall be authorized to remove the tag.

5.   It shall be mandatory to wear any personal safety equipment required for conducting operations safely in the laboratory.

6.   It is recommended that a laboratory safety committee review each new experiment planned for such a facility to determine if the experiment can be carried out in the facility. If the risk is such that experiments may affect the environment, or the surrounding community, it is recommended that the committee include at least one lay person from the community, not currently affiliated directly or indirectly with the institution or corporation. In this context, "new" is defined as being fundamentally different in character, scope, or scale from any experiment previously approved for the facility.

### C. Special Safety Equipment

1.   Any refrigerators or freezers shall be rated as acceptable for "Flammable Material Storage", except for ultra-low temperature units.

2.   A flammable material storage cabinet, either built-in or free standing, shall be used for the storage of flammable materials.

3.   The laboratory shall be equipped with a fume hood. The fume hood should meet any specific safety requirements mandated by the nature of the research

program. A discussion of hood design parameters will be found in a later section but, for high hazard use, the interior of the hood and the exhaust duct should be chosen for maximum resistance to the reagents used; the blower should either be explosion proof or have nonsparking fan blades; it should be equipped with a velocity sensor and alarm; the interior lights should be explosion proof and all electrical outlets and controls should be external to the unit. It may be desirable to equip the unit with an internal automatic fire suppression system.

4.    The laboratory shall be equipped with an eyewash station and a deluge shower.
5.    The laboratory shall be equipped with an appropriate fire suppression system and be provided with one or more 12BC, or larger, fire extinguishers or Class D units if reactive metals are in use.
6.    An emergency lighting system shall be provided.
7.    A first-aid kit shall be provided and maintained.
8.    Any special equipment mandated by the research program shall be provided. For example, electrical equipment other than refrigerators may need to be designed to be explosion-safe.

## D. Laboratory Facilities

1.    The floor of the laboratory is designed to be easily cleaned. Durable, seamless floors of materials that are substantially impervious to spilled reagents, are easily decontaminated, and which have curved junctures to walls aid in accomplishing this.
2.    Two well-separated exit doors shall be available to the laboratory which shall swing in the direction of exit travel.
3.    Bench tops should be resistant to the effects of acids, bases, solvents, moderate heat, and should not absorb water. To facilitate cleaning, the tops should have few seams or crevices.
4.    Furniture should be designed to be sturdy and designed for convenient utilization. Storage spaces should be designed to meet any special requirements and should be easily accessible.
5.    Aisle spaces should be 40 to 48 in. wide and shall not be constricted to less than 28 in. by any temporary obstacles. The aisles should lead as directly as possible toward a means of egress.
6.    The organization of the facility shall be such as to reduce the likelihood of having to pass an originating or secondary hazard to evacuate the facility in the event of an emergency.
7.    Electrical outlets shall be three-wire outlets with high quality, low resistance ground connections. Circuits should be clearly identified to correlate with labels in breaker panels.
8.    The laboratory shall be supplied with a sink. The trap shall be of corrosion-resistant material. The plumbing shall be sized to accommodate the deluge shower and eyewash station. With average water pressure, this would normally be a 1-in. line or larger.
9.    Six air changes per hour of 100% fresh air shall be supplied to the facility. No

air shall be recirculated. The ventilation system shall be designed such that the room air balance is maintained at a small negative pressure with respect to the corridors whether the fume hood is on or off. Where toxic and explosive gases and fumes are present, the system is to be designed to be efficient in exhausting these fumes, by locating the exhaust intakes either near the source of fumes or near the floor (except for lighter-than-air gases). Typical air flow patterns are to be such as to draw dangerous fumes away from the normal breathing zones of the laboratory's occupants.

10. The facility shall include a separation of work spaces and desk areas as well as a second exit, equivalent to the arrangement shown in the standard laboratory module, Figure 3.1 (see Section 3.1.1).

### 3.1.3.2.4. High-Risk Facility

A distinguishing feature of a high risk facility is that, in some important aspect, the operations of the laboratory pose an immediate and substantial danger to either the occupants, the general public, or the environment if not performed safely in a suitable facility. The users of the facility and those permitted access to it must be limited to those individuals of the highest competence, training, and character. The training must be specifically tailored to inform the personnel in the facility of the risks to which they are exposed, the mandatory preventative safety procedures which must be followed, and the measures which must be taken in an emergency. Because it is so difficult to guarantee the degree of safety which must be met, an academic building would not normally be suitable for such a facility, nor would most common industrial research facilities.

A second distinguishing feature of a high risk facility is the need for isolation. If, for example, unless specific exceptions are permitted under the building code, then a building of use group H (Hazard) shall not be located within 200 ft of the nearest wall of a building of use group A (Assembly), E (Education) or I (Institutional). There are other restrictions under the code for specific types of risk, but each of these again defines conditions of isolation. In some cases, this is achieved by distance as above. In other instances, isolation is achieved by building walls and other structural components to a higher than normal level of construction. In cases where the level of risk is not so much physical, as is basically the concern of most building codes, but involves toxic materials or biologically pathogenic organisms, isolation can be achieved by such devices as airlocks and hermetically sealed doors. Where the risk is biological, isolation may be achieved, in part, by autoclaving and/or treating and disinfecting all garments, waste, and other items leaving the facility. Personnel may be required to wear self-contained, air-supplied suits while inside the facility or, instead, conduct all operations inside glove boxes or enclosures, in extreme cases using mechanical and electrical manipulating devices. Exhaust air from such a facility may require passing through a flame to kill any active organisms. Where the risk is of this character, rather than representing a danger due to fire or explosion, it may be possible to accommodate the facility within a building of generally lower risk level.

It will be noted that the four sections following are similar to those for the Substantial Risk Facility. However, there are some significant differences.

## A. Standard Practices

1. Access to the laboratory is limited to authorized personnel only except at times and under such conditions as designated by written rules as established by the laboratory supervisor and when accompanied by an authorized individual. The doors shall be locked at all times, with a formal key (or equivalent) control program in place.
2. All chemicals must be stored properly, according to compatibility. All chemicals which pose a special hazard or risk shall be limited to the minimum quantities needed for the research program, and materials not in actual use shall be stored under appropriate safe conditions. For example, excess quantities of explosives shall be stored in magazines, away from the immediate facility.
3. An annual chemical inventory shall be performed and sent to a central data collection point. A continuous inventory log shall be maintained of chemicals posing special risks and a formal accountability program implemented. Outdated and obsolete chemicals will be promptly disposed of through a chemical waste disposal program. Ethers and other materials which degrade to unstable compounds shall be shelf dated for disposal 6 months after being opened, but no more than 12 months after purchase, even if unopened, unless processed to prevent dangerous levels of the unstable compounds from forming.
4. An MSDS file will be maintained for all chemicals purchased in the laboratory for which they are available. In some instances, these are not available. Where equivalent data exist in whole or in part, this information will be made part of the MSDS file. The file will be accessible to the employees in the laboratory at all times. All laboratory workers shall be trained in how to interpret the information in an MSDS.
5. All secondary containers, holding materials of more than 1% of a hazardous component or combination of hazardous components (0.1% for carcinogens), which will be used more than a single work day shall be labeled with a label listing the hazardous components.
6. All hazardous wastes are placed in appropriate and properly identified containers for disposal through a chemical waste disposal program. Any wastes which pose a special hazard, or fall under special regulations, and require special handling shall be isolated and a program will be developed to dispose of them safely and legally. Normal, nontoxic waste shall be disposed of according to standard practices appropriate to such wastes, subject to any restrictions needed to prevent breaching any isolation procedures.
7. The laboratory will be maintained in an orderly fashion. Any spills or accidents will be promptly cleaned up and the affected area decontaminated or rendered safe.
8. No food or drink can be brought into the operational areas of the laboratory, nor can anyone smoke or apply cosmetics.
9. Any required signage or posting mandated by any regulatory agency shall be posted on the outside of the door to the entrance to the laboratory. In addition, a placard or other warning device shall be placed on the door or on the wall immediately adjacent to the door identifying any other major classes of hazards in the laboratory (See Section 2.3.4). A sign shall be placed on the door stating

in prominent letters, meeting any regulatory standards, "AUTHORIZED ADMISSION ONLY".

10. The telephone numbers of the laboratory supervisor, any alternates, and the department head shall be posted on the outside of the laboratory door or the adjacent wall.

## B. Special Practices

1. Specific policies, depending upon the nature of the hazard, shall be adopted and scrupulously followed to minimize the risk to laboratory personnel, the general public, and the environment. Several examples of laboratory practices for various hazards are given below. This list is not intended to be comprehensive but instead represents some of the more likely special precautions needed for a variety of types of risks.

    a. All work with hazardous kinds or quantities of materials shall be performed in a fume hood, specifically designed to provide the maximum safety for the hazard involved, or in specially designed and totally enclosed systems. It may be desirable for the hood or enclosed system to be equipped with a permanent internal fire suppression system. If the work involves a material which could be hazardous to the public or to the environment if released, a High Efficiency Particulate Aerosol (HEPA) filter would probably be needed. If so, then a pressure sensor to measure the pressure drop across the filter would be required to ensure that the filter would be replaced as the air passages become clogged.

    b. Work with explosives shall be limited to the minimum quantities needed. For small quantities used in a hood, an explosion barrier in the hood, with personnel wearing protective eye wear, face masks and hand protection may be sufficient protection. Note that most hoods are not designed to provide primary explosion protection. For larger quantities, the facility must be specifically designed for the research program. It is strongly recommended that a formal hazard analysis be completed, following guidelines such as those given in NFPA 49, Appendix C, if explosives are a major factor in designating the facility as a high risk facility. During periods of maximum risk, occupancy of the facility shall be limited to essential personnel.

    c. Some gases, such as fluorine, burn with an invisible flame. Apparatus for work with such materials should be placed behind a barrier to protect against an inadvertent introduction of a hand or other part of the body, so as to prevent burns.

    d. Systems containing toxic gases or gases which allowed to escape could pose explosive or health hazard ratings of 3 or 4 (lesser ratings if they provide no physiological warning), shall be leak-tested prior to use and after any maintenance or modification which could affect the integrity of the system. Where feasible, the gas cylinders shall be placed external to the facility and the gases piped into the laboratory to help minimize the quantity of gas available to an incident. As few cylinders as feasible shall be maintained within a given facility, preferably three or less. Permanently installed gas sensors, capable of detecting levels of gas well below the danger limits may be needed in some cases, such as when escaping gas provides no physiological warning signal.

e.     Vacuum systems, capable of imploding, resulting in substantial quantities of glass shrapnel or flying debris, shall be protected with cages or barriers, or for smaller systems, shall be wrapped in tape.

f.     Systems representing other physical hazards, such as high voltage, radiation, intense laser light beams, high pressure, etc. shall be interlocked so as to prevent inadvertent injuries. The interlocks shall be designed to be fail-safe such that no one failure of a component would render the safety interlock system inoperative.

2.     Activities in which the attention of the worker is not normally engaged with laboratory operations, such as record maintenance, calculations, discussions, study, relaxation, etc. shall not be performed in the laboratory proper but shall be performed in an area isolated from the active work area. The segregated desk area of the standard laboratory module is specifically intended to serve this purpose. Depending upon the nature of the hazard, it is usually feasible to make at least a portion of the upper half of the barrier separating the two sections of the laboratory transparent so that operations can be viewed if necessary.

3.     Workers in the laboratory should participate in a medical surveillance program if they actively use materials, for a significant portion of their work week, which would pose a significant short- or long-term risk to their health. Employees shall be provided medical examinations if they work with any material, such as regulated carcinogens, requiring participation in a medical program by OSHA or other regulatory agency under conditions which do not qualify for an exemption. Employees shall notify the laboratory supervisor as soon as possible of any illness that might be attributable to their work environment. Records shall be maintained of any such incident.

4.     No safety feature or interlock of any equipment in the facility shall be disabled without written approval of the laboratory supervisor. Any operations which depend upon the continuing function of a critical piece of safety equipment, such as a fume hood, shall be discontinued should the equipment need to be temporarily removed from service for maintenance. Any such item of equipment out of service shall be clearly tagged with a signed "Out of Service" tag. Only the person originally signing the tag, or a specific alternate, shall be authorized to remove the tag.

5.     It shall be mandatory to wear any personal safety equipment required for conducting operations safely in the laboratory.

6.     It is recommended that a laboratory safety committee review each new experiment planned for such a facility to determine if the experiment can be carried out in the facility. If the risk is such that experiments may affect the environment, or the surrounding community, it is recommended that the committee include at least one lay person from the community, not currently affiliated directly or indirectly with the institution or corporation. In this context, "new" is defined as being fundamentally different in character, scope, or scale from any experiment previously approved for the facility.

**C. Special Safety Equipment**

1.  Any refrigerators or freezers shall be rated as acceptable for "Flammable Material Storage", except for ultra-low temperature units.
2.  A flammable material storage cabinet, either built-in or free standing, shall be used for the storage of flammable materials. Any other special storage requirements, such as for locked storage cabinets or safes for drugs or radioactive materials shall be available and used.
3.  If the nature of the research program is such as to require it, the laboratory shall be equipped with a fume hood. The fume hood shall meet any specific safety requirements mandated by the nature of the research program. A discussion of hood design parameters will be found in a later section but, for high hazard use, the interior of the hood and the exhaust duct should be chosen for maximum resistance to the reagents used; the blower should preferably be explosion proof or, minimally, be equipped with nonsparking fan blades; it shall be equipped with a velocity sensor and alarm; the interior lights shall be explosion proof, and all electrical outlets and controls shall be external to the unit. It may be desirable to equip the unit with an internal automatic fire suppression system.
4.  The laboratory shall be equipped with an eyewash station and a deluge shower.
5.  The laboratory shall be equipped with a fire alarm system connected so as to alarm throughout the building, an appropriate fire suppression system, and be provided with one or more 12BC, or larger, fire extinguishers or Class D units if reactive metals are in use.
6.  An emergency lighting system shall be provided.
7.  A first-aid kit shall be provided and maintained.
8.  Any special equipment mandated by the research program shall be provided. For example, electrical equipment other than refrigerators may need to be designed to be explosion-safe.
9.  Any special equipment needed to maintain the required isolation for materials in the laboratory shall be provided. Examples are specially labeled waste containers, autoclaves, other decontamination equipment, or disposable clothing.

**D. Laboratory Facilities**

1.  The floor of the laboratory is designed to be easily cleaned. Durable, seamless floors of materials that are substantially impervious to spilled reagents are easily decontaminated, and which have curved junctures to walls aid in accomplishing this. The walls are to be similarly painted with a tough substantially impervious paint to facilitate cleaning and decontamination.
2.  Two well-separated exit doors shall be available to the laboratory which shall swing in the direction of exit travel.
3.  Bench tops should be resistant to the effects of acids, bases, solvents, moderate heat, and should not absorb water. To facilitate cleaning, the tops should have few seams or crevices. Although not necessarily subjected to the same level of abuse, other surfaces of the furniture should be readily cleaned or decontaminated.

4. Furniture should be designed to be sturdy and designed for convenient utilization. Storage spaces should be designed to meet any special requirements and should be easily accessible.

5. Aisle spaces should be 40 to 48 in. wide and shall not be constricted to less than 28 in. by any temporary obstacles. The aisles shall lead as directly as possible toward a means of egress.

6. The organization of the facility shall be such as to reduce the likelihood of having to pass an originating or secondary hazard to evacuate the facility in the event of an emergency.

7. Electrical outlets shall be three-wire outlets with high quality, low resistance ground connections. Circuits should be clearly identified to correlate with labels in breaker panels. If the nature of the hazard is such to generate potentially explosive or ignitable aerosols, vapors, dusts, or gases the electrical wiring, lights, and electrical switches shall be explosive proof.

8. The laboratory shall be supplied with a sink. The trap shall be of corrosion-resistant materials. The plumbing shall be sized to accommodate the deluge shower and eyewash station. With average water pressure, this would normally be a 1-in. line or larger.

9. At least six air changes per hour of 100% fresh air shall be supplied to the facility. No air shall be recirculated. The ventilation system shall be designed such that the room air balance is maintained at a small negative pressure with respect to the corridors, whether the fume hood is on or off. Where toxic and explosive gases and fumes are present, the system is to be designed to be efficient in exhausting these fumes, by locating the exhaust intakes either near the source of fumes or near the floor (except for lighter-than-air gases). Typical air flow patterns are to be such as to draw dangerous fumes away from the normal breathing zones of the laboratory's occupants.

10. The facility shall include a separation of work spaces and desk areas as well as a second exit equivalent to the arrangement shown in the standard laboratory module, Figure 3.1 (see Section 3.1.1), unless the risk is so pronounced as to require complete separation of operational and nonoperational areas.

### 3.1.4. ACCESS*

Much of the present chapter has been spent on details directly concerning the laboratory itself. However, a laboratory is rarely an isolated structure; it is almost always a unit in a larger structure. It often appears that the typical laboratory manager or employee is insufficiently aware of this. If it is necessary to dispose of some equipment, it is often simply placed outside in the hall, where it is no longer of concern to him. The thought that it may reduce the corridor width to well below the required minimum width probably does not arise. A door swinging into the hall such that it may block the flow of traffic appears similarly unimportant if it preserves some additional floor or wall space within the facility. The use of the corridor as a source of makeup air often seems reasonable, yet the possibility of this permitting a fire or toxic fumes to spread from one laboratory to another is clear once it has been pointed

---

* BOCA Basic National Building Code 1984 edition, copyright 1983, Building Officials and Code Administrators International, Inc. Published by arrangements with the author. All rights reserved.

out. The natural inclination is to concentrate one's thoughts on the operations within a laboratory since, for most research personnel, this is where virtually everything important to them takes place. The ideas presented in the previous sections relating to optimizing safety within the facility are quickly grasped and accepted by most laboratory personnel, but extending these same concepts beyond the confines of their own laboratory frequently appears to be more difficult to communicate. However, due to the inherent risks in laboratory facilities, it is critical that sufficient safe means of egress are always available. Except for scale and specific code requirements, most of the principles used in the laboratory to allow safe evacuation extend readily to an entire building.

### 3.1.4.1. Exitways*

An exitway consists of all components of the means of egress leading from the occupied area to the outside of the structure or to a legal place of refuge. Included as exitway components are the doors, door hardware, corridors, stairs, ramps, lobbies, and the exit discharge area. The function of the exitway is to provide a protected way of travel to a final exit from the facility to a street or open area. Elevators are not acceptable as a required means of egress.

*3.1.4.1.1. Required Exits*

Any required exitway is required to be maintained available at all times, unless alternate means are approved in advance by a building official** which will provide equivalent protection. This is probably one of the most common code violations. An extremely serious violation was personally observed by the author while attending a *safety* conference at a major university. The meetings were held within a large multistory building containing meeting rooms, dining facilities serving up to 300 persons, and offices, but which had all but one small, poorly marked, out-of-the-way exit blocked. This condition existed for a period of several weeks during a renovation project, during which full operations continued in the building.

Another common violation of the same type is the chaining of exits for protection against theft during low-usage hours. It is common, however, for research buildings to be partially occupied at almost any time. All exits may not be required during periods of low activity, but enough legal exits must be available to serve the occupants. It is essential that occupants know which exits remain usable if some which are normally available are blocked during certain hours. Most of these problems arise because the persons making the decisions to eliminate or reduce the size or number of exits are not personally knowledgeable of the legal requirements and fail to check with those who are.

If sufficient legal exits cannot be maintained during renovations or at other times, the occupancy load must be reduced or perhaps those sections of the building served by the needed exits closed temporarily. As a minimum, every building with an occupancy load up to 500 must have at least two exits, between 500 and 1000, three exits, and above 1000, at least four exits.

---

* BOCA Basic National Building Code 1984 edition, copyright 1983, Building Officials and Code Administrators International, Inc. Published by arrangements with the author. All rights reserved.

** A building official in this context is a person or agency specifically authorized to administer and enforce the building code applicable to the building, not a person in charge of a building or facility.

### 3.1.4.1.2. Exit Capacity*

In designing the needed exits for a facility, it is necessary to consider (1) the number of occupants in the building, (2) the number that could be in the building if the maximum density of occupants allowed by the building code were present, or (3) the latter number plus any who might have to pass through the building from another space to reach an exit. The exits must be sufficient to accommodate the largest of these three numbers. The maximum floor area allowed per occupant under BOCA is (space occupied by permanent fixtures is not counted):

| | |
|---|---|
| Assembly without fixed seats, chairs only | 7 ft² net |
| Assembly without fixed seats, chairs and tables | 15 ft² net |
| Business areas | 100 ft² gross |
| Educational: classrooms | 20 ft² net |
| Educational: shops, etc. | 50 ft² net |

The exit capacity from an area must be sufficient for the number of occupants of the space involved. Each usage group has a different occupant load per unit of exit capacity under BOCA:

#### A. Without A Fire Suppression System (number of occupants per unit capacity)

| Use group | Stairways | Doors, ramps, corridors |
|---|---|---|
| Assembly | 75 | 100 |
| Business | 60 | 100 |
| Educational | 75 | 100 |
| Hazardous | — | — |
| Storage | 60 | 100 |

#### B. With A Fire Suppression System (number of occupants per unit capacity)

| Use group | Stairways | Doors, ramps, corridors |
|---|---|---|
| Assembly | 113 | 150 |
| Business | 90 | 150 |
| Educational | 113 | 150 |
| Hazardous | 60 | 100 |
| Storage | 90 | 150 |

Each unit of egress width is 22 in. Credit for one half unit is given for an amount 12 in. or more in excess of a multiple of 22 in., but none is given if the excess is less than 12 in. The required exit capacity governs the width of exitway corridors as well, but in most cases they shall not be less than 44 in. or, for educational occupancies with more than 100 occupants, not less than 72 in. However, for small buildings serving less than 50 occupants, the width of the corridors may be as little as 36 in.

### 3.1.4.1.3. Travel Distance*

The characteristics of the routes of egress to an exit are also important, especially

---

* BOCA Basic/National Building Code 1984 edition, copyright 1983, Building Officials and Code Administrators International, Inc. Published by arrangements with the author. All rights reserved.

in as critical a facility as a laboratory building. Care should be taken, just as within the laboratory, for the distances to be as short and direct as practicable. The location of hazardous areas should be chosen to eliminate or minimize the probability of the direction of travel being toward a likely hazard during an emergency. The maximum travel distances for the use groups of interest are

| | Distance (ft) | |
| --- | --- | --- |
| Use group | Without fire suppression system | With fire suppression system |
| Assembly | 150 | 200 |
| Business | 200 | 300 |
| Educational | 150 | 200 |
| Hazardous | — | 75 |
| Storage — moderate risk | 200 | 300 |

There is a further advantage in having a fire suppression system in determining the points from which you measure the distances in the table above. Without a fire suppression system, the distances are measured from the most remote point, except that if the space is divided into rooms and the travel distance in a room does not exceed 50 ft, the measurement can start at the room door. With an automatic fire suppression system, this intraroom travel distance can be extended to 100 ft. Again, as with the standard laboratory module, when a *building* requires more than one exit, these exits shall be as remote as practicable from each other. They must also be arranged so that access is available from more than one direction from the area served, so that it is unlikely that access from both directions will be blocked in an emergency.

If the laboratory building is classed as either an educational or assembly occupancy, there must be a main exit which, in addition to being large enough to serve as a sufficient discharge for all the exitways which lead to it, must also be at least large enough to accommodate half of the building's occupant load.

It is acceptable to use an adjacent room or space as a means of egress from a room, as indicated in the standard laboratory module, if the room provides a path of egress to an exit, is not of higher hazard than the original space, and is not subject to locking. Thus, the laboratory modules should be arranged in blocks of comparable level of risk if this concept is used to provide a second exit from each laboratory.

### 3.1.4.1.5. Corridors*

In the introduction to this section, two points were used as illustrations, one maintaining the corridors free of obstruction and the second concerning the undesirability of using halls as a source of makeup air. Both of these points are intended to insure that the corridors remain available for evacuation. A door may not swing into a hall such that it reduces the width of the corridor to less than one half the legally required width, nor can the door, when fully open, protrude into the hall more than 7 in. An obvious implication of the first provision, considering that most laboratory

* BOCA Basic/National Building Code 1984 edition, copyright 1983, Building Officials and Code Administrators International, Inc. Published by arrangements with the author. All rights reserved.

doors are 36 in. wide or wider, when combined with one half of the minimum legal width of 44 in. for a corridor (except for buildings occupied by 50 persons or less), would mean a minimum actual permissible width of 58 in., unless the doors are recessed into alcoves in the connecting rooms.

If the corridors were to serve as a plenum for return air, they could spread smoke and toxic fumes from the original source to other areas. Further, instead of being a protected exitway, they themselves would represent a danger. In many fires, it is the individuals trapped in smoke-filled corridors and stairs who often fail to survive. Laboratories, in general, need to be kept at negative pressure with respect to the corridors, but the 200 or 300 cfm through an open door needed to maintain a negative air pressure will not violate the prohibition of corridors scriving as a plenum. Space above a false ceiling in a corridor can be used as a plenum if one can justify the corridor not being of rated construction or if the plenum is separated by fire resistance-rated construction.

If a corridor serves as an exit access in the building occupancies which are being considered here, the corridor walls must have at least a 1-h fire resistance rating. The corridor walls must be continuous to the ceiling separation to ensure this rating. Cases have been observed where the wall was not taken above a suspended ceiling or continued into open service alcoves. Floors of corridors are to have a slip-resistant surface.

The eventual point of exit discharge must be to a public way, courtyard, or other open space leading to a public way which is of sufficient width and depth to safely accommodate all of the occupants. On occasion, during renovation or construction projects, the areas outside the exits are not maintained in such a way as to satisfy this condition.

A fairly common error that tends to creep into older buildings as renovations take place is the creation of dead-end corridors. Frequently, for other design reasons, more corridors are built than are actually needed. Later, as space becomes tighter or the space needs to be reconfigured, the corridors are modified in order to recoup this "wasted" space and dead-end corridors are created. These dead-end corridors cannot be longer than 20 ft. If they are of sufficient width, some of the dead-end corridors can be converted into offices or other uses, as long as they conform to code requirements for the class of occupancy.

Where there would be an abrupt change in level across a corridor (or across an exit or exit discharge) of less than 12 in., so that a stair would not be appropriate, a ramp is required to prevent persons from stumbling or tripping at the discontinuity. Other unnecessary obstacles should be avoided as well, such as low-hanging signs and similar devices which may protrude into a corridor or even safety devices such as deluge showers with low-hanging chains which could strike a person in the face in a partially dimmed or darkened corridor.

### 3.1.4.1.6. Stairs*

Stairwells are another exitway component that are frequently abused. They are an egress component providing a protected way of exiting a building. In order to provide additional ventilation or to avoid having to continually open doors, a very common

---

* BOCA Basic/National Building Code 1984 edition, copyright 1983, Building Officials and Code Administrators International, Inc. Published by arrangements with the author. All rights reserved.

practice is to use wedges of various types to permanently prop doors open. The result is to not only void the protection afforded by the required fire-rated enclosure, but also create a chimney through which fire and smoke on lower floors may rise, changing the stairway from a safety device to a potential deathtrap and providing a means for problems on lower floors to spread to higher levels.

In order to provide a protected means of egress, a required interior stair must be enclosed within a fire separation meeting the fire resistance ratings given in Table 3.3. The stair enclosure cannot be used for any other purpose, such as storage underneath the stairs or within any enclosed space under a required stair. Any doors leading into the stair enclosure must be exit doors. This also precludes creating closets underneath stairs for storage. The width of the stairs and landings at the head, foot, and intermediate levels must meet the minimum dimensions established by the calculated required exit capacity. All doors leading onto a landing must swing in the direction of egress travel. The restrictions on reduction of width of the landings due to doors opening are the same as for corridors.

Stairways which continue beyond the floor level leading to an exit discharge or to a basement level are common. In an emergency situation, unless the stairs are interrupted at this floor, it would be very likely that persons evacuating the building would continue downward, even though there is an additional requirement that each floor level be provided with a sign indicating the number of the floor above the discharge floor, for stairways more than three floors high. Since the location of these signs is often above the door, smoke may render the sign unreadable. The persons continuing downward might be sufficiently confused to reenter the building before they recognized their mistake or, in an even more serious situation, cause the lower exit to be blocked, preventing those from the lower floor from using the exit stair.

Handrails represent an exception to the type of materials represented by Table 3.3. Wooden handrails are permitted in all types of construction. Except for the smallest buildings requiring only one exit, there must be a handrail on both sides of a stair and if the stair is more than 88 in. wide, there must be additional handrails such that the maximum space between rails is no more than 88 in.

The treads and risers for all classes of occupancy likely to be involved in laboratory buildings shall be a minimum of 10 in. for a tread and a maximum of 7.5 in. for a riser. The maximum variation in the actual widths are to be no more than ± 3/16 in. This seems a trivial point at first glance, but the importance of it should be clear to anyone who has ever stumbled over uneven ground in the dark. As one goes up or down a flight of stairs, one quickly grows accustomed to the step configuration and a substantial unexpected change could easily lead a person in a hurry to stumble.

A similar rationale exists for the continuance of a handrail beyond the ends of a stairway to ensure that the person traveling the stair has something to grasp to help avoid a fall if they cannot see that the stairs have ended. At both the top and bottom, the handrail should turn to be parallel to the floor for at least 1 ft (plus a tread width at the bottom).

### 3.1.4.1.7. Doors*

Doors are perhaps the most abused exitway component. The fire separation they

---

* BOCA Basic/National Building Code 1984 edition, copyright 1983, Building Officials and Code Administrators International, Inc. Published by arrangements with the author. All rights reserved.

are intended to provide is defeated when they are wedged open in order to improve ventilation or to eliminate the inconvenience of having to open them every time the passageway is used, especially if it is frequently used for moving supplies and equipment. In some instances, doors which are required to be shut because they represent openings into a fire resistance-rated corridor are left open simply because individuals want to leave their office doors open to be easily accessible to persons wishing to see them, as openness and accessability are seen to be desirable behavioral traits. Doors are often damaged when they are prevented from closing by the use of the now ubiquitous soft drink can forced into the hinge opening. Even maintenance departments may not be aware that a required fire resistance rating is achieved by the entire door assembly, including the frame and hardware, not just by the door itself, so that a repaired door may no longer meet code specifications.

Conversely, doors which are required to be openable may be blocked or rendered inoperable for a variety of reasons, one of the most common being to increase security. Compact, easily portable, and saleable instrumentation, especially computers and computer accessories, represents a tempting target for theft. As a result, doors which should be readily openable are fitted with unacceptable hardware to provide additional security, in many cases by the occupants themselves. In other cases, doors are blocked simply because individuals are careless and do not consider the consequences of their actions, such as locating a piece of equipment in such a way that a rarely used door cannot swing open properly.

Doors which are required exits must be prominently indicated as such, while doors which do not form part of a legal exitway must not be marked as exits, even though they may, in fact, form an additional means of egress.

For the classes of occupancies being considered here, the width of a door used as an exit must be at least 32 in. and the maximum width of a single leaf of a side-hinged, swinging door must be no more than 48 in. (except for certain storage spaces). If a door is divided into sections by a vertical divider, the minimum and maximum widths apply to each section. A normally unoccupied storage space of up to 800 ft$^2$ can have a door of up to 10 ft in width. The minimum height of a door is 80 in. If two doors are to be placed in series, as might be the case where separation of a facility from a corridor must be maintained, the doors must be separated by a minimum of 7 ft.

In general, all doors for laboratory structures should be of the side-hinged, swinging type, opening in the direction of exit travel. For doors opening onto stairways and for an occupant load of ten or more, or for a high hazard occupancy, this type of door is required. However, a sliding door with a side-hinged, swinging door incorporated in it of the proper dimensions is acceptable. Revolving doors are not.

It must be possible to open a door when coming from the normal direction of egress without using a key, nor can draw bolts, hooks, bars, or similar devices be used. For assembly or educational occupancies with an occupant load of 100 or more, exit doors must be fitted with panic hardware. An essential element of a door is that it cannot be too difficult to open. Panic hardware must require no more than 15 lbs force to release and a door not normally provided with power assistance cannot require more than 30-lb force to swing. A power assisted door cannot require more than a 50-lb force to swing should the power fail. These restrictions on the force required to operate a door can easily be exceeded should the ventilation system be modified without taking this concern into account. A very moderate atmospheric pressure differential of just over 0.3 in. (water gauge) would result in a force of more than 30-lb force on a door of the minimum acceptable size. Addition of hoods to a laboratory, without provision of additional makeup air, could easily cause this limit to be exceeded on a more representative 3 by 7 ft door.

Doors opening from rooms onto corridors, into stairways, and forming part of a required fire resistance-rated assembly must be rated. Most doors, such as those from offices opening on a corridor, are required to have at least a 20-min fire rating, while doors leading from rooms of 2-h fire resistant construction, as determined from Table 3.3, must be at least 1.5-h fire doors, as should those entering stairways. Wired glass, 1/4 in. thick and specifically labeled for such use, may be used in vision panels in 1.5-h fire doors, provided the dimensions do not exceed 33 in. high and 10 in. wide with, however, a total area of no more than 100 in$^2$. If the potential injuries and damage resulting from dropping chemicals as a result of being struck by a swinging door are considered, there is clearly merit in taking advantage of the provision for vision panels in laboratory, corridor, and stair doors.

Doors opening onto fire resistant-rated corridors and stairways must be self-closing or close automatically in the event of a fire. The first of these requirements usually includes offices opening directly off corridors and is the case alluded to earlier as representing one of the most commonly violated fire regulations. For whatever reason, most individuals usually prefer to work with their office door open. Unfortunately, in an emergency evacuation, many do not remember to close their doors and so the integrity of the fire separation is breached at these points.

### 3.1.4.1.8. Exit Signs, Lights, Emergency Power*

The need for emergency lighting within laboratories has already been discussed, independent of code issues. However, the need for lighting of exitways and identification of exits in emergencies is as critical outside the laboratory proper as it is inside. It is essential in a laboratory building that evacuation not be hindered by lack of lighting, especially in multistory buildings where stairways and corridors typically do not provide natural lighting.

Internally illuminated exit signs are a key component of an evacuation system. In every room or space served by more than one exit, which is recommended for most laboratory rooms** all the required means of egress must be marked with a sign with red letters on a contrasting background at least 6 in. high, with a minimum width of 3/4 in. for each segment making up the letters. The light intensity at the surface of the sign must be at least 5 fc.

There are self-illuminated signs, containing radioactive tritium (an isotope of hydrogen with a half-life of 12.33 years) which are acceptable, both under usual fire codes and to the Nuclear Regulatory Commission. The radiation from tritium is exceedingly weak (18.6 keV beta) and since these signs are completely sealed, no

---

\* BOCA Basic/National Building Code 1984 edition, copyright 1983, Building Officials and Code Administrators International, Inc. Published by arrangements with the author. All rights reserved.

\*\* Remember that while two exits are recommended here for most laboratories in order to provide the maximum degree of safety, neither building code provisions nor OSHA regulations require two exits unless the laboratory represents a high hazard area or is occupied by more than 50 people.

radiation can be detected from them. The transparent enclosure completely absorbs the radiation. As long as they remain sealed, they represent no hazard. However, in order to provide the required level of illumination, the amount of radioactivity in each sign is substantial. If they were to be broken in an accident or in a fire, an individual handling them could inhale a quantity of radioactivity substantially in excess of the permissible amount. Therefore, as a precautionary policy, the radiation safety committees at a number of institutions have taken the position that these signs are not permissible at their facilities. It might be well to consider the risk versus benefit whenever the use of such units is contemplated. (Note that normal glow-in-the-dark signs do not contain radioactive material. They depend upon phosphorescence, a completely different physical phenomenon.)

In addition to signs at the exit, it may be necessary to install supplemental signs to help guide persons to an exit when the distance is substantial or the corridor curves or bends. If a sign incorporates an arrow, it should be difficult to modify the direction of the arrow. However, when a sign is damaged, maintenance personnel have been known to inadvertently install a replacement sign with the arrow pointing in an incorrect direction. It is well not to take anything for granted. Users of the building should check such work independently.

Exit signs and means of egress must be lighted whenever a building is occupied, even if the normal source of power fails. The level of illumination at the floor level must be at least 1 fc. There are a number of methods by which power can be provided to emergency lighting circuits. They all must provide sufficient power to the lights and paths of egress to meet the required lighting standards for at least 1 h so that the building occupants will have ample time to evacuate.

For relatively small facilities, battery-powered lights, continuously connected to a charging source and which automatically come on when the power fails, are often used. Units are available which have extended useful lifetimes of 10 years or more. As with any other standby device, it is necessary to test them on a definite schedule. There are battery-powered units designated as "uninterruptible power supplies" which switch over within milliseconds. These are often used to maintain power to computers or electronic equipment where a loss of power can cause data to be lost. Such units can be sized to also support emergency electrical lighting.

Standby generators are another alternative to provide energy to the required emergency lights and other equipment which may need to be supported during a power failure. For assembly and educational occupancies and for business buildings with more than 1000 occupants, an emergency electrical system is required. A generator is preferred over a battery system for other than small buildings to provide the fairly substantial amount of power needed. However, these generators should be checked once a month, under load, to ensure that they will come on within the required 10 sec for emergency lighting and within 60 sec for other loads. Architects often fail to consider all circumstances in designing such systems. In one case, the architect designed an excellent system, but for economic reasons, the exhaust of the generator was located immediately adjacent to the building air intake, on the premise that in an emergency the ventilation system would shut down. However, because exhaust fumes were drawn into the working ventilation system during tests, the scheduled operational tests could not effectively be performed until the problem was corrected. Failure to provide proper maintenance and tests can lead to embarrassing and costly incidents when outside power fails and stairways and corridors are not lighted. Academic institutions are more vulnerable than corporate facilities since, in a given building, there is more likely to be a higher percentage of individuals who are relatively unfamiliar with the evacuation routes.

A last option, but one which must be used with considerable caution, is to provide outside power from two completely separate utility power feeds. Such an arrangement can be approved by code authorities if it can be shown that it is highly unlikely that a single failure can disrupt both sources of power. For example, if the local distribution system is fed by several alternative power lines and has alternate local lines to provide power to a building, it is conceivable that local building officials would approve the system, but one must remember that entire states and even larger regions have suffered total power losses in recent years.

*3.1.4.1.9. Other Topics*

A number of other topics related to exitways have not been touched upon here for the same reason given before. This handbook is not intended to be comprehensive. The intention is to cover those topics which are most meaningful to a person working in a laboratory building with enough detail so that reasonable persons can evaluate their facilities to insure that their safety is not compromised by renovations or the actions of individuals during normal usage. The reader should also be able to follow the reasoning for many of the architectural decisions made during the planning of a facility and should be able to actively participate in the planning process. However, provisions and specifications for components such as exterior stairs, fire escapes, access to roofs, communicating floors, vestibules, and lobbies, which are all relevant topics under the general subject "means of egress", would be important to architects, but probably less so to laboratory personnel.

### 3.1.5. CONSTRUCTION AND INTERIOR FINISH*

Code specifications for the details of construction and the materials used as interior finish are intended to provide a degree of fire protection commensurate with the use of a building. Although, as has been noted, a laboratory building may be constructed under several different classifications, laboratory usage does imply that a certain amount of risk is involved. Therefore, care is needed to ensure that construction practices and materials used in the interior finish do not add to the risks or defeat the intended level of protection.

In addition to fire protection, there are other potential hazards which may also be reduced by construction details and choices of materials. In Section 3.1.3.2 , under the topic "Laboratory Facilities" for each class of laboratory, many of the features stipulated characteristics of finish materials. As a general principle, laboratory floor coverings, wall finish, and table and bench tops should be durable, easy to clean, and resistant to common reagents.

A normal tile floor meets many of these requirements, but the seams around each tile form cracks in which substances such as mercury and other materials can lodge. For example, mercury can remain *in situ* in these cracks for extended periods and

---

* BOCA Basic/National Building Code 1984 edition, copyright 1983, Building Officials and Code Administrators International, Inc. Published by arrangements with the author. All rights reserved.

create a substantial mercury vapor pressure when, ostensibly, all spilled mercury may have been "cleaned up". Radioactive materials and biological agents can similarly be trapped and pose a continuing problem unless very thorough cleaning is performed on a regular basis. One argument frequently used in favor of tile floors has been that damaged or contaminated tiles can be easily and cheaply replaced. Recently, however, common asbestos tile has been included within the category of asbestos materials by the EPA and, when removed, must be removed according to the procedures for removing asbestos, frequently a very expensive process.

Any material which is used must meet required standards for fire spread and smoke generation, in addition to having the other properties for which it is selected. When a material has been selected which has both the desired properties and the requisite ratings, the construction contract should contain peremptory language stipulating that no substitutes will be acceptable without specific approval. In too many instances where vague language that "equivalent materials may be substituted" is incorporated into a contract, substitutes have been introduced which do not meet the original specifications, possibly innocently, because the supervisor on site did not realize the difference. It usually is not possible to simply look at an item and determine what its properties will be from appearances alone.

Although not particularly attractive, a plain sealed concrete floor or one painted with a durable paint probably is the best for most laboratories, while wood and carpeting would be the worst. Many different kinds of floor finishes are available, designed to prevent slipping, generation of sparks, and resistance to corrosives or solvents. Simple concrete block walls are often used for interior partitions in buildings or, alternatively, dry wall on steel studs. Both of these are relatively cheap for original construction and can be modified easily as well. Concrete block is relatively porous, which can pose decontamination problems, but can be painted so as to eliminate this as a problem. Paints for interior surfaces are available which will provide waterproofing, resistance to corrosives and solvents, and enhance fire resistance. Where biological cleanliness is an important criteria, there are paints which have been approved by the EPA which will inhibit the growth of biological organisms. Incidentally, the sand used in concrete blocks in many parts of the country is a source of Radon, which is a concern to many persons.

### 3.1.5.1. Construction Practices*

The intent of the fire code as it applies to interior finish and acceptable construction practices is to prevent the spread of fire from one fire area to another, i.e., to make sure that the fire walls and other fire separation assemblies are constructed in such a way and of such materials as to maintain the fire resistance rating of the structure. In the absence of any special cases, the fire grading for the building classes most appropriate for laboratories are given by BOCA (in hours):

| | |
|---|---|
| Assembly, lecture halls | 2 |
| Business | 2 |
| Educational | 1.5 |
| High Hazard | 4 |
| Storage, moderate hazard | 3 |

* BOCA Basic/National Building Code 1984 edition, copyright 1983, Building Officials and Code Administrators International, Inc. Published by arrangements with the author. All rights reserved.

Architectural plans and specifications must include documentation for all required fire resistance ratings.

It is not possible to ensure total separation of fire areas and still provide for air intake and exhaust or return air plenums unless they are separated from the surrounding spaces by required fire resistant shaft and wall enclosures, plus properly engineered, labeled fire dampers meeting Ul 555 specifications, installed where ducts pass through fire separations. Unless the architect, a representative of the building owner, or a contractor's inspector provides careful supervision of the workmen during construction, dampers may be installed improperly, perhaps at the end of a convenient piece of duct work, even if this happens to be in the middle of room, far away from the fire separation wall. As noted earlier, it is possible to use the space between the ceiling of a corridor and the floor above if the space is properly separated. However, ceiling spaces used for this purpose cannot have fuel, fixed equipment, or combustible material in them. The requirement for a fire damper does not apply to ducts used for exhausting toxic fumes (e.g., from a fume hood) since, in a fire, it is usually desirable, as a means of protecting normal building occupants and fire fighting personnel, for toxic materials to continue to be exhausted. This requires that ducts carrying toxic fumes be continuously enclosed within a shaft of the proper rating to the point of exhaust.

It is essential that the integrity of the fire separation walls not be significantly diminished. For example, walls less than 8 in. thick are not to be cut into after they are constructed in order to set in cabinets or chases. Among the most common violations of the integrity of fire separations are penetrations for running utilities and, today, cable chases for electronic services such as video signals, data cables, and computer lines. Often, these penetrations are roughly done, leaving substantial unfilled gaps. Even in new construction, especially if the penetrations are in hard-to-inspect or otherwise awkward locations, the gaps around the ducts, pipes, and conduit are frequently left unfilled. Where retrofitting of spaces to accommodate such devices is done by maintenance and construction personnel, this deficiency is even more likely to occur. Whenever this error is found, the gaps must be filled with materials meeting fire resistance standards. If a renovation or new construction involves setting a structural member into a hollow wall, the space around the member must be filled in for the complete thickness of the wall and at least 4 in. on all four sides with approved fire stopping material.

Openings can exist in a fire wall, else how could doors and windows exist? However, there are limits on the size of the openings — 120 ft² (but no more than 25% of the length of the fire wall), except in buildings with an approved automatic sprinkler system. Larger openings (240 ft²) can exist on the first floor of a building for the passage of trucks, with the provision of a 3 h-rated protective assembly and a water curtain. The openings must be protected with an appropriately rated assembly, which may be a fire door. If the wall has a 3-h rating, the rating of the door must also be 3 h. For walls with 1.5- or 2-h ratings around exits, shafts, or elevators, the doors must be 1.5 h. If the fire separation requirement is only 1 h, then the fire door need only be a 3/4 h-rated door. Unless the interior space is rated, doors to rooms such as offices opening onto a 1-h corridor need only be rated at 20 min.

Fire walls shall extend completely from one rated assembly to another, such as floor to ceiling, extending beyond any false or dropped ceiling which may have been added. The joint must be tight.

This section has been primarily concerned with the interior of a building, but measures are required to prevent a fire from spreading due to the exterior design of a building as well. The exterior walls must be rated to withstand the effects of fires within the building. Windows arranged vertically above each other in buildings of three or more stories for business, hazardous or storage use shall be separated by appropriate assemblies of at least 30 in. in height from the top of a lower window to the bottom of the one above. Although there are a number of exceptions, if the exterior wall is required to have a fire rating of 1 h or more, then a parapet of 30 in. or greater in height above the roof is required for nonexempt structures.

### 3.1.5.2. Interior Finish*

Materials used for interior trim or finishing must meet standards for flame spread and smoke or toxic fume generation. For buildings of Type 1 or 2 construction, which includes most laboratory facilities in the second group, the materials must not cause the building to be declassified to a lower group.

Materials are rated in accordance to how well they perform on tests made according to the ASTM E84 procedure, with lower numbers corresponding to the better materials. Class 1 materials have a rating between 0 and 25, Class 2, 26 to 75, and Class 3, 76 to 200. As far as smoke generation is concerned, materials used for interior finish must not exceed a rating of 450, as tested according to the provisions of ASTM E84. Based on these ratings, the interior finish requirements for the categories of interest are

| Use group | Required vertical exits and passageways | Corridors for exit access | Rooms/enclosed spaces |
|---|---|---|---|
| Assembly | 1 | 1 | 2 |
| Business | 1 | 2 | 2 |
| Educational | 1 | 2 | 3 |
| High hazard | 1 | 2 | 3 |
| Storage, moderate hazard | 2 | 2 | 3 |

As usual, there are numerous exceptions, based on special circumstances, the most notable being (1) Class 3 materials may be used in rooms in places of assembly if the capacity is 300 or less and (2) if there is an automatic fire suppression system, Class 2 and 3 materials may be used instead of Class 1 and 2 respectively.

Materials used on floors are classified differently. Often, the propensity for materials to burn depends upon the physical configuration. For example, a match placed on a piece of carpeting lying on the floor may smolder and go out, while a match applied to the bottom of the same piece of carpeting, mounted vertically, may result in a vigorous fire. Most common floor coverings are exempt from being rated, but where this is required by the local building official, they must meet the standards given in the table below. In this table, Class 1 corresponds to a critical radiant flux of 0.45 W/cm$^2$, while Class 2 corresponds to 0.22 W/cm$^2$. The classification DOC FF- 1 is a legal "pill test" requirement for all carpet sold in the U.S. Where carpet is

---

* BOCA Basic/National Building Code 1984 edition, copyright 1983, Building Officials and Code Administrators International, Inc. Published by arrangements with the author. All rights reserved.

not used, and in most laboratories this would be the case, the alternative flooring must have a minimum equivalent critical radiant flux of 0.04 W/cm$^2$.

| Use groups | Required vertical exits and passageways | Corridors for exit access | Rooms/enclosed spaces |
|---|---|---|---|
| Assembly | 2 | 2 | DOC FF-1 |
| Business | 2 | 2 | DOC FF-1 |
| Educational | 2 | 2 | DOC FF-1 |
| High Hazard | DOC FF-1 | DOC FF-1 | DOC FF-1 |
| Storage, Moderate Hazard | DOC FF-1 | DOC FF-1 | DOC FF-1 |

Where interior finish materials are regulated, they must be applied such that they are not likely to come loose when exposed to temperatures of 200°F (93.3°C) for up to 30 min. The materials must be applied directly to the surfaces of rated structural elements or to furring strips. If either the height or breadth of the resulting assembly is greater than 10 ft, the spaces between the furring strips must be firestopped. Class 2 and 3 finish materials less than 1/4 in. thick must be applied directly against a noncombustible backing treated with suitable fire retardant material or have been tested with the material suspended from the noncombustible backing. This seems to be a fairly minor restriction, but most of the inexpensive paneling available today from builder supply houses is either 3/16 in. or 4 mm thick. At academic institutions, many departments have their own technicians who often do their own remodeling for economic reasons and build improper partitions of this noncomplying material. Rated material is available, 1/4 in. thick or more, which looks exactly the same on the surface. The only realistic options available to prevent violations of the code requirements is to totally prohibit purchases of building material, strictly enforce policies of no "home-built" structures to the extent of tearing down such constructions, or provide a source of rated material which must be used.

Roofing materials are not interior finishing material, but also must meet standards in order to maintain adherence to classes of construction. Class 1 roofing materials are effective against severe fire exposure and can be used on any type of construction. Class 2 materials are effective against moderate fire exposures and Class 3 materials are effective only against light fire exposures. Class 3 is the minimum acceptable material to be used on Type 2, 3, 4, and 5A construction. Typical materials meeting Class 1 requirements would be cement, slate, or similar materials, while metal sheeting or shingles would meet Class 2. Class 3 materials would be those that had been classified as such after testing by an approved testing agency.

## REFERENCES (SECTIONS 3.1.2 TO 3.1.5.2)

1. *Occupational Safety and Health Administration Proposed Performance Standard for Laboratories Using Toxic Substances,* 51 FR 26660, OSHA, Washington, D.C., July 24, 1986.
2. *Standard on Fire Protection for Laboratories Using Chemicals,* NFPA 45, National Fire Prevention Association, Quincy, MA, 1982.
3. *The BOCA Basic National Building Code,* 9th ed., Building Officials and Code Administrators International, Country Club Hills, IL, 1984, 29, 37, 42, 48—51, 58—71, 125—151, 269—300.
4. Biosafety in Microbiological and Biomedical Laboratories, U.S. Department of Health and Human Services, Public Health Service, Centers for Disease Control, and National Institutes of Health, HHS Pub. No. (CDC) 84-8395, U.S. Government Printing Office, Washington, D.C., 1984.

### 3.1.6. VENTILATION

Few research buildings at either corporate or academic institutions are constructed today without central air handling systems providing heating, cooling, and fresh air. Experience seems to indicate that relatively few of these are designed completely properly to provide suitably tempered air, where it is needed, and in the proper amounts at all times. High energy costs mandate that the energy expended in heating or cooling the air supplied to a facility be optimally minimized. Laboratory buildings, however, have highly erratic needs for tempered air. In academic buildings, for example, when both faculty and students cease working in the laboratory to meet classes or attend to other responsibilities, fume hoods, which typically exhaust around 1000 cfm/min, may individually be turned on and off. In a medium-sized research building containing 50 hoods, as a result of this factor, the required capacity for makeup air could theoretically vary as much as 50,000 cfm. The occupants rarely conform to a sensible daytime work regimen. In academic institutions especially, individuals are almost as likely to be working at 4:00 a.m. as at 4:00 p.m., or while the majority may be taking a Christmas vacation, there are always a few in the middle of a project that cannot be interrupted. Under such circumstances, it is very difficult to continuously provide the right amount of air all the time to every laboratory economically. Economy is the easiest parameter to forego since engineering technology is capable, at least technically, of maintaining proper ventilation under almost any circumstance, even though it may be expensive to do so. Further, health and safety of individuals should never be compromised for economic reasons.

Most written material on laboratory ventilation concentrates almost exclusively upon fume hoods. Ventilation does play an important part in the proper performance of hoods, and they, in turn, usually have the most significant impact of any piece of laboratory equipment on the design and performance of laboratory building air handling systems. However, there are many other aspects to laboratory ventilation. Hoods will be treated as a separate topic in Section 3.2.2 and some aspects of ventilation will be deferred to that section. Those portions of hood performance which concern the general topic of space ventilation will be covered in the following material.

Active laboratory areas should be provided with 100% fresh air. No air should be recirculated. There are laboratories for which this would not necessarily be essential, but as noted earlier, the character of research conducted in a given space may change. Ventilation, which depends upon supply and exhaust plenums to the space being built into the building structure, is one of the more expensive services to provide as a retrofit. It is better to design for the most demanding requirements, and use controls to modify the supply if the actual needs are less. If the active laboratory space can be adequately isolated from administrative, classroom, and service areas, the requirements for these other spaces may be met with a recirculating system, where the amount of fresh air introduced could be as little as 10%. For reasons associated with building air quality, it is often desirable to recirculate a large fraction of a building's air, as long as sufficient fresh air is provided to accommodate the basic needs of the occupants.

The amount of fresh air to be provided to a laboratory space should depend upon the activities within the facility, but there is little data to support a given amount. Epidemiological data, gathered by OSHA, indicates that there are health risks asso-

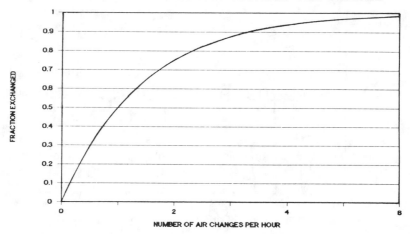

FIGURE 3.3.   The effect of increasing the rate of air change diminishes as the rate increases.

ciated with working in a laboratory. In five studies cited by OSHA, although the overall mortality rate appeared to be lower among chemists than in the general population, there was some evidence that indicated additional dangers from lymphomas and leukemia, development of tumors, malignancies of the colon, cerebrovascular disease, and prostate cancer, although virtually every study indicated a lesser rate of lung cancer. The general good health might be attributable to the generally high economic and educational status of the groups being studied, which probably translates into more interest in their health and being in a position to afford to maintain it. The general consensus that it is not a good idea to smoke in laboratories could impact on the number of cases of lung cancer. In a survey among the members of the California Association of Cytotechnologists, investigating the use of xylene in the laboratories in which they worked, out of the 70 who responded, 59% felt their ventilation was inadequate, 22.6% had no exhaust system, and 43% stated that their ventilation systems had never been inspected. In several recent health hazard evaluations conducted by NIOSH, it was found that ineffective exhaust ventilation was a major contributor to the hazardous conditions. If proper procedures are employed and all operations calling for the use of a fume hood are actually performed within a hood, the general room ventilation would be expected to have relatively little bearing on health of laboratory workers. However, sufficient hood space is not always available and, even where available, is not always used. Consequently, general laboratory ventilation should be sufficient to provide good quality air to the occupants.

In the absence of specific requirements, there are guidelines. The book, *Prudent Practices for Handling Hazardous Chemicals in Laboratories*,[8] recommends between 4 to 12 air changes per h. Guidelines for animal care facilities recommend between 10 to 20. Storage facilities used for flammables are required by OSHA to have six air changes per h. The latter number appears to be a reasonably satisfactory design target. If the air in the room is thoroughly mixed, six air changes per h would result in more than 98.4% of the original air being exchanged, as can be seen in Figure 3.3. Increasing this to seven would result in less than 0.8% additional gain,

FIGURE 3.4. Stratification of air due to poor ventilation system design.

at the expense of a further increase of 14.3% in the loss of tempered air. A critical consideration is whether the air does, in fact, become thoroughly mixed. This depends upon a great many factors, including the location of the room air intakes and exhaust outlets, the distribution of equipment and furniture, and the number, distribution, and mobility of persons in the room. Given time, any gases or vapors present in the air will eventually diffuse and attain a fairly uniform mix, even in an unoccupied space. Substantial amounts of movement in the room will tend to redistribute air within a room more quickly, but there will still be spaces and pockets in almost any room in which, because of the configuration of the furniture and the air circulation, mixing of the air will be slow.

Poor design of the air intake and exhaust system can negatively affect the needed air exchange significantly. In Figures 3.4 and 3.5, the results of tests of a particularly bad system are depicted. In this facility, fresh air is delivered from a unit ventilator mounted on a roof above a corridor and then ducted through the laboratory wall at a height of about 9 ft. Along the same wall, about 12 ft away, is an exhaust duct leading back to the roof. Air is blown into the room horizontally toward the opposite wall. It then, supposedly, traverses the room twice and leaves through the exhaust duct or a hood. Smoke bomb tests of this system, however, showed almost no vertical mixing of air in the room. Half an hour after the smoke was released, a clear line of demarcation about 8 ft above the floor between clear air and smoky air could still be discerned, the latter being partially replenished by exhaust air that had been recaptured and reentered the building. The occupants in this facility did not benefit at all from the air being introduced through the standard air intake, which also did not serve the fume hoods in the room. For the hoods to work properly, air had to be drawn in from either the doors or, when weather permitted, through open windows. Using the corridors as a source of air sufficient to supply even one hood violates code restrictions, and using open windows often results in an erratic air supply due to wind gusts.

Ideally, air entering the laboratory should enter gently and in such a way that the air in the breathing zones of the individuals working in the laboratory is maintained free of toxic materials and that the air flow into hoods in the room will not be

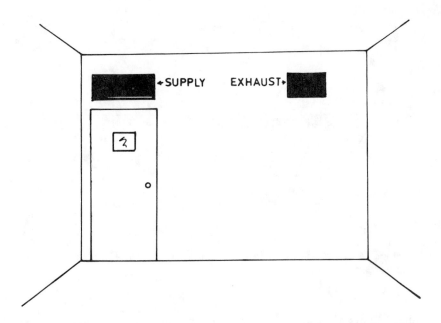

FIGURE 3.5. Head-on view of inlet and exhaust system for room shown in FIGURE 3.4.

interrupted or disturbed by traffic. Studies indicate that air directed toward the face of a hood or horizontally across its face will cause the most serious problems in meeting the latter condition, while air introduced through a diffuse area in the ceiling or from louvered inlets along the same wall on which the hood is situated will be affected the least by movement of personnel.

The majority of laboratory fumes and vapors are heavier than air and will preferentially drift toward the floor, although some will diffuse throughout the room air and some will be carried upward by warm air currents. Room exhausts should be located so as to efficiently pick up the fumes. Placing exhausts in the ceiling or high up on walls is not efficient and, as in the case described previously in this section, can serve to "short-circuit" the supply of fresh air to the room. Even if high air exhaust outlets were effective, they would tend to pull noxious fumes through the occupant's breathing zones. Exhausts placed near the floor or at the rear of workbenches would prove more effective as long as they remained unobstructed, and the direction of air movement from a source would be away from the occupant's face. Localized exhausts, using local pickups exhausting through flexible hoses, can be used to remove fumes from well-defined sources of fumes, but they must be placed close to the source. The air movement toward the nozzle is reduced to less than 10% of the original value within a distance equal to the nozzle diameter. Outside this distance, it is unlikely that a localized exhaust would be very effective in removing fumes. If all work with hazardous materials were done in hoods which ran continuously, it would be possible to rely on hoods to provide the exhaust ventilation to a room. However, this is not the case. Sometimes hood sashes are closed and the hoods used to store chemicals. On other occasions, hoods are turned off while apparatus is

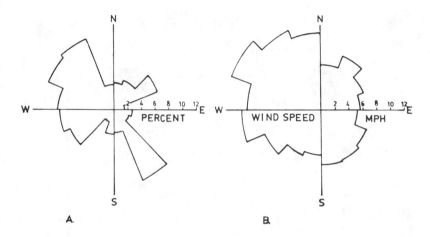

FIGURE 3.6.   (A) Wind direction, percent of time during year; (B) average wind speed vs. direction.

installed or they are off while being serviced. Therefore, the design of the air exhaust system from a laboratory must be done carefully to provide continuing replacement of fresh air in the room. The fume hood system and the supplementary exhaust system should be interlocked so as to insure a stable room air balance at all times.

If there are administrative, classroom, or service areas within the same building as laboratories, the entire laboratory area should be at a modest negative pressure with respect to these spaces so that any air flow that exists will be from the nonresearch areas into the space occupied by laboratories.

It is important that the source of air for a building be as clean as possible and that the chances for air to reenter the building be minimized. In most locations, there are preferred wind directions. In Figure 3.6A and B, directional and velocity wind data are shown, averaged over a year for a typical building site. Such data can be obtained for a region from airports and weather bureaus. However, the data are strongly affected by local terrain, other nearby buildings, trees, and other local variables (note the anomalously high percentage of time the wind comes from a sharply defined southeastern direction here). Where reliable data are available or can be obtained, the air intakes should be located upwind as much as possible with respect to the building. At this site, locating the air intakes at the northwest corner of the building would clearly be desirable because this is the predominant wind direction and the higher wind velocities from this direction would help minimize the recapture of fumes. Building exhausts, again from the data, should be to the south of the air intakes. Obviously, there will be periods when this configuration will lead to the exhaust fumes being blown toward the air intakes, but other measures can also be taken to aid in the reduction of recirculation. One situation which must be avoided is that shown in Figure 3.7. Here, the air intake is located on the lee side of a building elevation. As shown in the figure, the air moving over the top of the obstruction tends to be trapped and circulate downwind from an obstruction. If fumes are swept into this volume, either from the roof above or in the space contiguous to the obstruction

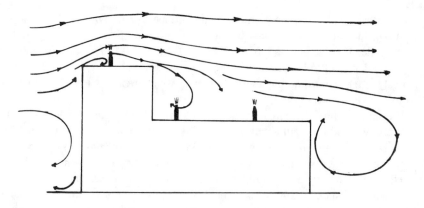

FIGURE 3.7.  Effect of obstructions on wind patterns.

slightly further downwind, they will tend to remain there. If an air intake were located in this space, the entrapped fumes would be returned to the building.

The exhausts from the building should discharge fumes outside the building "envelope", i.e., the air volume surrounding the building where air may be recaptured. Physically, this can be done with tall, individual hood exhaust ducts or the exhausts from individual hoods can be brought to a common plenum and discharged through a common tall stack. The needed height of individual stacks often make this alternative a physically unattractive concept. Two "rules of thumb" are employed to estimate needed heights. For one- or two-story buildings, the stack height above the roof should be about 1.5 times the building height. For taller buildings, this rule would lead to very high stacks, so a height equal to 0.5 times the building's width is often used in such cases. We will return to the latter concept again later because there are a number of design details which are required to insure that a common duct can be safely used. In the former case, if it is desired to have the air escape from the vicinity of a building, inverted weather caps above duct outlets clearly should not be used since they would direct the air back toward the building. Updraft exhaust ducts with no weather caps are preferable, in which the outlets narrow to form a nozzle, thereby increasing the exit velocity. Since the exhaust air has a vertical velocity, it will initially continue to move upward, so the effective height of the duct will be higher than the physical height. The gain in the effective height will depend upon a number of factors, viz., the duct outlet diameter, d; the exit velocity of the gas, v; the mean wind speed, $\mu$; the temperature difference between the exhaust gas and outside air temperatures $\delta T$; and the absolute temperature of the gas, T. The effective height gain is given by:

$$\text{Height gain} = d[v/\mu]^{1.4} [1 + \delta T/T]$$

For simplicity's sake, assume that the indoor and outdoor temperatures are the same. For a duct diameter of 8 in., a nozzle velocity of 4900 fpm, and a mean wind speed of 700 fpm (equal to the annual average of approximately 8 mph at the site for which data were given in Figure 3.6), the height gain would be about 10 ft. Under some weather conditions, the plume would continue to rise, while under others, it would fall. In gusty wind, it could be blown back down upon the building roof. In any case, the effective height above the roof of about 13 ft (duct height plus height gain) would be helpful in reducing the amount recaptured by the building and obviously is far better than the alternative using weather caps, where the fumes are directed down toward the roof.

An examination of the equation used to determine height gain shows that, if the exit velocity could be maintained, it would be advantageous to have a larger duct diameter. For example, if several hoods could be brought to a common final exhaust duct, 2 ft in diameter, so that the exit velocity remained the same, the net gain would be 30 ft instead of 10 ft, and it would be more acceptable to have a single tall chimney rather than a forest of exhaust stacks. With this arrangement, it would be possible to be reasonably certain that the fumes would not return to the level of the air intake until the plume left the vicinity of the building.

Although some concern is usually expressed about chemical reactions due to mixing the fumes from different hoods, generally the fumes from each hood are sufficiently diluted by the air through the faces of the hoods so that the reactions in the plenum will not be a significant problem. The most serious problem is maintaining the balance of the system as the number of hoods exhausting into the common plenum varies. If all the hoods ran continuously, this would not be a problem, but for energy conservation as well as other reasons, this mode of operations is not the most desirable. There are certain conditions that must be met. Each contributing hood exhaust must be kept at a negative pressure with respect to the building as a whole so that fumes will not leak into the building through a faulty exhaust duct. To ensure that no fumes from the common plenum are forced back into the laboratory, the plenum must always be at a negative pressure with respect to the individual ducts, so the plenum must be serviced by a separate blower system. It would be difficult to meet both the balance and energy conservation requirements simultaneously with a single plenum exhaust motor. Multiple motors are needed which go on and off line automatically as the number of hoods which are actually on vary, so that a reasonably constant negative pressure differential, as determined by pressure switches, between the plenum and the individual hoods is maintained. The negative pressure in the plenum would increase the effectiveness of the individual hood exhaust fans.

The risk in such a system is that the motors serving the plenum might fail, while those serving the individual hoods do not. In this event, the fumes in the common plenum would mix and the chances would be good that some would be returned to the laboratories, especially those in which the hoods had been turned off. Since the hoods would be exhausting into a volume at a higher than normal pressure, the effectiveness of the individual hood systems would be diminished and the probability of fumes spilling from the hoods would increase, even for those hoods which continued to operate. If multiple motors serve the common plenum, the problems would not be serious if an individual motor failed since the system should be designed to compensate until the motor is returned to service. However, if an electrical power failure caused the entire plenum system to go down, the potential would exist for serious problems within the laboratories. It is essential that such a system be provided with sufficient standby electrical power, as well as an alarm system, to permit the system to continue to serve all operations which cannot be temporarily terminated or reduced to a maintenance level.

### 3.1.6.1. Quality of Supplied Air

Quantity of air is important, but so is the quality. Humans and equipment work best within a fairly narrow range of temperature and humidity. The term "fresh air" implies that the air is free of noxious fumes, but says nothing about the temperature and humidity. In 1979, emergency building temperature regulations were imposed which required that the temperature set points be set at a minimum of 78°F in the summer and a maximum of 68°F in the winter in order to conserve energy. Although these temperatures were eminently satisfactory to some individuals, a large number were vociferously unhappy. Similarly, very low humidity in the laboratory is frequently encountered during the winter as outside air at low temperatures, containing very little moisture, is brought inside and heated, resulting in desert-like characteristics. As a result, many people develop respiratory problems. During the summer, unless sufficient moisture is wrung from the hot outside air while it is being reduced to more comfortable temperatures, the interior humidity may rise to very high levels. Persons will feel clammy and uncomfortable since they will be unable to perspire as readily. Moist warm air is also conducive to microbial growth. Under such conditions, workers are less likely to wear personal protective equipment such as lab coats, chemical splash goggles, and protective gloves. Unless the temperature and humidity stay within a relatively narrow range, people become less productive and make more errors. Laboratories are not work environments where error-producing conditions should be acceptable. Although individuals differ, a comfortable temperature to a large number of people seems to lie between 68 and 75°F and a comfortable humidity, between 40 and 60%.

Because of all the potential negative results of having poor quality air, it is clearly desirable to have a properly designed and maintained system to make the air as conducive to comfort as possible. In order to assure this, the initial building contract should contain clauses defining explicitly the specifications for the temperature, humidity, and volume of air for each space within the building, and the contractor should be required to demonstrate that the building meets these specifications before the owner accepts the building. This is as important, from both a usage and health and safety standpoint, as any other part of the design.

A laboratory building, with the need for 100% fresh air and due to the large quantities of air discharged by each fume hood, is an energy-inefficient building, almost by definition. Laboratory equipment in the building also is a very substantial heat load. Buildings housing laboratories, therefore, are logical candidates for energy recovery and management systems. In implementing an energy recovery system, care must be taken to ensure that the system is a true energy exchange system, where the incoming air characteristics are moderated by the air being exhausted from the building, but no air is recirculated. It is difficult to manage an energy system within a building if the occupants have the capability of modifying it locally. Thus, in a managed facility, it is likely that any windows will be permanently sealed so that they cannot be opened and disturb the local air balance. The use of ceiling spaces as return plenums is less desirable than the use of fixed ducts since the former permits an individual to modify the air circulation in his space by simply making an opening into the ceiling.

A building designed to meet all the requirements of a well-designed and managed building in terms of air quality is usually a "tight building", with few chances for air to leak into and out of the building. Experience has shown that the "tight building syndrome" occurs when a large fraction of the populace of an entire building appears prone to developing environmentally related illnesses, sometimes suddenly and acutely. Such problems may, on occasion, be triggered in such a building by an unfamiliar odor, by the overall air quality moving out of the comfort zone, by an individual suddenly and unexplainably becoming ill, or for no apparent reason. Laboratory buildings, with their common and frequently unpleasant pervasive chemical odors, are especially vulnerable to this problem. Where the building occupants have little or no means of modifying their environment, the frustration of having no control over the problem seems to exacerbate the likelihood of the problem developing and worsens the impact when it does. Of course, a bad odor does not always trigger a tight-building syndrome response from a building's occupants, but the inability to personally do something about it seems, at the very least, to increase a person's irascibility. Often, even after prolonged investigation, no real cause of a tight-building syndrome incident is ever discovered and it is attributed to stress or other psychological causes, especially when the problem seems to disappear when no corrective measures are taken.

Some tight-building syndrome incidents represent real health problems, with individuals persistently complaining of discomfort and showing evidence of physical distress. Most commonly, the symptoms are respiratory distress, headaches, fatigue, dizziness, nausea, skin and eye irritation, complaints regarding odors, or a chemical "taste" in the mouth. These problems are similar to allergenic reactions, and if some individuals exhibit the problems while others do not, this is typical of the widely varying sensitivity of individuals to allergens. If no other source of the problems can be located, an evaluation of the heating, ventilation, and air conditioning (HVAC) system is in order.

All HVAC systems include filtration systems and some provide humidification and dehumidification functions. The filters may become dirty and the amount of fresh air may decrease to an insufficient level or the filters may begin harboring dust mites or fungal growth. If the filters in different parts of the system become less able to pass air preferentially, the building air balance may shift so that areas which at one time had sufficient air may no longer have an adequate supply. This alone can cause problems. The moisture used for humidification or the drain pans into which moisture from the dehumidifying process goes may become contaminated with biological organisms. Many individuals are allergic to dust, dust mites, fungi, and other microorganisms which would be distributed by a contaminated air supply. In severe cases, the air handling system could harbor Legionella organisms, which would require massive and disruptive decontamination efforts, if these were possible at all.

In order to avoid contamination of HVAC systems and maintain a proper level of performance, a comprehensive and thorough maintenance program is required for all centralized HVAC systems. Filters should be changed or cleaned on a regular and frequent schedule. Water employed for humidification and chillers, or cooling towers, should be checked frequently for biological growth. Condensate pans should be cleaned regularly. Decontamination of an afflicted ventilation system is difficult, time consuming, and, consequently, very costly. Preventative maintenance is very cost effective for ventilation systems.

# REFERENCES (SECTIONS 3.1.6 TO 3.1.6.1)

1. **Li, F., et al.,** Cancer mortality among chemists, *J. Natl. Cancer Inst.,* 43, 1159, 1969.
2. **Olin, R.,** Leukemia and Hodgkin's disease among Swedish chemistry graduates, *Lancet,* 2, 916, 1976.
3. **Olin, R.,** The hazards of a chemical laboratory environment: a study of the mortality in two cohorts of Swedish chemists, *Am. Ind. Hygiene Assoc. J.,* 39, 557, 1978.
4. **Olin, R. and Anlbom, A.,** The cancer mortality rate among Swedish chemists graduated during three decades, *Environ. Res.,* 22, 154, 1980.
5. **Hoar, S. K. and Pell, S.,** A retrospective cohort study of mortality and cancer incidence among chemists, *J. Occup. Med.,* 23, 485, 1981.
6. Comment by the California Association of Cytotechnologists on xylene exposure regarding the OSHA Proposed Performance Standard for Laboratories Using Toxic Substances, 51 FR 26660, July 24, Cypress, CA, 1986.
7. U.S. Department of Health and Human Services, Centers for Disease Control, National Institute for Occupational Safety and Health, Health Evaluation Reports HETA 83-048-1347, HETA 830076-1414, and HETA 81-422-1387, Cincinnati, 1981, 1983.
8. *Prudent Practices for Handling Hazardous Chemicals in Laboratories,* National Academy Press, Washington, D.C., 1981, 193.
9. The Guide for the Care and Use of Laboratory Animals, Publ. No. 85-23, National Institute of Health, Bethesda, MD, 1985.
10. Occupational Safety and Health Administration General Industry Standards, 29 CFR, Part 1910, § 106.(d)(4)(iv), Washington, D.C.
11. **Caplan, K. J. and Knutson, G. W.,** The effect of room air challenge on the efficiency of laboratory fume hoods, in *ASHRAE Trans. 83,* Part 1, 1977.
12. **Fuller, F. H. and Etchells, A. W.,** Safe operation with the 0.3m/s (60 fpm) laboratory hood, *Am. Soc. Heat. Refrig. Air Cond. Eng. J.,* 49, October, 1979.
13. Laboratory Ventilation for Hazard Control, NCI Cancer Research Safety Symp., Fort Detrick, MD, 1976.
14. **Wilson, D. J.,** Effect of stack height and exit velocity on exhaust gas dilution, in *ASHRAE Handbook,* Am. Soc. Heating, Refrigerating and Air-Conditioning Eng., Atlanta, GA, 1978.
15. **Cember, H.,** *Introduction to Health Physics,* Pergamon Press, New York, 1969, 334.
16. Tight building syndrome; the risks and remedies, *Am. Ind. Hyg. J.,* 47, 207.
17. Energy conservation in new building design, in ASHRAE 90, Am. Soc. Heating, Refrigerating and Air Conditioning Eng., Atlanta, GA, 1980.
18. Fundamentals Governing the Design and Operation of Local Exhaust Systems, ANSI Z9.2-1979, American National Standards Institute, New York, 1979.

### 3.1.7. ELECTRICAL SYSTEMS

Electrical requirements for laboratories are relatively straightforward. The entire system must meet the NFPA Standard 70,[1] the National Electrical Code, and must be properly inspected before being put into service. As far as the laboratory worker is concerned, the details of the service to the building are relatively unimportant, but it is important to him that there are sufficient circuits of sufficient capacity to provide enough outlets for all of the equipment in the laboratory. In most new facilities, the designers usually provide enough, but in many older facilities where less use of electrically operated equipment was anticipated, the number of outlets is often inadequate. Many of the older electrical systems were originally designed based on two wire circuits instead of the three wire circuits that are currently required. Any circuits of the older type should be replaced as soon as possible, and laboratory activities should not be assigned to spaces provided with such electrical service until this is done. All circuits, whether original equipment or added later, should consist of three wires: a hot or black wire, a neutral or white wire, and a ground or green wire. The ground connection should be a high-quality, low-impedance ground (on the order of a few ohms) and all grounds on all outlets should be of comparable quality. A poor-quality ground, perhaps due to a workman failing to tighten a screw firmly, can result in a substantial difference of potential between the grounds of two outlet receptacles. This can cause significant problems in modern solid-state electronic equipment which typically operate at voltages of less than 24 V (frequently at 3, 5, or 6 V). A further problem which can result from a poor ground is that leakage current through the high-impedance ground connection can develop a significant amount of localized electrical heat. This is often the source of an electrical fire, rather than a "short-circuit" or an overloaded circuit.

Electrical circuits should be checked with a suitable instrument which is capable of providing a quantitative measure of the ground impedance. Commonly available, inexpensive plug-in "circuit-checkers" which indicate the condition of a circuit by a combination of lights can give a valid indication of a faulty circuit, but a "good" reading can be erroneous. If the distance to another connection is substantial, capacitive coupling of the ground wire can result in a false indication of a low-impedance reading.

The female outlets, as noted, have three connections with openings located at the points of an equilateral triangle. The ground connection is round, while the other two openings are rectangular. If the two rectangular openings are of different sizes, the neutral connection will be longer and the hot connection shorter. A male plug has matching prongs and should only fit in the outlet in a single orientation. A "cheater" or adapter can be used to allow a three-prong male plug to be plugged into a two-wire circuit, but this is not desirable. All connectors, switches, and wiring in a circuit must be rated for the maximum voltage and current they may be expected to carry.

There should be enough outlets appropriately distributed in a laboratory so that it should not be necessary to use multiple outlet adapters plugged into a single socket or to require the use of extension cords. Where it is necessary for additional circuits to be temporarily added, the circuits should be run in either conduits or metal cable trays, both of which should be grounded and installed by qualified technicians or electricians. Even though their use should be discouraged, extension cords will continue to be used. However, as a minimum, they must be maintained in good condition, include three wires of sufficient size to avoid overheating, preferably of 14 gauge wire or better for most common uses, and be protected against damage. They should not be placed under stress and should be protected against pinching, cutting, or being walked upon. Where abuse may occur, they must be protected with a physical shield sufficient to protect them from reasonably anticipated sources of damage.

Circuits must be protected by circuit breakers rated for the maximum current to be carried by the circuit. Normally, many breakers for a room or group of rooms are located together on a common breaker panel. All circuits should be identified, both within the facility and at the breaker panel, so that when required, the power supply to a given circuit may quickly and easily be disconnected. This is especially important when it is necessary to disconnect power due to an emergency. There should be no ambiguity about which breaker needs to be thrown to kill the power to a given receptacle.

Where breaker panels and electrical switches are placed in separate electrical closets or rooms, there is some question about the propriety of individual laboratory workers having access to the space. If there are electrically live parts with which individuals might accidentally come into contact, access to the spaces must only be by qualified, authorized individuals. Normally live components on a breaker panel are completely covered (if not, then prompt action to replace the cover should be initiated), so that if the panels are segregated by a locked or otherwise secure barrier from areas containing electrically active components, then laboratory workers should be allowed to enter these spaces for essential activities. However, access to these spaces must not be abused by considering them as extra "storage space". Access to the electrical panels, switches, and other electrical equipment in the space must not be blocked by extraneous objects and materials.

The location of electrical circuits and electrically operated equipment in a room should be such that they are unlikely to become wet or in an area susceptible to condensation or where a user might be in contact with moisture. As unlikely as it may appear, instances have been observed where equipment has been located, and electrical circuits installed, where water from deluge showers would inundate them. For some equipment, such as refrigerators, freezers, dehumidifiers, and air conditioning units, moisture is likely to be present due to condensation and these equipment items are required to be grounded.

### 3.1.7.1. Hazardous Locations

Most laboratories do not represent hazardous locations in the context of requiring special electrical wiring and fixtures, although there may be individual equipment items which may need to be treated as such. The classification of a facility as hazardous depends upon the type of materials employed in the facility and whether flammable fumes or gases, electrically conducting materials, or explosive dusts are present in the air within these facilities in the normal course of routine activities or only sporadically due to some special circumstance. Explosion-proof wiring and fixtures are substantially more expensive than ordinary equivalents and need be used only if there are no acceptable alternatives. Where the need does exist, however, they definitely should be used.

The National Electrical Code defines three different categories of hazardous locations, Classes 1, 2 and 3 (note that these are not the same as the classification of flammable liquids into Classes 1, 2 and 3). Class 1 represents locations where flammable vapors or gases may be in the air. Class 2 locations involve facilities where electrically conducting or combustible dusts may be found, and Class 3 locations contain ignitable fibers. Each class is split into two divisions. In Division 1 for each class, the hazardous conditions are present as a normal course of activities, are sufficiently common due to frequent maintenance, or may be generated due to equipment failure, such as emission of dangerous vapors by the breakdown of electrical equipment. Division 2 includes locations which involve hazardous materials or processes similar to those in Division 1, but under conditions where the hazardous gases, fumes, vapors, dusts, or fibers are normally contained or the concentrations maintained at acceptably low levels by ventilation so that they are likely to be present only under abnormal conditions. Division 2 locations are also defined to include spaces adjacent to, but normally isolated from, Division 1 locations from which problem materials might leak under unusual conditions.

Within Class 1, there is a further division by groups into A, B, C, and D, depending upon the materials employed, with the distinction between groups being based essentially on the flammable limits in air, by volume. A long list of chemicals, with the groups identified to which they may belong, is given in NFPA 497M.[3] Information on other chemicals should be available to any researcher in the material safety data sheets which chemical users should be maintaining or, at least, to which they should have access. However, in this listing, acetylene, with a flammability range of 2.5 to 81%, is the only chemical listed in Group A. Group B chemicals, with flammability ranges between about 4 to 75% and with flash points less than 37.8°C or 100°F, include acrolein, ethylene oxide, propylene oxide, hydrogen, and manufactured gas (>30% hydrogen by volume), although if equipment in the facility is isolated by sealing all conduit 1/2 in. or larger in diameter, according to specifications in the National Electrical Code, the first three of these may be placed in groups of lesser risk. Among those flammable liquids with flash points above 37.8°C or 100°F, but less than 60°C or 140°F, allyl glycidal ether and *n*-butyl glycidal ether would also be in Group B, but with the same exception as to conduit sealing.

Groups C and D have flammability ranges of about 2 to 30 and 1 to 17%, respectively. A few common chemicals falling into Group C are acetaldehyde, carbon monoxide, diethyl ether, ethylene, methyl ether, nitromethane, tetrahydrofuran, and triethylamine. A partial list of common Group D chemicals includes acetone, benzene, cyclohexane, ethanol, gasoline, methanol, methyl ethyl ketone, propylene, pyridine, styrene, toluene, and xylene.

Generally, if the location uses chemicals in any of these four groups which have flash points less than 37.8°C or 100°F and otherwise meets the specifications of a Class 1, Division 1 location, special electrical equipment would normally be required. Such equipment would normally be required for Class 2 flammables (flash points equal to or above 37.8°C or 100°F, but less than 60°C or 140°F) only if the materials are stored or handled above the flash points, while for Class 3A flammables (flash points equal to or above 60°C or 140°F, but less than 93.3°C or 200°F), special electrical equipment is needed only if there are spaces in which the temperature of the vapors may be above the flash points.

Group 2 includes conductive and combustible dusts, with some dusts falling into both categories. Dusts with resistivities above $10^5$ $\Omega$/cm are not considered conductive, while those with lesser resistivities are. If operations were such as to generate significant levels of dust in the ambient air in the work location, the decision as to whether special electrical equipment would be required would be based on whether a cloud of the dust in question would have an ignition sensitivity equal to or greater than 0.2 and an explosion severity equal to or greater than 0.5. Both of these are dimensionless parameters based on a comparison to a standard material, Pittsburg Seam Coal.

The definitions of these two parameters are

$$\text{Ignition Sensitivity} = \frac{(P_{max} \times P)_2}{(P_{max} \times P)_1}$$

$$\text{Explosion Severity} = \frac{(T_c \times E \times M_c)_1}{(T_c \times E \times M_c)_2}$$

where  $P_{max}$ = maximum explosive pressure
$P$ = maximum rate of pressure rise
$T_c$ = minimum ignition temperature
$E$ = minimum ignition energy
$M_c$ = minimum explosive concentration
Subscript 1 = standard dust
Subscript 2 = specimen dust

There are a number of metals, and their commercial grades and alloys, which could give rise to the need for special electrical equipment, some of which are listed in NFPA 497.[2] Some metals such as zirconium, thorium, and uranium, which would be found in some special laboratories, have both low ignition temperatures (around 20°C, 68°F) and low ignition energies, so work with these materials would require special precautions and safeguards.

A large number of nonconductive dusts have an ignition sensitivity of 0.2 or higher and an explosion severity of 0.5 or greater. Many agricultural products, such as grains, can form dusts in this category. Other nonconducting materials with similar characteristics would be many carbonaceous materials, chemicals, dyes, pesticides, resins, and molding compounds. A long but incomplete list is given in NFPA 497M[3]. OSHA lists many industrial operations which could be classified as Class 2, and in the context of laboratory safety, there are pilot or bench-scale research operations which would emulate these industrial locations.

Class 3 locations represent hazardous locations because of easily ignitable fibers. However, it is unlikely that they would be present in sufficient concentrations to produce an ignitable mixture. Thus, no special electrical wiring requirements would normally exist. However, research facilities which generate substantial airborne quantities of fibers of cotton, synthetic materials such as rayon, wood, or similar materials should take care to avoid the potential for fire.

The special electrical equipment or wiring procedures needed to satisfy the requirements of a hazardous location are specified in the National Electrical Code, NFPA 70. In general, fixtures suitable for use in hazardous locations will be rated and certified as safe by a nationally recognized testing laboratory such as Underwriters Laboratory or Factory Mutual Engineering Corporation. In some laboratory installations, equipment items may be found which are nonstandard and are not listed as acceptable by any appropriate organization. In such cases, the equipment may be certified as safe if an appropriate agency charged with enforcing the provisions of the National Electrical Code so as to ensure the occupational safety of users finds the equipment in compliance with the Code. In such cases, records need to be kept showing how this determination was made, from manufacturer's data or actual tests and evaluations. Special electrical equipment for use in Class 1 locations is not necessarily gas-tight, but if an explosion does occur within it, it should be contained and quenched so that it does not propagate further.

# REFERENCES (SECTIONS 3.1.7 TO 3.1.7.1)

1. *National Electrical Code,* NFPA 70-84, National Fire Protection Association, Quincy, MA, 1984.
2. *Recommended Practices for Classification of Class I Hazardous Locations for Electrical Installations in Chemical Plants,* NFPA 497, National Fire Protection Association, Quincy, MA, 1975.
3. *Manual for Classification of Gases, Vapors and Dusts for Electrical Equipment in Hazardous (Classified) Locations,* NFPA-497, National Fire Protection Association, Quincy, MA, 1983.
4. *Prudent Practices for Handling Hazardous Chemicals in Laboratories,* National Academy Press, Washington, D.C., 1981, 179.
5. The following ANSI Standards cover electrical apparatus for use in hazardous locations (by number only): C33.3-1973, 781-1977, 844-1978, C33.27-1974, C33.29-1972, 1002-1977, and C33.97-1973, ANSI, New York. These are issued jointly with Underwriter's Laboratories.

## 3.1.8. PLUMBING

There are two aspects of laboratory plumbing system design, as opposed to operations, which are relevant to safety and health. The first is the capacity of the system to withstand the waste stream which the system may be called upon to handle. The second is the need to prevent the operations within the laboratory facility from feeding back into and contaminating the potable water system which supplies the building.

### 3.1.8.1. Sanitary System Materials

Under the standards governing the disposal of toxic and hazardous chemicals into the sanitary sewage system and into the public waters, much smaller quantities of chemicals should be going into sink drains than in the past. A very large number of the chemicals used in laboratories are now classified as hazardous waste and should be collected and disposed of according to the provisions of the Resource Conservation and Recovery Act. However, a substantial amount of acids, bases, solvents, and other chemicals still go into the sanitary system, even if from no other source than the cleaning of laboratory glassware. These should be highly diluted in large quantities of water, but, given time, can corrode ordinary plumbing systems.

Plumbing materials used to service laboratory facilities must be resistant to a large range of corrosives, physically durable, relatively easy to install and repair, and relatively cheap. Metal plumbing materials are generally unacceptable because they are vulnerable to inorganic acids. Plastics such as PVC, which is used in residential systems, will not withstand many common organic solvents or will absorb other solvents and not remain dimensionally stable. Glass would serve very well, but is brittle and not inexpensive in large systems. Its desirable properties make it suitable and acceptably cost effective for the components which are the most vulnerable to the actions of chemicals, the sink trap and the fittings which connect the sink to the sanitary system. A plastic which is resistant to a large range of inorganic and organic waste streams, dimensionally stable, durable, and relatively inexpensive is polypropylene. Other materials with comparable physical properties may be either more expensive or more difficult to install and maintain.

One characteristic not mentioned in the preceding paragraph is the necessity for the material to have established fire resistance ratings as specific structural members, i.e., tested by appropriate testing laboratories in the configurations in which they might be used. Polybutylene plumbing components are available which meet this requirement.

### 3.1.8.2. Backflow Prevention*

Many jurisdictions have specific legal requirements that devices to prevent contamination of the potable water supply must be installed on each service line to a building's water system wherever the possibility exists that a health or pollution hazard to the waterworks system could exist. This includes any facility, such as a laboratory, where substances are handled in such a manner as to create a real or potential risk of contaminating the water supply external to the building. Although the regulations may only address the problems of cross contamination between buildings, the risk may exist as easily within a building. For example, if by some means contaminated water is drawn from a laboratory sink back into the potable water system, the contamination could easily express itself in the water available from the water fountain immediately outside in the hall if the latter were fed from the same line supplying the faucets in the sink. The contamination would not necessarily be only close to the contaminating source but could be anywhere in the system downstream from the origin of the pollution.

There are two possibilities for cross connections, the simplest being a direct connection between a contaminated system and a clean system, connected by a valve which could be opened as in Figure 3.8. Obviously, this is not a desirable situation, but can occur. The second possibility is that a pressure differential between the clean and contaminating systems may be established such that contamination may be forced or drawn into the clean system through some linkage. This linkage may not necessarily be an open or leaky valve, although these are two of the most likely sources. An example of an inadvertent cross connection would be a heat exchanger which over a period of time has developed small leaks between the primary and secondary loops. As long as the primary or clean side is at a higher pressure than the secondary or contaminated side, any water flow would be from the clean to the contaminated side and, as long as the leaks were small, would probably go undiscovered. If, however, the water supply on the primary side were to be reduced and pressure were to fall below the secondary, the flow would be reversed and the clean water supply would be polluted. This would be an example of a forced backflow problem and the connection could, as noted above, be due to human error — simply leaving a valve partially open that is normally used to add makeup water to the system.

A very common situation which could exist in a laboratory is to find a section of plastic tubing, draining the effluent from a piece of apparatus, lying in the bottom of a far-from-clean sink. Perhaps the sink is being used to wash dirty glassware and the plastic tubing is under several inches of water. If the water pressure should suddenly fail, perhaps due to a reduced supply because of maintenance, coupled with the

---

* Much of the material in this section is derived from Woodruff, P. H., Prevention of contamination of drinking water supplies, in *Handbook of Laboratory Safety*, 2nd ed., Steere, N. V., Ed., CRC Press, Boca Raton, FL, 1971, 587.

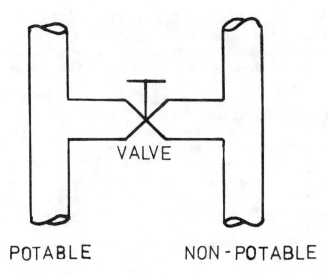

POTABLE                    NON - POTABLE

FIGURE 3.8.   Simple cross connection between liquid systems.

simultaneous flushing of several toilets, it would be possible for the water in the sink to be siphoned back through the system and reach the potable water supply. Any system in which a connection to the potable water supply can be flooded with contaminated water would be subject to the same type of problem. Inexpensive vacuum breakers for installation on sink faucets are available from most major laboratory supply firms.

The correction of these problems is not always obvious. In the first example, the problem with a heat exchanger, a pressure sensor with redundancy features should be installed between the primary and secondary loops such that when the positive pressure differential falls below a stipulated value (well above the point at which the pressures are reversed), operations should be shut down and the secondary loop emptied. A better solution would be for the heat exchanger to incorporate an intermediate loop so that the secondary and primary sides could not be directly coupled.

Figure 3.9 is a simple diagram of vacuum breakers used to prevent back-siphonage. In order to prevent backflow due to excess pressure, vacuum breakers cannot be used and an air gap provides the most protection. Figures 3.10 and 3.11 illustrate two versions of air gaps. The second of these is very straightforward: simply do not connect anything to the water source which could become flooded. Any formal backflow or antisiphonage device which is installed must be an approved type which has been tested by a recognized laboratory testing agency and be of satisfactory materials

## REFERENCES (SECTIONS 3.1.8 TO 3.1.8.2)

1. **Woodruff, P. H.,** Prevention of contamination of drinking water supplies, in *Handbook of Laboratory Safety,* 2nd ed., Steere, N. V., Ed., CRC Press, Boca Raton, FL, 1971, 587.

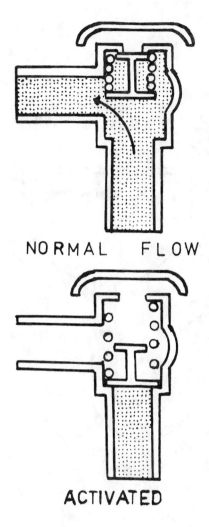

NORMAL  FLOW

ACTIVATED

FIGURE 3.9.    Vacuum breaker.

### 3.1.9. OTHER LABORATORY UTILITIES

There are a number of other possible utilities which may be provided to a laboratory. Among these are natural gas, compressed air, distilled water, vacuum, steam, refrigerated brine, and other gases. Although there are safety issues such as limitations on the pressure available from a compressed air line and the need to incorporate provisions for pressure relief, in some instances, to insure that personnel will not be injured by explosions due to excess pressure, the quality of the air is often more of a problem. The compressor supplying the system should be capable of supplying air which is clear of oil and moisture. Some facilities which have a large pressurized liquid nitrogen tank at hand use the vapors from the tank as a source of ultra-clean compressed "air" to clean work surfaces.

The dangers of natural gas are well understood. This does not prevent numerous

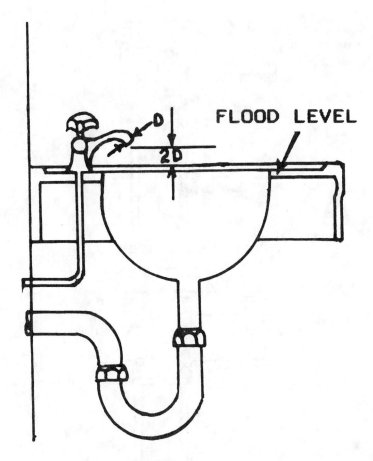

FIGURE 3.10. Minimum desirable vertical air gap on sink.

accidents each year due to gas explosions. Natural gas used to heat experimental devices should not be left unattended unless a heat-sensitive automatic cut-off device is connected to the system. Otherwise, if gas service were to be interrupted, gas escaping from burners or heaters left on could easily cause a major explosion. If the gas device is set up in a properly working fume hood, the gas would be exhausted through the fume hood duct system, but as noted earlier, fumes exhausted outside a laboratory building can be recaptured under some conditions. Whenever the odor of gas is perceived, under no circumstances should electric light switches, other spark-producing electric devices, or, for that matter, any other possible source of ignition be operated. If the gas concentration is low enough to make it safe for personnel to enter the room, all devices connected to the gas mains in the laboratory should be immediately turned off. Any windows in the room which are not fixed shut and are capable of being opened, should be opened. Consideration should be given to having all nonessential personnel evacuate the building while volunteers check the remainder of the building or facility for other systems which might also be leaking. If the concentration is already high, everyone should evacuate the building at once, the gas

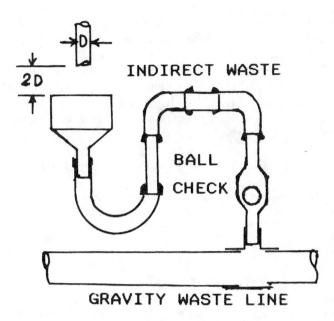

FIGURE 3.11.   Air gap in line subject to back pressure.

service turned off at the service entrance to the building, and the utility service company notified.

Even small gas leaks, barely perceptible by odor, should be sought out and repaired. Gas can seep slowly through cracks and seams and accumulate in confined spaces in which the air interchange is very slow and, if an ignition source presents itself, explode.

OSHA has specific regulations on four gases — acetylene, nitrous oxide, hydrogen, and oxygen. Generally, the requirements in the OSHA regulations are those provided for by the Compressed Gas Association (GSA), and for the first two of the gases mentioned above, the OSHA standards simply refer to the appropriate pamphlet issued by the GSA. For hydrogen and oxygen, the regulatory requirements are spelled out in detail in Sections 1910.103 and 1910.104, respectively. The requirements for oxygen generally pertain to bulk systems of 13,000 ft$^3$ or larger. Relatively few laboratory facilities would involve a system of this size.

For hydrogen, the regulations do not apply to systems of less than 400 ft$^3$, so laboratory systems involving a single, typical gas cylinder would not be covered. However, in the context of the present chapter, which is intended to involve the design of a building, it is entirely possible that a hydrogen gas system will be larger than 400 ft$^3$, and would fall under the provisions of the OSHA standard. All of the requirements of the standard will not be repeated here, but among these are provisions that (1) cast iron pipes and fittings shall not be used (note that for acetylene, brass and copper pipes may not be used. Each gas has specific properties for which provision must be made in the design), (2) the system shall be above ground, (3) the electrical system shall be a Class 1 system, (4) outlet openings shall be at the high points of the wall and roof, and that (5) explosion venting be provided.

Every utility that is provided should be properly identified with a clear, unambiguous label. Color-coded discs with engraved name labels which screw into each service fixture are available from at least one company for most common service utilities. All fittings and connectors should be provided according to appropriate standards. For example, all gas fittings should comply with the provisions of the Compressed Association, Standard V-1. Every utility, if used improperly, can cause safety problems, the only difference being that some require more of an effort on the part of a user than do others.

# REFERENCES (SECTIONS 3.1.9 TO 3.1.10)

1. Occupational Safety and Health Administration General Industry Standards, 29 CFR, Part 1910 § 101-104, Washington, D.C.
2. Standard V-1, Compressed Gas Association, New York.
3. *Prudent Practices for Handling Hazardous Chemicals in Laboratories,* National Academy Press, Washington, D.C., 1981, 75.

### 3.1.10. MAINTENANCE FACTORS

Every building and every piece of equipment installed in a building should be designed with the realization that each item of equipment and most building components will eventually need service. As noted at the very beginning of this chapter, the current modus operandi of most research facilities is to change the nature of the research in a given space fairly frequently. Thus, the space itself, treated as a piece of equipment, will need relatively frequent service. It should be possible to bring any needed services to virtually any position within the laboratory. Good access to services should be provided in the initial design, which will make it possible for maintenance personnel to work on the equipment conveniently and safely.

All ducts, electrical circuits, and utilities need to be clearly identified with labels or other suitable means. Fume hood ducts on the roof should be identified with at least the room number where the hood is located and a hood identification if there is more than one hood in the room. It would also be desirable for the general character of the effluents from the exhaust duct to be indicated by means of a written statement or a color code. Similarly, every electrical circuit should be clearly labeled so that there will be no confusion as to which breaker controls the power to the circuit.

A significant amount of space is set aside in every major building for mechanical service rooms, electrical closets, and other building services. They are rarely seen by most of the building occupants, but they are critical to the good operation of the building. Most of the major equipment used to provide building services is either located within these spaces or the control panels are located within them. These areas need to be maintained well and not be used as storage space. Unless a major component such as a compressor for the air conditioning system is out or it is necessary to turn off the water to an entire area, the operations within these rooms are usually outside the experience of the usual occupants of the building and access to these spaces should not normally be available to them. To most occupants of the buildings, maintenance involves work either in their area or directly affecting their area.

The building plans should show all utilities, ducts, and electrical circuits correctly, as they physically exist throughout the facility. During construction projects, it frequently is necessary to make some adjustments to the original plans. Sometimes these are documented with change orders, if they are sufficiently major, but at other times they are considered minor and are not well documented. In principle, all changes should be reflected on the final "as built" drawings and specifications provided to the building owner. This should include not only physical changes in locations, but any changes and substitutions made in materials as well. For example, a complicating factor in dealing with asbestos in existing buildings stems from the uncertainty as to whether asbestos was used under provisions in many contracts which allowed substitution of "equivalent" materials. Such uncertainty has resulted in delays and additional costs in renovation projects while insulating materials were tested for asbestos. Most asbestos was used when it was perfectly legal and desirable to do so, and it is necessary now to know where it is in order to remove or treat it to render it harmless. If such information were reliably available from building plans, it would be much easier to estimate removal costs and design asbestos abatement projects.

Unfortunately, after a period of occupancy, it is common for many changes to be initiated by the facility's users themselves. These are intended to serve a specific purpose and are rarely built with consideration for anything except this purpose in mind. As a result, maintenance may be significantly impaired. Perhaps the most common occupant-initiated changes involve electrical circuits, which may or may not be installed or labeled properly. Similarly, the interior configuration of a laboratory space may be changed, which can significantly affect the distribution of air from the ventilation system serving the area. It should be institutional or corporate policy to prohibit such modifications and to require that they be removed, once found. As a minimum, any such remodeling on the part of the occupants should be required to be reviewed by appropriate personnel and changed where necessary to comply with building codes and maintenance programs. They should not be allowed to be left in place where they are disruptive to good design and use practice.

It is necessary that appropriate provisions be made for unusually dangerous maintenance operations. A good example is the need to change contaminated high-efficiency particulate and aerosol (HEPA) filters, where the contaminant might be a carcinogen or other life-threatening biologically active agent. Ample access room to the exhaust duct must be provided so that bag-out procedures, where the contaminated filters are withdrawn into a sealed bag, may be done without difficulty and without risk of the agent of concern affecting the maintenance personnel or spreading to adjacent areas. Another example, representing an acute danger rather than a delayed one, would be repair to a perchloric acid ejector duct after a period of operation when the wash-down cycle failed to work. In this case, there would need to be provision for washing the interior of the duct for several hours prior to beginning maintenance operations as well as developing contingency plans to protect the workmen in the event flooding the duct was not totally effective in removing dried perchloric acid.

Maintenance operations often considered as routine, such as replacement of a fume hood motor or repair to a laboratory sink, must be considered hazardous operations if the potential exists for exposure of the workmen to toxic chemicals. An example is exposure to chemicals during repairs to a hood exhaust system because of fumes from the nearby exhaust ducts of operating hoods or chemical residues on the motor, fan, or other components of the unit under repair. Even maintenance activities totally unrelated to laboratory operations, such as patching a roof, can permit the workmen to be exposed to possibly toxic effluents from the air handling system or fume exhausts. The converse is also true; fumes from roofing materials, paints, welding, etc. can be drawn into a building via the air intake to the accompanying discomfort of the building's occupants. Under the new 1986 OSHA Hazard Communications Standard, it is required that any individual potentially exposed to dangerous chemicals must be informed of the risks and of the measures needed and available to protect them from the dangerous effects. Thus, when anyone is asked to perform maintenance on a piece of laboratory apparatus which could be contaminated or to work under such conditions that chemical exposures could result, it is now mandatory that the situation be evaluated for potential dangers. It may be necessary to plan for a period of time when operations can be postponed or modified to eliminate the release of toxic materials or, alternatively, the workmen provided with appropriate protective equipment which they must wear during the maintenance operations. Since workmen typically have a minimal chemical background, they tend to be more concerned about the dangers of exposure than an ordinary laboratory employee. It is important that an effective training program be established to ensure that they are informed of the actual risks to which they might be exposed, but which will neither exaggerate nor minimize them.

## 3.2. FIXED EQUIPMENT AND FURNITURE

In Section 3.1.3, requirements for furniture in the various classes of laboratories were briefly given, viz., "Furniture should be designed to be sturdy and designed for convenient utilization" and "Bench tops should be resistant to the effects of acids, bases, solvents, moderate heat, and should not absorb water". Flammable material storage cabinets were among some of the special items of equipment which were needed if solvents were to be used. This section will discuss these items and others in detail. The quality of the laboratory furniture should be as good as can be afforded and as versatile as possible, unless it is certain that usage will involve only a limited range of reagents in a few applications. Similarly, the specifications on equipment should be written to insure that it will perform well and *safely* . A refrigerator may be bought on sale for perhaps as little as one quarter of the price of a flammable material storage model. However, the possible consequences of storing solvents in the former make it a very poor bargain in most laboratories. Even if the immediate need is not there, few of us are so certain of the future that we can be confident that in the 15- to 20-year lifespan of a refrigerator, it will not become necessary to store flammable solvents in one.

### 3.2.1. LABORATORY FURNITURE

Just as ordinary furniture can be bought in many grades of quality, which are not always discernible to a casual examination, laboratory furniture also varies in quality of construction. The least expensive may look attractive and appear to offer the same features as better units, but it will not be as durable and will have to be replaced frequently. It is also likely to not be as safe. If shelves are not firmly attached, the weight of chemicals on them can cause them to collapse. If cabinets and drawers do not have positive catches, they can rebound or fail to close and lead to a spill if containers inside were to be struck through the opening or cause an accident to

### TABLE 3.5
### Work Surface Materials

| Material | Use level | Applications |
|---|---|---|
| Stainless steel | Heavy | Radioisotopes, perchloric acid, solvents |
| Modified epoxy resin | Heavy | Most applications |
| Natural stone, resin impregnated | Mod. heavy | Most applications, avoid severe heat stress |
| Furan-coated composition cement, chem. resistant filler | Mod. heavy | Solvents, most wet chemistry operations |
| CA. Al. silicate, chem. resistant filler, vinyl coated | Moderate | Not for corrosives |
| Portland cement silica mixture vinyl coated | Moderate | Academic laboratories, moderate duty hoods |
| Welded fiber, baked on resin coatings | Light | Work benches, general science facilities |
| Melamine laminate, bonded to phenolic flakeboard | Light | Low-abuse applications |

someone who does not observe the obstruction. Poorly protected work surfaces can corrode or become contaminated and be difficult to decontaminate.

Good quality modular laboratory furniture is available today in a variety of materials which can be installed in configurations to fit almost any needs. Units can be obtained precut to accommodate connections to utilities. In most cases, the utilities can be brought to the correct locations prior to installation of the furniture, which makes it simple to perform maintenance and make changes should a need arise to modify the existing configuration.

### 3.2.1.1. Base Units and Work Tops

Base units can be obtained in steel, wood, or plastic laminates. The steel in the steel units should be heavy gauge, with a pretreatment to reduce the corrosive effects of chemicals. Painting all surfaces with a durable, baked on, chemically resistant paint finish will also help minimize chemical effects. Many individuals continue to prefer wooden laboratory furniture. Because of cost, solid wood furniture is not an economic choice, but durable wood (or plastic) veneer furniture is available which can meet most safety requirements. Although wood may be more absorbent to liquids than steel, it is less reactive and more resistant to a very wide range of chemicals than many materials. There may be some surface degradation, but the furniture will remain usable. There are exceptions, such as facilities using perchloric acid.

Bench tops can be obtained in several different materials. Some are satisfactory for light duty, while others can be used for almost any purpose. Again, unless it is certain that both immediate and long-range usage will permit the use of a lesser material, it is suggested that the bench tops for a new facility be selected from among the more rugged and versatile materials. Table 3.5 lists several materials provided by one vendor for laboratory bench tops, an approximate level of duty for which they appear to be suitable, based on the characteristics of the materials and the vendor's own description, and a few representative applications.* Although this is only the range of materials offered by one supplier, it is a reasonable approximation of the types and ranges of materials in common use.

When using radioactive materials or perhaps some unusually toxic materials, where it would be necessary to decontaminate the work surface after a spill, the choice clearly should be limited to materials, such as stainless steel, least likely to absorb materials. However, if the use of such materials is minimal and limited to very low levels of activity or concentrations (as it should be for use on an open bench), the work surface can be protected with an absorbant paper with a chemically resistant backing. The higher cost of the premium materials can be avoided if the need does not otherwise exist.

Wherever possible, in order to avoid seams in which toxic materials could become trapped, the backsplash panel, if one is used, should be an integral part of the worktop. Service shelves above the backsplash panel can be of lesser-duty materials in most cases since they will be primarily used to store limited quantities of reagents intended for short-term usage in closed containers. Laboratory sinks incorporated in the worktops are usually made of stainless steel or materials equivalent to the top two or three other materials listed in Table 3.5 in order to provide the essential levels of resistance needed for corrosives, solvents, and other organic and inorganic materials. However, current restrictions on disposing of many chemicals into the sanitary system should reduce the burden on the laboratory sink and the remaining plumbing components.

In most instances, the tendency is to utilize all of the space underneath a work top for cabinetry and other forms of storage space. It is a good idea to leave at least one portion open and available for storage of movable carts and other items of equipment which must be left on the floor, in order to avoid reducing the aisle space below acceptable limits.

## REFERENCE (SECTION 3.2.1.1)

1. Fisher Scientific Company 1986 Catalog, pp. 825-842.

### 3.2.1.2. Storage Cabinets

Facilities for storage of research materials in the laboratory should be selected with as much care as any other item of equipment, although many chemicals may be stored on ordinary shelves or cabinets, with only common-sense safety provisions being necessary. Obviously, the shelves or cabinets must be sturdy enough to bear the weight of the chemicals. Storage should be such as to make it unlikely that the materials will be knocked off during the normal course of activities in the room. Shelves should not be overcrowded. It should not be necessary to strain to reach materials or to return them to their places. Finally, the amount of storage should not be excessive, in order to restrict the amount of chemicals not in current use that would otherwise tend to accumulate within the facility. Although these are, as noted,

---

* Table 3.5 is derived from material presented in the 1986 Fisher Scientific Laboratory Supply Catalog. Descriptive names are used instead of the trade names given in the catalog so that the materials can be compared to similar information provided by other vendors.

common sense procedures, it is surprising how frequently many laboratories violate one or more of the rules and how many accidents occur as a result.

A number of classes of materials should be stored in units designed especially for them because of their dangerous properties or because there are restrictions on the use of some materials which require that provisions be made for storing them securely. Among these are flammables, drugs, and radioactive materials. In addition, some materials, such as acids, are best stored in cabinets designed to resist the corrosive action of the materials. Most of the storage units sold specifically for this purpose are intended to be free standing, but they also may be purchased to be compatible with other modular components, equipped with worktops or as bases for fume hoods so that work space will not be lost.

### 3.2.1.2.1. Flammable Material Storage*

Flammable liquids are among the most universally employed chemicals in laboratories and represent one of the most significant hazards due to the possibility of ignition of the liquid and the vapors from the liquids. Even a small quantity of a very volatile flammable liquid can spread across a substantial surface if allowed to spill, and the vapors from the spilled liquid (or from an open container) can be ignited from a remote distance if the vapors spread and come into contact with an ignition source. In the latter case, the flame can flash back to the original source and result in a fire. Since the vapors of most flammable liquids are heavier than air and tend to remain as a relatively coherent mass as they spread, the risk of ignition by flash back is significant. If substantial amounts of liquid are stored on open shelves or workbenches, it is possible for a small spill to relatively quickly escalate into a large fire, possibly involving the entire facility due to the availability of the additional fuel. Flammable material storage cabinets are primarily designed to restrict the amount of flammable liquids available to a fire by providing protected storage for the amounts needed for routine laboratory operations. Except for the quantities required for the immediate work at hand, all the reserves should be stored in flammable material storage cabinets and should be returned promptly to these cabinets when the required amounts are removed from the containers.

Flammable liquids are divided into various classes, as given in Table 3.6, and the maximum amounts permitted to be stored in a single cabinet depends upon the class of flammable liquid involved.

Note that neither combustible 2 nor combustible 3A materials include mixtures in which more than 99% of the volume is made up of components with flash points of 93.3°C (200°F) or higher.

In Section 1910.106(d)(3) of the General Industry Standards, OSHA limits the amounts of Class 1 and Class 2 liquids in a single flammable material storage cabinet to 60 gal and the amount of Class 3 liquids to 120 gal. Thus, even if the integrity of a single storage cabinet were broached in a fire or if an accident occurred while the cabinet was open, no more than 60 gal of Class 1 and 2 liquids or 120 gal of a Class 3 liquid could become involved in the incident. However, even these quantities are

---

* Additional information on the properties and hazards associated with the use of flammable liquids may be found in Section 4.5.6, Flammable Liquids, and in Section 4.6.2.6.1, Flammable Hazards. However, for the convenience of the reader, the first two tables in this section will be repeated again in Section 4.5.6.

## TABLE 3.6
### Definitions and Classes of Flammable and Combustible Liquids

| Class | Boiling points[a] | Flash points[a] |
|---|---|---|
| Flammable | | |
| 1A | <37.8 (100) | <22.8 (73) |
| 1B | ≥37.8 (100) | <22.8 (73) |
| 1C | — | 22.8 (73) ≤ and < 37.8 (100) |
| Combustible | | |
| 2 | — | 37.8 (100) ≤ and < 60 (140) |
| 3A | — | 60 (140) ≤ and < 93.3 (200) |
| 3B | — | ≥93.3 (200) |

Note that neither Combustible 2 nor Combustible 3A materials include mixtures in which more than 99% of the volume is made up of components with flashpoints of 93.3 (200) or higher.

[a]  Temperatures in °C (°F).

## TABLE 3.7
### Maximum Allowable Size of Containers and Portable Tanks

| Container type | Class | | | | |
|---|---|---|---|---|---|
| | 1A | 1B | 1C | 2 | 3 |
| Glass or approved plastic | 1 pt | 1 qt | 1 gal | 1 gal | 1 gal |
| Metal (other than DOT drums) | 1 gal | 5 gal | 5 gal | 5 gal | 5 gal |
| Safety cans | 2 gal | 5 gal | 5 gal | 5 gal | 5 gal |
| Metal drums (DOT specs.) | 60 gal | 60 gal | 60 gal | 60 gal | 60 gal |
| Approved portable tanks | 660 gal | 660 gal | 660 gal | 660 gal | 660 gal |
| Polyethylene spec. 34 or as authorized by DOT exemption | 1 gal | 5 gal | 5 gal | 60 gal | 60 gal |

substantial and, unless the rate of use warrants maintenance of these amounts, the totals actually stored should be limited.

As noted above, the purpose of a flammable material storage cabinet is to postpone the involvement in a fire of the materials within the cabinet, should a fire develop in the room, long enough to allow persons in the immediate area to evacuate the area, or in some cases to permit the fire to be extinguished. Technically, if a 10-min fire test were to be run according to NFPA Standard 251-1969, for the cabinet to pass the test, the interior temperature of the cabinet should not exceed 164.1°C (325°F), all joints and seams should remain intact during the fire, and the cabinet doors should remain closed.

Even if flammable materials are kept within a suitable storage cabinet, there are also restrictions imposed by OSHA standards on the maximum sizes of containers, even if they are kept within flammable material storage cabinets or rooms designed or built for the purpose of flammable storage. These restrictions are given in Table 3.7.

Most commercially available flammable material storage cabinets are made of metal and would have to meet at least the following specifications:

1.    The bottom, top, and sides shall be of at least 18 gauge sheet iron and shall be double walled with a one and one half (1.5) in. air space.

2.    Joints shall be riveted, welded, or made tight by equally effective alternative means.

3.    The cabinet door would have to be provided with a three-point lock.

4.    The door sill would have to be raised at least 2 in. above the bottom of the cabinet.

5.    The cabinets must be labeled in conspicuous lettering "Flammable — Keep Fire Away".

Note that there are no requirements for automatic door closers provided with a fusible link. This is clearly a desirable feature since it ensures that the cabinet will close in the event of a fire, even if it were inadvertently left open. There are also no requirements for vent connections, but most flammable material storage cabinets have provisions for installing ventilation ducts. The value of these is not as universally accepted. If the containers within the cabinets are always tightly sealed, then there is no reason to make provisions for exhausting fumes. However, should the containers not be tightly closed, it would be possible for volatile fumes to accumulate within the cabinet, which it would be desirable to exhaust. Although it is likely that on some occasions containers will be returned to the cabinet improperly closed, the author's own preference is to not utilize the vents since they often are not properly ducted into an acceptable exhaust and volatile vapors could escape into the facility.

If the decision is made to vent the cabinets, the venting ducts should be made of materials such as steel which would withstand a fire so that the vent opening itself will not represent a point at which flames could ignite any vapors within the cabinet. Ducting into a fume hood would ensure that the fumes would be exhausted properly if released into the hood well back of the sash.

Commercial steel flammable material storage cabinets are available in a variety of sizes, the most common being 30, 45, and 60 gal units. However, other sizes are available. In addition, if a facility is short of space, cabinets are available with workbench surfaces. Flammable material storage cabinets are also available as fume hood bases, which facilitate the venting of the interior of the cabinet directly into a fume hood.

Wooden flammable material storage cabinets, if properly constructed according to the provisions of Section 1910.106(d)(3)(ii)(b) of the OSHA General Industry Standards, are also acceptable. These provisions are

> The bottom, sides, and top shall be constructed of an approved grade of plywood at least 1 inch in thickness, which shall not break down or delaminate under fire conditions. All joints shall be rabbited and shall be fastened in two directions with flathead wood screws. When more than one door is used, there shall be a rabbited overlap of not less than 1 inch. Hinges shall be mounted in such a manner as not to lose their holding capacity due to loosening or burning out of the screws when subjected to the fire test.

Although wood will eventually burn, thick sheets of plywood such as required by the standard can withstand a substantial amount of heat before becoming directly involved. The wood can also be treated with heat-resistant paint to further improve its usefulness. Wood of the required thicknesses is also a much better insulator than the thicknesses of metal normally used for cabinetry. Wooden cabinets can be readily fabricated on site by local personnel and be made to fit into nonstandard size spaces within existing laboratories.

In 1959, the Los Angeles Fire Department performed a number of comparative tests using various combinations of metal and wood to simulate the walls of a storage cabinet.* The experimental walls were fastened to the opening of a furnace operating in the range of 704 to 788°C (1300 to 1450°F). A thermocouple was attached to the opposite face of the simulated wall cross section. The experimental mockups were the following:

Metal

1.  A double wall, metal structure of 18 gauge CR steel, approximately 7 in. × 10 in. with 1.5 in. air space
2.  A cross section similar to number 1 made with a core of 5/8 in. sheetrock suspended midway in the 1.5 in. air space
3.  A cross section similar to number 1 with a core of untreated 1/2 in. Douglas Fir plywood suspended in the air space
4.  A metal-walled structure insulated with 1 in. of 1 lb-density fiberglass blanket in the 1.5 in. air space
5.  A metal-walled structure insulated with 1.5 in. of mineral rock wool (density unavailable)

Wood

1.  Two layers of 1-in. Douglas Fir plywood
2.  One layer of 1-in. Douglas Fir plywood
3.  A laminate formed of 1/2 in. plywood on each side of 1/2 in. sheetrock.

Although the experimental arrangement does not simulate a storage cabinet perfectly, the temperature data as a function of time, presented in Table 3.8 do show the rate at which heat is transmitted through the various combinations. They show that the poorest of the wooden panels transmitted heat at a slower rate than the best of the metal combinations. Thus, a wooden flammable material storage cabinet, although it probably would need to be custom constructed in a laboratory facility, would appear to offer superior fire protection. Eventually, however, the wood itself, even if protected by fire-retardant paints, would represent a source of fuel for a protracted fire.

### 3.2.1.2.2. Cabinets for Drug Storage

Security is the primary concern for the storage of controlled substances or drugs because of the potential for theft and misuse. The Drug Enforcement Agency, in CFR Title 21, Parts 1301.72 to 1301.76, delineates the security requirements for Schedule

---

* This material was taken from Fire-protected storage, in *Handbook of Laboratory Safety,* 2nd ed., Steere, N. V., Ed., CRC Press, Boca Raton, FL, 1971, 183—184.

TABLE 3.8
Simulated Storage Cabinet Wall Configurations
(Time [in Minutes] vs. Temperature Data)

| Sample | 5 | | 10 | | 15 | | 20 | |
|---|---|---|---|---|---|---|---|---|
| | °F | °C | °F | °C | °F | °C | °F | °C |
| Metal | | | | | | | | |
| 1 | 430 | 221 | 500 | 60 | 510 | 266 | 650 | 343 |
| 2 | 150 | 66 | 180 | 82 | 210 | 99 | 240 | 116 |
| 3 | 130 | 54 | 140 | 60 | 160 | 71 | 170 | 77 |
| 4 | 310 | 154 | 430 | 221 | 470 | 243 | — | — |
| 5 | 270 | 132 | 433 | 223 | 466 | 341 | — | — |
| Wood | | | | | | | | |
| 1 | 100 | 38 | 100 | 38 | 100 | 38 | 100 | 38 |
| 2 | 120 | 49 | 133 | 56 | 166 | 74 | 200 | 93 |
| 3 | 90 | 32 | 100 | 38 | 110 | 43 | 130 | 54 |

1 through 5 controlled substances, including provisions for both cabinets and storage vaults. For individuals who are practitioners (there are several different categories of practitioners, including physicians, dentists, pharmacists, and some institutional personnel, all of whom are legally allowed to dispense drugs), the requirements are simple: the storage cabinet needs to be a securely locked, substantially constructed cabinet. In general, most institutional research programs can be designed to meet this level of required security. However, wherever there is a question of security, it is essential that storage facilities be adequate to prevent the loss or theft of the materials. Where the programs would fall under the requirements for nonpractitioners, Parts 1301.72 to 1301.74 describe the needed security provisions. Although these are given in considerable detail, Section 1301.71 provides that actual security requirements which offer structurally equivalent protection would be acceptable. The simplest manner in which protection can be provided for small quantities is a steel safe or cabinet. Among other requirements, the safe or cabinet should be sufficiently durable to prevent forced entry for at least 10 min, either be sufficiently heavy (750 lb or more) or rigidly bolted to a floor or wall so that it cannot readily be bodily carried away, and should be equipped with an alarm which will sound in a central control station, manned at all times. All of these are not required for a practitioner's storage cabinet, but the implication of these provisions for tight security should serve as a guide in defining what is meant by a "securely locked, substantially constructed cabinet".

### 3.2.1.2.3. Storage of Radioactive Materials

Storage cabinets for radioactive materials need to meet multiple needs. As with drugs, there is the problem of security, although for a totally different reason. There is a widespread fear of radioactive materials among the general population. As a result, the public is extremely sensitive to the possibility that radioactive material could be lost or taken from the areas in which it is normally used, with the result that members of the general public could be exposed to radiation. Although the quantities of materials normally used in most research laboratories are often extremely small and, hence, would rarely cause any exposure to the public significantly above that due to natural radiation, in order to address the public's concern, the rules and regulations governing the use of radioactive materials are extremely strict, including those on security. The use of radioactive materials and radiation will be dealt with at length in Chapter 5, as well as the other major concerns involving storage of radioactive materials, shielding against radiation from the original research materials and of radioactively contaminated waste. Some of the concepts which will be developed more fully there will be used here without immediate explanation. In the following material, if there are any terms which are unfamiliar, definitions or explanations will be found in Chapter 5.

Title 10, Part 20.207 of the Code of Federal Regulations covers the basic security requirements very briefly, and this brevity may appear to reflect that security is not a major issue. However, directives have been issued by the NRC to emphasize the importance which the NRC places on security. All radioactive materials must be kept in a secure, restricted area to which access is controlled or in a securely locked cabinet or other type of storage unit unless a qualified, authorized user is in the immediate area. This means that if an experiment is in progress involving radioactive material, if the researcher were to leave the laboratory, he must either lock the facility behind him or return the radioactive material to a locked storage unit, should there be no other authorized user in the laboratory unit.

If an NRC inspector were to conduct an unannounced inspection of the facility and find radioactive material in use and unattended, it would be considered a violation of the radioactive material license and, depending upon the circumstances, could result in a fine or loss of the license to use radioactive materials. In an extreme case, if the institution or corporation were to be operating under a Broad License, the lack of compliance of a single laboratory, if it could be shown that the institution was not enforcing the rules, could lose a license for the entire corporation or institution. Clearly, then, it is essential that every laboratory using radioactive materials have a sturdy, lockable storage cabinet in which to keep the material. Frequently, a radioactive material is incorporated into a research material which would have physical or chemical properties that would require specialized storage, such as flammable materials, because of these properties as well.

In addition to security, the radiation emitted by some of the isotopes used in research requires that shielding must be incorporated into the storage cabinet. For the radiation from some isotopes, the choice of material used for shielding is critical because the wrong type, depending upon whether it is of a low atomic number or not, can exacerbate the radiation problem. Although almost any radioactive material can be used in some applications, in the majority of chemical and biological laboratories, a relatively small number of isotopes represent the major usage. Several of these are weak pure beta emitters and can be stored safely in any cabinet since the betas will be completely stopped by virtually any shielding. Among these isotopes, tritium ($^3$H), $^{14}$C, $^{35}$S, $^{33}$P, $^{45}$Ca and $^{63}$Ni (used primarily as a source in gas chromatographs) are used most frequently and are among the safest to use. On the other hand, $^{32}$P is also a beta emitter, but the betas are very energetic and shielding is required. As the energetic betas from this isotope are slowed down and brought to a stop in the shielding, the deceleration creates a penetrating type of electromagnetic radiation called "bremstrahlung". This effect is much more pronounced for shielding made of higher atomic number materials such as lead than in materials such as plastics, composed primarily of carbon and hydrogen. Therefore, for this particular beta emitter, which is one of the more frequently used isotopes because of its chemical and biological properties, the shielding in the storage cabinet should be of plastic or some similar material.

Several commonly used isotopes also emit a penetrating electromagnetic radiation called gamma radiation. Among the more commonly used isotopes in this category are $^{125}$I, $^{22}$Na, $^{51}$Cr, $^{65}$Zn, $^{60}$Co and $^{137}$Cs. The appropriate shielding material to be used in or around the storage unit in such a case would be lead.

One last type of radioactive material for which special storage criteria are needed is in the relatively rare instance where neutron sources are used. Neutron radiation is found primarily in reactor facilities, but neutron sources are also used in moisture density probes, which are used in a number of research areas. If the source were to be taken from the instrument, the shielding would need to consist of a layer of plastic or paraffin, 5 to 20 cm thick, either impregnated with boron or surrounded by cadmium. Some gamma shielding might be required in an unrestricted area, which could be provided by concrete blocks.

### 3.2.1.2.4. Corrosive Materials

Storage cabinets can be obtained which can serve as either flammable material storage cabinets or as cabinets for the storage of acids. In the latter case, the shelves must be provided with protection against the effects of corrosion. One of the more economical ways of achieving the required protection is to use polyethylene shelf liners. If they are not needed or need to be replaced, they can readily be taken out or replaced with stock polyethylene. The interior walls can be protected from the effects of acid vapors and fumes by being painted with an acid-resistant paint. An alternative interior finish which provides excellent protection against corrosion that has recently become available is glass reinforced cement.

### 3.2.1.2.5. Record Protection*

Research records are not adequately protected against fire damage in most laboratories and research facilities. Research notes, thesis materials, computer disks, and financial and academic records may be irreplaceable or more difficult to reproduce than equipment or the building. In addition, the need exists today to be able to provide proof, at the laboratory level, that the operations of the laboratory have been in accordance with the requirements of equal employment practices and training programs of the Hazard Communication Standard. Further, if the proposed OSHA Laboratory Safety Standard is published in approximately its present form, each laboratory will need to develop a performance plan to provide equal protection to the employees in the laboratory, as is provided by the specific standards in the remainder of the OSHA standard to other types of employees. Documentation will be needed to show that this is the case and this documentation might well be needed in the event of a laboratory fire.

---

* This section, for the most part, is taken directly from *Handbook of Laboratory Safety*, 2nd ed., Steere, N. V., Ed., CRC Press, Boca Raton, FL, 1971, 179, 180. It has been edited to reflect changes in standards.

There are several ways of providing fire-protected storage, ranging from inexpensive methods of gaining additional minutes of protection to built-in systems and construction for providing hours of fire protection. We believe the potential for fire loss in laboratories calls for (1) realistic evaluation of possible losses, (2) immediate steps to minimize losses, and (3) planning for systematic long-range improvements. While various innovations and expedients may be useful for temporary or partial protection, tested and approved equipment and construction will be more economical in the long run.

One method of providing protection for records is extremely simple. Maintain two sets of records in two different buildings. This is expensive because of the additional space required, and the labor involved in the duplication of records is certainly not negligible, but it is the only virtually certain way in which valuable records can be protected. However, as records become older and are needed only rarely, the set maintained in the laboratory can be eliminated. The amount of expensive laboratory space devoted to records can be minimized, and only that amount of space needed for records that have not been duplicated as yet need to be protected against fire damage.

Fire in a university laboratory building will usually result in an attempted influx of graduate students, research faculty, and post-doctoral associates trying to rescue their research records to avoid having to repeat a great deal of work, if this is feasible at all. To avoid loss of valuable research material and interference with fire fighting operations, research safety considerations should include fire-protected storage.

Ordinary desks and file cabinets provide almost no protection for records since unprotected wood will burn rapidly and metal will transmit enough heat in a few minutes to ignite or seriously char papers and other records.

### Improvement of Existing Record Storage

When funds and space are limited, ordinary desks and file cabinets can be improved to provide a measure of fire-protected storage by use of special fire-retardant paints. Intumescent paints foam when heated and form an insulating barrier against heat and flames. Several such paints have been tested and are listed as fire-retardant coatings by Underwriter's Laboratories, 207 East Ohio St., Chicago, IL 60611 in their Building Materials List.

Providing adequate fire protection for records will require a detailed analysis and a comprehensive plan based on determination of the value of various records, hazards to which the records are exposed, and the present protection afforded.

Acquisition of the latest version of the Standard for the Protection of Records, National Fire Protection Association Standard No. 232, is recommended for every research department and organization. The version available at the time this document was being prepared was adopted in 1980 and issued in 1981. The publication details the need for management and protection of records, gives standards for fire-resistive vaults and file rooms, and describes standards for fire-resistive safes and other protective equipment. It also provides some excellent references for methods of salvaging records damaged by fire suppression efforts.

Although some large laboratories may need file-resistive record vaults and file rooms, the discussion which follows will be limited to the standards for record protection equipment. The only way to be sure that record protection equipment will meet the need for which it is purchased is to see that it bears the label of Underwriter's Laboratories, or another nationally recognized testing laboratory. Chapters 4 and 5 and Appendix A of the NFPA Standard 232 would be useful in preparing purchase order specifications.

### Record Protection Equipment

Equipment should be selected which will protect records from fires of the most destructive intensity and duration which may occur. If stairwells are open, if there are no automatic sprinkler systems, and if the building is of combustible construction, the probable maximum condition is complete destruction of all combustible contents and portions of the building. It is hoped that if the building is built and equipped according to the principles previously discussed in this chapter, complete destruction will not occur. However, if fire fighting efforts are seriously hampered or prevented by the effects of the chemical contents of the building, complete destruction would still be a distinct possibility. Record protection equipment in older wood-joisted or other nonfire-resistive research buildings should be rated to withstand the impact following floor collapse as well as the fire exposure.

The method described in Appendix A-5-1.1(a) of NFPA Standard 232 for selecting record protection equipment for a building is based on estimating the combustible material to which the equipment may be exposed, assuming uniform distribution and dividing by the floor area to obtain the weight of combustibles per unit area. Flammable liquids are not included in this formula, but may be approximated, according to Steere, by multiplying their weights by a factor of two (2) for the purpose of approximating their fuel contribution equivalent to ordinary combustibles. The following formula provides the fuel loading per square foot and, if the liquid fuel factor L is deleted, is the same as that given in NFPA Standard 232.

$$\text{Adjusted fuel load } I = \frac{H}{\text{gross floor area in ft}^2}$$

Here, H = adjusted or derated fuel load in pounds, where $H = F - (C \times G + 0.25D)$.

  C = weight in pounds of books and paper in totally enclosed steel containers
      = volume × 28
  D = weight in pounds of books and papers in partially enclosed (five-sided) steel containers = volume × 28
  F = Fuel load = A + 2B + C + D + E + 2L, where
  A = Weight in pounds of all wood or cellulose furniture
  B = Weight in pounds of all plastic furniture
  E = Weight in pounds of all free combustibles (books and paper) located on horizontal surfaces 6.5 ft or less from the floor and in combustible containers and steel containers with less than five sides = volume × 28.

Factor G depends upon the ratio of factor C to factor F.

If C/F is less than 0.5, then G = 0.6
      equal to or between 0.5 and 0.8, then G = 0.8
      greater than 0.8, then G = 0.9

## TABLE 3.9
### Equipment for a Fire-Resistive Building

| Floor area exposed (lb/ft²) | Noncombustible desks, filing cabinets, lockers, and other closed containers (not over 30% of combustibles exposed) | Combustible desks, filing cabinets, shelving containers, etc. |
|---|---|---|
| <5 | 1-h device (without impact) or file room | 1-h device (without impact) or file room |
| 5—10 | 1-h device (without impact) or file room | 1-h device (with impact) or file room |
| 10—15 | 1-h device (without impact) or file room | 2-h device or file room |
| 15—20 | 1-h device (with impact) or file room | 2-h device or file room |
| 20—30 | 1-h device (with impact) or file room | 4-h device, file room, or vault |
| 30—35 | 2-h device or file room | 4-h device, file room, or vault |
| 35—45 | 2-h device or file room | 6-h vault or file room |
| 45—50 | 4-h device, file room, or vault | 6-h vault or file room |
| 50—60 | 4-h device, file room, or vault | 6-h device or file room with no combustible near door |

Once factor I is calculated, Table 3.9* may be used to determine the devices needed to provide protected storage in a fire-resistive building, which a research building should be. The required devices depend upon both the total combustible contents per floor and the percentage of combustibles that are in an exposed condition on any given floor.

Since each device which provides additional protection is more expensive than the lesser one, it can be seen from Table 3.9 that reducing the quantity of combustible contents in a laboratory and the amount exposed will reduce the cost of record protection. Replacing or the fire-retardant coating of combustible desks, filing cabinets, shelving, and containers will also reduce the cost of providing adequate record protection.

## REFERENCES (SECTIONS 3.2.1.2 TO 3.2.1.2.5)

1. Occupational Safety and Health Administration General Industry Standards, 29 CFR, Part 1910, § 106(d), Washington, D.C., 1988.
2. Flammable and Combustible Liquid Code, NFPA 30, National Fire Protection Association, Quincy, MA, 1981.
3. Fire Protection for Laboratories Using Chemicals, NFPA 45, National Fire Protection Association, Quincy, MA, 1981.
4. Standard Methods of Fire Tests of Building Construction and Materials, NFPA 251, National Fire Protection Association, Quincy, MA, 1985.
5. **Steere, N. V.,** Fire-protected storage for records and chemicals, in *Handbook of Laboratory Safety,* 2nd ed., Steere, N. V., Ed., CRC Press, Boca Raton, FL, 1971, 179.
6. Standard for the Protection of Records, NFPA 232, National Fire Protection Association, Quincy, MA, 1980.
7. *Prudent Practices for Handling Hazardous Chemicals In Laboratories,* National Academy Press, Washington, D.C., 1981, 226.

* This table is the same as that provided by Steere in his article, incorporating changes in NFPA 232.80.[6]

8. Drug Enforcement Agency, Title 21 CFR, Part 1301, § 72-76, Washington, D.C., 1988.

9. Nuclear Regulatory Commission, 10 CFR, Part 20, § 207, Washington, D.C., 1988.

### 3.2.2. HOODS

A key item of the fixed equipment in most research laboratories other than those employing only the least hazardous materials is a work enclosure, usually denoted, at least for chemicals, by the common name "fume hood". Some laboratory facilities have recently been constructed with no open bench space, with all work within the facility being done within hoods. Others are being built or planned around the same concept. It has been suggested that a reasonable rule of thumb to be used in designing laboratories is to provide one hood for every two research personnel. However, since the activities in a laboratory vary so widely, there can be no absolute standard for a suitable number for a given laboratory other than that there should be sufficient hoods available so that no work which should be performed in a hood need be done on an open bench. In the standard laboratory module described in Section 3.1.1, only one hood is shown. However, additional hoods could be added as needed, with these being placed as far back in the laboratory as possible in each case. The air supply intakes to the room should be designed to permit location of these additional hoods such that the flow of air into the hoods will be minimally disturbed. In general, the air speed at the face of the hood must be somewhat less than the face velocity of the hood, in the range of 20 to 30 fpm.

The purpose of a hood is to capture, retain, and ultimately discharge any noxious or hazardous vapors, fumes, dusts, and microorganisms generated within it. It is not intended to capture contaminants generated elsewhere in the room; however, since a hood exhausts so much of the laboratory air when operating, it does serve as a major component of the ventilation system for the room. A few specially designed hoods are intended to confine moderate explosions, but most are not. Unless it is designed, built, used, and maintained properly, a hood will not perform its intended function.

Some individuals have been observed to be so hypnotized by the concept of a hood that they continue to use hoods which are not functioning, still counting on them to provide a normal level of protection. It has actually been necessary on occasion to padlock the sashes of hoods closed to prevent this. Unless a hood is fully functional, it should not be used.

There are many different hood designs, even within a single category such as chemical or biological. Some configurations perform better than others, while the desirability of some features depend upon individual preferences. A good quality general-purpose hood should be able to withstand corrosion, be easily decontaminated (especially for some uses), suitable for the use of flammable materials, and capable of withstanding the effects of a fire for a reasonable period of time, sufficient to either allow an attempt to put out the fire or to initiate an orderly evacuation. The design should be such as to minimize the possibility of initiating a fire or explosion.

#### 3.2.2.1. Factors Affecting Performance

This has been discussed briefly before in discussing the location of the hood within a laboratory and the effects of doors, windows, air supply inlets, and traffic. Although there have been a number of advances in fume hood designs and standardization of commercial units since the article by Horowitz et al.[1] on science laboratory hoods, much of the material is still relevant. In addition, Steere[6] has a good article in the same Handbook and portions of the article, "Air Conditioning for Laboratories" provided by the American Society of Heating, Refrigeration, and Air Conditioning Engineers, also provided excellent information on factors affecting fume hood performance. Much of the information found in these earlier articles will be recognized in the material below, which has been revised, combined, and edited to meet current regulations. Some material is used without changes and is identified as quotations. In the next few sections, most of the material is oriented toward what has become the "conventional" hood — one with a vertical sash, with air entering through the face of the hood below the sash. Other types of hoods and some of the parameters affecting their performance will be discussed in Section 3.2.2, Types of Chemical Hoods.

### 1. Face Velocity

One of the major factors affecting the performance of a hood is face velocity. "Satisfactory performance of a fume hood requires that the airflow past the opening into the enclosure occur within minimum and maximum limits, since both are of great importance. They vary somewhat depending upon a number of factors relating to the design of the particular hood, its location in the laboratory and the degree of hazard of the experiments; therefore, the limits must be selected with judgment. The minimum face velocity must be great enough to ensure that the direction of air movement at any point in the area of the open face of the hood will always be into the hood. It is desirable that the lowest possible amount of air be exhausted, consistent with safety requirements, because of the economic advantage of reducing losses of heated air in winter and cooled air in summer. Factors influencing the minimum face velocity are numerous.

The upper limit of air velocity for a fume hood is related to the aerodynamic flow pattern created by the air stream flowing past the person in front of the hood, past the experiment itself, and out the exhaust opening. The person standing in front of the hood serves as a barrier to the air stream, and when the air velocity reaches a certain point, a low pressure or partial vacuum is created directly in front of him. The low-pressure zone extends into the fume hood increasingly as the face velocity is increased. A velocity can be reached where the low-pressure zone may extend into the area of the fume hood occupied by the experimental apparatus. When this condition is reached, the fumes generated by the experiment will fill the low-pressure zone and may contact the researcher's skin, or be inhaled unless he is protected by a portable shield, a horizontal-sliding sash,"[1] or the vertical sash is lowered well below the individual's face.

"Within the two limiting extremes of face velocity, the choice of face velocity is based on the desire to reduce to a minimum the total flow of air being exhausted and to maintain a safe environment."[1] There is no uniform agreement on what the minimum safe velocity is, with recommended minimums from various sources ranging from 60 to 150 fpm. Tests have shown that, although the face velocity can be reasonably uniform across the opening to the enclosure, the amount of spillage may vary widely from point to point, depending in part upon the external factors listed earlier, but also upon the internal design and set-up of the hood, and upon the manner in which it is used. It has been shown that such factors as the distance of the fume source from the face of the hood and the height of the sash opening (for a vertically sliding sash) can affect the performance especially significantly. The presence of a heat source within the hood can cause the spillage to increase. A depressed area of the base (to retain spilled liquids) can modify the air flow pattern into the hood. Generally, however, the influence of these factors on the spillage from the hood diminishes as the face velocity increases. As a result of these factors, although hoods can be operated safely at 60 fpm, they are likely to be more vulnerable to operating conditions. An individual walking by at about 2/3 mph (a comfortable walking speed is of the order of 2 or 3 mph) would exceed this speed. Too rapid an arm movement can cause pulses of spillage. Unless every other factor is optimally adjusted to prevent spillage, 60 fpm provides little safety margin. Although the actual operating velocity may be adjusted to a lower speed, such as 100 fpm, the system should be designed with a capacity to provide a somewhat greater speed, perhaps 125 fpm. This permits some error in the actual performance, compared to the design value, so that an operating value of 100 fpm can definitely be obtained. In addition, as the performance of the hood decreases with time due to losses in fan effectiveness from dirt in the exhaust duct or stretching of the drive belt from the motor to the fan and other factors, the design level capacity of 125 fpm provides some slack in the maintenance required. The air velocity can be adjusted in many hoods by a damper in the exhaust duct. In others, variable motor speed controls are provided. In both cases, the air balance to the room can be disturbed. It is preferable that adjustments be made at the fan location by maintenance personnel who should, at the same time, check the room air balance to insure that it remains about the same and slightly negative with respect to the corridor.

Most good quality commercial hoods today incorporate air foil features at the top, bottom, and sides to ensure an aerodynamic flow of air into the hood to minimize eddy currents and provide for maximum operating efficiency. Many are available which provide a reasonably constant volume of air through the hood, but split between an opening below the sash and a path over the top of the sash, eliminating excessive velocities at smaller sash openings (please refer to Section 3.2.2.2, Types of Chemical Hoods).

A common procedure, which frequently leads to low face velocities and poor performance, is for the scientist to personally select the exhaust motor from the catalog. Each hood installation needs to be configured by a ventilation engineer. The length and diameter of the duct, the number of bends and turns, the type of fan, and the termination of the duct will all influence the size of the blower motor required. Few research personnel are qualified to correctly select the size fan required.

As noted earlier, individuals will continue to use hoods that are not working because they appear to be working. The air speed of about 1 mph through the face is so slight that, unless there are visible fumes, it is not readily apparent whether the hood is working. A light found on the outside of many hoods usually only indicates that power is being supplied to the motor. It does not provide a positive indication of air flow. In at least one instance, a perchloric acid hood was checked and found to have zero face velocity. When the fan on the roof was examined, the motor was running, but the blades had corroded so badly that they had fallen off and were strewn across the roof. A positive velocity sensor of some type should be incorporated into every hood. It should provide a visual and audible alarm when the air velocity falls below a predetermined safe level and should give an indication should the sensor itself fail.

Hoods should be checked upon installation, when any maintenance is done on them, when there is any modification to the room which could affect air flow patterns in the vicinity of the hood, when any significant maintenance or modifications are made to the building HVAC system, and on a regularly scheduled basis. The last check should be done at least annually and preferably quarterly or semiannually. The anemometers or velometers employed for testing the performance of the hoods should be calibrated before use. Measurements should be made at a number of places over the sash opening to insure that the velocity does not vary by more than ± 25% at any point. For a conventional hood, it may be desirable to mark the sash heights on the side of the hood opening at the positions where the face velocity drops below the selected acceptable level and above a level of 150 fpm, where turbulence due to a person standing in front of the hood could lead to spillage. Smoke generators can also be used to check for spillage under various configurations of face velocity, baffle position and experimental apparatus positioning.

## 2. Dimensions

"The depth of a hood is the most important dimension with respect to satisfactory operation. In general the deeper the hood, the more satisfactory it will be in providing uniform suction across the open face. The depth is also of great importance with respect to the size of apparatus and experimental setups that can be accommodated."[1] A deeper unit will also allow fume-generating apparatus to be placed further back in the hood. "Generally, the lower the height of a fume hood (and the clear opening), the more satisfactory will be its airflow characteristics. A very tall hood with the exhaust duct connection near the top," as most are, "will present the greatest risk of escape for fumes because of the difficulty of securing uniform airflow at the base."[1]

## 3. Baffles

"The function of the baffles in a fume hood is to distribute the suction of the exhaust duct in such a way that uniform airflow through the face of the hood will result. The design and position of baffles and their openings are critical to satisfactory performance. Opinions on proper design of baffles vary widely among experts on the subject."[1] Most commercial hoods are provided with a single baffle which has adjustable slots at the top and bottom. The baffle and the slots should be adjusted after installation in the laboratory to provide the most uniform airflow across the face of the hood. Larger hoods may have more slots in the baffle to aid in adjusting the air flow.

## 4. Construction Materials

The three previous factors were concerned with the ability to capture and retain noxious and dangerous fumes and vapors within a stream of air passing through a hood. This section will be concerned with the physical ability of a hood to contain and withstand the corrosive actions of the materials used in it. In principle, the selection of the materials used in the fabrication of the hood should depend upon the types of chemicals intended to be used in the research program. However, hoods are major fixed pieces of equipment which would be difficult and expensive to change as the nature of a research program changes, so, as has been the general tenor of the recommendations in this chapter, the hood should be selected to be as versatile as funds will permit. If a hood is intended for a dedicated use which is not expected to change for an extended period, then it is only common sense not to spend more than necessary. The materials used for the walls and lining, base, sash, and some of the interior fittings will be discussed separately.

*Walls and Lining*

"A great many materials have been used for fume hood construction. Most of them have been proven to be satisfactory when used correctly and properly selected with regard to the requirements of the experiments." As noted above, "it makes good sense to use the least expensive material that will do the job. Unfortunately, however, materials which are most versatile tend to be the most expensive."[1] In the following list, the materials which will be discussed are the ones which are normally available from commercial vendors. It is rare for hoods to be fabricated by the user. The list is approximately in order of versatility.

**a. Transite** — Transite is a material in which asbestos fibers are bonded with a resin and, until recently, was probably the most popular lining material for general-purpose, heavy-duty hoods. Although it may become discolored with use, it is highly resistant to a large number of chemicals. In recent years, because of the concern for the health effects of airborne asbestos fibers, its use has decreased rapidly. Although the asbestos fibers are tightly bonded in the transite, some older hoods have been observed in which the fibers had become friable and liable to become detached. When transite is broken or cut, asbestos fibers may again become airborne and pose a potential health hazard to anyone in the vicinity at the time. In the near future, it is likely that the use of asbestos will be discontinued in most products used in the laboratory, including hoods.

**b. Stainless steel** — Stainless steel may be attacked by some chemicals, but type 316 stainless or its equivalent is commonly used for the lining of perchloric acid hoods and in radioisotope hoods which need to be easily decontaminated. Because of its vulnerability to some chemicals and its relatively high cost, it is not recommended for general-purpose fume hood use. Among the problem chemicals for stainless steel are acids and compounds containing halides.

**c. Fiberglass-reinforced polyester** — This material is a popular material for the lining of general-purpose hoods and is highly resistant to a large number of materials. However, as with stainless steel, there are some materials which may cause some problems with heavy usage. Care should be used in selecting this material to avoid applications involving chemicals for which it is not suited. Among those chemicals for which it is suitable for limited service are acetone, ammonium hydroxide, benzene, and hydrofluoric acid.

**d. Smooth-surfaced, glass-reinforced cement** — At least one major firm has substituted this material for transite as a general-purpose hood lining material. It is a relatively new material, but if it proves to be sufficiently durable, it will probably become a popular choice.

**e. Epoxy resin** — This material is comparable to fiberglass in versatility and is affected by some chemicals. If a wide range of chemicals are to be used, advice should be sought from the vendor to ascertain if there are any significant problems.

Among the chemicals for which it may be unsatisfactory are benzene, fatty acids, concentrated hydrochloric acid, and nitric acid.

**f. Polyvinyl chloride (PVC)** — PVC offers good protection for a wide range of chemicals, but is affected by some. Among these are liquid ammonia, amyl acetate, aniline, benzene, benzaldehyde, bromine, carbon disulfide, carbon tetrachloride, chloroform, ether, fluorine, nitric acid, and fuming sulfuric acid.

**g. Plastic-coated steel** — Plastic (usually vinyl)-clad steel linings are used for limited or light duty applications for materials which will not adversely affect the vinyl cladding.

**h. Cold-rolled steel** — This material is intended for light duty liner applications only. However, carefully prepared and treated to resist corrosion, or vinyl clad as in "g" above, it is the most popular material for the external components of fume hoods.

*Fume Hood Bases*

Hoods which need to be decontaminated frequently, such as radioisotope hoods, often come with a stainless steel base surface which forms an integral part of the hood interior. Other models are also available with integrated bases of other materials. However, many hoods are sold without bases and they can be selected separately. Base units and work tops are discussed in Section 3.2.1.1 and the same selection criteria discussed there apply here. The most popular material used for fume hood bases is a molded, modified epoxy resin. Since the material is molded, it is easy to incorporate a shallow depression in the work surface to function as a watertight pan to catch and retain spills.

*Sashes*

The most popular sash configuration is the vertically sliding type. The sash must be counterweighted, especially if the sash window is made of heavy glass. In order to avoid a "guillotine" effect should the counterweight cable break, a safety device must be incorporated in the sash. As for transparent materials used for the sash window, there are only a few kinds used for sashes in good quality, currently available commercial fume hood models.

**a. Laminated safety glass** — This is probably the best material for a sash because of the resistive properties of glass to most chemicals and the safety features provided by laminating tempered glass. However, it is not as effective for higher-temperature applications (good only up to about 70°C), as is tempered glass. In the event of a moderate explosion, it is possible that the sash will remain within the frame. If the sash remains essentially intact, the employees in the laboratory will continue to be afforded some protection against fires within the hood and the escape of hazardous fumes.

**b. Tempered glass** — Although tempered glass breaks into fragments that are not sharp in the event of an explosion, the result will be a loss of glass from the sash and the loss of protection afforded by the enclosure. Tempered glass will withstand more heat (up to about 200°C) than laminated glass.

**c. Clear, high-impact PVC** — PVC will provide impact protection comparable to laminated safety glass, but will not be as resistant to the effects of chemicals in the hood.

**d. Plexiglass** — Plexiglass is not as durable as the other materials and is primarily provided in economy models.

**e. Polycarbonate** — Polycarbonate sashes are recommended when a heavy use of hydrofluoric acid is involved.

*Internal Fixtures*

The lights in the hood should be shielded from the hood body in, as a minimum, a vapor-proof enclosure. These can also be selected as explosion-proof units should the usage include substantial amounts of very flammable materials. In fact, if a large portion of the work in the hood is expected to involve very volatile flammable materials, the hood should be designed as an explosion-proof unit. In any case, all electrical outlets and switches should be located outside of the hood on the vertical facia panels. Similarly, utilities should be by remotely controlled valves in the side panels operated by handles outside the hood. The connections to the utilities inside the hood should be chemically resistant and should be clearly identifiable as to the utility provided.

### 3.2.2.2. Types of Chemical Fume Hoods

There are several different types of fume hoods. Among these are (1) conventional hood, vertical sash, (2) conventional hood, horizontal sash, (3) bypass hood, (4) auxiliary air hood, (5) walk-in hood, and (6) self-contained hood. The differences in types 1, 4, and 6 are especially important in terms of the amount of tempered air lost during operations, while types 1, 2, and 3 differ primarily in the air flow patterns through the sash openings. Figures 3.12 to 3.17 illustrate each of these types and the air currents through them during typical operations. In addition, there are specialty fume hoods for perchloric acid and radioisotopes which will be treated separately. All of the hoods to be discussed in this section will be updraft units, where the exhaust portal is at the top of the hood.

One class of hood which will not be discussed here is the canopy hood. These have their uses, where it is desired to capture and exhaust hot fumes carried upward by convection currents until they come close enough to the canopy so that the fumes become entrained within the hood. The speed of the air movement in the vicinity of the hood face, due to the air flowing through the canopy, falls off very rapidly, dropping to about 7.5% at a distance equal to the effective size of the canopy opening. If the canopy is a reasonable distance away from the bench top, the airflow at the work surface due to the hood will be on the order of the average air movement speed within the room or less. A further disadvantage is that the fumes, if drawn upward, would pass through a worker's breathing zone. For these reasons, canopy hoods are not recommended as general-usage laboratory fume hoods.

*3.2.2.2.1. Conventional Fume Hood*

A conventional fume hood is illustrated with the sash open and closed in Figure 3.12. Note that a section of the internal baffle essentially remains in contact with the sash at all opening positions. It is equipped with a vertically opening sash and an interior baffle so arranged that some of the air sweeps the base of the hood and is directed up behind the baffle to the exhaust opening. The remaining air passes through the hood interior and is directed into the exhaust portal over the top of the interior baffle. The volume of air through the hood is relatively constant, although some losses occur as the sash opening becomes smaller. A more important conse-

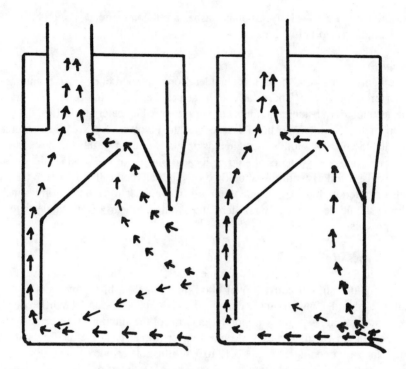

FIGURE 3.12.   Conventional hood, sash open and closed.

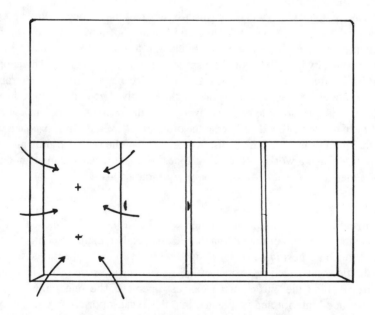

FIGURE 3.13.   Conventional hood, horizontal sash.

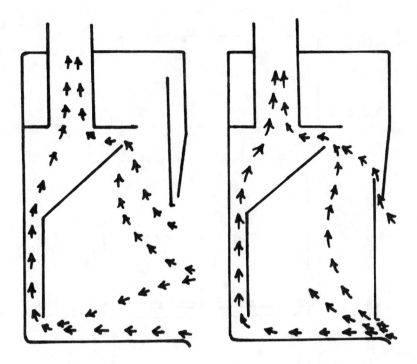

FIGURE 3.14.    Bypass hood, sash open and closed.

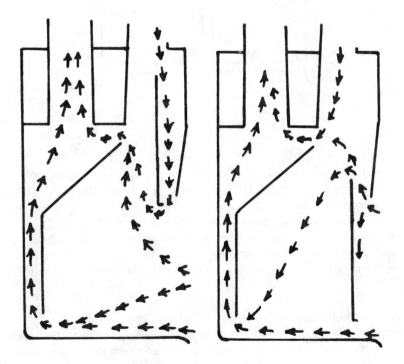

FIGURE 3.15.    Auxiliary or add-air hood.

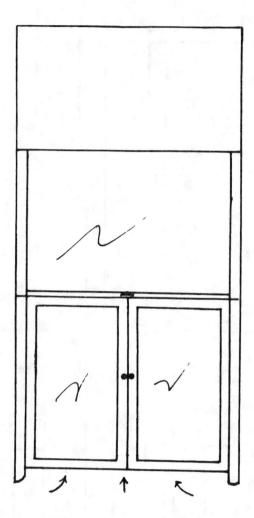

FIGURE 3.16.   Typical walk-in hood.

quence of the decreasing sash opening is the increasingly high velocity of the air
through the narrowing opening. The increased air speed could disturb the operation
of the experiment within the hood, and the effects of turbulence around apparatus
sitting on the bottom of the hood would be increased. A person representing an
external obstruction could also give rise to increased turbulence so that spillage from
the hood would occur, although with the bottom of the sash well below face height,
the possibility of toxic fumes entering directly into the breathing zone would not be
high.

### 3.2.2.2.2. Conventional Hood, Horizontal Sash

Some hoods are available with two or three horizontally sliding sashes instead of
a vertical sash. The maximum hood opening is greater with three sashes than with

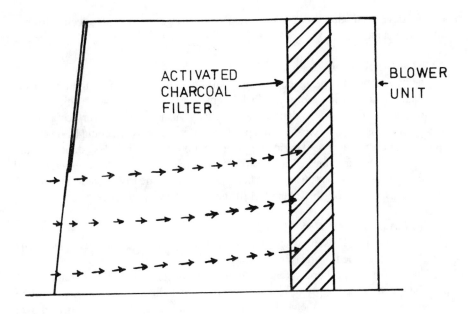

FIGURE 3.17.   Self-contained hood.

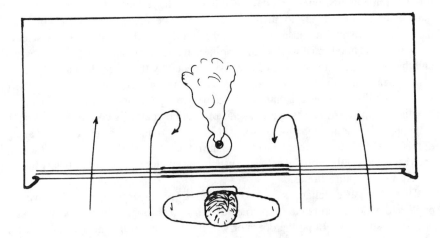

FIGURE 3.18.   Protection offered by overlapping horizontal sashes.

two. It is also possible, with three sashes, to have more flexibility in terms of where the sash opening would be located. Since it would be possible to have the center section of the hood opening blocked and the fume-generating apparatus placed behind it, fumes entrained in the eddy currents would have more difficulty reaching a person working in the position shown in Figure 3.13 than with a similar vertical sash hood. The section of sash between the operator and the experiment would also afford some protection from a moderate explosion or chemicals ejected from a runaway reaction.

### 3.2.2.2.3. Bypass Hood

The bypass hood is designed so that a portion of the air entering the face of the hood may pass over the top of the sash opening as well as below it. This has two consequences. The first is that the air velocity near the work surface remains reasonably constant, so excessive air speeds will not occur which could be detrimental to delicate apparatus or experiments. The second is that there is less static pressure and, hence, less frictional resistance to the flow of air than with the conventional hood, so the volume of air through the hood remains more nearly the same at different sash heights, permitting better control of the laboratory air balance.

### 3.2.2.2.4. Auxiliary Air Hoods

This type of hood is somewhat controversial. The original comments of Horowitz et al.[1] are still echoed in today's literature.

"The auxiliary air hood attempts to reduce air-conditioning requirements by providing a separate supply of air that has not been cooled and dehumidified in the summer or fully heated in the winter. The supply of air for such a hood may be drawn from outside or from the service chases within the building, which are, in turn supplied by air from attic or mechanical equipment rooms. Such hoods can substantially reduce the air-conditioning equipment capacity required to make up losses through fume hoods; operating costs can likewise be reduced. However, there are a number of disadvantages to such a hood. One type of auxiliary air-supply hood discharges untreated air just in front of the hood, usually at the head. A scientist working at the hood must work in unconditioned air. The disadvantages are obvious, and the annoyance of scientists has been evidenced by their very human attempts to invent means of foiling the intended mode of operation. One such effort consists of securing cardboard over the outlets with adhesive tape, thus closing or reducing the auxiliary supply.* Attempts to rectify this problem by partially cooling or heating the air supply, depending upon the season, substantially reduce any economic advantage of this type of hood. Another type of hood introduces the auxiliary air within the hood enclosure and is inherently unsafe because the face velocity is reduced below the rate necessary to capture fumes."

There have been changes in hood design since the article quoted above was published which permits the supplementary air to be brought down outside the sash, but not in such a way that the operator will be standing in the airflow, as shown in Figure 3.14. Specifications on this type of hood usually permit up to 70% of the air to be provided by the auxiliary air supply, with at least 85% of the supplied air passing through the hood face. Untempered air, especially during the winter when the outside temperature may be very cold, would mandate that this supplied air be heated to at least 10°C, which eliminates much of the savings for this type of hood. The installation and configuration of such hoods is critical and they are difficult to maintain

---

\* It should be noted that here is a good example where the users would continue to use a hood which they themselves had caused to function improperly. There is a serious attitude problem among many, although by no means a majority, of scientific workers which puts expediency and comfort ahead of safety.

in proper operating condition. Most authorities today still discourage selection of an auxiliary air hood.

### 3.2.2.2.5. Walk-In Hood

Walk-in hoods are those which usually rest directly upon the floor, or on a pad resting directly upon the floor. They are designed to accommodate tall apparatus which will not fit in a standard hood sitting upon a base unit or work bench. Because their height is usually somewhat out of proportion to their width, the air flow characteristics may not be as favorable for avoiding spillage as with a standard hood, as noted earlier in Section 3.2.2.1. Because their height would require an abnormally long sash travel, these hoods are often provided with dual sashes, each of which would cover half the opening, or with a single sash which would come down only about half way, with swinging doors being used to provide access to the lower portion. These doors may terminate a few inches above the floor to ensure that there is always some airflow through the entire length of the vertical space. As with other types of hoods, they may be obtained in different configurations, such as bypass or auxiliary air units.

### 3.2.2.2.6. Self-Contained Hoods

These are listed last because they are the least desirable alternative as chemical fume hoods. Self-contained chemical units should not be confused with biological safety cabinets, which also are often self contained. They are not recommended for general usage. However, there are circumstances where they can be used relatively safely. Good quality units can cost as much as a conventional hood, so their use is not mandated upon the basis of purchase price. Their use is usually contemplated when a hood is needed, but no exhaust duct is accessible and it is impractical to install one. The commercial units available are generally intended for use in histology and cytology procedures involving materials such as xylene, formalin, toluene, alcohol, etc.

A typical self-contained unit pulls room air through the face of the unit, over the work surface, and through an activated charcoal filter. The fan unit does not become contaminated because the air is filtered before it reaches it. A typical filter will absorb several pounds of solvent before it becomes saturated and must be replaced. A major problem with most of the units which are available is the inability to tell when the filter has become saturated. At least one vendor has solved this problem by placing a material at the back of the filter which reacts with the solvent when it is no longer being absorbed by the filter and which then emits a pungent odor. The unit provided by this same vendor probably comes about as close to the performance of a regular fume hood as any self-contained unit on the market, but it is also comparably expensive. The face velocity is set at 75 fpm, which is comparable to that of a standard hood. However, even this relatively well-performing unit is no real substitute for an actual fume hood and should be used only for light-duty applications.

### 3.2.2.2.7. Other Modes of Exhaust

The trend is to do more laboratory work in hoods, but the use of hoods is a substantial burden on the energy budget of a building. An alternative is to use spot ventilation. Many laboratory supply houses provide small exhausters where an adjustable inlet can be placed very close to well-defined spot sources of noxious

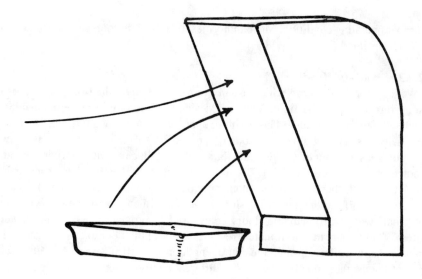

FIGURE 3.19.    Local fixed exaust.

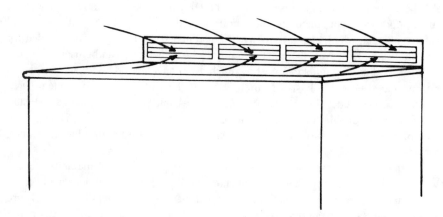

FIGURE 3.20.    Rear plenum exhaust system.

fumes. The fumes are then typically discharged into a fume hood. If the work is repetitive enough to warrant setting up a permanent spot exhaust, then the designs shown in Figures 3.19 and 3.20 might be usefully employed for limited-risk work. In the first design, perhaps a repetitive operation involving pipetting into test tubes or sample vials could be done virtually at the face of the cowl-shaped spot exhauster. With this physical relationship, the air intake should be very effective in exhausting any aerosols or fumes which might be generated. Note that the air flow would be, as with a hood, away from the research worker.

The second design would be more effective for a tray type operation. Here, one or more narrow slots, with air being drawn through them at a relatively high inlet velocity, are placed at the rear or, occasionally, to the side of the work bench close to the level of the bench top. This design takes advantage of the fact that most solvents are heavier than air. Air flowing across the surface would entrain the vapors from the tray and exhaust them through the slots at the rear of the work bench. The fumes would not have an opportunity to rise into the worker's breathing area, but would be pulled back and away from the worker. A modified version of this system is used for silk screening. In this version, the entire circumference has either an aerodynamic slot around the edge or the last several inches close to the edge of the table top are perforated with hundreds of holes through which, in both cases, air is pulled down prior to being exhausted. There are consumer range tops for cooking built on this principle which work very well. Smoke from the food being prepared rarely rises more than an inch or two above the cooking surface before being captured and exhausted.

### 3.2.2.2.8. Perchloric Acid Hood

Individuals working with perchloric acid and perchlorates must be trained in procedures which will let them conduct their research with maximum safety. These are extremely dangerous materials.

There is an extensive section later on the problems of working with perchlorates and perchloric acid. This section will be restricted to a discussion of the critical factors which are applicable to the proper performance of perchloric acid hoods. It will be sufficient to say at this point that perchloric hoods are designed to avoid accumulation of precipitates from perchloric acid or to avoid perchloric acid from coming into contact with materials with which it may react vigorously and explosively. Hoods used for hot perchloric acid use should not be used for research with other types of materials. The hood should be prominently labeled with a sign stipulating that it is for perchloric acid work only. Exhaust systems should not be manifolded into a common exhaust plenum.

For conventional hoods, and their variants which have been covered in Sections 3.2.2.2.1 to 3.2.2.2.6, the discussion of appropriate ducts and exhaust fans has been deferred to separate sections. However, perchloric fume hood systems are uniquely dangerous and will be treated as an integral concept.

Perchloric hoods are usually constructed with an integral liner of a single piece of stainless steel, such as 316 stainless, which will resist the effects of the acid, although PVC can also be used as a liner. The liner should have cover corners and as few seams as possible to allow ease of decontamination. In order to avoid the build up of perchloric precipitates in the hood and duct system, a hood intended to be used for perchloric acid work must be equipped with a rinse system which will make it possible to thoroughly flush the interior of the hood and ductwork with water. This may be done with a manual control system or an automatic system which at the end of a work session, will come on and rinse the system for 20 to 30 min. A combination of an automatic system which can be bypassed for additional rinses is preferable if the researcher feels they are needed.

The ductwork should also be stainless or PVC. Caution must be exercised to ensure that, during installation, workers do not use standard organic caulks to seal the joints. Organic materials, when contaminated with perchloric acid, are highly flammable and dangerous, and such joints will also tend to leak, allowing perchloric acid to escape outside the ductwork. Under circumstances which would allow this errant material to be exposed to heat or to receive a sudden shock, the result could be a fire or an explosion in the space outside the duct. The most desirable procedure for stainless steel ducts is to weld the sections of ducts together. Since this will require heliarc welding, it is a relatively expensive procedure compared to welding ordinary steel ductwork. Some fluorinated hydrocarbon materials can be used as a sealant if welding is not feasible.

It is recommended that the interior fittings of a perchloric acid hood should be nonsparking and that the lights should be explosion proof. This concept should be extended to any apparatus placed in the hood. With the dangers already represented by perchloric acid, there should be no contributory factors which could initiate an explosion. PVC ductwork can be employed instead of stainless steel, but it would be much less likely to remain intact in the event of a significant fire exposure. However, if the ductwork is enclosed within a 2-h fire-rated chase, as it usually should be, this would not be a serious drawback.

The ductwork for a perchloric acid hood should have as few bends as possible and be taken to the roof in the shortest, most direct path. No horizontal runs should be permitted, and even slopes of less than 60 to 70° should be avoided wherever possible. For aesthetic reasons, architects prefer to place exhaust ducts away from the edge of a building where they cannot be seen easily. As a result, in some older designs, horizontal runs of 100 ft or more of perchloric fume hood exhaust ducts have been observed. Even if a washdown mechanism were incorporated in the design, it would be unlikely to come into contact with and clean the upper portion of the duct in the horizontal section. In one instance where a perchloric hood was installed below grade in the basement of a building, the exhaust duct first was run horizontally for approximately 75 ft under the floor of an adjacent section of the building. An exhaust fan was installed in this horizontal run and then the duct was run vertically for three floors. When this was discovered, the horizontal section of the ductwork beyond the fan had corroded through and perchloric acid crystals were observed on the external surface of the duct and on the ground below it. This became a major and costly removal project.

As a minimum, the blades and any other portion of the exhaust fan which might come into contact with the perchloric fumes should be coated with PVC, teflon, or another approved material which will resist the effects of the perchloric acid. An induction exhaust fan, where none of the fumes actually pass through any part of the motor or fan, is recommended. Under no circumstances should the exhaust fumes be directed down upon the roof to be absorbed in the roofing material. The contaminated roofing material could itself constitute a danger. The exhaust point should be well above the roof to avoid the fumes readily reaching any portion of the roof prior to dilution by the outside air. The washdown mechanism should be capable of cleaning the entire duct, from the point of exhaust all the way back to the hood. The washdown system plumbing should automatically drain when shut off to avoid rupturing the supply lines due to freezing in the winter. The rinse water may be permitted to drain directly into the sanitary system, where it will be quickly diluted.

Labconco makes an excellent perchloric acid ejector duct which has all of these desirable features and exhausts the perchloric fumes at a point about 10 ft above the roof level. It is not inexpensive, but performs exceptionally well.

Maintenance personnel as well as laboratory employees should be trained in the dangers inherent in the material and the potential for injury represented by any residual material in crevices or other places where perchloric acid or byproducts might accumulate. This training should be done in a positive way to instruct individuals how to work with such material properly, and not in such a way as to unduly frighten anyone.

### 3.2.2.2.9. Radioisotope Fume Hood

Radioisotopes are frequently used in the life sciences and in nuclear medicine in diagnostic applications. Only rarely are the amounts employed large enough to be of immediate danger to the laboratory worker, if used properly. In addition, many of the more commonly used radioisotopes emit low-energy beta radiation only, which will not penetrate the skin. However, not all radioactive materials have this favorable property and many of even the relatively safe materials may cause delayed injury, perhaps 20 years later, if ingested and inhaled. The word "may" is not to be construed here in any definite sense of "they will cause injury". It is intended to indicate only that the possibility of a health effect may be increased. At very low levels, there is no direct evidence of either immediate or delayed injury, although an enormous number of studies have tried to resolve the issue. Most studies seeking to prove the point either way have typically been controversial and not accepted universally. The possibility of health effects is based on a conservative linear extrapolation from known detrimental effects at high levels to possible detrimental effects at the much lower levels usually encountered in research. However, because of the possibility of a finite risk and because of public concern, it is public policy that laboratory use of radioactive materials be stringently controlled to minimize exposures. A carefully monitored license is required. There will be a separate major section in Chapter 5 on radiation safety.

A radioisotope fume hood is designed to minimize risks of exposure to the laboratory worker by making it easier to maintain the hood in an uncontaminated condition. The liner is usually made of a single piece of stainless steel, as with perchloric acid units, and for the same reason, for ease of decontamination. There should be a minimal number of seams or hard-to-clean areas. The major classes of research employing radioisotopes are often intended to retain the compounds containing radioisotopes in the endproduct since a frequent purpose in employing radioactive materials is as a tracer. Therefore, although there will be some radioactive fumes generated, typically, there will be less than with some other dangerous materials. Even so, it is usually recommended that the duct work for a radioactive fume hood be of stainless steel since this is easily decontaminated.

Where relatively high levels of radioactive materials are used or where the levels of fumes generated containing activity may be substantial, it will probably be necessary to install a two-stage, high-efficiency particulate air (HEPA) filter unit, which will filter out 99.97% of all particles 0.3 μm in size or larger, in line in the exhaust duct to insure that the legal minimum concentrations of activity can be maintained at the point where the fumes are discharged to the outside. The most appropriate location for this filter is at the exit portal of the hood since this will prevent any of the ductwork from being contaminated and the location will make it convenient to service the filter. Where a HEPA filter is used in a fume hood involving chemicals, it is essential that a device to monitor the air velocity through the face of the hood be installed. The air passages in a HEPA filter used in a chemical system where particulates are generated will soon become clogged, increasing the difficulty of drawing air through the filter, so the efficacy of the fan unit will fall off rapidly. A less expensive prefilter will significantly extend the life of the HEPA filter. A device to measure the pressure drop across the filter unit may be used to monitor the condition of the filter, but this will not necessarily measure the velocity of the airflow through the hood and duct. Gaseous radioactive materials will not be stopped by a HEPA filter. An activated charcoal or alumina filter should be used for these materials.

Although the stainless steel liner is relatively easy to clean, there are a number of measures which will make it easier to keep it in an uncontaminated condition. Among these are the use of trays to contain any spills which might occur and the use of plastic-backed absorbent paper, which may be discarded as waste, on the work surfaces.

The property of emitting radiation which makes radioisotopes useful in research and potentially dangerous is also the property which makes their use relatively easy to control by research personnel who conscientiously follow good laboratory practice. The work surface, the hands and clothing of the persons performing the work, and the tools and equipment employed in the work can easily be checked for contamination by use of appropriate instrumentation. It is unfortunate that the dispersion of many dangerous chemical and biological agents in the laboratory cannot be monitored and controlled so readily.

The only other unique feature in a radioactive fume hood may be the need of the base to support shielding materials. The base may need to be stronger than usual to support the concentrated weight of the lead shielding which may be needed to protect the workers from radiation, such as when synthesizing a compound where relatively large amounts are employed and substantial amounts of shielding may be required. Most of the time, only small quantities of radioactive materials are in use at a single time, so personnel shielding needs would be small. However, substantial amounts of lead may be needed to provide adequate shielding against background radiation for the sensitive detectors used to detect minute traces of the experimental radioisotopes in the material being studied.

Fume hoods used for radioactive materials should be marked "RADIOISOTOPE HOOD" and, in addition, should be labeled with a "CAUTION-RADIOACTIVE MATERIALS" sign bearing the standard radiation symbol. The isotopes being used should be identified on the label. Under some circumstances, additional signs may be needed.

### 3.2.2.2.10. Carcinogen Fume Hood

Clearly, work with a carcinogen mandates a high-quality fume hood. The features discussed in the previous two sections which would minimize spaces for materials to be trapped and facilitate decontamination are strongly recommended. In the OSHA General Industry Standards, Subpart Z — Occupational Health and Environmental Control, many of the specific standards for the carcinogens regulated in this section contain the following standard paragraph relating to fume hoods:

> "Laboratory type hood" is a device enclosed on three sides and the top and bottom, designed and maintained so as to draw air inward at an average linear face velocity of 150 feet per minute with a minimum of 125 feet per minute; designed, con-

structed and maintained in such a way that an operation involving (name of regulated carcinogen) within the hood does not require the insertion of any portion of any employees' body other than his hands and arms.

### 3.2.2.3. Exhaust Ducts

Exhaust ducts are necessary to take the fumes from the hood to the point at which the fumes are to be exhausted. For the purpose of this section, it will be assumed that the duct will exhaust directly to the outdoors, rather than to a plenum (see Section 3.1.6).

It is recommended that, if it is desired to manifold more than one hood into a common duct, this practice be limited to hoods within the same room. Otherwise, it is less likely that individuals using different hoods would be aware of each other's activities, and one might make changes which would affect the performance of hoods in the other room. For example, a pressure differential might be established between one laboratory and another, so that fumes could be exchanged between the two areas. In addition, maintenance of the air balance may be made more difficult and there may be problems in complying with fire codes if fire walls are penetrated.

### Materials

Many of the comments in Section 3.2.2.1 regarding materials are relevant here as well. At one time, transite was a very popular duct material, but it is no longer recommended due to the concerns regarding asbestos. The joints between sections required cutting by the maintenance personnel preparing or installing the ductwork, with a consequent release of airborne asbestos fibers which could be inhaled. Stainless steel is used for special types of applications, such as for perchloric acid systems, but is not universally suitable for all chemicals. PVC can be used for many applications, is easy to install and custom fit, and is comparatively inexpensive. Steel ductwork coated with a chemically resistant material, such as an epoxy coating, is popular because it is relatively inexpensive and is especially adaptable for custom installations.

### Dimensions*

There are numerous sources of frictional losses for the airflow through the hood and ductwork which must be overcome by the exhaust fan. The air entering the hood must be accelerated from minimal velocity in the room to the velocity within the duct. Due to physical factors, there is always some turbulence created during this process, so the pressure difference created by the fan must be sufficient to provide the desired airspeed in the duct and overcome the losses due to turbulence. In the specifications for a hood, the hood static pressure data provided is a direct measure of the total energy needed for acceleration and overcoming turbulence losses. The static pressure will be proportional to the square of the velocity of the air.

There are significant losses along even a straight, relatively smooth section of duct due to air friction. This is due to the energy required to maintain a velocity gradient ranging from the essentially stagnant air in contact with the walls to the rapidly moving air near the center of the duct. For example, for a nominal 10-in. internal

---

\* For more thorough coverage of the material covered briefly in this section, the reader is referred to ANSI Z9.2-1979[9] or a superceding version.

diameter PVC duct through which 1500 cfm of air is passing, each 10-ft section contributes about 0.1 in., water gauge, pressure loss. This also varies rapidly with the volume of air movement; by increasing the speed by one third, the pressure loss is increased by about two thirds. The diameter of the duct is also important; for the same volume of air, 1500 cfm, the losses in a nominal 8-in. ID duct will be more than three times greater than in the 10-in. duct, while the losses in a 12-in. duct will only be about 40% of the amount in a 10-in. duct. In general, the total amount of duct friction in a round duct varies in direct proportion to the length, inversely to the diameter, and is proportional to the square of the velocity of the air moving through it. The equations below give an approximate value for the skin resistance of a duct.

$$wg = \frac{2LV^2}{CDT} \quad \text{for circular ducts}$$

where  wg = pressure loss in in water gauge
  L  = length of duct in feet
  V  = velocity of air in feet per second
  D  = diameter of the duct in inches
  T  = absolute temperature on Fahrenheit scale (460 + °F)
  C  = constant = 55 for new steel ducts and 45 for older steel ducts
Similarly, for a rectangular duct, sides A and B:

$$wg = \frac{LV^2(A + B)}{CABT} \quad \text{for rectangular ducts}$$

A shock loss occurs whenever there is a sudden change in the air velocity caused by a change in the direction of the air or a change in the diameter of the duct. Every bend in the duct dramatically decreases the efficacy of the fan motor; generally, the sharper the bend, the more severe the loss becomes. This loss may be estimated from the following equation:

$$wg = \frac{0.118V^2k}{T}$$

where k varies as follows:

| Mean radius of bead/duct diameter or width (width = side measured along radius of bend) | Circular ducts k | Rectangular ducts k |
|---|---|---|
| Right angle elbow | — | 1.25 |
| 0.50 | 0.75 | 0.95 |
| 0.75 | 0.38 | 0.33 |
| 1.00 | 0.25 | 0.17 |
| 1.50 | 0.17 | 0.09 |
| 2.00 | 0.15 | 0.08 |
| 3.00 | 0.13 | 0.07 |
| 6.00 | 0.10 | 0.05 |

The last equation can also be used to estimate the loss in changing duct sizes by substitution of $k_c$ for k in the equation. The ratio given below is the ratio of the smaller flow area to the larger.

| Ratio | 0.1 | 0.2 | 0.3 | 0.4 | 0.5 | 0.6 | 0.7 | 0.8 | 0.9 | 1.0 |
|-------|-----|-----|-----|-----|-----|-----|-----|-----|-----|-----|
| $k_c$ | 1.25 | 1.20 | 1.15 | 1.05 | 0.95 | 0.80 | 0.65 | 0.45 | 0.25 | 0 |

Similarly, if a secondary branch joins another duct, the angle at which it joins critically affects the fractional velocity pressure loss of the air through the secondary branch, ranging from a very small percentage at shallow angles to over 40% at 60°. Branches should not enter at right angles or opposite each other.

It has already been pointed out that a deflecting weather cap is inappropriate for a fume hood since it is highly undesirable for the exhaust fumes to be deflected back toward the roof. It would also cause significant pressure losses. The design shown in Figure 3.21 below would cause minimal pressure losses and yet would protect almost as well against rain, unless the rain were falling virtually straight down. Unless the fan were off, the normal speed of the discharged air would also keep any rain from entering the duct.

Most persons buying hoods for use in their facilities are not ventilation engineers. Even a cursory look at ANSI Standard Z9.2[9] should convince most individuals to assign the design of fume hood air systems to professional ventilation engineers. If the system is inadequate, the hood will represent a danger to persons using it and, if oversized, will be wasteful of energy and funds.

### Fan Selection

Centrifugal fans are the most commonly used type of fume hood exhaust fans. Within this category are several different variants, with choice depending upon the requirements of the individual installation (Figure 3.22).

Forward-curved or squirrel-cage fans are primarily suited for relatively low pressure applications, from 0 to 2 or 3 in. water gauge. They typically have a relatively large number of small blades set close together around the periphery of the wheel, each blade being curved forward in the direction of wheel rotation. Because the spaces between the blades in this type of fan are small, they are prone to collect dust and become clogged.

Paddle-wheel or radial-blade fans are used where high pressures of 15 in. or more water gauge are required. The radial blades, typically six, are relatively heavy and resist corrosion and abrasion well. Since they have large spaces between the small number of blades, they are the least likely to become clogged.

Fans with backward-curved blades are best for medium pressures, from approximately 1 to 8 in. water gauge. This type of fan resembles the forward-curved type in that the blades are placed around the periphery of the wheel, but they usually have fewer blades, typically less than 16, but more than radial-blade fans. This type of fan operates more efficiently than the other two types.

The inlets to the fans should be designed to take maximum advantage of the fan's performance capabilities. Either the duct should feed directly into the fan intake or be brought into it with a smooth bend to the duct. Connections which require the air to make a right-angle turn will have a significant adverse effect on the performance of the fan.

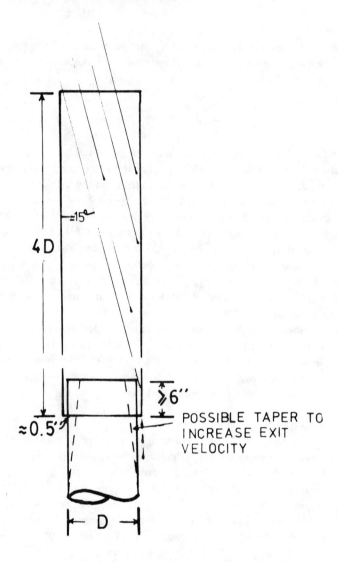

FIGURE 3.21.    Exhaust duct rain shield.

For general purpose fume hoods, as a minimum, the fan blades should be non-sparking. Usually this is achieved by using coated aluminum or stainless steel fan impellers. If the fume hood is certain to be used heavily for highly flammable solvents, the motor should be explosion proof as well. If the exposure to corrosion is expected to be severe, materials with good corrosion resistance such as PVC or fiberglass-reinforced polyester materials should be selected for the fans. If, for

| FOWARD | BACKWARD | AXIAL |

## CENTRIFUGAL FAN TYPES

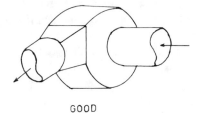

GOOD                                  POOR

## FAN CONNECTIONS

FIGURE 3.22.   (Top) Types of centrifugal fan blade designs; (Bottom) exhaust duct fan connections.

reasons of economy, ordinary steel fans are used, they should be coated with teflon or, if the corrosion problem is somewhat less severe, PVC or polypropylene.

For flexibility, the fan should be driven by a belt from the motor since, within limits, this will allow the speed of the fan to be changed to make up losses in efficiencies in the exhaust system over a period of time or to make planned design changes. If a deliberate change is to be made in the amount of air discharged through a given duct, the ramifications of the change on the room air balance and the performance of other hoods in the facility should be considered in advance.

## REFERENCES (SECTIONS 3.2.2 TO 3.2.2.3)

1. **Horowitz, H., Heider, S. A., and Dugan, C. N.,** Hoods for science laboratories in *Handbook of Laboratory Safety,* 2nd ed., Steere, N. V., Ed., CRC Press, Boca Raton, FL, 1971, 154.
2. *Applications Handbook,* American Society of Heating, Refrigerating, and Air Conditioning Engineers, Atlanta, GA, 1978, chap. 15.
3. Committee on Industrial Ventilation, *Industrial Ventilation,* Section 4, 15th ed., American Conference of Governmental Industrial Hygienists, Cincinnati, OH, 1978.
4. **Fuller, F. H. and Etchells, A. W.,** Safe operation with the 0.3 m/s (60 fpm) laboratory hood, *Am. Soc. Heat. Refrig. Air Cond. Eng. J.,* p. 49, October 1979.
5. **Caplan, K. J. and Knutson, G. W.,** The effect of room air challenge on the efficiency of laboratory fume hoods, in *ASHRAE Trans.,* 83, Part 1, 1977.
6. **Steere, N. V.,** Ventilation of laboratory operations, in *Handbook of Laboratory Safety,* 2nd ed., Steere, N. V., Ed., CRC Press, Boca Raton, FL, 1971, 141.
7. *Useful Tables for Engineers and Steam Users,* 8th ed., Babcock & Wilcox, 1963.
8. Laboratory Chemical Fume Hood Standards, Manual 232.1, U.S. Department of Agriculture, Administrative Services Division, Washington, D.C., December, 1981.

9.  *Fundamentals Governing the Design and Operation of Local Exhaust Systems,* ANSI Z9.2, American National Standards Institute, New York, 1979.
10.  Corrosion Resistance Chart, Sheldons Manufacturing Corporation, Elgin, IL.

### 3.2.2.4. Biological Safety Cabinets

The purpose of a biological safety cabinet is to protect the laboratory worker from particulates and aerosols generated by microbiological manipulations. HEPA filters are used to remove these sources of danger from the air passing through them. Since HEPA filters are ineffective against gaseous chemicals, biological safety cabinets are generally not intended to be used for protection against this type of chemical hazard. One type, in which the air is totally exhausted after a single pass through the work area, may be used to a limited degree for chemical applications.

There are basically three classes of biological safety cabinets, Classes 1, 2, and 3. Class 2 units are divided into two main subclasses, 2A and 2B. There are several versions of the Class 2B units.

*3.2.2.4.1. Class 1 Cabinet*

The Class 1 cabinet, illustrated schematically in Figure 3.23, is essentially a variation on the chemical fume hood. As with a chemical fume hood, worker protection is provided by air flowing inward through the work opening to prevent escape of biologically active airborne agents. Unlike a chemical hood, the view screen, equivalent to the sash in a chemical fume hood, is usually designed to provide a fixed opening, typically 8 to 10 in. (20 to 25 cm) high . The view screen is generally made so that it can swing upward to permit putting equipment into the unit, although some units have an air lock-type door on the end of the unit to serve the same purpose.

The airflow through the face may range between 75 and 100 fpm. An aerodynamic shape at the entrance or suction slots near the edge to ensure that air flows inward aids in the performance of the cabinet, while a baffle at the rear ensures that some of the inlet air will flow directly across the work surface, while the remainder is drawn upward through the cabinet to the exhaust portal. It is primarily the directional flow of air which ensures that the agents of concern stay within the cabinet.

As with the chemical fume hood, the performance of the cabinet can be adversely affected by a number of factors. The velocity of air movement within the room may result in a degradation of the performance of the hood, as can too rapid movements by the worker, location of the work too close to the entrance, or perhaps the effect of a thermal source, e.g., bunsen burner, within the unit. The effectiveness of the unit can be enhanced by placing a panel over the opening in which arm holes are cut. The airflow through these holes will be much higher than the designed air speeds through the front, especially with the worker using them, but the relatively small gaps normally should keep the turbulence caused by the higher speeds through the portals from causing materials to escape from the cabinet and creating a problem for the user.

A Class 1 cabinet, used properly, can provide excellent protection for the research worker, but it does not provide any protection for the work since the air flowing into the cabinet is "dirty", i.e., ambient air from the room that has not been specifically cleaned. This was not discussed for chemical work since it is rarely a problem, but with biological research, elimination of contamination of research materials may be important.

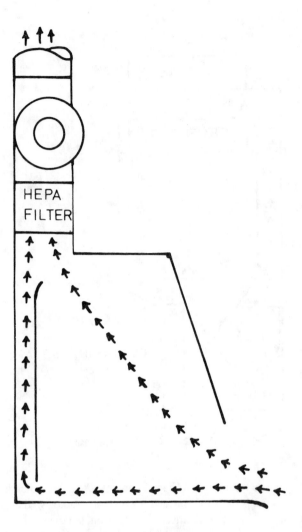

FIGURE 3.23.  Class 1 biosafety cabinet.

The air from the cabinet is usually exhausted through a HEPA filter placed above the exit portal and before the exhaust fan. If the discharged air does not contain any chemical agent or other material which could be expected to pass through a HEPA filter, it is not absolutely essential that the exhaust be to the outdoors. Absorbent filters, such as are used in the self-contained chemical hood, can be used to supplement the HEPA filter if there are possible chemical effluents. However, a HEPA filter can begin to leak and, hence, it is probably desirable to take the exhaust to the outside. It is not as critical that the entire ductwork be maintained at a negative pressure, as with a chemical hood, so the exhaust fan can be integrated into the cabinet if necessary. The exhaust can be integrated into a chemical air discharge system as well.

A unit sometimes confused by a novice purchaser with a Class 1 biological safety cabinet is a horizontal laminar-flow cabinet or work table. This type of unit serves

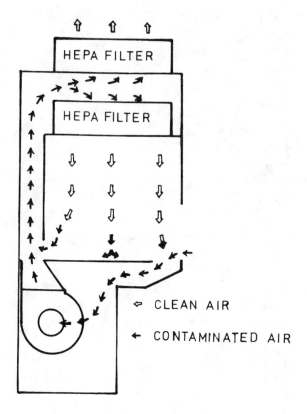

FIGURE 3.24.    Class 2 biosafety cabinet.

precisely the opposite function of a Class 1 cabinet. Clean air which has been HEPA filtered is blown across the work surface *toward* the worker so that the research or product materials are protected against contamination, but the worker is not protected at all. Such a unit is unsuited for microbiological work, except for applications which would cause no harm to the user.

### 3.2.2.4.2. Class 2 Cabinets

Class 2 cabinets provide protection for both the researcher and the research materials within the cabinet. Figure 3.24 shows an idealized version of a Class 2 cabinet. In this design, room air is drawn in the front opening, but, instead of passing over the work surface, is pulled down into a plenum by a fan unit under the work surface. Then the output of the fan is passed up through a channel at the rear of the cabinet into a space between two HEPA filters. A portion of the air is exhausted at this point through one of the filters and the other portion passes through the second filter and is directed down as clean air to the work surface. This stream of air is considered to be a laminar flow stream, which will resist encroaching air and remain clean, so the work surface is in a clean environment. Part of this air enters the intake air grill at the front of the cabinet and the remainder goes through an exhaust grill at the rear of the unit. Both of these air streams continue going through the cycle. The result is that "dirty" room air or air that has passed over the work surface and, hence, is also "dirty", is restricted to the air on its way to the filters. The work surface is in a clean air space and pathogens are blocked from escaping into the room by the inward flow of air at the work opening and by being removed by the HEPA exhaust filter. As noted at the beginning of this section, there are several variants on this concept.

### Class 2A Biosafety Cabinets

Class 2A units are very similar to the basic Class 2 biosafety cabinet just described. In order to keep the work area free of room air and air that has been contaminated by the work, the plenum containing the blower and the channel through which air is delivered to the HEPA filters under a positive pressure must be carefully sealed to be leak tight. The Class 2A unit is designed so that about 30% of the air is exhausted from the cabinet and about 70% recirculated each cycle. Thus, the make up air through the front must provide an amount of air equal to that exhausted, or 30%. The air being supplied through the front opening must be carefully balanced with the amount of air being exhausted. If too small an amount of air is exhausted to the outside, the air in the working volume could be at a positive pressure and pathogens could be forced out into the operator's area. If too small an amount of makeup air is provided at the face, the pressure in the working volume could become negative, allowing the work area to be contaminated by room air. This type of unit is especially sensitive to any thing in the work area that could perturb the laminar flow of air. Examples would be equipment blocking portions of the duct, rapid arm motions and gas flames.

In consequence of both a smaller front opening, 8 to 10 in., and the fact that only 30% of the air is being discharged at any one time, the amount of tempered air needed for a laboratory is much less than for a chemical fume hood, typically in the range of 250 cfm for a 4-ft unit as compared to close to 1000 cfm for the same size chemical hood. It could even be more favorable if the biosafety cabinet exhaust were to be discharged into the room, in which case no additional tempered air would need to be supplied to the room.

### Class 2B Biosafety Cabinets

Class 2B biological safety cabinets differ in several different ways from Class 2A cabinets. Two major differences are the amount of air recirculated, a much smaller proportion for a type B unit than for a type A, and the air in the plenums surrounding the work area is filtered, or clean air rather than contaminated air.

Shown in Figure 3.25 is a generalized drawing of a representative Class 2B biosafety cabinet, designed originally by the National Cancer Institute. As with other Class 2 units, air entering the cabinet is immediately drawn into an air intake. Also entering this intake is filtered air from above, aiding in blocking the entrance of room air further into the working volume. In this unit, 100 fpm of room air are provided as supply air. The air drawn through the front grill is sucked through HEPA filters below the work surfaces by blowers situated below the filters. The resulting clean air is forced up plenums on each end of the unit and into a diffuser area above the work area. The clean air is circulated into the work zone through a diffuser panel at a reduced speed of 50 fpm. Most of this air, 70%, is exhausted as contaminated air

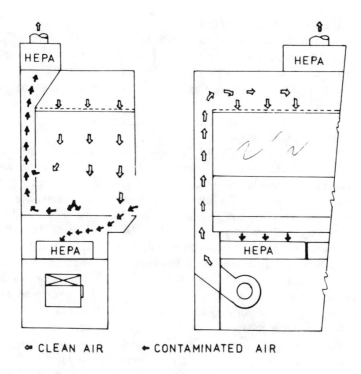

CLEAN AIR        CONTAMINATED AIR

FIGURE 3.25.    Class 2B biosafety cabinet.

through two rows of slots at the rear of the cabinet, through a HEPA filter, while 30% is recirculated through the front grill. Blowers external to the cabinet, which are typically a part of the laboratory system, provide the exhaust pressure.

Another version of the Class 2B biosafety cabinet is the total exhaust unit shown in Figure 3.26. All of the air entering the cabinet makes only one pass through the cabinet before being discharged through a HEPA filter. One hundred (100) fpm of room air enters through the front opening, providing protection for the operator against pathogens escaping from the cabinet, and is drawn down into a plenum below the work surface by a blower. All of this air is exhausted. None is recirculated. In order to provide clean air for the work zone and to block the entrance of room air into the cabinet interior, air is drawn in from above the cabinet by another blower, through a HEPA filter into the work zone. Part of this air passes through the inlet opening at the front of the cabinet and part through a slot at the rear, as contaminated air. All of the air in the exhaust plenum is discharged through another HEPA filter.

Since none of the air which has passed through the work zone is recirculated, at least in principle this type of cabinet could be used for moderate chemical applications. However, care would have to be taken to insure that the exhaust filter would not suffer from loading of the filter by chemicals. The different rate of loading of the inlet and exhaust filters is a problem for all types of Class 2 cabinets, but the more

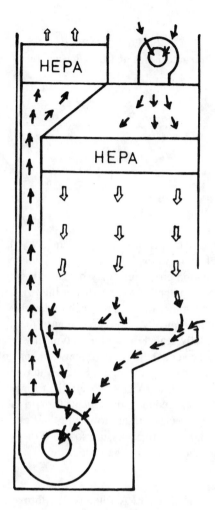

FIGURE 3.26.    Total-exhaust biosafety cabinet.

complicated air flow systems of Class 2B biosafety cabinets, where two or more blowers are involved, exacerbate the problems.

Table 3.10 provides a summary of the characteristics of the three types of Class 2 biosafety cabinets.

Class 2 biosafety cabinets are intended for low- to moderate-risk hazards. As a minimum, they should be required to meet the National Safety Foundation (NSF) Standard 49 for Class 2 (laminar flow) biohazard cabinetry. The working enclosures and plenums through which air moves should be constructed of materials that are easy to decontaminate, such as stainless steel or a durable plastic.

### 3.2.2.4.3. Class 3 Cabinets (Glove Boxes)

Class 3 units are totally sealed cabinets in which the user performs the manipulations with the research materials inside the chamber, using attached arm-length, impermeable gloves. Materials are usually placed in the cabinet prior to work

**TABLE 3.10**
**Basic Characteristics of Class 2 Biosafety Cabinets**

|  | Type | | |
|---|---|---|---|
|  | 2A | 2B | 2B (total exhaust) |
| Inlet air speed | 75 fpm | 100 fpm | 100 fpm (min) |
| Volume of air recirculated | 70% | 30% | 0% |
| Positive-pressure plenums | Contaminated | Filtered | Contaminated, but isolated |
| Exhaust air for a typical 4-ft unit | 225 cfm | 500 cfm | 700 cfm |

beginning or cycled into the interior through an air lock. Materials taken from the chamber are usually taken out through a double-door autoclave, through a second air lock where decontamination procedures may be carried out, or through a chemical dunk tank. These units are intended to be used for high-risk materials. They are, in a sense, the reverse of a laminar flow work bench in that they provide virtually total personnel protection, but, typically, little protection to the product.

Because most are totally sealed, glove boxes typically have minimal ventilation requirements. At least one vendor, however, Baker Company of Sanford, ME, makes a Class 3 unit in which air enters through a HEPA filter and is exhausted through ultra-high-efficiency HEPA filters (99.999% efficient for particle removal) in tandem. These filters, which exceed the requirements for normal HEPA filters by better than a factor of 10, are especially tested to insure their performance. The cabinets should be kept under slight negative pressure so that if there are any leaks at all, the leakage would be from the outside to the inside.

### 3.2.2.4.4. HEPA Filters

An accepted definition of a HEPA filter is: "A throwaway extended-pleated-medium dry-type filter with (1) a rigid casing enclosing the full depth of the pleats, (2) a minimum particle removal efficiency of 99.97% for thermally generated monodisperse DOP smoke particles with a diameter of 0.3 $\mu$m, and (3) a maximum pressure drop of 1 inch water gauge when clean and operated at its rated airflow capacity." A properly functioning HEPA filter is an essential component in both Class 1 and Class 2 units and may be important in some Class 3 biosafety cabinets.

HEPA filters used in Class 2 biosafety cabinets meet the requirements of Underwriters' Laboratory's UL 586 and mil-spec MIL-F-51068. As stated in the definition, they are made by folding a continuous sheet of filter paper back and forth over corrugated separators. The separators provide strength to the assembly and form air passages between the pleats. The filter paper is made up of submicron fibers in a matrix of larger, 1 to 5 $\mu$m fibers with a small admixture of organic binder. It is a fragile material and subject to puncture or cracking if abused. The filter is usually glued to the edge of the frame with a glue that hardens, thus making this connection a highly vulnerable stress point. The frame is seated on a gasket and this seat is also vulnerable to leaks. If there are any questions about the integrity of the HEPA filter, operations should cease until it has been tested and recertified.

A standard filter assembly of 24 × 24 in. and 5 7/8 in. deep has a rated capacity of 500 cfm. This filter will contain about 110 ft$^2$ of filter paper. Operating at rated capacity, the air will be moving through the filter at about 5 fpm and the speed of the air leaving the face will be about 125 fpm. The filter will offer, in clean condition, a pressure drop across the filter of no more than 1 in. water gauge. When the pressure drop reaches 2 in. water gauge, it is usually time to replace the filter.

Because the desired laminar air velocity inside a Class 2 unit is somewhat less than 125 fpm (in the unit designed by the National Cancer Institute, described in Section 3.2.2.4.2 , the downward design velocity is 50 fpm), most HEPA filters operate at well below their rated capacity. This results in a significantly increased lifetime. The decrease in resistance is inversely proportional to the velocity of the air through the filter, and the decrease in the amount of loading on the filter is directly proportional to the volume of air passing through the filter. The net effect is to greatly extend the life of the filter unit.

*3.2.2.4.5. Installation, Maintenance, and Certification*

The guidelines governing the most favorable location of laminar flow cabinets are essentially the same as for a chemical fume hood. Place them in the far end of a laboratory, in a low traffic area, and where there are no drafts. The environment of the hood should be checked to ensure that air speeds in the neighborhood of the cabinet opening are small, compared to the face velocities of the cabinet.

The installation of laminar flow cabinets is a specialized skill. An examination of a cabinet shows that the components are carefully sealed together, with a large number of bolts being used to maintain sufficient pressure on the seals between the individual sections of the cabinet to insure that they fit tightly together. Shipping or even moving the cabinet across the room may be sufficient to break some of the critical seals. Demonstration that the unit's seals are intact and that it meets all specifications at the factory, simply shows that it was in good condition at that time. It does not demonstrate that it is still in a similar condition after installation. The equipment needed to test the integrity of the cabinets is expensive and requires skills and training which relatively few laboratories, institutions, or corporations have in-house. However, in order to be sure that a cabinet is providing the designed protection for the operator, product, or both, the cabinet needs to be tested and certified at installation, after any relocation, and periodically, thereafter, such as annually or after 1000 h of use.

A purchase order for a cabinet should include provision and funds for testing and certification of the unit after delivery and setup, but before final payment is made. Few, if any, vendors routinely make provision for this service or have in-house personnel to do it. In order to avoid a potential conflict of interest, this should be arranged by the purchaser with an independent contractor. The cabinet vendor should be willing to delay payment of their invoice for a reasonable period to allow this to be done. This provision needs to be made in the purchase contract so that no misunderstanding will occur. Training programs are available for teaching how to perform the tests properly as are firms or individual consultants who will perform the tests at a reasonable fee.

It has been noted that it is not *essential* that the cabinets be exhausted out-doors, but there are several reasons why this is highly desirable. The most obvious reason is that no system is absolutely foolproof. A cabinet may be checked and certified in the morning and an accident occur during the afternoon which could cause a seal to fail or the HEPA filter to begin to leak. If the cabinet is exhausted outside the facility, no pathogens should be released into the facility. Decontamination with formaldehyde or a comparably undesirable chemical may be required, which should not be released into the laboratory. Some research may involve chemical carcinogens, radioisotopes, or other materials which would not be eliminated by the HEPA filters and should not be discharged indoors. In fact, it may be desirable to incinerate some organisms by passing them through a flame before evacuating them even to the outside. Finally, Class 2B cabinets typically do not provide an integral exhaust blower to discharge air from the unit, but depend upon the facility to provide the exhaust fan. If the cabinet is connected to the system in order for the suction on the exhaust portal to be provided, there is no reason for the system to exhaust back into the laboratory or building. Where cabinets are exhausted to the roof, the height of the exhaust stack should be sufficient to ensure that the effluents are discharged above the head height of any maintenance personnel who may be present on the roof. No weather cap should be provided for these exhaust ducts.

The removal and replacement of contaminated HEPA filters should be performed by trained personnel who take precautions against contaminating themselves and the facility. This will require advanced planning in order to provide sufficient access to the filters. For filters anticipated to be contaminated with human pathogens (or animal, where animal exposures are a matter of concern), provision for isolation bag-out procedures should be made in advance. In general, HEPA filters should be disposed of as contaminated biological waste, preferably by incineration.

Cabinets should be decontaminated periodically. A UV tube placed inside the cabinet will aid in disinfecting the surface, but will not significantly decontaminate the air passing through it. Cleaning after each day's use or at the end of a sequence of operations with a weak solution of household bleach is recommended, but other materials such as quaternary ammonium compounds may serve as well, be less irritating to the user, and cause less corrosion. There are procedures recommended by the National Cancer Institute and available on a slide cassette package for decontaminating cabinets. Basically, this consists of sealing the cabinet and vaporizing an amount of dry paraformaldehyde sufficient to provide a concentration throughout the cabinet of 7 to 8.5 mg/m$^3$. The cabinet should remain sealed for about 4 h. The temperature should be between 20 and 25°C and the humidity above 70%. The paraformaldehyde should then be exhausted to the outdoors and the cabinet ventilated for at least 8 h. It should be noted that since this procedure was recommended, questions about the carcinogenity of formaldehyde have arisen. It is likely that the OSHA permissible exposure level (PEL) will be lowered. Persons performing the decontamination must not exceed the occupational limits. There is no question that some individuals are irritated by levels somewhat lower than those now permitted.

## REFERENCES (SECTIONS 3.2.2.4 TO 3.2.2.4.5)

1.  Class II (Laminar Flow) Biohazard Cabinetry, Standard No. 49, National Sanitation Foundation, Ann Arbor, MI, 1983.

2. **Burchsted, C. A., Kahn, J. E., and Fuller, A. B., Eds.,** *Nuclear Air Cleaning Handbook,* Oak Ridge National Laboratory, ERDA 76-21, Oak Ridge, TN.

3. **McGarrity, G. J. and Coriell, L. L. ,** Modified laminar flow biological safety cabinet, *Appl. Microbiol.,* 28 (4), 647, 1974.

4. **Coriell, L.L. and McGarrity, G. J.,** Biohazard hood to prevent infection during microbiological procedures, *Appl. Microbiol.,* 16 (12), 1895, 1968.

5. *Selecting a Biological Safety Cabinet,* slide cassette, NAC No. 00709 and 01006, National Audio Visual Center (GSA), Washington, D.C., 1976.

6. *Effective Use of the Laminar Flow Biological Safety Cabinet,* slide cassette, NAC No. 00971 and 003087, National Audio Visual Center (GSA), Washington, D.C., 1976.

7. *Certification of Class II (Laminar Flow) Biological Safety Cabinets,* slide cassette, NAC No. 003134 and 009771, National Audio Visual Center (GSA), Washington, D.C., 1976.

8. *Formaldehyde Decontamination of the Laminar flow Biological Safety Cabinet* slide cassette, NAC No. 005137 and 003148, National Audio Visual Center (GSA), Washington, D.C., 1976.

### 3.2.3. BUILT-IN SAFETY EQUIPMENT

In a very real sense, the hoods and safety cabinets that have been the subjects of the last several sections can be considered built-in safety equipment. However, most persons would consider the items to be discussed in the following sections as fixed safety equipment.

#### 3.2.3.1. Eyewash Stations

One of the most devastating injuries a person can suffer is loss of eyesight. There are a number of protective measures which should be taken in the laboratory to prevent eye injury. However, should all of these measures fail and chemicals enter the eye, an effective eyewash station is an essential item of fixed equipment that should be immediately available. OSHA requires in Section 1910.151(c) of the General Industry Standards that:

> Where the eyes or body may be exposed to injurious corrosive materials, suitable facilities for quick drenching or flushing of the eyes and body shall be provided within the work area for immediate emergency use.

There are no fixed standards on the maximum distance of travel which would be acceptable to reach an eyewash station. The National Safety Council has recommended that the travel distance should be no more than 25 ft or take no more than 15 sec to reach. A more realistic criterion should probably be that no one should have to open a door to reach an eyewash station and that the distance *and* time of travel should be no greater than those recommended. No one in pain, and possibly blinded, should have to overcome any additional impediments or obstacles in seeking relief. An eyewash station should be centrally placed in a laboratory along a normal path of egress or in an equally logical location in a given facility.

Small squeeze bottles containing a pint or, at most, perhaps a quart of water are not acceptable as eyewash devices. The basic problem is lack of volume. As a minimum, eyes suffering even a light chemical burn need to be flooded with potable water for 15 to 20 min. The second problem is that the water in the bottle may become contaminated. Where water lines are not available, eyewash units connected to pressurized portable containers of water are acceptable substitutes if they contain sufficient amounts of water, preferably 15 to 20 gal.

Eyewash stations should provide an ample amount, at least 0.4 gal/min, of water at a relatively low pressure, 25 psi or less, in such a manner as to flood both eyes and the entire face with aerated, potable water. The most common type, with two nozzles facing upwards and aimed slightly inward, toward each other, is probably the best overall design. An alternative consisting of a spray nozzle connected to a flexible hose is not a bad supplement, but it should not be the only eye washing device available. An individual alone may be in too much pain to do much more than hold his face in flowing water and could certainly not simultaneously manipulate a hose and use his hands to hold his eyelids open.

Turning on the eyewash station should require minimal manual dexterity. Any number of mechanisms to turn it on are possible, but perhaps the most popular is a simple paddle which the injured person can push aside. The eyewash should remain on continuously, with no additional effort after the initial activation, but if an automatic cutoff is provided, it should not activate for at least 15 min or until 40 to 50 gal of water have been delivered. Many eyewash stations are mounted as part of the plumbing over a sink. This is convenient, but not essential. In the following section on deluge showers, it will be pointed out that in the case of eye injuries, safety is more important than spilled water, which can be mopped up, so floor drains are not required for safety shower installations. Preferably, eyewash stations and deluge showers should be installed as a package since it is likely that if the eyes and face have been exposed to chemicals, other portions of the body are likely to have been contaminated as well.

A major problem with most eyewash stations and deluge showers is that they are usually connected to the cold water line. Typically, tapwater temperatures are in the 60 to 70°F (15.5 to 21°C) temperature range and in colder climates can be much less during the winter. Water at these temperatures is painful itself and in extreme cases can cause the injured person to go into shock. Although relatively few eyewash installations are capable of conveniently providing it, lukewarm water between 90 to 95°F (32-35°C) would be ideal.

A permanently installed eyewash station is an essential component in or very near every laboratory, but if one is not available, any source of water, provided it is not too hot or extremely cold, should be used in an emergency — a sink faucet, shower, or even a large basin of water in which the injured person could immerse his eyes.

All eyewash stations should be checked under full-flow conditions on a definite schedule. Any deficiencies should be corrected immediately.

Brief instructions on how to activate and use an eyewash fountain and simplified instructions on how to help the patient keep eyes fully open so that the water will be able to reach the injured tissues, should be placed immediately adjacent to the unit.

### 3.2.3.2. Safety (Deluge) Showers

Safety showers should also be situated such that it is not necessary to travel more than 25 ft or for longer than 15 sec to reach one in the event of an accident, nor should there be any obstacles in the way. However, showers are often placed in hallways, usually where they can service more than one laboratory. This also avoids creating a massive water flood in a crowded laboratory in favor of a much more easily cleaned hallway. This latter point is not an insignificant advantage since it may be difficult to locate a shower in a small facility so that water will not splash into sophisticated and easily damaged equipment. Nevertheless, the potential of a serious life-threatening or maiming injury in a chemical spill is sufficiently likely that the concern for personal safety should override this and other factors. In addition, units located in public corridors are subject to vandalism. Therefore, it is recommended that each laboratory should be individually equipped with a combination safety shower and eyewash station. As noted in the previous section, accidents involving facial splashes are also likely to involve other parts of the body. Clearly, if the units are separated, it is not practical to travel from one to the other when both are needed. Both should be in the same location.

The demands of a safety shower on the water supply are much more severe than for an eyewash fountain. Their alternate name, deluge shower, is not idly applied. The water supply should be able to provide up to 30 gal/min. For normally available water pressures, this would usually indicate that the shower water supply be at least a 1-in. line. The problem of temperature is, again, a serious one, but even more so because of the greater area of the body involved. Usually, the showers are connected to a cold-water supply line. In colder climates, the stress of inundating the entire body with cold water at temperatures perhaps in the 50 to 60°F (10 to 15.5°C) range may be sufficiently severe to cause a person to go into shock. Ideally, water temperatures should be around 90 to 95°F (32 to 35°C), but not over 100°F (38°C).

Instructions for activating and deactivating a shower should be prominently posted near the shower. The mechanism for turning a shower on can be a paddle which is simply pushed out of the way or a chain which can be pulled. Both are simple and require minimal physical control or manual dexterity, which may be important in this type of an emergency. The shower should continue to run until it is deliberately turned off, but it should deliver at least 50 to 60 gal before any automatic cutoff activates. Showers should be checked at least once a year and is conveniently done by catching the flow in a large funnel connected to a fire hose which is discharged into a 55-gal drum. This provides both a rate and volume check simultaneously, while avoiding a mess to be cleaned up.

It should not be necessary to point out that safety showers should not be located near any source of electricity with which the flowing water from the shower could come into contact, but numerous instances have been observed where this has occurred. Usually, it has been due to the laboratory workers themselves moving portable equipment too near the shower, but occasionally one is found improperly installed by maintenance personnel or because of changes made by renovation crews.

Drains are unnecessary. Although a mess will be created which will have to be mopped up when a shower is used, this should be sufficiently rare that it does not justify the cost of installing additional drains to accommodate the safety showers, especially if a retrofit is necessary. However, there should be provision for interrupting the water flow after a reasonable interval of 15 to 20 min to avoid floods that could damage equipment.

## REFERENCES (SECTIONS 3.2.3.1 AND 3.2.3.2)

1. American National Standard for Emergency Eyewash and Shower Equipment, ANSI Z358.1, American National Standards Institute, New York, 1981.
2. Stearns, J. G., Safety showers, in *Handbook of Laboratory Safety,* 2nd ed., Steere, N. V., Ed., CRC Press, Boca Raton, FL, 1971, 121.
3. Srachta, B. J., Safety Showers & eyewashes, in *Safety and Health,* National Safety Council, Chicago, 1987, 50.

### 3.2.3.3. Fire-Suppression Systems

Although many laboratory buildings have automatic fire suppression systems, many do not, despite the fact that the most common type, a water sprinkler system, is effective well over 90% of the time in either extinguishing a fire or controlling it until fire fighting personnel arrive. In a very high percentage of the cases in which the sprinkler system was not effective, it was due to either poor installation or human error. Some scientists do not want a fire-suppression system in their laboratories because they feel that it will lead to problems. The possibility that a fire in one area will cause the entire system throughout the building to activate is one basis for concern, although in most systems, only those sprinkler heads in the vicinity of a fire activate. In over one third of the fires in which the sprinkler system was successful, only one sprinkler head was activated in controlling the fire. There is some concern that water will react with chemicals in the laboratory, spread the fire due to burning solvents being carried away by excess water, or damage sensitive equipment. If there are problem chemicals, there are alternative fire-suppression systems which do not use water and will not damage even delicate equipment.

The potential for fires is higher in most research facilities than in a typical building because of the variety and character of the chemicals employed. Often, much of the equipment is home-built or temporarily rigged, and has not been checked for safety by an accredited testing laboratory. Many of the operations involve heat. Some type of fire-suppression system is recommended for laboratory buildings, if for no other reason than that the presence of one in a building can lead to substantial savings in insurance costs. Currently, insurance premiums are rapidly escalating and do not appear to show signs of abating, although there is some reason to believe that government actions will eventually be taken to limit these costs.

The time to stop a fire is when it is very small. This is the purpose of portable fire extinguishers which laboratory personnel can use before a fire becomes too large to control. Unfortunately, most laboratories are not staffed at all hours, while some heat-generating equipment such as stills or heat baths may be left functioning continuously. A sprinkler system serves essentially the same purpose as a person with an extinguisher 24 hours a day, 365 days a year. It can put water, carbon dioxide, or other fire-suppression media directly onto a fire while it is still small enough to be easily extinguished. There are fires which expand so rapidly that a fire-suppression system will be overwhelmed, but, as noted above, the number of fires in this category are very small.

OSHA does not mandate automatic fire extinguishing systems, but the General Industry Standards, Subpart L, Sections 1910.155 to 1910.165, do provide regulations covering the essential requirements which installed systems must meet.

The principles of an automatic fire-suppression system are straightforward, but designing an actual system requires substantial engineering skills to be sure that the system will adequately and efficiently serve its purpose. The following brief sections on built-in fire-suppression systems are intended only to provide basic information on the essential features of the various types of fire-suppression systems so that concerned laboratory facility managers will have some insight into this critical area.

*3.2.3.3.1. Water Sprinkler Systems*

There are several essential components of an automatic water sprinkler system.

There must be a supply system to provide water to the system. Although in most cases this supply system need not be so large as to service a sprinkler system for an entire building, it still must be sufficient to supply water to a substantial fraction of the total suppression system, and it must have sufficient pressure to supply water to the units highest up in the building. There must be a control valve to serve the suppression system. A distribution system must carry the water to the spaces to be protected by the system, when needed, to respond to the presence of a fire. In each area where the sprinkler system is to provide protection, a carefully engineered pattern of sprinkler heads is required to distribute the water in such a fashion that complete coverage is obtained. The sprinkler heads themselves must be selected to meet the needs of the location. In many systems, the heads incorporate heat sensing devices, in the form of fusible links, which cause them to activate. The design of the system must take into account the maximum normal temperature that will be reached in the vicinity of the sprinkler heads and specify, for the fusible links, an operating temperature a reasonable amount above this temperature. In some systems, the sensing devices are separate from the sprinkler heads. Finally, when a sprinkler system is activated, a sensing device is needed which will transmit an alarm to the occupants of the building, preferably to a manned location which can immediately summon fire fighters, or directly to a fire station. Any automatic water sprinkler system should be installed according to the NFPA Standard 13, as most recently amended.

The water supply system may be from a water works system, a gravity tank, or a pressure tank. NFPA standard 13 provides specifications for each of these to ensure that they provide sufficient water volume and pressure to supply a sprinkler system. This standard provides for seven different categories of occupancy classifications ranging from light hazard to extra hazard. These levels are only for the purpose of designing an appropriate sprinkler system and do not correspond directly to the occupancy classes under the BOCA building code. The extra hazard, Group 2, represents a facility where, in part, there are moderate to substantial amounts of flammable and combustible liquids. As a minimum for a light hazard occupancy, the water supply must provide for a residual pressure of 15 psi at the level of the highest sprinkler and a duration of 30 min for the volume of water needed for the system. The requirements on duration and the amount of water needed are higher for other classifications. A gravity tank supply should be at least 35 ft above the level of the highest sprinkler head to provide an adequate pressure, while a pressure tank of sufficient size, filled two thirds full, must be pressurized to at least 75 psi.

There must be at least one connection through which a fire department can pump water into the sprinkler system. When a large number of sprinkler heads have become involved in a fire or the fire department has connected to the same supply to service their own hoses, the flow of water through the sprinkler pipes may be reduced so that the required volume and pressures cannot be met. In such cases, the fire department can boost the pressure and volume through this connection, preferably from separate water mains not being used in their own fire suppression efforts.

There are two different types of water sprinkler systems, wet pipe and dry pipe. In a wet-pipe system, the lines leading to the sprinkler heads are full of water so that water will be discharged immediately from an open sprinkler head, while in a dry-pipe system, the lines are full of air under pressure instead of water. The latter type of unit should be used where the temperature is not maintained above freezing at all times. Both types of systems include a main control valve which is designed to not only supply water to the sprinkler heads, but also to provide a mechanism which causes an alarm to sound. They also usually provide a visual indication of whether they are open or closed. Except during maintenance, they should be in the open position. There are a number of design features for these valves which are beyond the scope of this book, except to note that some are intended to avoid false alarms due to surges or variations in the water supply pressure to the system.

A variation of the dry-pipe system is the deluge system. The sprinkler heads are continuously open and water is prevented from entering the system by a deluge valve. When a fire is detected, the valve to the water supply opens and water flows into the system and out of ALL the sprinkler heads. This not only wets the immediate area of a small fire, but the entire area to which the fire may spread. This is usually done when the contents of the space are unusually hazardous. A variation on this system is the pre-action system in which the sprinkler heads are not open, but, when a fire is detected, the deluge valve opens and water is supplied to the sprinkler heads. The water entering the system causes an alarm to be sounded. When the heat causes the sprinkler heads to fuse, or open, water then is discharged onto the fire.

After the control valve is activated, water is distributed through one or more vertical risers to portions of the system. Smaller cross mains are connected to the risers which then service several branch lines which are still smaller. The sprinkler heads are connected to these branch lines. For laboratories, both the piping and sprinkler heads should be especially selected to prevent corrosion.

Sprinkler heads are very simple and rugged devices in which a valve is kept tightly closed by lever arms held in place by fusible links or other devices which fuse or open when they are heated to a predetermined level. When this occurs, the valve opens and water is discharged. There are many different versions of this simple device. Figure 3.27 shows a generalized drawing of an upright design.

In most cases, the water flow pattern is designed to obtain a uniform overlapping distribution similar to that shown in Figure 3.28. The water is intended to be in the form of a fine spray. In instances where there are expected to be strong, vertically rising convective air flow currents, the design may be modified to provide larger droplets which would be more likely to overcome the upward moving air.

The orifice within the sprinkler head is usually 1/2 or 17/32 in. in diameter. Depending upon the design and the level of risk of the occupancy, a sprinkler head with an orifice of these sizes is intended to protect between 90 and 130 ft$^2$ for the higher levels of risk.

The temperature at which the fusible link or other device fuses depends upon the normal maximum temperature to which it may be exposed. For maximum ceiling temperatures in the vicinity of 100°F or 38°C, which would be comparable to those found in most laboratories, the fusible links should be selected to open with a temperature rating in the range of 135 to 170°F (57 to 77°C).

Water is not dangerous to humans, and a water sprinkler system is comparatively inexpensive and easy to maintain. The major advantages of water are due to its physical properties. It has a high heat capacity and a large heat of vaporization. It provides cooling and wetting of the fuel, which aids in putting out the fire, and once the fire is subdued, the presence of water discourages reignition of the fire. Water does cause considerable damage to equipment and can react with many materials.

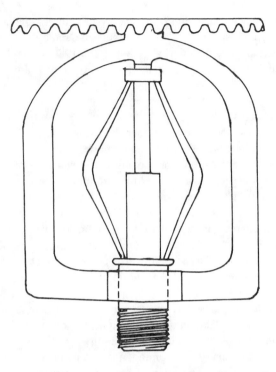

FIGURE 3.27.    Representative upright sprinkler head.

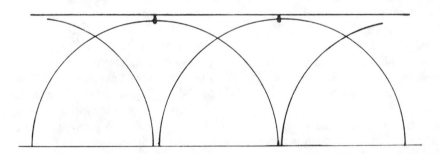

FIGURE 3.28.    Flow pattern from sprinkler system.

### 3.2.3.3.2. Halon™ Systems

Halon™ is a generic name for halogenated agents used in fire-suppression systems. There are two major agents of this type in common use, Halon™ 1301 and Halon™ 1211. The different numbers in the nomenclature correspond to the number of carbon atoms, fluorine atoms, chlorine atoms, and bromine atoms, in that order. If there were a fifth number in the designation, it would refer to the number of iodine atoms. Halon™ 1301 is the one most commonly used in occupied spaces, while Halon™ 1211 applications are typically in storage areas or other areas which are rarely or sporadically occupied.

There is some concern that fluorocarbons are causing a depletion of the ozone layer in the upper atmosphere. As a result, there are some proposals to freeze or reduce the amount of these materials at or below the 1986 levels. If such proposals were to go into effect, at some point in the future the use of Halon™ systems might be restricted.

### Halon™ 1301 Systems

Halon™ 1301 is a colorless, odorless gas that is noncorrosive and is electrically nonconductive. It is relatively nontoxic and at concentrations up to at least 7% by volume in air, personnel can perform a normal evacuation of an area without significant risk. It is especially useful for fires in normally occupied spaces, where flammable and combustible liquids are present, and where electrical devices such as computers, data terminals, and electronic instrumentation and control equipment are present. The latter characteristics make it very useful for laboratory applications.

The design of Halon™ 1301, bromotrifluoromethane ($CBrF_3$) systems is covered in NFPA Standard 12A.[4] This standard also contains much useful background material on the characteristics of the material and is the major reference for this section.

There are two basic types of systems, one providing total flooding of an area and one designed for local application of the extinguishing material directly to the fire. Fixed systems will consist of a container of liquified compressed gas pressurized with nitrogen, detectors, distribution piping, nozzles, manual releases, and a control panel incorporating an alarm. The control panel may also provide circuitry to cut off ventilation systems and close windows, louvers, and other openings to the space.

It is generally accepted that Halon™ agents interrupt the chemistry of a fire to extinguish it. Since fires involving the vapor phase of a material are most susceptible to contact with the extinguishing agent, Halon™ systems work best on this type of fire. In the case of a solvent fire, for example, if the fire can be detected quickly enough and Halon™ 1301 applied immediately, substantial fires can be put out in seconds. Gaseous systems such as Halon™ work well on liquid fires, in part because the temperature of the liquid generally will not exceed the boiling point of the liquid. Hence, if the availability of oxygen is eliminated, the liquid itself will not be hot enough to reignite the fire. Other types of fires, such as those on the surfaces of solid materials, will take longer. This agent may not be appropriate for deep-seated fires, although if a sufficient concentration can be maintained long enough, even these fires can be extinguished.

In a total flooding system, the required concentration of the extinguishing agent will depend upon what fuel is present and whether it is sufficient to only extinguish the fire or if it is desirable to "inert" the space, i.e., provide a sufficient concentration of Halon™ 1301 in the atmosphere so that the flammability range for the fuel-air combination will be nonexistent at the ambient temperature. The latter concentrations are substantially greater than required for extinguishing the fire. The concentrations needed at a temperature of 77°F or 25°C∗ to extinguish fires for a number of common flammable liquids are given in Table 3.11, as well as design concentrations to provide a reasonable safety factor and required inerting concentrations. Note that the design concentration was chosen to be at least 5%.

---

∗ These data are derived from MFPA 12A, Tables A-2-3.2.2 and A-2-3.2.3.

**TABLE 3.11**
**Halon™ 1301 Extinguishing Characteristics**

| Fuel | Extinguishing concentrations | Design concentrations | Inerting design concentrations |
|------|------------------------------|-----------------------|--------------------------------|
| Acetone | 3.3 | 5.0 | 7.6 |
| Benzene | 3.3 | 5.0 | 5.0 |
| Ethanol | 3.8 | 5.0 | 11.1 |
| Ethylene | 6.8 | 8.2 | 13.2 |
| Methane | 3.1 | 5.0 | 7.7 |
| N-heptane | 4.1 | 5.0 | 6.9 |
| Propane | 4.3 | 5.2 | 6.7 |

*Note:* Values expressed in percent by volume in air at 77°F (25°C) at 1 atm.

Note that if ethylene were the primary fuel present, a higher concentration of the extinguishing agent than the recommended 7% would be needed. This would not preclude the use of Halon™ 1301, but it would mandate an effective and rapid evacuation plan. The fact that only flammable liquids or gases are given in Table 3.11 is not to infer that Halon™ 1301 is only suitable for such materials. If a fire is confined to the surface of a solid material, chances are that Halon™ will be effective, i.e., a 5% concentration could possibly put it out if the fuel is allowed to "soak" in the design concentration for 10 or more min, and if the fuel temperature falls below the level required for spontaneous ignition after the Halon™ is dissipated. Deep-seated fires, which in this context are defined as ones which will not be put out at a concentration of 5% with a soaking time of 10 min, may still be extinguished at higher concentrations, longer soaking times, or a combination of both. Some of the advantages of Halon™ 1301 diminish if higher concentrations and longer times are required and some disadvantages arise. Most of these negative factors are related to health factors.

Halon™ 1301 appears to have very low toxicity as a pure material. At concentrations below 7%, little if any effect on humans has been noted for test periods up to 30 min. In order to provide a reasonable safety margin below 7% concentrations by volume, exposures of up to 15 min are permissible. Some individuals have experienced mild effects on the central nervous system, such as dizziness and a tingling sensation in the extremities at concentrations between 7 and 10%, so, to be conservative again, at these concentrations exposures should be limited to 1 min. Above 10%, the effects on the central nervous system are more pronounced, so between 10 and 15%, exposures should be limited to 15 sec. Personnel should not be exposed above 15%. Although a number of other physical effects have been investigated, the only other significant effect noticed in tests on animals has been to the cardiovascular system, where the heart has been made abnormally sensitive to elevated levels of adrenalin, as might be present during the stress of a fire, leading to possible cardiac arrhythmia. All of the observed effects have been shown to be transitory, disappearing after the exposure ceased.

The decomposition products of Halon™ 1301 are dangerous in sufficient concentrations. The chief ones are the acids of the halide components, HF, HBr, and $Br_2$.

Fluorine is too reactive to be present in substantial amounts alone. Small amounts of carbonyl fluoride and carbonyl bromide ($COF_2$ and $COBr_2$), respectively have been found as well. Because of the potential for these toxic materials being present after the intervention of a Halon™ 1301 system, fire personnel and others entering the area of a fire should exercise caution. The use of a positive-pressure, air-supplied breathing apparatus is recommended where the fire involves substantial amounts of solid fuel, in searching the area for injured personnel.

The disadvantages alluded to above when discussing the use of Halon™ 1301 for deep-seated fires in solid materials are the risks associated with the required higher concentrations of Halon™ 1301, especially if an individual is injured or unconscious and is unable to evacuate quickly, and due to the increased generation of decomposition byproducts with more extinguishing agent present. Cost is a further, and significant, disadvantage. Halon™ 1301 is expensive and most systems are designed to completely release the supply of the agent. As a result, fixed Halon™ systems are usually installed where they will be cost effective, such as in large computer installations where other expensive configurations of sensitive electronic devices are in use, or where substantial amounts of solvents are stored or in use.

To take advantage of the excellent extinguishing properties of Halon™ 1301, especially for very rapidly spreading fires from burning flammable liquids, a quick acting fire detector should be selected, such as one which detects either ultraviolet or infrared light from the flames. Although precautions must be taken for both types of detectors to avoid false activation of the system, the rapid response capability of these units means that the extinguishing agent can be applied to the fire almost instantaneously.

Local application systems work as precisely as the fixed systems, with the exception that the agent is applied directly to the immediate area of the fire. The primary need is for the concentration to remain sufficiently high in the area of the fire long enough to insure the extinction of the fire and to permit sufficient time to elapse to allow cooling of the fuel to avoid spontaneous reignition once the concentration decreases below the level required to extinguish a fire. Whereas in a fixed system it is possible to calculate the amount of Halon™ 1301 needed with some precision, each local application is a unique situation and must be evaluated individually.

As with other common types of fire-suppression systems, Halon™ 1301 is ineffective for fires involving a number of materials, including reactive metals, metal hydrides, materials which do not require the presence of air to burn, such as gunpowder, some organic peroxides, and hydrazine.

Although Halon™ 1301 is not ideal for every laboratory facility, it does offer advantages over many other systems. Its toxicity is low; it is very effective on many common types of laboratory fires; it acts rapidly, which is especially critical in minimizing the spread of a fire in a laboratory situation; and it will not damage the increasingly sensitive equipment found in modern laboratories. Its cost and propensity to be set off by minor flames when ultraviolet or infrared sensors are used are its major disadvantages.

### Halon™ 1211 Systems

Much of the material in the preceding section is equally relevant to systems using Halon™ 1211, bromochlorodifluoromethane ($CBrClF_2$), as far as effectiveness is

## TABLE 3.12
## Halon™ 1211 Extinguishing Characteristics

| Fuel | Extinguishing concentrations | Design concentrations | Inerting design concentrations |
|------|------------------------------|-----------------------|--------------------------------|
| Acetone | 3.6 | 5.0 | NA |
| Benzene | 2.9 | 5.0 | 5.0 |
| Ethanol | 4.2 | 5.0 | NA |
| Ethylene | 7.2 | 8.6 | 13.2 |
| Methane | 3.5 | 5.0 | 10.9 |
| N-heptane | 4.1 | 5.0 | NA |
| Propane | 4.8 | 5.8 | 7.7 |

*Note:* Values expressed in percent by volume in air at 77°F (25°C) at 1 atm.

concerned. There are some minor differences. Instead of being odorless, it has a faintly sweet smell, but otherwise many of the physical properties are similar. Table 3.12 gives the extinguishing characteristics for Halon™ 1211. Again, the two materials are very similar, as can be seen by comparing the data in Table 3.12 to the corresponding information in Table 3.11 for Halon™ 1301.

The fact that the design characteristics are not very different be taken to imply that the two materials could be used virtually interchangeably. However, whereas the onset of toxic effects on personnel for the most part did not become significant for Halon™ 1301 until concentrations of 7% or higher, which are above most of the required design concentrations, for Halon™ 1211, the onset of problems such as dizziness become definite within a few minutes at exposures above 4%, which is below the required extinguishing concentrations. For this reason, the use of Halon™ 1211 is not approved for occupied spaces or for normally unoccupied spaces where an evacuation time of more than 30 sec would be required.

Even though there are restrictions on its use, because of its excellent fire extinguishing properties, Halon™ 1211 is frequently used in flammable store-rooms, especially small stock rooms where self-contained units are available which can be simply hung from the ceiling. Most portable Halon extinguishers are filled with Halon™ 1211.

### 3.2.3.3.3. Carbon Dioxide Systems

Carbon dioxide is a colorless, odorless, nonconductive, chemically inert gas. It is commonly used in small portable fire extinguishers for putting out Class B fires, i.e., fires that involve solvents, petroleum products, grease, and gases. It can also be used for Class C fires where the nonconductivity of the extinguishing agent is important. It is not very effective against fires involving ordinary combustible Class A materials, such as wood and paper. Carbon dioxide is also used in total flooding systems. The characteristics of carbon dioxide as an extinguishing agent are included in NFPA Standard 12, which also provides information on the requirements of an approved system.

A major distinction between the use of carbon dioxide as an extinguishing agent and the Halon™ systems described in the last two sections is that the extinguishing

**TABLE 3.13**
**Carbon Dioxide Extinguishing Characteristics**

| Fuel | Extinguishing concentrations | Design concentrations |
|------|------------------------------|-----------------------|
| Acetone | 27 | 34 |
| Benzene | 31 | 37 |
| Ethanol | 36 | 43 |
| Ethylene | 41 | 49 |
| Methane | 25 | 34 |
| Propane | 30 | 36 |

*Note:* Values expressed in percent by volume in air.

mode is primarily simple smothering of the burning fuel, with no chemical action involved. There is little cooling action, with the effectiveness of carbon dioxide being about one tenth that of an equivalent amount of water. Hence, once the carbon dioxide has dissipated, the possibility of reignition exists if there are any sufficiently hot areas still present.

Table 3.13 gives the minimum extinguishing concentrations by volume and the design concentrations to provide a margin of safety. These should be compared to the equivalent data in Tables 3.11 and 3.12.

It is clear that far higher concentrations are required than for Halon™ systems — in fact, so high that an individual trapped in a space flooded by the minimum recommended amount of 34% would quickly become unconscious due to lack of oxygen since, in such a case, the oxygen concentration would fall to 13.8%.

When a fixed system is activated, a major portion of the rapidly expanding gas will become carbon dioxide vapor, while the remainder will become very fine particles of dry ice. There will also be condensed water vapor due to the cooling action of the expanding gas. As a result, visibility may be limited and individuals trapped in an area may have difficulty finding their way out. An area equipped with an automatic carbon dioxide extinguishing system should be posted with warning signs, such as:

**WARNING**
**AREA EQUIPPED WITH**
**CARBON DIOXIDE FIRE-SUPPRESSION SYSTEM**

**EVACUATE IMMEDIATELY**
**WHEN ALARM SOUNDS**

Predischarge alarms are essential and training should be provided for personnel in spaces where a carbon dioxide system is installed and for those individuals in adjacent spaces where the gas could flow. Any aisles providing a path of egress should be amply wide and kept clear at all times. Doors should swing in the direction of exit travel. Any automatic door-closing systems used to minimize leakage of the suppressant from the fire area should be equipped with delay circuits. Self-contained, positive-pressure breathing apparatus should be maintained nearby to make it safe for personnel to conduct rescue efforts for individuals who may be trapped in the area. Fire fighting and rescue forces entering an area where a total flooding system has been triggered should exercise caution and, unless they are confident that the gas has been dissipated, should wear breathing apparatus.

Because of the risk to personnel, carbon dioxide systems should not be used in most laboratory situations, although they could be used in storage areas or where electrical and electronic devices are employed. It does not leave a residue and will not damage equipment.

### 3.2.3.3.4. Dry Chemical Systems

Dry chemical fire extinguishing systems use a variety of dry powders as the fire fighting agent. They can be stored in pressurized containers and discharged, when needed, very much like the water and gaseous materials previously covered. Most of the agents are primarily effective against fires involving solvents, greases, and gases and can be used around active electrical circuits and electrical equipment since the chemicals are nonconductive. They do leave a residue which can be a problem for delicate electronic equipment. In general, however, the residue can be readily brushed off surfaces and vacuumed or swept up. There are multipurpose formulations which can be used on ordinary combustibles such as wood and paper. Most dry chemical formulations use monoammonium phosphate, sodium bicarbonate, potassium chloride, or potassium carbonate as their fire extinguishing agent.

Dry chemical agents are nontoxic, but personnel in the area when the agents are being discharged may have respiratory difficulties and vision problems due to the copious amounts of powder in the air.

The primary extinguishing mechanism is similar to that of the halogenated agents. They disrupt the chemistry of a fire so that it will not propagate. Other mechanisms include reducing the oxygen concentration in the flame zone, heat absorption by the chemical agent, and, for liquid fires, reducing the amount of vapor entering the air from the liquid by reducing the amount of energy radiated by the flame that reaches the surface and causes evaporation. The action of the dry chemicals is very rapid, as with the halogenated agents, which makes them desirable where it is essential to prevent a fire from spreading. If the problem with the residue is acceptable, a dry chemical system would appear to be a good choice for laboratories and for chemical stockrooms and store-rooms.

Monoammonium phosphate ($NH_4H_2PO_4$) is described as a multipurpose agent. As with other common dry chemical agents, it is effective on Class B and C fires, but unlike most others, which are not recommended for Class A fires, this compound does provide a mechanism for putting out fires in ordinary combustible materials. When heated, it decomposes and forms a solid residue which adheres to the heated fuel surface and excludes oxygen, thus eliminating an essential component for a fire, except for those materials which will sustain combustion without additional oxygen. Because of the formation of the solid residue which could adhere to equipment, it may be less desirable for laboratory applications, but it would appear to be a good choice for a stockroom, where materials are often stored in their original paper and wood shipping containers.

*3.2.3.3.5. Foam Systems*

Foam systems are usually intended to be used in extinguishing fires involving flammable liquids, rather than as general purpose fire extinguishing agents, although some versions are useful on Class A fires. The foam is intended to float on the surface of the liquid and extinguish the fire by excluding air and by cooling the fuel and hot objects that may have been heated by the action of the flames. It also prevents reignition by suppressing the production of flammable vapors, which could come into contact with heated solid objects. Most foam systems are not intended to be used on three-dimensional fires, i.e., fires involving solid materials of a significant height. High-expansion foam systems which are designed to smother a fire by flooding the area with a foam layer about 2 ft thick can be used in storage areas.

To be effective, the foams must remain intact and not mix with the burning liquids. Some foams work well on ordinary liquid hydrocarbons, but would readily mix with and lose their effectiveness on polar liquids such as alcohols, acetone, and ketones. However, there are different formulations which are effective for each type of liquid, and at least one, a synthetic alcohol foam, which may be used on either one.

Unless the laboratory is a very specialized one involving large quantities of flammable liquids, it is unlikely that a foam fire-suppression system would be selected. Foam-generating equipment could be useful in such a facility in covering large spills of flammable liquids to prevent their ignition.

### 3.2.3.4. Fire Detection and Alarm Systems

Every laboratory facility should be equipped with at least a manually activated alarm system, although an automatic system is preferable since it will continue to function when the facility is unoccupied. This is a special advantage in academic institutions since there are periods when the population of the campus is very low. In many cases, depending upon local code requirements and the occupancy classification, an automatic fire alarm system may be required. Every component should be approved by Underwriter's Laboratory or by the Factory Mutual System or other nationally recognized accrediting and testing organizations.

*3.2.3.4.1. Detectors*

The first essential component in any system is a device which can be used to detect and initiate an alarm. In a manual fire alarm system, this essential "device" is an individual who will recognize the probability of a fire. This is not a trivial point. If a fire is behind closed doors or in a concealed chase or plenum, it may not be readily apparent. Persons in buildings where the likelihood of a fire exists should be familiar with the normal state of affairs and be able to recognize discrepancies. For example, there are fairly common situations in chemistry laboratories and from operations such as welding where visible smoke and fumes may be generated. Certainly, no fire alarm should be sounded in such situations. However, where smoke is observed under unusual circumstances, it should be investigated. When starting to enter a room and the door or doorknob is felt to be warm, it can reasonably be assumed that a fire is burning on the other side of the door. Under suspicious circumstances which cannot readily be clarified, it is usually desirable to activate an alarm.

In a manual fire alarm system, the device used to activate the alarm in most cases is a pull box or pull station. The mechanism is very simple, pulling the switch either makes or breaks an electrical circuit which, in turn, causes an alarm to sound. As will be discussed more fully in the next section, the alarm may sound only in the individual building or it could initiate a signal at a remote location as well.

In most research buildings, vandalism is rarely a problem. If local circumstances are such that this is not the case, there are protective measures which can be taken. The use of a glass cover, to be broken by a small hammer, should be avoided since the person pulling the alarm can be cut by the broken glass. An alternative that has proven very effective is a cover for the pull station which, when removed, sounds a local alarm. Usually, this frightens away a person intending to initiate a false alarm.

In an automatic system, devices are used which depend upon physical phenomena which are uniquely characteristic of fires. These can be classified under four categories: (1) heat, (2) visible products of combustion, (3) invisible products of combustion, and (4) electromagnetic energy output.

Heat is the most obvious choice of a characteristic by which a fire can be automatically recognized. In the section on fire-suppression systems, the fusible links in the sprinkler heads represented one type of heat detector. Alloys can be developed which will have definite melting points. When the temperature at the detector site exceeds the melting point of the alloy, contacts are allowed to move so that the device can either make or break a circuit, just as with a manual alarm system. There are plastics which can perform in the same manner. Fixed temperature systems are very stable and not prone to false alarms, but are slow to respond. There are several other versions of these fixed temperature detectors, including bimetallic strips, where the differential rate of expansion of two different metals causes the strip to flex or bend to either make or break the contact. Others depend upon the expansion of liquids.

One version of a heat-sensing detector which depends upon the properties of materials changing with temperature that has recently had wide application involves cables which can be run in cable trays or conduit, along with data and video cables, to detect excessive heat or fires within the chases. These serve to isolate, within the length of each section of the heat-sensitive cable, where a problem has occurred or, perhaps, even provide warning of an impending problem. Two wires in the cable are normally insulated from each other, but heat causes the insulator to change so that a short develops between the wires. Transmittal of data and interlinkage of computers is such an important part of modern technology, that the use of these cables as part of a heat-sensing system is highly recommended.

Rate-of-rise heat sensors are devices which react more rapidly to rapid changes in temperature than the fixed-temperature units. Most of these units use the thermal expansion of an enclosed volume of air to activate a pair of electrical contacts. They include a small vent to allow some air to leak slowly in and out to avoid false alarms due to slow changes in the ambient temperature or to changes in barometric temperatures. The vent is designed to be too small to accommodate rapid changes in air volume due to a sudden rise in temperature. Some units use thermocouples, where advantage is taken of the property of two dissimilar metals in contact with each other to generate a small electric current if a temperature difference is established between one junction and another, to detect heat. A rapid change in the electric current can be detected and used to initiate an alarm, while slow changes in the ambient temperature can be discriminated against electronically.

Smoke is the obvious visible product of combustion that usually accompanies a fire, although the smoke-generating properties of fires vary greatly. In recent years, the use of local smoke detectors has become very widespread. Smoke detectors include two components, a light source and a light sensor. Because of the latter component, an alternate name is "photoelectric" detector. In most units of this type, the detector is usually shielded from the light source. The detector is activated by the light being scattered into the sensor by the smoke which enters the unit. The location of the unit should be such that there is some air movement in the area which will cause smoke to be carried into the detector. They should not be placed in areas of stagnant air.

A major failing of smoke detectors is that they can be triggered by extraneous light-scattering materials. These can be aerosols, such as generated by spray cans, or vapor from showers, dust from sweeping, maintenance operations such as welding or construction activities, or insects. They can be set off accidentally by cooking, persons smoking near them, or, unfortunately, by individuals deliberately introducing smoke into them. Some of these deficiencies can be avoided by careful location of the detectors or by training, where maintenance personnel are taught to cover the units when working nearby. The design should include a very fine grid over the passages by which smoke enters the unit so that only the smallest insects can enter them, or they can be constructed with dual chambers which will require both chambers to provide a positive indication. Pest strips on the outside can be used to kill any insect crawling into them. Finally, a frequent cleaning (at least twice per year) program will assist in reducing false activations.

The actions recommended above can keep the number of false alarms to a reasonably acceptable level, but deliberately initiated false alarms are a more difficult problem. Fortunately, at corporate facilities and in the academic and service buildings at academic institutions, deliberate alarms are not a major problem, although they are a major problem in dormitories at the latter class of institution. One of the few recourses available is to use verification circuits to test the detector a brief time (30 sec or less) after the initial triggering event. If the device has cleared, the alarm will not be activated; if it has not, an alarm will be sounded.

Ionization detectors are alternatives to smoke detectors. In this type of detector, a very weak radiation source, emitting beta particles, is placed near two plates, one of which is charged positively and the other negatively. As the beta particles pass through the air between the plates, they create ion pairs which are collected by the plates, creating a weak electric current. When ionized combustion products enter the air space between the plates, they neutralize the ionization current and cause it to decrease or cease. The lack of current then triggers an alarm circuit. Some of the same problems, such as cooking or sources of ionized particles from some laboratory operations, can also trigger these units and cause false alarms. Both this type of detector and the smoke detector have good sensitivity.

Sensors can be built which will detect the light from a fire. However, in occupied spaces, it would not be feasible to use the visible light region between about 4000 and 7000 Å. Sensors can be designed to detect energy generated by the fire in the ultraviolet region below 4000 Å and in the infrared region above 7700 Å.

Other sources of ultraviolet and infrared light must be prevented from entering these two types of light sensors. Welding generates ultraviolet light. Lightning reflected from a polished floor has been known to trigger an ultraviolet sensor. For infrared detectors, fairly sophisticated filters are needed to eliminate the background infrared radiation. There are electronic filtering circuits which can be used to help systems employing both of these types of detectors avoid triggering false alarms.

*3.2.3.4.2. Automatic Alarm Systems*

These systems can range from very simple to very sophisticated. In the simplest possible system, an alarm triggered by any of the automatic sensors in the previous section will sound in the building and nowhere else. It will continue alarming until shut off. At the other extreme, the system will not only sound a local alarm and initiate a number of local measures to promote life safety and to confine and extinguish the fire, but will also send a large amount of useful information to a central location where fire fighting resources are available. No matter what level of system is present, the control panel should be located at the most convenient point of entrance for fire fighting personnel so that they can quickly gain as much information as possible from the indicators on the panel. It should not be hidden away in a difficult location such as a locked electrical closet.

A significant improvement in the basic system is for the building to be divided into zones, with a separate module in the control panel for each zone. A zone will normally include a number of detectors within a fairly compact, contiguous area of a building. Firemen reporting to the scene can tell from the panel in which zone the alarm had been initiated. A map of the building indicating the boundaries of the individual zones should be located in the immediate area of the control panel. Although the individual zone where the alarm was initiated would be identified on the panel, there would still have been only a single alarm that sounded throughout the building. If provision has been made to transmit the signal to a remote location, this type of system can directly transmit the zone information to emergency personnel or it can be limited to a single indication of a problem in the building. In some systems, this signal may go instead to a locally manned station, perhaps a security area manned 24 hours per day, which will, in turn, notify a fire safety crew.

The alarm signal may be any number of different devices, such as a bell, horn, or recorded voice message. It should be different from any other similar signal. For example, in an academic institution, there should be no confusion between the fire alarm signal and class bells. In addition, requirements for the handicapped need to be met. In the case of an alarm signal, it is necessary to make provision for individuals with a hearing handicap by means of a visual signal, usually a strobe light.

In addition to initiating an alarm in the building, which may or may not be sent to a central station, the control system frequently will be capable of initiating a number of other actions. The system can activate mechanisms which can close fire doors, turn off ventilation systems, close dampers, send elevators to the ground floor and lock them out of service, and activate emergency lights.

The advent of small but increasingly powerful computers shows promise of revolutionizing automatic fire alarm systems. With a multiplexing system, if enough contact points are available and with the proper interface to the computer, it would be possible to identify the individual detector in alarm at a remote computer console. With enough memory, it would be possible to display a map of the building on the screen, locating the alarm source and if the information had been put into a database, listing the physical conditions or the types and amounts of hazardous materials expected to be present. Under current OSHA and EPA regulations, fire fighting groups must to be notified of hazardous materials they may encounter at facilities in their jurisdiction. In addition, they must be provided with Material Safety Data Sheets for these materials so that they may be informed of the risks to which they may be exposed and the appropriate emergency measures they may need to take.

Presently, the required hazardous material information is primarily being provided to fire departments as printed material. However, generic data bases containing thousands of Material Safety Data Sheets have become available recently on optical disks which provide encyclopedic amounts of data. Small computers have also developed the capability of placing information on these disks so that locally specific information can be added. It should not be too long before companies selling computerized alarm systems will incorporate these most recent advances into their products and be able to provide even small organizations with extremely sophisticated information resources to aid fire fighters to cope with fires in research buildings and other complex facilities.

Often, fire alarm system needs can be accommodated on computer or communication systems installed for other purposes. For example, several companies sell small systems primarily designed for energy management, but which are designed to accept input from other types of systems such as security and fire safety and provide software packages to support them.

The cost of laboratory facilities is so great and the progress in science is so rapid that no competitive corporate or academic research organization can afford the loss of a major facility. Even if insurance is available to cover the physical losses, how can the value of intellectual properties, perhaps irreplaceable, be determined? It takes a minimum of 2 to 3 years to construct a major facility. What would the experimental research personnel, formerly housed in the lost facility, do in the meantime? What would be the position of granting agencies that might be supporting the research? The value of a good fire safety system, including appropriate detectors, an alarm system, and a fire suppression system, is inexpensive compared to the potential losses, including, in addition to those cited, injuries to personnel.

## REFERENCES (SECTIONS 3.2.3.3 TO 3.2.3.4.2)

1. *Low Expansion Foam and Combined Agent Systems,* NFPA 11, National Fire Protection Association, Quincy, MA, 1983.
2. *Medium and High Expansion Foam Systems,* NFPA 11A, National Fire Protection Association, Quincy, MA, 1983.
3. *Carbon Dioxide Extinguishing Systems,* NFPA 12, National Fire Protection Association, Quincy, MA, 1985.
4. *Halon 1301 Fire Extinguishing Systems,* NFPA 12A, National Fire Protection Association, Quincy, MA, 1980.
5. *Water Spray Fixed Systems for Fire Protection,* NFPA 15, National Fire Protection Association, Quincy, MA, 1985.
6. *Sprinkler Systems, Installation of,* NFPA 13, National Fire Protection Association, Quincy, MA, 1985.
7. *Dry Chemical Extinguishing Systems,* NFPA 17, National Fire Protection Association, Quincy, MA, 1985.
8. **Laughlin, J. W., Ed.,** *Private Fire Protection and Detection,* International Fire Training Association, Fire Protection Publ., Oklahoma State University, Stillwater, 1979.
9. Occupational Health and Safety Administration, *General Industry Standards,* 29 CFR, Part 1910, Subpart L, Washington, D.C., §155—165.

### 3.2.4. OTHER FIXED EQUIPMENT

Buildings need to be equipped with two separate elevator systems, one for personnel and one for freight. In order to make spaces available to the handicapped, many buildings without elevators are being retrofitted with them and others are having former freight elevators converted to passenger use. Because it is often too expensive, or there may be no reasonable way to install two separate elevators, many of these elevators are being used for both purposes. This is not desirable, especially in research facilities where hazardous materials are in use. Passenger elevators should not be used to transport hazardous materials, nor should passengers, other than those essential to manage the materials, use a freight elevator at the same time it is being used to move hazardous items between floors.

Both types of elevators should be equipped with a means to signal to a manned location in the event of an emergency, preferably by telephone. In the event of a fire in a building, elevators should automatically be sent to the ground floor, or the floor representing the normal entrance level to the building, and should be constrained to remain there unless fire or other emergency personnel override the interlock.

## 3.3. CHEMICAL STORAGE ROOMS

The OSHA requirements for inside storage rooms for chemicals are given in 29 CFR, Part 1910.106(d)(4), and are based on NFPA Standard 30-1969. The latter standard has been amended since 1969, but at this time, the OSHA requirements have not been changed to reflect the later changes. The BOCA building code is based on the changed standards. The major difference in the more recent NFPA standard involves changes in the ventilation requirements, although there are minor differences elsewhere. In a few instances, additional safety precautions which are generally accepted practice have been added as recommendations.

### 3.3.1. CAPACITY

The amount of flammable and combustible liquids permitted in inside storage rooms depends upon the type of construction and whether an automatic fire suppression system is installed. The permissible amounts under both the OSHA and more recent NFPA standards are the same and are given in Table 3.14.

It should be noted that these limits are generous for most operations, permitting between 300 and 5000 gal of flammable and combustible liquids overall in an inside storage room. However, NFPA 30 currently does not permit more than 660 gal of 1A liquids, 1375 gal of 1B liquids, 2750 gal of 1C liquids, and 4125 gal of Class 2 liquids, so these limits should not individually be exceeded. Excess storage should be avoided, regardless of the legal maximums permitted. The OSHA standard permits other materials to be stored in the same space, provided they create no fire hazard to the flammable and combustible liquids. Materials which react with water must not be stored in the same room.

### 3.3.2. CONSTRUCTION FEATURES

In the earlier material (Sections 3.1.2.1 to 3.1.2.3) on the building code, the factors which govern the classification of the storage space as to the degree of hazard and the implications of the classification were covered in detail. However, there are

**TABLE 3.14**
**Storage in Inside Rooms**

| Fire protection provided | Fire resistance (h) | Maximum size (ft²) | Total allowable quantities (gal/ft² of floor area) |
|---|---|---|---|
| Yes | 2 | 500 | 10 |
| No | 2 | 500 | 4 |
| Yes | 1 | 150 | 5 |
| No | 1 | 150 | 2 |

*Note:*  1 ft² = 0.0929 m². 1 gal = 3.785 l.

minimum construction standards which are specified by OSHA. These basic requirements are

1.    Inside storage rooms must be constructed to meet the fire-resistive ratings for their use. Doors must be approved self-closing fire doors. Windows opening on the room, exposing other parts of the building or other properties, must be protected according to NFPA Standard 80, Standard for Fire Doors and Windows.
2.    The room must be liquid tight where the floor meets the walls. Spilled liquids must be prevented from running into adjacent rooms by one of three methods: (1) by having at least a 4 in. (10.16 cm.) high sill or ramp at the opening, (2) by recessing the floom at least 4 in., or (3) by cutting an open-grated trench in the floor within the room which drains to a safe location.
3.    If the room is to be used for Class 1 liquids, the wiring must be adequate for Class 1, Division 2, hazardous locations. If only Class 2 and Class 3 liquids are to be stored in the room, the wiring need only meet standards for general use.
4.    Shelves, racks, scuffboards, and floor overlays may be made of wood if it is at least 1 in. (nominal) thick.

### 3.3.3. VENTILATION

1.    Every inside storage room must be provided with ventilation, either gravity or mechanical. Either type system must be capable of six complete air changes per hour according to the OSHA standard. Instead of six air changes per hour, NFPA 30 specifies an exhaust ventilation rate of 1 cfm/ft² of floor space (0.305 m³/min), but not less than 150 cfm (45.7 m³/min). For a ceiling height of 10 ft (3.05 m), the two requirements are the same.
2.    If a mechanical system is used to provide the ventilation, OSHA requires that it be controlled by a switch on the outside of the door to the room. The lights in the room are to be operated from the same switch, and if Class 1 liquids are dispensed within the room, a pilot light must be installed adjacent to the operating switch. In order to accommodate hearing handicapped users, a strobe alarm light is recommended as well.

3. If the ventilation is provided by a gravity system, both the intake air inlets and the exhaust air outlet must be on the exterior of the building.
4. Exhaust air should be taken from a point no more than 1 ft (0.3048 m) from the floor and exhausted directly to the roof of the building. The air intake in the room should be on the opposite side of the room from the exhaust. Since, in general, the vapors are heavier than air, this design is intended to sweep the floor clean of vapors before they accumulate and pose a hazard. The aisles of the room should be such as to not block this sweeping action. If ducts are used, they must meet the requirements of NFPA 91, Standard for the Installation of Blower and Exhaust Systems, for Dust, Stock, and Vapor Removal or Conveying, and should not be used for any other purpose. It would be preferable for the air intakes to be from outside of the building and upwind from the most prevalent wind direction.

### 3.3.4. FIRE SAFETY

1. Smoking or open flames must not be allowed in a flammable or combustible material storage room. A prominent FLAMMABLE MATERIAL STORAGE/ NO SMOKING sign should be posted on the outside of the door to the facility.
2. At least one 12B or larger portable fire extinguisher must be located outside the door to a flammable material storage area, no more than 10 ft from the door.
3. Any fire suppression system installed in the storage room must meet the standards of 29 CFR, Parts 1910.155 to 1910.165.
4. There must be at at least one clear aisle at least 3 ft (0.9144 m) wide in every flammable and combustible material storage room. No container should be more than 12 ft (3.66 m) from an aisle. Containers of 30 gal (113.5 l) capacity or larger must not be stacked more than one layer high.

## REFERENCES (SECTIONS 3.3 TO 3.3.4)

1. *Flammable and Combustible Liquids Code,* NFPA 30, National Fire Protection Association, Quincy, MA, 1984.
2. *Prudent Practices for Handling Hazardous Chemicals in Laboratories,* National Academy Press, Washington, D.C., 1981, 218.
3. *Fundamentals Governing the Design and Operation of Local Exhaust Systems,* ANSI Z9.2, American National Standards Institute, New York, 1979.
4. *Standard for the Installation of Blower and Exhaust Systems, for Dust, Stock, and Vapor Removal or Conveying,* NFPA 91, American National Standards Institute, New York, 1973.

## 3.4. MOVABLE EQUIPMENT

Many items of movable equipment represent special safety problems in the laboratory. In some instances, it is the probability of initiating an explosion or fire that is the major concern. In other cases, the major problem may be generation of toxic fumes, and in still others, the equipment is inherently dangerous due to the physical injuries which improper maintenance or use may cause.

Many items of laboratory equipment include electric motors, switches, relays, or other spark-producing devices. In the presence of vapors from flammable materials, a spark can initiate a fire if the concentration of the vapor is between the upper and lower flammable limits and the temperature is above the flash point at which the given vapor can be ignited. Motors used in laboratory equipment should be equipped with induction motors which are nonsparking. Series-wound motors with graphite brushes that are used in many home appliances should not be used in laboratory equipment, and appliances designed for the home such as hot plates, vacuum cleaners, blenders, and power tools should not be brought into laboratories where flammable liquids are actively used. Switches and contacts for electrical controls should be located in flammable vapor-free areas whereever possible. Equipment should be purchased which is designed to minimize the possibility of flammable vapors entering internal spaces where sparks may occur or where the vapors may come into direct contact with heating elements.

Electric shock is another hazard common to many pieces of laboratory equipment. Any electrically powered item of laboratory equipment which is subject to spillage of chemicals or water or which exhibits signs of excessive wear should be used carefully. All equipment should be provided with three-wire power cords (some tools may be double insulated as an acceptable alternative) which should be replaced if the insulation is cracked or frayed. Metallic parts of the equipment should be grounded separately if necessary. Care must be taken to ensure that any ground is, in fact, a good one. An alligator clip on a water pipe is not sufficient. A poor ground connection can generate a high temperature if sufficient current passes through the high-resistance contact, and instead of being a safety feature, actually represent a fire hazard. The potential difference between two poorly grounded pieces of electronic equipment can be enough to damage sensitive electronic components.

Many devices are left on continuously or are left operating unattended for long periods. Any device which could overheat to such a degree that it could result in a fire within the facility should be equipped with redundant controls, heat sensors, or overload protection which will cause the equipment to shut off if excess heat is generated or to fail in such a way as to minimize heat generation. Although much concern about fires stem from the presence of flammable vapors, many fires start in overheated ordinary combustible materials, with flammable and combustible liquids becoming involved at a later stage.

Many items of equipment result in significant amounts of fumes being generated in laboratory operations, many of these typically being used outside of fume hoods. In general, these items of equipment, if used properly, do not cause fumes to be generated at levels exceeding or even approaching the allowable limits of exposure established by OSHA. However, these levels are subject to revision as more information becomes available, usually being lowered, and there are individuals of more than average sensitivity for whom even the original limits may be too high. Under these circumstances, it would appear prudent for laboratory managers to adopt, as an informal policy within their facility, a chemical policy similar to that used to minimize exposure to radiation. This policy, which is intended to achieve exposures as low as reasonably achievable (ALARA), would appear to be a good working policy as well as a good policy for equipment design.

In the following sections, attention is directed to a number of specific items of equipment for which explicit problems may arise.

**TABLE 3.15**

| Chemical | Flash point | | Ignition temperature | | Flammable limits (% by volume in air) | |
|---|---|---|---|---|---|---|
| | °F | (°C) | °F | (°C) | Lower | Upper |
| Acetaldehyde | −36 | (−37.8) | 347 | (175) | 4.0 | 60 |
| Carbon disulfide | −22 | (−30) | 176 | (80) | 1.3 | 50 |
| Diethyl ether | −49 | (−45) | 320 | (160) | 1.9 | 36 |
| Ethylene oxide | ←18 | (<0) | 804 | (429) | 3.6 | 100 |
| Ethyl nitrite | −31 | (−35) | 194 | (90) | 4.0 | 50 |
| Propylene oxide | −35 | (−37) | 840 | (449) | 2.8 | 37 |
| Vinyl ethyl ether | −50 | (−46) | 395 | (202) | 1.7 | 28 |

## 3.4.1. REFRIGERATION EQUIPMENT

Refrigeration units represent a hazard as an item of laboratory equipment for a number of reasons. For example, improper use of laboratory refrigerators for food to be consumed by the laboratory workers is a continuing problem, but which is readily solvable by firm enforcement of policies defining acceptable practices by managers. Under no circumstances should laboratory refrigerators and freezers used for toxic chemicals and pathogenic biological agents ever be permitted to be used for the storage of food. However, the major problem with refrigeration units is the tightly sealed space within them.

The confined space within refrigeration units permits vapors from improperly sealed containers to accumulate. In some instances, the vapors may be toxic, so an individual peering in to find the container desired has an opportunity to breathe in fumes which may substantially exceed acceptable safe levels. Unless the material has a distinctive or offensive odor, it may not even occur to the person using the unit that a problem may exist. It would be desirable to have more than one refrigerator in a laboratory, one of which would be designated and prominently labeled for the storage of dangerous materials only, which would encourage users to be especially careful in using it and to make sure that everything placed in it was properly sealed. Unfortunately, both space and funds are often limited in laboratories. Careful training in how to seal containers placed in refrigerators and freezers and insistence that these procedures be followed should be a part of every laboratory's management program.

Beakers, flasks, and bottles covered with aluminum foil or plastic wrap are unacceptable for storage in a refrigeration unit. Corks and glass stoppers also may not form a good seal. Screw-cap tops with a seal inside are much better, when screwed on firmly. No type of top is foolproof when used in haste.

Fortunately, the problem of confined flammable vapors is one for which there is an engineering solution since the consequences of an accumulation of flammable vapors in a normal refrigeration unit are potentially life threatening. The vapors of flammable liquids may be ignited by sparks (or other heat sources), but some are more easily ignited than others, over a wider range of concentrations. Data for some flammable liquids are given in Table 3.15.

The materials in the above list were selected as examples because they have a wide range of concentrations in which their vapors could be ignited and flash points* well under temperatures found in most household refrigerators and freezers. The reason this is important is that the interiors of household refrigerators contain a number of electrical contacts which could generate sparks to ignite the contained vapors. Among these are the light switch, temperature control, defrost heater (in "frost-free" models), and fan. Many frost-free models also have a drain which could allow the vapors to reach the space occupied by the compressor. Models are available in which all of these sparking devices have either been eliminated, modified to be explosion proof (such as the compressor), or moved to a safer location outside the refrigeration unit.

The confined vapors in a refrigerator or freezer, if ignited, can create a major explosion and fire within a facility. Anyone standing in front of an exploding unit would be in serious danger of losing their life or of being seriously injured. It is highly likely that any flammable liquids not directly involved in the explosion would catch on fire and the fire, in turn, could spread to other stored materials within the laboratory. In an unprotected facility or in an older structure, the final result could be the destruction of an entire building. There could also be other toxic or hazardous materials within the refrigerator or laboratory which could be spread by the incident even if the fire did not spread. In at least one instance, a refrigerator which blew up was in a very active radiochemical laboratory. An entire floor of the building had to be decontaminated.

Refrigerators which have been commercially modified to be safe for the storage of flammable materials are designated as "Flammable Material Storage Units" and meet NFPA Standard 56C and Underwriter's Laboratory standards. These units are not "explosion proof". They have only had components removed which could cause sparks within the interior of the unit. Units intended to be used in hazardous locations, where a spark inside or outside the refrigerator could cause a fire or an explosion, are designated as explosion proof. The electrical power wiring to the latter class must be installed in conformance with "Commercial Refrigerator/Freezer for Hazardous Locations" Class 1, Groups C and D, code requirements. Although in theory it is possible to modify an ordinary refrigerator to be acceptable, in practice it is difficult to be sure that it has been done properly. It is strongly recommended that all laboratory refrigerators and freezer units, with the exception of ultra-low temperature units which operate at temperatures lower than the flash points of any commonly used flammable liquids, be the flammable material-storage type.

The cost of flammable material-storage refrigerators and freezers is usually two to four times higher than comparable products used in the home. As a result, many individuals object to a blanket policy requiring the purchase of these safer units, especially those who are not using the refrigerators in their laboratories to store solvents or other flammable liquids. There are several reasons for overriding these objections. The most important ones are based on the exceptionally long useful life of refrigerators. Only rarely do they last less than 10 years and many continue to work well for more than 20 years. Few research programs endure for comparable periods and few individuals remain in the same position as long. Thus, although assurances

---

* The flash point is the lowest temperature at which a liquid gives off vapor in a high enough concentration to form an ignitable air-vapor mixture above the surface of the liquid.

can be given and signs can be placed on doors forbidding the use of the refrigerator in question for the storage of flammables, there is no feasible means of guaranteeing that they will not be used at some time in the useful life of the refrigerator or freezer for the storage of flammables. Although the initial cost is high, over the total life of a unit, it is an extremely inexpensive price to pay to totally eliminate a major source of fire and explosions.

Ordinary-sized refrigerators and freezers as well as combination units are available from a number of sources. However, as the size of the units becomes larger, only a few suppliers offer explosion-proof units, normally as ordinary refrigerators modified at the factory to be explosion safe at an additional cost.

There are reasons to make a few exceptions. If the use is for a basic departmental function which would never entail the use of flammable liquids and the department is a stable, established discipline or research field, then there is no reason to pay the additional costs involved. Units to be placed permanently in isolated, normally unoccupied locations also might be candidates for ordinary units. Refrigerators to be used only for the storage of food and beverages for the convenience of the employees should also be permitted.

Large walk-in refrigerators and freezers, or cold rooms, pose an additional problem when electrically operated equipment is placed inside them — the problem of condensation of water vapor due to the very high humidity usually present, on the equipment. Care should be taken to avoid shorts and electrical shocks to personnel. All of the equipment should be well grounded and any electrical cords should be insulated with waterproof insulation. A last precaution would be to have all of the electrical sockets in the interior wired with ground-fault interruptors or require that any equipment used inside must be connected through one.

## REFERENCES (SECTION 3.4.1)

1. *Fire Hazard Properties of Flammable Liquids, Gases, and Volatile Solids,* NFPA 325M, National Fire Protection Association, Quincy, MA, 1977.

### 3.4.2. OVENS

Electrically heated ovens are another device found in many laboratories in which the problem of ignition of flammable fumes may exist. They are used for baking or curing materials, out-gassing, removing water from samples, drying glassware, or in some cases providing a controlled, elevated temperature for an experiment. Very few are provided with any provision for preventing any of the materials evaporated from the samples from entering the laboratory. Ovens should be designed so that any fumes generated in the interior do not have an opportunity to come into contact with the heating elements or any spark-producing control components. A single passthrough design in which air is drawn in, heated, and then exhausted is a relatively safe design as long as nothing in the oven impedes the flow of air. Most ovens intended for home kitchens are not constructed in this manner and should not be used in the laboratory environment.

Every oven should be equipped with a back-up thermostat or temperature controller which will either control the unit should the primary one fail or shut the oven down. If the secondary unit permits the oven to continue to operate, it should provide a warning that the failure has occurred so that the researcher can decide whether to continue the operation. In any event, as soon as practicable and before beginning a new run, the oven should be repaired. No unit with only a single thermostat should be used for long, unattended programs.

Because most laboratory ovens exhaust directly to the laboratory, they should not be used to heat any material from which a toxic vapor or gas would be expected to evolve unless provisions are made to exhaust the fumes outdoors, as would be done with a fume hood.

Ovens can be purchased which are designed for heating materials which contain flammable liquids. One commercial model, designed to be used for small amounts of solvents, purges the interior with several complete air changes prior to turning the heat on in order to remove any residual gas which may be present. It also automatically turns off the heat if either the exhaust fan fails or the temperature rises above the maximum temperature for which the unit is designed. The door to the unit has an explosion-venting latch which allows it to blow open in an explosion. However, a recommended feature of ovens used for solvents are explosion vents on the rear of the unit, so that any explosion would be vented away from the laboratory and its occupants.

Where asbestos has been used as insulation in laboratory ovens, some concern has been voiced about the potential exposure of service personnel performing maintenance on the units. Although under normal circumstances such operations will involve minimal contact with the insulation, it is desirable to purchase units which use other insulating materials.

### 3.4.3. HEATING BATHS

Heating baths are used to heat containers partially immersed in them and to maintain them at a stable temperature, on some occasions for extended periods. Heating baths should be equipped, as in the case of ovens, with redundant heat controls or automatic cutouts should the temperature-regulating circuits fail. Often the material used in the bath is flammable, and excessive temperatures could result in a fire.

A number of materials are used in heating baths. Water can be used up to about 180°F (82°C). Mineral oil and glycerine are used up to about 300°F (about 150°C). Paraffin is employed in the range of up to about 400°F (about 200°C). The last three materials are flammable, although the NFPA rating of each is 1, on a scale of 0 to 4. Silicone oils are recommended at temperatures up to about 570°F (about 300°C). These are also moderately flammable and are more expensive than organic oils. In a 1957 tabulation of materials which were in use for heating baths, Egly listed tetracresyl silicate as an expensive material, but one which had very good characteristics. It was listed as nontoxic, noncorrosive, fire resistant, and was suitable for use from near room temperatures to approximately 750°F (about 400°C).

Heating baths should be in durable, nonbreakable containers and set up with a firm support so they will not be likely to tip over. They should not be placed near either flammable and combustible material, including wood and paper which, if exposed to continuing heat over a sufficient period of time, could reach kindling temperatures, or sources of water (particularly deluge showers), which could cause the bath liquid to splatter violently from the container. In most cases, the bath temperatures are high enough to cause severe burns. If it is necessary to move the full container, it should be done while the liquid is cool, again to avoid the risk of burns.

If the container itself does not include a heating element, any immersion heater should be insulated to avoid the potential of electrical shock and should include a cutout device if the temperature exceeds the set point. Alternatively, a second temperature sensor should be placed in the heat bath to act as a circuit breaker to cut off power to the heater if its thermostat fails. The thermostat should always be set well below the flash point of the heating liquid in use. A thermometer placed in the bath at all times it is in use is recommended to provide a visual indication of the actual temperature of the bath.

# REFERENCES (SECTION 3.4.3)

1. *Techniques in Modern Chemistry,* Vol. 3, Part 2, Weissberger, A., Ed., Interscience, New York, 1957, 152.

## 3.4.4. STILLS

Individual stills are frequently set up in laboratories to provide distilled water to the facility and are usually left running unattended for extended periods. The concern here, as with many of the other devices discussed in this section, is the possibility of overheating, with the subsequent initiation of a fire. Some units use two water sources, pretreated water for the boiler, from which various impurities are removed, and ordinary tap water for cooling, although some use tap water for both. The still should be equipped with an automatic cutoff should it overheat because of either the water pressure failing or the boiler becoming dry. Both the water supply and the heat should cut off if the collector bottle becomes full, and if the power fails, a valve should shut the water supplies off.

## 3.4.5. KJELDAHL SYSTEMS

Kjeldahl units and other digestor and distillation units used for nitrogen determinations and trace-element analyses can be sources of potentially unacceptable fume levels if not vented properly. Since larger units are constructed so that several digestions can take place simultaneously, substantial amounts of corrosive fumes can be generated. The fumes in such units are usually drawn through a manifold to a discharge point at one end of the unit. At this point, the fumes are either exhausted by an integral blower or drawn to an aspirator where they are diluted and condensed by the water spray and disposed of into the sanitary system. Because the blower is integral to the unit, the exhaust duct downstream from the unit would be at a positive pressure. If the duct were to corrode, fumes could leak into the surrounding spaces. The total volume of fumes disposed of into the sanitary system would normally be small enough to be well diluted in the sanitary waste stream, so the aspirator method has some advantages.

## 3.4.6. AUTOCLAVES

Pressurized sterilizing chambers or autoclaves are used primarily in the life sciences. Glassware, instruments, gloves, liquids in bottles, biological waste, dressings, and other materials are sterilized in them by steam under pressure, typically at a pressure of a little under two atmospheres, at temperatures of up to 275°F (135°C). Since they are heated pressure vessels, they should be checked periodically to ensure that the seals to the closures are in good condition, and they should be equipped with safety devices to prevent excessive temperatures and pressures. There are a number of potential problems associated with their use. All users should be thoroughly trained in safe techniques and acceptable practices.

Fortunately, most autoclaves are designed so that they cannot be opened while the chamber is under pressure. However, the materials inside will still be very warm, and removing them too hastily or forgetting to wear insulating gloves would be very likely to cause the item being handled to be dropped. In some cases, this would only cause a loss of sterility in the dropped material, but in other cases, a bottle containing a liquid would probably be broken.

Liquids placed inside in sealed bottles may explode, and liquids in ordinary glass bottles instead of pyrex containers which are designed for the temperatures and pressures, may rupture. If the unit is set to exhaust rapidly, as might be done for instrument sterilization, boiling may take place in bottles of liquids, with a consequent loss of liquid into the autoclave. Flammable liquids or chemicals which could become unstable at the temperatures reached in the autoclave should not be run through the sterilizing cycle.

Operating instructions, and a list of good safety practices, should be posted near any autoclave for ready reference.

## REFERENCES (SECTION 3.4.6)

1. **Miller, B. M., Gröschel, D. H. M., Richardson, J. H., Vesley, D., Songer, J. R., Housewright, R. D., and Barkely, W. E., Eds.,** *Laboratory Safety Practices,* American Society for Microbiology, Washington, D.C., 1986.

## 3.5. ANIMAL LABORATORY SPECIAL REQUIREMENTS*

### 3.5.1. FIXED EQUIPMENT IN ANIMAL HOLDING FACILTIES
### 3.5.1.1. Sanitization Equipment — Cage Washers

Most facilities have equipment for sanitization of cages and cage racks. There are three major types of mechanical cage washers:

**Rack washer** — This unit can hold one or more cage racks or racks containing cage boxes. It can have a single entrance or two entrances, allowing movement of cages/racks from a "dirty" processing area to a "clean" area, with the areas separated by a wall. Spray arms in the unit direct water at high pressure on all sides, and the unit reaches the recommended temperature of 180°F for a minimum of 3 min, sufficient to destroy pathogenic (disease-causing) microorganisms. It can handle a larger number of cages/racks than a cabinet washer and is thus less labor intensive.

**Cabinet washer** — The cabinet washer has smaller internal dimensions than the rack washer and can accommodate cages, but not racks. For larger, heavier cages (i.e., rabbit, nonhuman primate, dogs), this unit proves to be more labor intensive and

---

* This material was prepared by Dr. David M. Moore, D.V.M.

more time consuming. Racks would require some other method for sanitization: steam generators, high pressure spray units, or chemical disinfection.

**Tunnel washer** — "Shoebox" rodent cages, water bottles, cage pans, and other small equipment can be placed on this unit's conveyor belt. It is more efficient than the cabinet washer for sanitizing small items.

These units generate quite a bit of heat, and the ventilation for the cage wash area should be adjusted accordingly.

### 3.5.1.2. Sanitization Equipment — Autoclaves

Autoclaves provide support for animal surgical facilities and may be used in barrier facilities to sterilize food, bedding, water and water bottles, cages, and other equipment prior to entry into the barrier. Steam autoclaves can potentially dull sharp surgical instruments, and the heat can reduce vitamin levels in feed. Special autoclavable diets are manufactured with higher levels of vitamins to insure that appropriate levels remain after autoclaving. Ethylene oxide sterilizers are used for materials which might be damaged by the temperatures in steam autoclaves (i.e., surgical instruments, plastic tubing and catheters, electronic devices). However, ethylene oxide gas poses a health risk for humans, and materials should be allowed to "off gas" for 24 to 48 h before coming in contact with animals. Care should be taken in designing a gas scavenging system, and safety standard operating procedures should be established for use of the ethylene oxide sterilizer.

### 3.5.1.3. Incinerator

Most facilities dispose of solid wastes and animal carcasses by incineration. If chemical carcinogen-contaminated materials are to be incinerated, a unit designed to operate at 1800 to 1900°F with a retention time of 2 sec should be utilized.[1]

## REFERENCES (SECTIONS 3.5.1 TO 3.5.1.3)

1. Chemical Carcinogen Hazards in Animal Research Facilities, Office of Biohazard Safety, National Cancer Institute, Bethesda, MD, 1979, 15.

### 3.5.2. EQUIPMENT FOR ANIMAL LABORATORIES AND HOLDING AREAS

Caging for small and large laboratory animal species may be either fixed or movable. Sanitization of fixed caging and the room environment is less easily accomplished than is sanitization of movable racks and cages. Fixed caging might also provide safe haven for vermin and reduces the flexibility of use of that holding room.

Animal caging is designed for the convenience of the investigator and the husbandry staff, but more importantly, for the confort, safety, and well being of the animal. The Guide for the Care and Use of Laboratory Animals[2] provides recommendations for cage materials and sizes for a variety of species. Adherence to these recommendations will also assure compliance with the Federal Animal Welfare Act requirements regarding cage sizes. A description of various caging systems is given by Hessler and Moreland.[1]

# REFERENCES (SECTION 3.5.2)

1. **Hessler, J. R. and Moreland, A. F.,** Design and Management of animal facilities, in *Laboratory Animal Medicine,* Fox, J. G. et al., Eds., Academic Press, Orlando, FL, 1984, 517.
2. The Guide for the Care and Use of Laboratory Animals, National Institutes of Health Publ. No. 85-23, Bethesda, MD, 1985.

# Laboratory Operations 4

## 4.0. GENERAL CONSIDERATIONS

The attitude of laboratory personnel toward safety is the most important factor affecting the safe conduct of research. It is more important than the quality of the equipment, regulations, managerial policies, the inherent risks associated with the materials being employed, and the operations being conducted. If the attitude of everyone in the laboratory is positive, and this attitude is clearly supported by either the corporation or the academic institution, then it is highly probable that a strong effort will be made for the research program to be conducted safely because conscientious individuals will usually try to ensure for themselves that their operations are as safe as possible and attempt to comply with regulations and policies which have been established for their protection. On the other hand, no matter how strict management policies are and how many regulations have been established, individuals with an attitude that safety concerns are not important and that nothing will ever happen to them will manage, somehow, to circumvent any inconvenient restrictions.

Rarely do you have as black and white a situation as implied by the two extremes in the preceding paragraph. No one is so careful that they avoid taking any risks, nor is anyone totally unconcerned about their own safety. The goal should be to avoid taking unreasonable risks. It is the responsibility of laboratory managers to establish, by policy and example, reasonable standards of conduct to ensure that this goal is met.

A safe laboratory operation is usually a well-run operation. For example, labeling of secondary containers of reagents is not only a good safety practice to avoid accidental reactions leading to injuries, but it also serves to prevent errors which could negatively affect the research program.

The failure of a laboratory manager to establish the right atmosphere toward safety and to enforce established safety and health policies can render the manager vulnerable to litigation on the part of an injured employee, especially if it can be shown that the failure was due to willful negligence. Even if written policies have not been established, if a reasonable individual can be shown to have been likely to have anticipated a problem and due care to protect an employee under his supervision was not exercised, a civil court suit against the manager by the injured party could well be successful. On the other hand, an employee, at least in an academic institution, who deliberately does not comply with safety precautions of which he has been

201

informed, and which are normally expected to be followed, may weaken his case due to contributory negligence to the extent that his suit would not succeed or the award would be substantially diminished. In the corporate world, there are workman's compensation laws which usually provide for compensation to an injured employee regardless of who is at fault, although there are differences in coverage depending upon many factors in the different states. The whole concept of liability is constantly being modified by court actions. However, for financial as well as ethical reasons, the prudent manager or employer should be sure that the research programs for which he is responsible are conducted according to good safety practices, as defined by laws and regulations, corporate and institutional policies, and reasonableness.

It is symbolic of our society that this chapter, intended to provide guidelines to assist in making laboratory operations safer, should start with such a strong legal tone. Formal safety standards have had to be established because of concerns for the rights of individuals and society, and because of abuses by the very small minority that may place results ahead of the well being of the persons involved. Individuals are no longer willing to take what they believe to be excessive risks on behalf of their employer. Many are willing to go to court to protect themselves, to the extent that this prerogative is at risk of being abused. However, even without the need for laws and regulations, such a chapter in a book on laboratory safety would still be needed to provide guidelines to research personnel on how to avoid or minimize the risks associated with the conduct of research.

Much has been made of the professional expertise, experience, and judgment of scientists which should allow them to be the best judge of the safety program needed in their research. In chemistry laboratories in the academic world, however, where competent, enlightened scientists should be found, it has been estimated that the accident rate is 10 to 50 times higher than that in industrial laboratories. The broad range in the estimate is attributed to the reluctance of academic personnel, particularly students, to report accidents. The disparity between the two situations is partially explained by the greater likelihood in industry that scientists will be required to do a careful hazard analysis and follow strict safety precautions. In addition, the expertise cited is often confined to the scientific object of the research program. Very few scientists have taken formal courses in safety, health, and toxicology. Most of the relevant safety articles are published in journals devoted to topics outside of their major field of interest. They are likely to have no better judgment or common sense, on average, than any comparably well-educated and intelligent group. They may, in fact, because of the intensity of their interest in a very narrow field, have only a limited awareness of information extraneous to those interests which would assist them in making safety decisions. In the academic area, many profess to be concerned that academic freedom could be abridged by rules imposed from the outside. Academic freedom, however, should not be confused with issues which govern the health and safety of individuals and the environment, which transcend this desirable concept.

There are legitimate concerns that research laboratories may become overregulated by too specific a set of rules since they do not fit the standard mold for which the original OSHA and other regulatory standards were designed. Instead of working with a few chemicals, a single laboratory may work with hundreds over the course of time, often for limited periods. There may be extremely limited safety and health information (or none at all for newly synthesized substances) for many of the materials with which a scientific investigator may work. In general, research laboratory safety and health policies should not be regulated on a chemical-by-chemical basis except for specific known serious risks, but this does not mean that otherwise there should be no safety rules. Health and safety programs should be based on well-defined general policies, sufficiently broad in scope, which are conservatively designed to encompass any *reasonable* hazard to laboratory personnel. They should be administered uniformly as institutional or corporate policies, tempered by local circumstances, to assure that all laboratory workers, including students, are equitably treated.

## 4.1. OPERATIONAL PLANNING

A typical research proposal goes into great detail on the significance of the proposed research, the approach which is to be taken, and the results which are sought. The proposal always provides a thorough justification for the technical manpower and equipment resources needed to carry out the planned program. Occasionally, mention will be made of some of the hazards which will be encountered, if these are sufficiently dangerous or unusual, and the means by which they will be controlled. Except in these relatively infrequent instances, and in a few others where the research involves very stringently regulated materials, the reviewer of the proposal often must take on faith that a basic infrastructure has been established to ensure that the research can be carried out safely, in compliance with contemporary regulatory standards. This infrastructure does not just happen; it requires careful planning. It is the intent of this chapter to provide essential information to guide planning for safe operations in the laboratory.

If the basic design of the facility is satisfactory, the first order of priority, when initiating a research program, is to order all of the essential items of equipment which will be needed, if they are not already available. Orders for major items of equipment frequently take extended periods of time to be processed and delivered, 9 to 10 weeks being as short an interval as might reasonably be expected. If installation is required, such as in the installation of an additional hood, this period could be extended for months since the work will have to be carefully planned to ensure, among other things, that the air handling system has sufficient capacity and that fire code requirements can be met, especially if the ductwork must penetrate multiple floors. Scheduling and pricing of the actual work cannot be done in such instances without working plans. This delay is critical when the research is scheduled to be completed within a fixed contract period, with annual renewals depending upon progress, as are the majority of academic research contracts.

If new employees need to be hired, a number of factors need to be considered in addition to technical skills. As noted earlier, attitude is extremely important. A research laboratory is not the place for a casual attitude toward safety. One of the most important considerations should be personality. It is critical in any group effort for the personnel to be able to work together. It is not necessary to be "popular", but it is important for individuals to be receptive to the ideas of others and tolerant of differences in points of view. A group of persons working under the stress of strained relationships are likely to be an unproductive and unsafe group. A principal investigator needs to establish a clear line of authority for the laboratory personnel, both for day-to-day operations and for emergencies. These may not be the same. The individual trained to manage the scientific aspects of the research may not have as appropriate a background to handle an emergency situation as would a senior technician. The latter might have received special training in safety areas such as chemical spill control or emergency first aid. Where there is the possibility of ambiguity, responsibility for various duties, especially those associated with safety, needs to be clearly assigned. It would be well, for example, to designate a relatively senior person as a local laboratory safety officer and, if necessary, provide access to additional safety training to that individual. This individual could be responsible for such items as the safety orientation of new employees and the safety training of all employees when new materials or procedures are incorporated into the laboratory operations. The person might be asked to perform a hazard analysis of any new laboratory operation and secure any authorizations or clearances which might be needed. It might be this individual's duty to assign other persons responsibility for ensuring that chemicals are shelved according to compatibility, to maintain safety items such as first aid kit supplies, personal protective equipment, spill kit materials, and material safety data sheets, or to maintain equipment in safe condition. The safety officer and/or the laboratory supervisor needs to act as liaison with the safety department to provide access to any new information which might affect the laboratory's operations. A knowledgeable employee needs to be designated as the person responsible for the safe disposal of hazardous materials. This individual needs to be responsible for seeing that all surplus and waste materials are properly identified, and segregated if waste materials are combined into common containers.

An emergency plan needs to be developed for each laboratory which is consistent with and integrated into the plan for the entire building and that of the corporation or institution. It needs to take into account procedures for temporarily interrupting the research operations or for automating uninterruptible operations to allow employee evacuation during an emergency. This plan should be reviewed periodically to ensure that it is still appropriate. As has been noted, research programs, especially in academic institutions, tend to change rapidly, not only in the materials which are in use and the operations being conducted, but also in the participating personnel.

Every aspect of the laboratory operations should be evaluated to see if it could be made more efficient and safer. Purchasing of reagents, for example, should be reviewed to see how much is actually needed on hand at a given time. If all chemicals are ordered early in the program, program needs may shift and a substantial investment in surplus chemicals could result. Today, where disposal of waste chemicals has become such a major legal issue, the cost of disposing of surplus chemicals often will exceed their original costs. The quality of partial containers of chemicals may become dubious and, again, an investment in excess chemicals will represent a drain on available funds. Anticipation of needs is critical, especially where equipment is involved. As noted earlier, delivery of essential items of equipment may be delayed for extended periods. In such situations, the temptation is to make-do with equipment not specifically designed to meet the actual needs, with serious safety implications being involved on occasion.

The regulations, and the information on which they are based, change so frequently that it is unreasonable to expect every purchaser to be able to keep up with the current regulations. Further, the entire body of relevant information regarding laboratory safety has become so extensive and so complex that, again, it is unlikely that a single individual can be sufficiently knowledgeable to adequately consider every factor. For example, the review of the purchase of a fume hood is usually not based so much on the characteristics of the hood as it is on the installation. Has the location been reviewed for availability of sufficient make-up air? Has the path of the exhaust duct been selected and the exhaust blower sized appropriately? Will fire separations have to be penetrated? Flagging the purchase will permit these questions to be answered, and if they have not been considered, make sure that they are before the order is processed. It is highly likely that the order will have to be modified if these factors have not been addressed, and it is highly desirable that specifications be changed prior to ordering unsuitable equipment.

An evaluation of the potential exposures of individuals to hazardous materials should be made as soon as possible. It may be necessary to consider selectively placing individuals in work assignments, although one has to be very careful in such cases to avoid triggering charges of discrimination. Still, if there are known risks, for example, of teratogenic effects from a chemical, it would certainly be surprising if an expectant mother did not have some concerns about working in an area where it was in use, even if the levels were well below the acceptable OSHA limits for the average worker. Any work regimen would need to be fully discussed between the individual and the supervisor in such a case, based on knowledge, not speculation. Often, once the exposure potential or lack of one is clearly understood, concerns may disappear. Failure to consider the employee's rights to a working environment free of recognized hazards could lead to a complaint to OSHA or other regulatory agency which could, in extreme cases, cause the program to be interrupted, pending resolution of the safety issues.

Prior planning is needed, especially in facilities where students are expected to be working. Legal safety standards usually have been designed for employees, and although many graduate students and undergraduate students on wages or work-study programs receive stipends for their efforts, they are usually not considered or treated as employees. They typically have less experience and a different purpose in being in the laboratory than do permanent personnel. The pressures associated with completing the various hurdles of a degree program, especially those accompanying the completion of a research program within a tight schedule, often lead to students working long hours, going without enough sleep, and eating odd diets. The result may be their working without adequate supervision under conditions which could cause impaired judgment. The laboratory safety program should take these differences into account and make a special effort to see that these younger persons understand the goals of the safety program, relative to laboratory operations, and the need to comply with the safety procedures.

### 4.1.1. QUANTITIES

The recommendation that the volumes of reagents on hand be kept to the minimum needed for a reasonably short working period is found in virtually every safety manual. However, a visit to almost any laboratory will reveal many bottles and other types of containers which have accumulated substantial layers of dust. Many of the more recently acquired reagents will very likely be duplicates of these older materials. There must be good reasons for this apparently needless duplication.

It would appear to make a great deal of sense to order what you need and replenish the supply when it appears likely that more will be needed. There are two major reasons why this common-sense approach is so rarely followed, both of which are attributable to purchasing considerations.

1.    It takes time to process an order. Unless a central stores facility maintains a stock of chemicals at the research facility, the processing of a requisition, receipt of an order by the vendor, and delivery is unlikely to take less than a month, unless an alternative buying process has been established, such as a blanket order system or precleared requisitions for low-value purchases. Under these circumstances, a purchaser tends to buy more than is currently needed to avoid having to order frequently and to avoid delays in receipt of the needed material.
2.    Chemical costs decrease rapidly with the size of the container. For example, for one grade of sulfuric acid, the following pricing schedule has been established by one major vendor (note that these have been normalized to set the price per liter of the smallest size equal to 1).

| Container size | Cost/l |
|---|---|
| 1 l, each | $1.000 |
| 6 × 1 l, case | 0.558 |
| 4 l, each | 0.526 |
| 4 × 4 l, case | 0.359 |
| 10 l | 0.303 |
| 20 l | 0.225 |

Obviously, if the volume of usage justifies the purchase, the largest size is the most economical to buy. However, there are several reasons why such a purchase is probably unwise. It increases the potential risk to have more of such a corrosive material than is actually needed, and storage space will have to be found for the excess material. If it is not used relatively quickly, the quality may become suspect so that the user will be reluctant to employ it in the research program. The cost of disposal of any eventual surplus material is likely to wipe out any initial economic gain from buying in volume.

In addition to the two reasons given above, sometimes a researcher wants to be sure of the consistency of the reagent, so he buys enough for his needs from one lot. However, some chemical firms will, upon request, set aside an amount of a given lot and maintain it at their regional warehouse to accommodate a large user.

An examination of the purchases of research reagents by most university or corporate research facilities will probably reveal that relatively few of them are bought in substantial quantities. At the author's institution, fewer than 75 of the more than 1200 different chemicals purchased during a typical year exceeded 50 kg. Where this is true, it would appear desirable to set up a central stores, at least for these frequently used chemicals. Stocking of these materials should probably emphasize the middle ranges of sizes. If, in the example given above, case lots of 4-l containers were the primary sizes purchased from the vendor for stocking, most of the cost savings of volume purchases could be passed on to the local purchaser. It is likely that purchaser wastage of larger sizes would make up for the remaining cost differential. At most institutions, there are some high-volume users who could economically use the larger sizes. It might be desirable to restrict the purchases of larger sizes to those who can establish a need or for those items for which it is feasible to disburse chemicals from drums into smaller containers by stores workers.

Except for the high-volume materials, most remaining chemicals are bought in relatively small quantities to meet specific needs of individual programs. Some chemicals pose unusual hazards and it is essential to keep track of which group is ordering them and where they are to be found. A central stores area would make a convenient distribution center for these special materials and would facilitate maintenance of records of their use.

In mid-1987, the Community Right-To-Know standard became law, requiring users of hazardous materials to inform nearby communities when they had significant holdings of any of several hundred hazardous chemicals. The definition of significant holdings ranges from 1 lb (0.454 kg) to 10,000 lb (4539 kg), depending upon the chemical. Where the amount exceeds another, usually larger threshold, the law requires that emergency planning programs be established. It is also required to report within 60 days any time these two levels are exceeded. Clearly, it is desirable to maintain in storage amounts less than the trip-point levels. The need to comply with this standard virtually mandates establishing a system to keep records of the various materials purchased, as a tentative running inventory. A major problem in maintaining these records is identification of the various materials. For the 406 chemicals on the initial EPA list of hazardous chemicals (currently reduced to 366) covered by the Right-To-Know standard, there are more than 5400 different synonyms. Without a sophisticated computer data entry and analysis system, it is almost an impossibly time consuming job to check each chemical purchase to see if a given material is on the list. One way to facilitate the task would be to require the CAS registry number to be placed on the initial requisition for each item ordered. This would eliminate the problem of synonyms or trade names.

In summary, it is desirable to order and maintain in stock as small amounts of chemicals as is practicable in order to (1) minimize the risks in the event of an incident, (2) reduce the overall expense by reducing the amount requiring disposal as hazardous waste, and (3) minimize the problem of meeting the limits imposed by the Community Right-To-Know standard. However, in order to encourage a laboratory manager to buy and stock smaller containers, purchasing procedures need to be established to conveniently provide smaller sizes at a reasonable cost.

### 4.1.2. SOURCES

One of the more difficult tasks associated with the purchase of equipment and materials meeting acceptable safety standards is to do so in a system which requires acceptance of the low bid. Many of the safety standards or guidelines are minimal standards, which it is often desirable to exceed. Usually, chemicals from any major company or distributor will be acceptable, but the same is not necessarily true of equipment. In order to obtain the quality desired, purchase specifications must be carefully written to include significant differences which will eliminate marginally acceptable items. In some cases, it is virtually impossible to write such a specification, and it is necessary to include a performance criterion. This often requires considerable effort on the part of the purchaser. For example, chemical splash goggles are sold by many companies at prices which differ by an order of magnitude or more. All of these will usually meet ANSI standard Z-87 for protective eye wear, but many are so uncomfortable or fog up so rapidly that they will not be worn by laboratory workers. Thorough comparative testing under actual laboratory conditions will identify a handful of the available models which offer superior performance. With documented data, it is usually possible to obtain permission from the purchasing department to limit purchases to sources meeting acceptably high safety *and* performance criteria, rather than minimal standards. This not only applies to smaller items, but to major ones, such as fume hoods, as well. Where there is a significant difference in quality which will enhance the performance and/or the safety of any unit at a reasonable price, a cooperative effort should be made by the purchaser, the purchasing department, and the safety department to obtain needed items from these sources.

### 4.1.3. MATERIAL SAFETY DATA SHEETS

The federal government enacted a Hazard Communication Standard in 1984. Chemical manufacturers, importers, and distributors were required to comply with the standard by November 25, 1985, and affected employers, by May 25, 1986. Originally, the standard applied only to Standard Industrial Code Classifications 20 through 39. As a result of litigation, OSHA extended the coverage to a much broader range of employees. After September 23, 1987, it required that material safety data sheets (MSDS) be provided to nonmanufacturing employees and distributors with the next shipment of chemicals to these groups. After May 23, 1988, all employers in the nonmanufacturing sector had to comply with all provisions of the standard. However, prior to this decision, many states had enacted similar standards which extended the coverage within their own jurisdiction. Some specifically extended coverage to public employees, which included individuals at public universities and colleges.

Under the standard, chemical manufacturers and importers must obtain or develop a MSDS for each hazardous chemical they produce or import. These MSDSs must reflect the latest scientific data. New information must be added to the MSDS within 3 months after it has become available. The manufacturer or importer must provide an MSDS to a purchaser the first time a given item is purchased, and an updated version must be provided after the information becomes available. Distributors of chemicals must provide MSDSs to their customers.

The MSDS can be in different formats as long as the essential information is included, although a standard format eventually may be adopted. The minimal information to be provided, which must be in English, is

1.  The identity of the chemical, as used on the label of the container.
    a.  For a single substance, the chemical name and other common names.
    b.  Mixtures tested as a whole: the chemical and common names of all ingredients which contribute to known hazards and common names of the mixture itself.
    c.  Mixtures untested as a whole: chemical and common names of all ingredients which are health hazards and which are in concentrations of 1% or more, or carcinogens in concentrations of 0.1% or more. Carcinogens are defined as those identified as such in the latest editions of (1)

the National Toxicology Program (NTP) *Annual Report on Carcinogens,* (2) International Agency for Research on Cancer (IARC) Monographs, or (3) 29 CFR 1910, Subpart Z, Toxic and Hazardous Substances, OSHA.

If any of the ingredients which do not exceed the concentration limits in the previous paragraph could be released from the mixture such that they could exceed an established OSHA permissible exposure limit or an American Conference of Governmental Industrial Hygienists (ACGIH) threshold level value, or could represent an occupational health hazard, their chemical and common names must be given as well. The same information is also required for any ingredient in the mixture which poses a physical hazard (as opposed to a health hazard).

2. Physical and chemical characteristics of the hazardous chemicals.
3. Physical hazards of the hazardous chemical, specifically including the potential for fire, explosion, and reactivity.
4. Known acute and chronic health effects and related health information.* This information is to include signs and symptoms of exposure and any medical conditions which are generally recognized as being aggravated by exposure to the chemical.
5. Primary routes of entry into the body.
6. Exposure limits data.
7. If the hazardous material is considered a carcinogen by OSHA, IARC or NTP (see 1.c above).
8. Precautions for safe handling, including protective measures during repair and maintenance of apparatus employed in using the equipment and procedures for clean-up of spills and leaks.
9. Relevant engineering controls, work practices, or personal protective equipment.
10. Emergency and first aid procedures.
11. Date of MSDS preparation or latest revision.
12. Name, address, and telephone number of the entity responsible for preparing and distributing the MSDS.

Although this list appears to be straightforward, the MSDSs provided by different companies vary significantly in quality. Many are very incomplete, not necessarily always due to lack of information.

Provision of an MSDS at the time of the initial purchase of a chemical is the responsibility of the chemical vendor, and if the vendor fails to provide it, it is the responsibility of the purchaser to take the necessary steps to require the vendor to do so. A typical MSDS can be up to several pages long, and a comprehensive file of hundreds of these, which might be required in a typical laboratory, will be bulky and difficult to maintain.

Both the distributor and the purchaser of a chemical have a major problem in complying with the requirement that an MSDS be provided to the user where purchasing authority is widely distributed, as it often is on a university campus. Many

---

* It is possible that future usage will substitute "immediate" for "acute" and "delayed" for "chronic".

institutions permit direct delivery to the actual location ordering a given material, while, in others, there is a central receiving point. In the former situation, a chemical vendor may supply an MSDS to the first institutional purchaser of a chemical, but subsequent purchasers may not receive one, even though they did not receive a copy of the first one sent to the initial purchaser. Where all the separate purchasers of a chemical are part of the same institution and located within the contiguous confines of a single site, it is probable that the vendor technically can meet the legal requirement of furnishing an MSDS to the institution as an entity by providing a single MSDS to the individual laboratory first ordering a substance. In a large research institution, this would result in a very incomplete distribution of MSDSs. Designation of a single department, such as the safety department, to receive all MSDSs from the chemical vendors and to establish a master file of them, perhaps with some partial or complete duplicate files at other locations, will partially alleviate the problem. These files would need to be in places that are easily accessible to the users for a large portion of the day in order to approximate compliance with the requirement of being readily available to the employees. Unless the information as to which unit actually ordered the material accompanies the MSDS, it would be impossible to distribute them further internally unless an individual department requested a specific MSDS which it wished to maintain in its local file. However, unless each department received a notice of the receipt of any revised MSDS and took the initiative to upgrade its own files, the local files would soon become obsolete. This could lead to possible liability problems if an employee assumed that the local files were current.

Some firms have avoided the entire problem, as far as they are concerned, by sending an entire set of MSDSs for all of their products to corporations or institutions with whom they do a substantial business. It is then up to the university or corporation to decide how to distribute them properly so as to comply with the regulatory requirement that any needed MSDS be readily available to employees.

If all chemicals are delivered to a central receiving location, a fairly straightforward, but potentially expensive, solution to the problem exists. A master file of all MSDSs can be maintained at the central receiving location, as well as a list of all departments or other definable administrative units which have previously ordered each chemical. If a department is not on the latter list for a given chemical, then a copy of the MSDS can be made and sent along with the material when it is delivered. A revised MSDS would be sent to every department listed as having the specific chemical in their possession. Although this sounds relatively easy, the amount of record maintenance required and the time spent in checking the files would be substantial. For a major research institution, the amount and variety of materials ordered, coupled with the large number of independent administrative units, would probably mandate a full-time-equivalent employee for the program.

As noted, the quality of MSDSs from different firms varies and it is cumbersome to try to keep separate ones for every individual company. An alternative would be for a commercial organization to develop generic MSDSs for every hazardous chemical used in any significant volume in research and sell these to be used in lieu of the ones provided by a chemical vendor. Any MSDS needed at a given location could be copied from the master file. The problem of getting revised MSDSs to individual laboratories would still exist, but it would be no worse than with the ones provided by the chemical manufacturers, importers, or distributors. There are, however, serious liability questions for the firm doing this. A company wishing to sell a chemical product is compelled to prepare an MSDS, and the cost is absorbed as part of its business expenses. However, a firm selling generic MSDSs would have to recover its costs entirely from sales of the product, and these costs would have to cover the cost of any liability insurance needed to pay for any awards for injuries claimed to be caused by erroneous information in an MSDS, as well as the production expenses. There are compilations of MSDSs for sale for a limited number of chemicals, and some firms have established MSDS databases to which access can be had for a fee. At least one firm has made available a microfiche copy of virtually every MSDS developed by any commercial vendor. As photographic copies, any data errors would still be the responsibility of the chemical vendor, not the seller of the microfiche.

Recently, one firm has made available over 10,000 MSDSs on an optical disk, which can be processed by microcomputers. Much of the cost of these generic MSDSs is due to the constant effort needed to keep them up to date, as required by the standard, and to cover the cost of liability insurance. If an organization has the ability to transmit data electronically to all of its individual locations, a computerized database of all MSDSs used by the organization is the best approach to take. The database is the legal file and it is equally up to date for everyone. If someone wishes to print a copy of an individual MSDS, they may do so, but it must be with the understanding that a revision in the database supercedes it. The computer system at most research institutions is normally intended to be available almost 100% of the time. Any alternative procedure based on receiving and distributing printed copies of MSDSs to everyone will be very difficult to maintain properly in a research-oriented university and in most large corporate research laboratory facilities.

## 4.2. PURCHASE OF REGULATED ITEMS

There are a number of classes of items for which purchases must be carefully monitored for compliance with safety and security regulations. Several of these can be purchased only if a license is held by the individual or by the corporation or institution. There also are many restrictions on the transportation of hazardous materials. Usually, the purchaser will expect the vendor to be responsible for meeting these shipping requirements. However, there will be occasions when the institution or corporation will initiate a shipment. It is recommended that a subscription to a hazardous material transportation regulatory advisory service be taken out by anyone who ships any hazardous material frequently, due to the relatively rapid changes in shipping regulations.

### 4.2.1. RADIOISOTOPES

With certain exceptions, the purchase of radioactive materials is generally restricted to those persons who are licensed to own and use the materials under 10 CFR, 30 or 33. In this context, the word "person" is used quite broadly. In Part 30, which provides the rules for domestic licensing of byproduct material, "person" is defined as "any individual, corporation, partnership, firm, association, trust, public or private institution, group, Government agency other than the Commission or Department..., any State, any foreign government or nation or any political subdivision of any such

**TABLE 4.1**

**Exempt Quantities of Some of the Most Often Used Radioisotopes**

| Isotope | Quantity (μCi) | Isotope | Quantity (μCi) |
|---|---|---|---|
| Calcium-45 | 10 | Iodine-131 | 1 |
| Carbon-14 | 100 | Iron-59 | 10 |
| Cesium-137 | 10 | Mercury-203 | 10 |
| Cobalt-60 | 1 | Molybdenum-99 | 100 |
| Chromium-51 | 1000 | Nickel-63 | 10 |
| Hydrogen-3 | 1000 | Phosphorus-32 | 10 |
| Iodine-125 | 1 | Sulfur-35 | 100 |

government or nation, or other entity; and any legal successor, representative, agent or agency of the foregoing." Clearly, virtually any assemblage of persons can be licensed to own and use radioactive byproduct materials if they can fulfill the licensing conditions provided by Part 30 and have an approved radiation management program meeting the standards of Part 20. In approximately half of the states, the oversight function to ensure compliance with the standard is done by the state rather than the NRC. These are "agreement states".

There are a few more definitions which will be useful. The federal regulations in Part 30 usually apply only to "byproduct material". This refers to "radioactive materials, other than special nuclear material, yielded in or made radioactive by exposure to the radiation incident to the process of producing or utilizing special nuclear material." The NRC definition of special nuclear material is lengthy, but essentially it means plutonium or uranium enriched in the fissionable isotopes U-233 or U-235. There are naturally occurring radioactive materials which are mostly unregulated. There also are radioactive materials made radioactive by using accelerators. The latter materials are regulated by the states independently and not by the NRC. Exposure to some naturally radioactive materials, such as radon, is regulated under some circumstances.

A number of classes of radioactive materials do not require a license. If the amount is less than the exempt quantity for a given material, as listed in Paragraph 30.71, Schedule B, of the regulations, a license is not required. The amounts meeting these criteria are given in Table 4.1 for a few of the radioisotopes most commonly used in research. The units are in microcuries (μCi), with 1 μCi equal to 37,000 disintegrations per second, since this is the way they appear in the regulations. A set of units different from these has been recommended by the International Commission on Radiological Protection. In the international system of units (SI units), the unit of activity is the becquerel (Bq) and is equal to 1 disintegration per second. A microcurie, therefore, equals 37,000 Bq.

There are a number of other classes, described in paragraphs 30.15 to 30.20 of 10 CFR, where the persons purchasing certain items containing radioactive materials are exempt from having a license, although the original manufacturer must have had a specific license to allow production of the units. Among these are self-luminous devices and gas and aerosol detectors.

The amounts in Table 4.1 are very small and are usually exceeded in most research applications. For practical research using radioactive materials, it is necessary to obtain a license. A discussion of this will be deferred to Chapter 6. However, assuming that a license has been obtained and a radiation safety program has been established satisfying the NRC (or its equivalent in an agreement state. Henceforth, when the NRC is mentioned, it will be understood to include this addendum), there are still formal steps to go through in purchasing and receiving radioactive materials.

In a research institution, it is common practice to establish a license to cover all users of radiation at the institution. This is called a broad license and provides limits on the total amount of each isotope that can be in the possession of the licensee at any specific time. These limits are normally chosen by the institution and approved by the NRC. If there are several separate users, the sum of all their holdings for each isotope, including unused material, material in use, and material as waste, must not exceed these limits. Since each individual user cannot keep track of the holdings of other independent users, it is essential that all purchase orders as well as all waste materials be cleared through a radiation safety specialist whose responsibility (among many others) is to ensure that the license limits are not violated. Adherence to this and all other radiation safety regulations is essential. At one time, the primary threat in the event of a violation was the possible suspension of a license. This was such a severe penalty that it was invoked very infrequently. In recent years, substantial fines have been levied against universities and other users who violated the regulations and terms of their licenses. On March 12, 1987, a city attorney filed 179 *criminal* charges against a major university within the city's jurisdiction and ten individual members of its faculty for violations of the state standards. No matter how this case is resolved, it will have established a major precedent as a conceptual possibility. Even more recently, another university reached a settlement with the surrounding community to conduct a $1,300,000 study of the possible dispersion of radioactive materials into the community, in addition to a substantial fine, because of their management of the use of radioactive materials.

Without a valid copy of the license of a person ordering radioisotopes, vendors are not allowed to fill an order. Since the radiation safety specialist is such a key person in the process, this person should be responsible for providing current copies of licenses, including any amendments, to prospective suppliers of radioactive materials. At many facilities, the radiation safety specialist has been assigned virtually all responsibility for the ordering and receipt of radioactive materials. Section 20.205 requires each licensee to establish safe procedures for the receipt and opening of radioactive packages. Although mistakes are rare, they do happen, so it is highly desirable that the radiation safety specialist directly receive each package of radioactive materials, verify that the paperwork is correct, and check the external radiation levels and containers for damage. There have been instances when all of the paperwork conformed to the expected material, but the wrong material or the wrong amounts of the ordered material were shipped. Where it is impossible for the radiation specialist to receive all packages, provision needs to be made for the temporary secure storage of packages until they can be checked.

Many radioactive materials are used in the form of labeled compounds, often prepared specifically to order. In some of these, the half-life of the isotope used in the compound is short, so procedures need to be established to ensure prompt handling and delivery to the user. In other cases, the compound itself will deteriorate at ordinary temperatures. These materials are usually shipped in dry ice and must be delivered immediately upon receipt or stored temporarily in a freezer until delivery. If it is necessary to ship radioactive material, the material must be packaged according to 49 CFR 173. Again, the radiation safety specialist is the individual who normally would be expected to be familiar with all current standards affecting shipment and to arrange for transportation according to the regulations.

### 4.2.2. CONTROLLED SUBSTANCES (DRUGS)

The purchase, storage, and use of many narcotic, hallucinogenic, stimulant, or depressive drugs are regulated under 21 CFR, 1300 to the end. In addition, these substances are usually regulated by state law, which in many cases is much more stringent than federal law. The substances covered by the Controlled Substances Act are divided into five schedules. Schedule I substances, which have no accepted medical use in the U.S. and a high potential for abuse, are the most tightly controlled, while Schedule V substances contain limited quantities of some narcotics with limited risk. In this case, the federal Drug Enforcement Agency (DEA) does not permit a broad agency license, but, rather, requires a single responsible individual in each functionally independent facility to obtain a separate license which spells out which schedules of controlled substances the facility may possess. This individual can permit others to use the controlled substance under his direction or issue it to specific persons for whom he will assume responsibility, but there is no required equivalent of the RSO to monitor programs internally. Thus, the individual license holder is responsible for ordering, receiving, and maintaining an accurate, current inventory of the drugs used in his laboratory.

One institutional responsibility that should be assigned to an individual or department is monitoring the expiration dates of licenses. Although the DEA has a program which should remind each licensee in ample time that their license is about to expire, experience has shown that the program has not been wholly successful. An individual should maintain a file of all licenses held by employees of the institution and take appropriate steps to see that applications for renewals are filed in a timely manner to avoid purchasing controlled materials on expired licenses.

Packages containing controlled substances must be marked and sealed in accordance with the provisions of the Controlled Substance Act. Every parcel containing these sensitive materials must be placed within a plain outer container or securely wrapped in plain paper through which no markings indicating the nature of the contents can be seen. No markings of any kind which would reveal the nature of the contents are permitted on the parcel.

### 4.2.3. ETIOLOGIC AGENTS

Hazardous biological agents are classified as etiological agents. An etiologic agent is more specifically defined as a viable (can live in the environment which it is in) microorganism or its toxin (a poisonous substance produced by an organism or microorganism) which may cause human disease. The importation or subsequent receipt of etiologic agents and vectors of human diseases is subject to the regulations of the Public Health Service given in 42 CFR 71.156. The Centers for Disease Control (CDC) issues the necessary permits authorizing the importation or receipt of regulated materials and specifies the conditions under which the agent or vector is shipped, handled, and used. The interstate shipment of indigenous etiologic agents, diagnostic specimens, and biological products is subject to the applicable packaging, labeling, and shipping requirements of 42 CFR 72, Interstate Shipment of Etiologic Agents. There is a current proposal to eliminate the shipping of dangerous etiologic agents by mail because of problems arising from the shipment of defense-related materials.

In addition to the regulations of the Public Health Service, the U.S. Department of Transportation (DOT) has additional regulations in 49 CFR 173.386 to 173.388. The U.S. Postal Service provides regulations covering the mailability of biological materials in the Domestic Mail Manual, Section 124.38. All of these provide explicit instructions on how etiologic agents can be shipped. Additional restrictions for international shipments are covered by the International Mail Manual. The ability to make foreign shipments is restricted to laboratories by approval of the General Manager, International Mail Classification Division, USPS Headquarters, Washington D.C. 20260-5365.

Whether a person or laboratory purchases a given etiologic agent should depend upon the available facilities for the research program, the training and experience of the laboratory employees, and the type and scale of the operations to be conducted. If, as discussed in Section 5.5, Microbiological and Biomedical Laboratories, the etiologic agent is one that would require planned operations to be conducted in a laboratory meeting Biological Safety Level 3 or 4 standards, the purchase should require the prior approval of the institutional biosafety committee or the biosafety officer.

Comparable restrictions for the importation, possession, use, or interstate shipment of certain pathogens of domestic livestock and poultry are administered by the U.S. Department of Agriculture (USDA).

For additional information regarding etiologic agents of human diseases and related materials, contact:

> Centers for Disease Control
> Attention: Office of Biosafety
> 1600 Clifton Road, N.E.
> Atlanta, GA 30333
> (404) 329-3883

For additional information regarding animal pathogens, contact:

> Chief Staff Veterinarian
> Organisms and Vectors
> Veterinary Services
> Animal and Plant Health Inspection Service
> U.S. Department of Agriculture
> Hyattsville, MD 20782
> (301) 436-8017

### 4.2.4. CARCINOGENS

There are no restrictions on ordering known carcinogenic materials. However, for carcinogenic materials covered by the regulations in 29 CFR 1910, Subpart Z,

**TABLE 4.2**
**Regulated Carcinogenic Materials**

Asbestos, tremolite, anthrophyllite, and actinolite
4-Nitrobiphenyl
α-Naphthylamine
Methyl chloromethyl ether
3,3′-Dichlorobenzidine (and its salts)
bis-Chloromethyl ether
β-Naphthylamine
Benzidine
4-Aminodiphenyl
Ethyleneimine
β-Propiolactone
2-Acetylaminofluorene
4-Dimethylaminoazobenzene
*N*-Nitrosodimethylamine
Vinyl chloride
Inorganic arsenic
Coke oven emissions
1,2-dibromo-3-chloropropane
Acrylonitrile
Ethylene oxide
Benzene

purchases for research should be limited to individuals who formally commit themselves to complying with the terms and conditions of the standards. To ensure that this is done, every requisition for purchase of one of the regulated carcinogens should be referred to the institutional safety department for review. This will normally involve a review of the research protocols to ascertain whether the planned use will meet any criteria exempting the proposed program from some of the more stringent, and often expensive, regulatory requirements. If the program does not appear to qualify for exemptions, then the investigator and the safety reviewer should go through each of the requirements under the standard to confirm that they can be met. Although this will seem excessive to some users, it is necessary not only to protect the employees, but also to minimize the possibility of litigation for the research director and the academic institution or corporation.

There are a number of known carcinogenic materials, and the list is growing as the necessary studies of suspected carcinogens are completed. It is recommended that purchases of these be limited and exposures minimized as much as possible to promote the safety of everyone exposed to the materials and in consideration of potential future regulatory restrictions. As discussed in Section 4.3.3, for the purpose of the MSDSs, a listing as a carcinogen by either the NTP or IARC is sufficient to be considered as one for the purpose of the Hazard Communication Standard. However, the only carcinogens which are specifically regulated as such in 29 CFR 1910, Subpart Z, are those for which individual regulatory standards have been issued. Materials currently regulated as carcinogenic by OSHA are given in Table 4.2.

Note that coke oven emissions on this list are the result of processes and not chemicals to be used in the laboratory.

In addition to the list above, a so-called "California" list of chemicals has been developed as a result of the passage of Proposition 65 in California, which requires the governor of the state to publish annually a list of chemicals known to cause cancer or reproductive toxicity. The California list, as of January 1, 1988, is given in Tables 4.3A and 4.3B.

## TABLE 4.3A
### "California" List of Carcinogens

| Chemical | CAS number | Chemical | CAS number |
|---|---|---|---|
| 2-Acetylaminofluorene | 53963 | Certain combined chemotherapy for lymphomas | — |
| Acrylonitrile | 107131 | | |
| Adriamycin | 23214928 | Chlordecone (kepone) | 143500 |
| AF-2;[2-(2-furyl)-3-(5-nitro-2-furyl)] acrylamide | 3688537 | 1-(2-Chloroethyl)-3-cyclohexyl-1-nitrosourea (CCNU) | 13010474 |
| Aflatoxins | — | Chlorambucil | 305033 |
| o-Aminoazotoluene | 97563 | Chloroform | 67663 |
| 4-Aminodiphenyl | 92671 | Chloromethyl methyl ether (technical grade) | 107302 |
| 4-Amino-5-(5-nitro-2-furyl)-1,3,4-thiadiazole | 712685 | 4-Chloro-o-phenylenediamine | 95830 |
| Amitrole | 61825 | Chromium (hexavalent compounds) | 7440473 |
| o-Anisidine and o-Anisidine hydrochloride | 90040 | Coke oven emissions | — |
| | | Conjugated estrogens | |
| Analgesic mixtures containing phenacetin | — | p-Cresidine | 120718 |
| | | Cupferron | 135206 |
| Aramite | 140578 | Cycasin | 14901087 |
| Arsenic (inorganic arsenic compounds) | — | Cyclophosphamide | 50180 |
| Asbestos | 1332214 | Dacarbazine | 4342034 |
| Auramine | 492808 | Daunomycin | 20830813 |
| Azaserine | 115026 | DDT (1,1,1-Trichloro-2,2-bis [p-chlorophenyl] ethane) | 50293 |
| Azathioprine | 446866 | | |
| Benz(a)anthracene | 56553 | 2,4-Diaminoanisole sulfate | 39156417 |
| Benzene | 71432 | 4,4'-Diaminodiphenyl ether | 101804 |
| Benzidine (and its salts) | 92875 | 2,4' Diaminotoluene | 95807 |
| Benzo(b)fluoranthene | 205992 | Dibenz(a,h)acridine | 226368 |
| Benzo(j)fluoranthene | 205823 | Dibenz(a,j)acridine | 224420 |
| Benzo(k)fluoranthene | 207089 | Dibenz(a,h)anthracene | 53703 |
| Benzo(a)pyrene | 50328 | 7H-Dibenzo(c,g)carbazole | 194592 |
| Benzotrichloride | 98077 | Dibenzo(a,e)pyrene | 192654 |
| Benzyl violet 4B | 1694093 | Dibenzo(a,h)pyrene | 189640 |
| Beryllium and beryllium compounds | — | Dibenzo(a,i)pyrene | 189559 |
| N,N-bis-(2-chloroethyl)-2-naphthylamine (chlornapazine) | 494031 | 1,2-Dibromo-3-chloropropane (DBCP) | 96128 |
| | | 3-3'-Dichlorobenzidine | 91941 |
| bis-Chloroethyl nitrosourea (BCNU) | 154938 | 3,3'-Dichloro-4-4'-diaminodiphenyl ether | 28434868 |
| bis-Chloromethyl ether | 542881 | | |
| 1,4-Buanediol dimethanesufonate (myleran) | 55981 | Dichloroethane (ethylene dichloride) | 107062 |
| | | Diepoxybutane | 1464535 |
| β-Butyrolactone | 3068880 | Di(2-ethylhexyl)phthalate | 117817 |
| Cadmium and cadmium compounds | — | 1,2-Diethylhydrazine | 1615801 |
| Carbon tetrachloride | 56235 | Diethyl sulphate | 64675 |
| Carrageenan (degraded) | 9000071 | Diethylstilbestrol | 56531 |

## TABLE 4.3A (continued)
## "California" List of Carcinogens

| Chemical | CAS number | Chemical | CAS number |
|---|---|---|---|
| Dihydrosafrole | 94586 | 2-Naphthylamine | 91598 |
| 3,3'-Dimethoxybenzidine | 119904 | Nickel refinery dust from the | — |
| 4-Dimethylaminoazobenzene | 60117 | pyrometallurgical process | |
| *trans*-2-[(Dimethylamino)-]methylimino- | 55738540 | Nickel carbonyl | 13463393 |
| 5-[2-5-nitro-(2-furyl)vinyl]- | | Nickel subsulfide | 12035722 |
| 1,3,4-oxadiazole | | Nitrilotriacetic acid | 139139 |
| 3,3'-Dimethylbenzidine (*o*-tolidine) | 119937 | Nitrofen (technical grade) | 1836755 |
| Dimethylcarbamoyl chloride | 79447 | Nitrogen mustard | 51752 |
| 1,2-Dimethylhydrazine | 540738 | 2-Nitropropane | 79469 |
| Dimethyl sulfate | 77781 | *N*-nitrosodi-*n*-butylamine | 924163 |
| 1,4-Dioxane | 123911 | *N*-Nitrosodiethanolamine | 1116547 |
| Direct black 38 (technical grade) | 1937377 | *N*-Nitrosodiethylamine | 55185 |
| Direct blue 6 (technical grade) | 2602462 | *N*-Nitrosodimethylamine | 62759 |
| Epichlorohydrin | 106898 | *p*-Nitrosodiphenylamine | 156105 |
| Estradiol 17B | 50282 | *N*-Nitroso-*n*-propylamine | 621647 |
| Estrone | 53167 | *N*-Nitroso-*N*-ethylurea | 759739 |
| Ethinylestradiol | 57636 | *N*-Nitroso-*N*-methylurea | 684935 |
| Ethylene dibromide | 106934 | *N*-Nitrosomethylvinylamine | 4549400 |
| Ethyleneimine | 151564 | *N*-Nitrosomorpholine | 59892 |
| Ethylene oxide | 75218 | *N*-Nitrosonornicotine | 16543558 |
| Ethylene thiourea | 96457 | *N*-Nitrosopiperdine | 100754 |
| Ethyl methanesulfonate | 62500 | *N*-Nitrosopyrrolidine | 930552 |
| Formaldehyde (gas) | 50000 | *N*-Nitrososarcosine | 13256229 |
| Formylhydrazino-4-(5-nitro-2-furyl) | 3570750 | Oxymetholone | 434071 |
| thiazole | | Panfuran S (dihydroxymethylfuratrizine) | 794934 |
| Glycialaldehyde | 765344 | Phenazopyridine | 94780 |
| Gyromitrin (acetaldehyde | 16568028 | Phenazopyridine hydrochloride | 136403 |
| methylformyl-hydrazone) | | Phenytoin | 57410 |
| Hexachlorobenzene | 118741 | Phenytoin, sodium salt of | 630933 |
| Hexachlorocyclohexane (technical | — | Polybrominated biphenyls | 59536651 |
| grade) | | Polychlorinated biphenyls (those | — |
| Hexamethylphosphoramide | 680319 | containing 60% or more chlorine | |
| Hydrazine | 303012 | by mol wt) | |
| Hydrazine sulfate | 10034932 | Procarbazine | 671169 |
| Hydrazobenzene | 122667 | Procarbazine hydrochloride | 366701 |
| Indeno[1,2,3-*cd*]pyrene | 193395 | Progesterone | 57830 |
| Iron dextran complex | 9004664 | 1,3-Propane sulfone | 1120714 |
| Lead acetate | 301042 | β-Propiolactone | 57578 |
| Melphalan | 148823 | Propylthiouracil | 51525 |
| Methoxsalen with ultraviolet A | 298817 | Sodium saccharin | 128449 |
| therapy (PUVA) | | Safrole | 94597 |
| 2-Methylaziridine (propylene amine) | 75558 | Soots, tars, and lubricant base oils and | — |
| 4,4'Methylene-*bis*-(2-chloroaniline) | 101144 | derived products; specifically, vacuum | |
| 4,4'-Methylenedianiline | 101779 | distillates, acid-treated oils, aromatic | |
| 4,4'-Methylenedianiline dihydrochloride | 13552448 | oils, mildly solvent-refined oils, mildly | |
| Metronidazole | 443481 | hydrotreated oils, used engine oils, and | |
| Michler's ketone | 90948 | mineral oils when used in occupations | |
| Mirex | 2385855 | such as mulespinning, metal machining, | |
| Mustard gas | 505602 | and jute processing | |

## TABLE 4.3A (continued)
## "California" List of Carcinogens

| Chemical | CAS number | Chemical | CAS number |
|---|---|---|---|
| Streptozocin | 18883664 | Toxaphene | 8001352 |
| Sulfallate | 95067 | 2,4,6-Trichlorophenol | 88062 |
| 2,3,7,8-Tetrachlorodibenzo-*p*-dioxin (TCDD) | 1746016 | Treosulfan | 299752 |
| | | Tris (1-aziridinyl) phosphine sulfide (thiotepa) | 52244 |
| Thioacetamide | 62555 | | |
| Thiourea | 62566 | Tris (2,3-dibromopropyl) phosphate | 126727 |
| Thorium dioxide | 1314201 | Urethane | — |
| *o*-Toluidine | 95534 | Vinyl chloride | 75014 |
| *o*-Toluidine hydrochloride | 636215 | | |

## TABLE 4.3B
## Reproductive Toxicants

| Chemical | CAS number | Chemical | CAS number |
|---|---|---|---|
| Aminopterin | 54626 | Etretinate | 54350480 |
| Chlorcyclizine hydrochloride | 82939 | Isotretinoin | 4759482 |
| 1,2-Dibromo-3-chloropropane (DBCP) | 96128 | Lead | 7439921 |
| Diethylstilbestrol (DES) | 56531 | Methyl mercury | — |
| Diphenylhydantoin | 630933 | Thalidomide | 50351 |
| Ethyl alcohol in alcoholic beverages | 64175 | Valproate | 99661 |
| Ethylene oxide | 75218 | Warfarin | 81812 |

The following four lists of chemicals are being considered for addition to the California list of carcinogens given in Table 4.3A above.

Materials identified by the IARC as reasonably anticipated to be carcinogens or as probable carcinogens are

| | |
|---|---|
| Lasiocarpine | 303344 |
| Merphalan | 531760 |
| Mestranol | 72333 |
| 5-Methylchrysene | 3697243 |
| Methylazoxymethanol and methylazooxymethyl acetate | 592621 |
| 4,4'-Methylene-bis(2-methylaniline) | 838880 |
| Methyl methanesulfonate | 66273 |
| 2-Methyl-1-nitroanthraquinone of uncertain purity | 129157 |
| *N*-Methyl-*N*'-nitro-*N*-nitrosoguanidine | 70257 |
| Mitomycin C | 50077 |
| Monocrotaline | 315220 |
| 5-(Morpholinomethyl)-3([5-nitrofurfurylidene]-amino)-2 oxalolidinone | 139913 |
| Nafenopin | 3771195 |

| | |
|---|---:|
| Niridazole | 61574 |
| 4-Nitrobiphenyl | 92933 |
| 5-Nitroacenaphthene | 602879 |
| 1-[(5-Nitrofurfurylidene)-amino]-2-imidazolidinone | 555840 |
| N-(4-[5-nitro-2-furyl]-2-thiazolyl) acetamide | 531828 |
| Nitrogen mustard and its hydrochloride | 51752 |
| Nitrogen mustard N-oxide and its hydrochloride | 302705 |
| N-Nitroso-N-methylurethane | 615532 |
| Orange oil SS | 2646175 |
| Phenoxybenzamine and its hydrochloride | 59861 |
| Ponceau MX | 3761533 |
| Ponceau 3R | 3564098 |
| Sterigmatocystin | 10048132 |
| Testosterone and its esters | 58220 |
| 4,4′-Thiodianiline | 139651 |
| Tobacco smoke, tobacco, and oral use of smokeless tobacco | |
| o-Toluidine and its hydrochloride | 636215 |
| Trp-P-1 (tryptophan-P-1) | 62450060 |
| Trp-P-2 (tryptophan-P-2) | 62450071 |
| Trypan blue (commercial grade) | 72571 |
| Uracil mustard | 66751 |

Chemicals listed by the NTP as reasonably anticipated to be carcinogens are

| | |
|---|---:|
| Lead phosphate | 7446277 |
| Methyl iodide | 74884 |
| 5-Nitro-o-anisidine | 99592 |
| Phenacetin | 62442 |
| Reserpine | 50555 |
| Selenium sulfide | 7488564 |

Chemicals listed by the EPA Carcinogen Assessment Group as probable carcinogens are

| | |
|---|---:|
| Acetaldehyde | 75070 |
| Aldrin | 309002 |
| Allyl chloride | 107051 |
| 1,3-Butadiene | 106990 |
| Chlordane | 57749 |
| Dichloromethane or methylene chloride | 75092 |
| Dieldrin | 60571 |
| 2,4-Dinitrotoluene | 121142 |
| Diphenylhydrazine | 122667 |

Also under consideration by the California state group are

| | |
|---|---:|
| bis(2-Chloroethyl) ether | 111444 |
| Heptachlor | 76448 |

| | |
|---|---|
| Heptachlor epoxide | 1024573 |
| Hexachlorodibenzodioxin | 34465468 |
| N-Nitroso-diphenylamine | 86306 |
| Tetrachloroethylene or perchloroethylene | 127184 |
| Trichloroethylene | 79016 |
| Unleaded gasoline vapor | |

Additional data concerning tests and publications on chemicals are published frequently. In order to remain current, it is almost essential that facilities subscribe to reference publications which are published frequently such as the Bureau of National Affairs (BNA), the Commerce Clearing House (CCH), and a number of others which seek out and publish relevant information. For example, since the material listed above was tabulated, the National Institutes for Occupational Safety and Health (NIOSH) issued a warning (late January, 1988) of importance to individuals working on solar energy devices that gallium arsenide may be a carcinogen because it releases inorganic arsenic and gallium in the body.

### 4.2.5. EXPLOSIVES

A substantial amount of laboratory research involves materials considered, in the legal sense of the term, as explosives rather than simply chemicals which can explode under appropriate conditions. The term "explosive" in this relatively narrow sense is defined as any material determined to be within the scope of Title 18, United States Code , Chapter 40 — Importation, Manufacture, Distribution and Storage of Explosive Materials — and any material classified as an explosive by the DOT in the Hazardous Material Regulations (49 CFR 100 to 199). A list of the materials which are within the scope of 18 USC 40 is published periodically by the Bureau of Alcohol, Tobacco, and Firearms, U.S. Department of the Treasury.

The classification of explosives by the DOT 49 CFR Chapter I, is as follows (the wording follows that used by OSHA in 29 CFR 1910.109[a][3]):

1. Class A — "Possessing detonating or otherwise maximum hazard, such as dynamite, nitroglycerin, picric acid, lead azide, fulminate of mercury, black powder, blasting powder, blasting caps, and detonating primers."
2. Class B — "Possessing flammable hazard, such as propellant explosives (including some smokeless propellants), photographic flash powders, and some special fireworks."
3. Class C — "Includes certain types of manufactured articles which contain Class A or Class B explosives, or both, as components but in restricted quantities."
4. Forbidden or Not Acceptable Explosives — "Explosives which are forbidden or not acceptable for transportation by common carriers by rail freight, rail express, highway, or water in accordance with the regulations of the U.S. Department of Transportation, 49 CFR, Chapter I."

Some activities involving explosives require a federal license or permit under Title XI, 18 USC, Chapter 40. Those activities not covered in these regulations are covered by NFPA 495, where the latter standard has been adopted as a legal requirement. Under NFPA 495, no explosive materials are to be sold or transferred in any way to a person without a valid permit to have them, and no one is to conduct any operations involving explosives without an appropriate permit. Laboratories that are engaged in research with an explosive in areas that have adopted NFPA 495 would require a "Permit to Use" unless test blasts are involved, in which case an additional "Permit to Blast" would be required.

As noted, the OSHA standards are based on the 1970 version of NFPA 495, and the requirements differ in some respects from the current version. However, they still require stringent safety precautions which must be followed by any research facility employing or investigating explosive materials. Consequently, any acquisition of explosive materials should be internally reviewed to insure that the intended recipient can provide adequate facilities and safeguards to comply with the standards.

### 4.2.6. Equipment

In Chapter 3 a large number of items of equipment were discussed in terms of the requirements which should be met in order to ensure that the equipment meets appropriate safety standards. These standards change as technological improvements occur, as new information becomes available, and as new or revised regulatory standards become effective. Acquisition of many items of equipment affecting the safe operation of a facility should be internally reviewed, in part to ensure that the equipment being acquired meets current safety specifications and to alert units within the organization which need to know of the purchase. In the latter case, it may be necessary to confirm that sufficient attention has been paid to the installation. Among those items of equipment which need to be routinely reviewed — usually by the safety department, sometimes by buildings and grounds, and, if separate from the latter, planning and engineering (or equivalent groups) — are

- Biological safety cabinets
- Electron microscopes
- Chemical fume hoods
- Lasers
- Equipment incorporating radioactive sources
- Refrigeration equipment
- All safety equipment
- Water stills
- X-ray equipment

## 4.3. FREE MATERIALS

It is tempting to accept free materials. However, with the advent of regulations providing for the safe and environmentally responsible disposal of hazardous chemicals under the Resource Conservation and Recovery Act (RCRA), the acceptance of free materials may eventually generate unexpected and substantial disposal costs. In many cases involving the study of an experimental material where free materials are the basis for the intended research program, the amount proffered often is substantially more than is actually needed. The excess amount is typically unusable for other purposes and will require legal disposal. Unfortunately, a few organizations have shifted disposal costs to the recipient organization by giving inexperienced persons unwanted surplus or out-of-date materials. Even if this is not the case, bypassing normal purchasing procedures could increase the possibility of unrecorded dangerous materials arriving on site. All of these negative possibilities can be avoided by adopting a policy similar to the following for regulating the receipt of free materials.

1. The amount of free material which may be accepted by an individual, laboratory, or other administrative unit to be used in a program of research must be limited to the amount which is likely to be actually needed in the proposed program.
2. The donor must agree in writing to accept the return of any unused amounts or pay for the legal and safe disposal of the material. The recipient may agree to waive this requirement if he is prepared to pay for the disposal from his own funds.
3. If the utilization or storage of the free material is likely to pose any substantive risk to personnel or property, the safety department and the risk management (insurance) office must be informed prior to completion of the agreement to accept the material, to allow time to determine if the risks are acceptable and if adequate facilities are available, and to insure that the proposed research has been reviewed for health and safety implications.

In some cases, undesirable materials may be a part of, or within, an item of equipment. In an example at the author's institution, an engineering department accepted a large "free" transformer, which eventually turned out to be unusable. The transformer contained 474 gal of high purity polychlorinated biphenyl (PCB) insulating oil. The total disposal cost of the transformer and its contents was well over $10,000 because of the PCB. Recently, the state in which our institution is located issued a directive requiring any state agency to audit any real estate the agency might acquire, either as a purchase or as a gift, for evidence of any prior hazardous waste dumping.

These policies may appear unnecessarily formal to many individuals. However, the cost of disposing of hazardous materials and the rate of increase of these costs has reached the point that hazardous waste disposal is becoming a major problem for many corporations and institutions. Most cannot afford to incur additional costs by accepting free materials without considering the future obligations which the gift may engender.

## REFERENCES (SECTIONS 4.0 TO 4.3)

1. **Bilsom, R. E.,** Torts among the ivy: some aspects of the civil liability of universities, University of Saskatchewan, Saskatoon, Saskatchewan, Canada.
2. Legal liability, in *Better Science Through Safety,* Gerlovich, J. A. and Downs, G. E., Eds., Iowa State University Press, Ames, 1981, 17.
3. **Kaufman, J.,** Laboratory Safety Workshop, Curry College, Milton, MA.
4. Occupational Safety and Health Administration, Hazard Communication; Final Rule, 29 CFR Parts 1910, 1915, 1917, 1918, 1926, and 1928, *Fed. Reg.,* 52(163), 31852, 1987.
5. Occupational Health Services, Inc., New York, NY, 10123.
6. Nuclear Regulatory Commission, Title 10, Code of Federal Regulations; Parts 20, 30 and 33, Washington, D.C., 1988.
7. Food and Drug Administration, Title 21, Part 1300, Washington, D.C.
8. U.S. Public Health Service, Title 42, Parts 71.156 and 72, Washington, D.C.
9. Department of Transportation, Title 49, Code of Federal Regulations, Sections 173.386 to 173.388, Washington, D.C.
10. Occupational Safety and Health Administration, Title 29, Code of Federal Regulations, Subpart Z, Washington, D.C.
11. Title 18, United States Code, Chapter 40, Importation, Manufacture, Distribution, and Storage of Explosive Material, 1988.

12. Department of Transportation, Title 49, Code of Federal Regulations, Parts 100 to 199, Washington, D.C., 1988.
13. Occupational Safety and Health Administration, Title 29, Code of Federal Regulations, Part 1910, §109, Washington, D.C., 1988.
14. *Code for the Manufacture, Transportation, Storage and Use of Explosive Materials,* NFPA 495, National Fire Protection Association, Quincy, MA, 1982.

# 4.4. PURCHASING OF ANIMALS*

## 4.4.1. INTRODUCTION

Nothing can sabotage an animal-related experiment quicker than the use of animals with latent (hidden) or overt signs of disease. The use of healthy, unstressed animals is critical to obtaining quality experimental results. The following information will aid in the selection of appropriate vendors and provide criteria for selecting "clean" animals.

### 4.4.1.1. Selection Criteria for Rodents and Rabbits

Animals purchased from commercial vendors could have latent bacterial, viral, or parasitic diseases which can radically affect experimental results.[1-3] To avoid the use of infected ("dirty") animals as well as the contamination and subsequent infection of "clean" animals in a facility, selection criteria for the purchase of animals should be established.

Commercial vendors should be required to submit copies of their monthly or semiannual quality assurance health testing reports for review prior to purchase. Testing includes serological evaluation for the presence/absence of latent viral infections, culturing of various anatomical sites to assess the presence/absence of microbial pathogens (disease-causing bacteria), and evaluation for internal and external parasites. Small[4] lists the organisms which should *not* be present in mice, rats, hamsters, guinea pigs, and rabbits prior to purchase.

Since the vendor animal health report represents a snapshot in time and does not guarantee avoidance of subsequent contamination in the vendor's facility or during transport to the research facility, it would be wise to test a representative sample of the animals after receipt while they are housed in the quarantine area. Should serological testing reveal a latent viral infection, the entire group of animals should be destroyed and the quarantine area thoroughly disinfected.

Studies using inbred strains of rats and mice can be decimated by genetic impurity. Vendors also test inbred strains for genetic purity, and a copy of their most recent test report can be requested prior to animal purchase.

### 4.4.1.2. Laws Affecting Animal Purchasing

The Federal Animal Welfare Act (PL 89-544) governs the purchase of dogs and cats for use in research and teaching. Under this law, these species may be purchased only from (1) a dealer licensed with the U.S. Department of Agriculture Animal, Plant Health Inspection Service (USDA/APHIS), (2) a commercial breeder licensed with USDA/APHIS, (3) local or county animal shelters (recognizing that, at this time, 13 states prohibit the sale of pound animals for research or teaching), or (4) another research institution which has obtained the animals from any of these sources.

---

* This section was written by Dr. David M. Moore.

Research facilities must keep the following records on dogs and cats, usually on the USDA Individual Health Certificate and Identification Form (VS Form 18-1):

1.    The name, address, and license number of the dealer from whom the animal was purchased
2.    The date of acquisition of each live dog and cat
3.    The official USDA tag number or tattoo assigned to each animal
4.    A description that includes species, sex, date of birth or approximate age, color and distinctive markings, and breed or type
5.    Any identification number assigned to that animal by the research facility

### 4.4.1.3. Transportation of Animals

Transport of animals from the vendor to the research facility is also regulated by the Animal Welfare Act. Regulations on ambient temperature limits might prohibit shipment of animals on days that are too cold or too hot. Since most shipments are by common carrier (airplane, truck), one cannot be sure that animals were not subject to environmental stressors (heat or cold) while in loading areas or during transport. Heat stress can cause debilitation or death in rodents and other species. Additionally, animal transport boxes from several vendors might be shipped together, with the potential for cross-contamination of "clean" animals by "dirty" animals during shipment. Some vendors ship animals in their own environmentally controlled vehicles to preclude this problem, but this service is not available to all parts of the country.

### 4.4.1.4. Additional Laws Affecting Animal Purchase

Some states, such as California, prohibit entry of certain species (i.e., gerbils, ferrets) into the state without appropriate approval forms. Contact the state veterinarian in your state to ascertain whether similar regulations/restrictions exist.

## REFERENCES (SECTIONS 4.4 TO 4.4.1.4)

1.    **Bhatt, P.,** Virus infections of laboratory rodents, *Lab. Anim.,* 9(3), 43, 1980.
2.    **Orcutt, R. P.,** Bacterial diseases: agents, pathology, diagnosis, and effects on research, *Lab. Anim.,* 9(3), 28, 1980.
3.    **Hsu, C. K.,** Parasitic diseases: how to monitor them and their effects on research, *Lab. Anim.,* 9 (3), 48, 1980.
4.    **Small, J. D.,** Rodent and lagomorph health surveillance-quality assurance, in *Laboratory Animal Medicine,* Fox, J. G. et al., Eds., Academic Press, Orlando, FL, 1984, 709.
5.    **Smith, K. P., Hoffman, H. A., and Crowell, J. S.,** Genetic quality control in inbred strains of laboratory rodents, *Lab. Anim.,* 11(7), 16, 1982.
6.    **Weisbroth, S. H.,** The impact of infectious disease on rodent genetic stocks, *Lab. Anim.,* 13(1), 25, 1984.

## 4.5. STORAGE

Laboratory storage practices may enhance or diminish overall laboratory safety. There are many factors to be considered in addition to those concerned with flammable materials touched upon in Chapter 3 in the design and selection of facilities and equipment for those specific substances. Among these are the amount, location, and organization of the stored chemicals. The types of vessels in which they are

## TABLE 4.4
## Compatible Chemical Groups

| Number | Chemical group | Do not store with group numbers |
|---|---|---|
| 1 | Inorganic acids | 2—8, 10, 11, 13, 14, 16—19, 21, 22, 23 |
| 2 | Organic acids | 1, 3, 4, 7, 14, 16, 17—19, 22 |
| 3 | Caustics | 1, 2, 6, 7, 8, 13—18, 20, 22, 23 |
| 4 | Amines and alkanolamines | 1, 2, 5, 7, 8, 13—18, 23 |
| 5 | Halogenated compounds | 1, 3, 4, 11, 14, 17 |
| 6 | Alcohols, glycols, glycol ethers | 1, 7, 14, 16, 20, 23 |
| 7 | Aldehydes | 1—4, 6, 8, 15—17, 19, 20, 23 |
| 8 | Ketones | 1, 3, 4, 7, 19, 20 |
| 9 | Saturated hydrocarbons | 20 |
| 10 | Aromatic hydrocarbons | 1, 20 |
| 11 | O'efins | 1, 5, 20 |
| 12 | Petroleum oils | 20 |
| 13 | Esters | 1, 3, 4, 19, 20 |
| 14 | Monomers, polymerizable esters | 1—6, 15, 16, 19—21, 23 |
| 15 | Phenols | 3, 4, 7, 14, 16, 19, 20 |
| 16 | Alkylene oxides | 1—4, 6, 7, 14, 15, 17—19, 23 |
| 17 | Cyanohydrins | 1—5, 7, 16, 19, 23 |
| 18 | Nitriles | 1—4, 16, 23 |
| 19 | Ammonia | 1—2, 7, 8, 13—17, 20, 23 |
| 20 | Halogens | 3, 6—15, 19, 21, 22 |
| 21 | Ethers | 1, 14, 20 |
| 22 | Elemental phosphorus | 1—3, 20 |
| 23 | Acid anhydrides | 1, 3, 4, 6, 7, 14, 16—19 |

contained and the information on the container labels are important. Some types of materials represent special hazards for which specific protective measures may be indicated or perhaps mandatory due to regulations. The following sections will address these topics.

### 4.5.1. COMPATIBLE CHEMICAL STORAGE

Many laboratories, if not the majority, find it easiest to store a large portion of their chemicals alphabetically, although there are often partial exceptions even in these facilities. The most frequently used flammable liquids, for example, usually are placed in a common area. Other, less frequently used flammable liquids may still be stored on shelves with other chemicals. Comparable usage may occur for acids, bases, or other heavily used reagents. For the most part, however, general chemical storage in a laboratory is often done without regard to compatibility. Determination of compatibility does require some effort, as reflected by Table 4.4, adapted from material in the *Journal of Hazardous Materials.*[1]

There are longer and more detailed lists of compatible chemicals which could be used to determine appropriate storage. However, even the system represented by Table 4.4 may be too elaborate to encourage individuals to use it. Several of the major chemical firms have developed less complicated systems using a color code to define the groups which should be stored together. Unfortunately, although there are some similarities, the schemes of the different companies are not wholly compatible. The color code systems of two major chemical companies are shown below for comparison with each other and with Table 4.4. The color codes used by chemical vendors to define the major groups are prominently incorporated into the label. This makes it easy to decide where to place and return a new container after use. Although providing less selectivity in segregating different materials than a system listing individual chemicals which should not be stored together or a list of classes of materials as in Table 4.4, the ease of use should make color coding acceptable to most laboratory managers and employees.

Both of these systems provide an alternative for the color blind. In the Fisher system, the first letter of the color is displayed prominently in the color bar, while the Mallinckrodt system spells the color out.

| **Fisher** | | **Mallinckrodt** | |
|---|---|---|---|
| RED | Flammable | RED | Flammable |
| BLUE | Health hazard | BLUE | Health hazard |
| YELLOW | Reactive and oxidizing agents | YELLOW | Reactivity hazard |
| WHITE | Corrosive | WHITE | Contact hazard |
| GRAY | Moderate hazard, general storage | GREEN | Minimum or no hazard, general storage |
| STOP | Exception, incompatible with reagents of same color code, store separately | NAVY BLUE | Band at bottom of label, compatible with reagents with same color code, store separately |

Other firms have similar systems which also vary slightly in detail. Usually, they agree for chemicals coded red, blue, yellow, and white, but differ in the way they denote exceptions or indicate a material for which there are few or no storage problems. For example, the J. T. Baker Company, one of the pioneers in developing a color-coded storage system, uses striped colors to denote exceptions and orange to denote chemicals which may be stored in the general storage area.

Until an accident occurs in which the contents of the different containers come into contact with each other, failure to store materials according to compatibility may have few, if any, repercussions. In some instances, mixing of spilled chemicals will result in no reaction or in relatively nonviolent reactions which can be controlled easily. However, in other cases, the result could be the insidious release of deadly toxic fumes or perhaps a violent explosion or fire. In any event, the failure to store chemicals according to their properties is too much of a risk to personnel, to property, and possibly even to the intellectual value of accumulated research data files that may represent the product of years of effort. It is inexpensive insurance to make the relatively modest effort to segregate chemicals according to a color-coded system or even to follow a more complicated program using groups such as those defined in Table 4.4.

### 4.5.2. LABELING

There are two types of labels which are important in laboratories. The labels on commercial containers are usually extremely comprehensive, providing not only information on the nature, amount, and quality of the product but also a very large amount of safety-related data. Typically, a commercial label will readily meet the requirements of the Hazard Communication Standard. On the other hand, labels placed on secondary containers in the laboratory by employees may be something such as "soln. A" or even less. This may be sufficient if all of the material is to be promptly used by the individual placing the label on the container, but otherwise it is not. In most instances, secondary containers of hazardous chemicals must be marked with labels identifying the chemical in the container and providing basic hazard warnings. In order to meet the requirements of the Hazardous Communication Standard, the secondary label must be affixed before the container is put into use. The label, with the hazard information, must be listed in English as the primary language. As long as an English version is on the label, the same information may be provided in other languages to meet the needs of the personnel in the area. The intent of the labeling requirements under the Hazard Communication Standard is primarily to protect the immediate users of the material by ensuring that they have access to the identity of the material with which they are working. However, superficial and uninformative labels cause a major problem for the legal disposal of containers possibly holding hazardous chemicals. It is difficult to dispose of unknown materials under RCRA rules. This is a serious problem in academia due to the rapid turnover in research personnel, especially students, as they finish their degree programs. Full compliance with the Hazard Communication Standard should greatly alleviate this problem.

Laboratory employees should familiarize themselves with the commercial labels on the chemical containers they are using. The labels will reflect the information available to the manufacturer at the time the material was packaged. The information will not be as current as that provided by up-to-date MSDSs, which are revised as often as significant new information becomes available. However, in most cases the information for a specific chemical will change sufficiently slowly so that the information on the label may be used with reasonable confidence.

An examination of typical chemical labels indicates that the following safety-related information is included on the labels of virtually all manufacturers:

- Name of compound
- Impurities, other components
- Flash point (if applicable)
- Storage color code
- Risk descriptor (danger, warning, caution)
- Risk descriptive statement
- Handling advice
- First aid emergency medical advice
- NFPA hazard diamond
- Recommended fire extinguisher class
- CAS number
- UN number

Many labels contain additional useful information. The following items also are found frequently on chemical container labels:

- DOT symbols
- Hazard ratings (may be different from NFPA ratings for laboratory applications)
- Recommended protective equipment (some do this in words, some by means of stylized pictographs)

- Location for dating receipt of material
- Bar code

In the near future, it is likely that a greatly revised American National Standards Institute (ANSI) labeling standard, ANSI Z129.1-1987, will be published. This standard, as with other ANSI standards, will not become a regulatory standard unless adopted by a federal, state, or local jurisdiction. However, even if not adopted in total, it may assist in rectifying the differences between the labels of different chemical vendors and extend the scope of the information presented upon the container labels.

### 4.5.3. REGULATED MATERIALS

Many regulated materials require special storage facilities, primarily for security reasons, although the security requirements frequently are based upon safety or environmental concerns. The following materials were discussed previously in Section 4.2.4 concerning restrictions on their purchase:

- Explosives
- Controlled substances (drugs)
- Radioisotopes
- Etiologic agents
- Carcinogens

This list is approximately in order of the security required, although there is considerable overlap.

#### 4.5.3.1. Explosives

Explosives pose the most immediate danger to individuals of any of the items on the list in the preceding section. OSHA covers the storage requirements for all Class A, B, and C explosives, any special industrial explosives, and any newly developed, and hence unclassified, explosives in 29 CFR 1910.109(c). All of these materials must be kept in magazines which meet the specified requirements in the same section. As noted earlier, the OSHA standard is based on a 1970 version of NFPA Standard 495, although it does include some amendments which were added in 1978. Where the OSHA requirements do not cover a specific point, the most recent version of NFPA 495 may be consulted. Some local jurisdictions may have adopted later versions as part of their codes.

The OSHA storage requirements differentiate between storage of less than 50 lb ($\approx$22.7 kg) and more than 50 lb. Some large research programs may require more than 50 lb to be available at a given time, but most programs typically involve much smaller amounts, often on the order of a few grams. Class 1 magazines, which are required for the larger quantities, have structural requirements appropriate to room size spaces, while Class 2 magazines, appropriate for smaller operations, may be mounted on wheels and be mobile. This section will be limited to a discussion of Class 2 magazines, which are most suitable for typical laboratory-scale use of explosives.

Normally, explosive magazines are expected to be located outdoors. However, a Class 2 magazine may be permitted within a building (the OSHA standard specifically references warehouses and wholesale and retail establishments) when it is located on a floor which has an entrance at outside grade level and is not situated more than 10 ft from the entrance. Also, it is not normally expected to have two magazines within the same building, but if one is used solely to store no more than 5000 blasting caps and is at least 10 ft from the other, it is permissible to do so.

Class 2 magazines may be constructed primarily of wood or metal and will normally be a combination of the two. The wood of the bottom, sides, and cover of a primarily wooden magazine must be of 2-in. hardwood. The corners must be well braced and the magazine must be covered with sheet metal of not less than 20-gauge thickness. In order to avoid contact of the stored explosive with metal, any exposed nails in the interior must be well countersunk.

Primarily metal magazines must have their bottom, top, and cover constructed of sheet metal and be lined on the interior with $^3/_8$-in. plywood or the equivalent. The edges of the metal covers must overlap the sides by at least 1 in. Both metal and wood magazines must be lockable. The covers must be attached securely with substantial strap hinges.

Class 2 magazines must be painted red and labeled on each of the four sides and the top with white letters at least 3 in. high:

## EXPLOSIVES — KEEP FIRE AWAY

When Class 2 magazines are kept inside a building, they must be equipped with substantial wheels or casters to make it possible to remove them from the building in the event of a fire. When necessary due to climate, Class 2 magazines must be ventilated.

The general OSHA standards for the method of storage of explosives within magazines are primarily intended to apply to larger units, but a number are appropriate for smaller Class 2 units. Among the more relevant are

1.    Packages must be stored lying flat, with the top side up. Piles of packages or containers must be stable.
2.    The oldest material of a given type of explosive must be used first to minimize the risks associated with instability upon aging.
3.    Packing and unpacking of explosives must not be done within 50 ft of a magazine.
4.    Except for metal tools to cut open fiberboard boxes, all other tools must be nonsparking.
5.    The magazine must not be used for the storage of metal tools or for other general storage.
6.    Smoking, matches, open flames, spark-producing devices, firearms (other than those in the possession of guards), and combustible materials must not be permitted within 50 ft of the magazine.
7.    A competent individual must be charged with the care of the magazine at all times and is responsible for enforcement of all safety precautions. It is most important that access to the explosives in the magazine be limited to those who have demonstrated or documented experience in the safe use and handling of the explosive materials stored in the magazine. It is also important that an accurate record of the contents be maintained. This record should include an identification of each separate package in the magazine, when it was placed

inside, the contents of the package, and the date and amounts of any materials that have been removed. The log should also identify individuals by name, not initials, who have been permitted to remove explosives. The log should be audited periodically by an independent person.

In addition to the materials which are specifically classified as explosives, there are many chemicals which, under appropriate conditions, can act as explosives, i.e., react or decompose very rapidly, accompanied by a large release of energy. The violence of an explosively rapid reaction is largely dependent upon the gas pressures produced in the reaction, enhanced by any thermal energies produced.

The best precaution in working with materials which are potential explosives is to minimize the amounts actually present. Some of this is simply a matter of maintaining proper inventory control in order to dispose of chemicals which tend to form unstable materials with age such as ethers, in which peroxides form, or perchloric acid or picric acid, which become dangerous if they are allowed to become dehydrated.

Other safety measures which should be taken in storing these laboratory materials are

1.  Keep the minimum quantities needed in a cool, dry area, protected from heat and shock.
2.  The materials should be segregated during storage from materials with which they could react as well as flammables, corrosives, and other chemicals which are likely to interact with each other.
3.  Potentially explosive materials should be stored and used in an area posted with a sign in prominent letters:

**CAUTION**
**POTENTIAL EXPLOSIVE HAZARD**

4.  If the material is being kept because of its potentially explosive properties, it should be treated as an explosive of the appropriate class and kept in a magazine or the equivalent.
5.  Make sure that all occupants of the laboratory are aware of the potential risks and trained in emergency procedures, including evacuation procedures, fire containment, and emergency first aid for physical injuries that might result from an explosion.

As noted, many laboratory chemicals may cause an explosion under appropriate conditions. Table 4.5 presents a brief list of chemicals which have a reactivity rating of 3 or 4 (mostly 4) according to the NFPA No. 704M system for the identification of hazardous chemicals and are sensitive to shock, heat, and/or friction. Materials on this list and those with comparable properties should always be treated with extreme care.

### 4.5.3.2. Controlled Substances (Drugs)

As noted in Section 3.2.1.2.2, the major concern regarding controlled substances is security or control of the materials used in the research program. This is particu-

### TABLE 4.5
### Highly Reactive Shock/Heat-Sensitive Materials

| | |
|---|---|
| Ammonium perchlorate | Dibenzoyl peroxide |
| Ammonium permanganate | Diisopropyl peroxydicarbonate |
| Anhydrous perchloric acid | Dinitrobenzene (ortho) |
| Butyl hydroperoxide | Ethyl methyl ketone peroxide |
| Butyl perbenzoate | Ethyl nitrate |
| Butyl peroxyacetate, tert | Hydroxylamine |
| Butyl peroxypivalate, tert | Peroxyacetic acid |
| 1-Chloro-2,4-dinitrobenzene | Picric acid |
| Cumene hydroperoxide | Trinitrobenzene |
| Diacetyl peroxide | Trinitrotoluene |

larly important for those materials which can be used by individuals or sold as narcotics. The storage unit must be sufficiently strong to prevent forced entry for at least 10 min or more and either sufficiently heavy (750 lb or more) or rigidly bolted to a floor or wall to prevent the entire storage unit from being carried away. This is relatively easy to provide, but, as with explosives, it is equally important that a complete current record be maintained of all materials in the storage cabinet. The log should contain the date and amounts of each substance placed in the storage cabinet and disbursed from it. The amounts must be accurately quantified. The name, not initials, of the person to whom the materials are issued should be recorded. Distribution of the material should either be under the direct control of the principal investigator identified as the holder of the license allowing possession and use of controlled substances or an alternate designated by that person in writing.

Unfortunately, narcotics are occasionally abused by persons who may legally possess them. Provision should be made for a periodic audit of the utilization of controlled substances by each licensee to ensure that any such behavior is quickly detected, for the licensee's own protection and for that of the organization with whom he is affiliated.

### 4.5.3.3. Radioisotopes

Radioactive byproduct materials are probably the most closely regulated research materials in wide use. It is rare for any organization involved to a significant degree in research, especially in the life sciences, not to have a license to use radioactive materials, which commits the organization to establishing a radiation safety committee and designating a RSO. Among the duties of these officials, under the terms of the license, are to make sure that each person authorized to have or use radioisotopes or sealed sources of radiation has no more than he is authorized to have in his possession at one time and that the total amount of each radioisotope does not exceed the overall limits provided for under the organization's broad license.* It is necessary for each authorized user to make sure that all radioactive materials in their possession

---

\* Most larger organizations and many smaller ones have a license to use radioactive materials covering the entire organization or at least that part of it at a single location. However, it is possible to have multiple individual users at the same facility directly licensed by the NRC. In the latter case, it is still required to establish a committee and identify a safety officer to monitor the use of radioisotopes.

are securely stored. Unlike controlled substances, most radioisotopes normally have no practical use outside the research laboratory and, hence, the strength of the storage facilities are not comparably demanding, but whenever a legitimate user is not physically present, even for short periods of time, the radioactive material must be under lock and key. It is necessary to keep track of the amount of materials used, although some types of use make it difficult to do so with high accuracy. For example, if the use of carbon-14 results in the generation of carbon dioxide, a somewhat uncertain amount may escape into the exhaust system of a fume hood.

Many radioactive materials used in the life sciences are incorporated in materials which require refrigeration to prolong their usefulness. Not only must a refrigerator or freezer used for radioactive storage be lockable, but care must also be taken to avoid contaminating the unit and its contents by spillage or leakage of material from a container due to freezing of the contents. In such a case, the lost material will usually be trapped in the ice or frost within a freezer. The trapped material could represent a personnel problem or a possible uncontrolled release of radioactivity into the sanitary system when the freezer is defrosted.

Not only must the experimental radioactive materials be stored properly, but all of the waste products which could contain any residual active materials must be kept and stored within the laboratory until they can be disposed of safely, usually by a radiation safety specialist. The temporary storage of radioactive waste gives rise to the possibility of accidental removal of the material as ordinary trash. Any trash containers containing radioactive waste should be distinctively marked, and custodial personnel should receive special training to recognize the containers and instructions so as not to combine the contents with ordinary trash. Laboratory personnel must be sure to place radioactive waste and potentially contaminated waste in these special receptacles. Written labels or signs are not sufficient to prevent accidental losses of radioactive waste as ordinary trash.

An inability to account for the radioactive material in the possession of a licensee is likely to be taken as a serious event by the NRC or its local surrogate in an agreement state. Each loss of radioactive material, if in such quantity and under such circumstances as to pose a hazard to persons in unrestricted areas, must be reported immediately by telephone and telegraph to the director of the NRC for the region in which the facility is located. Even if the amounts lost do not appear to pose a significant risk, operational procedures for the laboratory in question should be reviewed by the internal radiation safety committee and the NRC should be notified as an item of information. Operations of facilities where continuing problems of accountability occur, even for minor problems, should be carefully monitored because they can lead to a loss of credibility of the oversight program for the entire institution or commercial research laboratory. A pattern which appears to reflect poor governance of the facilities using radiation, with a subsequent failure of the radiation safety program to take prompt, effective, corrective actions, has resulted in substantial fines by the NRC in recent years.

Many laboratories conduct operations which do not require the use of radioactive materials in the same space as those in which radiation research is employed. In many of these instances, different personnel are employed in the two programs. Nonusers of radioisotopes should be made sufficiently aware of the procedures required for the safe, legal use of radiation so that they will neither inadvertently violate any safety requirements for the use of byproduct materials nor misunderstand any actions of the employees involved with radiation. While the latter are present, they can and must take precautions to avoid unnecessarily exposing the other persons in the laboratory to radiation, but when they are not present, nonusers need to be aware of the areas where radioactive sources and waste are stored and areas which they should avoid if there is any possibility of contamination. Any area containing radioactive materials should be clearly marked with signs bearing the radiation symbol and the label:

**CAUTION**
**RADIOACTIVE MATERIALS**

### 4.5.3.4. Etiologic Agents

Laboratories which employ etiologic agents as well as chemicals in their research differ from ordinary chemical laboratories in a number of ways, primarily because the agents which are involved may be infectious to humans. Although some data indicate a higher incidence of diseases among this class of laboratory worker that is associated with the organisms used in the research, other data show that there are virtually no secondary infections from the primary laboratory infections. There is a wide variation in the level of professionalism among employees in laboratories using biological agents, so it is extremely important that the laboratory manager and all employees set and maintain high standards of safety.

There is less agreement in this area than in most as to the actual risks posed by a number of operations and specific etiologic agents. In many cases, the scale of the operation determines the level of safety required. In other cases, the determining factor might be the availability of a safe, effective vaccine. The CDC has published a safety guide to assist the laboratory director in making appropriate decisions as to the level of biosafety precautions required and has defined sets of standard practices, special practices, containment equipment, and laboratory equipment appropriate for each level. A condensed version of the salient features of these guidelines will be found in Chapter 5. Note that the word "guidelines" was used. The CDC manual is a set of recommended practices which do not have the force of law. However, they do form a body of information which has been carefully reviewed by a large number of highly trained, experienced, professionals. It is strongly recommended that the recommendations be incorporated into a formal biosafety program at any academic or corporate research institution which engages in a substantial level of research in the life sciences.

Although a comprehensive biosafety committee is not required by law, as it is for radiation safety, the creation of one to monitor the use of etiologic agents within an organization is highly desirable. For specific areas such as recombinant DNA and research using animals, a regulatory committee is required. If the committee is provided with an appropriate charge, it can be of considerable help in justifying expenditures to provide safety resources and services for the biological laboratories as well as providing a uniform set of safety and performance standards for the laboratories.

In the context of this section, any biologically active materials which could result in human infections should be stored and used so as to preclude or minimize the probability of infections for all laboratory employees as well as any other employees such as custodial workers, maintenance personnel, and visitors to the laboratory. This can best be done by limiting access to the facility to those who have a specific need to be present, keeping materials in active use within the appropriate enclosure, keeping materials not currently required in storage, and decontaminating work surfaces frequently.

### 4.5.3.5. Carcinogens

Cancer, in its many forms, is the second largest cause of death in the U.S. and most other developed countries. Anything which increases the likelihood of initiating the eventual onset of cancer is a matter of significant concern to those working with such materials. Although there are a few specific materials which have been known or suspected for some time to be carcinogenic agents, the long latency period, ranging from as few as 5 to as many as 30 years, has delayed identification of causal relationships in many cases. OSHA has, in the past, regulated possible carcinogens on a case-by-case basis and continues to do so. This is the basis of the list in Section 4.2.4.4. As a result of the difficulty of generating a standard for each chemical and taking the standard through the entire regulatory process, OSHA proposed and adopted a generic carcinogen standard, 29 CFR 1910.101 to 1910.152. In this standard, which is intended to shorten the regulatory process, OSHA proposed standards and procedures for evaluating human epidemiological data and animal studies so that materials which met the standards would be considered to be carcinogenic. At this time, disagreements on the validity of the criteria have resulted in no new materials being considered carcinogens under the new standard; the only ones currently regulated as carcinogens are materials which have individually gone through the standard-setting process.

Before a research program is begun using any of the currently regulated carcinogens or any which may become regulated in the future, either by the individual standard-setting process or the generic standard, there are a number of factors which should be considered:

1.  Are there suitable alternative, noncarcinogenic materials which could be used?
2.  Are there acceptable permissible exposure and ceiling limits below which it would be acceptable to use the material, with resultant exemptions from some restrictions upon the use of the material?
3.  If the answer to question 2 is no, can the provisions of the standard be met? An examination of the standards for the currently regulated carcinogens, as well as the provisions of the generic standards, reveals a number of restrictions upon the spaces in which the material is to be stored and used. The research investigator and employer must commit to meeting all of these conditions.

It would be inappropriate in this section, which is concerned primarily with the storage of materials, to pursue the subject of research with carcinogens further. However, unless there is a firm commitment to compliance with all regulations pertaining to carcinogenic research, it makes very little sense to acquire a carcinogenic chemical, especially if it is regulated or appears likely to become regulated, so that it cannot be legally used. The responsibility of assuring that it is not used, and the potential liability if someone exposed to the material were to eventually develop cancer, would appear to preclude acquiring the material unless a definite research program of sufficient merit exists which would justify the risk and the compliance effort.

If a program requiring the use of a carcinogen does exist, the materials should be kept in either a secure cabinet (in sealed, unbreakable containers) or a sealed, protected system. A current inventory of the quantities held must be maintained. The containers should be marked:

### DANGER, CONTAINS _____
### CANCER HAZARD

Containers for a potential carcinogen in which the evidence of carcinogenicity is based on limited, but suggestive, evidence should be labeled:

### POTENTIAL CANCER HAZARD

If the material is such that the research is not exempt from the provision requiring the establishment of a regulated area where the material is stored or used and in which it is possible for a person to be exposed to concentrations above the permissible levels (assuming that such have been established), then the entrances to the area are to be clearly marked with signs stating:

### DANGER, _____
### CANCER HAZARD
### AUTHORIZED PERSONNEL ONLY

Any employee expected to handle a carcinogen must be informed of the potential risk and the procedures to be followed in an emergency. An emergency plan specific for the area and for the research program must be developed in order to prevent exposures of the normal occupants of the area and to prevent the accidental exposures of others outside the immediate research facility. The training program should include everyone who may work within the regulated area and who handles the materials, including persons who may only stock the shelves and those individuals performing custodial and maintenance services, unless all such services are performed only with operations suspended and in the presence of a qualified laboratory employee who is fully trained.

## REFERENCES (4.5.3 TO 4.5.3.5)

1. *J. Hazardous Mater.*, 1, 334, 1975.
2. Allied Fisher Scientific, Fairlawn, NJ, 1986.
3. *LabGuard Safety Label System,* Mallinckrodt, Paris, KY, 1986.
4. *Baker Saf-T-Data Guide,* J. T. Baker Chemical Company, Phillipsburg, NJ, 1987.
5. *Chemical Labeling Standard,* ANSI Z129.1 (pending), American National Standards Institute, New York, 1987.
6. Occupational Safety and Health Administration, General Industry Standards, Title 29, Code of Federal Regulations, Part 1910.109(c), Washington, D.C.
7. *Hazardous Chemical Data,* NFPA 49, National Fire Protection Association, Quincy, MA, 1975.
8. Nuclear Regulatory Commission, Title 10, Code of Federal Regulations, Part 20, §207, Washington, D.C.

9. Biosafety in Microbiological and Biomedical Laboratories, HHS Publ. No. (CDC) 84-8395, Department of Health and Human Services, Washington, D.C., 1984.

10. **Favero, M. S.,** Biological hazards in the laboratory, in *Proc. Institute on Critical Issues in Health Laboratory Practices,* Richardson, J. H., Schoenfeld, E. S., Tulis, J. J., and Wagner, W. W., Eds., DuPont, Wilmington, DE, 1985.

11. **Pike, R. M.,** Laboratory-associated infections: incidence, fatalities, cases and prevention, *Annu. Rev. Microbiol.,* 33, 41, 1979.

12. Occupational Safety and Health Administration, General Industry Standards, Title 29, Code of Federal Regulations, Parts 1910.101 to 1910.152, Washington, D.C., 1988.

## 4.5.4. ETHERS

Ethers represent a class of materials which can become more dangerous upon prolonged storage because they tend to form explosive peroxides with age. Exposure to light and air enhances the formation of the peroxides. A partially empty container increases the amount of air available and, hence, the rate at which peroxides will form in the container. It is preferable, therefore, to use small containers which can be completely emptied rather than to take the amounts needed for immediate use from a larger container over a period of time, unless the rate of use is sufficiently high so that peroxides will have a minimal time to form.

The following material is taken directly from the second edition of this handbook from the article by Steere, Control of Peroxides in Ethers.[25] It has been edited slightly so as to conform to the format of the current edition. The sections on Detection and Estimation of Peroxides and Removal of Peroxides have been substantially shortened, in line with the philosophy espoused elsewhere in this section to keep on hand only amounts that will be quickly used.

"Ethyl ether, isopropyl ether, tetrahydrofuran, and many other ethers tend to absorb and react with oxygen from the air to form unstable peroxides which may detonate with extreme violence when they become concentrated by evaporation or distillation, when combined with other compounds that give a detonatable mixture, or when disturbed by unusual heat, shock or friction. Peroxides formed in compounds by autoxidation have caused many laboratory accidents, including unexpected explosions of the residue of solvents after distillation, and have caused a number of hazardous disposal operations. Some of the incidents of discovery and disposal of peroxides in ethers have been reported in the literature, some in personal communications, and some in the newspapers. An "empty" 250-cc bottle which had held ethyl ether exploded when the ground glass stopper was replaced (without injury), another explosion cost a graduate student the total sight of one eye and most of the sight of the other, and a third explosion killed a research chemist when he attempted to unscrew the cap from an old bottle of isopropyl ether.

Appropriate action to prevent injuries from peroxides in ethers depends on knowledge about formation, detection and removal of peroxides, adequate labeling and inventory procedures, personal protective equipment, suitable disposal methods, and knowledge about formation, detection and removal of peroxides.

## FORMATION OF PEROXIDES

Peroxides may form in freshly-distilled and undistilled and unstabilized ethers within less than two weeks, and it has been reported that peroxide formation began in tetrahydrofuran after three days and in ethyl ether after eight days. Exposure to air, as in opened and partially emptied containers, accelerates formation of peroxides in ethers, and while the effect of exposure to light does not seem to be fully understood, it is generally recommended that ethers which will form peroxides should be stored in full, air-tight, amber glass bottles, preferably in the dark.

Although ethyl ether is frequently stored under refrigeration, there is no evidence that refrigerated storage will prevent formation of peroxides, and leaks can result in explosive mixtures in refrigerators since the flash point of ethyl ether is –45°C (–49°F).

The storage time required for peroxides as $H_2O_2$ to increase from 0.5 ppm to 5 ppm has been reported to be less than two months for a tin-plate container, six months for an aluminum container, and over 17 months for a glass container. The same report stated that peroxide content was not appreciably accelerated at temperatures about 11°C (20°F) above room temperature. Davis has reported the formation of peroxides in olefins, aromatic and saturated hydrocarbons, and ethers particularly, with initial formation of an alkyl hydroperoxide which can condense on standing or in the presence of a drying agent to yield further peroxidic products. Davis refers to reports that the hydroperoxides initially formed (e.g., from isopropyl ether and tetrahydro-furan) may condense further, particularly in the presence of drying agents, to give polymeric peroxides and that cyclic peroxides have been isolated from isopropyl ether.

The literature contains an extensive report on autoxidation of ethyl ether.

Isopropyl ether seems unusually susceptible to peroxidation and there are reports that a half-filled 500-ml bottle of isopropyl ether peroxidized despite being kept over a wad of iron wool. Although it may be possible to stabilize isopropyl ether in other ways, the absence of a stabilizer may not always be obvious from the appearance of a sample, so that even opening a container of isopropyl of uncertain vintage to test for peroxides can be hazardous. Noller comments that "neither hydrogen peroxide, hydroperoxide nor the hydroxyalkyl peroxide are as violently explosive as the peroxidic residues from oxidized ether."[1]

## DETECTION AND ESTIMATION OF PEROXIDES

Appreciable quantities of crystalline solids have been reported as gross evidence of formation of peroxides, and a case is known in which peroxides were evidenced by a quantity of viscous liquid in the bottom of the glass bottle of ether. If similar viscous liquids or crystalline solids are observed in ethers, no further tests are recommended, since in four disposals of such material there were explosions when the bottles were broken.

The potassium iodide method of testing for peroxides in ethyl ether, as described by the Manufacturing Chemists Association, is as follows:

Add 1 cc of a freshly-prepared 10% solution of potassium iodide to 10 cc of ethyl ether in a 25-cc glass-stoppered cylinder of colorless glass protected from light; when viewed transversely against a white background, no color is seen in either liquid.

If a yellow color appears when 9 cc of ethyl ether are shaken with 1 cc of a saturated solution of potassium iodide, according to one organization, there is more than 0.005% peroxide and the ether should be discarded.

## INHIBITION OF PEROXIDES

No single method seems to be suitable for inhibiting formation in all types of ethers, although storage and handling under an inert atmosphere would be a generally useful precaution.

Some of the materials which have been used to stabilize ethers and inhibit formation of peroxides include the addition of 0.001% of hydroquinone or diphenylamine, polyhydroxylphenols, aminophenols, and arylamines. Addition of 0.0001 g pyrogallol in 100 cc ether was reported to prevent peroxide formation over a period of two years. Water will not prevent formation of peroxides in ethers, and iron, lead and aluminum will not inhibit the peroxidation of isopropyl ether, although iron does act as an inhibitor in ethyl ether. Dowex-1© has been reported effective for inhibiting peroxide formation in ethyl ether, 100 ppm of 1-naphthol for isopropyl ether, hydroquinone for tetrahydrofuran, and stannous chloride or ferrous sulfate for dioxane. Substituted stilbene-quinones have been patented as a stabilizer against oxidative deterioration of ethers and other compounds.

### REMOVAL OF PEROXIDES

Reagents which have been used for removing hydroperoxides from solvents are reported to include sodium sulfite, sodium bisulfite, stannous chloride, lithium tetrahydroalanate (caution: use of this material has caused fires), zinc and acid, sodium and alcohol, copper-zinc couple, potassium permanganate, silver hydroxide, and lead dioxide.

Decomposition of ether peroxides with ferrous sulfate is a commonly used method; 40 g of 30% ferrous sulfate solution in water is added to each liter of solvent. Caution is indicated since the reaction may be vigorous if the solvent contains a high concentration of peroxide.

Reduction of alkylidene or dialkyl peroxides is more difficult but reduction by zinc dissolving in acetic or hydrochloric acid, sodium dissolving in alcohol (note ease of ignition of hydrogen) or the copper-zinc couple might be used for purifying solvents containing these peroxides.

Addition of one part of 23% sodium hydroxide to 10 parts of ethyl ether or tetrahydrofuran will remove peroxides completely after agitation for 30 minutes; sodium hydroxide pellets reduced but did not remove the peroxide contents of tetrahydrofuran after two days. Addition of 30% of chloroform to tetrahydrofuran inhibited peroxide formation until the eighth day with only slight change during 15 succeeding days of tests; although sodium hydroxide could not be added because it reacts violently with chloroform, the peroxides were removed by agitation with 1% aqueous sodium borohydride for 15 minutes (with no attempt made to measure temperature rise or evolution of hydrogen).

A simple method for removing peroxides from high quality ether samples without need for distillation apparatus or appreciable loss of ether consists of percolating the solvent through a column of Dowex-1© ion exchange resin. A column of alumina was used to remove peroxides and traces of water from ethyl ether, butyl ether, dioxane and petroleum fractions and for removing peroxides from tetrahydrofuran, decahydronapthalene (decalin), 1,2,3,4-tetrahydronaphalene (tetralin), cumene and isopropyl ether."

Because they have a limited shelf life, as noted earlier, ethers should be bought in the smallest practicable containers appropriate to the rate of usage within the facility, preferably in 500-ml containers. In Section 4.2.1, it was acknowledged that

## TABLE 4.6
## Some Materials which Tend to Form Peroxides

| | | |
|---|---|---|
| Acrolein | Cyclooctene | *p*-Dioxane |
| Aldehydes | Decahydronaphthalene | Divinyl ether |
| Allyl ethyl ether | Diacetylene | Ethyl methyl ether |
| Allyl phenyl ether | Dibutyl ether | Methyl acetylene |
| Benzyl ether | Dicyclopentadiene | *o*-Methylanisole |
| Benzoyl-*n*-butyl ether | Diethylene glycol | *m*-Methylphenetole |
| Bromophenetole | Diethylene glycol diethyl ether | Phenetole |
| Butadiene | Diethylene glycol mono-*o*-butyl ether | Tetrahydrofuran |
| *p*-Chloroanisole | Diethyl ether | Tetrahydronaphthalene |
| Cumene | Dimethyl ether | Vinyl acetate |
| Cyclohexene | Dimethyl isopropyl ether | Vinylidene chloride |

buying in small sizes invokes a heavy financial penalty, but some of this price disadvantage can be eliminated by buying in case lots or having a central stock room buy in multiple case lots. It also eliminates much of the need for frequently checking the condition of the contents of a large container. No matter what size container is purchased, each container should be dated when it is received and placed in stock. For isopropyl and diethyl ethers, it is recommended that even unopened containers be disposed of after 1 year, while opened containers should be discarded 6 months after they are first used if they have not been tested periodically during the interval. Opened containers should be tested after 1 month, and continue to be tested until emptied, or at frequent intervals. If only modest amounts of peroxides are found, the ethers can be decontaminated. If an alumina column is used, the contaminated alumina can be treated with an aqueous solution of ferrous sulfate and discarded as chemical waste. However, it may be difficult to get a hazardous waste disposal firm to accept it. They are very reluctant to accept any waste material if there is any possibility of an explosion. The concern about possible explosions, the cost of the manpower involved in the needed frequent checking, and the possibility of having to pay a premium price for disposal of excess materials as a potential explosive should obviate the argument of a lower unit cost for the larger sizes.

Table 4.6 contains a brief list of some the more common chemicals in which peroxides may form.

## ACKNOWLEDGMENT

[From original article]. Reprinted with permission from *The Chemistry of Ether Linkage,* 1967, Saul Patai, Editor, Interscience Publishers, Inc., Division of John Wiley & Sons, Inc.

## REFERENCES (SECTION 4.5.4)

1. **Douglas, I. B.,** *J. Chem. Educ.,* 40, 469, 1963.
2. **Steere, N. V.,** Control of hazards from peroxides in ethers, *J. Chem. Ed.,* 41, A575, 1964.
3. Accident Case History 603, Manufacturing Chemists Association (reported in part in Reference 2).
4. **Fleck, E.,** Merck, Sharp & Dohme Company Memo, Rahway, NJ, May 11, 1960.

5.  **Noller, C. R.**, *Chemistry of Organic Compounds,* W. B. Saunders, Philadelphia, 1951.
6.  **Rosin, J.**, *Reagent Chemicals and Standards,* 4th ed., D Van Nostrand, Princeton, NJ, 1961.
7.  **Brubaker, A. R.**, personal communication, December 4, 1964.
8.  **Davies, A. G.**, Explosion hazards of autoxidized solvents, *J. R. Inst. Chem.,* 386, 1956.
9.  **Lindgren, G.**, Autoxidation of diethyl ether and its inhibition by diphenylamine, *Acta Chir. Scand.,* 94, 110, 1946.
10. **Pajaczkowski, A.**, personal communication, September 29, 1964.
11. **Davies, A. G.**, *Organic Peroxides,* Butterworths, London, 1961.
12. **Brasted, H. S.**, personal communication, June 1, 1964.
13. **Dugan, P. R.**, *Anal. Chem.,* 33, 1630, 1961.
14. **Dugan, P. R.**, *Ind. Eng. Chem.,* 56, 37, 1964.
15. **Feinstein, R. N.**, Simple method for removal of peroxides from diethyl ether, *J. Org. Chem.,* 24, 1172, 1969.
16. *Encyclopedia of Chemical Technology,* Vol. 5, Kirk, R. E. and Othmer, D. F., Eds., Interscience, New York, 1950, 142, 871.
17. *Encyclopedia of Chemical Technology,* Vol. 6, Kirk, R. E. and Othmer, D. F., Eds., Interscience, New York, 1950, 1006.
18. **Jones, D. G.**, British Patent 699,179, 1953; *Chem. Abstr.,* 49, 3262f, 1955.
19. **Moffett, R. B. and Aspergren, B. D.**, Tetrahydrofuran can cause fire when used as solvent for LiAlH$_4$, *Chem. Eng. News,* 32, 4328, 1954.
20. Chemical Safety Data Sheet — SD 29, Ethyl Ether, Manufacturing Chemists' Association, Washington, D.C., 1956.
21. **Dasler, W. and Bauer, C. D.**, Removal of peroxides from ethers, *Ind. Eng. Chem. Anal. Ed.,* 18, 52, 1946.
22. **Birnbaum, E. R.**, personal communication, August 11, 1964.
23. Manuals of Techniques, Inc., Beverly, MA, 1964.
24. **Ramsey, J. B. and Aldridge, F. T.**, Removal of peroxides from ethers with cerous hydroxide, *J. Am. Chem. Soc.,* 77, 2561, 1955.
25. **Steere, N. V.**, Control of peroxides in ethers, in *Handbook of Laboratory Safety,* 2nd ed., Steere, N. V., Ed., CRC Press, Boca Raton, FL, 1971, 250.

## 4.5.5. PERCHLORIC ACID*

### Hazards of Perchloric Acid

Hazards of perchloric acid are described in MCA SD-11 as follows.

1.  Perchloric acid is strong, and contact with the skin, eyes, or respiratory tract will produce severe burns.
2.  Perchloric acid is a colorless, fuming, oily liquid. When cold, its properties are those of a strong acid but when hot, the acid acts as a strong oxidizing agent.
3.  Aqueous perchloric acid can cause violent explosions if misused, or when in concentrations greater than the normal commercial strength (72%).
4.  Anhydrous perchloric acid is unstable even at room temperatures and ultimately decomposes spontaneously with a violent explosion. Contact with oxidizable material can cause an immediate explosion.

The following are listed among the causes of fires and explosions involving perchloric acid.

---

* The material in Section 4.5.5 is taken from the article by Everett, K. and Graf, F. A., Jr., Handling perchloric acid and perchlorates, in the *Handbook of Laboratory Safety,* 2nd ed., Steere, N. V., Ed., CRC Press, Boca Raton, FL, 1971. A revised version of this article will be found in Section 4.6.2.4.

1.  The instability of aqueous or of pure anhydrous perchloric acid under various conditions.
2.  The dehydration of aqueous perchloric acid by contact with dehydrating agents such as concentrated sulfuric acid, phosphorous pentoxide, or acetic anhydride.
3.  The reaction of perchloric acid with other substances to form unstable material.

Combustible materials, such as sawdust, excelsior, wood, paper, burlap bags, cotton waste, rags, grease, oil and most organic compounds, contaminated with perchloric acid solution are highly flammable and dangerous. Such materials may explode on heating, in contact with flame, by impact or friction, or may ignite spontaneously.

**Perchloric Acid Storage**

Within the Laboratory: The maximum advisable amount of acid stored in the main laboratory should be no more than two 8-lb (3.6-kg) bottles. A 450-g (1-lb) bottle should be sufficient for individual use. Storage of perchloric acid should be in a fume hood set aside for perchloric acid use. The acid should be inspected monthly for discoloration; if any is noted, the acid should be discarded.

Outside of the Laboratory: Perchloric acid should be stored on an epoxy painted metal shelf, preferably in a metal cabinet away from organic materials and flammable compounds. Discolored acid should be discarded.

**Storage of Anhydrous Perchloric Acid**

The anhydrous acid is much more dangerous than the concentrated 72% acid. If stored for more than 10 d, the acid is likely to develop a discoloration and be capable of spontaneously exploding. A principal hazard in the use of the anhydrous acid is the breakage of containers. Anhydrous perchloric acid will explode in contact with wood, paper, carbon, and organic solvents. Anhydrous perchloric acid should only be made as required and should never be stored.

The building in which perchloric acid is stored as well as the building in which it is used, should present few opportunities for the perchloric acid to encounter organic materials. It would be desirable for the floors to be of epoxy-painted concrete. The bench tops should be of materials which will not react with the perchloric acid, such as epoxy composites. Painted metal furniture should be used instead of wood. None of the furniture should be screwed into the floor, forming openings into which acid could enter and form dangerous metal perchlorates, which could then explode if the bolt is removed. In general, the laboratory furniture should have as few seams, in which the acid could enter and become dry and more dangerous, as possible.

## REFERENCE (SECTION 4.5.5)

1.  Chemical Safety Data Sheet — SD-11, Perchloric Acid, Manufacturing Chemists' Association, Washington, D.C.

## 4.5.6. FLAMMABLE LIQUIDS

Should a fire or an explosion occur in a laboratory, a major concern is to reduce the amount of fuel available to support the fire. Many solvents commonly used in laboratories are highly flammable, and should even a small quantity become involved in the fire, it would have the capacity of significantly increasing the probability of the fire spreading.

The OSHA Act, in Section 1910.106(d)(3), places restrictions on the maximum amounts of flammable liquids allowed to be stored, depending on class, in flammable material storage cabinets within a room, and defines in Section 1910.106(d)(2) the maximum size of individual containers for the various classes of flammables. NFPA Standard 45 provides guidelines for the maximum amounts of flammable liquids that should be allowed in the three classes of facilities — A, B, and C — defined in that standard. This standard has been mentioned, but the three laboratory classes have not been stressed in this volume, in favor of the proposed four-level standard, paralleling legally defined classes of risk for biological facilities and involving wider classes of hazards than that due to the amount of flammable materials in the laboratory alone. For the purposes of this section, the low- and moderate-risk facilities described in Sections 3.1.3.2.1 and 3.1.3.2.2, respectively, may be taken to be approximately equivalent to an NFPA Standard 45 Class C facility, a substantial-risk facility (Section 3.1.3.2.3) to be roughly equivalent to a Class B facility, and a high-risk facility (Section 3.1.3.2.4) to include a Class A facility, but with more restrictions than the latter would require. Note that OSHA has not adopted the restrictions of NFPA Standard 45 and does not address the issue of the total amount of flammables permitted in a laboratory area, although the amount permitted in an interior storage room is defined. OSHA regulations regarding container sizes are based on sections of the 1969 NFPA Standard 30. Before returning to the topic of flammable material storage cabinets, the following two tables define the various classes of flammable and combustible liquids and the maximum container sizes permitted by OSHA for each class. The third table provides the maximum amount of flammable liquids which would be permitted in the three classes of laboratories defined by NFPA 45 (*outside* flammable material storage cabinets and not in safety containers) if the latter standard were to be adopted.

The definitions depend upon the flash points and, in some cases, the boiling points of the liquids. The flash point of a liquid is legally defined in terms of specific test procedures used to determine it, but conceptually is the minimum temperature at which a liquid forms a vapor above its surface in sufficient concentration to be ignited.

In Table 4.7, the first temperature is in degrees Celsius and that in parentheses is the equivalent Fahrenheit temperature. Note that neither Combustible 2 nor 3A materials include mixtures in which more than 99% of the volume is made up of components with flash points of 93.3°C (200°F) or higher.

Table 4.8 is equivalent to Table H-12 from the OSHA General Industry Standards. Polyethylene containers have become widely available, so information on this type of container has been appended to the end of the table.

Several exceptions to Table 4.8 permit glass or plastic containers of no more than 1-gal capacity to be used for Class 1A and 1B liquids: (1) if a metal container would be corroded by the liquid, (2) if contact with the metal would render the liquid unfit for the intended purpose, (3) if the application required the use of more than 1 pt of a Class 1A liquid or more than 1 qt of a Class 1B liquid, (4) an amount of an analytical

## TABLE 4.7
### Definitions and Classes of Flammable and Combustible Liquids

°C (°F)

| Class | Boiling points | Flash points |
|---|---|---|
| Flammable 1A | <37.8 (100) | <22.8 (73) |
| Flammable 1B | ≥37.8 (100) | <22.8 (73) |
| Flammable 1C | — | 22.8 (73) ≤ and <37.8 (100) |
| Combustible 2 | — | 37.8 (100) ≤ and <60 (140) |
| Combustible 3A | — | 60 (140) ≤ and <93.3 (200) |
| Combustible 3B | — | ≥93.3 (200) |

## TABLE 4.8
### Maximum Allowable Size of Containers and Portable Tanks

| Container type | 1A | 1B | 1C | 2 | 3 |
|---|---|---|---|---|---|
| Glass or approved plastic | 1 pt | 1 qt | 1 gal | 1 gal | 1 gal |
| Metal (other than DOT drums) | 1 gal | 5 gal | 5 gal | 5 gal | 5 gal |
| Safety cans | 2 gal | 5 gal | 5 gal | 5 gal | 5 gal |
| Metal drums (DOT specs) | 60 gal | 60 gal | 60 gal | 60 gal | 60 gal |
| Approved portable tanks | 660 gal | 660 gal | 660 gal | 660 gal | 660 gal |
| Polyethylene spec 34 or as authorized by DOT exemption | 1 gal | 5 gal | 5 gal | 60 gal | 60 gal |

## TABLE 4.9
### Amounts of Flammable Liquids Permitted in Laboratory Units Outside Flammable Material Storage Units and Safety Cans (in gal)

| Laboratory class | Flammable/combustible liquid class | Max. quantity per 100 ft$^2$ | Max. quantity per laboratory unit | |
|---|---|---|---|---|
| | | | Unsprinklered | Sprinklered |
| A | 1 | 10 | 300 | 600 |
| | 1, 2, 3A | 20 | 400 | 800 |
| B | 1 | 5 | 150 | 300 |
| | 1, 2, 3A | 10 | 200 | 400 |
| C | 1 | 2 | 75 | 150 |
| | 1, 2, 3A | 4 | 100 | 200 |

standard of a quality not available in standard sizes needed to be maintained for a single control process in excess of one-sixteenth the capacity of the container sizes allowed by the table, and (5) if the containers are intended for export outside the U.S.

Table 4.9 is derived from data in NFPA Standard 45 and gives the amounts permitted by that Standard in laboratories, excluding quantities in storage cabinets and safety cans. The reader is referred to the original standard for the total amounts which would be permitted in laboratories.

**TABLE 4.10**
**A Brief List of Common Class 1A Liquids**

| | | |
|---|---|---|
| Acetaldehyde | Furan | Methyl sulfide |
| 2-Chloropropane | Isoprene | *N*-Pentane |
| Collodian | Ligroine | Pentene |
| Ethyl ether | Methyl acetate | iso-Propylamine |
| Ethanethiol | Methylamine | Propylene oxide |
| Ethylamine | 2-Methylbutane | Petroleum ether |
| Ethyl vinyl ether | Methyl formate | Trimethylamine |

The amounts for instructional laboratories would only be one half of those shown in the table, and an instructional laboratory could not be a Class A laboratory.

In the author's opinion, the amounts permitted by Standard 45 would be excessive for most laboratories, especially academic research facilities. Under these limits, the standard laboratory module, used as the basis for most of this chapter, would be permitted to have 80 gal of mixed flammables or 40 gal of Class 1A flammables in unprotected storage cabinets or containers if the facility were rated as a high-hazard operation. Except in unusual cases where some very active laboratories could use such large quantities in a sufficiently short period to justify their presence, flammable materials in these amounts, other than those contained in research apparatus, should be kept in protected storage.

In Section 1910.106(d)(3) of the General Industry Standards, OSHA limits the amounts of Class 1 and 2 liquids in a single flammable material storage cabinet to 60 gal and the amount of Class 3 liquids to 120 gal. Thus, even if the integrity of a single storage cabinet were breached in a fire, or if an accident occurred while the cabinet was open, no more than 60 gal of Class 1 and 2 liquids or 120 gal of a Class 3 liquid could become involved in the incident. Even this amount is substantial.

### 4.5.7. REFRIGERATION STORAGE

Two of the most dangerous storage units in any laboratory are the ordinary refrigerator and, to a lesser extent, the freezer. This is primarily due to the storage of flammable materials within them, although there are also problems due to individuals using them as a place to store food as well as for their intended purpose. Refrigerators intended for the storage of laboratory chemicals and biological materials should not be used for personal items, particularly food and beverages.

Table 4.11 lists the flash points of a number of common solvents which are below or close to the normal operating temperature (about 38°F or 3.3°C) of a common refrigerator. The flammable limits, in percent by volume in air, are also given for these same solvents. Most of these solvents evaporate rapidly, so they would quickly reach equilibrium concentrations in a small confined space.

Storage of flammable materials in refrigerators or other confined spaces in which the vapors can be trapped and which also contain sources of ignition represent a potential explosion hazard. Carelessly closed containers, e.g., screw caps that are not firmly tightened or beakers containing solvents covered only with aluminum foil or plastic wrap, will allow vapors to escape from the container and, given sufficient time, build up in the confined space until they could reach a concentration in excess

## TABLE 4.11
### Flammability Characteristics of Some Common Solvents

| Chemical | Flash point (°C) | Flammable limit (%) | |
|---|---|---|---|
| | | Lower | Upper |
| Acetaldehyde | −37.8 | 4 | 60 |
| Acetone | −17.8 | 2.6 | 12.8 |
| Benzene | −11.1 | 1.3 | 7.1 |
| Carbon disulfide | −30.0 | 1.3 | 50 |
| Cyclohexane | −20.0 | 1.3 | 8 |
| Diethyl ether | −45.0 | 1.9 | 36 |
| Ethyl acetate | −4 | 2.0 | 11.5 |
| Ethyleneimine | −11 | 3.6 | 46 |
| Gasoline (approximate) | −38 | 1.4 | 7.4 |
| n-Heptane | −3.9 | 1.05 | 6.7 |
| n-Hexane | −21.7 | 1.1 | 7.5 |
| Methyl acetate | −10 | 3.1 | 16 |
| Methyl ethyl ketone | −6.1 | 1.8 | 10 |
| Pentane | −40.0 | 1.5 | 7.8 |
| Toluene | 4.4 | 1.2 | 7.1 |

of the lower flammable limit. A spark may then cause ignition, and, because the reaction is temporarily constrained, very high pressures can build up until the refrigerator door latch fails and a powerful explosion ensues. Many such cases have been observed to occur, and in most cases, if there were workers in the vicinity, in front of the refrigeration unit, it is likely that they would have sustained serious, if not fatal, injuries. Fortunately, it appears in many cases, that the propensity of laboratory workers to place improperly sealed containers in refrigerators is greatest at the end of the work day, when they may be in a hurry to leave. This treatment, coupled with the materials remaining undisturbed for extended periods of time after normal working hours, tends to make night hours the most likely time for a refrigerator explosion to occur.

A normal refrigerator has many sources of ignition, the thermostat, interior light, the light switch on the door, the defrost heater, the defrost control switch, the compressor unit, and the fan. Most of these are located within the space being maintained cool, but self-defrosting units containing an internal drain can permit the internal vapors to flow into the compressor space below the usable space.

It is possible to modify a normal home refrigeration unit to remove the internal sources of ignition, but unless it is done by a person who knows precisely what to do and does it very carefully, it is unlikely that the result will be as safe as one initially designed to be used for flammable material storage. The liability which could result from the failure to prevent an explosion of an improperly locally modified unit makes this an imprudent economy measure. In most cases, refrigeration units need only be rendered safe for prevention of ignition by components of the refrigerator themselves, i.e., be designated as safe for the storage of flammable materials, instead of meeting standards for total explosion safety, which would permit them to be operated in locations where flammable vapors and gases exist outside the refrigeration units.

The additional cost of the latter units, plus the cost of making the proper electrical explosion proof connections, is an unnecessary expense for most laboratories.

In virtually all cases, refrigeration units which operate as "ultra-lo" units (i.e., internal temperatures on the order of –60°C to –120°C) need not be flammable material storage units or explosion safe. Note that none of the materials in Table 4.11 have flash points in this range.

As noted earlier, refrigerators last for as many as 20 to 30 years. It is not feasible to accept assurances by laboratory managerial personnel that no flammable materials will ever be placed in an ordinary refrigerator because neither the individual making the promise nor the program for which the refrigerator is purchased are likely to occupy the space for such an extended period of time. It is also not reasonable to depend upon marking laboratory refrigerators, no matter how prominently, as not to be used for flammable material storage and count on compliance with the restriction. If there is an ordinary consumer-quality refrigerator in the laboratory, it is virtually certain that someone will eventually use it improperly. Therefore, it is recommended that all refrigerators purchased for use in laboratory areas be originally constructed for flammable material storage and bear an appropriate label on the front of the unit indicating that it meets such standards. Exceptions should be very few and restricted to those programs that are, in fact, essential to the basic operations of a stable department and not dependent upon the program of an individual or a limited number of persons constituting a temporary research group. Older units not meeting the standards for flammable material storage should be phased out as rapidly as possible, or moved from laboratories in which usage of flammables is a normal activity to noncritical areas where current usage is not likely to involve the storage of flammable liquids, and should be replaced with a suitable refrigerator or freezer. This procedure will enable all laboratories to be equipped with safe units relatively economically over a period of time.

## REFERENCES (SECTIONS 4.5.6 TO 4.5.7)

1. Occupational Safety and Health Agency, Title 29, Code of Federal Regulations, Section 1910.106(d)(3), Washington, D.C. General Industry Standards, 1988.
2. *Flammable and Combustible Liquids Code,* NFPA-30, National Fire Protection Association, Quincy, MA, 1981.
3. *Standard on Protection for Laboratories Using Chemicals,* NFPA-45, National Fire Protection Association, Quincy, MA, 1982.
4. Flammability Ratings of Flammable Liquids, Fisher Chemical Company, Chicago, IL.
5. **De Roo, J. L.,** The Safe Use of Refrigerator and Freezer Appliances for Storage of Flammable Materials, Union Carbide, South Charleston, WV,
6. **Langan, J. P.,** Questions and Answers on Explosion-Proof Refrigerators, Kelmore, Newark, NJ.
7. Properties of Common Flammable and Toxic Solvents, Division of Industrial Hygiene, New York State Department of Labor, Albany, 1977.

### 4.5.8. GAS CYLINDERS

Gas cylinders are used for many purposes in the research laboratory. Most individuals usually think of gas cylinders in the context of the standard industrial gas cylinder, which is 23 cm (9 in.) in diameter and 140 cm (55 in.) high. However, many other sizes are available. All of these cylinders can represent a significant hazard. A standard cylinder weighing about 64 kg (140 lb) often contains gas at pressures of about 21 MPa (3000 psi) or more. Should the valve connection on top of the cylinder be broken off, the cylinder would correspond to a rocket capable of punching a hole through most laboratory walls and would represent a major danger to all occupants in any area where such an incident occurred. The contents of cylinders also frequently represent inherent hazards. These pressure-independent hazards associated with the contents include flammability, toxicity, corrosiveness, excessive reactivity, and potential asphyxiation if the volume of air displaced by the contents of the cylinder is sufficient. Obviously, measures need to be taken to insure that the integrity of the cylinder is totally maintained. Compressed gas cylinders can be used safely if due care is taken with them and the accessories and systems with which they may be combined.

A compressed gas is defined by the DOT as "any material or mixture having in the container either an absolute pressure greater than 276 kPa (40 lbf/in.$^2$) at 21°C (70°F), or an absolute pressure greater than 717 kPa (104 lbf/in.$^2$) at 54°C (129.2°F) or both, or any liquid flammable material having a Reid vapor pressure greater than 276 kPa (40 lbf/in.$^2$) at 38°C (100.4°F)".

The actual pressure will depend upon the type of gas in the cylinder and the physical state of the contents. Cylinders containing gases which are gaseous at all pressures that are practicable for the cylinder, such as nitrogen or helium, will provide a pressure reading which reflects the amount of material in the cylinder, while those in equilibrium with a liquid phase, such as ammonia, carbon dioxide, or propane, will be at the pressure of the vapor as long as any of the material remains in the liquid phase, provided the critical temperature is not exceeded. The weight of the cylinder is used to measure the amount of the material in the cylinder for the latter type of gas.

Compressed-gas cylinders will usually be stamped near the top of the cylinder with the DOT code appropriate to the specification under which the cylinder was manufactured and with the pressure rating at 21°C, usually in lbf/in.$^2$. The last date on which it was tested will usually be stamped near the upper end of the cylinder. In most cases, the test interval for a steel cylinder is 10 years. It is the responsibility of the company distributing the cylinders to be sure that they are within the appropriate test span. Unfortunately, it appears that this is not always done. A spot-check of a large group of cylinders at one facility revealed that more than 10% were significantly beyond the required test date.

Very few laboratory facilities have the capability to refill cylinders, nor in most cases is the use heavy enough to make it economical to acquire the capability to do so safely. Most users of compressed-gas cylinders have an arrangement with a vendor to periodically replace empty cylinders with full ones of the same type on a regular basis, paying a demurrage charge on the number which they maintain in use. As a result, those on hand are continually changing and care must be taken to ensure that the identity of the gas in the cylinder is known. Color codes are unreliable, especially for the caps since these are always taken off in use. Tags attached to the cap are not appropriate for the same reason. A stenciled name or an adhesive label placed on the side of the cylinder would constitute a satisfactory system, but the contents of the replacement cylinders should be confirmed upon each delivery. A cylinder for which the contents are not certain should not be accepted. Any units for which the identification label or stencil has become defaced should be returned to the supplier.

OSHA at one time had specific standards for the inspection of compressed gas cylinders and their safety relief devices. However, these standards were revoked as of February 10, 1984. OSHA still has general standards on gas cylinders incorporated into Section 1910.101(a) which require that visual and other inspections be conducted as prescribed in the hazardous materials regulations of the DOT (49 CFR 171 to 179 and 14 CFR 103). Where those regulations are not applicable, visual and other inspections shall be conducted in accordance with Compressed Gas Association (CGA) pamphlets C-6-1968 and C-8-1968. Section 1910.252, which concerns welding, cutting, and brazing, also contains regulations for compressed gas cylinders used in these operations. Users should seek further guidance from the most current edition of these publications and from the most current version of the CGA publication, Safety Relief Device Standards.

### 4.5.8.1. Bulk Storage

Most building codes have restrictions on the location and arrangement of bulk storage facilities for compressed gas cylinders. Storage areas for gas cylinders should not intrude on a required path of egress, such as stairs and hallways, or be in an outside area where occupants evacuating a building would be required to pass them or congregate. Cylinders in storage containing oxidizing agents should either be at least 25 ft from those containing reducing agents or the two different types should be separated by a fire wall at least 5 ft high with at least a 30-min fire rating. If possible, flammable gas cylinders should be stored separately from other cylinders, even those containing inert gases, so that in the event of a fire, their contribution of additional fuel would not increase the possibility of other nearby cylinders rupturing. Nominally empty cylinders should not be stored in the same location as full ones and all "empty" cylinders should be clearly marked as empty. If separate storage is provided, simply marking a cylinder with the letters "MT" is usually sufficient. Empty oxidizing- and reducing-agent cylinders should be separated as if they were full since they should never be emptied to less than about 172 kPa (25 lbf/in.$^2$) in order to prevent contamination of the interior of the cylinders.

In addition to any code restrictions, there are a number of common-sense guidelines which should be followed in providing a storage location. Cylinders should be stored in well-ventilated areas, and air should be able to circulate freely around them, so that any leakage of gases will be quickly eliminated. The area in which they are stored should not be damp, and the surface on which they sit should be dry to minimize corrosion of the steel. Outdoor storage areas should be protected from the weather, and cylinders should not be stored in areas where heavy objects will fall on them or in areas with heavy vehicular traffic where they might be struck by a moving vehicle. Indoor storage areas should be of fire-resistant construction. Since many compressed gases are heavier than air, the storage location should preferably be above grade and should not readily connect to spaces where it would be likely for escaping gases to flow and collect. Cylinder caps should always be on the cylinders while they are in storage and at any time they are moved. Signs should be prominently displayed at the storage area listing all gases stored at the location. All cylinders containing a specific variety should be grouped together.

Compressed-gas cylinders should be handled carefully when being moved to and from the bulk storage area and during normal use. Although they appear sturdy, the cylinders are designed as shipping containers for the gases within them and, hence, are designed to be as light as possible, consistent with reasonable margins of safety for pressure and physical handling. Cylinders should always be moved using a transporting device (e.g., strapped to a common hand dolly) or any number of commercial units specially designed to transport cylinders. They should never be moved by supporting the valve cap with one hand while rolling the base of the cylinder along the floor with the other. There are also manual hand trucks specially designed for moving heavy objects up and down stairs. If a freight elevator is not available when moving a cylinder from one floor to the next, one of these devices should be employed. A cylinder should never be hoisted by attaching a cable to the cylinder cap.

Except for those designed to hold toxic gases, most cylinders incorporate a rupture disk as an over-pressure safety device, which will melt at the relatively low temperatures of 70 to 95°C (158 to 203°F). Because of these safety devices, the temperature in an area where cylinders are stored should not exceed 52°C (125°F), nor should the cylinders be exposed to localized heating.

Cylinders should be stored in an upright position (i.e., with the valve end up), never on their side. The storage area should contain facilities which would make it possible to firmly secure the cylinders in an upright position and prevent them from falling or being knocked over. Parallel bars with space between them just sufficient to accommodate a cylinder, with a chain holding them in place, makes a secure storage area; a group of cylinders stored compactly with a chain drawn tightly around them would also be satisfactory. In the latter case, however, the length of the chain holding them upright should be adjustable so that if the number stored decreases substantially, the chain will not become so slack as to permit cylinders to fall over.

### 4.5.8.2. Laboratory Storage

Storage of cylinders in a laboratory at a given time should be restricted to those in actual use or attached to a system ready for use. If this is not feasible, the actual number of cylinders present should be maintained at an absolute minimum. No cylinder should be in a laboratory unless it is securely fastened to a support so that it cannot fall over. No free-standing cylinder should be allowed in a laboratory, even a nominally "empty" one. There is always a possibility that the empty cylinder has been mislabeled. A good rule of thumb is to treat a compressed gas cylinder as you would a gun; unless confirmed otherwise, always assume that it is loaded and treat it as such.

As long as cylinders in a laboratory are not connected to a system and potentially in use, the policies in Section 4.5.8.1, Bulk Storage, should apply. Such restrictions as the physical separation of oxidizing and reducing agents should not be abrogated unless circumstances are appropriate.

Additional information on operations involving gas cylinders will be presented in Section 4.6.2.8.

## REFERENCES (SECTIONS 4.5.8 TO 4.5.8.2)

1. *Prudent Practices for Handling Hazardous Chemicals in Laboratories,* National Academy Press, Washington, D.C., 1981, 75, 221.
2. Specialty Gas Data Sheets, Air Products and Chemicals, Emmaus, PA.

3. *Safety In Academic Chemistry Laboratories,* 4th ed., American Chemical Society, Washington, D.C., 1985.
4. **Pinney, G.,** Compressed gas cylinders and cylinder regulators, in *Handbook of Laboratory Safety,* 2nd ed., Steere, N. V., Ed., CRC Press, Boca Raton, FL, 1971, 565.
5. *Safety Relief Device Standards — Cylinders for Compressed Gases,* Compressed Gas Association, New York, 1965.
6. *Standard Compressed Gas Cylinder Valve Outlet and Inlet Connections,* V-1, Compressed Gas Association, New York, 1965.

## 4.5.9. ANIMAL FOOD AND SUPPLY STORAGE*

### 4.5.9.1. Animal Food

The Federal Animal Welfare Act (PL 89-544) requires that animal food be stored in facilities which protect it from infestation or contamination by vermin (wild rodents, birds, and insects). Food can be stored in individual animal rooms in verminproof containers with lids, such as plastic garbage containers. Ideally, bulk-food shipments should be stored in a room or warehouse where the temperature can be maintained at less than 70°F and the relative humidity at 50% or less. The room should have doors which prevent the entry of rodents or birds. Vermin control is important since wild rodents, birds, and insects can contaminate stored feed with bacteria, viruses, or parasites, which could adversely affect laboratory animal health. Pesticides should not be used to control vermin in this area while food supplies are present; contamination of food with pesticides can seriously affect experimental results in animals. Boric acid powder can be placed along the walls to control cockroaches, without the negative experimental impact of organophosphate insecticides.

Most commercial laboratory diets contain preservatives and stabilizers which maintain nutrient quality in the diet for up to 6 months. However, diets containing vitamin C (i.e., guinea pig chow and nonhuman primate chow) have a limited shelf life of 90 days because of the instability of vitamin C in the diet. Feed sacks are coded at the manufacturer with the date of milling, and this date should be recorded upon receipt of the shipment, recognizing that the food should be used within 90 days after the milling date to avoid vitamin C deficiency problems in guinea pigs and monkeys. Most facilities use the first in-first out method of warehousing feed, stacking the pallets/bags such that the oldest feed is most accessible for transport to the individual animal rooms.

The ingredients in purified or chemically defined diets are not as stable as those in most commercial diets, and the NIH Guide for the Care and Use of Laboratory Animals recommends that these diets be stored at 39°F or colder.

Diets which contain potential or known hazardous compounds (carcinogens, mutagens) should not be stored in the same area as control diets.

### 4.5.9.2. Supply Storage

Potentially hazardous compounds such as detergents, chemical disinfectants, and insecticides should not be stored in the same area with bulk-feed stores to prevent contaminating the latter. The storage area should be clean and orderly, with appropriate precautions taken to keep it free of vermin.

* This section was written by Dr. David M. Moore.

### 4.5.9.3. Animal Carcass Storage

Animal carcasses that are not immediately incinerated should be kept refrigerated at 44°F or lower. Those to be kept for an extended period should be frozen. Refrigerator units should not be used to store food if used for carcass retention.

## REFERENCES (4.5.9 TO 4.5.9.3)

1. Guide for the Care and Use of Laboratory Animals, NIH Publ. 85-23, Nuclear Regulatory Commission, Washington, D.C., 1985, 22.
2. Animal Welfare Act (PL 89-544), Title 9, Subchapter A, Subpart B, Section 3.25 — Facilities, General, paragraph C — Storage.
3. **Hessler, J. R. and Moreland, A. F.,** Design and management of animal facilities, in *Laboratory Animal Medicine,* Fox, J. G. et al, Eds., Academic Press, Orlando, 1984, 509.

## 4.6. HANDLING AND USE OF CHEMICALS: LABORATORY OPERATIONS

Laboratory personnel work in a *potentially* extremely hazardous and unforgiving environment. The substances with which they work may be toxic, flammable, explosive, carcinogenic, pathogenic, or radioactive, to mention only a few unpleasant possibilities. The hazards may cause an immediate or acute reaction, or the effects may be delayed for several years. A worker may be lulled into a false sense of security because of the seeming safety of a material, according to current knowledge, but eventually evidence may develop indicating that continued exposure may cause unexpected or cumulative and irreversible effects.

The equipment in the facility, if used improperly or if it becomes defective, could represent physical hazards which could result in serious injuries or death. Electric shock, cuts, explosions due to the rupture of high-pressure systems (or implosion of large vacuum systems), exposure to cryogenic materials, excessive levels of exposure to ionizing and nonionizing radiation, heat, mechanical injuries due to moving systems, equipment or supplies simply falling on a person, among many other possibilities, may occur in a laboratory environment.

Of course the laboratory environment is not the only place an injury can occur. Scoffers who do not put a high probability on the possibility of an accident happening in the laboratory often point out that they could have been injured while driving to work. Of course this is possible, as is being struck by lightning or any number of other possibilities. If we were to brood about all the things which could happen each day, we might choose to not get up each morning. However, it is necessary that we do so and that some risks be taken. It is prudent, however, to follow practices which will minimize the risks. Most of us would not choose to drive with faulty brakes, on the wrong side of the road, or through red lights and stop signs, nor would most of us deliberately violate similar common-sense rules governing practices which would lead to ill health. It is impossible to achieve absolute safety, but in the presence of hazards it is only reasonable to take those steps which will efficiently and effectively reduce the risks to acceptable levels. Laboratory workers should follow ALARA principles (using the parlance of radiation safety practice) and reduce the risks to a level as low as reasonably achievable.

Laboratory operations are so varied that it would be totally impractical to attempt to exhaustively cover the topic. There are, however, some basic considerations which should be used to enhance the safety of laboratory operations. Some of them are common sense and some have been mandated by regulatory requirements because some safety-related practices are too important to be left to choice. Prior to addressing specific topics, the following list of simple rules, if followed faithfully, would dramatically reduce the number of laboratory accidents which occur or would diminish the consequences of those which do occur.

1. Plan the work carefully. At the beginning of an extended project, formally analyze the proposed program for possible hazards and consider the consequences of possible failures or errors. Ask a colleague to review the hazard analysis with you. Being too close to a subject often leads to overlooking potential problems.

2. Make sure the right equipment is available and in good condition. All too often, makeshift equipment or equipment which has deteriorated is the cause of an accident. Rarely is it worth the risk to take chances. Most persons with more than a few years of experience can think of a number of examples where this has proven true, sometimes tragically.

3. Make sure all systems are assembled in a stable and solid manner that accommodations for the specific limitations or failure modes of the individual components are factored into the operation of the total system.

4. If the release of a toxic or hazardous substance may occur, the work should be done in a fume hood appropriately designed for the operation.

5. Use an explosion shield or other protective enclosure if there is a possibility of a violent reaction. Do not overlook the possibility that scaling up a process will change the safety parameters.

6. Chemicals should be handled carefully at all times, using appropriate containers and carrying devices. Open containers should be closed after use and unneeded reagents should be returned to storage.

7. Do not hurry unnecessarily or compromise on safety. Take the time to label temporary containers.

8. Do not work alone. As a minimum, a second person should be aware of an individual working alone in a laboratory and definite arrangements should be made for periodic checks. Excessively long working hours increase the likelihood of mistakes due to fatigue.

9. Follow good housekeeping practices. Maintain the work area in a reasonably orderly fashion.

10. Do not set up equipment so as to block means of egress from the work area. Consider the activities of others sharing the facility with you in establishing your own work space.

11. Conscientiously use any protective equipment required and wear appropriate clothing.

12. Make sure that you are familiar with and conscientiously follow all safety and emergency procedures.

13. The work area in a laboratory is not a restaurant or a place to socialize. Coffee and meal breaks should be taken at a desk outside the work area or in a lounge set aside for the purpose.

14. Anyone indulging in horseplay or practical jokes within a laboratory should be excluded from the facility.
15. Never work while under the influence of drugs or alcohol. This should include prescribed drugs for legitimate problems, e.g., codeine, antihistamines, etc.

## REFERENCE (SECTION 4.6)

1. *Safety in Academic Laboratories,* 4th ed., American Chemical Society, Washington, D.C., 1985.

### 4.6.1. PHYSICAL LABORATORY CONDITIONS

Many of the points to be made in this section have been alluded to earlier in Chapter 3, where the factors that should be considered in laboratory design were discussed. If the layout of the laboratory is similar to that of the standard laboratory module shown in Figure 3.1, and repeated here as Figure 4.1, many safety practices which depend upon the physical configuration of the facility will almost automatically follow. However, in many cases, laboratories are often placed in structures originally designed for other purposes and ill adapted for the intended use. Even in the latter case, safety can be significantly enhanced by following a few straightforward guidelines as closely as the available space permits.

#### 4.6.1.1. Organization of the Laboratory

The basic premise in laying out the interior design of a laboratory facility or allocating space for the various activities within an existing facility is to separate areas of high risk from those of low risk as much as possible and to place high risk operations where there will be the least traffic and the least probability of blocking escape from the laboratory in case of an accident. Escape routes should, wherever possible, lead from high- to low-risk areas. A high-risk component may not always be obvious. For example, storage of chemicals in appropriate cabinets does not represent a high risk under most circumstances, but if left open, a flammable material storage cabinet along a path of egress can become a major danger if the liquids stored inside become involved in a fire. If the configuration of the laboratory permits, the laboratory furniture should be selected to permit two alternative paths from any point in the room. One of these, constituting a secondary escape path, may not necessarily lead directly from a high- to a low-hazard area, but even a poor alternative is better than none at all.

Fume hoods are intended to be used to house activities which should not be done on an open bench because of the potential hazard which they represent, usually the generation of noxious fumes. The ability of fume hoods to capture and retain fumes generated within them is especially vulnerable to air movement, due to either traffic or other factors such as the location of air system ducts, windows, doors, or fans. Clearly, they should be located, as in the standard laboratory module, in a remote portion of the laboratory selected for low traffic and minimal air movement. Other fume-generating apparatus such as Kjeldahl units should also be placed in out-of-the way places where errant air motion will not result in dispersion of the fumes into more heavily occupied areas of the room. A point that needs to be considered is the work habits of laboratory employees. Data on the possible health effects of long-term exposures to the vapors from most laboratory chemicals are relatively scant, although

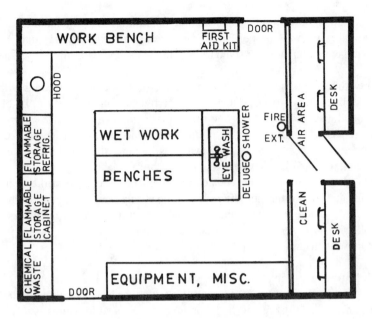

FIGURE 4.1    Standard laboratory module.

there are some epidemiological data for some types of exposures indicating some general problems. There are very little data on the synergistic effects of combinations of general laboratory chemicals. It is known, however, that the sensitivity of individuals to materials varies widely. In the spirit of ALARA as applied to chemical usage, when the evidence is missing, a conservative approach should be taken. Research personnel should spend no more time than is essential to the work in progress in areas where the generation and concentration of chemical vapors is likely to be high, compared to other spaces in the facility. The practice of allowing or requiring laboratory technicians to have desks in the work area, either for convenience or to monitor the work in progress, should be discontinued for both health and safety reasons.

The location of the various items of equipment in a laboratory should depend upon a number of factors such as frequency of use, distances to be traveled, and the need to transport chemicals to and from the work location and storage cabinets. The distances traveled to and from the most heavily used apparatus should be minimized, as should the distances and frequency with which chemicals are moved. Specialized work, such as that involving the use of radioactive materials, should be isolated from the other activities in the laboratory, especially if only some of the employees are involved in the work.

Dangerous apparatus should be placed in areas where protection can be afforded for the maximum number of laboratory employees. For example, a temporary glass system containing a highly reactive material under pressure might be located to one side of the laboratory, with explosive barriers placed on either side of the system so that if the system did rupture, the flying particles of glass would be directed toward a normally unoccupied area of the laboratory. Of course, if the probability of an explosion is a significant rather than remote possibility, the work should take place in a laboratory with the proper explosion venting and explosion-resistant barriers. Note that the hood area in the standard module is located with this consideration in mind.

Safety showers and eyewash stations should be conveniently located within the facility so that the approach to them is uncomplicated and unlikely to be blocked. They should also be located close to the primary entrance to the laboratory so that persons rendering assistance to an injured individual would not have to enter the laboratory any further than necessary, which will reduce the possibility of having to enter a contaminated area. Note that, in the standard laboratory module, both the shower and eyewash station are located at the end of the central workbench immediately inside the door through the partition separating the desk area from the active work area. Also note that the first aid kit and the portable fire extinguisher are on a side wall as close to the combination eyewash station and deluge shower as possible, so that these items are close at hand and available if needed.

### 4.6.1.2. Eating, Studying, and Other Social Activities

Unless provision is made for acceptable alternatives or extremely tight discipline is maintained, the work area of the laboratory will be used for eating, studying, and social activities. Such activities should not take place, and, as a minimum, the laboratory areas where these activities will not be allowed under any circumstances should be clearly marked. The standard laboratory module provides a convenient and acceptable location within the laboratory for eating, record and lab-book maintenance, studying, and social activities by placing desk spaces immediately inside the laboratory facility, separated from the rest of the laboratory by a partition. The location of this office class space also provides direct access to the exitway corridor and to the remainder of the building without having to reenter the laboratory. The upper half of the partition separating the laboratory and desk spaces in the laboratory module is intended to be transparent, made either of tempered safety glass or a plastic material such as Lexan™. This makes it possible to keep an eye on laboratory operations while taking a coffee or lunch break in safety or to perform any other desired activity without interfering with laboratory operations or being disturbed by others still in the laboratory. Since the laboratory, in most instances, should be at a modest negative pressure with respect to the corridor, the passage of vapors from the laboratory into the desk area should be inhibited, reducing the exposure of those at their desks significantly. The low speed of the air through the door, on the order of 10 to 20 fpm, will still make it possible for traffic through the door to carry some laboratory odors into the office compartment. The two doors in sequence also serve an additional safety function, representing a simplified airlock separating the corridor from the laboratory, thus adding some stability to the HVAC demands within the room. If they are closed following an evacuation, they would provide an additional barrier to any fire or noxious gases spreading from the laboratory to the remainder of the building.

### 4.6.1.3. Maintenance

Topics generally overlooked in laboratory safety are safety factors involved in providing needed maintenance and custodial services. Access to equipment needing service must be provided to service personnel under conditions which will make it possible for them to perform their work safely. Generally, equipment maintenance in the laboratory by support personnel should be coordinated by an individual who is familiar with current and recent research programs so that he can advise the workmen of possible risks in handling the various components. An example is maintenance on fume hood exhaust fans. Instances are known of workmen servicing fume hood exhausts who suffered severe reactions to contaminants on the equipment and the roof in the vicinity of the exhaust duct, even though the hood was not in use at the time. It is not enough to warn the maintenance department upon the initial request for services. Direct information needs to be provided to service persons on the scene. Some materials remain a problem for extended periods of time. In such cases, the workmen need to be protected while doing the work, using appropriate items of protective clothing such as gloves, respirators, coveralls, and goggles or full-face respirators, depending on the level of risk.

Where fume hood exhausts are brought to the roof through individual ducts, the area in which maintenance is needed may be surrounded by exhaust ducts in use since, in most cases, it is impractical to shut off operations for an entire building or even a significant portion of one because it is too disruptive of the research programs. In such cases, it is probably desirable to have a standard personnel protective equipment package for the maintenance workers to use, consisting of half- or full-face respirators that provide protection against solvents, particulates, and inorganic acids; chemically resistant coveralls; and gloves selected to provide a broad spectrum of protection against chemicals. Requiring personnel to wear these may appear to be excessively cautious, but, as noted earlier, there have been instances where unanticipated severe and long-lasting health effects have occurred.

Fume hood maintenance is one of the more active areas where maintenance personnel have concerns and where both support and laboratory personnel need to assume responsibility for seeing that the work is properly coordinated. A simple suggestion that has been found useful is to make sure each exhaust duct on the roof is properly labeled with the location of the hood itself. Workmen have been known to turn off power to motors on hoods in active use. Where hoods are dedicated to special uses which represent unusual hazards, such as radioactive materials, perchloric acid, toxic gases, or any other especially unusual risk, the duct should also be labeled with the application involved or a color code employed to identify these unusual risks. The latter program would alert maintenance personnel to definitely contact the laboratory from which the duct came before working in the vicinity of the duct. Power to the motors on the roof should be provided in such a way as to ensure that maintenance personnel on the roof can completely control the circuits while working to avoid accidental activation of the circuits from the laboratory. However, should they turn off the exhaust motor without notifying laboratory personnel, an alarm should sound in the laboratory, warning that the hood is not functional. A tagging and lockout procedure should also be employed during the maintenance operation.

Once hoods are removed from service to perform maintenance, they should not be returned to use until it is verified that they are performing according to required standards. It is easy to miswire a three-phase motor so that the fan rotates opposite to the desired direction. Belts may need to be tightened or a pulley size changed to achieve the proper face velocity.

Fume hoods have been used to illustrate some of the problems that can arise from lack of coordination among maintenance and laboratory personnel, but there are many other possible problems. Explosions can occur if gas service is turned off without everyone being aware of it; gas jets may be open, which could flood a facility with gas when service is restored. Stills can overheat if condenser water supplies are interrupted. Electrical service to an area should be discontinued and restored only with full prior notification to all persons who might be affected. Individual laboratory technicians or students often modify their facility without informing the groups responsible for maintenance, thereby raising the possibility of an injury to an unsuspecting serviceman or a repair which would not be based on an accurate assessment of the conditions or load present.

### 4.6.1.4. Housekeeping

Another problem related to maintenance is what should reasonably be expected of custodians. Experience has shown that there is a tremendous variation in the level of expectations and wishes among laboratory supervisors. Some supervisors do not want custodians to enter their laboratory at all, while others have no qualms in asking them to clean up a hazardous chemical spill. Most safety and laboratory personnel would agree that the latter is asking too much; most would also agree that, if they wish, facility personnel should be allowed to take care of their own housekeeping, as long as reasonable standards of cleanliness are maintained. Most laboratory groups, however, fall somewhere between these two extremes.

The salary levels of most custodial positions are usually among the lowest in most organizations and, hence, limit the skill levels one can expect from the persons filling the positions. Unfortunately, literacy rates are less than average and, in many cases, it certainly would be unrealistic to expect a significant level of technical training which would permit an adequate level of understanding of the problems they might encounter in a laboratory. As a result, they are often afraid of the laboratory environment. However, alternative positions are also usually hard to find for these employees, so they frequently are very concerned about losing their jobs. Most cannot afford to do so. As a result of these conflicting pressures, they may attempt to do things they really do not understand and are afraid to ask about, thus making mistakes. It is the responsibility of the laboratory supervisor, working with custodial management, to carefully establish the areas of responsibility for custodians in the laboratory.

Among the tasks a custodian can reasonably be expected to do in most laboratories are

1.  Clean and maintain the floor area.
2.  Dispose of ordinary trash. However, if other than ordinary solid waste is generated in the laboratory, it should be placed in distinctively shaped and/or colored containers. If the custodians are still expected to handle it, then the circumstances and procedures should be carefully delineated and training given.
3.  Wash windows. If they are expected to wash bench tops or other laboratory furniture, it should only be when additional supervision is provided by laboratory personnel.

Among the tasks which they should not be expected to do are

1. Clean up chemical spills. They are not trained to do it according to established regulatory guides, nor to do it in such a way that they do not expose themselves to potential injury.

2. Dispose of broken glass, syringes, or "empty reagent containers". They can dispose of these items if they are carefully prepared by the laboratory workers in advance. For example, broken glass can be disposed of if it is placed in a sturdy kraftboard box, sealed, and labeled as "broken glass". Syringes should be placed in a sturdy, narrow-mouth plastic container in the laboratory and then placed in a sturdy kraftboard container for disposal. It would be preferable for syringes to be incinerated, rather than placed in ordinary trash. Empty reagent containers should be triple-rinsed and then placed in a box labeled "triple-rinsed reagent containers". Unless these restrictions are met, the waste should be left alone, and the facility reported to the custodian's supervisor so the problem can be resolved.

3. Handle special wastes in any way, including radioactive materials, chemical wastes, or contaminated biological materials. All of these require special handling by specialists, and precautions must be taken to ensure that these materials are not accidentally collected by custodians. In some cases, such as with radioactive wastes, failure to provide adequate controls could result in fines or loss of a radioisotope use license for flagrant and repeated violations.

4. Clean the work surfaces and equipment in the laboratory, except in special circumstances and under the direct supervision of a responsible laboratory employee. Even in this case, a preparatory program should have been carried out in advance by laboratory personnel to remove or secure items which could be dangerous in the area being cleaned.

Housekeeping also means maintaining the laboratory in a reasonably organized fashion on a day-to-day basis. This is the responsibility of all laboratory personnel, but individuals will follow the laboratory manager's own performance as a guide. Reagents not in use should be returned to proper storage. Secondary containers should be labeled according to the requirements of the Hazard Communications Standard. Glassware should be cleaned and put away. Trash should not be allowed to accumulate. Equipment should not be allowed to encroach upon aisles. Cables and temporary electrical extensions should not become a tripping hazard. Periodically, refrigerators and other storage units should be inspected and cleaned out. An audit of materials should be made periodically to get rid of old, degraded, and obsolete materials before they become a hazard. Chemicals stored inappropriately outside of their hazard class should be returned to their proper locations. Bottles heavily covered with dust, indicating a lack of use for an extended period, will probably be considered unusable and should be eliminated. No one should expect a busy laboratory to be spotless, but neither should it be a disaster area. Unless a concerted effort is made from time to time, housekeeping problems will slowly accumulate. An effective mechanism used by the author to combat this erosion of order was to schedule a quarterly "field day" on which all personnel, including faculty, staff, and students, ceased research and returned everything to reasonable order. This rarely took more than a few hours and furthered a sense of cooperation between the various groups of people.

### 4.6.1.5. Signs and Symbols

In many situations, a person needs to be made aware that a hazard exists in the area he is entering or that restrictions are placed on persons entering the area. In addition, there are signs which are intended to provide information to individuals in an emergency. There are literally hundreds of specialized safety signs and symbols which can be purchased for the laboratory. Given below is a partial list of some of the more important ones, along with a brief description of the types of applications where they would be needed. In many cases, the signs are mandated by regulatory requirements, while in other cases they represent common safety practices. In most cases, the hazard signs will be prefaced by a risk descriptor defining the level of risk represented in the specific instance. The three cautionary words in normal use, in decreasing order of risk, are DANGER, WARNING, and CAUTION.

1. **AREA UNSAFE FOR OCCUPANCY.** This is used to indicate a contaminated area or an area otherwise rendered unsafe, temporarily or otherwise, for normal use.
2. **AIRBORNE RADIOACTIVITY AREA.** Some applications involving radioactive materials result in the generation of airborne radioactive materials in excess of those permitted by the standards of the NRC or the equivalent state agency in an agreement state. Should such an operation exist, the boundaries of the room, enclosure, or operating area where the airborne material may exist must be posted with this sign. The legend will be accompanied by the standard radiation symbol shown earlier in Figure 25.
3. **ASBESTOS.** Areas known to contain asbestos such that airborne fibers could be in the air must be marked with an appropriate warning sign. A typical sign would be

**CAUTION**
**ASBESTOS-CONTAINING MATERIAL PRESENT**
**AVOID CREATING DUST**

Note that a sign such as this does not say that there are asbestos fibers in the air. The sign is intended to alert people that their actions could result in the generation of airborne asbestos fibers. If asbestos fibers were known to be present, more restrictive signs would be used, but in such cases the facility should not be used until the problem is abated.

4. **AUTHORIZED ADMISSION ONLY.** This sign may accompany many other signs or it may be used alone in restricting access to an area to those who have legitimate reasons to be there or who are aware of the risks within the area to which they may be exposed.
5. **BIOLOGICAL HAZARD.** This sign will be accompanied by the standard biological hazard symbol, indicating that an agent which may prove infectious to human beings is present within the area.
6. **CARCINOGENIC AGENT.** The entrances to research areas in which car-

cinogens are used must, according to the OSHA standards governing carcino-gens, be marked with a warning sign such as the one below and a carcinogenic agent symbol.

### CANCER-SUSPECT AGENT
### AUTHORIZED PERSONNEL ONLY

Under some circumstances where exposure is more probable, the sign could indicate that protective gear is required to enter the area.

7. **CHEMICAL SPLASH GOGGLES REQUIRED WHILE WORK IN PROGRESS.** It is recommended that this sign be used at the entrances to all active laboratories where chemicals are employed and actively being used. In order to enforce the requirement, care must be taken to select goggles which resist fogging, do not become oppressively warm while being worn at comfort-able room conditions, and do not exert uncomfortable pressure on the face. They should also accommodate wearing normal-size prescription glasses at the same time. Many goggles which meet the minimum regulatory standard, based on ANSI Z87.1, for impact protection do not meet all of these practical considerations, but there also are several that do. When work is not in progress or when a person is in an area well separated from the active work area, it may be permissible, for reasons of comfort, to allow goggles to be removed.*

8. **CRYOGENIC LIQUIDS.** All containers which contain cryogenic liquids — most commonly liquid nitrogen, as in the example below — but also other gases maintained at very low temperatures, should be prominently labeled:

### CAUTION
### LIQUID NITROGEN

The container, usually a large flask with walls separated by a vacuum called a dewar, will also usually be labeled with the cautionary information:

### FRAGILE CONTAINER UNDER VACUUM
### MAY IMPLODE

9. **EMERGENCY INFORMATION SIGNS.** Prominent signs such as **EXIT, EMERGENCY SHOWER, EYEWASH STATION, FIRST AID KIT, FIRE EXTINGUISHER**, etc. should be posted near the safety device men-tioned to aid in locating it in an emergency. Symbols can be used in place of or in addition to some of these.

10. **EXPLOSIVES.** If explosives are stored in Class 1 magazines or in outdoor Class 2 magazines, the property must be posted with signs stating **EXPLO-SIVES — KEEP OFF.** Class 2 magazines must have labels on all sides except the bottom in letters at least 3 in. high, **EXPLOSIVES — KEEP FIRE AWAY**.

---

* A point was made here, which will not be repeated for reasons of brevity, that signs and rules must take into account human factors. Otherwise, they are likely to be ignored and weaken the safety program.

11. **FLAMMABLE MATERIALS.** Cabinets containing flammable materials and areas or rooms where flammable materials are stored or used must be posted with this sign, which may also be indicated by a symbol. This sign should always be accompanied by the **NO SMOKING** sign, which may be augmented by a standard no smoking symbol.

12. **HIGH VOLTAGE — DANGER.** Spaces which contain accessible high voltage panels, such as switch rooms and electrical closets, should be locked and provided with these signs to warn persons who lack training and experience in high voltage circuits not to enter. Equipment containing high voltage circuits should also bear this warning label.

13. **HYDROGEN — FLAMMABLE GAS, NO SMOKING OR OPEN FLAMES.** This sign must be posted in all areas where hydrogen is used or stored.

14. **INTERLOCKS ON.** Equipment with internal hazards, such as X-ray diffraction cameras, or areas where the space is rendered unsafe to enter by the presence of a hazard are often provided with a fail-safe circuit or interlock, which will turn off the equipment creating the problem if the circuit is broken. The sign provides a warning that the interlock is on to prevent access to the hazard.

15. **LASERS.** Labeling of lasers should follow 21 CFR 1040, the federal Laser Product Performance Standard. The spaces in which lasers are located should also have a similar warning at the entrance. The label will depend upon the class of laser involved. All of the labels will include a stylized sunburst symbol, with a tail extending to the left. The signal word **CAUTION** is to be used with Class 2 and 3A laser systems, while the signal word **DANGER** is to be used for all Class 3B and 4 systems.

16. **MACHINE GUARDS IN PLACE.** OSHA requires that many machines, such as vacuum pumps or shop equipment, be provided with guards over the moving parts. Signs should be posted near these machines to remind employees not to use the equipment if the guards are not in place.

17. **MICROWAVES.** This sign must be posted in any area where it is possible to exceed the current occupationally legal limit of exposure to microwave electromagnetic radiation.

18. **NO EATING, DRINKING, SMOKING, OR APPLYING COSMETICS.** This sign should be posted wherever toxic materials are used, in the working areas of wet chemistry laboratories, and in biological laboratories using pathogenic substances.

19. **NO SMOKING.** A no smoking sign must be posted wherever flammables are in use, where there is a risk of explosion due to the presence of explosives, gases, vapors, or dusts, and should be posted where toxic materials are in use.

20. **RADIATION AREA.** Areas where the radiation exceeds a level established by the NRC must be posted with this sign. If the level exceeds a higher level set by the NRC, the area must be posted with a **HIGH RADIATION AREA** sign. Most of these areas will be within an area posted with a **RESTRICTED AREA — AUTHORIZED ADMISSION ONLY** sign. Specifics on these requirements will be given in Chapter 5.

Signs defining radiation areas should be used to post areas where radioactive materials are stored unless radiation levels equal or exceed the stipulated limits. Areas where radioactive materials are stored should be posted with a **CAUTION — RADIOACTIVE MATERIALS** sign.

21. **RADIOACTIVE WASTE.** This is not a sign specifically required by the NRC, but is recommended to denote areas within a laboratory where radioactive waste is temporarily stored, prior to being removed for permanent disposal, in order to help avoid accidental removal of radioactive waste as part of normal laboratory waste. Much radioactive waste resembles ordinary trash, such as paper and other trash.

22. **REFRIGERATOR (FREEZER) NOT TO BE USED FOR STORAGE OF FLAMMABLES.** All refrigerators or freezers not meeting the standards permitting the storage of flammable materials (see Section 4.5.7.4) must be marked with this sign.

23. **REFRIGERATORS NOT TO BE USED FOR STORAGE OF FOOD TO BE USED FOR HUMAN CONSUMPTION.** Laboratory refrigeration units used for the storage of chemicals and biological materials must be posted with this sign to prevent the use of units to store food and beverages.

24. **RESPIRATORY PROTECTIVE EQUIPMENT REQUIRED.** Wherever airborne pollutants are present which exceed the permissible exposure limits (PEL) established by OSHA, respiratory protection is required. In many cases, ACGIH threshold limit values (TLV)® are lower than the OSHA PEL, and respiratory protection is recommended when the levels approach these lower limits. In most cases, it is recommended that an action level of half or less of the TLV® values be set to accommodate in part the different sensitivity of individuals to materials.

25. **SAFETY GLASSES REQUIRED.** This sign is to be posted wherever there is a risk of eye injury due primarily to impact.

26. **TOXIC GAS.** Areas where toxic gases are used or stored must be posted with this warning sign.

27. **ULTRAVIOLET LIGHT, EYE PROTECTION REQUIRED.** This sign should be posted wherever there is a risk of eye injury due to ultraviolet light emission.

There are many other signs and symbols identifying hazards or denoting specific requirements to aid in reducing a specific risk. The following generic signs are representative of many of these.

28. **(SPECIFIC ITEM) PERSONAL PROTECTIVE EQUIPMENT REQUIRED.** Many other risks exist which would require specific items of protective equipment. Where these items are needed, the area should be appropriately posted.

29. **(SPECIFIC) TOXIC OR HAZARDOUS MATERIAL.** There have been a number of materials that pose known risks and the areas in which these materials are used should be posted with an appropriate sign.

30. **(SPECIFIC) WASTE CHEMICAL ONLY.** Disposal of waste chemicals according to RCRA standards requires that wastes be identifiable, in some cases by class only, but in most cases it is desirable that wastes not be mixed.

Posting of areas or containers with this sign, where several waste streams of different character are located, will aid in the legal disposal of waste.

## 4.6.2. WORKING PROCEDURES

This section can only touch upon the broad topic of safe laboratory working procedures because of the immense scope of the subject. Some of the more common areas which offer the potential for mishaps will be covered in some detail, but there will undoubtedly be areas that many persons consider comparably important which will be touched upon lightly or not at all. Sections will be devoted to a few of the more hazardous chemicals to illustrate the precautions that need to be taken when working with such materials. In addition to physical hazards such as fire and explosions, health risks will be emphasized since in many cases these are more insidious and less often recognized by many laboratory workers. The next several sections will be concerned primarily with physical hazards and the latter part of the chapter will be devoted to short- and long-term aspects of laboratory operations on the workers' health.

### 4.6.2.1. Protection Against Explosions

Unusually careful planning must take place whenever there is any reason to suspect that work to be undertaken may involve the risk of an explosion. However, not all potentially explosive situations are recognized in advance. Letters from experimenters describing work in which unexpected explosions occurred can be found in a substantial proportion of the issues of *Chemical and Engineering News*. Because these incidents were unanticipated, sufficient protective measures were often not employed. Consequently, injuries which could have been avoided are often reported in these letters.

Explosions may occur under a variety of conditions, the most obvious being a runaway or exceedingly violent chemical reaction. Other situations could include the ignition of escaping gases or vapors, ignition of confined vapors with the subsequent rupture of the containment vessel, rupture of a system due to over-pressure caused by other mechanisms, or a violent implosion of a large vessel operating below atmospheric pressure. Partial confinement within a hood can enhance the dangerous effects of an explosion; areas in front of the open face may be damaged more severely than if the explosion were not confined.

Injuries can occur due to the shock wave from a detonation (if the release of energy occurs at supersonic speeds) or deflagration (if the energy release occurs at subsonic speeds). Most laboratory reactions are the latter class. Hearing loss may result if the shock wave causes a substantial over-pressure on the eardrums. According to Table C-3.1(a) of NFPA 45, Appendix C, the equivalent of as little as 1 g of TNT can rupture the eardrum of a person within 0.75 m ($\approx$2.4 ft), while 10 g is likely to rupture the ears of 50% of persons within 67 cm (2.2 ft) of the explosion. The shock wave, as a wave, can "go around" barriers or be reflected and reach areas that would be shielded from direct line-of-sight interactions. Injuries can occur due to the heat or flames from the explosion. Fume hood materials should be selected to contain fires occurring within them. However, if the sash is severely damaged, flames or burning material can escape through the front opening and the flames may spread to other fuels in the vicinity. Due to this possibility, flammable materials should not be stored in close proximity to fume hoods. Respiratory injuries can occur due to inhalation of fumes and reaction products. However, the most serious hazard is flying debris, including fragments of the containment vessel, other parts of the experimental apparatus, or nearby materials or unreacted chemicals, which can inflict physical injuries. The risk of the latter type of injuries can be reduced by eliminating the possibility of line-of-sight or single-ricochet paths for missiles from likely sources of an explosion to workers or to equipment which could be damaged and result in secondary harmful events. The possibility of extraneous material becoming involved in an explosion is a powerful argument in favor of not using a hood as a storage area, especially in experimental activities. The reflected shock wave can act in much the same way as a piece of physical debris in causing damage external to a fume hood. Overreaction of a worker or involuntary reflexive actions to even minor explosions can also lead to quite serious secondary injurious incidents.

In addition to immediate injuries, an insufficiently contained explosion can lead to fires or cause damage sufficient to wipe out expensive apparatus, destroy months or years of research effort, or even destroy an entire facility. Conservative precautionary measures to reduce the likelihood of these repercussions are worthwhile from this aspect alone.

Ordinary fume hoods offer a fair amount of protection to the sides and rear of the hood, if they are of good quality with substantial walls. However, most fume hoods are not intended to provide really significant explosion protection against a major explosion for a user standing immediately in front of the hood, although sash materials are usually designed not to contribute to the hazard. This is accomplished by having the sash material made of either laminated or tempered glass so that the broken sash will not cut persons standing in front of the hood. A hood with a three-section horizontal sash, where the user stands behind the central section, provides superior protection to the more common vertical sash hood. If the work to be done involves a known explosive risk, certainly a hood specifically designed to contain any explosion which could be anticipated, or to provide safe explosion venting, should be employed.

For the majority of laboratories, which are equipped only with ordinary fume hoods, supplementary measures should be taken to minimize the type of risks described above if a careful analysis of the planned operation reveals any likelihood of a potential explosion.

A simple way to reduce the potential risks is to minimize the amount of material involved in the experiment. The smallest amount sufficient to achieve the desired result should be used. Care should be taken in scaling up from a preliminary trial run in which minimal quantities were employed. Increasing the amount of material in use could significantly change the physical parameters so that insufficient energy removal, inadequate capacity for the reaction products, or excessive pressures could develop in the scaled-up version of the work and lead to a dangerously unsafe condition.

A number of other measures can be taken to enhance the protection of workers against explosions. Provision of barriers is a straightforward measure. The selection of an appropriate barrier will depend upon the circumstances. A variety of factors should be considered.

The strength of the barrier material is clearly an important factor. Tests have been

**TABLE 4.12**
**Shock Tests on Transparent Shields**

| Material | Thickness | | Drop ball | | ASTM D 256 (ft-lb) |
|---|---|---|---|---|---|
| | mm | in. | kg/m | in.-lb | |
| Double-strength glass | 3.2 | 0.125 | 446 | 25 | |
| Laminated glass | 6.4 | 0.25 | 1964 | 110 | |
| Plate glass | 6.4 | 0.25 | 1964 | 110 | |
| Wired glass | 6.4 | 0.25 | 2000 | 112 | |
| Tempered glass | 6.4 | 0.25 | 10393 | 582 | |
| Methyl methacrylate | 6.4 | 0.25 | 19400 | 1086 | 0.4 to 0.5 |
| Polycarbonate | 6.4 | 0.25 | | | 12.0 to 16.0 |

made of many materials commonly used in laboratory protective barriers and available either in commercial units or readily amenable to fabrication of custom shielding. Table 4.12 is adapted from a study by Smith[2] in which each material tested was 0.25 in. or 6.4 mm thick. The relative susceptibility to fracture was measured by either the ASTM D 256 test method or by dropping balls from various heights. It required 12 to 16 ft-lb of energy to fracture the polycarbonate material in the ASTM D 256 test. Additional protection can be obtained by increasing the thickness of the materials used in fabricating the shield, approximately proportional to the thickness.

An equal thickness of steel would have a relative effectiveness on this scale of about 40.

Resistance to fracture is not the only consideration. Wired glass, for example, may represent an additional hazard due to the presence of the wire, if shattered. Ordinary glass should not be used due to the danger of cuts from the flying debris. Methyl methacrylate is not suitable where high temperatures may occur. However, sheets of methyl methacrylate are commonly available at moderate cost and can readily be fabricated into custom shields. Polycarbonate obviously offers considerable strength, but can be damaged by organic solvents. Steel is resistant to both heat and solvents, but does not offer the desired transparency. However, there are alternatives to this deficiency, such as mirrors, optical devices, or closed-circuit television. Remotely controlled manipulating devices can be used to control apparatus behind any shield material.

The simplest types of supplementary protection suitable for moderate risks are commercial shields, which are available from most laboratory supply firms. Shields usually found in catalogs today are curved sheets of transparent material, most commonly polycarbonate, weighted at the bottom to increase their stability. Since these are free standing, they often will not remain upright in explosive incidents and, if the explosion is large enough, may actually be hurled through the air and cause injury themselves. Since the scale of an explosion cannot always be accurately estimated, it would be desirable to secure these shields firmly to the work surface. For small-scale reactions, they offer a worthwhile degree of added protection. If used, the shields should be located so as to provide the maximum protection against flying debris, chemicals, or, as noted earlier, external shock wave interactions. Individuals in the laboratory should be trained to use the shields correctly and not to move or modify them to increase their convenience in performing tasks, if these changes could reduce the level of protection.

For larger-scale experiments, especially if the operation is to continue for a significant length of time, the small commercial shields should be replaced with custom-fabricated shields. Because of its availability in a wide range of thicknesses, relatively high impact resistance, moderate cost, and ease of fabrication, methyl methacrylate is a convenient material to use in constructing custom transparent shields. The same essential design considerations that apply to commercial shields also apply to custom shields: (1) select the material that has the appropriate mechanical and physical properties and (2) place the shield so as to provide the maximum protection against flying debris and other external effects, both for direct injuries to personnel and minimize secondary events. As the size or hazard of the experiment increases, more consideration should be given to relocating the work to a more appropriate containment facility, preferably one specifically designed and engineered to limit damage, should an explosion occur. If it is essential that the work be done in a specific location which is not explicitly designed for work with potentially explosive situations, it is strongly recommended that the design of the proposed renovation or modification be done by a qualified engineer and fabricated by professionals, not graduate students.

There are differences in the cost of different types of shielding materials and in the approach that is taken to provide adequate protection. A cost-benefit analysis is always appropriate in selecting or designing any experimental system. On the other hand, selecting too inexpensive an approach can be a false economy. The person making the decision may be the one injured or killed if the protection is insufficient. There is always the question of liability if others are injured and, finally, there is always the ethical question of what should have been done if one only did what was required to be done and a person was injured as a result.

**Personal Protective Equipment**

Personal protective equipment will be covered in more detail in a later chapter, but an important aspect of enhancing the safety of workers where the potential for explosions exists is to provide and *require the use of* protective equipment. Every laboratory worker in a facility where any potentially injurious chemical is being used should always wear protective eyewear. Common spectacles with side shields are not nearly as effective as properly selected chemical safety goggles in protecting workers' eyes. The latter not only fit all around the eyes and thus protect against direct impact, but also protect the eyes from flying liquids. Since a properly fitting pair of goggles offers a snug fit to the face, it generally provides superior protection from lateral impacts which could knock off an ordinary pair of safety spectacles.

Additional protection to the eyes, face, and throat should be provided, as circumstances warrant, by the use of face masks in addition to goggles. A mask which protests the throat is preferable to one which does not, due to the vulnerability of the carotid arteries on each side of the neck.

In addition to eye protection, sturdy laboratory smocks, preferably made of a flame-retardant or flame-proof material, should be worn. Short-sleeved shirts, T-shirts, shorts, or sandals should not be allowed in any laboratory where the potential for exposure of the skin to chemicals exists or where even minor explosions can occur. Sturdy gloves or gauntlets, selected for the immediate requirement of manual dexterity, should be worn if manual manipulation of apparatus or materials with the potential for explosion is needed. Although the work should be done in a specially equipped facility if the potential for a major explosion exists, in principle, even bullet-proof vests could be used in certain situations. The potential for hearing damage can be greatly reduced if research personnel use good quality earmuffs. If these are to be issued, the workers should be provided an audiometric hearing evaluation beforehand to determine if a hearing loss has actually occurred subsequent to an accident.

### Summary

Laboratory explosions occur frequently in situations where they are least expected. Simple caution should dictate a conservative estimate of the probabilities of such an event and encourage the use of appropriate preventative measures in all laboratory work. It is far better to prevent an explosion than to attempt to confine one or reduce the severity of the damage resulting from one. However, appropriate barriers and personal protective devices can aid in reducing the seriousness of consequences when explosions do occur, if these ameliorative measures are used properly. Training all personnel to understand the risks associated with their work, coupled with encouragement by management and the cooperation of all laboratory workers to follow safe procedures, can significantly reduce the risks of explosions.

## REFERENCES (SECTION 4.6.2.1)

1. *Fire Protection for Laboratories Using Chemicals,* Appendix C, NFPA Standard 45, National Fire Protection Association, Quincy, MA.
2. **Smith, D. T.,** Shields and barricades for chemical laboratory operations, in *Handbook of Laboratory Safety,* 2nd ed., Steere, N. V., Ed., CRC Press, Boca Raton, FL, 1971, 113.

### 4.6.2.2. Corrosive Chemicals

The definition of corrosive chemicals is very broad. However, in the sense that the action of the chemical will result in an immediate, acute erosive effect on tissue as well as other materials, strong acids and bases, dehydrating agents, and oxidizing agents are commonly considered to be corrosive materials. These names may not be mutually exclusive.

Accidents with corrosive materials in which the material may splash on the body are very common in the laboratory. The eyes are particularly vulnerable to injury, and injuries to the respiratory system may range from moderate irritation to severe injury. Skin injuries may be very slow to heal. Ingestion can cause immediate injury to the mouth, throat, and stomach and, in severe cases, can lead to death. Some common household chemicals that are equivalent to laboratory corrosives, such as drain cleaners, are common causes of child fatalities. Work with these corrosive materials should be done in a fume hood. Personal protective equipment such as gloves, laboratory aprons, and chemical splash goggles should be used and, if the possibility of inhalation is significant, appropriate cartridge respiratory equipment should also be used. Every laboratory, especially those using these materials, should be individually equipped with deluge shower/eyewash fountain combinations.

Corrosive chemicals should always be handled with the greatest care. Where available, they should be purchased in containers coated with a protective plastic film so that if they are dropped, the probable result should at most be a leak through the film instead of a potentially dangerous splashing of the chemical onto the skin of the person transporting the material and possibly others. Although there is a surcharge in most cases for ordering containers with protective films, the additional protection afforded is substantial. Even with the film, the material leaking out of the container can still represent a nasty mess which must be cleaned up promptly.

Safety carriers are available for use in transporting containers of dangerous chemicals. These should be used especially if the chemical containers are unprotected breakable bottles. Most persons will not take the trouble to use them for the movement of chemicals within a laboratory, which is much more common than transporting materials from one laboratory to another or from the stockroom to the laboratory. Just as most car accidents occur within a few miles of home, most chemical accidents occur where the workers spend most of their time. The use of protected containers can largely ameliorate the seriousness of these accidents since it does not require an extra effort to use them.

When it is necessary to move chemical containers a significant distance, the use of safety carriers, even for protected containers, is recommended. Although the result is likely to be much less serious when coated bottles are used, accidents can still occur and it is always desirable not to have to clean up after one. If several bottles are to be moved at once, the bottles should be moved on a low cart with a substantial rim around the edge so that, if the cart is struck, the chemicals will be likely to stay on it or, if not, have only a short distance to fall.

Wherever possible, chemicals should be moved from one floor to another on a freight elevator, rather than carrying them manually up and down a flight of stairs. If only a passenger elevator is available, the use of the elevator should be delayed until no passengers are using it or passengers desiring to use it should be courteously asked to wait until the chemicals have been moved. Although elevator accidents are rare, should a dangerous material be released in one while the passengers were trapped inside, the results could be catastrophic.

Laboratory personnel should be trained to quickly limit the area affected by a spill, and the necessary supplies to enable them to do so should be immediately available within the laboratory. Kits containing absorbent pillows, neutral absorbent materials, or neutralizing materials are commercially available which will enable knowledgeable persons to safely contain most small accidents, such as spills of the contents of a single bottle of reagent, until a final cleanup of the materials in the container as well as those used to contain it can be disposed of, usually as hazardous waste.

If the average person who has taken chemistry remembers one safety rule, it is probably the one about always adding acid to water, never the reverse (usually brought up in the context of sulfuric acid, which is a strong dehydrating agent). This precaution is taken to avoid splashing the acid due to the localized generation of excessive heat as the two substances mix.

There are a number of other basic safety procedures involving corrosives. Keep the container sizes and quantities on hand as small as possible, consistent with the rate of use. Store each class by itself. Keep containers not in use in storage, and store the containers either in cabinets or on low shelves. Remember that reactions involving these substances will usually generate substantial heat, so closed containers could rupture due to excessive pressures.

Some brief comments about some of the classes of corrosives are given below. In the sections which follow, some of these as well as others are covered in more detail.

**Strong acids** — Concentrated strong acids can cause severe and painful burns. The pain is due in part to the formation of a protein layer which resists further penetration of the acid. In general, inorganic acids are more dangerous than organic acids, although the latter can cause deep-seated burns on extended contact with the skin. Leakage from containers and material remaining on the outside of the containers following a sloppy transfer can cause corrosion of the shelving, and if the acids are stored with materials with which they may react, accidents can occur from resulting reactions.

**Strong alkalis** — Alkali metal hydroxides are very dangerous when allowed to come into contact with tissue. Contact with the skin may be less painful than a comparable exposure to acid because the protective protein barrier is not formed. Damage may extend to greater depths as a result of the lesser pain because the injured person may not be as aware of the seriousness of the incident. Any area exposed to a strong alkaline material should be flooded with water for at least 15 min or longer. This is particularly important in eyes since the result of an exposure can be global rupture.

**Nonmetal chlorides** — Compounds such as phosphorous trichloride and corresponding bromides react violently with water and are a common cause of laboratory accidents.

**Dehydrating agents** — Strong dehydrating agents such as sulfuric acid, sodium hydroxide, phosphorous pentoxide, calcium oxide, and glacial acetic acid can cause severe burns to the eyes because of their strong affinity to water. When they are added to water too rapidly, violent reactions, accompanied by spattering, can occur.

**Halogens** — Halogens are corrosive on contact with the skin, eyes, and the linings of the respiratory system as well as being toxic. Because they are gases, they pose a greater danger, especially by inhalation, of coming into contact with tissue.

### 4.6.2.3. High-Energy Oxidizers

Oxidizing agents such as chlorates, perchlorates, peroxides, nitric acid, nitrates, nitrites, and permanganates represent a significant hazard in the laboratory because of their propensity under appropriate conditions to undergo vigorous reactions when they come into contact with easily oxidized materials such as metal powders and organic materials such as wood, paper, and other organic compounds. Elements from group 7A of the Periodic Table — fluorine, chlorine, bromine, and iodine — react similarly to oxygen and are classified as oxidizing agents as well.

Most oxidizing materials increase the rate at which they decompose and release oxygen with temperature. The rate of decomposition of hydrogen peroxide goes up by a factor of about 1.5 with each 10°F (5.6°C). Because of this ability to furnish increasing amounts of oxygen with temperature, the reaction rate of most oxidizing agents is significantly enhanced with increasing temperature and concentrations. The hazard associated with these agents increases as well. For example, cold perchloric acid at a concentration of 70% or less has little oxidizing power, but at concentrations above 73%, it has significant oxidizing power at room temperatures and increases still further at higher concentrations. Hot, concentrated perchloric acid is a very strong oxidizing agent. Containers of oxidizing agents may explode if they are involved in a fire within a laboratory.

The quantities of strong oxidizing agents within the laboratory should be minimized and these materials should be rigorously segregated from materials with which they could react. The containers should be protected glass, with inert stoppers instead of rubber or cork.

Quantities of potentially vigorously reacting materials, such as strong oxidizing agents, used in a given research operation or evolution should be kept to the minimal quantities needed. The work should always be performed in a hood, with appropriate safety features (such as the wash-down system recommended for research involving hot perchloric acid digestions). Oxidizing agents should be heated with fiberglass heating mantles or sand baths. The use of personal protection devices, including sturdy gloves and eye protection which provides both chemical splash and impact protection, should be mandatory. If the potential for explosions is determined to be significant, the operator as well as others within the facility should be protected with explosion barriers. If the risk is sufficiently high, the research should be performed in an isolated facility specially designed for the program, which would include explosion venting and explosion-resistant construction.

Listed below are brief comments regarding the hazardous properties of a number of common, powerful oxidizing reagents. Most form explosive mixtures with combustible, organic or easily oxidizable materials and most yield toxic products of combustion. Current MSDSs will provide additional data on each of these materials.

**Ammonium perchlorate ($NH_4ClO_4$)** — Similar in explosion sensitivity to picric acid. Explosive when mixed with organic powders or dusts. Highly sensitive to shock and friction when mixed with powdered metals, carbonaceous materials, and sulfur.

**Ammonium permanganate ($NH_4MnO_4$)** — May become shock-sensitive at 60°C (140°) and may explode at higher temperatures. Avoid contact with readily oxidizable, organic, or flammable materials.

**Barium peroxide ($BaO_2$)** — Combinations of this compound and organic materials are sensitive to friction. Sensitive to contact with small quantities of water.

**Bromine (Br)** — Highly reactive material. Causes serious burns to tissue; toxic; when inhaled, can cause serious damage to respiratory system.

**Calcium chlorate ($Ca[ClO_3]_2 \cdot 2H_2O$)** — Explosive mixtures are ignitable by heat and friction.

**Calcium hypochlorite ($Ca[ClO]_2$)** — Ignites easily when in contact with organic and combustible material. Chlorine evolves at room temperatures when mixed with acids.

**Chlorine trifluoride ($ClF_3$)** — Vapor at room temperature and is dangerously reactive. Most combustible materials ignite spontaneously on contact. This is an exception to the recommended use of glass containers. The material reacts strongly with silica, glass, and asbestos. Extremely toxic; causes severe burns to tissue.

**Chromium anhydride or chromic acid ($CrO_3$)** — Ignites on contact with acetic acid and alcohol and may react sufficiently vigorously with other organic materials to ignite.

**Dibenzoyl peroxide ($[C_6H_5CO]_2O_2$)** — Extremely explosion sensitive to shock, heat, and friction. Comparatively low toxicity.

**Fluorine ($F_2$)** — Extremely reactive gas, reacting vigorously with most oxidizable materials at normal room temperatures, often vigorously enough to ignite. Causes severe burns to tissue. Severe danger of damage to respiratory tract.

**Hydrogen peroxide ($H_2O_2$)** — Commercial products usually sold inhibited against decomposition. At concentrations between 35 and 52%, shares hazards of other oxidizing agents associated with coming into contact with easily oxidizable materials, but also may violently decompose when coming into contact with many common metals and their salts, e.g., brass, bronze, chromium, copper, iron, lead, manganese, silver, etc. At higher concentrations, most combustible materials will ignite on contact. Mixing of organics with concentrated hydrogen peroxides may create very sensitive, explosive combinations.

**Magnesium perchlorate (Mg [$ClO_4$]$_2$)** — Sensitive to ignition by heat or friction.

**Nitric acid ($HNO_3$)** — Explosively reactive with carbides, hydrogen sulfide, metallic powders, and turpentine. Causes severe burns to tissue.

**Nitrogen peroxide (in equilibrium with nitrogen dioxide) $N_2O_4$;$NO_2$** — May cause fire on contact with clothes and other combustible materials. Reactions with other fuels and chlorinated hydrocarbons may be violent. Vapors are life threatening at very low concentrations. Severely dangerous to tissue.

**Nitrogen trioxide ($N_2O_3$)** — May cause fire on contact with combustible materials. Very damaging to tissue, especially respiratory tract, where fatal pulmonary edema may result, although onset of symptoms may be delayed for several hours.

**Perchloric acid ($HClO_4$)** — Very dangerous oxidizing agent at high concentrations and elevated temperatures (see Section 4.6.2.4).

**Potassium bromate ($KBrO_3$)** — Sensitive to ignition by heat or friction. Relatively moderate health hazard.

**Potassium chlorate ($KClO_3$)** — Explosive properties similar to potassium bromate, but is toxic and fumes liberated by combustion are toxic.

**Potassium perchlorate ($KClO_4$)** — Similar to potassium chlorate. Yields toxic fumes in fires and is an irritant to eyes, skin, and respiratory system.

**Potassium peroxide ($K_2O_3$)** — Reacts vigorously with water. Mixtures with combustible, organic, or easily oxidizable materials are explosive. They ignite easily with heat, friction, or small quantities of water. Toxic if ingested.

**Propyl nitrate (normal) ($CH_3[CH_2]_2NO_3$)** — Very dangerous material. Forms explosive mixtures with air. Very wide flammable limits (2 to 100%); flash point = 20°C (68°F); very low energy required for ignition; comparable with acetylene and hydrogen. Vapors are heavier than air and may travel some distance to ignition source and flash back. Material itself is toxic by either inhalation or ingestion and combustion products highly toxic.

**Sodium chlorate ($NaClO_3$)** — Properties similar to potassium chlorate.

**Sodium chlorite ($NaClO_2$)** — Releases explosive, extremely poisonous chlorine dioxide gas upon contact with acid.

**Sodium perchlorate ($NaClO_4$)** — Properties similar to potassium perchlorate.

**Sodium peroxide ($Na_2O_2$)** — Properties similar to potassium hydroxide.

## REFERENCES (SECTIONS 4.6.2.1 TO 4.6.2.3)

1. *Hazardous Chemical Data,* NFPA 49, National Fire Protection Association, Quincy, MA, 1975.
2. **Armour, M. A., Browne, L. M., and Weir, G. L.,** *Hazardous Chemicals, Information and Disposal Guide,* 2nd ed., University of Alberta, Edmonton, Canada, 1984.
3. **Sax, N. I.,** *Dangerous Properties of Industrial Materials,* 5th ed., Van Nostrand Reinhold, New York, 1979.

### 4.6.2.4. Perchloric Acid*
### INTRODUCTION

Considerable interest has been taken in the explosion hazards to be encountered in the use of perchloric acid since a mixture of perchloric acid and acetic anhydride exploded in a Los Angeles factory in 1947, killing 15, injuring 400, and causing $2 million damage. On a smaller scale, Robinson reported a detonation of 3 g of a perchlorate salt of a rhodium-polyamine complex undergoing an evaporation step in a rotary evaporator[1] A violent explosion destroyed the evaporator, smashed a lab jack, cracked the bench top, and chipped walls over 15 ft away. Fortunately, this happened in an empty laboratory. Literature surveys reveal that descriptions of explosions in laboratories using perchloric acid have been reported over a period of more than a century.

The most detailed available account of the chemistry of perchloric acid and a reference highly recommended to everyone who will be working with perchlorates is given by J. S. Shumacher in the American Chemical Society monograph *Perchlorates, Their Properties, Manufacture, and Uses*.[2] Cummings and Pearson have reviewed the thermal decomposition and the thermochemistry, and in a later report Pearson reviewed the physical properties and inorganic chemistry of perchloric acid.[3,4] Shorter, but very useful, reviews of the chemistry of perchloric acid also have been published,[5,6] and laboratory accidents which have occurred in France have been reviewed.[7] A review of the circumstances leading to the Los Angeles explosion and of five laboratory incidents involving perchloric acid has been published,[8] and a résumé of five serious accidents has been reported.[9] A summary of a number of additional accidents reported in the literature is appended to this section.

### GENERAL DISCUSSION

Perhaps the most disturbing features of accidents involving perchloric acid are (1) the severity of the accidents and (2) the fact that the persons involved are, in the majority of cases, experienced workers. Harris concludes that the basic cause of accidents involving perchloric acid is due to contact with organic material or to the accidental formation of the anhydrous acid.[8] Smith emphasizes the hazard of allowing strong reducing agents to come into contact with concentrated (72%) perchloric acid.[6]

*Properties of Perchloric Acid Solution*

Review of the MSDSs prepared to meet the OSHA hazards standard by each manufacturer is strongly recommended.

The important physical and chemical properties of perchloric acid are that it is a water white liquid, it has no odor, the boiling point (of constant boiling mixture at atmospheric pressure) is 203°C (397°F), and while under high vacuum, a 73.6% composition can be produced. Perchloric acid can be dangerously reactive. At ordinary temperatures, 72% perchloric acid solution reacts as a strong nonoxidizing acid. At elevated temperatures (approximately 160°C [320°F]), it is an exceedingly strong and active oxidizing agent as well as a strong dehydrating reagent. Contact

---

* This material is taken from an updated (by Graf) version of an article by Everett and Graf[22] which appeared in the second edition of the *Handbook of Laboratory Safety*. The remaining portion of this article forms part of Section 4.5.5.

with combustible material at elevated temperatures may cause fire or explosion. Review of the safety data sheets prepared to meet the OSHA hazards standard by each manufacturer is strongly recommended.

*Relative Oxidizing Power of Perchloric Acid*

Although no data are known to describe the change in the oxidizing power of perchloric acid with temperature and/or concentration, some observations describe the phenomenon sufficiently.

Cold perchloric acid, 70% or weaker, is not considered to have significant oxidizing power. The oxidizing power, however, increases rapidly as the concentration increases above 70%. Acid of 73+% (which gives off fumes in even relatively dry air) is a fairly good oxidizer at room temperature. The monohydrate of perchloric acid (85% acid strength and a solid) is, indeed, a very good oxidizer at room temperature as it will even react with gum rubber, whereas the 73% acid does not.

Temperature increases will also increase the oxidizing power of perchloric acid solutions; therefore, hot, strong perchloric acid solutions are very powerful oxidizing agents.

*Explosive Reactions*

A chemical explosion is the result of a very rapid increase in volume due to the evolution of gas or vapor, the reaction normally being exothermic. The force of the shock wave is governed by the rate at which the reaction takes place. The point is made by Burton and Praill[5] that, apart from the thermochemical aspects of explosions, it is the velocity of the decomposition which determines whether a reaction is explosive, and that the power of the explosive is governed largely by the pressure of the gases produced in the decomposition. Where the temperature of the explosion is several thousand degrees centigrade, the power of the explosion is increased further by the thermal expansion of the gases.

When considering the hazards involved in the use of perchloric acid, this point should be clearly recognized: many of the reported serious laboratory accidents involved only small quantities (<1 g) of reactant. (See Appendix included in this section.)

*The System: Perchloric Acid-Acetic Anhydride-Acetic Acid*

Both Shumacher[2] and Burton and Praill[5] examined the perchloric acid-acetic anhydride system in some detail, the former reproducing a triangular diagram for the system. Burton and Praill quote the equation:

$$2.5\ Ac_2O + HClO_4 \cdot 2.5\ H_2O = 5\ AcOH + HClO_4 + 18.4\ kcal$$

showing that there is a considerable evolution of heat when mixing the reagents, and that if excess acetic anhydride is present, the solution may be considered as a solution of anhydrous perchloric acid in acetic acid. The most explosive mixture is given as the one in which complete combustion occurs.

$$CH_3COOH + HClO_4 = 2\ CO_2 + 2\ H_2O + HCl$$

These authors give the explosion temperature as 2400°C and calculate that 1 g of the mixture produces, instantaneously, about 7 l of gas at the explosion temperature. Finally, they state that, for this system, all the investigated explosions have been due to the use of potentially dangerous mixtures, together with faulty equipment or technique.

In 1961, in a university laboratory, a young metallurgy student lost the sight of both eyes when an explosion took place while he was preparing an acetic anhydride — perchloric acid — water electropolishing mixture. Turner and Bartlett[12] investigated the process and discovered that the reagents should be mixed in the following order.

1. Add the perchloric acid to the acetic anhydride.
2. Add water to the mixture, slowly and in small portions.

*The Use of Magnesium Perchlorate as A Drying Agent*

Several cases are on record in which magnesium perchlorate (anhydrone) has exploded while being used as a desiccant (Appendix 1). Smith[6] regards the use of magnesium perchlorate for the drying of alcohol vapors as permissible, but Burton and Praill[5] warn that if magnesium perchlorate is to be used for drying organic liquids, the purity of the drying agent should be determined since the preparation may have left traces of free perchloric acid in the salt. Explosions involving magnesium perchlorate may have been caused by the formation of perchloric esters in the system. It should be noted that methyl and ethyl perchlorate are violently explosive compounds.

*Miscellaneous Reactions*

Burton and Praill[5] state that it is impossible to overemphasize that the simple alkyl esters are extremely dangerous. They note that many documented explosions resulted from the standard method of determining perchlorates or potassium as potassium perchlorate, during which an ethyl alcohol extraction is used. The same authors state that most organic perchlorate salts, with the exception of the diazonium salts, are safe unless they are overheated or detonated, but mention that pyridine perchlorate can be detonated by percussion.

*Reducing Agents*

In general, mixtures of strong reducing agents and concentrated 72% perchloric acid should be regarded as very hazardous[6] (Appendix 1).

**SAFE HANDLING OF PERCHLORATES**

Shumacher[2] states that:

> Perchlorates appear to fall into two broad categories: those (1) more or (2) less sensitive to heat and shock. Included in the group of those qualitatively less sensitive are pure ammonium perchlorate, the alkali metal perchlorates, the alkaline earth perchlorates, and perchloryl fluoride. Among the more sensitive compounds are the pure inorganic nitrogenous perchlorates, the heavy metal perchlorates, fluorine perchlorate, the organic perchlorate salts, the perchlorate esters, and mixtures of any perchlorates with organic substances, finely divided metals, or sulfur. Any attempt to establish a more precise order

of the degree of hazard to be expected from any given perchlorate seems unwarranted on the basis of available data. Each perchlorate system must be separately (and cautiously) evaluated.

There do not appear to be any uniform recommendations for the safe handling of perchlorates which are generally applicable. A number of heavy metal and organic perchlorates, as well as hydrazine perchlorate (hydrazinium diperchlorate) and fluorine perchlorate, are extremely sensitive and must be handled with great caution as initiating explosives. Mixtures of any perchlorates with oxidizable substances are also highly explosive and must be treated accordingly. For all of these, it is essential to avoid friction, heating, sparks, or shock from any source (as well as heavy metal contamination), and to provide suitable isolation, shielding, and protective apparel for personnel. However, the more common ammonium, alkali metal, and alkaline earth perchlorates are considerably less hazardous.

Synthesis of new inorganic or organic perchlorates should only be undertaken by an experienced, cautious investigator who is familiar with the literature.

A simple test to evaluate impact sensitivity can be conducted by placing a crystal or two of the perchlorate on a steel block and striking it with a hammer. The degree of noise and relative impact to produce an explosion can be roughly correlated with the impact sensitivity.

A simple thermal stability test can be conducted by placing a crystal or two on a hot plate and observing the time required to create a violent decomposition reaction. A gram or less of the material can be heated slowly in a loosely closed vial for a more exact determination of thermal stability.

## RECOMMENDATIONS FOR THE SAFE HANDLING OF PERCHLORIC ACID

Several organizations have drawn up recommendations for the safe handling of perchloric acid, among them the Association of Official Agricultural Chemists, the Factory Mutual Engineering Division, and the Association of Casualty and Surety Companies. Graf updated these recommendations in a paper.[13] The recommendations from these and other sources are combined and summarized below.*

### Walk-In Hood Design

Perchloric acid should be handled in a masonry building with concrete or tile floors. Handling acid on wooden floors is dangerous, especially after the acid has dried. The wooden floor will then become sensitive to ignition by friction.

For conventional wooden wall construction, which is not desirable, it is highly recommended that a 6-in. concrete curb be provided for the walls to rest on. In this way, acid seepage under the wall is minimized.

**A. Floors** — Concrete, of course, is not resistant to acids, and thus should be covered. Epoxy paints in general are resistant to room temperature perchloric acid spills; however, epoxy paint will peel off concrete if pools of water stand for several days. Therefore, the floor should have a gentle slope to a drain and contain no low spots.

---

* Note that since the original preparation of this article, commercial equipment specifically designed to be used with perchloric acid has become generally available — specifically, hoods and ejector ducts to exhaust perchloric fumes well above roof level.

## TABLE 4.13
## Materials Compatible or Incompatible with 72% Perchloric Acid[13]

### Compatible

| Material | Compatibility |
|---|---|
| Elastomers | |
| Gum rubber | Each batch must be tested to determine compatibility |
| Vitons[a] | Slight swelling only |
| Metals and Alloys | |
| Tantalum | Excellent |
| Titanium (chemically pure grade) | Excellent |
| Zirconium | Excellent |
| Niobium | Excellent |
| Hastelloy C[a] | Slight corrosion rate |
| Plastics | |
| Polyvinyl Chloride | |
| Teflon[a] | |
| Polyethylene | |
| Polypropylene | |
| Kel-F[b] | |
| Vinylidine fluoride | |
| Saran[c] | |
| Epoxies | |
| Others | |
| Glass | |
| Glass-lined steel | |
| Alumina | |
| Fluorolube | |

### Incompatible

Plastics
  Polyamide (Nylon)
  Modacrylic ester, Dynel (35-85% Acrylonitrile)
  Polyester (Dacron)
  Bakelite
  Lucite
  Micarta
  Cellulose based lacquers
Metals
  Copper
  Copper alloys (brass, bronze, etc.) for very shock sensitive
    perchlorate salts
  Aluminum (dissolves at room temperature)
  High nickel alloys (dissolve)
Others
  Cotton
  Wool
  Wood
  Glycerin-lead oxide (letharge)

[a]  du Pont Trademark.
[b]  3M Company Trademark.
[c]  Dow Chemical Corporation Trademark.

No equipment of any kind should ever be bolted to a floor with bolts that screw into the floor. Perchlorates can enter and form hazardous metallic perchlorates that can initiate a detonation when the bolt is removed. Studs, firmly and permanently set into the floor, to which the equipment can be bolted, are far safer. The nuts can then be flushed with water and sawed off with a hack saw under a constant water spray to remove that equipment.

*Building Equipment*

**Laboratory benches**. Laboratory benches should be constructed of resistant materials, and not wood, to prevent acid absorption, especially at the bottom surface which rests on the floor and would be subject to the greatest exposure from acid spills. Bench tops of resistant and nonabsorbent materials such as chemical stoneware, tile, epoxy composites, and polyethylene are recommended.

**Shelves and cabinets**. Shelves and cabinets of epoxy-painted steel are highly recommended rather than wood.

*Laboratory Equipment*

**Heating source.** Hot plates (electric), electrically or steam-heated sand baths, or a steam bath are recommended for heating perchloric acid. Direct flame heating or oil baths should not be used.

**Vacuum source.** Smith describes a simple apparatus, using a water aspirator or pump, for drawing fumes from a reaction vessel. The use of this apparatus is to be commended in that the contamination of the fume hood duct with a dust/perchloric acid layer is avoided and the vapors are drawn into water and discharged safely to the drain.* A similar apparatus is marketed for carrying out Kjeldahl digestions. Vacuum pumps from which all traces of petroleum lubricants have been flushed and refilled with halocarbon, Kel-F, or fluorolube are recommended.

Silverman and First have described a self-contained unit, which has been developed and field tested, for collecting and disposing of chemical fumes, mists, and gases.[15] It is portable and compact and, when assembled, only requires connection to an electrical receptacle and water tap to be completely operational. Although originally designed for use in filter-type radiochemical laboratory hoods for safe disposal of perchloric acid fumes arising from acid digestions, it may be used as a substitute for a permanent hood in a variety of (low risk) locations. Collection and disposal of the acid at the source of emission is the guiding design principle.

The conventional vacuum pump has been used both in the laboratory and in a pilot plant handling large quantities of 72% perchloric acid. The pumps were protected by the use of well-designed cold traps and desiccant columns as well as by maintaining the practice of changing petroleum-based oil daily. The desiccant was also routinely changed, along with the frequent thawing and removal of the cold trap contents.

**Glassware.** The hazards that may ensue if an apparatus cracks or breaks due to thermal or mechanical shock are sufficient to make it desirable that quartz apparatus be considered, especially as it is necessary in many experiments to chill rapidly from the boiling point.

---

* For an individual laboratory, such a practice should pose no problem. However, if the facility is big enough, such practices may eventually become regulated, especially if the water treatment works is marginal. The same concern may eventually become involved in maintaining air quality near facilities with a large number of hood exhausts.

Glass-to-glass unions, lubricated with 72% perchloric acid, seal well and prevent joint freezing arising from the use of silicon lubricants. Rubber stoppers, tubes, or stopcocks should not be used with perchloric acid due to incompatibility.

**Stirrers.** Pneumatically driven stirrers are recommended rather than the electric motor type. Repeated exposure of the copper motor windings to perchloric acid vapor could result in a fire, or detonation unless the motor is an explosion-proof type, which would be unlikely.

**Sundry items.** The choice of tongs for handling hot flasks and beakers containing perchloric acid mixtures should be given due thought. Since the use of radioactive materials has become commonplace, much thought has been put into the design of indirect handling equipment. The cheap, commonly used crucible tongs are most unsuitable for picking up laboratory glassware. If possible, tongs with a modified jaw design should be used to insure that a safe grip is obtained.

*Acid Handling in the Laboratory*
**73% perchloric acid or less**
1. Use chemical splash and impact-rated goggles for eye protection whenever the acid is handled.
2. Always transfer acid over a sink in order to catch any spills and afford a ready means of disposal.
3. In wet combustions with perchloric acid, treat the sample first with nitric acid to destroy easily oxidizable matter.
4. Any procedure involving heating of perchloric acid must be conducted in a perchloric acid fume hood with the sash down.
5. No organic materials should be stored in the perchloric acid hood.
6. Do not allow perchloric acid to come into contact with strong dehydrating agents (concentrated sulfuric acid, anhydrous phosphorous pentoxide, etc.).
7. Perchloric acid should be used only in standard analytical procedures from well-recognized analytical texts. (This does not apply to analytical research workers.)
8. Keep the quantities of perchloric acid handled at the bare minimum for safety.

**73 to 85% perchloric acid**
1. Same precautions as above.

**Anhydrous perchloric acid (greater than 85%)**
1. Only experienced research workers should handle anhydrous perchloric acid. These workers must be thoroughly familiar with the literature on the acid.
2. A safety shield must be used to protect against a possible exlosion, and the acid must be used in an appropriate hood with a minimum of equipment present. No extraneous chemicals should be present in the hood.
3. A second person should be informed of the intended use of anhydrous acid and be in the same room with the research worker using this extremely strong oxidizer.
4. Safety goggles, a face shield, thick gauntlets, and a rubber apron must be worn.
5. Only freshly prepared acid should be used.
6. Do not make any more anhydrous perchloric acid than is required for a single day's work.

7.   Dispose of any unused anhydrous acid at the end of each day by dilution and neutralization.
8.   Contact of the anhydrous acid with organic materials will usually result in an explosion.
9.   Any discoloration of the anhydrous acid requires its immediate disposal.

*Acid Disposal*

**Spills.** Perchloric acid spilled on the floor or bench top represents a hazard. It should not be mopped up, nor should dry combustibles be used to soak up the acid. The spilled acid should first be neutralized and then soaked up with rags or paper towels. The contaminated rags and paper towels must be kept wet to prevent combustion upon drying. They should be placed in a plastic bag and sealed and then placed in a flammable waste disposal can. If the spill can be rinsed down a chemical drain, neutralization of the wetted area is recommended, followed by additional rinsing.

[Other recommendations in the literature are to wear a face shield and gloves while working on the spill. Cover the spill with a weak solution of sodium thiosulfate, and then transfer the slurry into a large container of water, where it should be neutralized with soda ash. After neutralization, it can be drained into the sewer, accompanied by abundant water.]

**Disposal.** Stir the acid into cold water until the concentration is less than 5%, followed by neutralization with aqueous sodium hydroxide, and then dispose of the resulting mixture in the sanitary system, accompanied by abundant water. [Larger quantities in the original unopened containers may be acceptable to a commercial hazardous waste vendor. If it is potentially explosive, the best option available is to hire a firm specializing in disposal of exceptionally hazardous materials. This will be expensive.]

## DISMANTLING AN EXHAUST VENTILATION SYSTEM SUSPECTED OF CONTAMINATION WITH PERCHLORATES

Dismantling a laboratory exhaust system contaminated with shock-sensitive perchlorates is a hazardous operation, as evidenced by published and unpublished case histories. The procedures used by one university to reduce the hazards were described by Peter A. Breysse in the *Occupational Health Newsletter* (15[2, 3] 1, 1966) published by the Environmental Health Division, Department of Preventive Medicine, University of Washington. The problem, procedures, and confirmation of perchlorate contamination were reported as follows.

A short time ago, the manager of Maintenance and Operations was requested to dismantle and relocate six laboratory exhaust systems. The possibility of perchloric acid contamination of these systems was considered. An investigation indicated that several laboratories serviced by the exhaust systems were utilizing, or had in the past used, perchloric acid for wet washing of tissues. Furthermore, the exhaust hoods were constructed with sharp corners and cracks, permitting the accumulation of contaminants not readily noticed or easily removed. The ducts were made of ceramic material and contained numerous joints as well as a number of elbows — areas conducive to perchlorate build-up. Organic compounds were also used — to pack the duct joints, as an adhesive for the flexible connectors, and as a sealing compound for the fan.

Recognizing the potential dangers of dismantling these systems, the following procedures were established and successfully carried out.

1.  It was deemed desirable to dismantle the systems on the weekend when occupancy would be at a minimum.
2.  The entire system was washed for 12 hours, just prior to dismantling, by introducing a fine water spray within the hoods, with the fans operating.
3.  The fans were then hosed down.
4.  Fan mounting bolts and connectors were carefully removed. Nonsparking tools were and should be used throughout.
5.  The fans were immediately removed to the outdoors. As an added precaution during removal, the fans were covered with a wet blanket.
6.  After all the fans were taken outdoors, one fan at a time was placed behind a steel shield for protection during dismantling. This fan was again washed down.
7.  Plate bolts were evenly loosened to remove the plate without binding. If a fan puller is necessary, it should be nonsparking.
8.  All disassembled parts were washed and cleaned. The gasket material contained on the flanges was scraped off with a wooden scraper.
9.  Ordinarily, the ceramic ducts would be removed by breaking them apart with a sledge hammer. In this instance, the ducts were washed down again just prior to and during dismantling. A high-speed saw was used to remove the duct work.

One of the flexible connectors and a piece of duct-joint sealing compound were collected and taken to the laboratory for examination. Qualitative analysis by X-ray fluorescence and chemical tests indicated the presence of perchlorates in both samples. While these procedures for dismantling and decontamination seem unduly severe, the uncertainty requires that they be followed.

[One of the problems in conducting such an operation is providing appropriate liability insurance. This question should be considered before doing such work in-house, and contracts for work done by outside vendors should have the provision of liability insurance as a specific requirement.]

Some additional words of caution include: (1) do not unscrew any nuts or bolts — cut them off after water flushing, (2) do not produce any impact upon the hood parts, and (3) do not cause any friction between parts of the hood or ducts during dismantling.

## SUMMARY AND CONCLUSIONS

The use of perchloric acid is becoming increasingly widespread and the properties of both the acid and its derivatives make it likely that the trend will continue. Safety hazards associated with the use of perchloric acid may be reduced, provided that its hazardous properties are clearly recognized, the purpose of the acid in a process is fully understood, and measures are taken to avoid known possibilities.

It is, however, clear that no one should attempt to use perchloric acid who is not fully conversant with the chemistry of the material who has not made a careful appraisal of his operating conditions and techniques, and who exhibits an unsafe attitude about his work.

## APPENDIX 1 — SOME ACCIDENTS INVOLVING PERCHLORIC ACID

1.  Explosions may occur when 72% perchloric acid is used to determine chromium in steel, apparently due to the formation of mixtures of perchloric acid vapor and hydrogen. These vapor mixtures can be exploded by the catalytic action of steel particles.[2]

2.  Two workers are reported to have dried 11,000 samples of alkali-washed hydrocarbon gas with magnesium perchlorate over a period of 7 years without accident. However, one sample containing butyl fluoride caused a purple discoloration of the magnesium perchlorate, with the subsequent explosion of the latter.[2]

3.  A worker using magnesium perchlorate to dry argon reported an explosion and warned that warming and contact with oxidizable substances should be avoided.[2]

4.  An explosion was reported when anhydrous magnesium perchlorate used in drying unsaturated hydrocarbons was heated to 220°C.[2]

5.  An explosive reaction takes place between perchloric acid and bismuth or certain of its alloys, especially during electrolytic polishing.[2,5]

6.  Several explosions reported as having occurred during the determination of potassium as the perchlorate are probably attributable to heating in the presence of concentrated perchloric acid and traces of alcohol. An incident in a French laboratory is typical: an experienced worker, in the course of a separation of sodium and potassium, removed a platinum crucible containing a few decigrams of material and continued the heating on a small gas flame. An explosion pulverized the crucible, and a piece of platinum entered the eye of the chemist.[7]

7.  A violent explosion took place in an exhaust duct from a laboratory hood in which perchloric acid solution was being fumed over a gas plate. It blew out windows, bulged the exterior walls, lifted the roof, and extensively damaged equipment and supplies. Some time prior to the explosion, the hood had been used for the analysis of miscellaneous materials. The explosion apparently originated in deposits of perchloric acid and organic material in the hood and duct.[8]

8.  A chemist was drying alcohol off a small anode over a bunsen burner in a hood reserved for tests involving perchloric acid. An explosion tore the exhaust duct from the hood, bent a portion of the ductwork near the fan, and blew out many panes of window glass.[8]

9.  An employee dropped a 7-lb (3.2-kg) bottle of perchloric acid solution on a concrete floor. The liquid was taken up with sawdust and placed in a covered, metal waste can. Four hours later, a light explosion blew open the hinged cover of the can. A flash fire opened three sprinklers, which promptly extinguished the fire.[8]

10. A 7-lb bottle of perchloric acid solution broke while an employee was unpacking a case containing three bottles. The spilled acid instantly set the wood floor on fire, but it was put out quickly with a soda-acid extinguisher.[8]

11. At a malleable iron foundry, perchloric acid had been used for about 4 years in the laboratory for the determination of the silicon contents of iron samples.

A cast iron, wash-sink drain at the bench used for this purpose had corroded and the leaking acid had soaked into the wood flooring, which was later ignited while a lead joint was being poured. This fire was extinguished and part of the wood flooring was removed. Later in the day, at a point slightly removed from the location of the first fire, a similar fire occurred when hot lead was again spilled. This time, the fire flashed with explosive violence into the exhaust hood and stack above the work bench. Laboratory equipment and records were wet down extensively, and damaged.

12.  A stone table of a fume hood was patched with a glycerin cement and several years later, when the hood was being removed, the table exploded when a workman struck the stone with a chisel. The hood had been used for digestions with perchloric acid and, presumably, acid spills had not been properly cleaned up.[9]

13.  A conventional chemical hood normally used for other chemical reactions, including distillation and ashing of organic materials, was also used during the same time for perchloric acid digestion. During a routine ashing procedure, the hot gases went up the 12-in. tubular transit exhaust duct and one of a series of explosions occurred that tore the duct apart at several angles and on the horizontal runs.[9]

14.  During routine maintenance involving partial dismantling of the exhaust blower on a perchloric acid ventilating system, a detonation followed a light blow with a hammer on a chisel held against the fan at or near the seal between the rear cover plate and the fan casing. The intensity of the explosion was such that it was heard 4 miles away. Of the three employees in the vicinity, one sustained face lacerations and slight eye injury, the second suffered loss of four fingers on one hand and possible loss of sight in one eye, and the third was fatally injured when the 6-in. chisel entered below his left nostril and embedded in the brain.[9]

15.  A 6-lb (2.7-kg) bottle of perchloric acid broke and ran over a fairly large area of a wooden laboratory floor. It was cleaned up, but some ran down over wooden joists. Several years later, a bottle of sulfuric acid was spilled in this same location and fire broke out immediately in the floor and the joists.[9]

16.  A chemist reached for a bottle of perchloric acid stored on a window sill above a steam radiator. The bottle struck the radiator, broke, and the acid flowed over the hot coils. Within a few minutes, the wooden floor beneath the radiator burst into flame.[9]

17.  An explosion occurred when an attempt was made to destroy benzyl celluloses by boiling with perchloric acid.[11]

18.  An explosion occurred as anhydrous perchloric acid was being prepared via sulfuric acid dehydration and extraction with methylene chloride when a stopper was removed from the separatory flask.[14]

19.  A rat carcass was dissolved in nitric acid, the fat skimmed off, and perchloric acid added. The mixture was heated to dryness and touched, setting off an explosion that cracked the fume hood and nearly blew out the sash.[16]

20.  A perchlorate-doped polyacetylene film was prepared and stored under argon in a sealed vessel. Two weeks later, the film detonated when the vessel top was being removed. Earlier safety testing failed to show any reaction to flame or impact.[17]

21. Perchlorate-doped polyacetylene samples combusted violently in the oxygen atmosphere of a Schoniger flask.[18]

22. An explosion occurred in a fume hood upon ether drying of a second crop of crystals of hexamine chromium (111) perchlorate that were washed with absolute ethanol and anhydrous diethylether. Following aspiration of the ether wash, the ether damp filter cake was agitated with a glass stirring rod and the mass detonated.[19]

23. Perchlorate-doped, highly conducting polythiophene Pt-C104 exhibits excellent ambient stability, but the film should not be heated about 100°C. Touching an extremely dry Pt-C104 film (kept in a desiccator over $P_2O_5$) with tweezers might cause an explosion of the film.[20]

24. Some samples of rare earth organic fluoride were re-ashed with perchloric, sulfuric, and nitric acids in 1-l beakers. One of the beakers started foaming, turned yellow, and then exploded. The surface of the hot plate was bent downwards and the imprint of the beaker was left in the metal surface of the hot plate by the force of the explosion.[1]

25. Drying an acetonitrile adduct of neodymium perchlorate at 80°C in vacuum apparently produces a compound that can detonate on mechanical contact.[21]

## ACKNOWLEDGMENT

This chapter is based on and developed from a safety assessment prepared for use in the University of Leeds, England, and is reprinted with permission, as is the work of Graf and associates at Morton-Thiokol Corporation, Brigham City, UT. Revised in 1989.

## REFERENCES (SECTION 4.6.2.4)

1. **Robinson, W. R.,** Perchlorate salts of metal ion complexes: potential explosives, *J. Chem. Ed.,* 62(11), 1001, 1985.
2. **Shumacher, J. C.,** *Perchlorates: Their Properties, Manufacture and Uses,* American Chemical Society Monogr. Ser. No. 146, Reinhold, New York, 1960.
3. Perchloric Acid: A Review of its Thermal Decomposition and Thermochemistry, Rocket Propulsion Establishment Tech. Note No. 224, Ministry of Aviation, London, 1963.
4. Perchloric Acid: A Review of the Physical and Inorganic Chemistry, Rocket Propulsion Establishment Tech. Note No. 352, London, Ministry of Aviation, 1965.
5. **Burton, H. and Praill, P. F. G.,** Perchloric acid and some organic perchlorates, *Analyst,* 80, 4, 1955.
6. **Smith, G. F.,** The dualistic and versatile reaction properties of perchloric acid, *Analyst,* 80, 16, 1955.
7. **Moureu, H. and Munsch, H.,** Sur Quelches Accidents Causes par la Manipulation de l'Acide Perchloriques et des Perchlorates, *Arch. Mal. Prof. Med. Trav. Secur. Soc.,* 12, 157, 1951.
8. **Harris, E. M.,** Perchloric acid fires, *Chem. Eng.,* 56(1), 116, 1949.
9. **Scheffler, G. L.,** University Health Service, University of Minnesota, private communication to safety officer, Leeds University, London.
10. **Jacob, K. D., Brabson, J. A., and Stein, C.,** *J. Assoc. Off. Agric. Chem.,* 43, 171, 1960.
11. **Sutcliffe, G. R.,** *J. Textile Ind.,* 41, 196T, 1950.
12. **Bartlett, R. K. and Turner, H. S.,** An unappreciated hazard in the preparation of electropolishing solutions: the investigation of an explosion. *Chem. Ind.,* 1933-1934, November 20, 1965.
13. **Graf, F. A.,** Safe handling of perchloric acid, *Chem. Eng. Prog.,* 62(10), 109, 1966.
14. Explosion Involving Anhydrous Perchloric Acid, Letter WLC-151/63, Thiokol Chemical Corporation, April 11, 1963.

15. **Silverman, L. and First, M. W.,** Portable laboratory scrubber for perchloric acid, *Ind. Hyg. J.*, p. 463, 1962.

16. **Muse, L. A.,** Letter to editor, *Chem. Eng. News,* 51(6), 29, 1973.

17. **Elsenbanner, R. L. and Miller, G. G.,** Letter to editor, *Chem. Eng. News,* 63(25), 4, 1985.

18. **Varyu, M. E., Schenoff, J. B., and Dabkowski, G. M.,** Letter to editor, *Chem. Eng. News,* 63(27), 4, 1985.

19. **Pennington, B. E.,** Letter to editor, *Chem. Eng. News,* 62(32), 55, 1982.

20. **Osterholm, J. and Passiniemi, P.,** Letter to editor, *Chem. Eng. News,* 64(6), 2, 1986.

21. **Raymond, K. N.,** Letter to editor, *Chem. Eng. News,* 63(27), 4, 1985.

22. **Everett, K. and Graf, F. A., Jr.,** Handling perchloric acid and perchlorates, in *Handbook of Laboratory Safety,* 2nd ed., Steere, N. V., Ed., CRC Press, Boca Raton, FL, 1971, 265.

### 4.6.2.5. Ethers

Most of the concern in working with ethers is due to the problem of peroxide formation, already dealt with at some length in Section 4.5.4, which discussed this problem in the context of the storage of ethers. If ethers are bought in small sizes so that the containers in use on a given day are emptied, the concerns of peroxide formation in partially full containers in storage should not arise. Peroxides can form in unopened containers as well, although due to the absence of excess available air in the restricted empty space above the liquid level, the rate should be much slower. However, the inventory of unopened containers should be maintained at a reasonable level so that no ethers are kept for extended periods. All containers should be marked with the date received and a schedule of disposal or testing established. If a partially empty container is kept, a target date for testing for peroxides should also be placed on the container as well as subsequent dates for the tests to be repeated. If this is necessary, it possibly reflects poor management of resources because the effort needed to maintain a reliable program to check for the presence of peroxides could almost certainly be used to greater advantage on the basic research program. Unused portions of ethers should not be returned to the original container. Small quantities can be allowed to evaporate in a fume hood.

Because of the explosion risk associated with peroxides, older containers of ethers are not normally accepted as part of an ordinary hazardous waste shipment by a commercial chemical waste disposal firm. They would have to be disposed of as unstable explosive materials, which is extremely expensive. Any savings in buying the "large economy size" would be more than compensated for by the additional disposal costs. Attempting to treat the ethers to remove the peroxides or to have laboratory personnel dispose of them carries with it the risk of an explosion and the subsequent liability for injuries.

Although the major risk usually associated with ethers is, as noted, the problem of peroxides, they also pose additional problems because of their properties as flammable solvents. Many of them have lower explosion limits in the range of 0.7 to 3% and flash points at room temperature or, in several cases, much lower. Ethyl ether, the material which most frequently comes to mind when "ether" is mentioned, has a lower explosion limit of 1.85%, an upper explosion limit of 36%, and a flash point of $-45°C$ ($-49°F$). The vapors of most ethers are heavier than air and, hence, can flow a considerable distance to a source of ignition and flash back. Because of their flammable characteristics, many ethers, especially ethyl ether, placed in improperly sealed containers in an ordinary refrigerator or freezer release vapors which represent a potential "bomb" that can be ignited by sparks within the confined space and explode with sufficient force to seriously injure or even kill someone in the vicinity. A fire is virtually certain to result should such an incident occur, which, depending upon the type of construction and the availability of fuel in the laboratory area, could destroy an entire building. Because ethers as well as other flammable solvents are used so frequently in laboratory research, all refrigeration units in most laboratory environments should be purchased without ignition sources within the confined spaces and with explosion-proof compressors, i.e., meet standards for "flammable material storage", even if the current program does not involve any of these materials. At some time during the effective lifetime of most refrigeration units, it is likely that flammable liquids will be used in the facility where they are located.

Should a spill occur, it should be promptly cleaned up using either a commercial solvent-spill kit material to absorb the liquid or a preparation of equal parts of soda ash, sand, and clay cat litter, which has been recommended as an absorbent. Since the lower explosion limit concentrations are so low for so many of the commonly used ethers, all ignition sources should be promptly turned off following a spill and all except essential personnel should be required to leave the area. Personnel performing the cleanup should wear half-face respirators equipped with organic cartridges. The resulting waste mixture from the cleanup can be placed in a fume hood temporarily until removed from the laboratory for disposal as a hazardous waste.

Ether fires as well as other fires involving flammable liquids should be extinguished with Class B fire extinguishing agents. Usually, the most effective are dry chemical extinguishers which interrupt the chemistry of a fire, while carbon dioxide units can be used to smother small fires. Portable halon extinguishers are also usable if the fire is such that the fuel has time to cool before the concentration of the halogenated agents falls below the critical concentration at which it is effective. It is again worth pointing out that unless there is a reasonable chance of putting out a fire with portable extinguishers, it is preferable to initiate an evacuation as quickly as possible and to make sure that everyone can safely leave the building, rather than engage in a futile attempt to put out the fire.

In general, the toxicity of ethers is low to moderate, although this generalization should be confirmed for each different material to be used by information obtained either from the container label or the MSDS. Prolonged exposure to some ethers has been known to cause liver damage. Many have anesthetic properties and are capable of causing drowsiness and eventual unconsciousness. In extreme cases of exposure, death can result. Recently, four ethylene glycol ethers — 2-methoxyethanol, 2-ethoxyethanol, 2-methoxyethanol acetate, and 2-ethoxyethanol acetate — have been identified as causing fetal developmental problems, including fetal malformations and resorptions and testicular damage, in several animal species. Studies have also shown adverse hematologic effects and behavioral problems in the offspring of animals.

### 4.6.2.6. Flammable Solvents

Much of the concern in the literature is centered on the flammable characteristics of flammable liquids, much of which has already been addressed in Section 4.5.7. There are many other issues relating to the health effects of these solvents which also need to be considered.

### 4.6.2.6.1. Flammable Hazards

The fire hazard associated with flammable liquids should more appropriately be associated with the vapors from the liquid. It is the characteristics of the latter which determine the seriousness of the risk posed by a given solvent. In previous sections on the storage of liquids, two of the more important properties of flammable liquids were mentioned in terms of defining the classes to which a given solvent might belong. Definitions of Class 1A, 1B, 1C, 2, 3A and 3B were based on the boiling point and flash point of the solvents. The formal definition of these two terms will be repeated below as well as three other important parameters relevant to fire safety, the ignition temperature and the upper and lower explosion limits.

**Boiling point (bp)** — This is the temperature at which the vapor of the liquid is in equilibrium with atmospheric pressure (defined at standard atmospheric pressure of 760 mm of mercury).

**Flash point (fp)** — This is the minimum temperature at which a liquid gives off vapor in sufficient concentration to form an ignitable mixture with air near the surface of a liquid. The experimental values for this quantity are defined in terms of specific test procedures which are based on certain physical properties of the liquid.

**Ignition (autoignition) temperature** — This is the minimum temperature which will initiate a self-sustained combustion independent of the heat source.

**Lower explosion (or flammable) limit (lel)** — This is the minimum concentration, by volume percent in air, below which a flame will not be propagated in the presence of an ignition source.

**Upper explosion (or flammable) limit (uel)** — This is the maximum concentration, by volume percent of the vapor from a flammable liquid, above which a flame will not be propagated in the presence of an ignition source.

For a fire to occur involving a flammable liquid, three conditions must be met: (1) the concentration of the vapor must be between the upper and lower flammable limits, (2) an oxidizing material must be available, usually the air in the room, and (3) a source of ignition must be present. The management strategy is usually to either maintain the concentration of the vapors below the lower flammable limit by ventilation (such as setting the experiment up in an efficient fume hood) or eliminate sources of ignition. The latter is easier and more certain because the ventilation patterns, even in a hood, may be uneven or disturbed so that the concentrations may locally fall into the flammable range. While some materials may require an open flame to ignite, it is much easier to ignite others. For example, the ignition temperature for carbon disulfide is low enough (80°C or 176°F) that contact with the surface of a light bulb may ignite it.

In order to work safely with flammable liquids, there should be no sources of ignition in the vicinity, either as part of the experimental system or simply nearby. Use nonsparking equipment. When transferring a flammable liquid using metal containers, both containers should be grounded as the flowing liquid can itself generate a static spark. Flammable materials should be heated with safe heating mantles (such as a steam mantle), heating baths, or explosion-proof heating equipment. Many ovens used in laboratories are not safe for heating flammables because the vapors can reach the heating element, or either the controls or the thermostat may cause a spark. Any spark-emitting motors should be removed from the area. Flammable materials should never be stored in an ordinary refrigerator or freezer, which have numerous ignition sources in the confined volume. Placing the entire system in a hood where the flammable vapors will be immediately exhausted aids in limiting the possibility of the vapors coming into contact with an ignition source.

The vapors of flammable liquids are heavier than air and will flow away from the source for a considerable distance. Should they encounter an ignition source while the concentration is in the flammable range, a flame may be initiated which may flash back all the way to the source. In at least one instance, a fire resulted when a research worker walked by a fume hood carrying an open beaker of a low ignition point volatile solvent. There was an open flame in the hood, and when the fumes were pulled into the hood, they ignited, and flashed back to the container, which was immediately dropped, resulting in the feet and lower legs of the worker and the entire floor of the laboratory becoming engulfed in flame. Fortunately, a fire blanket and fire extinguisher were immediately available, so the worker escaped with only minor burns and the fire was extinguished before anything else in the laboratory became involved. This incident clearly illustrates the need to consider all possibilities of fire when using solvents, and that a hood does not totally isolate a hazardous operation.

# REFERENCES (SECTIONS 4.6.2.5 TO 4.6.2.6.1)

1. **Armour, M. A., Browne, L. M., and Weir, G. L.,** *Hazardous Chemicals Information and Disposal Guide,* 2nd ed., University of Alberta, Edmonton, Canada, 1984.
2. *Fed. Reg.,* 10586, April 2, 1987.
3. **Langan, J. P.,** *Questions and Answers on Explosion-Proof Refrigerators,* Kelmore, Newark, NJ.
4. *Prudent Practices for Handling Hazardous Chemicals in Laboratories,* National Academy Press, Washington, D.C., 1981, 57.
5. **Sax, N. I.,** *Dangerous Properties of Industrial Materials,* 5th ed., Van Nostrand Reinhold, New York, 1979.
6. *Hazardous Chemicals Data,* NFPA 49, National Fire Protection Association, Quincy, MA, 1975.

### 4.6.2.7. Reactive Metals

Lithium, potassium, and sodium are three metals that react vigorously with moisture (lithium to a lesser extent than the other two, except in powdered form or in contact with hot water) as well as many other substances. In the reaction with water, the corresponding hydroxide is formed along with hydrogen gas, which will ignite. Lithium and sodium should be stored under mineral oil or other hydrocarbon liquids that are free of oxygen and moisture. It is specifically recommended that potassium be stored under dry xylene.

In many ways, the chemical hazards of the three metals are similar. All three form explosive mixtures with a number of halogenated hydrocarbons, all three react vigorously or explosively with some metal halides, although potassium is significantly worse in this respect, and the reaction of all three in forming a mercury amalgam is violent. They all react vigorously with oxidizing materials. Potassium will form the peroxide and superoxide when stored under oil at room temperature and may explode violently when cut or handled. Sodium reacts explosively with aqueous solutions of sulfuric and hydrochloric acids. The literature provides a number of other potentially violent or vigorous reactions for each of these materials. It is not the intent here to list all of the potentially dangerous reactions that may occur, but, rather, to point out that there are many possibilities for incidents to occur. No one should plan to work with these materials without carefully evaluating the chemistry involved for potential hazards. The materials should be treated with the care which their properties demand at all times.

In one instance, a very old can of sodium was determined to have "completely" reacted to form sodium hydroxide, and the worker decided to flush it with water to dispose of the residue. The bottom two inches were still sodium metal and, consequently, the can exploded. The only entrance to the laboratory was blocked by an ensuing fire, so the occupants had to escape through windows. Fortunately, they were on the first floor and the windows were not blocked. The latter point is worth noting because shortly before this incident, bars which had been put over the windows to prevent break-ins had been removed at the insistence of the organization's safety department. Unsubstantiated assumptions or misplaced priorities are a major cause of injuries.

Since they all react vigorously with moisture, care should be taken to avoid skin and eye contact, which could result in burns from the evolved heat and direct action of the hydroxides. The materials should always be used in a hood, and, as a minimum, gloves and chemical splash and impact-resistant goggles should be worn while working. If the risk of a violent reaction cannot be excluded, additional protection such as an explosion barrier or a face mask should be considered.

If a fire should result, appropriate Class D fire extinguishers should be available within the laboratory. These vary somewhat in their contents, which usually are specific to a given material. Appropriate material suitable for extinguishing fires involving these reactive materials are dry graphite, soda ash, and powdered sodium chloride. Water (obviously), carbon dioxide, or halogenated units should not be used. There are a number of other reactive metals which, while not as active chemically as these three, once ignited, require Class D fire extinguishing agents as well. These include magnesium, thorium, titanium, uranium, and zirconium. Other materials for which Class D units should be used include metal alkyds and hydrides and organometallic compounds. Other materials which might be used in these Class D units are pulverized coke, pitch, vermiculite, talc, and sand. These materials will usually be mixed with various combinations of low-melting fluxing salts, resinous materials, and alkali-metal salts which, in combination with the other material in the extinguisher, form a crust to smother the fire.

As with many other materials, one of the major considerations in using these reactive metals today is the problem associated with disposal of unneeded surplus quantities or waste materials. There are suggestions in the literature for treating waste for each of these three materials. For example, small amounts of potassium residues from an experiment should be treated promptly by reacting them with tert-butyl alcohol because of the danger that they will explode (even if stored as potassium should be stored). This is appropriate for small quantities in the laboratory, but disposal of substantial quantities of unwanted material is a different matter. Recycling or transfer to another operation needing the material should be investigated, but disposal by local treatment should be avoided. Local treatment of these materials is limited under the Resource Conservation and Recovery Act (RCRA) without getting a permit to become a treatment facility. In addition, there are safety and liability risks associated with processing dangerously reactive materials which must be considered. Reactive metals are among those materials which require special handling by commercial waste disposal firms. The cost is very high. Quantities purchased and kept in stock should be limited.

# REFERENCES (SECTION 4.6.2.7)

1. *Hazardous Chemical Data,* NFPA-49, National Fire Protection Association, Quincy, MA, 1975.
2. **Armour, M. A., Browne, L. M., and Weir, G. L.,** *Hazardous Chemicals Information and Disposal Guide,* 2nd ed., University of Alberta, Edmonton, Canada, 1984.
3. **Sax, N. I.,** *Dangerous Properties of Industrial Materials,* 5th ed., Van Nostrand Reinold, New York, 1979.
4. **Sax, N. I. and Lewis, R. J., Sr.,** *Rapid Guide to Hazardous Chemicals in the Workplace,* Van Nostrand Reinhold, New York, 1986.
5. *Portable Fire Extinguishers,* NFPA-10, National Fire Protection Association, Quincy, MA, 1984.

## 4.6.2.8. Mercury

Mercury and its compounds are widely used in the laboratory. As metallic mercury, it is often used in instruments and laboratory apparatus and, in the latter application especially, is responsible for one of the more common types of laboratory accidents, mercury spills. Thermometers containing mercury are frequently broken; mercury is often spilled in working with mercury diffusion pumps or is lost when cleaning the cold traps associated with high-vacuum systems. Over a period of time, the small amounts of mercury lost each time can add up to a substantial amount. In one instance, the cold traps were always cleaned over a sink. After a few years of this, a large amount of mercury accumulated in the sink trap and finally eroded the metal and spilled on the floor. Over 15 lb of mercury were recovered. Mercury is frequently ejected from simple manometers consisting of mercury in plastic tubing connected to a system under vacuum. As an example of the consequences of this last type of accident, an employee working in a room previously used for years as an undergraduate biology laboratory, was diagnosed as having a fairly severe case of mercury poisoning, although he did not use mercury. Upon investigation, more than 50 lb of mercury was retrieved from under the wooden floor boards in the room. Although the instructors had "cleaned up the mercury" when spills had occurred, over the years, a substantial amount had worked its way through the cracks in the floor. No measurements of the airborne concentration were made at the time, but the normal equilibrium vapor pressure of mercury in air at normal room temperatures is between 100 and 200 times the current permissible levels of mercury in the workplace. It should also be noted that the vapor pressure rises rapidly with temperature. At the temperature of boiling water at standard atmospheric pressure, the vapor pressure is more than 225 times higher than at 20°C (68°F) (0.273 mmHg) and reaches 1 mmHg at 126.2°C (259.2°F). Clearly, mercury should always be heated in a functioning fume hood instead of an open bench.

Mercury poisoning has been known to affect many individuals, among them such prominent scientists as Pascal and Faraday as well as workers in various industries, such as those exposed to mercury as an occupational hazard while using mercuric nitrate in making felt in the hat industry. A frequently cited example of the effects of mercury poisoning was the "Mad Hatter" in Lewis Carrol's "Alice in Wonderland". In the 1950s, many Japanese in a small fishing village suffered serious permanent damage to their central nervous system, in many cases resulting in death or permanent disability, due to eating fish containing methyl mercury as a result of the industrial discharge of mercury compounds into the sea near their village. In the 1970s, fish taken from some of the common waters of Canada and the U.S. were found to be contaminated with mercury, and fishing was banned in some areas. Some commercial swordfish and tuna were found to be contaminated with mercury and had to be withdrawn from the market. In at least one instance, fish in a river in the southeastern U.S. were found to be contaminated by mercury-containing waste from a chemical plant. Eventually, the plant closed down its operations. Mercury compounds were at one time used as fungicides. Individuals died from mistakenly eating seed corn treated with these materials. Mercury, once it enters the ecology, is slow to biodegrade. As a result of the dangers inherent in materials containing mercury, the use of mercury for most agricultural purposes has been banned, and dumping wastes containing mercury compounds in such a way as to be able to contaminate the environment is no longer permitted.

Elemental mercury is probably not absorbed in the gastrointestinal tract, but many of its compounds are. Poisoning due to inhalation and absorption of mercury vapors results in a number of symptoms. Among these are personality and physiological changes such as nervousness, insomnia, irritability, depression, memory loss, fatigue, and headaches. Physical effects may be manifested as tremors of the hands and general unsteadiness. Prolonged exposure may result in loosening of the teeth and excessive salivation. Kidney damage may result. In some cases, the effects are reversible if the exposure ceases, but, as noted in the previous paragraph, ingestion of some organic mercury compounds may be cumulative and result in irreversible damage to the central nervous system. Alkyl mercury compounds have very high toxicity. Aryl compounds and, specifically, phenyl compounds are much less toxic (in the latter case, comparable to metallic mercury), and therapeutic compounds of mercury are less toxic still. In the case of the Minimata Bay exposure in Japan, the fetus was especially vulnerable to exposure. In recognition of the seriousness of the potential toxic effects, the permissible ceiling exposure level for metallic mercury is currently set by OSHA at 0.01 mg/M$^3$, and for alkyl organomercury compounds, an 8-h time-weighted average of 0.01 mg/M$^3$ has been established by OSHA, with a ceiling average of 4 (current ACGIH recommendation is 3 times that level).

The following four paragraphs are taken directly from an article by Steere[1] from the second edition of this handbook.

### Absorption of Mercury Vapor

The metabolism of inhaled mercury in the rat was reported in 1962 by Hayes and Rothstein in studies using a special exposure unit, $^{203}$Hg as a tracer, and whole-body counting techniques which allowed frequent estimation of the body burden. They found that 86% of the mercury was cleared from inhaled air, deposited in the lung, oxidized to the ionic form, and absorbed by the blood within a matter of hours. They report a study by Clarkson in 1961 which showed mercury in the blood to be bound 50% to the red blood cells and 50% to serum protein, with less than 0.1% filterable.

### Distribution of Mercury in the Body

In their study of the metabolism of inhaled mercury in the rat, Hayes and Rothstein found after exposures of 5 hours to a mercury level of 1.4 mg/M$^3$ that the metal was generally distributed in the body, but became highly localized in the kidney with an accumulation after 15 d of 70% or more of the body burden, or 150 times as high a concentration of the metal on a per gram basis as the other tissues. Assuming filterable mercury at 0.1%, they calculate the maximum amount of mercury that could be filtered by the kidneys in 24 hours after exposure at 0.86% of the body burden, and in comparison with the actual deposition of 16.7% of the body burden conclude that the extraordinary affinity of the kidney for mercury is not related to filtration phenomena.

### Excretion of Mercury

In the rat inhalation studies, about 30% of the body burden of mercury cleared from the body rapidly with a half-time of 2 days, associated with a rapid fecal excretion, and the rest moves more slowly with a half-time of 20 days, with about equal rates of fecal and urinary excretion. The study showed that excreta during 15 days accounted for 42% of the initial body burden, and suggested that an additional reduction of 14% of initial body burden may in large part be accounted for by loss of mercury in the form of vapor from the animals.

### Pharmacology of Mercury

The pharmacology of mercury and other heavy metals was reviewed in 1961 by Passow, Rothstein and Clarkson in an article which mentions that mercury in the body has high affinity for sulfhydryl groups, chloride ions and amines or simple amino acids, and that mercury inhibits urease, invertase and other enzymes carrying SH groups. Mercury can also block glucose uptake by erythrocytes and muscle, produce K+ loss from the cells of all species, cause lesions of the central nervous system, and influence bioelectric phenomena by altering transmembrane potentials and by blocking nerve conduction.

Mercury is dense (sp gr of just under 13.6 at 4°C [39.2°F]) and has high surface tension and low viscosity. As a result, it tends to break up into small droplets when it is poured or spilled. Anyone who has tried to pick up small droplets using a stiff piece of paper can attest to the appropriateness of the alternate name "quicksilver". As the droplets are disturbed, e.g., walked upon on the laboratory floor, they tend to break up into smaller and smaller droplets, eventually becoming too small to see. In a laboratory where mercury has been in use for an extended period of time, it is instructive to run a pen knife in the cracks in a tile floor or in the seams where cabinets and bench tops fit together. Invariably, small droplets of mercury will be found.

Although a thin film of oxide will form on the skin of mercury droplets, the film is very fragile and will break. Similarly, sprinkling flowers of sulfur on the location of a mercury spill has been suggested as a control measure, but the surface film which forms also apparently is very fragile and will allow the mercury underneath the film to be readily exposed.

### *4.6.2.8.1. Control Measures*
### Exposure Reduction

One of the simplest control measures is to reduce the amount of mercury used in instrumentation and equipment. Mercury thermometers can be replaced with alternatives. Vacuum gauges can be used to replace manometers and oil diffusion pumps can replace mercury diffusion pumps.

Wherever possible, work with mercury should be done in a fume hood, preferably one that has a depressed surface so that a lip will aid in preventing mercury spills from reaching the floor, and with a seamless interior, as recommended for radiological work and work with perchloric acid. As noted earlier, heating mercury causes it to emit fumes at concentration levels two to three orders of magnitude above the permissible exposure levels. Heating of mercury should never be done on the open bench.

The general restriction upon eating, drinking, and smoking in the laboratory should be strictly enforced in laboratories where mercury compounds are commonly used. Depending on the level of use and the availability of fume hoods, the use of personal protective equipment is recommended in addition to goggles and laboratory aprons. Respiratory protection, consisting of half-face respirators fitted with a cartridge which will absorb mercury, should be considered if there is any potential for mercury exposure. A number of mercury compounds are absorbed through the skin and are strong allergens. In such cases, the skin should be protected with gloves covering the forearms as well as the hands.

Tile is a commonly used material for the floors in a laboratory because it is "easy" to maintain and inexpensive to install. Ease of maintenance is no justification for a tile floor in a laboratory using mercury because of the propensity of the extremely small (20 μm or less) mercury droplets to collect in the cracks. A seamless vinyl or poured epoxy floor should be used instead, with the joints of the floor with the wall being curved or "coved". Similarly, the benchtop should be curved where it joins the back panel.

**Monitoring**

The fumes from mercury provide no direct sensory evidence that they are present. Where use is substantial, monitoring information should be readily available. The least expensive means to detect mercury levels is a detector tube in which a given volume of air is pulled through a glass tube containing a material which undergoes a color change when mercury comes into contact with the material. Normally, the air is drawn through by means of a hand-operated pump. This method is reasonably accurate, but any finding within approximately 25% of the permissible exposure level should be considered sufficient warning of a possible overexposure of personnel working in the area. Each measurement requires a new tube. A popular instrument used to provide a direct reading of mercury concentrations in the air is a hand-held atomic absorption spectrometer in which air contaminated with mercury is drawn through the instrument and the degree of interference with ultraviolet light of the wavelength corresponding to a characteristic line of the mercury spectrum is translated into a numerical reading of the concentration of mercury in the air. The calibration of these instruments must be carefully maintained. Another instrument which provides an accurate, rapid reading of the level of mercury in the air depends upon the property of mercury to amalgamate with a thin gold film. This type of instrument is probably the most accurate and reliable, but is also the most expensive. Unless the work with mercury compounds is quite heavy, the laboratory may be unable to afford either of the last two types of monitoring devices. In such a case, the safety department should be provided with an instrument to be used at all locations within an organization.

**Spill Control Measures**

Large globules of mercury can be cleaned up mechanically by carefully brushing them onto a dust pan or a stiff piece of paper. Another simple device is the use of a small, mechanical, hand-held pump to suck the globules into a small container. This is a tedious procedure which is limited to small spills and, of course, to droplets big enough to be seen. Bulk mercury recovered by these procedures can be recovered and purified for reuse.

Mercury spill kits are available commercially which usually include a small pump, sponges impregnated with a material to absorb mercury and which can be used to wipe up the area of a small spill, and a quantity of an absorbent powder that reacts with mercury to form a harmless amalgam. The latter can be spread on cracks and seams in the floor and furniture and is effective in collecting mercury from otherwise inaccessible places. After leaving the material on the floor or contaminated surfaces for several hours to allow the amalgam to form, the powder can be swept or brushed up and the waste material disposed of as a hazardous waste.

Ordinary vacuum cleaners *MUST NOT* be used to used to clean up a mercury spill. An ordinary vacuum filter bag will not stop an appreciable fraction of small particles in the region of several microns or less and, more importantly, the mercury globules pulled into the bag will be broken up into even finer droplets and spewed out of the vacuum's exhaust into the air, substantially increasing the surface area of mercury exposed to the air and enhancing the rate at which mercury vapor will be generated. There are commercial vacuum cleaners which are specifically designed to pick up mercury, however. One such unit, sold by Nilfisk of America, Inc., which also makes specialized units for other toxic materials, first draws the mercury into a centrifugal separator and collects the bulk of the material into an airtight plastic bottle. The contaminated air then is passed into a collection bag which collects bulk solid waste and then through an activated charcoal filter. Additional filters (some optional) follow the charcoal filter collector. This unit can be used to clean up virtually any spill alone, but can also be used with other control measures to insure a complete cleanup of the spilled material. The hose in the Nilfisk vacuum cleaner has an especially smooth surface to prevent mercury particles from adhering to the inside of the hose. As with most specialized units, the vacuum cleaner is not inexpensive. In some instances, a special purpose mercury vacuum cleaner is virtually indispensible, e.g., for use on a spill on a porous, rough material such as carpeting. The use of carpeting as a laboratory floor covering is very rare, but in at least one instance where this was done, a large area of the carpet was thoroughly contaminated by an extensive mercury spill.

**Ventilation**

The ventilation system in a laboratory using mercury or mercury compounds should conform to the general recommendation that wet chemistry laboratories involving any hazardous material be provided with 100% fresh air instead of having a portion of the air recirculated. Local systems, such as the exhausts of mechanical pumps servicing mercury diffusion pumps, should be collected with a local exhaust system and discharged into the fume hood exhaust system in the room or to a separate exhaust duct provided to service such units.

**Medical Surveillance**

As a minimum, it is recommended that permanent employees working with mercury or mercury compounds be provided periodic physical examinations with a test panel selected specifically for mercury poisoning.

# REFERENCES (SECTIONS 4.6.2.8 TO 4.6.2.8.1)

1. **Steere, N. V.,** Mercury vapor hazards and control measures, in *Handbook of Laboratory Safety,* 2nd ed., Steere, N. V., Ed., CRC Press, Boca Raton, FL, 1971, 334.
2. **Vostal, J. J. and Clarkham, T. W.,** Mercury as an environmental hazard, *J. Occup. Med.,* 15, 649, 1973.
3. **Armour, M. A., Browne, L. M., and Weir, G. L.,** *Hazardous Chemicals Information and Disposal Guide,* 2nd ed., University of Alberta, Edmonton, Canada, 1984.

### 4.6.2.9. Hydrofluoric Acid

Anhydrous hydrofluoric acid (HF) (CAS 7664-39-3) is a clear, colorless liquid. Because it boils at 19.5°C (67.1°F) and has a high vapor pressure, it must be kept in pressure containers. It is miscible in water and lower concentration aqueous solutions are available commercially. It is an extremely dangerous material and all forms, including vapors and solutions, can cause severe, slow-healing, burns to tissue. At concentrations of less than 50%, the burns may not be felt immediately, and at 20%, the effects may not be noticed for several hours. At higher concentrations, the burning sensation is noticeable much more quickly, in a matter of minutes or less. Fluoride ions readily penetrate skin and tissue and, in extreme cases, may result in necrosis of the subcutaneous tissue, which eventually may become gangrenous. If the penetration is sufficiently deep, decalcification of the bones may result. The current OSHA PEL 8-h, time-weighted average to HF is set at 3 ppm (2.5 mg/M$^3$), which also is the ceiling TLV currently recommended by the ACGIH. Chronic exposure to even lower levels may irritate the respiratory system and cause problems to the bones. Even brief exposures to high levels of the vapors may cause severe damage to the respiratory system, although the sharp, irritating odor of the acid usually is sufficient to avoiding inhalation in normal use. Contact with the eyes could result in blindness. If eye exposure occurs, it is urgent to flush the eyes as quickly as possible. Every laboratory using HF should have both an eyewash station and a deluge shower within the laboratory. Dilute solutions and vapors may be absorbed by clothing and held in contact with the skin, which will probably not result in an immediate sensation of pain as a warning, but may eventually lead to skin ulcers which, again, may take some time to heal. A generalization might be made here about absorbent clothing. In many instances, as in this case, absorbent clothing which can retain toxic materials and maintain them in close contact with the skin may be worse than no protection at all, changing the exposure from a transient phenomenon to a persistent one. This is not always a problem, but it should be kept in mind as a possibility when choosing protective apparel. All work with HF should be done in a fume hood.

HF attacks glass, concrete, and many metals (especially cast iron). It also attacks carbonaceous natural materials such as woody materials, animal products such as leather, and other natural materials used in the laboratory such as rubber. Lead, platinum, wax, polyethylene, polypropylene, polymethylpentane, and teflon will resist the corrosive action of the acid. In contact with metals with which it will react, hydrogen gas is liberated and hence the danger exists of a spark or flame resulting in an explosion in areas where this may occur.

The following article, Treatment of Severe Hydrofluoric Acid Exposures, is taken with permission from *Journal of Occupational Medicine*, 25(12), 861, 1983.

### Treatment of Severe Hydrofluoric Acid Exposures

Hydrogen fluoride (HF) or its aqueous solution is capable of producing severe damage to the lungs, skin, eyes, and other soft tissues from brief acute exposures. It can also produce changes in bone, teeth, muscle, and connective tissue in chronic overexposures. These chronic changes are well described in recent monographs[1,2] and will not be discussed here.

Du Pont experience with hydrofluoric acid exposures has resulted in previous publications on treatment of the resulting burns.[3,4] More recent experience has led to revisions in treatment methods that have appeared to improve results. This report will update the recommendations of the 1965 and 1966 reports.

The treatment methods we have found to be the most successful are summarized.

### Principles

The two basic tenets of the treatment program we have developed are the following:

1.  Speed is vital in initiating both first aid and definitive treatment. The exposed individual must remove contaminated clothing immediately, get under a safety shower, and begin flushing exposed skin areas with water.
2.  Evaluation and monitoring must always extend to the entire individual.

Delayed systemic poisoning may appear after relatively small areas of skin have been burned.[5]

Calcium glutonate 2.5% gel has proved a distinct advance in local treatment. It is packaged in 25-g tubes, which may be given to the patient with instructions to rub it on the burned areas whenever pain recurs. This method was described in 1974 by Browne[6] and has gained worldwide acceptance.

Eye exposures should also be treated with calcium gluconate by irrigation with a a 1% solution of calcium gluconate in physiologic saline solution, followed by 1% calcium gluconate in saline as eye drops. For infiltration of deeply burned skin, 10% calcium gluconate is diluted with an equal amount of physiologic saline solution to yield a 5% solution. For inhalation exposures, 2.5% to 3% solution is given by nebulizer with 100% oxygen, preferably by intermittent positive pressure breathing (IPPB). Calcium gluconate is also added to the intravenous solution for management of extensive skin or inhalation exposures to prevent potentially lethal hypocalcemia.[7]

### Skin Burns

Skin contacted by concentrated HF vapor or aqueous solution rapidly becomes erythematous with a white or gray color at the surface due to the coagulation of tissue. Treatment begins with immediate copious washing with water for five minutes. All clothing must be removed. Leather gloves and shoes cannot be decontaminated and must be destroyed.

All burned areas should have calcium gluconate 2.5% gel (HF burn jelly) applied to them as a first-aid measure. The application and massaging into the skin of HG burn jelly should be continued for three to four days, four to six times daily. Care should be taken to see that the personnel who apply the jelly, especially on the initial application, wear rubber gloves to prevent skin contamination with HF and the development of hand burns.

The attending physician should carefully estimate the total area of skin involved and the depth of the burns. Burn areas greater than 50 to 100 cm$^2$ should be of great concern and the patient should be hospitalized immediately. Involvement of areas greater than 100 to 150 cm$^2$ probably will require the facilities of an intensive care unit. All patients with extensive skin burns will have some fume inhalation unless the individual was wearing a proper respiratory protection apparatus. Each patient's pulmonary status must be carefully and rapidly evaluated.

Blood for a complete blood cell (CBC) count and for determining liver function, blood urea nitrogen or creatinine, electrolytes, serum calcium, magnesium, and fluoride levels should be drawn immediately. Intravenous fluid therapy should be initiated as soon as possible. Twenty milliliters of calcium gluconate should be added to the first liter of intravenous solution to raise the amount of available calcium in the blood. Frequent measurement of serum calcium and magnesium levels should guide further administration. The electrocardiogram should be monitored continuously for prolongation of the QT interval.[7,8]

The maintenance of an adequate amount of calcium ion in the blood and tissue is essential in the treatment of extensive HF exposures, whether the exposures occurred by skin, inhalation, or ingestion. Fluoride ion is rapidly absorbed and depletes the tissue calcium rapidly. Blood calcium must be frequently and carefully monitored and sufficient calcium given intravenously, as gluconate, to maintain the calcium level at or slightly above the upper limit of normal. This will encourage movement of the ion into areas of lesser concentration in the extracellular fluid, and then into the tissues.

Deep burns with HF solutions of concentrations greater than 20% may be treated by the use of subcutaneous infiltration of calcium gluconate. Ampules of 10% calcium gluconate should be diluted with physiologic saline solution to obtain a 5% concentration of calcium gluconate. The solution may be infiltrated under the burned area using a 25-gauge × 1$^1$/$_2$-in. or 24-gauge × 3$^1$/$_2$-in. spinal needle, if the area is extensive. Infiltration should be avoided on the digits and undertaken with great care on the hands, feet and face. The administration of calcium gluconate in concentrations above 5% tends to produce severe irritation of the tissues that may cause keloids and intractable scarring. Burns with solutions of HF less than 20% do not usually require infiltration.

System therapy in the severely burned patient is generally supportive: maintaining electrolyte balance and carefully monitoring the patient for signs of renal or hepatic toxicity, and maintaining respiratory and cardiac functions. ECG monitoring is necessary for the early detection of arrhythmias that may result from the alterations in calcium metabolism and for observing prolongation of the QT interval.

The use of adequate doses of carticosteroids to maintain the blood pressure and exert an anti-inflammatory action in the injured area is strongly advised. Hydrocortisone sodium succinate (Solu-Cortef) 500 mg, four to six times a day in the intravenous bottle is usually adequate.

Antibiotics by the systemic or local route are usually unnecessary. Our observation has been that HF burns seldom become infected. Since the tissue is digested, enzymes are unnecessary. One can use silver-sulfadiazine cream for local treatment after the initial several days of treatment with calcium glutonate or infiltration.

### Inhalation Therapy

Persons exposed to HF by inhalation should immediately be given 100% oxygen by mask or catheter. As soon as possible, they should be given 2.5% to 3% calcium gluconate solution by inhalation, preferably by IPPB utilizing a nebulizer alone. The patient should be carefully watched for edema of the upper airway with respiratory obstruction, and the airway should be maintained by tracheostomy or endotracheal intubation if necessary. All those with a history of exposure who experience respiratory irritation should be immediately admitted to an intensive care unit and carefully watched for 24 to 48 hours. Delayed pulmonary edema is likely in patients with burns of the skin of the face or neck.

If pulmonary edema develops, the patient should be placed on IPPB with positive end-expiratory pressure (PEEP). The administration of respiratory care should be very closely supervised, including the continuing administration of calcium gluconate by inhalation. The general supportive principles given in the discussion of skin burns should be observed, with the addition of the use of antibiotics to prevent pulmonary infections and the use of arterial blood gases for control of respiratory therapy.

Toxicity from pulmonary or dermal absorption of fluoride ion may develop rapidly in the liver and kidneys and may require more energetic measures of control, up to and including hemodialysis, if the blood urea nitrogen and potassium levels rise. Supportive care is necessary for all organ systems.[7]

Pulmonary reaction to HF continues for up to 3 weeks, and the patients must be closely monitored. The great solubility of HF in water will limit mild exposures to the bronchial tree, but severe exposures have been found on autopsy to result in bronchiolar and alveolar damage.

Radiological studies and pulmonary function tests are a great assistance in evaluation of long-term effects of HF exposure. Dyspnea during mild or heavy physical activity may occur as long as seven to nine months after exposure. The need for corticosteroid administration on every third day may be necessary periodically for as long as three months to control inflammatory episodes.

### Eye Exposure

The cornea and conjunctiva can be extensively damaged by vapor exposure alone. More severe damage is caused by liquid HF, which rapidly penetrates the outer layer of epithelium of the eye.

Immediate, rapid, copius washing of the eyes should be followed by ice packs. The ice pack should be used until a medical facility is reached. Here the eyes should be washed thoroughly with 1% calcium gluconate in normal saline for 5 to 10 minutes; thereafter, calcium gluconate in normal saline should be instilled every 2 to 3 hours for 48 to 72 hours. No oils or ointments should be used. Inflammation may be decreased by the use of corticosteroid solutions for ophthalmic use.

The cornea may cloud rapidly and obscure vision when exposed to HF. Desquamation may occur in four to 24 hours. Recovery usually takes four to five days, and scarring may occur, as may corneal perforation. An ophthalmologist should always be consulted and should be instructed that long-term monitoring may be required. Surgical procedures may be needed in correcting defects caused by HF exposure. Severe exposures to anhydrous HF have led to destruction of the eye and denucleation to protect surrounding tissues.

## Summary

In all HF exposures, time is of the utmost importance. One must help the patient wash with copius amounts of water immediately to minimize the extent and depth of burns, and then minimize the time needed to institute adequate medical treatment. The amount of tissue injury depends upon the concentration and amount of HF encountered and the duration of the exposure. Low concentrations may lead to delayed effects.

## REFERENCES (SECTION 4.6.2.9)

1. **Proctor, N. H. and Hughes, J. P.,** *Chemical Hazards of the Workplace,* Lippincott, Philadelphia, 1978, 290.
2. **Knight, A. L.,** *Occupational Medicine: Principles and Practical Applications,* Zenz, C., Ed., Year Book Medical Publishers, Chicago, 1975, 649.
3. **Wetherhold, J. M. and Shepherd, F. P.,** Treatment of hydrofluoric acid burns, *J. Occup. Med.,* 7, 193, 1965.
4. **Reinhardt, C. F., Hume, W. G., Linch, A. L. et al.,** Hydrofluoric acid burn treatment, *Am. Ind. Hyg. Assoc. J.,* 27, 166, 1966.
5. **Gosselin, R. D., Hodge, H. C., Smith, R. P. et al.,** *Clinical Toxicology of Commercial Products: Acute Poisoning,* 4th ed., Williams & Wilkins, Baltimore, 1976, 159.
6. **Browne, T. D.,** The treatment of hydrofluoric acid burns, *J. Occup. Med.,* 24, 80, 1974.
7. **Tepperman, P. B.,** Fatality due to acute systemic fluoride poisoning following a hydrofluoric acid skin burn, *J. Occup. Med.,* 22, 691, 1980.
8. **Abukurah, A. R., Moser, A. M., Baird, C. L. et al.,** Acute sodium fluoride poisoning, *JAMA,* 222, 816, 1972.
9. **Trevino, M. A., Herrmann, G. H., and Sprout, W. L.,** Treatment of severe hydrofluoric acid exposures, *J. Occup. Med.,* 25(12), 861, 1983.

### 4.6.2.10. Hydrogen Cyanide

Hydrogen cyanide (HCN) (CAS 74-90-8), also called hydrocyanic acid or prussic acid, is an extremely dangerous chemical that is toxic by ingestion, inhalation, or absorption through the skin. The current OSHA 8-h PEL to the vapors from this chemical is 10 ppm, as is the current ACGIH ceiling limit (with a cautionary note that skin absorption could be a contributory hazard). The material has a characteristic odor of bitter almonds, but the odor is not usually considered to be sufficiently strong to be an adequate warning of the presence of the vapors at or above the PEL. Not only is HCN toxic, it has a very low flash point –17.8°C [0°F], a lower explosion limit of 6%, and an upper explosion limit of 41%, so it also represents a serious fire and explosion hazard. It has a boiling point of 26°C (79°F), so it is normally contained in cylinders in the laboratory. Heating of the liquid material in a pressure-tight vessel to temperatures above 115°C (239°F) can lead to a violent heat-generating reaction. The material is usually stabilized with the addition of a small amount (0.1%) of acid, usually phosphoric acid, although sulfuric acid is sometimes used. Samples stored more than 90 d may become unstable.

HCN can polymerize explosively when amines, hydroxides, acetaldehyde, or metal cyanides are added to the liquid material, and it also may do so above 184°C (363°F).

Although there will be variations among individuals, a concentration of 270 ppm in air is usually considered fatal to humans. A few breaths above this level may cause nearly instantaneous collapse and respiratory failure. Exposures at lesser levels may be tolerated for varying periods, e.g., 18 to 36 ppm may be tolerated for several hours before the onset of symptoms. Initial symptoms of exposure to HCN include headache, vertigo, confusion, weakness, or fatigue. Nausea and vomiting may occur. The respiratory rate usually increases initially and then decreases until eventually it becomes slow and labored, finally ceasing. The symptoms reflect the mechanism by which the toxic action occurs. The chemical acts to inhibit the transfer of oxygen from the blood to tissue cells by combining with the enzymes associated with cellular respiration. If the cyanide can be removed, the transfer of oxidation will resume. On average, absorption of 50 to 100 mg of HCN, directly by ingestion or through the skin as well as by inhalation, can be fatal.

Treatment of a person poisoned by HCN is based on the introduction of methemoglobin into the bloodstream to interact with the cyanide ions to form cyanmethemoglobin. In any area where HCN is being used, a special emergency kit should be provided, containing an ample supply of ampules of amyl nitrite, a solution of 1% sodium thiosulfate solution, and an oxygen cylinder, accompanied by a facepiece and tubing to permit administering the oxygen. This kit should be labeled **FOR HCN EMERGENCIES ONLY**. Sodium nitrite might also be kept in the kit if there is someone available qualified to administer drugs intravenously. Introduction of sodium nitrate directly into the bloodstream has been suggested as a means of increasing the rate of conversion of cyanide to thiocyanate, which is less toxic. Treatment should begin as soon as possible after an acute exposure, and after recognition of the symptoms in less intense exposures. If the exposure has been due to contamination in the air, the patient should be removed from the area (if the source of vapor is from a cylinder, the valve on the cylinder should be closed). Any contaminated clothes should be removed and the skin flooded with water.

If the patient is not breathing, resuscitation should begin. As soon as the patient is breathing, an open amyl nitrite ampule should be held under the patient's nose for 15 sec/min, with oxygen being administered during the remaining 45 sec. Medical aid should be called for immediately. If there is an available person qualified to administer drugs intravenously, injection of sodium nitrite while administering amyl nitrite should prove beneficial. Subsequent intravenous injection of sodium thiosulfate also has been suggested as an ameliorative action. Rescue squad teams usually have at least one member qualified to administer drugs while under the direction (by radio) of an emergency room physician. If the patient has swallowed HCN, the recommended treatment is to get the patient to swallow 1 pt of the sodium thiosulfate solution, followed by soapy water or mustard water to induce vomiting. Vomiting should not be induced in an unconscious patient. Application of amyl nitrite may restore consciousness.

All work with HCN must be done in a fume hood operating with a face velocity of at least 100 fpm and with the apparatus set well back from the face of the hood to insure that all vapors will be captured and discharged by the exhaust system. The hood should comply in every respect with recommended good practices for the location, design, and operations of hoods in Chapter 3. The hood should have its own duct to the roof. If the work is a continuing program, the exhaust duct should be labeled: **DANGER, DO NOT SERVICE OR WORK IN THE VICINITY WHILE UNIT IS OPERATING.**

Protective gloves and chemical splash goggles should be worn while working with HCN. No one should work alone with this material. The laboratory entrance(s) should be posted with a warning sign, **DANGER, HCN, AUTHORIZED PER-SONNEL ONLY.** As noted above, care needs to be taken to be sure that persons outside the area, such as workmen on the roof, are not inadvertently exposed. All work with HCN should be done in trays or other shallow containers of sufficient capacity to retain any spill from the apparatus.

Cleaning up spills represents a serious problem with a chemical as dangerous as HCN, so extra care should be taken to avoid accidents with the material. If a spill occurs outside a hood, the laboratory should be evacuated as quickly as possible. All ignition sources and valves to cylinders of HCN should be turned off. If any individuals are splashed with the compound in an accident, they should immediately remove their contaminated clothes and step under a nearby deluge shower, preferably located in a space outside the laboratory in which the accident occurred. Occupants of the latter space should be warned of the accident and encouraged to evacuate as well. If the spill is substantial, the evacuation of all the contiguous spaces or the entire building might be considered. The evacuation of additional spaces is especially important in facilities in which the reentry of fumes exhausted from the building is known to be a problem. Unless, as described earlier, immediate on-site treatment is needed due to inhalation or skin absorption, medical care for any exposed persons should be obtained as soon as possible. It would be desirable to have sufficient self-contained escape-type breathing devices on hand to equip every occupant of the laboratory. Unless laboratory personnel have received specific training in handling hazard material incidents, the nearest hazardous material emergency response center should be called for assistance for a substantial spill.

HCN is categorized as a chemical which is immediately dangerous to life and health (IDLH). As such, the cleanup of spills should be handled very carefully. Individuals performing the cleanup should wear a self-contained, positive-pressure breathing apparatus equipped with a full face piece, rubber or neoprene gloves, and chemically protective outer wear. A type C supplied-air respirator unit operated at a positive pressure can be used as well, but a self-contained unit should be available as a backup. Anyone asked to wear this equipment must have received prior training in the proper use of the equipment. The material can be cleaned up using absorbent pillows or other absorbent materials. Waste should be placed in double heavy-duty plastic bags, which are then tightly closed by twisting the top, folding the top over, and wrapping it securely with duct tape. The sealed plastic bags should then be placed in heavy plastic containers or steel drums which can be tightly sealed. Waste material should not be placed in fume foods to evaporate or disposed of in drains. In the latter case, the possibility of fumes collecting in sections of the drain piping and reentering a building thorough a dry sink trap is too great. The waste should not go to a normal landfill. Incineration is the preferred means of disposal.

A leaking cylinder which cannot be readily repaired should be taken to a remote location where the gas in the cylinder can be released safely. Cylinders which are damaged, but not leaking, should be returned to the vendor for disposal wherever possible. Disposal of gas cylinders by commercial hazardous waste firms can be very expensive.

## REFERENCES (SECTION 4.6.2.10)

1. Occupational Health Guidelines for Chemical Hazards, DHHS (NIOSH) Publ. No. 81-123, Mackison, F. W., Stricoff, R. S., and Partridge, L. J., Jr., Eds., U.S. Department of Health and Human Services and U.S. Department of Labor, Washington, D.C., 1981.
2. *Prudent Practices for Handling Hazardous Chemicals in Laboratories*, National Academy Press, Washington, D.C., 1981, 45, 133.
3. **Chen, K. K. and Rose, C. L.,** Nitrite and thiosulfate therapy in cyanide poisoning, *JAMA*, 149, 113, 1952.
4. **Hirsch, F. G.,** Cyanide poisoning, *Arch. Environ. Health*, 8, 622, 1964.

### 4.6.2.11. Fluorine Gas

Fluorine (CAS no. 7782-41-4) is an extremely reactive gas which will react violently with a wide variety of materials, a representative sample of which are most oxidizable substances, most organic matter, silicon-containing compounds, metals, halogens, halogen acids, carbon, natural gas, water, polyethylene, acetylides, carbides, and liquid air. Many of these reactions will initiate at very low temperatures. Because it will react with so many materials, extreme care must be taken when working with fluorine. The work area should be very well ventilated and free of combustible materials which would act as fuel in the event of a fire. A written hazard analysis should be prepared for the research program prior to beginning work, and an emergency contingency plan developed.

The OSHA PEL is 0.1 ppm or 0.2 mg/M$^3$. However, the 1987-88 ACGIH TWA levels are ten times higher with short-term exposure limits (STEL) another factor of two higher. An exposure to 25 ppm for 5 min has caused severe eye symptoms. The LC$_{50}$ for a 1-h exposure for rats and mice is 185-150 ppm, respectively. It is highly irritating to tissue.

Fluorine will react with brass, iron, aluminum, and copper to form a protective metallic fluoride film. Circulating a dilute mixture of fluorine gas and inert gases through a system will passivate the surfaces and render them safe to use, provided the film remains intact. However, it is recommended that an inert gas be circulated through any fluorine system before the fluorine is introduced.

All systems containing fluorine should be checked frequently for leaks. Filter paper moistened with potassium iodide can be used to perform the tests. The paper will change color when any escaping gas comes in contact with it.

Systems using fluorine should be located within a fume hood. The research worker should be protected by an explosive shield. The worker should wear protective goggles and a face mask. Unless the cylinder valve is operated through a remote control device, the user should wear sturdy gloves with extended cuffs to protect his hands and arms while manipulating the valve. A protective apron should be worn as well. However, all of these may only give limited protection in the event of an accident since fluorine may react with many common items of personnel protective gear.

Self-contained escape breathing apparatus should be available for all occupants of a laboratory in which fluorine is in active use. In the event of an accident, the area should be evacuated immediately and doors closed as personnel leave to isolate the problem as much as possible. Evacuation of nearby areas should be considered or, depending upon the scale of the accident, perhaps the entire building, especially, as noted elsewhere, if there are known problems with exhausted materials reentering the building. No remedial measures should be attempted under most circumstances; the incident should be allowed to proceed until the fluorine supply is exhausted. Fire fighting efforts should be aimed at preventing a fire resulting from an incident from spreading. Applying water directly to the leak could intensify the fire. Obviously, it would be desirable to use smaller cylinders, if practicable, for the research program to reduce the scale of any incident.

Cylinders with valves that cannot be dislodged without application of sufficient force to damage the valve or the connection to the cylinder should be returned to the vendor for repair rather than taking a chance on a massive rupture and release of the contents of the cylinder. Ordinary maintenance personnel should not be asked to attempt to free the valve. Although the consequences here would be exacerbated by the extremely hazardous properties of the contents of the cylinders, the same recommendation would apply to virtually any cylinder containing a substantial volume of gas under high pressure.

## REFERENCES (SECTION 4.6.2.11)

1. Occupational Health Guidelines for Chemical Hazards, DHHS (NIOSH) Publ. No. 81-123, Mackison, F. W. Stricoff, R. S., and Partridge, L. J., Jr., Eds., U.S. Department of Health and Human Services and U.S. Department of Labor, Washington, D.C., 1981.
2. *Hazardous Chemical Data*, NFPA-49, National Fire Protection Association, Quincy, MA, 1975.

### 4.6.2.12. General Safety for Hazardous Gas Research

The previous section was devoted to fluorine as an example of an exceptionally hazardous gas. However, many others are commonly used in the laboratory which pose comparable risks, some because of their own toxicity or that of gases or vapors evolved as a consequence of their decomposition, others because of their flammable or explosive properties or due to reactions with other chemicals, or a combination of all of these characteristics. Even relatively innocuous gases such as nitrogen, carbon dioxide, argon, helium, and krypton can be a simple asphyxiant if they displace sufficient air, leaving the oxygen content substantially below the normal percentage. If the concentration of these inert gases approaches 33% or higher, symptoms of oxygen deprivation begin to occur and at concentrations of around 75%, persons will only survive for a brief period of time. Any gas under high pressure in a cylinder poses a problem if the cylinder is mishandled so as to rupture the containment of the gas. In such a case, the cylinder can represent an uncontrolled missile with deadly consequences for anyone in the vicinity. The ability of escaping gas to move readily throughout a volume greatly enhances the likelihood that a flammable gas will encounter a source of ignition. This problem is shared by the vapors of many volatile liquids. Many gases are heavier than air, so they may collect in depressions or areas with little air movement and can represent a danger to unsuspecting persons. Many gases do not have a distinctive odor or are not sufficiently irritating to warn you of their presence, and some which do, such as hydrogen sulfide, desensitize the sense of smell at levels which would be dangerous.

Cylinders connected to systems in the laboratory must always be strapped firmly to a support to insure that they do not fall over. Not only is there a risk of breaking the connection on the cylinder side of the regulator valve, with the concomitant risk of the cylinder becoming a missile, but the connection to the system also may be broken, allowing gas to escape from the low-pressure side of the regulator. If the amount of gas to be stored in the laboratory is substantial, it may be preferable to pipe the gas in from a remote outside storage area, with control valves located both outside and within the laboratory.

Where the explosion risk is substantial, the facility may need to be designed with explosion venting so that the force of any explosion and the resulting flying debris can be released in a relatively safe direction, minimizing the risk to the occupants. Systems for smaller-scale operations should be placed within a hood, perhaps one specially designed to provide partial protection against explosions. Explosion shields are available to aid in protecting the research worker. Wherever the possibility of an explosion exists, laboratory personnel should wear impact-resistant goggles, possibly supplemented by a face mask to protect the lower part of the face and throat. Gauntlets should be worn if there is risk to the hands and forearms while conducting experimental evolutions.

Not all work with high pressures involves gas cylinders. Reaction bombs are commercially available which operate at pressures up to 2000 psi (13.8 MPa) and temperatures up to 350°C (662°F). An error on the part of the research worker could permit these design parameters to be exceeded, and although these units are designed with a substantial safety factor, the failure of one of them could lead to disastrous consequences. Not only is there the immediate danger of injury due to flying components of the system, there also is the risk of reactions involving reagents from broken bottles. These secondary events could escalate the consequences far beyond the original scope of the incident. Any device in which the potential for a high-pressure accident exists should be set up in a hood to provide some explosion protection and explosion barriers should be used to provide additional protection for the occupants of the room. Personnel working in this type of research should be especially careful to not work alone. They should wear goggles, a face mask, and sturdy gloves to protect their hands and forearms.

Systems involving toxic gases should be adequately ventilated. If possible, the systems should be set up totally within a fume hood. Large walk-in hoods are often used for this purpose. All systems should be carefully leak-tested prior to introducing toxic materials into the system, periodically thereafter and after any maintenance or modifications to the system which could affect the integrity of the system.

Many gases are potentially so dangerous that access to the laboratory should be limited to essential personnel who are authorized to be present. No one should work with such materials alone. It may be desirable to have one person somewhat removed from the immediate area of operations, but a second person should be within the working area. Entrances to a high-hazard gas research facility should be marked with a **DANGER, SPECIFIC AGENT, AUTHORIZED PERSONNEL ONLY** or comparable warning sign. Hood exhausts also should bear a comparable warning legend. In some cases, automatic alarm sensors have been developed to detect the presence of gas approaching dangerous levels. It is recommended that warning trip points on these devices be set at no more than 50% of either the OSHA PEL or the

ACGIH TLV value, whichever is lower. If an automatic sensing device is available, circuitry can be devised to activate a valve to cut off the gas supply as well as to provide a warning. The latter is especially important if the operation is left unattended and no signal is transmitted to a manned location, rendering an alarm ineffective. The growing application of programed personal computers or laboratory computer work stations dedicated to experimental control has increased the amount of sophisticated experimentation that can be automated.

In any laboratory using highly hazardous materials, an emergency plan should be developed prior to initiating any major project or any major modification to an ongoing project, based on a thorough hazard analysis. Because of the special problems associated with gases, the emergency contingency plan should provide for the rapid evacuation of the laboratory using short-duration, self-contained breathing apparatus and for initiating evacuation from other areas of the facility. Provision of detailed information to emergency response groups is now required under the Community Right-To-Know law for many hazardous substances when the amount involved exceeds a prescribed threshold amount. It is also recommended that employees engaged in any research involving hazardous materials participate in a medical surveillance program consisting of a comprehensive prior screening exam, acquisition of a serum sample for comparison with a sample following a possible incident, and a complete medical history, so that baseline information on individuals will be available to medical personnel called upon to treat personnel that may have been exposed for example, to toxic gases. If specific organs or bodily functions might be affected, the examination may need to include special tests in these areas.

The most common problem associated with the use of gas cylinders is that of leaks. The cylinder can develop leaks at any of four points, assuming no gross rupture of the cylinder wall itself has occurred: (1) valve threads, (2) valve safety device threads, (3) valve stem, and (4) valve outlet. Repairs of leaks in the first two would require repairs done at high pressure and should not be attempted in most laboratory facilities. It may be possible to incorporate some adjustments into the design of the cylinder to stop leaks in the latter two areas. In either case, it is best to contact the manufacturer for advice before attempting any repair. Forced freeing of a "frozen" or corroded valve should not be attempted.

Leaks involving cylinders containing corrosive materials may increase in size with time as the corrosive gas interacts with atmospheric moisture. Removal of leaking cylinders containing corrosives from an occupied facility to a remote location should be done as quickly as possible. The vendor should be called for advice or assistance. If time permits, a solution would be to slowly exhaust the leaking cylinder into a neutralizing material. If the protective cap has corroded into position so that no bleeder hose can be attached, two heavy-duty plastic bags can be placed over the leaking end of the cylinder and the gases conducted from the bags through a hose into a drum of neutralizing solution. If the leak is too large for easy handling, a commercial, state, or local hazardous material response group should be called upon for assistance. The wearing of protective suits and self-contained breathing apparatus may be required, which requires special training.

Leaks involving toxic and flammable gases pose the risk of personal injury to individuals attempting repairs, and repairs should not be attempted by local personnel unless it is certain that they can be done safely. Evacuation is usually recommended.

Where flammable gas leaks are concerned, all ignition sources must be turned off — prior to evacuating a facility, if time permits. It is foolish to risk life or personnel injury unless, by doing so, it may be possible to save others.

In some cases, where the melting point or boiling point of the contained gas is sufficiently high, the leak rate may be substantially reduced or virtually stopped by putting the body of the cylinder into a cooling bath while deciding upon the appropriate corrective measures or waiting for assistance. This should not be done if the contained material will react with the coolant.

## REFERENCES (SECTION 4.6.2.12)

1. *Handling and Storage of Flammable Gases,* Air Products and Chemicals, Allentown, PA.
2. *Prudent Practices for Handling Hazardous Chemicals in Laboratories,* National Academy Press, Washington, D.C., 1981, 75, 221, 223, 229.

### 4.6.2.12.1. Some Hazardous Gases

The properties of a number of hazardous gases will be given in this section, along with a few brief comments on noteworthy problems associated with these materials. The common chemical name, formula, and CAS number will be given for each substance, followed by a definition of the primary hazard class(es) represented by the material, the boiling point in degrees Celsius (degrees Fahrenheit)*, the explosive range (lower explosive limit [lel] — upper explosive limit [uel]) in percent by volume in air (NA where not applicable or not available), the vapor specific gravity referred to air as 1, OSHA 8-h, time-weighted average PEL in ppm** and, finally, salient comments about the material.

**Acetylene** — $C_2H_2$; 74-86-2; explosive, flammable, asphyxiant; bp = –84.0°C (–119°F); er = 2 to 82%; sp gr = 0.91; PEL = 2500 ppm (10% of lel). Acetylene in cylinders is usually dissolved in acetone and relatively safe to handle, but purified acetylene has a very low ignition energy and a relatively low minimum ignition temperature, 300°C (571°F). It forms explosive mixtures with air over a wide range. Utilization of acetylene should be at a pressure of 15 psi gage or less. Piping for acetylene systems should be steel, wrought iron, malleable iron, or copper alloys containing less than 65% copper. May form explosive compounds with silver, mercury, and unalloyed copper.

**Ammonia** — $NH_3$; 7664-41-7; flammable gas, causes tissue burns, strong respiratory irritant; bp = –33.3°C (–28°F); er = 15 to 28%; sp gr = 0.6; PEL = 50 ppm. May cause severe injury to respiratory system and eyes, common 35% laboratory solution can cause severe skin burns. High concentrations may cause temporary blindness. Baseline physical should stress respiratory system and eyes. Skin should be examined for existing disorders. Tests should include pulmonary function and chest X-ray. Should wear self-contained breathing apparatus and rubber shoe covers when cleaning up a spill (by dilution with ample amounts of water and mop to a drain). On EPA list of extremely hazardous substances, 40 CFR 302.

---

* In most cases, the Celsius temperature is from the literature, while the Fahrenheit temperature is calculated from the former.
** The ceiling value is given if the level should not exceed this value at any time or for a limited period. If a level has not been set by OSHA and an ACGIH TLV value is available, the latter will be given.

**Arsine** — $AsH_3$; 7784-42-1; deadly poison by inhalation, fire and explosion hazard; bp = –62.5°C (–144.5°F); er = NA; sp gr = 2.66; PEL = 0.05 ppm. Recognized carcinogen; causes pulmonary edema; primarily poisonous due to interaction with hemoglobin, causes anemia; early symptoms are headache, dizziness, nausea, and vomiting. Severe exposures result in kidney damage, delirium, coma, and possible death. In increased use, due to applications to semiconductor research. Should be used in a sealed system or the system should be set up in an efficient fume hood. Leaking cylinders can be handled by allowing the leaking gas to interact with a 15% aqueous solution of sodium hydroxide.[8] The arsine forms a water soluble precipitate. After neutralization with sulfuric acid, the precipitate is filtered out to be disposed of as hazardous waste, and the neutral solution can be disposed of into the drain, diluted by large amounts of water. Individuals doing this should wear self-contained breathing apparatus and chemically resistant gloves. Arsine will ignite in contact with chlorine and undergoes violent oxidation by fuming nitric acid. On EPA list of extremely hazardous substances, 40 CFR 302.

**Boron trifluoride** — $BF_3$; 7637-07-2; inhalation poison, irritant, nonflammable; bp = –100°C (–148°F); er = NA; sp gr = 2.3; ceiling PEL = 1 ppm. Irritating to eyes and respiratory system. Animals have been shown to have kidney damage after high exposures. Baseline physical should stress respiratory system, eyes, and kidneys. Tests should include a chest X-ray and pulmonary function test. Cylinders with a slow leak can be allowed to leak into an efficient fume hood for disposal or the cylinder can be sealed and returned to the vendor. Produces a thick, white smoke by interaction with moisture in humid air, but otherwise insufficient data on warning properties. On EPA list of extremely hazardous substances, 40 CFR 302.

**1,3-Butadiene** — $C_4H_6$; 106-99-0; flammable, irritant to eyes and mucous membranes, suspect carcinogen; bp = –4.5°C (23.9°F); er = 2 to 11.5%; sp gr = 1.9; PEL = 1000 ppm (ACGIH = 10 ppm). Suspect human carcinogen; narcotic at high concentrations. May form peroxides on exposure to air; at high temperatures, may self-polymerize exothermally; forms explosive mixtures with air. Participation in a medical surveillance program is recommended for users. On EPA list of extremely hazardous substances, 40 CFR 302.

**Carbon dioxide** — $CO_2$; 124-38-9; nonflammable, asphyxiant; sublimes at –78.5°C (–109°F); er = NA; sp gr = 1.53; PEL = 5,000 ppm. Very common laboratory gas; also used as "dry ice". Causes problems primarily by displacement of air. At 5% concentration, respiratory volume quadrupled. Heart rate and blood pressure increases reported at 7.6%. At 11%, unconsciousness typically occurs in 1 min or less. No warning other than symptoms — dizziness, headaches, shortness of breath, and weakness — because it is colorless and odorless.

**Carbon disulfide** — $CS_2$; 75-15-0; flammable liquid (vapor pressure 400 mm @ 28°C (82.4°F), poisonous; bp = 46.5°C (115°F); er = 1.3 to 50% (flash point = –30°C (–22°F) and ignition temperature is only 90°C (194°F); sp gr of vapor = 2.64; PEL = 20 ppm. Central nervous system poison. Extended exposure can cause permanent damage to the CNS in severe cases. Numerous other physiological problems to heart, kidneys, liver, and stomach. Has strong narcotic and anesthetic properties. Poisonous if inhaled, ingested, or prolonged contact with skin. Baseline physical should stress central and peripheral nervous system, cardiovascular system, kidneys, liver, eyes, and skin. Tests should include urinalysis for kidney function, liver panel, electrocardiogram, and ophthalmic exam. Vapors can be ignited by contact with an incandescent light bulb. Air-carbon disulfide mixture can explode in the presence of rust. Do not pour down a sink. Do not use where electrical sparks are possible. Use nonsparking tools. Explosion-proof wiring and fixtures not necessarily effective in arresting flame. Use dry chemical or carbon dioxide to fight fires. On EPA list of extremely hazardous chemicals, 40 CFR 302.

**Carbon monoxide** — CO; 630-08-0; flammable, poisonous, experimental teratogen; bp = –191.1°C (–311.9°F); er = 12.5 to 74%, sp gr = 0.9678; PEL = 50 ppm. Combines highly preferentially with hemoglobin to the exclusion of oxygen and, hence, is a chemical asphyxiant. Pregnant women and smokers more susceptible to risk. Baseline physical recommended. Medical history taken to discover history of problems involving heart, cerebrovascular disease, anemia, and thyroidtoxicosis. A complete blood count should be taken. Insidious poisonous gas. Does not have adequate warning properties; odorless and nonirritating. Contact with strong oxidizers may cause fires and explosions. Dangerous fire hazard.

**Chlorine** — $Cl_2$; 7782-50-5; nonflammable gas, but supports combustion of other materials, toxic; bp = –34.5°C (–30.1°F); er = NA; sp gr = 2.49; ceiling PEL = 1 ppm. Forms explosive mixtures with flammable gases and vapors. Reacts explosively with many chemicals and materials such as acetylene, ether, ammonia gas, natural gas, hydrogen, hydrocarbons, and powdered metals. Many incompatibles; carefully review the literature before working with this material. Systems should be set up in an efficient fume hood. Strong odor, noticeable well below acute danger levels. Inhalation can cause severe damage to lungs. Baseline physical exam should stress eyes, respiratory tract, cardiac system, teeth, and skin. Pulmonary function test and chest X-ray recommended. Warning of the presence of gas well below PEL due to odor and irritating properties. On EPA list of extremely hazardous chemicals, 40 CFR 302.

**Cyanogen** — $C_2H_2$; 460-19-5; highly flammable, toxic; bp = –21°C (–5.8°F); er = 6.6 to 32%, sp gr = 1.8; ACGIH TLV = 10 ppm. Odor resembles almonds, poisonous by inhalation and through skin. Symptoms are headache, dizziness, nausea, vomiting, and rapid pulse; severe exposures lead to unconsciousness, convulsions, and death. Strong eye irritant; fire hazard when exposed to oxidizers, flame, and sparks. Reacts with fluorine (ignites) and oxygen (combination of liquid cyanogen and liquid oxygen will explode).

**Cyanogen chloride** — CCIN; 506-77-4; nonflammable, inhalant poison, eye irritant; bp = 13.1°C (55.6°F); er = NA; sp gr = 1.98; ACGIH ceiling level = 0.3 ppm. Toxic properties similar to hydrogen cyanide; heat causes it to decompose and emit highly toxic and corrosive fumes.

**Diazomethane** — $CH_2NO_2$; 334-88-3; explosion hazard, irritant to respiratory system and eyes, suspect carcinogen; bp = –23°C (–9.4°F); er = NA; sp gr = 1.4; OSHA PEL = 0.2 ppm. One of the most dangerous chemicals used in chemical laboratories. Strong allergen, irritating to eyes and respiratory system. May sensitize as well as irritate. Recommend initial medical history, full chest X-ray, and pulmonary function tests to individuals planning to work with this material. Explosively sensitive to shock, heat (about 100°C [212°F]), exposure to rough surfaces (e.g., ground glass joints), alkali metals, and calcium sulfate. Do not store; prepare fresh when needed. Always use in an efficient fume hood, use an explosive shield, and wear impact and chemical splash-resistant goggles, possibly supplemented by a face mask. Respiratory protection recommended if levels may approach PEL.

**Diborane** — $B_2H_6$; 19287-45-7; flammable, poison, suspect carcinogen; bp = –92.5°C (–134.5°F); er = 0.9 to 98% (ignites at temperatures of 38 to 52°C [100 to 125°F] or less in humid air); sp gr = 0.96; OSHA PEL = 0.1 ppm. Respiratory irritant, causes pulmonary edema. Strong irritant to skin, eyes, and other tissues. Keep cool and away from oxidizing agents. Baseline physical should stress lungs, nervous system, liver, kidneys, and eyes. Pulmonary function test and chest X-ray recommended. Reacts with aluminum and lithium to form hydrides which may explode in air. Use only in an efficient fume hood. Use an explosion shield, impact and chemical splash-protecting goggles supplemented with a mask, and hand and forearm protection. On EPA list of extremely hazardous chemicals, 40 CFR 302.

**Dimethyl ether** — $C_2H_6O$; 115-10-6; flammable, inhalation and skin irritant, narcotic properties; bp = –23.7°C (–10.7°F); er = 3.4 to 27%; sp gr = 1.62; OSHA PEL = NA. Explosion hazard when exposed to flames and sparks, forms peroxides, sensitive to heat.

**Ethylene** — $C_2H_4$; 74-85-1; simple asphyxiant, flammable, poisonous to plants; bp = –103.9°C (–155°F); er = 2.7 to 36%; sp gr = 0.98; OSHA PEL = NA. Dangerous when exposed to heat and flames.

**Ethylene oxide** — $C_2H_4O$; 75-21-8; flammable, strong irritant, inhalation and oral poison, carcinogen, bp = 10.7°C (51.3°F), er = 3 to 100%; sp gr = 1.52; OSHA PEL = 1 ppm. Regulated carcinogen, 29 CFR 1910.1047. Increases rate of miscarriages. Very commonly used as sterilizing agent. Should be used in a sealed system and exhausted outdoors. Personnel exposures should be monitored. It is a very strong irritant to eyes, skin and the respiratory tract. Can cause pulmonary edema of respiratory system at high levels. Forms explosive mixtures with air. Very reactive when in contact with alkali metal hydroxides, iron and aluminum oxides and anhydrous chlorides of iron, aluminum, and tin. Also reacts with acids, bases, ammonia, copper, potassium, mercaptans, and potassium perchlorate. On EPA list of extremely hazardous chemicals, 40 CFR 302.

**Formaldehyde** — $CH_2O$; 50-00-0; irritant to eyes, skin, and respiratory system; oral poison, allergen, suspected carcinogen, flammable; bp = –19.4°C (–3°F); er = 7 to 73%; sp gr = slightly greater than 1; OSHA PEL = 3 ppm (reduction to 1 ppm pending). Preceding physical properties are for the gas. It is normally sold as an aqueous solution of 37 to 52% formaldehyde by weight. Other solvents are also used. Liquid formaldehyde, when heated, can evolve the gas, which will burn. Ingestion of the solution will cause stomach pain, nausea, and vomiting; can result in loss of consciousness. Severe eye irritant. Extended exposures can cause skin and bronchial problems. The gas is on the EPA list of extremely hazardous materials, 40 CFR 302.

**Hydrogen** — $H_2$; 1333-74-0; flammable, explosive, asphyxiant; bp = –252.78°C (–422.99°F); er = 4.1 to 74.2%; sp gr = 0.0695; OSHA PEL = NA. Gaseous hydrogen systems of 400 ft³ (11.35 m³) and containers of liquid hydrogen of more than 150 l (39.63 gal) are regulated by OSHA, 29 CFR 1910.103. Systems to be used for hydrogen should be purged with inert gas prior to use. Consideration should be given to incorporation of safety systems required for larger systems, depending on the research program and facilities available. Hydrogen will burn with virtually an invisible flame. Care should be used in approaching a suspected hydrogen flame; holding a piece of paper in front of you is recommended to detect the flame. Because it is so light, it tends to escape from rooms where leaks occur.

**Hydrogen chloride (anhydrous)** — HCl; 7647-01-0; nonflammable, toxic gas by all routes of exposure and intake; bp = −84.8°C (−121°F); er = NA; sp gr = 1.27; OSHA ceiling PEL = 5 ppm. Gas combines with moisture to become corrosive to eyes, skin, and respiratory system. Baseline physical exam should stress respiratory system, eyes, and skin. Pulmonary test and chest X-ray recommended. Exposure to airborne concentrations above 1500 ppm can be fatal in a few minutes. Irritating properties detectable at about PEL. Should only be used in an efficient fume hood. Workers should wear gas-tight goggles and acid-resistant aprons, gloves, and outerwear.

**Hydrogen fluoride** — HF; 7664-39-3; strong irritant and corrosive to eyes, respiratory system, and internal tissue skin via contact, inhalation, and ingestion; noncombustible; bp = 19.5°C (67.2°F); er = NA; sp gr = 0.7; OSHA PEL = 3 ppm. Readily dissolves in water to form hydrofluoric acid. Prolonged exposure can cause bone changes. Baseline physical should stress eyes, respiratory tract, kidneys, central nervous system, skin, and skeletal system. Special tests should include urinalysis, pelvic X-ray (use shielding to protect genitals as much as possible), and an ophthalmic examination. Provides warning at levels near PEL due to irritant effects. Gas on EPA list of extremely hazardous chemicals, 40 CFR 302. For other comments, see Section 4.6.2.7, Hydrofluoric Acid.

**Hydrogen selenide** — $H_2Se$; 7783-07-5; flammable, very toxic via inhalation, also toxic by contact with eyes and skin; bp = −41.3°C (−42°F); er = NA; sp gr = 2.1; OSHA PEL = 0.05 ppm. Very offensive odor, but threshold for detection above dangerous levels. Recommend initial medical screening for existing respiratory problems and impaired liver function prior to work with this substance. Causes irritation of eyes, nose, throat, and lungs. Symptoms of exposure include nausea, vomiting, followed by a metallic taste, garlic odor to breath, dizziness, and fatigue. Can cause eye, liver, spleen, and lung damage. Use only in an efficient fume hood. Additional respiratory and eye protection recommended if potential exposure problem. Contact with acids, halogenated hydrocarbons, oxidizers, and water may result in fire or explosion. On EPA list of extremely hazardous chemicals, 40 CFR 302.

**Hydrogen sulfide** — $H_2S$; 7783-06-4; flammable, irritant, asphyxiant; bp = −60°C (−76°F); er = 4.3 to 46%; sp gr = 1.19; OSHA ceiling PEL = 20 ppm (single 10-min peak PEL = 50 ppm). Strong odor of rotten eggs, but sense of smell desensitized by gas after short interval (minutes) at high levels. Preliminary medical exam recommended prior to work with this material, with stress on eyes and lungs, including chest X-ray and pulmonary function test. Severe eye and respiratory irritant at moderate concentrations. Rapidly acting systemic poison which causes respiratory paralysis at high levels. Exposures at 1000 to 2000 ppm can cause immediate death. Prolonged exposure above 50 ppm can damage eyes and cause respiratory problems. Susceptibility increases with repeated exposures. Symptoms caused by low concentrations are headache, fatigue, insomnia, irritability, and gastrointestinal problems. Highly reactive with strong nitric acid and oxidizing agents. On EPA list of extremely hazardous chemicals, 40 CFR 302.

**Methane** — $CH_4$; 74-82-8; flammable, simple asphyxiant; bp = −161.4°C (−258.6°F); er = 5 to 15%; sp gr = 0.52; OSHA PEL = NA. Keep away from sources of ignition.

**Methyl acetylene** — $C_3H_4$; 74-99-7; flammable, anesthetic; bp = −23°C (−10°F);

er = 1.7 to 11.7%; sp gr = 1.4; OSHA PEL = 1000 ppm. Sweet odor. Overexposure causes drowsiness. Reactions with chlorine and strong oxidizing agents may result in fires or explosions. It forms very shock-sensitive compounds with copper. Equipment components containing more than 67% copper should not come into contact with the compound.

**Methyl acetylene propadiene mixture (MAPP)** — $C_3H_4$ isomers; no CAS number; flammable, anaesthetic; bp = −34.5°C (−30°F); er = 3.4 to 10.8%; sp gr = 1.5; OSHA PEL = 1000 ppm. Foul odor. Reactivity similar to methyl acetylene (MA). Overexposure can cause drowsiness and unconsciousness with MA. Odor detectable well below PEL. Will attack some plastics, films, and rubber.

**Methyl bromide** — $CH_3Br$; 74-83-9; fumigant, toxic by inhalation, contact with eyes and skin, and ingestion; cumulative poison; bp = 3.6°C (38.4°F); er = 13.5 to 14.5% (requires high-energy ignition source); sp gr = 3.3; OSHA ceiling PEL = 20 ppm (skin). Severe respiratory irritant, neurotoxin, narcotic at high concentrations. Symptoms of overexposure include headache, visual disturbances (blurred or double vision), nausea, vomiting, vertigo in some cases, tremors of the hand, and, in more severe cases, convulsions. Persistent depression, anxiety, hallucinations, inability to concentrate, and vertigo may follow severe exposures. Kidney damage may occur. Contact with skin can cause skin rash or blisters. Complete preuse physical recommended, stressing nervous system, lung function, and skin condition. Chest X-ray and pulmonary test recommended. Possible carcinogen. Contact with aluminum and strong oxidizers can lead to fires or explosions. On EPA list of extremely dangerous chemicals, 40 CFR 302.

**Methyl chloride** — $CH_3Cl$; 74-87-3; flammable, moderate irritant, suspect carcinogen, poison; bp = −23.7°C (−10.7°F); er = 8.1 to 17%; sp gr =1.78; OSHA PEL = 100 ppm (ceiling = 200 ppm, peak = 300 ppm for 5 min in any 3-h period). Dangerous fire hazard from heat and flame oxidizers. Prolonged exposures can cause psychological problems due to damage to central nervous system. Also can damage liver, kidneys, bone marrow, and cardiovascular system. Faint sweetish odor does not provide adequate warning of overexposure. Use only in an efficient fume hood.

**Methyl mercaptan** — $CH_4S$, 74-93-1; flammable, inhalant poison, possible carcinogen; bp = 5.95°C (42.7°F); er = 3.9 to 21.8%; sp gr = 1.66; OSHA ceiling PEL = 10 ppm. Warning odor of rotten cabbage. Reacts vigorously with oxidizing agents. Will decompose to emit toxic and flammable vapors when it reacts with water, steam, and acids.

**Nitric oxide** — NO; 10102-43-9; noncombustible, strong oxidizing agent, inhalant poison; bp = −152°C (−241°F); er = NA, sp gr = 1.0; OSHA PEL = 25 ppm. Causes narcosis in animals, drowsiness. Sharp, sweet odor provides warning well below dangerous levels. Changes to nitrogen dioxide in air (see following material), although conversion is slow at low concentrations. Reacts vigorously with reducing agents. Will attack some plastics, films, and rubber. On EPA list of extremely hazardous chemicals, 40 CFR 302.

**Nitrogen dioxide** — $NO_2$; 10102-44-0; poisonous by inhalation; bp = −21°C (−69.8°F); er = NA; sp gr = 2.83; OSHA ceiling PEL = 5 ppm. Brown, pungent gas. Odor detection threshold approximately the same as the OSHA PEL. Strong respiratory irritant. Exposure to 100 ppm for 1 h will normally cause pulmonary edema and possibly death; 25 ppm will cause chest pain and respiratory irritation. Onset of symptoms may be delayed. Recovery from an overexposure may be slow and may result in permanent lung damage. Recommend a baseline physical examination, with emphasis on respiratory and cardiovascular systems. Special tests recommended are chest X-ray, pulmonary function test, and electrocardiogram. Reacts vigorously with chlorinated hydrocarbons, ammonia, carbon disulfide, and combustible materials, possibly resulting in fires and explosions. High-temperature glass blowing operations may generate significant levels of this material. On EPA list of extremely hazardous chemicals, 40 CFR 302.

**Nitrogen trifluoride** — $NF_3$; 7783-54-2; poisonous by inhalation; bp = $-12°C$ ($-200°F$); er = NA; sp gr = 2.5; OSHA PEL = 10 ppm. May affect capacity of blood to carry oxygen; Reacts vigorously with reducing agents. Baseline physical should stress examination of blood, cardiovascular, and nervous systems, and liver and kidney functions. A complete blood count should be taken. Persons with a history of blood disorders should take special care to avoid exposures.

**Oxygen difluoride** — $OF_2$; 7783-41-7; poison via inhalation, strong respiratory irritant, noncombustible, strong oxidizing agent; bp = $-145°C$ ($-229°F$); er = NA; sp gr = 1.86; OSHA PEL = 0.05. Foul odor, but detection threshold too high to provide adequate warning. Sense of smell fatigues rapidly. Inhalation of less than 1 ppm causes severe headaches. Very corrosive to tissue. Strong irritant to respiratory system, kidneys, and internal genitalia. Baseline medical exam recommended, with particular emphasis on affected systems. Use only in efficient hood system.

**Ozone** — $O_3$; 10028-15-6; strong irritant to eyes and respiratory system; bp = $-112°C$ ($-169.6°F$); er = NA; sp gr = 1.65; OSHA PEL = 0.1 ppm. Sharp, distinctive odor; odor detection level about the same as the PEL. Affects central nervous system. May be mutagen. Powerful oxidizing agent for both organic and inorganic oxidizable materials. Some reaction products are very explosive. Baseline physical recommended, with emphasis on heart and lungs. Chest X-ray and pulmonary function test recommended. On EPA list of extremely hazardous materials, 40 CFR 302.

**Phosgene** — $CCl_2O$; 75-44-5; inhalant poison, nonflammable; bp = $8.2°C$ ($46.7°F$); er = NA; sp gr = 3.4; OSHA PEL = 0.1 ppm. Odor of "moldy hay"; sense of smell desensitized quickly. Irritating properties well above PEL, so does not provide adequate warning. Must be used carefully in an efficient fume hood. Paper soaked in a 10% mixture of equal parts of *p*-dimethylaminobenzaldehyde and colorless diphenylamine in alcohol or carbon tetrachloride, then dried, makes a good color indicator. Color changes from yellow to deep orange at about the maximum allowable concentration. Severe respiratory irritant, but irritation does not manifest itself at once, even at dangerous levels. Decomposes in presence of moisture in lungs to HCl and CO. Baseline medical examination recommended, with emphasis on respiratory system. Chest X-ray and pulmonary function test recommended. On EPA list of extremely hazardous chemicals, 40 CFR, 302.

**Phosphine** — $PH_3$; 7803-51-2; flammable, inhalation poison; bp = $-87.8°C$ ($-126°F$); er = 1%-?; sp gr = 1.17; OSHA PEL = 0.3 ppm. Fishy odor detectable well below PEL. Severe pulmonary irritant and systemic poison. Results of severe overexposure are chest pains, weakness, lung damage, and, in some cases, coma and death. Persons with prior history of respiratory problems should not work with this material without precautions. Baseline physical should include pulmonary function test. On EPA list of extremely hazardous chemicals, 40 CFR 302.

**Propane** — $C_3H_8$; 74-98-6; flammable, asphyxiant; bp = –42.1°C (–43.7°F); er = 2.3 to 9.5%; sp gr = 1.6; OSHA PEL = 1000 ppm. High exposures (100,000 ppm) cause dizziness after a few minutes. Odorless and nonirritating, so commercially sold propane usually has a foul-smelling odorant added as a warning device. Reacts vigorously with strong oxidizing agents. Explosion hazard from heat and flames.

**Propylene** — $C_3H_6$; 115-07-1; flammable, simple asphyxiant; bp = –47.7°C (–53.9°F); er = 2 to 11.1%; sp gr = 1.5; OSHA PEL = Simple asphyxiant; dangerous when exposed to heat and flames. Can react vigorously with oxidizing agents.

**Silane** — $SiH_4$; 7803-62-5; flammable, respiratory irritant; bp = –112°C (–169.6°F); er = NA; sp gr = NA; ACGIH TLV = 5 ppm. Moderate respiratory irritant. Repulsive odor. Easily ignites in air. Reacts vigorously with chlorine, bromine, and covalent chlorides.

**Stibine** — $SbH_3$; 7803-52-3; flammable, poison by inhalation; bp = –17°C (1°F); er = NA; sp gr = 4.34; OSHA PEL = 0.1 ppm. Odor similar to hydrogen sulfide, but data not available as to whether the odor is an adequate warning of the PEL. Toxic hemolytic agent which causes injury to liver and kidneys. Probable lung irritant. Symptoms of overexposure may be delayed for up to 2 d and would include nausea, headache, vomiting, weakness, and back and abdominal pain. Death could result from renal failure and pulmonary edema. Baseline medical exam should stress blood, liver, and kidneys. Special tests should include complete blood count, urine analysis, and liver panel.

**Sulfur dioxide** — $SO_2$; 7446-09-5; nonflammable, strong respiratory, eye, and skin irritant; bp = –10.05°C (13.9°F); er = NA; sp gr = 2.26; OSHA PEL = 5 ppm. Sharp, irritating odor. Detectable well below OSHA PEL. Reacts rapidly with moisture to form corrosive sulfurous acid ($H_2SO_3$). Mostly absorbed in upper respiratory tract. High concentrations can cause pulmonary edema and respiratory paralysis. Reacts vigorously with water, some powdered metals, and alkali metals such as sodium and potassium. Baseline medical exam should emphasize eyes and respiratory tract. Recommend chest X-ray and pulmonary function tests. Some individuals (10 to 20% of young adults) may be hypersensitive to the material. On EPA list of extremely hazardous chemicals, 40 CFR 302.

**Sulfur tetrafluoride** — $SF_4$; 7783-60-0; powerful irritant, poisonous by inhalation; bp = –40°C,F; sp gr = NA; ACGIH ceiling TLV = 0.1 ppm. Reacts with water, steam, and acids to produce toxic and corrosive fumes. On EPA list of extremely dangerous chemicals, 40 CFR 302.

**Trifluoromonobromomethane (Halon™ 1301)** — $CBrF_3$; 75-63-8; affects heart at high levels; bp = –57.8°C (–72°F); er = NA; sp gr = 5; OSHA PEL = 1000. Popular fire extinguishing agent used for solvent fires and delicate electronic equipment because of its nondamaging properties. No observed effects at design-use range, but can cause cardiac arrhythmias at high concentrations. Individuals with heart problems should use with caution. Can emit dangerous gases and vapors on decomposition by heat.

**Vinyl chloride** — $C_2H_3Cl$; 75-01-4; flammable, dangerous irritant, carcinogen; bp = –13.9°C (7°F); er = 4 to 33%; sp gr = 2.15; OSHA PEL = 1 ppm, ceiling = 5 ppm. Regulated under 29 CFR 1910.1017. Vinyl chloride *monomer* has been shown to cause a rare liver cancer, angiosarcoma. Latency period of 20 years or more. Participation in medical surveillance program required by OSHA standard if use meets prescribed conditions. Dangerous irritant to respiratory system, skin, eyes, and mucous membranes. Reacts vigorously with oxidizers. Decomposes when heated to generate phosgene.

# REFERENCES (SECTION 4.6.2.12.1)

1. Occupational Health Guidelines for Chemical Hazards, DHHS (NIOSH) Publ. No. 81-123, Mackison, F. W., Stricoff, R. S., and Partridge, L. J., Jr., Eds., U.S. Department of Health and Human Services and U.S. Department of Labor, Washington, D.C., 1981.
2. **Sax, N. I. and Lewis, R. J., Sr.,** *Rapid Guide to Hazardous Chemicals in the Workplace,* Van Nostrand Reinhold, New York, 1986.
3. **Armour, M. A., Browne, L. M., and Weir, G. L.,** *Hazardous Chemicals Information and Disposal Guide,* 2nd ed., University of Alberta, Edmonton, Canada, 1984.
4. **Sax, N. I.,** *Dangerous Properties of Industrial Materials,* 5th ed., Van Nostrand Reinhold, New York, 1979.
5. TLVs® , *Threshold Limit Values for Chemical Substances in Workroom Air,* American Conference of Governmental Industrial Hygienists, Cincinnati, OH, 1986.
6. Occupational Health and Safety, General Industry Standards, Subpart Z, Occupational Health and Environmental Control, 1910.1000 to 1910.1500, Washington, D.C., 1988.
7. *Hazardous Chemical Data,* NFPA-49, National Fire Prevention Association, Quincy, MA, 1975.
8. *Prudent Practices for Handling Hazardous Chemicals in Laboratories,* National Academy Press, Washington, D.C., 1981, 81.

### 4.6.2.13. Cryogenic Safety*

Cryogenics may be defined as low temperature technology, or the science of very low temperatures. To distinguish between cryogenics and refrigeration, a commonly used measure is to consider any temperature lower than –73.3°C (–100°F) as cryogenic. Although there is some controversy about this distinction, and some who insist that only those areas within a few degrees of absolute zero may be considered as cryogenic, the broader definition will be used here.

Low temperatures in the cryogenic area are primarily achieved by the liquefaction of gases, and there are more than 25 which are currently in use in the cryogenic area. However, the seven gases which account for the greatest volume of use and applications in research and industry are helium, hydrogen, nitrogen, fluorine, argon, oxygen, and methane (natural gas).

Cryogenics is being applied to a wide variety of research areas, a few of which are food processing and refrigeration, rocket propulsion fuels, spacecraft life support systems, space simulation, microbiology, medicine, surgery, electronics, data processing, and metalworking. [Recent advances in high temperature superconductors will further increase the use of liquid nitrogen.]

Cryogenic fluids (liquified gases) are characterized by extreme low temperatures, ranging from a boiling point of –78.5°C (–109°F) for carbon dioxide to –269.9°C (–453.8°F) for an isotope of helium, helium-3. Another common property is the large ratio of expansion in volume from liquid to gas, from approximately 553 to 1 for carbon dioxide to 1438 to 1 for neon. Table 4.14 contains a more complete summary of the properties of cryogenic fluids.

---

* This section, except for a short amount of material appended at the end, is taken directly from the article, Cryogenic Safety, by Spencer,[15] with minor editing in the second edition of this handbook.

**TABLE 4.14**
**Properties of Cryogenic Fluids**

| Gas | Normal boiling point °C | °K | Volume expansion to gas | Flammable | Toxic | Odor |
|---|---|---|---|---|---|---|
| Helium-3 | −269.9 | 3.2 | 757—1 | No | No | No |
| Helium-4 | −268.9 | 4.2 | 757—1 | No | No | No |
| Hydrogen | −252.7 | 20.4 | 851—1 | Yes | No | No |
| Deuterium | −249.5 | 23.6 | — | Yes | No | No |
| Tritium | −248.0 | 25.1 | — | Yes | Radioactive | No |
| Neon | −245.9 | 27.2 | 1438—1 | No | No | No |
| Nitrogen | −195.8 | 77.3 | 696—1 | No | No | No |
| Carbon monoxide | −192.0 | 81.1 | — | Yes | Yes | No |
| Fluorine | −187.0 | 86.0 | 888—1 | No | Yes | Sharp |
| Argon | −185.7 | 87.4 | 847—1 | No | No | No |
| Oxygen | −183.0 | 90.1 | 860—1 | No | No | No |
| Methane | −161.4 | 111.7 | 578—1 | Yes | No | No |
| Krypton | −151.8 | 121.3 | 700—1 | No | No | No |
| Tetrafluromethane | −128 | 145 | — | No | Yes | No |
| Ozone | −111.9 | 161.3 | — | Yes | Yes | Yes |
| Xenon | −109.1 | 164.0 | 573—1 | No | No | No |
| Ethylene | −103.8 | 169.3 | — | Yes | No | Sweet |
| Boron trifluoride | −100.3 | 172.7 | — | No | Yes | Pungent |
| Nitrous oxide | −89.5 | 183.6 | 666—1 | No | No | Sweet |
| Ethane | −88.3 | 184.8 | — | Yes | No | No |
| Hydrogen chloride | −85.0 | 188.0 | — | No | Yes | Pungent |
| Acetylene | −84.0 | 189.1 | — | Yes | Yes | Garlic |
| Fluoroform | −84.0 | 189.1 | — | No | No | No |
| 1,1-Difluoroethylene | −83.0 | 190.0 | — | Yes | No | Faint ether |
| Chlorotrifluoromethane | −81.4 | 191.6 | — | No | Yes | Mild |
| Carbon dioxide | −78.5 | 194.6 | 553—1 | No | Yes | Slightly pungent |

## HAZARDS

There are four principal areas of hazard relating to the use of cryogenic fluids or in cryogenic systems. These are: flammability, high pressure gas, materials, and personnel. All categories of hazard are usually present in a system concurrently, and must be considered when introducing a cryogenic system or process.

The flammability hazard is obvious when gases such as hydrogen, methane, and acetylene are considered. However, the fire hazard may be greatly increased when gases normally thought to be non-flammable are used. The presence of oxygen will greatly increase the flammability of ordinary combustibles, and may even cause some non-combustible materials like carbon steel to burn readily under the right conditions. Liquified inert gases such as liquid nitrogen or liquid helium are capable, under the right conditions, of condensing oxygen from the atmosphere, and causing oxygen enrichment or entrapment in unsuspected areas. Extremely cold metal surfaces are also capable of condensing oxygen from the atmosphere.

The high-pressure gas hazard is always present when cryogenic fluids are used or stored. Since the liquified gases are usually stored at or near their boiling point, there is always some gas present in the container. The large expansion ratio from liquid to gas provides a source for the build-up of high pressures due to the evaporation of the liquid. The rate of expansion will vary, depending on the characteristics of the fluid, container design, insulating materials, and environmental conditions of the atmosphere. Container capacity must include an allowance for that portion which will be in the gaseous state. These same factors must also be considered in the design of the transfer lines and piping systems.

Materials must be carefully selected for cryogenic service because of the drastic changes in the properties of materials when they are exposed to extreme low temperatures. Materials which are normally ductile at atmospheric temperatures may become extremely brittle when subjected to temperatures in the cryogenic range, while other materials may improve their properties of ductility. The American Society of Mechanical Engineers' *Boiler and Pressure Vessel Code, Section VIII — Unfired Pressure Vessels* may be used as a specific guide to the selection of materials to be used in cryogenic service. Some metals which are suitable for cryogenic temperatures are stainless steel (300 series and other austenitic series), copper, brass, bronze, monel, and aluminum. Non-metal materials which perform satisfactorily in low-temperature service are Dacron, Teflon, Kel-F, asbestos impregnated with Teflon, Mylar, and Nylon. Once the materials are selected, the method of joining them must receive careful consideration to ensure that the desired performance is preserved by using the proper soldering, brazing, or welding techniques or materials. Finally, chemical reactivity between the fluid or gas and the storage containers and equipment must be studied. Wood or asphalt saturated with oxygen has been known to literally explode when subjected to mechanical shock. When properties of materials which are being considered for cryogenic use are unknown, or not to be found in the known guides, experimental evaluation should be performed before the materials are used in the system.

Personnel hazards exist in several areas where cryogenic systems are in use. Exposure of personnel to the hazards of fire, high-pressure gas, and material failures previously discussed must be avoided. Of prime concern is bodily contact with the extreme low temperatures involved. A very brief contact with fluids or materials at cryogenic temperatures is capable of causing burns similar to thermal burns from high-temperature contacts. Prolonged contact with these temperatures will cause embrittlement of the exposed members because of the high water content of the human body. The eyes are especially vulnerable to this type of exposure, so that eye protection is necessary.

While a number of the gases in the cryogenic range are not toxic, they are all capable of causing asphyxiation by displacing the air necessary for the support of life. Even oxygen may have harmful physiological effects if prolonged breathing of pure oxygen takes place.

There is no fine line of distinction between the four categories of hazards, and they must be considered collectively and individually in the design and operation of cryogenic systems.

## GENERAL PRECAUTIONS

Personnel should be thoroughly instructed and trained in the nature of the hazards and the proper steps to avoid them. This should include emergency procedures, operation of equipment, safety devices, knowledge of the properties of the materials used, and personal protective equipment required.

Equipment and systems should be kept scrupulously clean and contaminating materials avoided which may create a hazardous condition upon contact with the cryogenic fluids or gases used in the system. This is particularly important when working with liquid or gaseous oxygen.

Mixing of gases or fluids should be strictly controlled to prevent the formation of flammable or explosive mixtures. As the primary defense against fire or explosion, extreme care should be taken to avoid contamination of a fuel with an oxidant, or the contamination of an oxidant by a fuel.

As a further prevention, when flammable gases are being used, potential ignition sources must be carefully controlled. Work areas, rooms, chambers, or laboratories should be suitably monitored to automatically warn personnel when a dangerous condition is developing. When practical, it would be advisable to provide for the cryogenic equipment to be shut down automatically as well as to sound a warning alarm.

When there is a possibility of personal contact with a cryogenic fluid, full face protection, an impervious apron or coat, cuffless trousers, and high-topped shoes should be worn. Watches, rings, bracelets, or other jewelry should not be permitted when personnel are working with cryogenic fluids. Basically, personnel should avoid wearing anything capable of trapping or holding a cryogenic fluid in close proximity to the flesh. Gloves may or may not be worn, but if they are necessary in order to handle containers or cold metal parts of the system, they should be impervious, and sufficiently large to be easily tossed off the hand in case of a spill. A more desirable arrangement would be hand protection of the potholder type.

When toxic gases are being used, suitable respiratory protective equipment should be readily available to all personnel. They should thoroughly know the location and use of this equipment.

## STORAGE

Storage of cryogenic fluids is usually in a well-insulated container designed to minimize loss of product due to boil-off.

The most common container for cryogenic fluids is a double-walled, evacuated container known as a Dewar flask, of either metal or glass. The glass container is similar in construction and appearance to the ordinary "Thermos" bottle. Generally, the lower portion will have a metal base which serves as a stand. Exposed glass portions of the container should be taped to minimize the flying glass hazard if the container should break or implode.

Metal containers are generally used for larger quantities of cryogenic fluids, and usually have a capacity of 10 to 100 liters (2.6 to 26 gallons). These containers are also of double-walled evacuated construction and usually contain some absorbent material in the evacuated space. The inner container is usually spherical in shape because that has been found to be the most efficient in use. Both the metal and glass Dewars should be kept covered with a loose-fitting cap to prevent air or moisture from entering, and to allow built-up pressure to escape.

Larger capacity storage vessels are basically the same double-walled containers, but the evacuated space is generally filled with powdered or layered insulated material. For economic reasons, the containers are usually cylindrical with dished ends, which approximates the shape of the sphere, which is expensive to build. Containers must be constructed to withstand the weights and pressures that will be encountered, and adequately vented to permit the escape of evaporated gas. Containers should also be equipped with rupture discs on both inner and outer vessels to release pressure if the safety relief valves should fail.

Cryogenic fluids with boiling points below that of liquid nitrogen (particularly liquid helium and hydrogen) require specially constructed and insulated containers to prevent rapid loss of product from evaporation. These are special Dewar containers which are actually two containers, one inside the other. The liquid helium or hydrogen is contained in the inner vessel, and the outer vessel contains liquid nitrogen which acts as a heat shield to prevent heat from radiating into the inner vessel. The inner neck should be kept closed with a loose fitting, non-threaded brass plug which prevents air or moisture from entering the container, yet is loose enough to vent any pressure which may have developed. The liquid nitrogen fill and vent lines should be connected by a length of gum rubber tubing with a slit approximately 2.54 cm (1 in.) long near the center of the tubing. This prevents the entry of air and moisture, while the slit will permit release of the gas pressure. Piping or transfer lines should be double-walled evacuated pipes to prevent the loss of product during transfer.

Most suppliers are now using a special fitting to be used in the shipment of Dewar vessels. Also, there is an automatic pressure relief valve, and a manual valve to prevent entry of moisture and air, which will form an ice plug. The liquid helium fill (inner neck) should be reamed out before and after transfer, and at least twice daily. Reaming should be performed with a hollow copper rod, with a marker or stop to prevent damaging the bottom of the inner container. Some newer style Dewar vessels are equipped with a pressure relief valve, and pressure gauge for the inner vessel.

Transfer of liquids from the metal Dewar vessels should be accomplished with special transfer tubes or pumps designed for the particular application. Since the inner vessel is mainly supported by the neck, tilting to pour the liquid may damage the container, shortening its life, or creating a hazard due to container failure at a later date. Piping or transfer lines should be so constructed that it is not possible for fluids to become trapped between valves or closed sections of the line. Evaporation of the liquid in a section of line may result in pressure build-up and eventual explosion. If it is not possible to empty all lines, they must be equipped with safety relief valves and rupture discs. When venting storage containers and lines, proper consideration must be given to the properties of the gas being vented. Venting should be to the outdoors to prevent an accumulation of flammable, toxic or inert gas in the work area.

## ACKNOWLEDGMENT

Reprinted with permission from the *Journal of the American Society of Safety Engineers,* 11(8), 15, 1963.

**Addendum to Section 4.6.2.13 (1988)**

There are some applications in which dewars with wide mouths are used, such as storage of certain biological materials. These come with a loosely fitting cap to prevent absorption of air and moisture into the liquid nitrogen, the refrigerant most frequently used in these Dewars. The problem briefly alluded to in the preceding article of a buildup of oxygen in liquid nitrogen containers over a period of time can become a problem if care is not taken to keep the cap on or to change the entire volume occasionally. If the liquid takes on a blue tint, it is contaminated with oxygen and should be replaced. The contaminated liquid should be treated as a dangerous, potentially explosive material. Most users fill Dewars from larger ones, usually by pressurizing the larger one with nitrogen from a cylinder, thereby forcing the liquid into the smaller one. In order not to waste liquid nitrogen by evaporation in a warm container, neither Dewar is usually allowed to become totally empty, again leading to possible oxygen contamination. If these practices are continued for a sufficiently long time, the oxygen content of the cryogenic liquid may become dangerously high.

There are two relatively common ways to maintain a supply of liquid nitrogen at a facility, one being to have a large reservoir of several thousand liters capacity from which individual users fill their smaller Dewars. The boil-off from a large reservoir can be used to provide a supply of ultra clean "air" to laboratories to clean surfaces. Liquid nitrogen is also usually available, if reasonably close to a distributor, in 160-l pressurized containers delivered directly to the laboratory. In either case, the quantities actually needed for most small laboratories can be obtained frequently enough to avoid having an elaborate piping and control system from a large central reservoir, with the associated problems of avoiding blockage of the system by ice plugs. There are, of course, applications where automatically controlled systems are necessary, which must be provided with safety relief and warning devices.

## REFERENCES (SECTION 4.6.2.13)

1. *Cryogenics,* Marsh & McLendon, Chicago, 1962.
2. *Industrial Gas Data,* Air Reduction Sales Company, Acton, MA,
3. *Matheson Gas Data Book,* 47th ed., The Matheson Company, East Rutherford, NJ, 1961.
4. *Precautions and Safe Practices for Handling Liquid Hydrogen,* Linde Company, New York, 1960.
5. *Precautions and Safe Practices for Handling Liquified Atmospheric Gases,* Linde Company, New York, 1960.
6. **Braidech, M. M.,** *Hazards/Safety Considerations in Cryogenic (Super Cold) Operations,* Conference of Special Risk Underwriters, New York, 1961.
7. **Hoare, Jackson, and Kurti,** *Experimental Cryophysics,* Butterworths, London, 1963.
8. **MacDomald, D. K. C.,** *Near Zero, An Introduction to Low Temperature Physics,* Doubleday, New York, 1961.
9. **Neary, R. M.,** *Handling Cryogenic Fluids,* Linde, New York, 1960.
10. **Scott, R. B.,** *Cryogenic Engineering,* D Van Nostrand, Princeton, NJ, 1959.
11. **Timmerhaus, K. D.,** Ed., *Advances in Cryogenic Engineering,* Vol. 7, Plenum Press, New York, 1961.
12. **Vance, R. W. and Duke, W. M.,** Eds., *Applied Cryogenic Engineering,* John Wiley & Sons, New York, 1962.
13. **Zenner, G. H.,** Safety engineering as applied to the handling of liquified atmospheric gases, in *Advances in Cryogenics Engineering,* Vol. 6, Plenum Press, New York, 1960.
14. Cryogenic Safety, A Summary Report of the Cryogenic Safety Conference, Air Products, Allentown, PA, 1959.
15. **Spencer, E. W.,** Cryogenic safety, in *Handbook of Laboratory Safety,* 2nd ed., Steere, N. V., Ed., CRC Press, Boca Raton, FL, 1971.

### 4.6.2.14. Cold Traps*
### COLD TRAPS

Cold traps are used in instrumentation and elsewhere to prevent the introduction of vapors or liquids into a measuring instrument from a system, or from a measuring instrument (such as a Mcleod gauge) into the system. A cold trap provides a very-low-temperature surface on which such molecules can condense, and improves pump-down [the achievable vacuum] by one or two magnitudes.

However, cold traps improperly employed can impair accuracy, destroy instruments or systems, and be a physical hazard. For example, many of the slush mixtures used in cold traps are toxic or explosive hazards, and this is not indicated in the literature.

The authors became aware of the deficiencies in tunnel instrumentation, where it was necessary to measure pressures in the micron to 760-torr** region. The instrumentation system used Stratham gauges for ambient pressure down to 100 to 150 torr or about 2 to 3 psia and NRC alphatron gauges for pressures to $5 \times 10^{-2}$ torr. To prevent calibration shifts and contamination of the NRC transducers by oil fumes from the vacuum pump and possible wind tunnel contaminants, a cold trap was placed in the line.

The cold trap was filled with liquid nitrogen, and the valve to the tunnel line shut off. When the valve was opened, cold gas shot out, shown by the condensation; the over-pressure developed in the system destroyed the stratham straingage bridge, although it was not sufficient to rupture the transducer diaphragm. As no satisfactory explanation was forthcoming, a glass cold trap was procured and set up in a dummy system. The cause of the phenomenon soon became apparent: air in the trap and system lines was liquified in the trap. When the valve was opened, this liquid air was being blown into the warmer lines by atmospheric pressure: the resultant volatilization of liquid into gas was practically an explosion.

Nevertheless, cold traps are often the only satisfactory means of removing contaminants, although in ordinary experimental work the charcoal trap is occasionally acceptable. A charcoal trap will remove oil and condensable vapors so that pressures to $10^{-8}$ torr or better may be secured, but it presents a serious restriction on pumping speed and requires bake out when it has become charged with oil and vapors. Molecular sieve traps place similar restrictions on pumping speed.

The errors introduced by the water vapor, when measuring low pressures, depend on the vacuum gauge used. The presence of water vapor also affects the magnitude of vacuum that can be achieved. The equilibrium point of a dry-ice-acetone slush is $-78°C$ ($-108.4°F$), which although sufficient to trap mercury vapor effectively, does not remove water vapor; a temperature of at least $-100°C$ ($-148°F$) is required to eliminate water vapor or, alternatively, exposure to anhydrous phosphorous pentoxide ($P_2O_5$). This material is usually rejected for field use because of possible biological, fire and explosive hazards: in absorbing water it produces heat, and reacts vigorously with reducing materials.

Slush mixtures using liquid air and liquid oxygen were considered and dropped, either because of the explosive hazard or toxicity of the vapors, or because they were

---

\*   This section is taken directly from the article Cold Traps, by Kaufman and Kaufman[4] in the second edition of this handbook.

\*\* The torr is equal to a pressure of 1 mmHg at standard conditions.

not cold enough. Table 4.15 lists many common thermal transfer and coolant fluids with their hazards and limitations.

### Virtual Leaks

If the cold trap is chilled too soon after the evacuation of the system begins, gases trapped will later evaporate, when the pressure reaches a sufficiently low value. The evaporation of the refrigerated and trapped gases is not rapid enough to be evacuated by the system, but is enough to degrade the vacuum, producing symptoms very similar to those of a leak.

To avoid these virtual leaks, keep the trap warm until a vacuum of about $10^{-2}$ torr is obtained. The tip of the trap is then cooled until ultimate vacuum is reached, at which time the trap may be immersed in the coolant to full depth.

### Safety Precautions

If liquid nitrogen is the coolant, liquid air can condense in the trap, inviting explosion. Liquid air, comprising a combination primarily of oxygen and nitrogen, is warmer than liquid nitrogen. Depending on the nitrogen content, air liquifies anywhere from $-190°C$ ($-310°F$) ($5°C$ warmer than liquid nitrogen) to $-183°C$ ($-297.4°F$) (liquid oxygen). If liquid nitrogen is used, the trap should be charged only after the system is pumped down lest a considerable amount of liquid oxygen condenses, creating a major hazard.

Handle any liquid gas carefully; at its extremely low temperature, it can produce an effect on the skin similar to a burn. Moreover, liquified gases spilled on a surface tend to cover it completely and intimately, and therefore cool a large area.

The evaporation products of these liquids are also extremely cold and can produce burns. Delicate tissues, such as those of the eyes, can be damaged by an exposure to these cold gases which is too brief to affect the skin of the hands or face.

Eyes should be protected with a face shield or safety goggles (safety spectacles without side shields do not give adequate protection). Gloves should be worn when handling anything that is or may have been in contact with the liquid; asbestos gloves are recommended. [This is no longer true because of the concern about airborne asbestos fibers from products containing asbestos. Gloves made of an artificial material such as Kevlar™ or Zetex™ are recommended as an alternative], but leather gloves may be used. The gloves must fit loosely so that they can be thrown off quickly if liquid should spill or splash into them. When handling liquids in open containers, high-top shoes should be worn with trousers (cuffless if possible) worn outside them.

Stand clear of boiling and splashing liquids and their issuing gas. Boiling and splashing always occur when charging a warm container or when inserting objects into the liquid. Always perform these operations SLOWLY to minimize boiling and splashing.

Should any liquified gases used in a cold trap contact the skin or eyes, immediately flood that area of the body with large quantities of unheated water and then apply cold compresses. Whenever handling liquified gases, be sure there is a hose or a large open container of water nearby, reserved for this purpose. If the skin is blistered, or if there is any chance that the eyes have been affected, take the patient immediately to a physician for treatment [call for emergency medical aid; normally,

**TABLE 4.15**

**Thermal Transfer Fluids[a] Used with Instrumentation-Type Cold Traps**

| Element[b] | Temperature[c] | | Hazard[d] | | | Remarks |
|---|---|---|---|---|---|---|
| | °C | °F | Inhalation toxicity | Skin toxicity | Explosive or fire | |
| Glycerine, 70% by weight Pressure | −38.9 | −38.0 | None | Slight | Slight | Vapor |
| Water 30% | | | | | | $2.5 \times 10^{-3}$ torr @ 50°C (122°F) |
| Ethyl alcohol, dry ice | −78 | −108 | Moderate | Slight | Dangerous | — |
| Ethylene glycol, 52.5% by volume; water 47.5% | −40 | −40 | None | Slight | Slight | — |
| Chloroform, dry ice | −63.5 | −82 | Extreme | Slight | Slight | Vapor pressure 100 torr @ 22°C (71.6°F) |
| Liquid SO₂ | −75.5 | −103 | Extreme | Extreme | None | Very dangerous |
| Methanol (methyl alcohol), dry ice | −78 | −108 | Slight | Slight | Dangerous | Vapor pressure 100 torr @ 22°C (71.6°F) Ingestion very dangerous |
| Acetone, dry ice | −78 | −108 | Moderate | Slight | Dangerous | Vapor pressure 400 torr @ 39°C (102.2°F) |
| Methyl bromide, dry ice | −78 | −108 | Extreme | Extreme | Moderate | Very dangerous |
| Fluorotrichloromethane (Freon 11), dry ice | −78 | −108 | Slight | Slight | None | — |
| Methylene chloride, dry ice | −78 | −108 | Moderate | Moderate | Slight | Very dangerous to eyes, vapor pressure 380 torr @ 22°C (71.6°F) |
| Calcium chloride | −42 | −44 | Slight | Slight | None | — |
| Ethyl methyl ketone | −78 | −108 | Moderate | Moderate | Dangerous | — |

[a]  Transfer fluids will freeze solid and become colder if subject to temperatures lower than their freezing point. A slush mixture is secured by lowering the temperature, such as by the introduction of limited quantities of dry ice until the mixture is quasi-frozen.

[b]  These materials are often sold under trade names (listed in the CRC *Handbook of Chemistry and Physics*). In general, any combination of elements shown was selected for the coldest slush mixture obtainable.

[c]  If the refrigerant is dry ice, the transfer fluid will not go below −78°C (−18.4°F), the temperature of solid $CO_2$.

[d]  The consensus is that many of these liquids, while hazardous at room temperature, are not hazardous when cooled since their evaporation at low temperatures is fairly low. For utmost safety, those noted dangerous should not be employed unless venting or other special precautions are taken. For greater detail, see Reference 2.

rescue squads can be in immediate contact with an emergency room physician by radio].

Oxygen is removed from the air by liquid nitrogen exposed to the atmosphere in an open Dewar. Store and use liquid nitrogen only in a well ventilated place; owing to evaporation of nitrogen gas and condensation of oxygen gas, the percentage of oxygen in a confined space can become dangerously low. When the oxygen concentration in the air becomes sufficiently low, a person loses consciousness without warning symptoms and will die if not rescued. The oxygen content of the air must never be allowed to fall below 16%.

The appearance of a blue tint in liquid nitrogen is a direct indication of its contamination by oxygen, and it should be disposed of, using all the precautions generally used with liquid oxygen. Liquid nitrogen heavily contaminated with oxygen has severe explosive capabilities. In addition, an uninsulated line used to charge Dewars will condense liquid air; liquid air dripping off the line and revaporizing causes an explosive hazard during the charging operation.

If the cold trap mixture is allowed to freeze, and the cold trap becomes rigid, slight movement in other parts of the apparatus could result in breakage of the trap or other glassware.

If a gas trap has to be lifted out of the Dewar cold bath for inspection, it will be difficult to reinsert into the slush. Therefore, it is preferable to use a liquid that will not freeze at $-78.5°C$.

## ACKNOWLEDGMENT

Reprinted with permission of Rimbach Publications, Pittsburgh, *Instrument and Control Systems,* 36, 109, 1963.

## REFERENCES (SECTION 4.6.2.14)

1. **Strong, J., Neher, H. V., Whitford, A. E., Cartwright, C. H., and Hayward, R.,** *Procedures in Experimental Physics,* Prentice-Hall, New York, 1938.
2. **Sax, N. I.,** *Dangerous Properties of Industrial Materials,* Reinhold, New York, 1961.
3. **Dushman, L.,** *Scientific Foundation of Vacuum Techniques,* John Wiley & Sons, New York, 1962.
4. **Kaufman, A. B. and Kaufman, E. N.,** Cold traps, in *Handbook of Laboratory Safety,* 2nd ed., Steere, N. V., Ed., CRC Press, Boca Raton, FL, 1971, 510.

### 4.6.2.15. Care and Use of Electrical Systems

Some of the problems associated with electrical systems have been covered in previous sections, such as Section 3.1.7. There may be some unavoidable repetition in this section, which will be primarily concerned with the safe use of electricity rather than characteristics of individual items, although there will be a brief discussion of generic problems associated with the design of equipment. Most of the hazards associated with the use of electricity stem from electrical shock, resistive heating, and ignition of flammables, and most of the actual incidents occur because of a failure to anticipate all of the ways in which these hazards may be evoked in a laboratory situation. This lack of recognition of the possible hazards may be reflected in the original choice of suitably safe electrical equipment or improper installation of the equipment. In some instances, the choice of equipment may simply involve a continued use of equipment on hand under conditions for which it is no longer suitable, so safety specifications are not really considered. Often, this is due to a familiarity with the existing resources rather than a deliberate choice.

Part of the problem may be a feeling that questionable electrical practices rou-

## TABLE 4.16
### Effects of Electrical Current in the Human Body

| Current (mA) | Reaction |
| --- | --- |
| 1 | Perception level, a faint tingle |
| 5 | Slight shock felt; disturbing, but not painful. Average person can let go. However, vigorous involuntary reactions to shocks in this range can cause accidents. |
| 6—25 (women) 9—30 (men) | Painful shock; muscular control is lost. Called freezing or "let-go" range.[a] |
| 50—150 | Extreme pain, respiratory arrest, and severe muscular contractions. Individual normally cannot let go unless knocked away by muscle action; death is possible. |
| 1,000—4,300 | Ventricular fibrillation (rhythmic pumping action of heart ceases); muscular contraction and nerve damage occur; death is most likely. |
| 10,000— | Cardiac arrest, severe burns, and probable death |

[a] The person may be thrown away from the contact if the exterior muscles are excited by the shock.

tinely followed at home as well as in the laboratory are actually safe unless you do something "really bad", such as standing in water while in contact with an electrically active wire or a similar feeling about wiring, "Just hook it up with that extension cord I bought on sale at the department store." When asked about a number of similar practices involving multiple connections to a single outlet or the use of extension cords, most persons will say they know that they shouldn't do it, but see no real harm.

Two major electrical factors need to be considered in the choice of most electrical items of equipment. The equipment needs to be selected so that it will not provide a source of ignition to flammable materials and it should be chosen so as to minimize the possibility of personnel coming into contact with electrically live components. The latter problem will be addressed first.

### 4.6.2.15.1. Electrical Shock

OSHA has included the relevant safety portions of the National Electrical Code in 29 CFR 1910, Subpart S. This regulatory standard, like many other sections of the OSHA regulations, is primarily oriented toward industrial applications, but it does speak directly to the problem of preventing individuals from coming into contact with electricity. Live parts of electrical equipment operating at 50 V or more must be guarded against accidental contact, while indoor installations that are at 600 V or more, and open to unqualified persons must be made with metal-enclosed equipment or within a space controlled by a lock. The higher-voltage equipment also must be marked with appropriate warning signs. Access points to spaces in which exposed, electrically live parts are present must be marked with conspicuous warning signs which forbid unqualified persons to enter.

The effects of electricity on a person depend upon the current level and, of course, on physiological factors unique to the individual. Table 4.16 gives typical effects of various current levels for 60-Hz currents for an average person in good health.

Several things affect the results of an individual incident. The duration of the current is important. In general, the degree of injury is proportional to the length of time the body is part of the electrical circuit. A suggested threshold is a product of time and energy of 0.25 Wsec for an objectionable level. The voltage is important because, for a given resistance, R, the current, I, through a circuit element is directly proportional to the applied voltage, V.

$$I = V/R \tag{1}$$

If the contact resistance to the body is lowered so that the total body resistance to the flow of current is low, then even a modest applied voltage can affect the body. The condition of the skin can dramatically alter the contact resistance. Damp, sweaty hands may have a contact resistance which will be some orders of magnitude lower than dry skin. The skin condition is more important for low-voltage contacts than for those involving high voltages since, in the latter case, the skin and contact resistance break down very rapidly. The remaining resistance is the inherent resistance of the body between the points of contact, which is of the order of 500 to 1000 $\Omega$. As shown in Table 4.16, the difference in a barely noticeable shock and a potentially deadly one is only a factor of 100. For an individual with cardiac problems, the threshold for a potentially life-threatening exposure may be even lower. The major danger to the heart is that it will go into ventricular fibrillation due to small currents flowing through the heart. In most cases, once the heart goes into ventricular fibrillation, death follows within a few minutes.

Even if an individual survives a shock episode, there may be immediate and long-term destruction of tissue, nerves, and muscle due to heat generated by the current flowing through the body. The heat generated is basically resistive heating such as would be generated in heating coils in a small space heater, with the exception that the resistive elements are the tissues and bones in the body. The power, P, or heat is given by

$$P = I^2R \tag{2}$$

The scope of the effects of external electrical burns is usually immediately apparent, but the total effect of internal burns may be manifested later by losses of important body functions due to destruction of critical internal organs, including the nervous system.

Several means are available to prevent individuals from coming into contact with electricity in addition to the exclusion of unqualified personnel mentioned in the introduction to this section. These include insulation, grounding, good wiring practices, and mechanical devices. Before addressing the latter options, it might be well to briefly discuss the concept of a qualified person.

Certainly, a licensed electrician would, in most cases, be a qualified person, and a totally inexperienced person would just as clearly not be a qualified individual. There is no clear definition of the training required to be "qualified" to perform laboratory electrical and electronic maintenance. As a minimum, such training should include instruction in the consequences of electrical shock, basic training in wiring color codes (so as to recognize correct leads), familiarity with the significance of ratings of switches, wiring, breakers, etc., simple good wiring practices, and recognition of problems and poor practices such as frayed wiring, wires underfoot, wires in moisture, overloaded circuits, poor grounding procedures, and improper defeat of protective interlocks. This would not make an individual a licensed electrician, which would require extensive training and experience, but would reduce the number of common electrical errors.

Insulation is an obvious means of protecting an individual against shocks. In general, good wiring insulation is the most critical, particularly that of extension cords, which is often abused. Insulation must be appropriate for the environment, which may involve extremes of temperature or exposure to corrosive vapors or solvents. The insulation itself may need to be protected by a metal outer sheath or the wire may need to be installed in conduit.

As it ages, insulation may become brittle and develop fine cracks through which moisture may seep and provide a conductive path to another component or to a person who simply touches the wire at the point of failure. Many plastic or rubber insulating materials will soften with heat and, if draped over a metal support, may eventually allow the wire to come into contact with the metal, thus rendering it electrically active if it does not first cause other problems. Extension cords, as noted above, are particularly susceptible to abuse. They are often carelessly strewn across the floor or furniture. On the floor, they may be walked upon, equipment may be rolled across them, or they may become pinched between items of furniture. First, extension cords should only be used as temporary expedients, but if they are used, they should be treated as any other circuit wiring, put out of harms way and properly supported on insulators separated by distances not to exceed 10 ft. Defective extension cords with badly deteriorated insulation should be discarded.

Insulation is not used solely to protect wiring. Insulation in the form of panels may break if excessive force is applied. If an arc temporarily flashes across an insulating surface, a carbonized conducting path may be permanently established on the surface which could render an external component, such as a chassis mounting screw, "hot". Care should be taken with all electrical equipment, especially older items, to ensure that the integrity of the insulation has been maintained.

Proper grounding of equipment is another requirement to insure that components are not electrically live. Most equipment today for use with 120-V circuits comes with a three-wire power cord, requiring a mating female connector at the power source, many of which are designed so that the neutral, hot, and ground connections can be readily identified and matched. The ground wire, which is either green or perhaps green with yellow stripes, is always connected to the female socket, which accommodates the round prong on the male connector. The neutral circuit wire, which normally completes the circuit for the equipment, is usually white or a natural gray. The socket and corresponding male connector are often wider than the connections for the hot wire. The hot wire is usually covered with a black insulator, although red may also be used. When there are both red and black wires, both usually will be hot wires. The male and female connectors are sometimes narrower than the neutral leads. Some equipment is double insulated and does not have the third ground wire in the power connector. Usually, these will have a polarized connector so that the neutral and hot wires will be properly oriented. Unfortunately, older circuits do not always provide the proper connections and should be replaced. If this is not feasible, the third wire on the power connector, if one is present, should be directly connected to a good-quality ground.

Autotransformers, which may be used to supply variable voltages to heating devices, may be connected in such a way that either outlet line may be high with respect to ground. They should be purchased with a switch which breaks the connection of both outlet sockets to the power input line or they should be rewired with a double-pole power switch to accomplish this.

The quality of all ground connections (and of all connections) needs to be good. This is often taken for granted, but the connection may vibrate or work loose, or a careless worker may fail to tighten a connection. In such cases, a significant difference of potential may arise between two different items of equipment. This can be enough to give rise to a discernible shock for a person coming into contact simultaneously with both pieces of equipment and, in some cases, cause damage to the equipment if they are interconnected. A careful researcher should have the electrical circuits checked for the resistance to ground of all the wiring in his facility. A ground with a resistance of 100 $\Omega$ will be at a difference of 10 V with respect to ground if a current of 100 mA were to flow through the ground connection. Good quality grounds with resistances of a few ohms are easily achievable with care.

Adapters or "cheaters" can be used to plug three-wire power cords into wiring providing only two wires or to avoid connecting the third wire to ground. There are only a few exceptions which would make this an acceptable practice; one is where an alternative direct connection to ground is provided for the equipment. Another would be those rare occasions where even the difference of a few millivolts between the separate ground connections would affect the experimental data signals. In the latter case, ground connections can be made directly between the components, with a single connection being made to the building ground. Except in clearly defined situations where their use is clearly made safe, these adapters should not be used.

Simple devices such as fuses, circuit breakers, and ground fault interrupters are available to cut off equipment when they overload, short out, or an imbalance develops between the input and output current from a device or circuit. More sophisticated devices can also be used to determine a problem, such as a redundant heat detector for deactivating a circuit serving a still, condenser, or heat bath when the temperature becomes too high.

Fuses and circuit breakers are the simplest devices used to shut off a circuit drawing too much current. A fuse inserted in one of the circuit legs functions by melting at a predetermined current limit and, in the breaker, by mechanically opening the circuit. The latter device is more flexible in that it can be reset, while the fuse must be replaced. A ground fault interrupter (GFI), on the other hand, specifically can protect an individual who comes into contact with a live component. The individual's body and the wires become parallel circuits through which a fraction of the total current flows. The amount flowing through the body makes the two normal halves of the circuit out of balance, which the GFI detects and causes it to break the circuit. A GFI can detect a difference of the order of 5 mA and break the circuit in as little as 25 msec. A comparison with Table 4.16 would show that the contact might be barely noticeable, but would cause no direct harm because of the short duration of the current flow. Although GFIs are generally used in the construction industry, they would serve a useful purpose, for example, in laboratories where moisture would be a problem.

The best defenses against electrical shock injuries are good work practices, as invoked by using good judgment and exercising care appropriate to the risk. Maintenance of electrical equipment or wiring should be done only with the system deenergized, unless it is essential that the circuit be active for the required maintenance. Procedures should be followed to confirm that power to the system has been disabled and remains so during the duration of the maintenance activity or, alternatively, if the circuit must remain powered, that a second person is available to disable the circuit and assist in the event of an incident. Formal lockout procedures are recommended where high-voltage circuits are involved. The tools used to perform maintenance should be in good condition. Barriers may be needed to isolate live circuits in the maintenance area. Good judgement should be used to determine safe distances, and metal ladders, etc. should not be used where it would be possible to contact a hot circuit. In some cases, it might be necessary to use rubber gloves and gauntlets, insulating mats, and hardhats certified for electrical protection.

A relatively simple protective strategy which should be followed by anyone working with or handling live electrical circuits is to remove all conducting jewelry, specifically items on or near the hands such as rings, watches, and bracelets or necklaces which may dangle from the neck and complete a circuit to the neck. If work activities involving direct contact with electrical components are infrequent, removing these items at the time may suffice. If the work is a normal activity, the practice of not wearing metallic objects should be routine to avoid having to remember to remove them. Avoiding the use of conductive items in the vicinity of electricity should extend to any object which might come into contact with the circuit. Many tools must be metallic, but any tool used in electrical work should have insulating materials in those areas normally in contact with the hands.

Interlocks should never be bypassed by the average laboratory worker. If it is necessary to do so, the decision to do so and the maintenance work should be done by a trained maintenance person. Bypassing should not be a decision permitted for an inexperienced graduate student. Whenever an interlock is bypassed, a definite procedure requiring positive confirmation that all personnel are no longer at risk must be adopted and in place. This may involve actual locks for which only the responsible person has a key, tags which cannot be removed without deliberately breaking a seal or an alarm, or a combination of the above. However, no preventive procedure should depend upon the continued functioning of a single device, such as a microswitch, which may fail in such a way as to defeat the alarm or interlock.

Each circuit should be clearly identified and labeled to correspond to a circuit breaker in a service panel. Access to these panels should be provided to most laboratory employees, but they should not be permitted to remove the covers protecting the wiring. No closet containing an electrical service panel should be used for storage purposes. Access to the panel should not be blocked by extraneous items, and accidental contact with the wires should not be possible.

One of the most effective safety practices, as well as one highly conducive to productivity, is a definite scheduled program of preventive maintenance. Each item of equipment should be periodically removed from service, carefully inspected, calibrated, any faults or indications of deterioration repaired, and tagged with the date of review and the name of the maintenance person, if more than one technician could have been responsible. A permanent file or maintenance log on each major item of equipment is useful for identifying trends or weak components.

Finally, in a facility in which electrical injuries are a reasonable possibility, it is strongly recommended that at least some permanent personnel be trained in CPR and the measures to be taken should a person receive a severe shock. Individuals should also be trained to effect a rescue without themselves becoming a casualty. If, for example, a live wire is lying across a person and the circuit cannot be readily broken, they should be instructed to find a meter stick or some other insulated device to lift the wire from the victim or use rubber gloves or other insulators in attempting to free a person from a circuit.

### 4.6.2.15.2. Resistive Heating

This is one of the two major electrical sources of ignition of flammable materials in a laboratory, the other being sparks. Electrical heating can occur in a number of ways — poor connections, undersized wiring or electrical components (or, alternatively, overloaded wiring or components), or inadequate ventilation of equipment.

Equation 2 in the previous section shows that the power or heat released at a given point in a circuit is directly proportional to the resistance at that point. A current of 100 mA through a connection which has a resistance of 0.1 Ω would generate a localized power dissipation at that point of only 10 mW while a poor connection of 1000 Ω resistance would result in a localized power dissipation of 100 W. The former would probably cause little problem, while the latter might raise the local temperature high enough to exceed the ignition temperature of materials in the vicinity. Poor or loose connections have, in fact, caused many fires due to just such localized heating. An alligator clip used to attach a grounding wire is a good example of a poor connection. Similarly, a contact which has been degraded by a chemical, a wire that has been insecurely screwed down, or the expansion and contraction of a wire such as aluminum may, in time, result in this kind of problem.

Overheating of switches, fixtures, and other electrical components can be avoided very simply by reading the limitations usually embossed on the item and complying with the limitations. If a switch is rated to carry 7 A at 120 V, it will not survive indefinitely in a circuit in which it is carrying 30 A.

Each size or gauge wire is designed to carry a maximum amount of current. This is based on the voltage drop per unit length and the amount of power dissipated in the wire. The voltage drop should not exceed 2 to 5% due to wiring resistance. A 5% drop in a 120-V circuit would mean an actual voltage at the item of equipment of only 114 V. The heat developed in an overloaded circuit may heat the wiring to a point where the insulation may fail or, in extreme cases, actually catch on fire. Even moderate overheating, continued long enough, will probably cause an eventual breakdown in the insulation. In addition, any energy dissipated in the wiring is wasted energy.

Table 4.17 gives the maximum current for copper wire of various sizes, the resistance, voltage drop, and power loss per 50 ft (about 15.25 m) of line (the latter two values are computed for a wire carrying the maximum rated current). Most inexpensive extension cords purchased at a department store are made of either 16- or 18-gauge wire. As can be seen from the table, inexpensive extension cords do not carry sufficient current to provide power to more than a few instruments at most when properly used. Overloading them will cause a larger voltage drop and power dissipation (heating) in the wire. Although extension cords made of wire which is too

TABLE 4.17
Electrical Characteristics of Wire per 50 Feet

| Wire size | Resistance ($\Omega$) | Maximum amperes | Voltage drop (V) | Power loss (W) |
|-----------|----------------------|-----------------|------------------|----------------|
| 18 | 0.3318 | 7 | 4.6 | 32 |
| 16 | 0.2087 | 10 | 4.2 | 42 |
| 14 | 0.1310 | 15 | 3.9 | 59 |
| 12 | 0.0825 | 20 | 3.3 | 61 |
| 10 | 0.0518 | 30 | 3.1 | 93 |
| 8 | 0.0329 | 40 | 2.6 | 105 |
| 6 | 0.0205 | 55 | 2.3 | 124 |
| 4 | 0.0129 | 70 | 1.81 | 127 |
| 3 | 0.0103 | 80 | 1.64 | 131 |
| 2 | 0.00809 | 95 | 1.53 | 138 |
| 1 | 0.00645 | 110 | 1.42 | 156 |
| 0 | 0.00510 | 125 | 1.27 | 159 |

small probably will not fail immediately in most applications, they are not suitable for continued use. The lower available voltages can result in damage to equipment or failure of relays in control circuits as their magnetic fields become weaker.

Additional electrical load is a problem for extension cords as well as permanent wiring if multiple outlet plugs are used in a socket. Virtually every safety professional has at least one photograph or slide of several multiple plugs plugged into each other, all drawing current from a single socket. The result will usually be an overheated fixture and wiring as well as a lower voltage. This process can continue all the way back to service panels and to the power supply to the facility. Examples of breaker panels almost too hot to touch are, unfortunately, fairly common. Low overall supply voltages are becoming common in older facilities as the larger electrical loads of laboratories replace the lesser loads of classrooms in many academic institutions. Table 4.18 lists a number of common laboratory devices with the current and power requirements for representative units. Note that some of these normally require 208- to 240-V circuits. Some may also require connection to three-phase current, which, if not done correctly, can result in an accident or, at best, poor performance (a relatively common mistake is to wire a three-phase motor so that it operates backwards). A fume blower miswired in this way would result in minimal exhaust velocity.

Devices with resistive heating elements such as furnaces, heat guns, hot plates, and ovens should be configured in such a way that personnel cannot come into contact with an electrically active element, nor should volatile solvents be used in the proximity of such devices (or in them, as in an oven), where the temperature will exceed the ignition temperature of the solvent.

### 4.6.2.15.3. Spark Ignition Sources

Induction motors should be used in most laboratory applications instead of series-wound electric motors, which emit sparks from the contacts of the carbon brushes. Sealed explosion-proof motors can also be used, but are expensive. It is especially important to use nonsparking motors in equipment which result in substantial amounts of vapor, such as blenders, evaporators, or stirrers. Equivalent ordinary

**TABLE 4.18**
**Electrical Requirements of Some Common**
**Laboratory Devices**

| Instrument | Current (A) | Power (W) |
|---|---|---|
| Balance (electronic) | 0.1—0.5 | 12—60 |
| Biological safety cabinet | 15 | 1,800 |
| Blender | 3—15 | 400—1,800 |
| Centrifuge | 3—30 | 400—6,000 |
| Chromatograph | 15 | 1,800 |
| Computer (PC) | 2—4 | 400—600 |
| Freeze dryer | 20 | 4,500 |
| Fume hood blower | 5—15 | 600—1,800 |
| Furnace/oven | 3—15 | 500—3,000 |
| Heat gun | 8—16 | 1,000—2,000 |
| Heat mantle | 0.4—5 | 50—600 |
| Hot plate | 4—12 | 450—1,400 |
| Kjeldahl digester | 15—35 | 1,800—4,500 |
| Refrigerator/freezer | 2—10 | 250—1,200 |
| Stills | 8—30 | 1,000—5,000 |
| Sterilizer | 12—50 | 1,400—12,000 |
| Vacuum pump (backing) | 4—20 | 500—2,500 |
| Vacuum pump (diffusion) | 4 | 500 |

household equipment or other items such as vacuum cleaners, drills, rotary saws, or other power equipment are not suitable for use in laboratories where solvents are in use. Blowers used in fume exhaust systems should at least have nonsparking fan blades, and in critical situations with easily ignitable vapors being exhausted, it may be worth the additional cost of a fully explosion-proof blower unit.

Any device in which an electrically live circuit makes and breaks, as in a thermostat, an on-off switch or other control mechanism is a potential source of ignition for flammable gases or vapors. Special care should be taken to eliminate such ignition sources in equipment in which the vapors may become confined — as already discussed for refrigerators and freezers, but also possible in other equipment such as blenders, mixers, and ovens — or the units should not be permitted to be used with or in the vicinity of materials which emit potentially flammable vapors.

## REFERENCES (SECTIONS 4.6.2.15 TO 4.6.2.15.3)

1. Occupational Safety and Health Administration, General Industry Standards, Subpart S — Electrical, 29 CFR 1910.301 to 1910.339.
2. *National Electrical Safety Electrical Code,* ANSI C2, American National Standards Institute, New York, 1981.
3. *Prudent Practices for Handling Hazardous Chemicals in Laboratories,* National Academy Press, Washington, D.C., 1981.
4. **Lockwood, G. T.,** Protective lockout and tagging of equipment, in *Handbook of Laboratory Safety,* 2nd ed., Steere, N. V., Ed., CRC Press, Boca Raton, FL, 1971, 511.
5. Electronic Industries Association, Grounding electronic equipment, in *Handbook of Laboratory Safety,* 2nd ed., Steere, N. V., Ed., CRC Press, Boca Raton, FL, 1971, 516.
6. **Dalziel, C. F.,** Deleterious effects of electrical shock, in *Handbook of Laboratory Safety,* 2nd ed., Steere, N. V., Ed., CRC Press, Boca Raton, FL, 1971, 521.

7. **Ehrenkranz, T. E. and G. W. Marsischky,** Electrical equipment, wiring, and safety procedures, in *Handbook of Laboratory Safety,* 2nd ed., Steere, N. V., Ed., CRC Press, Boca Raton, FL, 1971, 528.

8. **Ehrenkranz, T. E.,** Explosion-proof electrical equipment, in *Handbook of Laboratory Safety,* 2nd ed., Steere, N. V., Ed., CRC Press, Boca Raton, FL, 1971, 540.

9. Controlling Electrical Hazards, OSHA-3075, Occupational Safety and Health Administration, U.S. Department of Labor, Washington, D.C., 1983.

10. Electrical Hazard, Safety Manual No. 9, National Mine Health and Safety Academy, U.S. Department of the Interior, Washington, D.C., 1976.

## 4.6.2.16. Glass*

Glass is involved in a large percentage of laboratory accidents. Glass reagent bottles may be dropped and shattered. Flasks may explode or implode due to pressure differentials. Tubing may break while being handled, causing cuts which can be severe. Glass systems may fail due to stress. Individuals may burn themselves handling hot beakers or flasks, or while trying to fabricate glass items.

Nearly all the glasses used in the laboratory are based on silica as the structure-determining oxide. The silica tetrahedron has four atoms of oxygen and one atom of silicon; the oxygen atoms are shared with, and bonded to, neighboring silicon atoms. We can classify silica glasses into several large groups, according to their composition: silica glass, soda-lime glass, lead alkali glass, borosilicate glass, and aluminosilicate glass[o].

### Silica Glass

Silica glass is the most important of the single-oxide glasses. In many respects, pure vitreous silica can be considered to be an ideal glass. It has a high use-temperature, a very low coefficient of thermal expansion, and very low ultrasonic absorption. It is an excellent dielectric. Its chemical durability and stability to weathering are very good. It is resistant to radiation. Silica glass is used in ultrasonic delay lines, as supersonic wind tunnel windows, in various optical systems, and as crucibles for growing germanium or silicon crystals. Its high use-temperature results from its extremely high viscosity at high temperatures, which makes it a difficult glass to produce or to fabricate.

Some of the difficulties of manufacture of silica glass — fused silica — have been overcome in two ways: (1) a method of producing $SiO_2$ glass, other than of melting selected crystals of quartz, was devised — a vapor deposition of the reaction product of $SiCl_4$ and water; and (2) a glass was developed, containing at least 96% $SiO_2$, whose properties are comparable in many respects to those of $SiO_2$ itself. Products made from this material are sold under the trademark VYCOR, and are made by leaching a glass which is primarily $SiO_2$ and $Ba_2O_3$, after fabrication, to remove all but the last traces of $Ba_2O_3$. The new glass is nearly as refractory and nearly as low in expansion as $SiO_2$.

### Soda-Lime Glass

For nearly all commercial glasses, fluxes are added to reduce the high viscosity

* Much of this section was excerpted from an article by Smith[7] in the second edition of this handbook.

inherent in silica glass and to bring glass manufacture into the range of industrially accessible temperatures and refractories. The most usual flux is soda, $Na_2O$. The addition of alkali "softens" the glass, i.e., reduces its viscosity at high temperatures, by breaking Si-O bonds. The addition of alkali also increases solubility, and sodium silicate glasses form the basis of the soluble silicate industry. Stabilizing oxides are added to decrease greatly the solubility of the sodium silicates. Calcium oxide is a cheap and effective stabilizer. In the soda-lime-silica system, the optimum glass with respect to cost, durability, and ease of manufacture is the composition silica — 72%, soda — 15%, lime and magnesia — 10%, alumina — 2%, and miscellaneous oxides — 1%. The miscellaneous oxides result from, and enable the use of, cheap raw materials. With only slight variations, this composition is used throughout the world to produce flat glass, containers, and incandescent lamp envelopes. It accounts for about 90% of all glass made.

**Borosilicate Glass**

Addition of fluxes, as discussed above, to the silica network always increases its expansion coefficient. This effect is reduced by the addition of $Ba_2O_3$, another glass former. Thus, stable and workable glasses can be produced which retain enough of the low-expansion characteristic of $SiO_2$ to be classed as heat resisting. This fact, together with their generally good chemical durability, makes borosilicates desirable for such uses as cooking or baking ware, laboratory glassware, reagent bottles, and telescope mirrors. In the making of complicated laboratory apparatus, the low expansion coefficient reduces strain development and risk of breakage during fabrication. Code 7740 is an example of a borosilicate glass: products made from it are sold under many trademarks, including HYSIL, PHOENIX, DURAN 50, SIMAX, KIMAX, and K-33 as well as the familiar PYREX.

### 4.6.2.16.1. Glassware

Laboratory glassware should never be used for beverages or food containers, even if properly marked. The risk of confusion and the potential for serious injury is far too great for this to be an accepted practice. This should be a firm rule of good laboratory practice.

Virtually all glass laboratory containers should be made of borosilicate glass, except containers required for unusual applications which may require glass with different characteristics. Glass containers or systems which may undergo especially demanding heat stress should be fabricated of the more expensive silica glass. Reagent bottles, tubing, stirring rods, and similar glass items which usually will not be exposed to extremes of temperature are more commonly made of soft or soda-lime glass. Many laboratory items which were formally made of glass are now available in various plastics, and they are often used where the application permits. They have the distinct advantage of being unbreakable.

Spherical containers are stronger than glass items of comparable wall thickness which have different shapes. Larger items, especially if they are nonspherical, should be made with heavy walls to withstand the pressures to which they might be exposed. It may be desirable to tape glass containers or portions of equipment under pressure so that, should they implode or explode, the glass pieces will not fly away as dangerous shards of glass. Metal enclosures or screens will also protect nearby personnel. Reagent bottles of most hazardous materials can now be bought in most popular sizes clad in a thin film of plastic to protect users from the more serious consequences of dropping a glass container. Containers not so protected should be transported only in resilient rubber or plastic carriers.

Broken glass is a common result of many laboratory accidents. Generally, custodians should not be expected to clean up broken glass, especially if it is contaminated with chemicals. Custodial workers are rarely trained to clean up a chemical spill, and such responsibilities should not be assigned to them. Common practice today is for custodians to transfer ordinary trash into plastic bags, which are then taken to a centralized collection point. This type of container provides no protection to a person handling the bag, and a person not expecting sharp glass in the trash could be cut badly. All broken glass which is suitable, i.e., uncontaminated, should be placed in a sturdy corrugated cardboard box, taped shut, clearly labeled "Broken Glass", and placed beside the ordinary trash for the custodians to pick up.

Cuts from mishandling glass tubing, especially while attempting to insert tubing into a rubber stopper, hose, or piece of plastic tubing, are common. Unfired broken glass is liable to be extremely sharp and it is all too easy to cut one's hand badly when tubing breaks while attempting to force it through a stopper. In order to safely insert a piece of tubing (or a glass thermometer) into a hole, the procedures given below should be followed.

1.   The hole should not be too small. The hole should be just large enough to grip the tubing. If a new hole is to be cut in a stopper, the borer should be one size smaller than one which will just barely slip over the tubing. The hole should be clean and regular. Cutting of good holes is facilitated by using a well-sharpened borer, slicing through the material with a smooth rotary motion, while exerting a firm steady pressure on the borer. A borer can cut the hands as well as the stopper, so care needs to be exercised. Hold the stopper with the fingers only, and do not wrap the entire hand around it. Lubrication of the borer helps cut a clean hole.

2.   The edges of the glass tubing being inserted should be fire polished. An unpolished edge will tend to dig into the sides of the stopper. It will be more difficult to push through, and may damage the sides of the hole sufficiently to cause a poor seal.

3.   Lubricate the glass tubing with water, glycerol, stopcock grease, or other available lubricant.

4.   Wrap a cloth around the glass. Either wrap the hand holding the stopper or hose with another cloth, such as a piece of toweling, or put on a leather work glove.

5.   Grasp the glass tubing at a point within 1 to 2 in. of the end to be inserted into the hole.

6.   Push the end of the glass into the hole, while exerting moderate pressure, with a slight twisting motion. Do not attempt to push or twist too vigorously since this may lead to exerting sufficient lateral force on the glass to break it.

7.   The distance through which the glass must be inserted will vary according to the circumstances. If the object penetrated is a stopper, the distance needed beyond the stopper will be governed by the apparatus. If inserting a piece of glass tubing into a small piece of plastic tubing, a few centimeters may be sufficient, while a larger-diameter tube may need to be inserted up to 10 cm or

more. Until the job is completed, continue protecting your hands and exerting the force in the same manner.

It is also possible to cut oneself while deliberately breaking glass. In order to break a section of tubing, a nick extending about a third of the way around the circumference should be scored with a sharp file, preferably in a single smooth stroke. Multiple scribe marks will often lead to a jaggedly broken piece of tubing. To break the tubing, a cloth should be wrapped around the tubing, the thumbs placed against the covered glass on the side opposite the cut, and then pressure exerted on the glass with the thumbs.

It is relatively easy to heat and bend pieces of glass tubing to achieve a desired shape or to draw out a piece of tubing to diminish the internal diameter of the tube. Working glass to create complicated shapes for laboratory applications, however, is best left to the professional glassblower. For the shapes to be free of strains which could cause the glass to fail under stress, all of the components should be properly annealed, which requires experience and the correct equipment to do properly.

Two special health problems are associated with glass. When a glassblower is configuring a glass component, he frequently employs a glass lathe which has some asbestos components. The asbestos tends to become quite friable with time, and individuals working near the lathe can be exposed to airborne asbestos fibers. Alternative insulating materials such as Zetex™ should be used or the glass worker and nearby personnel should wear appropriate respiratory protection. The use of a localized exhaust to remove the asbestos fibers is usually not practical since the moving air will cause problems with the work.

The second problem is primarily associated with the construction of relatively large components of silica glass. The temperature at which fused silica can be conveniently worked is about 1580°C (2876°F), while ordinary borosilicate glass can be worked at a temperature nearly 500°C (900°F) lower. The formation of nitrogen dioxide from the air is significant at the higher temperatures involved in fabricating silica components.

Both of these problems can be alleviated by ensuring that the glass-blowing shop or work area is well ventilated. Because heat rises, canopy hoods placed a few feet directly above the work stations will be effective in trapping the gases, fumes, and even particulates entrained in the rising hot air. This is a major exception to the previous recommendation against canopy hoods.

Glass under strain is very vulnerable to sharp blows, but strains can also release spontaneously, especially with rising temperature. In some instances, the release will cause the glass to crack. Systems should be put together to minimize additional mechanical strains placed on the components. The strains in glass can also release spontaneously for no apparent reason. In one instance in which the author was personally involved, a gallon jug of a hazardous chemical was delivered to the secretaries in a research-oriented department. No one was at the laboratory to receive it, so it was left in the departmental mail room. About three hours later, the jug spontaneously split and the chemical ran out, dissolved the plastic protective material, ran into the carpet and floor underneath, and dripped through into an office area below, damaging the floor there as well. The material was toxic, so evacuation was necessary; the cleanup required working in protective clothing and the use of respiratory protection. The direct cost to repair the damage was around $2000 to $3000, with an additional, comparable cost in manpower. Fortunately, only a handful of persons were in the building at the time, and no one suffered an acute exposure.

Glass systems under either positive or negative pressure are very dangerous due to the inability of the glass to withstand impact. The force on the surface of an evacuated container may be very large. For example, the net force on an evacuated spherical container 10 cm (4 in.) in diameter is equal to the weight of an 81-kg (179-lb) mass. There is essentially no difference in this force for a container at a good vacuum and a poor one. The force exerted on a container with a poor vacuum equivalent to 1 mmHg is within 0.13% of the force on a container with a vacuum several orders of magnitude higher. The forces on high-pressure systems can, of course, be much higher than at atmospheric pressures. Pressures in gas cylinders can be 200 times atmospheric pressure or more and, hence, should never be attached to most glass research systems, except through a regulator valve. The existing forces on pressurized glass systems make them unusually vulnerable to other factors, such as a sharp blow, which they might otherwise withstand. A warning sign should be placed on systems under pressure to alert personnel of the unusual danger.

Anything that weakens the strength of the glass in a system under pressure increases the risk of either an implosion or explosion. Some laboratories use metal evaporating systems under a large bell jar to deposit metal films. A current is passed through a small crucible or "boat", raising its temperature to a level sufficient to evaporate a metal placed within it. To obtain a good deposition of the metal, an appropriate vacuum needs to be established. Although there usually is not sufficient air in the bell jar to transfer heat by either conduction or convection, radiative transfer does occur. To obtain good evaporation, in most cases the metal must be heated to a brilliant white-hot temperature — approximately 1500°C (2700°F) or better, well above the point at which borosilicate glass begins to soften. If the temperature is maintained at this level for a substantial length of time, the strength of the glass bell jar can be impaired, with a resulting implosion. Evaporating systems should always be enclosed within a sturdy wire cage during use.

## REFERENCES (SECTIONS 4.6.2.16 TO 4.6.2.16.1)

1.  **Parr, L. M.,** *Laboratory Glass-Blowing,* Chemical Publishing Company, New York, 1957.
2.  **Robertson, A. B. J. et al.,** *Laboratory Glass-Working for Scientists,* Academic Press, New York, 1957.
3.  *Guide for Safety in the Chemical Laboratory,* D Van Nostrand, New York, 1954.
4.  **Lewis, E. J.,** Proper care will prolong the life of chemical glassware, *Chem. Eng. News,* 21, 552, 1943.
5.  Laboratory Glassware, Data Sheet No. 23 (revised), National Safety Council, Chicago, 1964.
6.  *Safety in Academic Chemistry Laboratories,* 4th ed., American Chemical Society, Washington, D.C., 1985, 21.
7.  **Smith, G. P.,** Glass, in *Handbook of Laboratory Safety,* 2nd ed., Steere, N. V., Ed., CRC Press, Boca Raton, FL, 1971, 544.

### *4.6.2.16.2. Glassware Cleaning*

Cleaning laboratory glassware is an essential part of laboratory procedures. In some cases, a simple cleaning with soap and water is sufficient, but in many cases chemical cleaning is necessary. In the life sciences, contact with contaminated glassware could allow an individual to contract a contagious disease. Biologically contaminated glassware should be autoclaved and/or dunked in a sterilizing bath before further cleaning. In some larger laboratories, glassware is washed in a central location, and in such facilities, it is the responsibility of the individuals using the glassware to insure that workers in the washing facility are protected from materials still in or on the containers and utensils.

Most smaller chemical laboratories still clean their own glassware in the laboratory sink. The following short article (with very minor editing) by Tucker,[7] Acid Cleaning of Glassware (second edition) provides useful guidelines on acid cleaning of glassware. It also provides a useful example of how to prepare a formal hazard analysis for a simple, but potentially hazardous, laboratory procedure.

## ACID CLEANING OF GLASSWARE

The clinical or research laboratory has many complex safety problems, including the common practice of safety factors involved with procedures used in the acid cleaning of glassware. The major component in most acid cleaning solutions is sulfuric acid, and many laboratory personnel have experienced at least one minor acid burn. One survey of 3 year's experience in a university hospital clinical laboratory showed that 25% of the laboratory injuries were caused by acid or alkali, and a Manufacturing Chemists' Association publication lists 14 case histories of injuries caused by sulfuric acid.

It has been suggested that a job hazard checklist be used to determine the safe performance of any job by an individual, and this technique has been applied to acid cleaning of laboratory glassware. This job hazard analysis consists of a detailed listing of key job steps, the tools or equipment used, the potential health and injury hazards which may be involved, and the safety practices or equipment which may be needed to control the hazards.

**Job Hazard Analysis for Sulfuric Acid Cleaning of Laboratory Glassware**
1. Job description: Provide clinically clean glassware.
2. Job location: A sink within the laboratory area.
3. Key job steps:
    a. Prepare acid cleaning solution — sodium dichromate and concentrated sulfuric acid.
    b. Prepare glassware for acid cleaning by rinsing, soaking, or other means.
    c. Immerse glassware in acid or provide contact of the acid by rotation on the interior of the glassware.
    d. Rinse glassware with tapwater.
    e. Rinse glassware with deionized or distilled water.
    f. Oven dry the glassware.
    g. Dispose of contaminated acid.
4. Tools or equipment used:
    a. Carrier for acid bottle.
    b. Safe container for acid solution.
    c. Carrier for pipettes, other containers.
    d. Basket or rack for test tubes.
    e. Porcelain or stainless steel dipper for acid solution.
    f. Washer for rinsing pipettes.
5. Potential health and injury hazards:

a.  Sulfuric acid can splash or spill in routine handling, or if containers break and cause severe burns on contact with body tissues.

b.  Sulfuric acid mist can present an inhalation hazard if inhaled in concentrations greater than one milligram/cubic meter (current OSHA PEL) for an eight-hour exposure.

c.  Sulfuric acid can cause a fire hazard if hydrogen is liberated by the action of the acid on metals.

d.  Sulfuric acid can be highly corrosive to many metals and alloys, so sink traps and waste lines can create a safety hazard to workers and plumbers.

e.  Sudden dumping of several liters of concentrated sulfuric acid into the drain lines can cause the release of toxic hydrogen sulfide gas into laboratories through dry traps in waste lines or floor drains.

6.  Safety practices, apparel, and equipment:

a.  Safety goggles

b.  Face shield

c.  Rubber gloves

d.  Rubber apron

The job hazard analysis outlined above provides a basis for a thorough study of the entire procedure. Several physical modifications could be introduced which would facilitate safer performance by the worker.

The first of these physical modifications would be to provide a lead-lined or stainless steel double sink. One half of this double sink should be considerably deeper to house the acid container and automatic pipette washer, as shown in Figure 4.2. A perforated copper tubing connected to a cold water supply is fastened several inches below the top of the sink to provide a constant stream of water to flush away any excess acid. The trap and waste line should be of a borosilicate glass drain line construction material.

The second physical modification to make acid cleaning of glassware safer would be the use of the "Movable Safety Shield" as shown in Figure 4.3.

The shield is mounted on a track and is designed to be of any appropriate linear measurement. The entire track can be permanently mounted to a bench or semipermanently secured by the use of clamps. The worker should roll the safety shield between himself and the acid cleaning solution at all times.

Additional physical requirements should be as follows:

1.  The suggested level of ventilation for all laboratory areas should be a minimum of ten changes of air per hour. [Elsewhere in this volume, six air changes per hour of fresh air is recommended as a compromise between economy and air quality for most laboratories, with high-hazard or unusual hazard facilities requiring more.]

2.  The lower level of light available for all laboratories is 200 foot candles (2152.8 lux). [Building code requirements are for a minimum of 6 fc or 64.58 lx 30 in. or 76.2 cm above the floor.]

3.  A safety shower and floor drain should be located immediately behind the operator. [A safety shower and easy access to it should not be obstructed.

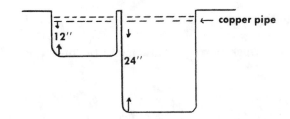

FIGURE 4.2

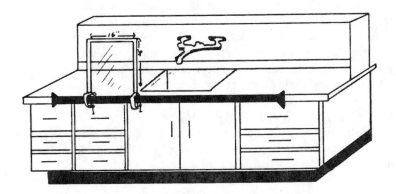

FIGURE 4.3

However, it does not have to be immediately behind the worker. If the sink is equipped with an eyewash station, the shower should be over the worker's head. Also, a floor drain while desirable, is not essential. Water can be mopped up.]

4.    A spray or hose should be available, with easy access, at the sink area.

5.    An eyewash fountain should be readily available. [See comment on 3 above.]

## REFERENCES (SECTION 4.6.2.16.2)

1. **Ederer, G. M. and Tucker, B.,** Accident surveys and safety programs in two hospital clinical laboratories, *Am. J. Med. Technol.,* 26, 219, 1960.
2. *Case Histories of Accidents in the Chemical Industry,* Vol. 1, Manufacturing Chemists' Association, Washington, D.C., 1962.
3. **DeReamer, R.,** *Modern Safety Practices,* John Wiley & Sons, New York, 1958, 47.
4. Chemical Safety Data Sheet SD-20, Manufacturing Chemists' Association, Washington, D.C., 1963.
5. **Rappaport, A. E.,** *Manual for Laboratory Planning and Design,* College of American Pathologists, Chicago, 1960, 75.
6. **Steere, N. V., Ed.,** Laboratory design considerations. II. Safety in the chemical laboratory, *J. Chem. Ed.,* 42, A666, 1965.
7. **Tucker, B.,** Acid cleaning of glassware, in *Handbook of Laboratory Safety,* 2nd ed., Steere, N. V., Ed., CRC Press, Boca Raton, FL, 1971, 557.

### 4.6.2.17 Distillation Units*

### The Safe Operation of Laboratory Distillations Overnight

#### INTRODUCTION

Scientific laboratories differ greatly in the extent to which laboratory distillations are permitted to run unattended during the evening hours. Most scientists consider the familiar "distilled water" still to be free of trouble. Many laboratories have had such stills operating day and night for years with the very minimum of attention. However, other distilling units are often regarded as hazardous either because of their construction or because the material being distilled is toxic or flammable.

If these other stills are examined carefully, the sources of potential trouble can be eliminated, and these stills too can be operated into or through the evening hours. [The location, at least, of "trouble-free" water stills also needs to be reviewed in many laboratories. A serious fire occurred in the editor's institution due to a water still overheating and setting some nearby combustible materials on fire.]

#### FAILURE OF THE COOLING WATER

Most distillation units operate with water-cooled condensers; therefore, it is essential for safe operation that the water supply be dependable. However, water pressures change:

1.  There are decreases in water pressure which may be caused by failure of a pump, by partial blocking of the mains, by increased consumption by other laboratories on the same water main, by water main breaks, by maintenance personnel working on the system, lawn sprinkling, or water department use.
2.  There are increases in water pressure which may occur when auxiliary pumps are being switched on, usually at the beginning of the working day; when other laboratories turn off their water valves, usually at the end of the day; or at other times during the day or evening when water system pressures increase because of decreased consumption by many or major users.

Either type of pressure variation can cause trouble. Too low a pressure results in an inadequate water flow to the cooling condenser, thereby allowing distillate vapors to escape. Too high a pressure can cause the tubing connections to the condensers to expand and burst, which is doubly undesirable: the laboratory is flooded, and vapors may be released since there is no flow of cooling water to the condenser.

It is not difficult to protect against both hazards if the following four-point approach is used:

1.  Install a simple pressure regulator in the water main ahead of the valve that is used to adjust the flow to the condenser. The regulator can be adjusted to maintain an intermediate water pressure; it will then minimize the effect of both increases and decreases in the supply pressure.

---

* Section 4.6.2.17 is a slightly edited version of an article that appeared in the *Handbook of Laboratory Safety,* 2nd edition.[7]

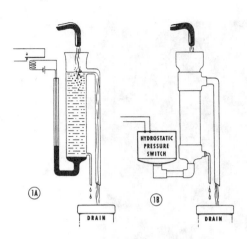

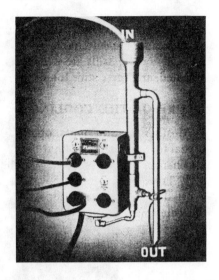

FIGURE 4.4. Cooling water-flow monitors (1A glass, 1B metal). When the flow ceases, the column drains through the capillary, and the mercury contact or the pressure switch opens.

FIGURE 4.5. Water-flow monitor equipped with alarm and convenience outlets.

2.  Protect the pressure regulator with a suitable water filter, which will prevent particles of rust and other foreign materials from affecting its operation. Commercial filters with replaceable cartridges made of cotton or polypropylene twine are readily available.

3.  Monitor the flow of water flowing from the cooling condenser to the drain. Monitoring this emergent water flow ensures that water has flowed through the condenser. Thus, it is inherently a more reliable approach than monitoring the main pressure.

4.  Install a solenoid valve in the water supply line and connect it electrically to the water-flow monitor so that electrical power to this valve and to the still-pot heater will be turned off when the water-flow monitor responds to a water failure.

A number of water-flow monitoring devices have been described in the literature. A modification of Houghton's device is shown in Figure 4.4A, made of glass, and in 4.4B, made of metal. This design has the advantage that it imposes no backpressure on the apparatus being monitored; it possesses a short time delay so that adjustments in the flow can be readily made; and its flow-restricting capillary can be readily inspected. [A number of commercial units are now available in virtually any general scientific supply catalog.]

Figure 4.5 shows a commercially available water-flow monitor based on this same principle. The commercial unit has a built in alarm plus a number of electrical receptacles, several of which are automatically turned off when the cooling water fails. The four-point approach recommended above is shown in Figure 4.6, items A, B, C, and D.

### FAILURE OF THE ELECTRICAL SUPPLY

It is important that the line voltage used for the distillation remain relatively

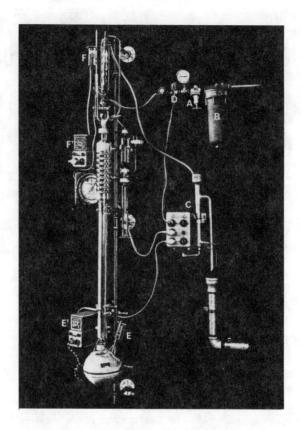

FIGURE 4.6.   Distillation column showing (A) water pressure regulator, (B) water filter, (C) emergent water-flow monitor, (D) solenoid valve in water line, and (EE′) (FF′) capacitance-actuated controllers for monitoring still-pot and still-head thermometers. (Glass equipment: Courtesy of Ace Glass Co., Vineland, New Jersey.)

constant since even moderate changes will affect the rate of distillation. However, electric voltages, like water pressures, change:

1.   There are decreases in line voltage due to increased consumption by other laboratories, usually early in the day. This will be most noticeable if the lines are inadequate. [Voltage levels from the power distributor may drop significantly due to excess demand placed on the generating and distribution system. This normally occurs during extremely hot or cold weather when either the air-conditioning or space-heating demand loads the system.]
2.   There are increases in line voltage when additional generating equipment is turned on, or when other laboratories turn off their electrically operated equipment at the end of the day.

The following two-point approach will guard against difficulties from these variations.

1.  As with water pressure, there is a convenient device for overcoming source fluctuations: the constant-voltage transformer. A 500-, 1000-, or 2000-W unit would be suitable. [Use the size which is appropriate to the electrical load. Constant-voltage transformers work best when nearly fully loaded.] Although bulky, the insurance it provides against fluctuations in line voltage greatly outweighs any inconvenience. These units are inherently reliable and trouble free.

2.  A constant-voltage transformer cannot cope with line voltage failure. Such failure in itself is not serious: the distillation merely stops. However, when the power is turned on again, "bumping" may well take place in the still-pot. If "bumping" is a serious problem, one should consider wiring a suitable relay into the still-pot heating circuit. Failure of the electrical supply should cause the relay to "drop out" and stay out until it is manually reset. [An alternative that is now available at a reasonable cost is the uninterruptible power supply. These devices switch over to a battery-operated supply until the power failure is cleared. The capacities available vary, but there are sizes which would carry even a heat-pot for periods from a few minutes to perhaps an hour. Most power interruptions are relatively short.]

An alternative way to achieve this same result is to plug the solenoid valve referred to previously into the water-flow monitor provided. A failure of the electric supply will cause the solenoid valve to close, shutting off the flow of cooling water and hence causing the water-flow monitor to de-energize its outlets. Since the solenoid valve is connected to one of these outlets, both it and the water flow-monitor are unable to reset themselves should the power failure be corrected. This approach has the advantage of having an inherent time delay — it will not be affected by momentary electrical cessations.

## TERMINATION OF DISTILLATION

A question that often arises in deciding whether to leave an unattended distillation running during the night is how to stop the distillation if it is not to run until morning. There are several methods; the choice depends upon the nature of the distillation.

*Timer Terminations.* If the distillation is similar to the distillation of a solvent, the still-pot and the still-head temperature will not change greatly during the course of a distillation. Consequently, if, as has been recommended, a constant-voltage transformer is used to guard against variations against line voltage, the distillation rate will be reasonably constant. With experience, one can, therefore, estimate rather closely the amount of distillate that should accumulate during the night, and the distillation can be safely terminated with an electric timer. Such timers are familiar tools in most laboratories.

*Temperature Terminations.* If the boiling range of the material is wide rather than narrow, the rate of distillation will not remain constant. In this case, it will be better to terminate the distillation at a predetermined temperature rather than at a predetermined time. This is a common situation, not only with distillations left running for days on end, but also with one-day runs that are not quite finished by "quitting time".

The scientist can readily terminate such a distillation through the use of a capacitance-actuated controller which senses the position of the mercury in a thermometer

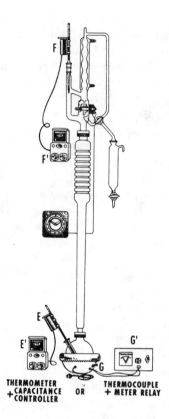

FIGURE 4.7.   Distillation column with still-head temperature being monitored for over-
temperature or under-temperature by FF′, and still-pot being monitored for over-temperature by either EE′ or GG′. Note: If FF′ is not used for safety monitoring, it may be used
to control the reflux ratio automatically by turning the reflux timer off whenever the still-
head temperature exceeds the set point.

located either in the still-pot or in the still-hood. This is shown in Figures 4.6 and 4.7,
EE′ and FF′. When the still-pot heater is plugged into the controller, the heat to the
still-pot will be cut automatically when the temperature rises to the desired value. (A
limit switch in the controller, which must be manually reset, keeps the power from
turning on again when the temperature drops.)

*Weight and Volume Terminations.* If one prefers to terminate the distillation on the
basis of the weight of the volume of the distillate, several other approaches may be
used:

1.   If the system need not be airtight and the product receiver can be supported on
     a pan balance, movement of the balance can be used to open an electrical circuit
     and terminate the distillation. When the balance moves, the heater is turned off
     (Figure 4.8). The weights used on the balance should allow for the force used
     to actuate the switch.
     This approach can be modified if the system must be airtight; flexible tubing
     can be run from the condenser to the product receiver mounted on the balance.

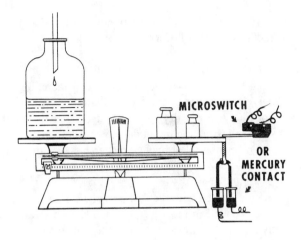

FIGURE 4.8. Termination of distillation by weight of product.

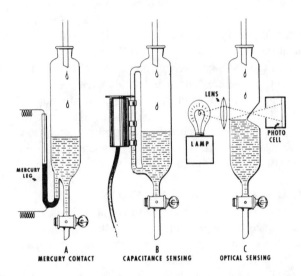

FIGURE 4.9. Termination of distillation by volume of product.

It may be necessary to run two flexible lines — one for liquid flow, the other for vent. Care should be used in selecting the tubing used. It must not be attacked by the materials being distilled.

If a volume termination is preferred, there are several choices — these include the use of mercury contacts, capacitance-actuated controllers, and optical means of detecting liquid level. Figure 4.9 shows three examples schematically.

3.  It is quite likely that automatic fraction collecting devices that have been developed for chromatographic purposes can be adapted for use with distilla-

tion columns. Termination of the distillation would be done on the basis of the number of fractions collected of a fixed volume.

## POSSIBLE OVERHEATING OF THE STILL-POT

Once the foregoing factors have been taken into consideration, the more subtle hazards should receive attention. Laboratory distillations involving many liters of raw material are not likely to run dry overnight. However, it is conceivable that sooner or later this unlikely situation might develop because of some oversight. The distilling apparatus can be monitored by an over-temperature or an under-temperature approach.

1a.  If a heating mantle is used with the still-pot, one can make use of the thermocouple that is embedded in most heating mantles. This can be attached to a suitable millivoltmeter equipped with an adjustable over-temperature sensing pointer (Figure 4.7, CC'). These meters, known as meter-relays, are commercially available. They should be ordered complete with auxiliary relays capable of interrupting heater currents of several amperes.

1b.  The still-pot itself can be monitored for over-temperature by the use of a thermoregulator plus an electronic relay or by a thermometer with a capacitance-actuated controller similar to that described above, but for monitoring rather than for termination (Figure 4.7, EE').

2.  In some cases, it is convenient to monitor the still-head temperature against a decrease in temperature. A decrease should occur if the liquid in the still-pot becomes quite small in volume; the *rate* of distillation will then decrease greatly and the still-pot temperature will drop. This monitoring of the still-head temperature can also be accomplished readily with either the thermocouple plus meter-relay or with the still-head thermometer plus a capacitance-actuated controller (Figure 4.7 FF').

## BREAKAGE AND PHYSICAL SEPARATION

In any safety discussion, it is important to consider the possible breakage of the glass equipment. Such breakage is highly unpredictable since it may result either from residual internal stresses in the glass, from stresses generated by improper external supports, or from an accidental blow.

Internal stresses in the glass apparatus may arise from inexperienced glass blowing or from careless commercial manufacture. *It is strongly recommended that all parts of the glass distillation equipment be carefully annealed and be checked for residual stress by means of polarized light.* Commercial polarized light units are available for this purpose, or a stress-analysis unit can be improvised from two sheets of Polaroid and a light source. Some large research laboratories inspect with polarized light all glass apparatus which they receive.

The method used to support the distillation column is of greatest importance in minimizing stress. Thought must be given to the design of the supporting framework, the method of clamping, and the design and location of the individual clamps.

A complete distillation column is usually rather tall and narrow. The column, head, and receivers can therefore be supported by a single support rod. The use of a rigid support rod and heavy-duty clamps and clamp holders is important in order to prevent one part of the apparatus from shifting with respect to the other parts. It is recommended that the rod be at least 1.90 cm ($^3/_4$ in.) in diameter and securely fastened to a wall. The following procedures are suggested:

1.  The column itself is first firmly clamped either at its top, bottom, or at some enlarged portion. The clamp used must be strong enough to carry the full weight of the column and head.
2.  A second clamp is then placed on the column. This is tightened only slightly; it is intended as a guide against lateral movement.
3.  When the still-pot is attached to the bottom of the column, it can be held in place by a cradle supported by springs. The still-pot will then be free to align itself with the column.
4.  The distillation head is placed on the top of the column and is held loosely in position by a clamp which allows it to move vertically, but largely restricts lateral movement.
5.  The receivers are then supported with care in order to prevent them from putting strain on the distillation head. Careful support of this part as well as of the rest of the apparatus is essential in order to prevent physical parting of the ground joints.
6.  If the apparatus is extensive or if the receivers, when loaded, will be heavy, then it may be desirable to introduce semi-flexible connecting links between the column and the receivers. One way of doing this is shown in Figure 4.10. The links illustrated are easily fabricated from ground-glass ball-and-socket joints.

In passing, mention should be made of an alternative method of supporting distillation equipment: namely, to mount the assembly on a rigid panel made of metal or other non-combustible material. In this case, a large number of supporting brackets may be used to clamp all parts of the assembly. A glassblower may seal the parts together after they are supported, thus eliminating any residual stresses due to the clamping.

Finally, breakage due to accidental blows should be prevented by (1) locating the apparatus in a corner of the laboratory, out of the main line of traffic, and by (2) the use of adequate safety shields. Such shields may be flat, rectangular, or semi-cylindrical. Those shown in Figure 4.11 are hinged and particularly convenient to use.

**OVERFLOW**

Once a laboratory distillation is set up and allowed to run overnight, the operation may prove so convenient and satisfactory that the scientist forgets his initial caution. He might, at this time, forget to empty the receiver at the proper time, or might underestimate the distillation rate. A simple safeguard is to use a product receiver that is large enough to contain the entire charge to the distillation. If this is undesirable, one can use a smaller receiver, but connect it to a large overflow bottle. A 5- or 10-liter bottle equipped with a two-hole rubber stopper and a drying tube in the vent line is a convenient safeguard.

Even after these precautions have been taken, it is worthwhile to mount the entire

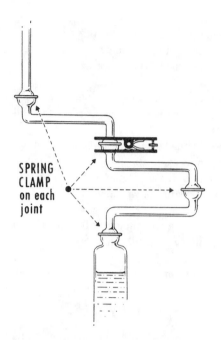

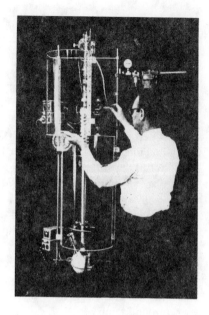

FIGURE 4.10.   Ball-and-socket linds in transfer lines can flex and hence eliminate any stress present.

FIGURE 4.11.   Modern acrylic shields protect apparatus against external blows, yet are easily opened for manipulation.

distillation apparatus in or over a metal tray large enough to contain all the liquid being distilled. A tray 8 cm (3.25 in.) deep is usually adequate and is not too cumbersome.

## MONITORING BY THE WATCHMAN

In large, well-staffed laboratories, the safety committee may insist that the night watchman, on his periodic trips through the laboratory, glance at any equipment that is running. Some scientists may feel that their "guard" has no feeling for scientific apparatus. Nevertheless, whenever apparatus is running at night, it is wise to have the apparatus so clearly labeled that an inexperienced, scientifically untrained guard could turn it off — preferably by turning no more than two or three conspicuously labeled switches and valves. [Most universities with large numbers of buildings and hundreds of laboratories usually lock the doors to the laboratories when unoccupied to reduce the chance of theft. In most cases, the watchman does not enter laboratories unless specifically instructed to do so, and few have the time to make many such inspections.]

## CONCLUSION

Many laboratories are quite cost conscious; other laboratories are hampered by shortages of personnel, while some are under great pressure to achieve results in a short period of time. Whatever the motivation, there are real advantages in being able to run laboratory operations, including distillations, in an unattended condition, provided this can be accomplished in a safe manner.

In safety programs, it is often difficult to agree on the exact sources of potential trouble and their relative importance. Hence, it is difficult at times to agree on which precautions are necessary and which are simply advisable. Nevertheless, there are few distillations that cannot operate unattended — not only during the night, but during the day as well — if careful plans are worked out along the lines suggested here.

## ACKNOWLEDGMENT

Reprinted from the *Journal of Chemical Education*, Vol. 43, A589 and A652, July and August, 1966, with permission.

Note: The original article provided a list of specific devices, identified by brand name and catalog number, which has been omitted here. Because of the length of time since the original article was published, many of these items have been superceded by newer models and companies have changed hands or ceased operations, so the reader is referred to catalogs of scientific supply houses to obtain equivalent equipment for the recommended devices.

## REFERENCES (SECTION 4.6.2.17)

1.  **Burford, H. C.,** Improved flow-sensitive switch, *J. Sci. Instrum.*, 37, 490, 1960.
2.  **Cox, B. C.,** Protection of diffusion pumps against inadequate cooling, *J. Sci. Instrum.*, 33, 148, 1960.
3.  **Houghton, G.,** A simple water failure guard for diffusion pumps and condensors, *J. Sci. Instrum.*, 33, 199, 1956.
4.  **Mott, W. E. and Peters, C. J.,** Inexpensive flow switch for laboratory use, *Rev. Sci. Instrum.*, 32, 1150, 1961.
5.  **Pike, E. R. and Price, D. A.,** Water flow controlled electrical switch, *Rev. Sci. Instrum.*, 30, 1057, 1959.
6.  **Preston, J.,** Bellows operated water switch, *J. Sci. Instru.*, 36, 98, 1959.
7.  **Conlon, D. R.,** The safe operation of laboratory distillation overnight, in *Handbook of Laboratory Safety*, 2nd ed., Steere, N. V., Ed., CRC Press, Boca Raton, FL, 1971, 227.

### 4.6.2.18. Control of Laboratory Processes*

### RUNNING LABORATORY REACTIONS UNDER SAFE CONTROL

### INTRODUCTION

Laboratory apparatus varies greatly in complexity — from the relatively simple thermostated ovens used for physical tests to quite complicated setups used for chemical operations and chemical reactions. Both scientists and safety engineers differ greatly in their attitudes toward apparatus and in their willingness to have apparatus running unattended at any time. Therefore, in some laboratories, practically every piece of apparatus is shut down at the end of the working day. In others, numerous pieces of equipment and even chemical reactions run unattended for either a few minutes, hours, or days.

* Section 4.6.2.18 is a slightly edited version of an article that appeared in the *Handbook of Laboratory Safety*, 2nd edition.[1]

The determining factors in deciding whether it is worthwhile to develop a set of conditions so that a laboratory apparatus can operate safely in the absence of the scientist [more likely a technician or graduate student at an academic institution] are

1. Is the operation repetitive and so demanding of close personal attention that it is quite tedious, although not lengthy? (Exothermic reactions and vacuum-stripping operations are examples.)
2. Is it a lengthy experiment requiring many hours or days to run?
   a. If the experiment is lengthy, are there real disadvantages in interrupting the operation?
   b. If it were to be left running, could it be terminated automatically at some *desired* point?
   c. Could it be terminated at any point should some malfunction develop?

If the answer to these questions is yes, then one should examine the apparatus, determining which parts should be "watched" or monitored automatically. In such an examination, one usually finds there are very few units so complicated that they cannot safely be left running under automatic control; but also one finds there are few units so inherently safe that they do not need some safety monitoring.

This section considers both the simple apparatus and the complicated apparatus, both the endothermic reaction and the exothermic reaction, both the short-term operation and the long-term operation, both "control monitoring" and "safety monitoring".

## SIMPLE ENDOTHERMIC EXPERIMENT

Figure 4.12 shows three apparatus types which from a safety point of view may be considered together: (A) the laboratory oven, (B) the constant temperature bath, and (C) the glass reaction flask. The assumptions made in grouping these together are: (1) only the variable of temperature need be "watched"; (2) should the temperature exceed a desired value, it can be detected by a monitor which automatically turns off the electric power; (3) since the operation is not exothermic, once the power is turned off, the apparatus will cool safely.

Apparatus such as ovens and constant temperature baths (Figure 4.12, A and B) are not usually considered to be hazardous. Therein lies the danger: they are overlooked in safety inspections. This is particularly dangerous in the case of oil baths. If the oil bath thermostat fails, the oil will overheat and a fire may start. (Even if no fire results, valuable apparatus may be damaged and irreplaceable samples may be lost.)

Figure 4.12D represents schematically any experimental apparatus which is heated by an electric heater of either an external or an immersion type. Electrical power is supplied to the heater cyclically through a temperature controller connected to a sensing probe in the apparatus.

As shown in Figure 4.13, to detect thermal malfunction of the apparatus in Figure 4.12, one may use a monitor to watch:

1. The temperature of the apparatus
2. The temperature controller

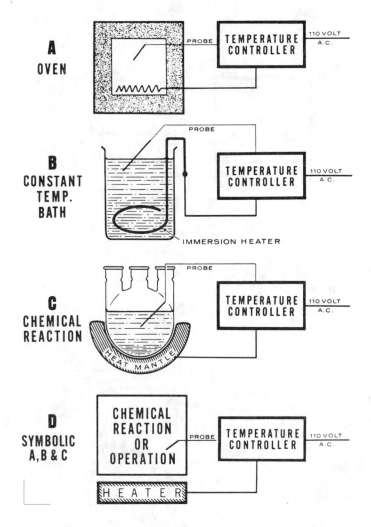

FIGURE 4.12. Examples of thermostated endothermic apparatus.

3. The temperature sensing probe
4. The power supplied cyclically to the heater
5. The temperature of the heater

One might reason that since the experiment is primarily concerned with the temperature of the apparatus (Item 1) this is what should be monitored. A good way of so monitoring is to use a secondary controller and probe that are independent of the primary temperature controller. If space is available, a relatively rugged bimetallic thermoregulator is ideal for the purpose. It should be set to operate 5° to 10° above the normal temperature of the apparatus and can be electrically connected to take over control of the temperature if the primary controller fails. In that event, it would hold the apparatus temperature at the 5° to 10° higher level. It may, on the other hand, be wired to a relay circuit (Figure 4.14) designed so that once the **THER-**

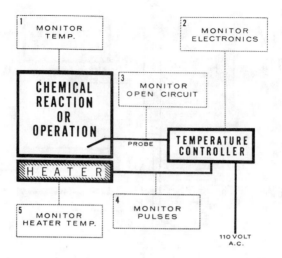

FIGURE 4.13.    Five possible ways to monitor an apparatus for malfunction of the temperature controller.

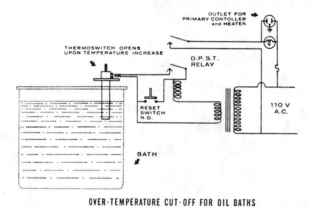

OVER-TEMPERATURE CUT-OFF FOR OIL BATHS

FIGURE 4.14.    Wiring diagram of an oven-temperature relay circuit. Reset switch must be pushed to re-energize circuit.

**MOSWITCH** opens, the relay will keep the power off until a push-button switch is reset. Figure 4.15 shows a commercially available "over-temperature guard" based on this design.

With smaller apparatus it may not be convenient to use a bulky bimetallic thermoregulator (although some are fairly small). If such is the case, one may use a thermocouple connected either to a pyrometer controller or to a meter relay. A thermistor plus a bridge-type controller may also be used.

If one prefers to insert no secondary probe into the apparatus, then one should either use an external sensor or should monitor items 2, 3, 4, or 5. In some cases (Item 2) the primary controller may be partially able to monitor itself. Pyrometer control-

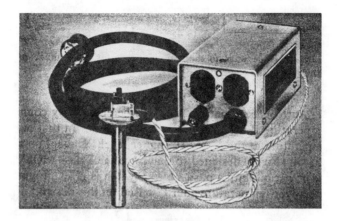

FIGURE 4.15. Compact relay unit based on circuit shown in Figure 4.14.

lers are often designed so that if certain internal components fail, the controller will cut off the heater power. (This is known as *Fail Safe* design.) A somewhat different approach is to make use of a pyrometer that has two independent control points. The second control point can be set at a higher level to sound an alarm, open a circuit, or take remedial action. Admittedly, the pyrometric approaches are not 100% *Fail Safe*, but they do help to guard against trouble. This is also the case when one monitors Item 3 — the sensing probe. This, again, is common practice with pyrometer controllers; they are usually designed so that the controller will turn off the heat if there is a thermocouple break.

Item 4 calls for monitoring the ON and OFF power cycles to the heater. This can be accomplished with a so-called "reset," or "interval," timer or a "time delay" relay. Such units are designed to open or close a circuit after a period of seconds, minutes, or hours. When connected parallel with the heater, one of these can sound an alarm or turn off the heat if for any reason the electric power remains continuously ON. This approach has the advantage in that it monitors the apparatus without physical contact with the apparatus.

Item 5 is sometimes the most convenient to monitor — particularly when one is using heating mantles into which the manufacturer has built a thermocouple. Again, one can attach the thermocouple either to a conventional pyrometer or to a meter relay.

Once the choice is made as to which approach (Items 1 to 5) one prefers, one then proceeds to the next decision: What action should the monitor take if an excessive temperature occurs?

1. Should the monitor permanently shut down the apparatus until an operator pushes a reset button?
2. Should the monitor allow the apparatus to continue to operate but at some slightly higher value?
3. Should the monitor give the disturbance a short time to disappear and take no action if the disturbance is only momentary?
4. Should the monitor make any record that a disturbance occurred?

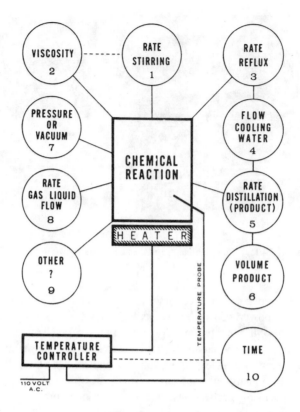

FIGURE 4.16. Schematic illustration of a complex reaction apparatus.

The above choices apply not only to the monitors described above but also to many of the monitors described in the following sections. For some situations, the monitor should shut down the apparatus rather than exert corrective action because normal operating conditions *cannot* possibly be restored. In other cases, however, the monitor should take corrective action because normal operating conditions *can* readily be restored.

## COMPLEX APPARATUS

While the scientist is almost always concerned with temperature, the apparatus is often more complicated and involves still more variables than the examples shown in Figures 4.12 and 4.13. Figure 4.16 illustrates schematically a more complicated setup. The ten circles symbolize the components and/or variables the scientist would naturally watch if he were personally present. They represent, in some cases, malfunction of the apparatus and, in other cases, the progress of the reaction. They are as follows:

1. The rate of the stirring
2. The viscosity of the reaction mixture
3. The rate at which reflux is rising into the reflux condenser
4. The flow of cooling water to any water-cooled part

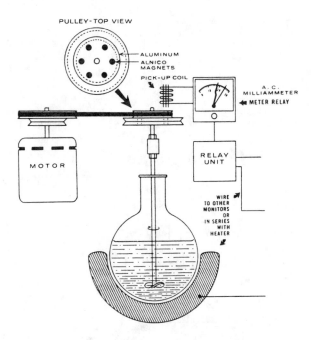

FIGURE 4.17.   An improvised stirrer monitor which uses Alnico magnets in an aluminum pulley to generate a rotating magnetic field. This changing magnetic field generates a current which is sensed by a meter relay connected to a pickup coil.

5.   The rate of distillation of product or by-product out of the reaction flask
6.   The volume of product or by-product in the distillate receivers
7.   The pressure in the apparatus: either hydrostatic pressure, gaseous pressure, or vacuum, depending upon the nature of the apparatus
8.   The rate of flow of gas (or liquid) to or from the reaction
9.   Other indicators of the progress of the reaction, depending upon the specific reaction
10.  The time

It should be remembered that with each variable one usually has the choice of several different approaches. To illustrate:

1.   The rotation of the stirrer may be monitored by (a) a meter relay which measures the electric current to the stirrer motor; or (b) a tachometer with either a switch output or a meter relay output; or (c) an improvised tachometer such as shown in Figure 4.17.
2.   The viscosity of the mixture is often more difficult to monitor. Three possibilities can be suggested: (a) Sometimes a commercially available viscometer can be adapted to the apparatus either by immersion into the reaction area or into a side stream. (b) If the rate of stirring is sensitive to the viscosity (and it should be if one selects an A.C. motor having "poor regulation") then one can monitor viscosity (1) as discussed above. (c) If the reactor is a continuous rather than

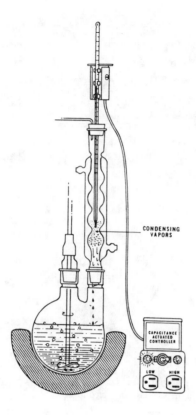

FIGURE 4.18.    Reflux rate monitoring with a thermometer located in condensing zone
of reflux condenser.

batch-type reactor, the viscosity may be monitored by watching the back
pressure as the reaction mixture is pumped.

3.  The rate of reflux can be monitored quite readily by sensing the vapor temperature in the reflux condenser either in the region of the condensing zone or
slightly above this zone. The temperature in these regions gives an indication
of the reflux rate because an increase in the rate causes the condensing zone to
move further up into the condenser. This results in a temperature increase at
any given point in the condensation zone. Such an increase is easy to monitor
using either a thermoregulator, a thermocouple plus pyrometer, or a thermometer with a capacitance-actuated controller attached (Figure 4.18). The monitor
selected should be set 5° to 10° *below* the normal condensing temperature of
the reflux vapors. The exact location of the sensing unit with respect to the
condensing zone (and the manner in which the monitor is connected electrically) will depend on whether one wishes the monitor to turn off the heat
cyclically whenever the rate of reflux gets too high, or to terminate the
distillation should the rate drop too low or decrease too much.

4.  The flow of cooling water to the condensers or to any other part of the
apparatus can be monitored with either a commercially available *"WATER-*

*FLOW CONTROLLER"* or the glass apparatus also described in Figure 4.4A.

5.  The rate of distillation is more difficult to monitor. One approach is discontinuous: to monitor the time required for successive samples of a given volume to accumulate. This could be done with either a relatively simple sample collector based on a syphon or one of the fixed-volume sample collecting and distributing devices which have been developed for chromatographic purposes. To monitor the time required for successive samples to accumulate, one would again use one of the time-delay relays or timers. The timer, in this case, could monitor the time elapsed per sample as being too long or too short, depending on how the timer was connected. (To monitor both high and low rate would require two timers). [The development of solid-state timers, digitizing circuits, and digitally based data acquisition and storage units since the original article was written provides enormous flexibility on the type of monitoring described here since it is now very easy to compare and manipulate data from different channels.]

6.  The amount of product in a distillate receiver can be monitored either on a volume or weight basis as described in [Section 4.6.2.15]. It might also be monitored by counting the number of fixed-volume fractions — if one were using a fraction collector as mentioned above.

7.  The pressure within the apparatus can be monitored by use of either (a) commercially available pressure switches of the bourdon, bellows, or diaphragm type; or (b) mercury monitors with contact probes; or (c) mercury manometers to which have been clipped capacitance-actuated controllers which sense movement of the mercury levels. The two ways to use the mercury monitors are shown in Figure 4.19A.

8.  The rate of gas or liquid flow either *to* the reaction or *from* the reaction can be a very important variable. The monitoring of a clean flow can usually be accomplished with a flow meter which develops a pressure differential across an orifice or capillary. Monitoring can be based on adjustable or fixed position contacts in a mercury *U*-tube connected across the capillary. It also can be monitored with a capacitance-actuated controller attached to one leg of the *U*-tube (Figure 4.19B).

9.  This variable includes pH, color, presence of precipitate, etc. In general, it represents those variables that can only be monitored with rather sophisticated electronic devices. These may be either quite specialized units designed for a specific apparatus or may be general-purpose laboratory tools that the scientist uses quite frequently.

10.  Time, the final variable on the list, can be monitored either (a) on the basis of the time of day or (b) elapsed time after a given stage of the reaction is attained. Scientists are quite accustomed to using electric clocks and timers to shut down apparatus at a predetermined time. They are less likely to be accustomed to the many possible ways of using the time-delay timers. Such a device, if interconnected to one of the other monitors listed above, can terminate the reaction *n* hours after the other monitor has indicated progress of the reaction. It is possible, for example, to connect a time-delay relay to the primary temperature controller and have the timer triggered by the first OFF cycle of the temperature

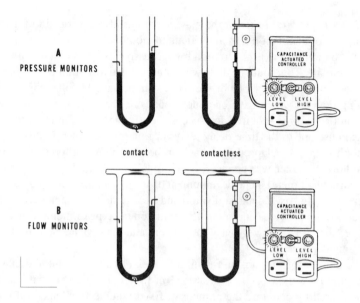

FIGURE 4.19.    Pressure and flow monitors based on manometers equipped either with contacts or capacitance-actuated controllers.

controller. This will enable it to control the time that the reaction was actually *at* temperature.

## COUPLING OF MONITORS

With a complex apparatus, it may be desirable to interconnect a number of monitoring devices.

In general, the monitoring units referred to above fall into several categories: mercury contact systems, mechanical-electrical devices (such as pressure switches), and electronic devices. These units generally have different electrical outputs.

1.    The mercury contact systems are either "normally open" or "normally closed," and may be able to carry only a few milliamperes of current.
2.    The electromechanical devices often have a microswitch with a single-pole, double-throw output that can be wired either "normally open" or "normally closed" and is able to carry at least 5 amperes of current. [Microswitches, although simple and durable devices, are prone to eventual failure. No safety or critical control function should depend on a single microswitch.]
3.    The electronic devices terminate in a variety of ways: (a) low voltage (or low current) control circuit, (b) an indicating meter that should be changed to a meter relay, (c) a 120-volt power circuit either normally ON or normally OFF, (d) a single-pole, double-throw relay equivalent to the microswitch. [Solid-state electronics usually depend on a voltage being "normally high", often around 5 V or less, or "normally low", usually meaning 0 volts.]

The output of any *single* monitor may be wired so that it interrupts the heat to the

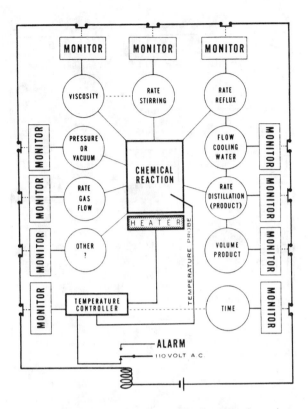

FIGURE 4.20. A number of normally closed circuit monitors may be readily connected in series.

reaction (or causes some other type of corrective action to take place). When a number of monitors are being used, they can be wired together so that if any one monitor is actuated, the heater can be turned OFF. A series circuit can be used if the output of every monitor is a "normally closed" relay or switch which opens upon malfunction (Figure 4.20). A parallel circuit would be used when every monitor has a "normally open" output which closes upon malfunction (Figure 4.21).

If one has a hybrid system of monitors — some "normally open," some "normally closed," some supplying a low-voltage output, and some supplying 120-volt output — then in order to combine them, one may need to add relays to at least some of the monitors or groups of monitors so that they can be connected in a series or parallel circuit. [Again, solid-state control equipment can be used to simplify the problem of working with a variety of inputs to make logical comparisons, i.e., "AND", "OR", "NAND", "NOR", etc. for control purposes.]

Mention should be made of a more elaborate method of coupling monitors, the "sampling" technique in which the monitors are periodically rather than continuously checked. This may be accomplished with a stepping-type switch wired to the monitors. As it steps, it connects each monitor in turn to the alarm and control circuits.

## EXOTHERMIC REACTIONS

Exothermic reactions, while usually not long-time reactions, at least so far as the

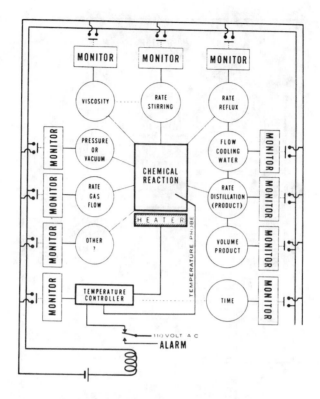

FIGURE 4.21.   A number of normally open circuit monitors may be connected in parallel.

exotherms are concerned, are particularly demanding of attention. When such reactions are running, the scientist may well feel he should not leave the laboratory for even a few minutes. This need not be, for many families of exothermic reactions are not dangerous and, once they are understood, can be entrusted to control by suitable monitors.

The monitors discussed under endothermic reactions may be used with exothermic reaction apparatus, both simple and complex, but with this difference: since the exothermic reaction evolves heat, the monitors may have to cause *positive* cooling to be applied to the apparatus. Thus, instead of wiring the monitor (or monitors) in such a way that it merely interrupts power to the heater, *it should be connected to turn on a cooling device.*

The nature of the cooling device will depend upon the nature of the exothermic reaction. Mildly exothermic reactions can be controlled by blowing air at the reaction flask. This is accomplished quite readily by having the temperature monitor turn OFF the heater and turn ON an electric fan, or open a solenoid valve in an air tube connected to the laboratory air supply.

More active, but still "mild," exothermic reactions may require that the heater be removed from contact with the apparatus (so that any residual heat stored in the heater does not reach the reaction) and that, simultaneously, cooling air be turned ON. Still more active reactions require still more positive cooling. This can be

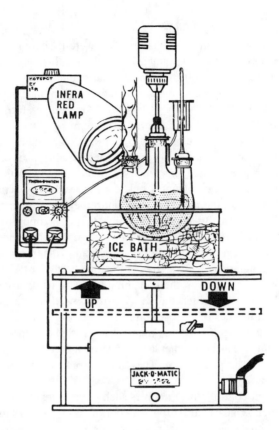

FIGURE 4.22. Jack-O-Matic used to control exothermic reactions by
raising and lowering ice baths.

accomplished if the cooling is applied in the form of an ice bath that can be raised
automatically around the bottom of the reaction. (Heat to the reaction is applied from
the top of the flask, by radiant heating.)

The steps described above can be applied automatically with a device known as
a Jack-O-Matic that is now commercially available. When used in connection with
a temperature-sensing system, Jack-O-Matic can either lower a heating bath from
around a reaction flask and apply positive air-blast cooling, or turn off a radiant
heater and raise an ice bath (Figure 4.22). It controls temperature by repeating these
operations cyclically as long as cooling is required. Thus it is able to control
automatically reactions that chemists have previously watched very closely. It is no
longer necessary for the chemist to watch the thermometer, to lower the heating
mantle, and to raise an ice bath around the reaction flask.

## OTHER FACTORS

The foregoing notes discuss many of the points that often have to be considered
in adding monitors and accessories to apparatus so that it can run unattended. The
following are miscellaneous items that are less frequently encountered:

*Programmed Control.* Sometimes the reaction, instead of being carried out at constant temperature (or constant pressure, flow, etc.), should be carried out at increasing or decreasing levels. A suitable controller for this purpose can usually be designed with the assistance of the laboratory instrumentation group and a research machinist. For example, a programmed temperature control that step-wise increases the temperature can be built around a group of constant-temperature regulators, each accurate to a few hundredths of a degree. The regulators would be selected in sequence by a stepping switch controlled by a time-delay timer. This arrangement would make it possible to maintain an apparatus at each of a series of fixed temperatures for a predetermined time. (This type of programming is particularly useful in connection with physical test apparatus.) [The same result may be achieved with a computer interfaced with the monitors and control devices, to change or control a given parameter over virtually any desired range, based on input from sensors. The computer can adjust a device, such as the heater voltage, to achieve the desired result. The major advantage of a computer-controlled system is that one can easily change the program by changing the software rather than the hardware. A similar comment would be appropriate for each of the remaining topics in this particular section.]

Programmed control in which the variable is allowed to increase or decrease gradually is sometimes desired. Programming controllers based on both linear and non-linear cams have been used for many years for plant-type process control. Although these units are usually rather large physically, they can be and are used with laboratory ovens and in-plant work. They may be also be used with laboratory reactions.

One can program-control other variables such as pressure or vacuum with plant-type cam controllers. One can also program-control these variables in an improvised manner by coupling a slow-speed motor to one of the commercially available pressure or vacuum regulators. The monitor will slowly drive the control point upscale. If a limiting value is desired, it can be achieved by adding a suitable cam which activates a microswitch and turns off the motor.

*Sequential Control.* Occasionally it may be desirable to program a sequence of operations. Thus, as an example, a reaction can be programmed so that after a predetermined number of hours a charge of reagent is added to the reaction flask; and after still another time period, still another charge is added. A different basis for sequential control is to monitor the reaction and use a given point in the development of the reaction, rather than time, to trigger a desired operation.

*Foam-Level Control.* In several rather different fields of laboratory work, excessive foaming in a laboratory operation can necessitate close personal attention and be a vexing problem. Vacuum stripping, although not a very long-time operation, is one example. As the lighter fractions are stripped from a mixture in a flask, foaming often occurs. If the stripping operation is not being watched quite carefully, part of the liquid in the still-pot may be swept over as foam into the condensing apparatus.

These boiling-type foams in glass flasks can usually be sensed either by probes coupled to capacitance-actuated controllers or by light beams using a light beam and a light beam photocell. In either case, the foam-sensing device can be connected to a solenoid valve which opens to admit a small amount of air or nitrogen to the apparatus (Figure 4.23). The resulting pressure pulse will cause the foam to subside almost immediately. As a result, vacuum stripping can be carried out much more rapidly and efficiently.

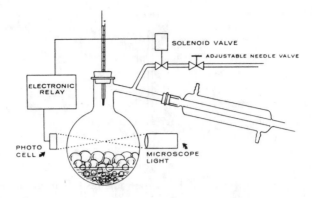

FIGURE 4.23.   Photoelectric monitoring and control of stripping foams.

A second field for foam monitoring and control is in microbiological preparations such as fermentations. Such processes are relatively slow compared with the foaming referred to above. Moreover, the fermentation foams are not as clean as the boiling foams. Finally, the fermentation apparatus is often made of stainless steel and not readily adaptable to optical sensing. Figure 4.24 shows foam-sensing, using a Teflon-insulated probe and a capacitance-actuated control. The controller could either open a solenoid valve and admit a small volume of "antifoam" or could turn on a small pump and inject "antifoam".

## MONITORING THE MONITOR

One can safely consider two different categories of monitoring — "control monitoring" and "safety monitoring." In designing apparatus that is to run unattended, one is concerned with both. The following notes illustrate the differences:

1.  "Safety monitoring" is always an addition to "control monitoring". In some cases, the "safety monitors" will be EXTRA monitors added to guard against failure of the "control monitors". In other cases, the "safety monitors" will be extra monitors added to watch variables that were not being controlled.
2.  "Safety monitoring" systems usually need not be as precise nor as fast in response as "control monitoring" systems. For example, in "control monitoring" the temperature of a bath, one may use a sensitive mercury thermoregulator and hold the temperature constant to a few hundredths of a degree. For "safety monitoring" the same bath it would be quite satisfactory to add the relatively insensitive, slow responding, bimetallic thermoregulator.
3.  Whenever possible, "safety monitors" should be unsophisticated devices whose mode of operation is easily visible, easily understood by the scientist, and readily checked. The mercury manometer is almost ideal from these points of view. It can be made into an excellent monitor for pressure, rate of pressure rise, rate of gas or liquid flow (or even flow of cooling water.) The electronic viscosity-measuring devices are examples of the opposite extreme of complexity. Admittedly, one is grateful for any method of measuring viscosity, but one wishes for devices that would be simple and readily checked.

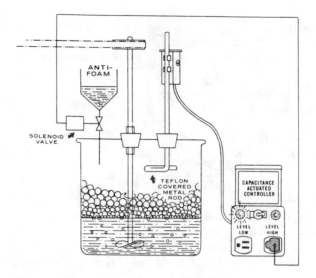

FIGURE 4.24. Monitoring and control of fermentation foams with a probe and a capacitance-actuated controller.

4. Although the "safety monitor" will usually be a device that is inherently simple and insensitive, there are some situations, particularly involving exothermic reactions, where the "safety monitor" must be a highly sensitive device with a fast response time. Under such conditions, one may well be willing to use as the "safety monitor" a device of greater complexity. In fact, it can be a duplicate of the primary temperature controller.

5. One may be tempted, at times, to consider using the "safety monitor" as a "control monitor". This may even, at times, be quite reasonable, particularly when used to terminate a reaction. As an example, one can use a "safety monitor" attached to the stirrer motor to terminate a reaction when the reaction mixture has reached the desired viscosity.

   However, one must guard against the opposite approach: considering that a "control monitor" eliminates the need for a "safety monitor." The five possible monitors shown earlier in Figure 4.13 were really "safety monitors" added to guard against failure of the primary controller.

6. Finally, this philosophy of guarding against all possible malfunctions brings up the following question: If one always has to assume that the primary controllers can fail and therefore need monitoring, will the "safety monitor" sooner or later fail? Should one monitor the "safety monitor"? This is the age-old problem: just how far should one go in monitoring the monitor, in guarding the guard, in policing the policeman? [Although not a complete answer to this question, a built-in or periodic check using an artificial signal to simulate a response of each "safety monitor" to an unsafe condition is highly recommended. This should be initiated as close to the original signal generation point as possible to test as much of the circuit as possible.]

   Sometimes the solution to this problem is not difficult since the variables are not independent. For example, if the temperature increases, other variables such as pressure or reflux rate will probably increase, and several different monitors will respond. [In effect, the concept here is to supply sufficient redundancy to insure safety. The system should be carefully analyzed to insure that the failure of no single component would defeat the safety systems. Sometimes the analysis is subtle and all failure modes are not fully considered. At least two competent persons should independently do a system analysis.]

In considering monitors and their design, one should remember that the degree to which the monitors will be able to cope with both the "likely" and the "unlikely" malfunctions of the apparatus will depend upon the thought that has been given to their design and installation. In the final analysis, it is the good judgment of the scientist that is involved.

## OTHER SAFETY RECOMMENDATIONS

Section 4.6.2.17 made a number of specific recommendations relative to safe operation of glass distillation equipment. Some of these recommendations also apply to the simpler physical testing apparatus and some to the more complex reaction apparatus. To review the recommendations briefly:

1. If electrical power fluctuations would affect the apparatus, these effects would be minimized by use of constant-voltage transformers.
2. Breakage of glass apparatus can be minimized by careful annealing, by inspecting the glass for strain, by proper support of the equipment, and by use of adequate safety shields.
3. Whenever liquid overflow of a liquid product is a possibility, oversize receivers should be used. It sometimes is advisable to place large trays under the apparatus.
4. The apparatus should be conspicuously labeled, so that an untrained person would have no trouble shutting it down in an emergency.

## CONCLUSION

Although the suggestions made in this chapter are necessarily limited and will have to be modified to fit various local conditions, certainly there are great advantages in adapting laboratory apparatus so that it can run safely unattended or with minimal attention. The advantages are: (a) apparatus that is run continuously produces the desired results much more rapidly than would otherwise be the case. (b) The quality, such as color, of the experimental samples produced when running without interruption is often better. Likewise, the quality of physical tests that are run continuously rather than interruptedly is better. (c) Whenever the experiment can be run continuously, the scientist will have greater assurance that the run is valid and reproducible. In fact, when interrupted runs are made, the scientist may even have reservations about their validity. (d) Finally, whenever one is able to make use of the evening hours, one can greatly reduce the cost per experiment. [(e) There is one other benefit. Provision of automatic controls eliminates or greatly reduces the opportunity for human error, and although there may be personnel in the laboratory, the automatic controls will mean that they may not need to be in the immediate vicinity of the apparatus. Both of these factors should reduce the number of injuries.]

The benefits of safe unattended operation of laboratory equipment are so great that creative efforts to achieve these are most worthwhile. In fact, such efforts by individual scientists and their supporting instrumentation personnel have been extremely profitable and are one of the most promising fields for the improvement of research.

## ACKNOWLEDGMENT

Reprinted from the *Journal of Chemical Education,* Vol. 43, A589 and A652, July and August 1966. With permission.

## REFERENCE (SECTION 4.6.2.18)

1. **Conlon, D. R.,** Running laboratory reaction under safe control, in *Handbook of Laboratory Safety,* 2nd ed., Steere, N. V., Ed., CRC Press, Boca Raton, FL, 1971, 235.

## 4.7. HAZARD AWARENESS (RIGHT-TO-KNOW)

In recent years, a movement has been growing to legally require that individuals exposed to chemicals in the workplace be informed about the risks which these chemicals pose to their health. A number of states passed what are commonly referred to as "Right-To-Know" laws applicable to various groups of employees in their states. Eventually, after extended hearings, OSHA published a national version of this requirement as a Hazard Communication Standard, 29 CFR 1910.1200. Comments were received from many sources, including industrial and trade associations, labor unions, academic groups, individuals, and others as to the content of the standard and which employees should be covered by the standard. As originally issued, the standard covered only the employers and employees in the Standard Industrial Classification (SIC) Codes 20 through 39. These classifications are given below:

| | | | |
|---|---|---|---|
| 20 | Food & kindred products | 30 | Rubber & plastic products |
| 21 | Tobacco manufacturers | 31 | Leather & leather products |
| 22 | Textile mill products | 32 | Stone, clay & glass products |
| 23 | Apparel & other textile products | 33 | Primary metal industries |
| 24 | Lumber & wood products | 34 | Fabricated metal products |
| 25 | Furniture & fixtures | 35 | Machinery, except electrical |
| 26 | Paper & allied products | 36 | Electrical equipment & supplies |
| 27 | Printing & publishing | 37 | Transportation equipment |
| 28 | Chemicals & allied products | 38 | Instruments & related products |
| 29 | Petroleum & coal products | 39 | Miscellaneous manufacturing products |

Laboratories in these industries were partially covered by some of the more critical portions of the standard. Laboratories in other industries and in academic institutions were not covered by the standard. Comments received during the hearings had revealed marked differences of opinion on how the coverage should be applied to research laboratories. Some comments reflected an opinion that research laboratories, particulary academic laboratories, were among the most critical areas where a definitive, explicit standard was needed, while others felt that as knowledgeable, responsible individuals, research scientists needed, at most, only a voluntary standard. The actual need is undoubtedly somewhere between these two disparate views. Research laboratories are extremely difficult operations to which to apply tightly worded standards, but, unfortunately, not all research personnel set as high a priority on safety as they should.

A large proportion of scientists agree on the need for some form of safety information program. Even those who argue for a voluntary program will usually concede that an individual does have the right to know the risks to which he is being exposed. Their objections to a formal program are usually based on the premise that intelligent and knowledgeable scientists, if not already informed, can be depended upon to take the time to make themselves so, and, in the academic area, some argue that the standard is still another infringement upon academic freedom. The mechanics required to demonstrate compliance are another factor to which many object, simply because they require considerable time and effort, at least initially, which must necessarily detract from the time available for research.

The standard, as finally written, is a performance standard. The required program is not spelled out except in very broad strokes, but the goals and objectives to be achieved are well defined.

Since many employees were omitted in the original version of the standard, court action followed almost immediately to force OSHA to cover all employees using or exposed to chemicals. After some delay in the court system, the legal issue was resolved in favor of OSHA being required to extend the coverage of the standard. The final rule was published in the *Federal Register* on August 24, 1987, to go into effect on May 23, 1988.

In a few areas, the standard was applied to some research laboratories outside of the original list of industries. Virginia, for example, in adopting the federal standard, added all public employees in the state to the list of those covered, which included laboratory personnel in the state's public universities and colleges as well as state-supported laboratories. Eventually, unless the standard is preempted by an explicit laboratory standard, it should apply to public and private laboratories. Even if the laboratory standard is adopted, the hazard communication standard possibly will become the basis of comparison for the written plans currently being proposed for individual laboratories.

The following sections will cover the major requirements of the standard which apply to research facilities and will offer suggestions on how to comply with the requirements. A facility with only a few laboratories and a limited number of persons involved should not have a significant problem with compliance. A major research institution, especially an academic institution, with hundreds of laboratories under approximately the same number of managers and using thousands of different chemicals, is faced with an enormously complicated task to fully comply with the standard.

### 4.7.1. BASIC REQUIREMENTS

Portions of the first few sections of 29 CFR 1910.1200, given below, spell out the basic concept of the standard.

*a. PURPOSE*

1.    The purpose of this section is to ensure that the hazards of all chemicals produced or imported are evaluated, and that information concerning their hazards is transmitted to employers and employees. This transmittal of information is to be accomplished by means of comprehensive hazard communication programs, which are to include container labeling and other forms of warning, material safety data sheets and employee training.

2.    This occupational safety and health standard is intended to address comprehensively the issue of evaluating and communicating the potential hazards of chemicals, and commuicating information concerning hazards and appropriate protective measures to employees, and to preempt any legal requirements of a state, or political subdivision of a state pertaining to this subject....

*b. SCOPE AND APPLICATION*

1.    This section requires chemical manufacturers or importers to assess the hazards of chemicals which they produce or import, and all employers to provide information to their employees about the hazardous chemicals to which they are exposed, by means of a hazard communications program, labels and other forms of warning, material safety data sheets, and information and training. In addition, this section requires distributors to transmit the required information to employers.

2.    This section applies to any chemical which is known to be present in the workplace in such a manner that employees may be exposed under normal conditions of use or in a foreseeable emergency.

3.    This section applies to laboratories only as follows:
     i.     Employers shall ensure that labels on *incoming* containers of hazardous chemicals are not removed or defaced;
     ii.    Employers shall maintain any material safety data sheets that are received with *incoming* shipments of hazardous chemicals, and ensure that they are readily accessible to laboratory employees; and,
     iii.   Employers shall ensure that laboratory employees are apprised of the hazards of the chemicals in their workplace in accordance with paragraph (h) of this section.

*c. EMPLOYEE INFORMATION AND TRAINING*

Employers shall provide employees with information and training on hazardous chemicals in their work area at the time of their initial assignment, and whenever a new hazard is introduced into their work area.

1.    *INFORMATION.* Employees shall be informed of:
     i.     The requirements of this section;
     ii.    Any operations in their work area where hazardous chemicals are present; and
     iii.   The location and availability of the new written hazard communication program, including the required list(s) of hazardous chemicals, and material safety data sheets required by this section.

2. *TRAINING*. Employee training shall include at least:
   i. Methods and observations that may be used to detect the presence or release of a hazardous chemical in the work area (such as monitoring conducted by the employer, continuous monitoring devices, visual appearance, or odor of hazardous chemicals when being released, etc.);
   ii. The physical and health hazards of the chemicals in the work area;
   iii. The measures employees can take to protect themselves from these hazards, including specific procedures the employer has implemented to protect employees from exposures to hazardous chemicals, such as appropriate work practices, emergency procedures, and personal protective equipment to be used; and
   iv. The details of the hazardous communication program developed by the employer, including an explanation of the labeling system and the materials safety data sheet, and how employees can obtain and use the appropriate hazard information.

## 4.7.2. WRITTEN HAZARD COMMUNICATION PROGRAM

Based on the requirements outlined above, employers must develop a written program. The program must include provisions to ensure proper labeling of chemicals, to maintain a chemical inventory, to maintain a current and accessible material safety data sheet file for all incoming chemicals, and to provide training and information to the employees in a number of relevant safety and health areas.

In order to develop an effective hazard communication program, it is desirable to have input from all of the groups that need to cooperate to make the program work. A straightforward way of accomplishing this is to establish a hazards communications committee to help define the program and to monitor the performance of the program once it has become operational. There should be representatives on the committee from the administrative departments, including health and safety, personnel, purchasing, and physical plant and from the research divisions. In an academic institution, having a representative from every affected department probably would create too large a committee, but each college should be represented by a member from a major chemical-using department in the college. One department should be assigned the leadership role in developing and implementing the program, probably the health and safety department, but it should be clear that all major constituencies share in the responsibility of formulating the program and making it work.

A number of things must be in a written program:

1. Although not explicitly required, all employees exposed to chemicals in the organization during the normal course of their employment have to be identified so that they can participate in the program. Note that this does not necessarily mean that all the individuals identified use chemicals. An electronics technician might not use chemicals himself, but could be considered to be exposed to those chemicals used by others working in the same room. In addition, an employee in this context would include both salaried employees and workers paid an hourly wage. On the other hand, a custodial worker would not be considered to have a significant exposure to a laboratory's chemicals if he only took out nonhazardous trash, but would have to be included if the

cleaning materials he uses contain hazardous chemicals. However, it probably would be desirable for a custodial employee to be informed of some of the general risks associated with chemicals in laboratories in his work area so he would understand the need to be careful while in the laboratory.

2. A list of hazardous chemicals in the work place must be compiled. The provisions in the standard only require laboratories to keep track of incoming chemicals since the effective date of the standard for employers, May 25, 1986. Existing inventories were, in a sense, "grandfathered". Eventually, as older stocks are disposed of or used, the list will come to reflect the actual holdings in a facility. The list should be kept as current as possible. If one person is assigned the responsibility of maintaining the list and the data are kept in a personal computer database, it is only necessary to keep track of additions and deletions to maintain a complete, current list. For purposes of complying with this portion of the standard, the quantities of each chemical in the laboratory are not needed, although these data would be important for a sound management program and would be helpful in planning a safety program. The list of chemicals, in combination with the list of employees, will serve to help define the training program.

3. The written program must define how the employees are to be informed of the requirements of the standard. This will include (1) details of how the employees are to be informed about the contents of the standard and (2) the contents of the written plan, (3) how they are to meet the labeling requirements, (4) how they are to learn of the methods available to them to warn them of exposures, (5) how to obtain and interpret a material safety data sheet for a given chemical, (6) the hazards associated with the chemicals to which they are exposed, (7) how they are to be trained in procedures which will eliminate or reduce these chemical hazards, and (8) how they are to react in an emergency.

4. Many laboratory uses of chemicals involve repetitive tasks, while others do not. Employees must be made aware of the risks associated with the latter type of activities as well as those accompanying the more routine uses of chemicals, and the same basic type of information must be provided as in item 3.

5. Although pipes are not considered containers for the purpose of this standard and need not be labeled, the plan must include education of employees about the hazards associated with any unlabeled pipes containing chemicals in their work area and how to deal with these hazards.

6. There must be a procedure or statement in the plan as to how transient employees, such as persons working on contract, are to be informed of the chemical hazards to which they may be exposed, and for provision of information on protective measures for these transient employees. Although not specifically spelled out in the standard, there is a need for the converse as well. Contractors are often called in to do renovations, perform an asbestos abatement project, or to conduct a pest control program, as examples, and use hazardous chemicals in the process or expose personnel to airborne hazards. Provision should be made in the contracts for these groups for them to provide information to the occupants of the spaces where their work is being done.

### 4.7.2.1. Personnel Lists

This appears to be relatively straightforward, but, in fact, can be rather complicated. In a large institution, the actual duties associated with a given job classification often become blurred over a period of time. For example, a job title of laboratory technician might appear to logically relate to chemical exposure, but the duties of the individual may have changed so that the job may never bring the individual into contact with chemicals at all. It is not possible to simply have the personnel department list all persons in specific job classifications as well as professional staff or faculty in research areas.

As an initial step to determine which employees need to participate in a formal hazard communication program, a questionnaire can be sent to each department or other internal division (1) asking them to define those areas in which chemicals are used in their department, (2) asking them to list each employee in those areas, with their job title, and (3) asking them for their appraisal of the involvement of these individuals with chemicals. This should be followed up with a second questionnaire to the managers of the individual areas, asking for the same information, and then the area should be visited to confirm the data provided. This sounds unnecessarily involved but experience has shown that all three steps are necessary. In many cases, through oversights, individuals are not identified who should have been included and, occasionally, someone is listed who has no exposure, usually because it was easier to list everyone rather than consider each individual case.

It is important to identify the position (most positions now have internal identification codes) so that when it becomes vacant, a mechanism can be established to ensure that the new person filling the position receives a proper orientation program. Often, such a position might qualify for participation in a preemployment medical screening examination as well, so the effort to correlate positions with exposure to chemical hazards might be justified for more than one purpose.

All persons being considered for positions covered by the hazard communication program should receive a brief written statement concerning the program so that they may ask appropriate questions at the time of their job interview. If they are selected, a more extensive document should be provided so that they will be aware of the explicit requirements of the standard.

### 4.7.2.2. Chemical List

A list of all hazardous chemicals in the workplace is required as part of the standard. It may be difficult to convince many research scientists to take the time to go through their stocks of chemicals to prepare a list for their laboratories. This is one of the more burdensome tasks associated with the standard. However, it is essential that an effort be made. Not only is it necessary in order to determine the training requirements, but it also is needed to prepare a MSDS file for the facility, although the MSDS file is only required for incoming chemicals for laboratories. In practice, however, it is difficult to justify not having an MSDS for a hazardous chemical in use in a laboratory, based on a technicality. There is no real alternative for the initial survey as a basis for defining the scope of the program.

It would be desirable if the problem of maintaining the list of incoming chemicals could be centralized, perhaps as the purchase order is being processed. However, although surprisingly few chemicals are bought in quantity in most research institutions, even very large ones, there may be more than a thousand different substances bought during the course of a single year, and several thousand purchased over a number of years. Many of these are bought under a number of synonyms, or as components of brand-name formulations. These two factors complicate capturing the data at a central point, where clerical personnel may not be qualified to properly identify a given chemical. If orders can be placed electronically, software can be written to compare the chemical name on the order against a list of synonyms to properly identify the chemical. There are so many synonyms that it is impractical to do the comparison manually when handling thousands of orders. The purchaser can be determined by means of budgeting or charge codes on the order. The problem of synonyms can be minimized if a standard identifier, such as CAS numbers which are provided in most general chemical supply catalogs, are included in all chemical purchase orders. Programs have already been written to perform this data acquisition, but most depend upon manual entry of the data, which is very slow. It is, however, practical for individual laboratories to use microcomputers to perform this task, using commercial programs such as Lotus 123©, dBase III+©, or a number of others, and it is at the individual laboratory where this information is most essential.

### 4.7.2.3. Labeling
Current labels on original containers of chemicals as purchased from the distributor or manufacturer will almost certainly exceed the requirements of the Hazard Communication Standard (see Section 4.5.2). These requirements are

1. The identity of the hazardous chemical
2. Appropriate hazard warnings
3. The name and address of the chemical manufacturer, importer, or other responsible party

Item 2 is the only ambiguous requirement. Most commercially sold chemicals provide this sort of information on the label in a number of ways, such as:

1. A risk descriptor (i.e., Danger, Warning, Caution)
2. The NFPA hazard diamond
3. A descriptive statement of the hazards
4. Stylized symbols, such as a radiation or biohazard symbol

Other useful safety-related information is normally provided as well, such as the flash point (if applicable), fire extinguisher type (if applicable), first aid and medical advice, a color code to aid in avoiding incompatible storage, and standard identifiers (such as a CAS number which can aid in referring to a MSDS database, and a UN number which is needed in disposing of the chemical as a hazardous waste).

Laboratories are specifically enjoined by the terms of the standard from removing or defacing the labels on incoming containers of chemicals. However, during the course of research, it is relatively common to transfer a portion of the contents from the original container to a secondary container. If this material remains under the control of the individual responsible for the transfer and is to be used during a single work session, then it is not necessary to label the secondary container. However, if it is not to be used under these conditions, then the secondary container must be marked with the identity of the chemical(s) in the container and with "appropriate" hazard warnings for the protection of the employee. These "appropriate" warnings need not be as comprehensive as the original label, but must be enough to provide adequate safety information.

The most likely occasions when secondary containers are used without proper labeling would be when chemicals are disbursed from a larger container into a smaller one at a central stockroom and when containers are taken from the laboratory into the field. Personnel must be sure to label the secondary containers in these cases and in any other comparable situation. If secondary containers are labeled properly, it also will help remedy one of the more troublesome problems associated with hazardous waste disposal, inadequately identified containers of chemicals.

The warning labels must be in English, although they may be provided in other languages *in addition,* if appropriate. In many academic institutions, graduate students who routinely use a language other than English as a primary language are becoming numerous, and, in some cases, consideration may be given to supplementing the commercial label with warnings in other languages, although the great majority of these graduate students can be expected to understand written English satisfactorily. All employees using chemicals must be instructed in how to interpret the hazard information on the labels.

A specific part of the written plan must address how the employees are to be made aware of the labeling requirements and how they are expected to comply with this standard. It would be highly desirable to develop a uniform program across an organization, particularly as to labeling of secondary containers, to avoid unnecessary confusion.

### 4.7.2.4. Material Safety Data Sheets

Since the receipt of MSDSs is tied so strongly to the purchase and receipt of chemicals, they were discussed earlier in some detail in Section 4.3.3 which defined the basic information to be provided by an MSDS. Firms must meet this requirement in different ways. The exact form of the MSDS is not mandated by the standard as long as the proper information is provided. A standard format is under consideration, but one has not been adopted as yet.

There are two basic requirements associated with MSDSs in the Hazard Communication Standard. Employees must be trained in how to use the information in them and they must be readily available to the employees.

As discussed in the earlier section, a major problem in a research institution, where the chemical users operate independently of each other and are likely to be in a number of different buildings, is to ensure that all users of a given chemical have ready access to a copy of the most recent version of the MSDS. The distributor is only required to send one copy of an MSDS to a purchaser, and an updated version when a revision is necessary because of new information. Where several different components of an organization order independently, one laboratory may receive an update while the others do not since the vendor may think it has fulfilled its obligation by sending the MSDS to the first unit making the purchase. If a centralized mechanism for tracking chemical purchases has been established, then all MSDSs could be sent to a single location, from which copies can be forwarded to all groups within the organization which need them. Another alternative, which does not provide as ready access to all users, but does not require the tracking mechanism referred to above, would be to have all MSDSs be received at one location and maintained in a master file, with copies placed in several secondary master files at locations which would be reasonably convenient to the users. A third alternative, but still less accessible to users, would be for a single master file to exist, with copies of individual MSDSs provided upon request. This might not be considered to meet the accessibility requirement if the delay in receiving the MSDS is more than one or two working days. All of these mechanisms are, at best, cumbersome and manpower intensive.

A better alternative is available if laboratories at the research institution have access to a centralized mainframe computer or can communicate to a dedicated microcomputer. Comprehensive generic MSDSs are now becoming commercially available on optical compact discs which can be processed by a computer and accessed at any time by the users. These are typically updated quarterly so that they can satisfy the need for the MSDS file to remain current. They are not inexpensive, but the cost is much less than the manpower which would be needed to maintain an equivalent hard-copy file and provide a comparable level of access to it. There also are firms which maintain computerized MSDS files available to subscribers as a database service. These are accessed by users from their terminals via modems. However, line and use charges for the service can become heavy.

No matter how a facility or organization sets up an MSDS file, one component of the employee training program must include an explanation of how an individual employee can obtain access to the file. It should be possible for an employee to obtain copies of an MSDS for a given chemical upon request. In addition to providing access to the MSDS file, part of the program training must also include training in how to review an MSDS to obtain appropriate hazard information.

The information presented in the various categories in an MSDS should pose no real difficulty to most technically trained persons. Some definitions of terms may need to be provided, such as $LD_{50}$ (lethal dose, 50% of the time for the test species), if an individual is not accustomed to using such terms, but even these are straightforward. However, some persons will not be as scientifically sophisticated and the training program for these individuals will need to be more thorough. In an instance at the editor's institution, grasping the distinction between a monomer and a polymer was a major problem for some clerical personnel who felt that they had been exposed to dangerous levels of a chemical due to some activities in the building where they worked. The health hazard data and the TLV values given in the MSDS for the chemical stated that they were for the monomer only and that the effects of exposure to the monomer would be serious at a few parts per billion in air. The employees' exposure was not to the monomer, but to very small quantities of the stable polymer, for which the health hazards were minimal. Extensive (and expensive) tests had to be run before the exposed personnel were convinced (some perhaps still have doubts) that they had not been unduly exposed. During training, an effort needs to be made to ensure that understanding has been achieved.

Compliance with the training requirements for utilization of an MSDS as a source of hazard information can be readily achieved for technically trained personnel. For example, a written handout informing the employees how they can obtain access to a needed MSDS and a short videotape explaining the contents might be all that is needed. An individual capable of explaining any confusing points should administer such a program and be available to answer questions. A statement affirming that the training had been received should be signed (and dated) by the employee after any questions had been resolved. For less knowledgeable employees, a formal training session should be set up and an instructor-student format used. The handout and videotape can still be part of the instructional program, but the instructor should go over each of the categories in an MSDS and encourage the employees to ask questions. The employees also should be provided with a written version of the concepts covered, for later reference.

Some organizations, in order to document that an employee has been exposed to the information and understands it, require that each employee take a very simple written quiz on the material covered, instead of signing a simple statement. This is not required to comply with the standard, but it does provide stronger documentation of an effective training program. Individuals should not resent the imposition of such a requirement, but some scientists feel, rightly or wrongly, that they have demonstrated sufficient proficiency in their area by fulfilling the requirements for a doctorate. If a quiz is made part of a program, it should be expected that a number of individuals will object to taking the quiz. Such a program can be made to work, with patience and the support of the organization.

### 4.7.2.5. Employee Training and Information

Portions of the training program which are common to every employee, the basic concepts of the organization's program, how it is administered, the requirements of the standard, and how to read and understand information labels and material safety data sheets have already been covered in the previous sections. These can be given by the lead department in the organization's hazard communication program, but a number of other areas will require a cooperative effort between the administrative department in overall charge of the program and the individual laboratories and the departments of which they are a part. In an organization which conducts research in a wide variety of disciplines and divisions of these disciplines, the managers of the local laboratories should be the primary parties responsible for providing much of the required training relating to operations specific to their program and facility.

The following topics need to be covered routinely in a training program in order to comply with the standard, in addition to those already discussed.

1. The physical and health effects of the hazardous chemicals which the employees may use or to which they may be exposed.
2. Means of detecting the presence of toxic materials in the workplace. This should include methods directly available to the employee, such as odor, presence of a respiratory irritant, visual means or various symptoms, such as dizziness, lassitude, etc. It also should include types of monitoring that can be done, either by laboratory personnel, by the organization's safety and health department, or by outside public and private agencies.
3. Means of reducing or eliminating the exposure of the employee to the risks associated with the hazardous chemicals in the workplace. This will include work practices (such as doing all work generating toxic fumes within an efficient fume hood instead of on an open bench) which will reduce the exposures or the use of personal protective equipment.
4. Actions the organization has taken to minimize the exposure of employees to

the chemical hazards. This can include the engineering controls which have been implemented, e.g., provision of fume hoods and biological safety cabinets, ventilation, shielding, and other safety and health-related features of the facility. It can also include policy positions which encourage or require that employees and their supervisors follow good safety practices at all times as well as programs which penalize those who do not follow safe work practices and which provide incentives for them to do so.

5.    Emergency procedures to follow in the event of an accidental exposure to a hazard material.

6.    Procedures to warn nonorganizational personnel working in the area of potential exposures. Generally, this will mean persons working under contract to the organization.

7.    Measures to provide information as to the hazards and the protective measures which both the employer and employee can take to reduce or eliminate the hazards associated with a nonroutine task involving chemicals.

8.    Measures to inform personnel of the hazards associated with unlabeled pipes carrying chemicals in their work area and the safety precautions which should be taken. This type of situation could arise fairly often, even in academic laboratories, in certain engineering disciplines such as chemical engineering.

The responsibility for training in these topics should be shared between the local administrative unit, usually the individual laboratory, the department of which the laboratory is a component, and the department assigned the lead in implementing the organization's program.

There are a number of generic chemical topics which would be essentially the same regardless of the discipline involved or the character of the research. Among these are

- Bases
- Flammable and combustible liquids
- Mercury
- Oxidizing and reducing agents
- Peroxides

- Explosive materials
- Gases and gas cylinders
- Mineral acids
- Perchloric acid
- Reactive metals

There also are a number of topics on protective measures to minimize exposures which could be made the topic of standardized presentations, such as:

- Biosafety cabinets
- Fire safety
- Personal protective equipment
- Safe working practices

- Electrical safety
- Fume hoods
- Pressurized systems
- Storage of chemicals

Standardized programs can be developed for these topics, as well as others, and video tapes made which can be used virtually anywhere. If the latter course is taken, much of the Hazard Communication Standard training requirements could be met by requiring a new employee to view selected tapes, supplemented by an opportunity for the employee to ask questions. Standardized programs are especially useful where there is a continuing turnover, with customized training being required for single individuals on a sporadic basis.

A large number of companies offer hazard communication training programs, many of which provide video tapes, as discussed above. However, many of these are based so specifically on the industrial environment that they are not very suitable for use in training laboratory employees. Others, in order to demonstrate compliance with one aspect of a safety program, do not follow good safety practices in other areas. Commercial programs are usually expensive, but they normally can be previewed, for a small charge, to see if they are suitable.

Although much of the training obligation can be met with standardized programs, the primary responsibility for training must be borne by laboratory managerial personnel. The actual exposures vary from laboratory to laboratory and even within a laboratory. No set of generic programs of reasonable size can accommodate all of the various chemicals that are in actual use in all of the laboratories in a research-oriented corporation or academic institution, nor can generic programs cover all the intricacies of the research techniques used in different areas. Further, the laboratory director, or a person delegated by the laboratory director, is the one who defines the research program, the one to whom the laboratory employees report, and to whom they look for guidance. It is impossible for any outside individual to be fully cognizant of all of the activities which take place within a facility. An outside person could not possibly know when procedural changes take place or when a different chemical might be used. It is also highly unlikely that an individual can be found with a background sufficiently comprehensive to respond to questions in all of the areas covered by the standard. Because of these factors, the laboratory director must be ultimately responsibile for the actual implementation of the factors under his control which are covered by the hazard communication program.

When training has been completed for a given area, or even for a single chemical, if this is all that is necessary when a new chemical is brought into the workplace, the employee should be asked to document that the training has been provided by signing a dated statement to that effect. These documents should be placed on file and retained for the duration of the individual's employment, and beyond if the materials to which he may have been exposed are known to pose problems under chronic exposure conditions or are known to cause delayed health effects.

## REFERENCE (SECTIONS 4.7 TO 4.7.2.5)

1. Hazard Communication, Final Rule, Department of Labor, Occupational Safety and Health Administration, 29 CFR, Parts 1910 (Section 1200), 1915, 1917, 1918, 1926, and 1928, *Fed. Regist.*, 52(163), 31852, 1987.

## 4.8. HEALTH EFFECTS

In recent years, there has been increasing emphasis placed on the health effects of chemical exposures. However, as has been frequently noted, health effects are much more difficult to quantitatively characterize than most physical safety parameters. It is straightforward to define with reasonable accuracy a number of physical hazards, such as the upper and lower explosive limits of the vapors of a flammable material. However, the exposure levels (see Figure 4.25, taken from the *Federal Register*, 53

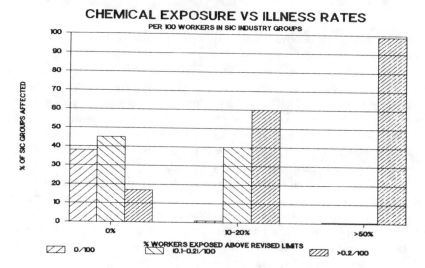

FIGURE 4.25.    Effect of increasing chemical exposure on illness rates.

(No. 109), 21342, 1988) which will cause a given physiological effect in humans are not nearly as precise, especially if the effect of interest is delayed, or is due to prolonged exposure to low levels of a toxic material.

Even individual reactions to low levels of those materials which cause serious immediate or acute effects at high doses are strongly dependent upon the inherent susceptibilities of individuals. Some will exhibit a reaction at extremely low levels, while others show no signs of responding at all to relatively high levels. Part of this is due to the natural range of sensitivity in a population, but part is also due to contributory effects. There are synergistic effects. For example, a heavy smoker may have developed emphysema due to the effects of inhaling smoke for an extended period of time. Such an individual would be affected by toxic materials which reduce pulmonary function before a person with healthy lungs. The sensitivity of individuals can change with time. An example is the common oleoresin allergen, poison ivy. The sensitivity of individuals is usually small to an initial exposure, but with successive exposures, the sensitivity increases. Similarly, the sensitization of individuals to bee stings is well known. The effects of medication can also modify the sensitivity of individuals. The serious problems associated with taking tranquilizers and drinking alcohol represent a well-known example of this phenomenon.

The natural differences in individuals due to genetic factors, age, sex, lifestyle, etc. make evaluation of laboratory tests quite subjective. On examination of the results of a typical blood panel, one notes that the patient's results are given for each parameter measured as well as the range to be expected for a typical healthy person. An individual can have results somewhat outside the normal range due to hereditary or environmental conditions and still be perfectly healthy. Fasting and foregoing any medication prior to laboratory tests is an attempt to eliminate as many variables as possible which would affect the tests. Unless a baseline series of tests are available from a time prior to an exposure when, presumably, the quantities to be measured are "normal" for the individual, then it is often difficult to determine with certainty whether a given result is due to an exposure. Even here, the results could be distorted by extraneous factors occurring earlier.

Dependence on observed symptoms to indicate an exposure to a toxic material is also very subjective. Again, the wide variation in the tolerance of individuals to concentrations of toxic agents causes a corresponding wide variation in the responses to an exposure. Many of the common symptoms associated with occupational exposures — shortness of breath, headaches, nausea, dizziness, etc. — often can also be the result of other problems — illnesses such as flu, lack of sleep, psychological problems such as stress due to personal problems or personality conflicts with a supervisor, overindulgence, etc. An individual needs to be aware of the symptoms which could result from an exposure and to try to determine when these are likely to be due to an occupational exposure and when they are likely not to be. If, for example, an individual normally enjoys good health, has not done anything which might result in any of the symptoms which he is experiencing, and there are no "bugs" going around, he might well suspect that he has suffered an exposure to some environmental hazard. In such a case, he should mention it to his co-workers, report it to his supervisor, and leave the work area. Especially if others are experiencing similar symptoms, it is very likely that an exposure has occurred and appropriate steps should be taken to seek medical aid for the exposed personnel, limit the exposure of others, and to correct the situation. If only one person is having difficulties, but the immediate work environment differs for each person, then the exposure may have been limited to the individual and precautionary steps such as leaving the area, lying down, and observation by a colleague should take place. Often, prompt recognition of a problem is critical in minimizing the consequences, especially when the possible culprit is a material which provides no other warning signs. Whenever an exposure has occurred which has resulted in physical effects, an evaluation by a physician should be obtained promptly. Even if no exposure has occurred, an individual complaining of an illness should be taken seriously. There could be a medical disorder requiring care, intervention, or at least documentation. Even if malingering is suspected, evaluation by a physician can help to confirm that the claim of illness is or is not valid.

Delayed effects due to prolonged exposures to relatively low levels of toxic materials or radiation rarely are reflected in immediate sensations of malaise, which can trigger concern about possible consequences of the exposures. If a material does not have any warning properties, then exposures may exist at unsafe levels indefinitely without the occupants of the area being aware of the exposure. The eventual consequences may be masked by naturally occurring illnesses of the same type. Lung cancer is, unfortunately, common and so the occurrence of lung cancer might not be recognized as due to an occupational exposure. Birth defects occur in about 3 to 6% of natural births (depending upon how birth defects are defined). What percentage might be due to an occupational exposure of the mother or father? Similarly, infertility is a problem for about 15% of married couples. What is the role of occupational exposures for the afflicted couple? Some neurotoxins cause deterioration of the central nervous system, but age and other illnesses may do the same. It is often difficult to establish a correlation between an occupational exposure and an illness, even statistically for a group, because it is difficult to isolate the effect from the interference of other variables or to define an equivalent control group. Often, unless the illness is rare, such as the angiosarcomas caused by exposure to vinyl chloride, it is impossible to definitely verify a causal relationship between an occupational exposure and a disease.

Not all delayed effects are due to low levels of exposure. The onset of cancer, which may occur due to exposure to asbestos or radiation, is often delayed for periods of 15 or more years or, and this is a key point, they may not occur at all. By no means do all individuals exposed to even high levels of such hazards suffer the consequences.

There are basically three mechanisms by which health hazard data may be acquired: (1) epidemiological studies of groups of exposed individuals, (2) human experimentation, and (3) animal studies. There are problems associated with each of these three sources.

The major problem with epidemiological studies is that usually one does not have a controlled experiment; the studies involve data either generated by an ongoing work situation or extracted from past medical records. In some instances, case reports are sufficiently unusual that they call attention to themselves, e.g., a reduction in fertility in a group of workers is so large that only a simple study to determine the cause-effect relationship between a common exposure factor and the resulting fertility depression is required, the effect being known; it only remains to determine what experience the workers have in common. Rarely are situations as simple as this, although they do occur.

In most instances, epidemiological studies to determine if an exposure to a substance results in a given effect take the form of cohort studies, in which two separate groups composed of exposed and unexposed individuals are studied. It is critical that the study be unbiased by either the way the participants are selected or the manner in which the outcome is tested. Another critical factor is whether the two groups are, in fact, similar in all respects which could affect the outcome of the study, or that the differences are such that they can be taken into account either in the design of the experiment or in the analysis of the data. In order to judge the validity of a study, all of the relevant factors must be completely documented and available for review.

Most of the epidemiological studies concerning exposure to toxic substances are from the industrial sector since only in such an environment is it likely that exposures would be limited to a single chemical or class of chemicals, and where the exposures would be relatively stable over a prolonged period of time. Most of the studies that are available tend to come from Scandinavian countries, where, for example, Finland maintains a computerized database of the health records of all its citizens. Similar records do not exist in this country, although some categories of specialized health data are maintained. Many epidemiological studies of exposures in this country depend upon records maintained by corporations or equivalent public agencies such as the national laboratories, which are managed by industrial firms and have similar medical surveillance programs. The limited range of chemicals for which such work situations provide the basis for valid epidemiological studies limit the scope of this approach.

Human experimentation is limited by statute to studies in which there is no prospect of permanent harm to the volunteers participating in the study. This obviously limits the scope of the results which can be obtained, although it can be employed to determine the onset of early symptoms or to determine threshold levels for detection of odors or irritation as a potential warning mechanism. Any experiment of this type must be carefully reviewed by a human subject review committee of the institution or corporate research facility where the research is being contemplated. Any subject of such experimentation must be fully informed of any risks and must, in most cases, be given an opportunity to withdraw at any point. However, many experiments using volunteers have been conducted with this restriction and significant data have been obtained on symptoms initiated by modest levels of exposure. However, it should also be remembered that "fully informing the volunteer of all known risks" can, in and by itself, skew the results. Therefore, published results on subjective symptoms using a small sample must be viewed as inconclusive or suspect.

Data have been obtained from direct human exposures due to accidental exposures to high levels of a number of hazardous chemicals. These, of course, are not controlled experiments and the dose levels have to be inferred from the circumstances of the incident, but, as direct evidence of the results of high exposures, they are extremely useful.

Since data obtained from direct human exposures are limited, much of the available data on the toxic effects of chemicals are obtained from animal data. The easiest data to obtain are the median dose or the median concentration in air which is fatal to an animal under a standardized experimental protocol. The most common animals used for this purpose are strains of rats and mice because they can be obtained relatively inexpensively with uniform characteristics, and the cost of housing and feeding them is small compared to most other species. In addition to rats and mice, many other animals are used such as primates (monkeys, chimpanzees), guinea pigs, rabbits, dogs, cats, chickens, etc. in efforts to obtain a model which would parallel the effect on humans. In the generic carcinogen standard, relevant animal studies are intended to specifically involve mammalian species.

The median lethal dose, $LD_{50}$, is given in mg/kg and the species is given. The median lethal concentration, $LC_{50}$, may be given in $mg/M^3$ or ppm for a given species, and the exposure time interval is usually specified. Lesser amounts of data are given in the literature at other survival fractions such as 25% or 75%. These data must be obtained under rigorously controlled conditions to be useful. The experimental protocol must be totally documented. Among other things, enough animals must be used to provide statistical accuracy. Where the effect to be measured is less well defined than lethality, the number of animals needed to obtain the data may become quite large. Even after the animal data have been obtained, in many cases the question remains of whether the animal model is sufficiently close to that of a human response to be used in determining the equivalent human response. Much of the controversy over using animal data to establish human exposure effects revolves around this question.

Another procedure which leads to varying interpretations on the health effects of tested materials is the practice of using large, nonlethal doses to reduce the number of animals required when studying other effects such as carcinogenicity. The premise is that administration of a large dose to a small number of animals is experimentally equivalent to small doses given to a large number of animals. A linear extrapolation hypothesis usually is used to estimate the effects at low exposures. This practice is not uniformly accepted and is often used as an argument to discredit the results which are obtained, but is the basis of much of the data on these nonacute effects. Another possibility for estimating the effects of low doses would be to assume that the response will approach zero more, or less, rapidly than the dose. A linear extrapolation generally is considered conservative. Data from experiments such as these are frequently used in arriving at health standards by pharmaceutical manufacturing companies and the media.

The use of animals has provided the greatest amount of health hazard data, but the practice has come under increasing attack by animal rights activists. Much of the public support for this movement originated from widely publicized instances where animals were not well treated and undoubtedly suffered more than was necessary. As a result of public pressure and a concern on the part of many scientists, many new safeguards have been instituted to minimize the amount of pain and suffering experienced by laboratory animals. Animal care committees are now required to review experimental protocols and must approve the procedures in order to qualify the research for federal support. The number of animals involved in the research is limited to the number required to achieve meaningful results and the pain experienced by the animals must be no more than is absolutely necessary. These committees must include persons not affiliated either directly or indirectly with the institution and who might be expected to care about the well-being of the animals.

Although conceding that improvements have been made in the care of experimental animals, the goal of animal-rights activists is to prevent the use of animals in any research which would adversely affect the animals. There have been instances where animals have been "liberated" from facilities and released into the environment. There are two practical problems with such actions, regardless of ethics: (1) the animals usually are not accustomed to surviving in the wild, and most often do not and (2) there generally have been no efforts to insure that the animals are healthy and, hence, a disease could be introduced into the environment.

The argument that there is no alternative to using animals to gain knowledge to prevent or cure human illnesses, since experimentation on humans cannot be done, is rebutted in two ways by the animal activists, the first being a purely moral stance of "why do we assume that it is morally right to cause pain to animals to help humans?". This is an issue that each individual must answer for himself, unless a legal restriction is imposed. The second argument is that animal experimentation is no longer necessary. Computer modeling can provide equivalent information. Relatively few scientists accept this argument as a generalization, although it is agreed that computer modeling can be used in some cases.

There is merit on both sides, although the extremists of both groups are undoubtedly too extreme, and some middle position will eventually become acceptable practice. However, animal health hazard data may be less available in the future.

A newer, but effective, modality which can be statistically relevant is to use cell cultures. Direct effects can be seen and measured as to benefits or toxicity. It is more complicated and has seen limited use to date.

A recent report by the Environmental Protection Agency indicates that the results of animal studies suggest that some toxic substances are less dangerous than formally supposed, specifically with regard to carcinogenicity. This finding, according to EPA spokespersons, is based on a better knowledge of how the metabolism of chemicals differs in various species, a better knowledge of how much of a chemical which has been taken into the body actually reaches an organ where it may do harm, and a better understanding of how the chemical influences the mechanisms that cause cancer.

This is a controversial position since, in general, it leads to higher acceptable exposure levels of the chemicals under discussion, such as dioxin and arsenic. Some scientists feel that the level of scientific knowledge does not yet justify moving away from a very conservative approach. Both sides should avoid treating the issue as one that can be settled by politically biased discussion, and the data on which the findings are based should be evaluated, as should any other hypothesis, on the basis of their scientific merit.

As stated in Section 4.7.2, OSHA defines health effects, for the purposes of the Hazard Communication Standard, in Appendix A to 29 CFR 1910.1200. The definitions given below are from that Appendix. They originated in 21 CFR Part 191.* A few modifications have been made to number seven.

For the purposes of this section any chemicals which meet any of the following definitions, as determined by the [following] criteria, are health hazards:

Criteria:

1. *Carcinogenicity:...* a determination by the National Toxicology Program, the International Agency for Research on Cancer, or OSHA that a chemical is a carcinogen or potential carcinogen will be considered conclusive evidence for purposes of this section.
2. *Human data:* Where available, epidemiological studies and case reports of adverse health effects shall be considered in the evaluation.
3. *Animal data:* Human evidence of health effects in exposed populations is generally not available for the majority of chemicals produced or used in the workplace. Therefore, the available results of toxicological testing in animal populations shall be used to predict the health effects that may be experienced by exposed workers. In particular, the definitions of certain acute hazards refer to specific animal testing results.
4. *Adequacy and reporting of data:* The results of any study which are designed and conducted according to established scientific principles, and which report statistically significant conclusions regarding the health effects of a chemical, shall be a sufficient basis for a hazard determination.

Definitions:

1. *Carcinogen.* A chemical is considered to be a carcinogen if:
   a.  It has been evaluated by the International Agency for Research on Cancer (IARC), and found to be a carcinogen or potential carcinogen; or
   b.  It is listed as a carcinogen or potential carcinogen in the *Annual Report on Carcinogens* published by the National Toxicology Program (NTP) (latest edition); or
   c.  It is regulated by OSHA as a carcinogen.
2. *Corrosive.* A chemical that causes visible destruction of, or irreversible alterations in, living tissue by chemical action at the site of the contact. For example, a chemical is considered to be corrosive if, when tested on the intact skin of

---

\* Some of the quoted material is rearranged and some nonessential wordage is deleted. The essential information is not changed.

albino rabbits by the method described in the U.S. Department of Transportation in Appendix A to 49 CFR Part 173, it destroys or changes irreversibly the structure of the tissue at the site of contact following an exposure period of four hours. This term shall not refer to action on inanimate surfaces.

3.   *Highly toxic.* A chemical falling within any of the following categories:

   a.   A chemical that has a median lethal dose ($LD_{50}$) OF 50 mg or less per kilogram of body weight when administered orally to albino rats weighing between 200 and 300 g each.

   b.   A chemical that has a median lethal dose ($LD_{50}$) of 200 mg or less per kilogram of body weight when administered by continuous contact for 24 hours (or less if death occurs within 24 hours) with the bare skin of albino rabbits weighing between 2 and 3 kg each.

   c.   A chemical that has a median lethal concentration ($LC_{50}$) in air of 200 parts per million by volume or less of gas or vapor, or 2 milligrams per liter or less of gas or vapor, or 2 mg per liter or less of mist, fume, or dust, when administered by continuous inhalation for one hour (or less if death occurs within one hour) to albino rats weighing between 200 and 300 grams each.

4.   *Irritant.* A chemical, which is not corrosive, but which causes a reversible inflammatory effect on living tissue by chemical action at the site of contact. A chemical is a skin irritant if, when tested on the intact skin of albino rabbits by the methods of 16 CFR 1500.41 for 4 hours exposure or by other appropriate techniques, it results in an empirical score of five or more. A chemical is an eye irritant if so determined under the procedure listed in 16 CFR 1500.42 or other appropriate techniques.

5.   *Sensitizer.* A chemical that causes a substantial proportion of exposed people or animals to develop an allergic reaction in normal tissue after repeated exposure to the chemical.

6.   *Toxic.* A chemical falling within any of the following categories:

   a.   A chemical that has a median lethal dose ($LD_{50}$) of more than 50 mg per kilogram but not more than 500 mg per kilogram of body weight when administered orally to albino rats weighing between 200 and 300 g each.

   b.   A chemical that has a median lethal dose ($LD_{50}$) of more than 200 mg per kilogram but not more than 1000 mg per kilogram per kilogram of body weight when administered by continuous contact for 24 hours (or less if death occurs within 24 hours) with the bare skin of albino rabbits weighing between two and three kilograms each.

   c.   A chemical that has a median lethal concentration ($LD_{50}$) in air of more than 200 parts per million by volume of gas or vapor, but not more than 2000 parts per million by volume of gas or vapor, or more than two milligrams per liter but not more than 20 mg per liter of mist, fume or dust, when administered by continuous inhalation for one hour (or less if death occurs within 1 hour) to albino rats weighing between 200 and 300 g each.

7.   *Target organ effects.* The following is a target organ categorization of effects which may occur, including examples of signs and symptoms and chemicals

which have been found to cause such effects. These examples are presented to illustrate the range and diversity of effects and hazards found in the workplace, and the broad scope employers must consider in this area, but are not intended to be all-inclusive.

a.  Hepatotoxins_____ Chemicals which produce liver damage.
    Signs and Symptoms _____ Jaundice, liver enlargement.
    Chemicals_____ Solvents such as toluene, xylene, carbon tetrachloride, nitrosoamines.

b.  Nephrotoxins_____ Chemicals which produce kidney damage.
    Signs and Symptoms_____ Edema, proteinuria, Hematuria, casts.
    Chemicals_____ Halogenated hydrocarbons, uranium.

c.  Neurotoxic _____ Chemicals which produce their primary effect on the nervous system.
    Signs and Symptoms_____ Narcosis, behavioral changes, coma, decrease in motor functions.
    Chemicals_____ Mercury, carbon disulfide, lead.

d.  Agents which act on the ____ Decrease hemoglobin function, deprive blood or hematopoietic system the body tissue of oxygen.
    Signs and Symptoms _____ Cyanosis, anemia, immune function depression.
    Chemicals _____ Carbon monoxide, cyanide.

e.  Agents which damage the ___ Chemicals which damage the pulmonary lung function.
    Signs and Symptoms_____ Cough, tightness in chest, shortness of breath.
    Chemicals_____ Silica, asbestos, organic fibers such as celluose-cotton.

f.  Reproductive toxins_____ Chemicals which affect the reproductive capabilities including chromosomal damage (mutations) and effects on fetuses (teratogenesis).
    Signs and Symptoms _____ Birth defects, sterility, functionality.
    Chemicals _____ Lead, DBCP, some blood pressure medications.

g.  Cutaneous hazards _____ Chemicals which affect the dermal layer of the body.
    Signs and Symptoms _____ Defatting of the skin, rashes, irritation, discoloration.
    Chemicals_____ Ketones, chlorinated compounds, soaps, solvents.

h.  Eye hazards_____ Chemicals which affect the eye or visual capacity.
    Signs and Symptoms _____ Conjunctivitis, corneal damage, blephaharitis.
    Chemicals _____ Organic solvents, acids, alkalis.

### 4.8.1. EXPOSURE LIMITS*

Consideration and attention to the concentrations of chemicals within any worker's environment is mandatory. This must include chemicals in respired air, water (used at work or for drinking), food contamination, contact with skin or eyes either directly or by vapor, and possible radiation from specific chemicals used. The garnering of meaningful data has been slow, and changes in acceptable levels occur as new resources and studies are produced and as policy formulations are agreed upon.[1]

The American Conference of Industrial Hygienists (ACGIH) has published threshold limit values (TLV)© for decades, and recent international meetings have formulated and categorized the values into what are called occupational exposure limits (OEL).[1,2] These limits refer to airborne values only. Where skin is involved, an "s" may be appended. The latter values refer to time-weighted averages (TWA) over a normal 8-h working day in a 40-h week, extrapolated for the worker's lifetime.

The exposure limits should not be used to define a boundary on one side of which working conditions are acceptable and on the other side, unacceptable. First, they are for a typical person, and individuals may not react in a typical way. Low levels for which a prolonged exposure could give rise to a delayed effect would not necessarily provide any warning for individuals who might be abnormally sensitive. Second, levels should be kept as low as possible, preferably well below the exposure limits. Maintenance of air concentrations below the TLV©, PEL, or OEL does not guarantee that a worker will not be affected at any given concentration. There is a very small percentage of workers who are so sensitive to particular chemicals that even minute quantities will precipitate severe reactions. It is not possible or feasible to try to make exposures acceptable for these individuals in most cases. Removing them from the environment and avoiding contact to the material causing the reaction is best for both the worker and the facility. The essential OSHA goal that "no worker will be affected" is unattainable in cases involving large numbers of persons. The variability of innate propensities as well as such factors as daily diet variations, stress levels, and circadian rhythms often relegates efforts to provide a "safe" environment to statistics and chance.

Dutch authorities have formulated a new means of evaluating exposures, the maximal accepted concentration (MAC), which helps to assess the available data to derive a more meaningful OEL. They state[3,4] that this formulation permits a better accommodation between the employee who does the work and the employer who orders or requires certain working conditions, procedures, and chemical usage. After deliberation by a working group of experts (WGE), a Dutch national MAC commission was founded (NMC) to promulgate the resultant report. This addresses the health-based OEL (HB-OEL), as recommended by WGE, and the socioeconomic and technological constraints. The NMC then issues a recommendation for an "Operational OEL" or MAC. This allows input from both employees and employers, and deliberations and reports are open to the public. The adoption of this procedure by the Dutch has proven workable, but a recommendation for international adoption has not occurred as yet. A very important concept must be stated — OELs are derived so as to maintain and preserve health, not just prevent disease or death.[5-7]

---

* This section has been co-authored by Richard F. Desjardins and A. Keith Furr.

One must differentiate between somatic effects and nuisance effects because:

1.  Such a distinction is based on the nonexistent dichotomy between soma and psyche.
2.  With adequate testing, a nuisance may be shown to actually be a somatic problem.
3.  The nuisance may be the premonitory signs and symptoms of a slowly developing somatic problem.
4.  The nuisance perception can be a statistically significant factor in diminished productivity.

Stokinger[7] appropriately considers nuisance perceptions and signs and symptoms as "co-inducers" of disease. Efforts to specifically categorize these areas might well be a fruitless endeavor. If they are considered mutually inclusive and are managed appropriately, the effect will be salubrious.

It would be valuable to keep in mind the effects of synergism in the workplace. In considering the many exposure/response relationships, extraneous factors must be considered as background variables which could skew results. Such factors include smoking, hypertension, heart disease, asthma or pulmonary fibrosis, diabetes, and many more. The preemployment examination made by the health facility has a responsibility through a good present and past health history, a work history with attention to types of exposures, a perceptive physical examination, and appropriate testings to determine whether an individual can healthfully be employed in a given area. This must be done at that time to (1) avoid harmful exposures which would compromise the person's health, and (2) avoid even the slightest suspicion that a disease was evolved within the facility.

The committee for the working conference on principles of protocols for evaluating chemicals in the environment in the U.S.[9] defined adverse effects as: those changes in morphology, growth, development and life span that:

1.  Result in impairment of functional capacity or in a decrement of the ability to compensate additional stress.
2.  Are irreversible if such changes cause detectable decrements in the ability of the organism to maintain homeostasis.
3.  Change the susceptibility of the organism to the deleterious effects of other environmental influences.

The "hygienic rule" states that exposure to chemicals at work should always be minimized as far as possible. Ultimately, the goal is zero. Pragmatically, this must at this time be interpreted as at the lowest detection achievable. The airborne exposure limits, regardless of their source — ACGIH, NIOSH, MAC, or OSHA regulatory limits — are intended to represent time-weighted average levels at which it should be acceptable for most workers to work normal 8-h, 5-d work weeks without suffering any ill effects. As noted earlier, individuals vary in terms of susceptibility, and there may be contributory effects which would make these levels inappropriate for a specific person. The values are intended to be based on the best available data from the sources described earlier, as evaluated by committees composed of professionals with appropriate expertise. The OSHA values are an exception, to a degree, in that they are adopted as part of a legal standard, and changing them is often a long, involved process which frequently includes legal contests. Therefore, changes in OSHA values tend to lag somewhat behind those data such as the limits suggested by the ACGIH, which are reviewed and revised annually, if the data warrant a revision. However, even the OSHA values were originally based on the same evaluation procedure and were, in most cases, actually taken from the other sources. (OSHA is currently proposing a massive revision of its standards, based primarily, but not exclusively, on ACGIH values.)

Many of the newer legal limits established by OSHA include an action-level concept of 50% of the OSHA PEL, at which point an employer is to take action to reduce the level or to ensure that the 8-h time-weighted PEL is not exceeded.

Although the 8-h, time-weighted average, by definition, includes periods during which the level exceeds the exposure limits, the departure should not be gross. In some instances, ceiling levels are recommended or mandated that exceed the 8-h, time-weighted-average exposure limits, which should never be exceeded. In other instances, short-term exposure limits are also set higher than the time-weighted average limits, usually for durations of no more than 15 min for a limited number of occasions (ACGIH uses four as a limit) and spaced widely throughout a workday. These are usually set for processes which may be sporadic or have occasions when higher levels are normal. Sometimes these higher numbers are based on toxicological data, and sometimes they are indirectly related to typical maximum excursions which might occur in an industrial environment. The TLV, along with the TWA, establish the guidelines for exposure control for both the employee's health as well as for possible corporate or institutional responsibility.

The responsibility for maintaining levels of toxic vapors and gases below acceptable limits is usually assigned to management, preferably by engineering controls unless these can be shown to be impractical, and otherwise by procedural controls and the use of personal protective devices. However, it is equally important for employees to properly use equipment in laboratories so as to not defeat engineering controls, to report promptly any safeguards needing repairs, to follow procedures which reduce safeguards, and to use and maintain personal protective equipment provided them. Neither a "macho" nor a "scoff-law" attitude is appropriate for laboratory activities as regards compliance with acceptable exposure levels.

## REFERENCES (SECTION 4.8.1)

1. **Zielhuis, R. L.,** Occupational exposure limits for chemical agents, in *Occupational Medicine,* Zenz, C., Ed., Year Book Medical Publ., Chicago, 1988.
2. Methods Used in Establishing Permissible Levels in Occupational Exposure to Harmful Agents, Tech. Rep. 601, World Health Organization, Geneva, 1977.
3. Principles and methods for evaluating the toxicity of chemicals, in *Environmental Health Criteria,* Vol. 6, Part 1, World Health Organization, Geneva, 1978.
4. **Zielhuis, R. L. and Notten, W. R. F.,** Permissible levels for occupational exposure: basic concepts, *Int. Arch. Occup. Environ. Health,* 42, 269, 1978/79.
5. **Zielhuis, R. L.,** Standards setting for work conditions as risky behavior, in *Standard Setting,* Grandjean, P. H., Ed., Arbejosmiljofondet, Copenhagen, 1977, 15.
6. **Sherwin, R. P.,** What is an adverse health effect?, *Environ. Health Perspect.,* 52, 177, 1983.
7. **Stokinger, H. E.,** Modus operandi of Threshold Limits Committee of the ACGIH, *Am. Ind. Hyg. Assoc. J.,* 25, 589, 1964.

8. **Stokinger, H. E.,** Criteria and procedures for assessing the toxic responses to industrial chemicals, in *Permissible Levels of Toxic Substances in the Working Environment,* International Labor Organization, Geneva, 1970.

9. Principles for Evaluating Chemicals in the Environment: A Report of the Committee for the Working Conference on Principles of Protocols for Evaluating Chemicals in the Environment, National Academy of Science, Washington, D.C., 1975.

## 4.8.2 ENVIRONMENTAL MONITORING*

Since many toxic materials provide no sensory warnings and, in many other cases, materials which have an odor or cause irritation (the two most common warning properties) do not result in a recognizable warning at or near the recommended limits or even at levels of immediate danger, the establishment of limits is of no practical value in many cases unless the capability of monitoring exists. Standard monitoring programs, based on 8-h, time-weighted average measurements are rarely appropriate for the laboratory environment, where a given material is rarely used this long. However, average and peak levels for the various activities should be determined.

Monitoring programs are usually performed by an industrial hygienist instead of the laboratory employees, except where permanently installed fixed monitors are used. Ideally, monitoring should measure the levels of chemical substances in the breathing zone of each employee. As will be discussed below, monitoring near the breathing zone is practical with small sampling systems which can be worn in a shirt pocket or attached to a lapel. Preferably, the employee should wear the monitoring device during an entire workday and, again preferably, over a period of several days to ensure that the results obtained will be truly representative, although in many research laboratories the variety of chemicals change so rapidly that there may not be a truly "typical" day. It would be good if this could, indeed, be done for every employee. The amount of time required would, in all likelihood, be prohibitive in most large research institutions to do this for every laboratory. However, for laboratories working with regulated carcinogens, for example, or where other highly toxic materials are in use, it may be necessary or desirable to establish such an intensive and thorough program.

Where the activities of many employees are essentially similar, the exposures of a typical, representative person may serve to reflect the exposures of all the persons doing the same tasks. In some instances, the areas where exposures to chemical hazards are relatively small and well defined, sampling devices may be placed at the location used to determine the average exposures per unit time of persons working in those areas. If these devices do not show concentrations above the acceptable limits, then individuals can be assured that their exposures do not as well.

In many instances, the rate of release of toxic vapors and gases varies widely, and an extended sample taken for 8 h may show a very low average level of exposure, but for brief intervals, the actual levels may actually be dangerously high. In such a case, a monitoring program should include "instantaneous" or "grab" samples taken during these periods of higher-than-normal levels. All of the organizations which publish levels recognize that, for some materials, even short-term high levels should not be permitted. This concept is embodied in the ceiling (C) or short-term exposure levels (STEL) incorporated in the standards for some materials. The monitoring

* This section was co-authored by A. Keith Furr and Richard F. Desjardins.

program should address the need to document compliance with these short-term excursions.

All data accumulated in the monitoring program should be maintained in a file, preferably for each employee, but definitely for each facility. These records should be dated, signed by the individual responsible for the data, and documented as to the measurement parameters, i.e., instrument used, duration of sample collection, characteristics of the detection device, location or person wearing the device, and, if possible, supporting data on the operations being conducted at the time of data acquisition.

Many different types of sampling instruments are used, and it is not the purpose of this article to dwell on the characteristics of each. Some are direct reading, while others require sophisticated followup procedures to analyze collected material. There is some legal merit in acquiring material for later analysis by accredited commercial laboratories, to provide additional credibility in case the results are contested at some later date. The delay in receiving results following this practice, which may vary from a day or so to weeks, may be unacceptable, and either in-house analysis or alternative methods may be chosen to obtain results more rapidly.

Direct reading instruments, by definition, provide "instantaneous" or "grab" samples which may be entirely appropriate, as when the need is present to determine if an IDLH situation exists. A good example is the need to use an oxygen meter when entering any confined space. A combustible gas meter is often used at the same time to determine the presence of organic gases, and these two devices are often combined in the same instrument, using different sensors. Other instances are reentry to areas where the presence of a highly toxic gas may be suspected. Direct reading instruments also are useful in conducting surveys within a facility to determine the relative levels of pollutants in different parts of a space. There may be areas, for example, where air exchange is minimal and toxic levels of gases can accumulate. In such instances, the data can be used to help determine what remedial actions can be taken. If previous measurements have not been taken, such as when an operation is to be performed for the first time, a direct reading instrument, perhaps incorporating an alarm, might well be used. In some cases, direct reading instruments may be incorporated into fixed monitoring devices to provide a warning and to automatically shut an operation down. If, for example, experimentation with explosive gases is being done, a sensor may be used to determine when the concentration of the gas reaches a predetermined level relative to the lower explosion limit of the gas and to automatically close a valve in the gas supply.

Some direct reading instruments are extremely sophisticated and others are very simple. Some instruments include a sensor for a single substance, while others can serve to detect and provide quantitative information on a number of materials. Among the more sophisticated units which provide direct readings are compact, portable infrared spectrometers, atomic absorption units, and gas chromatographs. These may be set up to provide a high sensitivity for a single element, such as the familiar Bachrach atomic absorption unit for mercury detection, while others such as the Miran infrared spectrometer, can provide accurate readings for hundreds of organic solvents. Individuals using these instruments need to be properly trained to obtain accurate data. Interferences are a major problem where two or more possible chemicals may contribute to the same "window" or peak area. If the potential presence of interfering chemicals is not recognized, erroneously high readings may be reported. If the problem is recognized, these same high readings may still be interpreted as an upper limit for the level of the chemical of concern, and if the combined contribution is still well below acceptable limits, then exposures may be no problem to the employees.

All of these instruments must be well maintained and kept properly calibrated in order for the data to be meaningful. In some instances, as with radiation measuring instruments, the calibration may be done by a commercial facility to provide traceability to a National Bureau of Standards standard in order to be certified, while in other cases, calibration can be done locally. Documentation of all maintenance and calibrations should be maintained in a log. Assurance of accuracy is especially important if the levels are near the acceptable limits. As a minimum, the accuracy should be such as to ensure that the results are within plus or minus 25% of the actual value, with a confidence level of 95%. It would, of course, be desirable to exceed this minimal goal.

A simple direct reading device is the familiar detector tube, intended to provide a measurement of the airborne concentration of a specific chemical, in which a known volume of air is drawn through a tube containing a material with which the chemical of concern will react. The known volume of air is usually provided by a manually activated pump attached to one end of a glass tube, and air is drawn in through the other. The amount of air may be provided by a single cycle of the air pump, or several cycles may be required. The reaction of the chemical in the air and the material in the tube will begin at the end away from the pump, and the length of the stain will indicate the level of the airborne contaminant. There are now hundreds of tubes available for individual chemicals, which provides a great deal of flexibility for a relatively modest cost. The units do have a limited shelf-life, so it is generally not feasible to keep tubes on the shelf for every possible contaminant. However, they can be the basis for a monitoring program for a chemical which is in relatively steady use. Although they are, as noted, relatively inexpensive per tube, if a large number of measurements are required and if a direct reading instrument is available, it may be a more economical long-term investment to buy a device which does not consume an expendable tube for each measurement. The detector tubes, although intended for a specific chemical, may, in fact, react with more than a single chemical, so interferences may be a problem. These interferences will generally be identified in the literature accompanying each tube. A major use of detector tubes, other than being an alternative for a more expensive device, is for testing atmospheres for entry into an area where high levels of a specific contaminant are expected to exist.

The devices previously mentioned are not suitable for taking extended measurements since they provide only an instantaneous measurement, unless the instrument has an output which can be connected to a device which will integrate the data over a period of time. There are a number of ways in which this can be done, from a relatively crude system based on a chart recorder where the area under the trace can be translated into an integrated reading, to a digital sampling device which transfers the data to a computer with an appropriate interface and software, or a dedicated multichannel scaler in which the data occurring in successive intervals are accummulated in successive channels.

Most long-term measurements depend upon a constant-volume air pump which pulls in a given volume of air over several hours, and the contaminating material in the air is collected within a collecting device. The type of contaminant will determine the choice of the collector. Many organic gases and vapors are adsorbed readily on the surface of materials such as activated charcoal, activated alumina, or silica gel from which they can be eluted within a laboratory or driven off by heat and the adsorbed amounts quantified using a gas chromatograph. In other instances, the materials are bubbled through a liquid into which they go into solution. The collection efficiency of such a unit can approach 100%. Particulates also may be absorbed into a liquid or, more frequently, collected by impact onto various types of filters, after which a number of techniques are available to quantify the amount of pollutant which has been collected. The efficiency of the filter will depend upon the type of filter used.

In some cases a collecting device called a cascade impactor is used, in which the rapidly moving contaminated air passes through a number of stages. There are slits between each stage of decreasing size. At each stage some of the particles, because of their inertia tend to continue in a straight line, impact on a collecting plate and adhere to it. In the first stage the larger particles which have the most inertia are collected, while smaller ones are drawn through the exit slit into the next slit and are collected there. This continues until the smaller particles are collected at the last stage on a permeable membrane.

Thus, not only are the contaminants collected, but they are also sorted into approximate sizes. Since the quantity of air which has been drawn through the collector is known and the amount on the filter or collectors can be measured, it is a simple matter to compute the concentration of the particulates in terms of milligrams per cubic meter.

In some instances, where there is only a single possible contaminant in the air, no further processing of the collected material is necessary, but in most cases, analysis of the collected material is necessary to separate and quantify the amounts present. NIOSH has published explicit analytical techniques to be used for a large number of airborne pollutants, and for meaningful results, these procedures should be followed exactly. In many cases, the commercial laboratory will provide the appropriate collectors as well as explicit instructions on how they should be used, including, for example, the air-flow rate and the length of time they should be used.

Each time a pump is used, it should be calibrated to insure that the volume of air is accurately known. Most companies that sell sampling pumps also sell convenient calibration units, but if one is not available, the pumps should be calibrated using a graduated cylinder and measuring the time a soap film across the cylinder moves a measured distance. If sampling pumps are to be mounted in a fixed location, they can generally be plugged into ordinary electrical power outlets, but if they are to be worn by an individual, they will have to depend on battery packs. Battery packs are usually nickel-cadmium and these require some care to be able to run for long periods. Unless a nickel-cadmium battery is frequently taken through a major portion of its discharge cycle, it will lose capacity. It is said to "remember" the range over which it is used, and if it is typically drawn down only one third, for example, during each use, eventually this will be its capacity. Nickel-cadmium batteries are best suited for frequent, heavy use.

A simple sampling device, well suited for personnel monitoring, uses a permeable membrane over an activated charcoal collector, in the form of a badge which may be worn over the period of interest. Normal movement by the wearer and the inherent motion of molecules in air are sufficiently characteristic to bring a definite quantity of the air bearing the contaminant into contact with the membrane through which the contaminant passes and is adsorbed by the charcoal. From this point on, the adsorbed material is eluted as with any other charcoal collector and analyzed. Normally, the company which sells the badge offers the analytical service as well, but the badges also may be processed locally. If care is taken, several organic solvents (if present in the air) can be identified with a single badge. Badges are available at this time for a relatively small number of materials, but the convenience of wearing a badge instead of a pump, no matter how quiet and lightweight, makes them desirable.

Sampling should be done as close to the breathing zone as practical. The results obtained as the worker moves around (and the air being sampled is representative of the individual's actual exposure) generally do not agree well with samples taken by fixed collectors.

For various reasons, individuals asked to wear monitoring devices, such as a small sampling pump, are not always cooperative and do things which invalidate the data which are obtained. They may feel that they are being "checked on", perhaps for careless or sloppy work. They may simply feel that the pump is too noisy or that it gets in their way. They may not want to know the levels to which they are being exposed for fear that they may be too high and that they could be in danger of losing their job. In any event, if the individual responsible for the monitoring program feels that the data obtained are not reliable, then steps should be taken to investigate and correct the problem. A direct person-to-person explanation of the reason for the monitoring program may be all that is required, persuasion may be effective, or in an extreme case of noncooperation where there are legal issues at stake, it might be necessary to make it a disciplinary matter. It would be hoped that the latter could be avoided since it would create the potential for a future adversarial relationship.

Any monitoring program only provides (hopefully valid) data for the exposure levels present in the area where the program was carried out and under the circumstances as they existed at the time the program was conducted. If the circumstances change, then it may need to be repeated, extended to other materials, or modified because procedures have changed. Most laboratory organizations do not have sufficient manpower to monitor laboratories often enough to catch every variation. It is incumbent upon laboratory directors or managers, or other employees, to notify the industrial hygiene specialist of internal operational changes which would affect the airborne levels of any hazardous material.

It also should be remembered that acceptable levels are those which would be appropriate for an average person not to experience any adverse health effects. There are many persons who are hypersensitive to a specific substance and such levels may not be suitable for them. It also is possible for persons not to be initially sensitive to a given material and become more sensitive to it over a period of time. Sometimes this sensitization process may carry over to additional materials. We know or can infer a substantial amount about the acute effects of many chemicals at relatively high levels, primarily because of animal experimentation, but we do not know a great deal about the delayed effects of a single exposure to toxic material or the delayed effects of prolonged exposure to low levels of contaminants, nor do we know very much about the synergistic effects of combinations of chemical exposures. It also is obvious that some individuals cannot tolerate particular environments because of heat, vibration, lighting conditions, ergonomic requirements, or because of the presence of chemicals or odors. However, most people can develop an equilibration with the environment, tolerating ambient conditions without any untoward effects. Occasionally a new element such as a change in the process, additional or different chemicals, heat, vibration, etc. occurs, resulting in a hypersensitization wherein the previously acceptable pollutants are no longer tolerable. This is not a synergism, but a new condition. Precipitating concentrations of any pollutant may be below the TLV or TWA. Good housekeeping measures may be of some value, but when hypersensitization has occurred, there might be no alternative but to remove the employee from the locus. For how long, only time will answer. Some individuals will gradually lose the sensitivity over 3 to 6 months, but some may retain it for life. Another condition can exist. Once the individual has acquired the immune response or developed a lower threshold and is removed from the situation, he may become asymptomatic. The immune system, however, may remember and, upon reexposure, the previous allergic response can reoccur in full force, just as it did at the time of primary acquisition. This is called the anamnestic response. Although it would be desirable to return an afflicted employee to his previous job, one must keep anamnesis in mind and proceed with caution. When reentry fails, one must avoid the tendency to become angry, frustrated, or impatient with the employee. The reaction is not imaginary, it is more real and discomforting to the employee than it is to management. Therefore, the legal limits or the recommended levels should be considered as upper limits, not to be exceeded or approached if reasonable measures are available to reduce the actual levels in the workplace. There is no reason to excessively fear the laboratory environment, but it is foolish to scoff at or ignore reasonable measures to reduce levels of exposure.

### 4.8.3 MODES OF EXPOSURE

In order to better understand how exposure to hazardous materials in the laboratory enters into operational safety, a brief article from the 2nd edition of the *Handbook of Laboratory Safety,* pp 314—316 by Herbert E. Stokinger will be used to illustrate this point.

### MEANS OF CONTACT AND ENTRY OF TOXIC AGENTS

Of the various means of body exposure to toxic agents, skin contact is first in the number of affections occupationally related. Intake by inhalation ranks second, while oral intake is generally of minor importance except as it becomes a part of the intake by inhalation or when an exceptionally toxic agent is involved. For some materials, as might be inferred, there are multiple routes of entry.

### SKIN CONTACT

Upon contact of an industrial agent with the skin, four actions are possible: (1) the skin and its associated film of lipid and sweat may act as an effective barrier which the agent cannot disturb, injure, or penetrate; (2) the agent may react with the skin surface and cause primary irritation; (3) the agent may penetrate the skin, conjugate with tissue protein, and effect skin sensitization; and (4) the agent may penetrate the skin through the folliculi-sebaceous route, enter the bloodstream, and act as a systemic poison.

The skin however is normally an effective barrier for protection of underlying body tissues, and relatively few substances are absorbed through this barrier in dangerous amounts. Yet serious and even fatal poisonings can occur from short exposures of skin to strong concentrations of extremely toxic substances such as parathion and related organic phosphates, tetraethyl lead, aniline, and hydrocyanic acid. Moreover, the skin as a means of contact may also be important when an extremely toxic agent penetrates body surfaces from flying objects or through skin lacerations or open wounds.

## INHALATION

The respiratory tract is by far the most important means by which injurious substances enter the body. The great majority of occupational poisonings that affect the internal structures of the body result from breathing airborne substances. These substances lodging in the lungs or other parts of the respiratory tract may affect this system, or pass from the lungs to other organ systems by way of the blood, lymph, or phagocytic cells. The type and severity of the action of toxic substances depend on the nature of the substance, the amounts absorbed, the rate of absorption, individual susceptibility, and many other factors.

The relatively enormous lung-surface area (90 square meters total surface, 70 square meters alveolar surface), together with the capillary network surface (140 square meters) with its continuous blood flow, presents to toxic substances an extraordinary leaching action that makes for an extremely rapid rate of absorption of many substances from the lungs. Despite this action, there are several occupationally important substances that resist solubilization by the blood or phagocytic removal by combining firmly with the components of lung tissue. Such substances include beryllium, thorium, silica, and toluene-2,4-disocyanate. In instances of resistance to solubilization or removal, irritation, inflammation, fibrosis, malignant change, and allergenic sensitization may result.

Reference is made in the following material to various airborne substances, and to some of their biologic aspects.

### Particulate Matter: Dust, Fume, Mist, and Fog

Dust is composed of solid particulates generated by grinding, crushing, impact, detonation, or other forms of energy resulting in attrition of organic or inorganic materials such as rock, metal, coal, wood, and grain. Dusts do not tend to flocculate except under electrostatic forces; if their particle diameter is greater than a few tenths micron, they do not diffuse in air but settle under the influence of gravity. Examples of dust are silica dust and coal dust.

Fume is composed of solid particles generated by condensation from the gaseous state, as from volatilization from molten metals, and often accompanied by oxidation. A fume tends to aggregate and coalesce into chains or clumps. The diameter of the individual particle is less than 1 micron. Examples of fumes are lead vapor on cooling in the atmosphere; and uranium hexafluoride ($UF_6$) which sublimes as a vapor, hydrolyzes, and oxidizes to produce a fume or uranium oxyfluoride ($UO_2F_2$).

Mist is composed of suspended liquid droplets generated by condensation from the gaseous to the liquid state as by atomizing, foaming, or splashing. Examples of mists are oil mists, chromium trioxide mist, and sprayed paint.

Fog is composed of liquid particles of condensates whose particle size is greater than 10 μm. An example of fog is supersaturation of water vapor in air.

### Gas and Vapor

A gas is a formless fluid which can be changed to the liquid or solid state by the combined effect of increased pressure and decreased temperature. Examples are carbon monoxide and hydrogen sulfide. An aerosol is a dispersion of a particulate in a gaseous medium while smoke is a gaseous product of combustion, rendered visible by the presence of particulate carbonaceous matter.

A vapor is the gaseous form of a substance which is normally in the liquid or solid state and which can be transformed to these states either by increasing the pressure or decreasing the temperature. Examples can include carbon disulfide, gasoline, naphthalene, and iodine.

### Biologic Aspects of Particulate Matter

The size and surface area of particulate matter play an important role in occupational lung disease, especially the pneumoconioses. The particle diameter associated with the most injurious response is believed to be less than 1 μm; larger particles either do not remain suspended in the air sufficiently long to be inhaled or, if inhaled, cannot negotiate the tortuous passages of the upper respiratory tract. Smaller particles, moreover, tend to be more injurious than larger particles for other reasons. Upon inhalation, a larger percentage (perhaps as much as tenfold) of the exposure concentration is deposited in the lungs from small particles. This additional dosage and residence time act to increase the injurious effect of a particle.

The density of the particle also influences the amount of deposition and retention of particulate matter in the lungs upon inhalation. Particles of high density behave as larger particles of smaller density on passage down the respiratory tract by virtue of the fact that their greater mass and consequent inertia tend to impact them on the walls of the upper respiratory tract. Thus, a uranium oxide particle of a density of 11, and 1 μm in diameter will behave in the respiratory tract as a particle of several microns in diameter, and thus its pulmonary deposition will be less than that of a low-density particle of the same size.

Other factors affecting the toxicity of inhaled particulates are the rate and depth of breathing and the amount of physical activity occurring during breathing. Slow, deep respirations will tend to result in larger amounts of particulates being deposited in the lungs. High physical activity will act in the same direction, not only because of the greater number and depth of respirations, but also because of the increased circulation rate, which transports the toxic amounts of certain hormones that act adversely on substances injurious to the lung. Environmental temperature also modifies the toxic response of inhaled materials. High temperatures in general tend to worsen the effect, as do temperatures below normal, but the magnitude of the effect is less for the latter.

### Biologic Aspects of Gases and Vapors

The absorption and retention of inhaled gases and vapors by the body are governed by certain factors different from those that apply to particulates. Solubility of a gas in the aqueous environment of the respiratory tract governs the depth to which a gas will penetrate in the respiratory tract. Thus very little if any of the inhaled, highly soluble ammonia or sulfur dioxide will reach the pulmonary alveoli, depending on concentration, whereas relatively little of insoluble ozone and carbon disulfide will be absorbed in the upper respiratory tract.

Following inhalation of a gas or vapor, the amount that is absorbed into the bloodstream depends not only on the nature of the substance but more particularly on the concentration in the inhaled air, and the rate of elimination from the body. For a given gas, a limiting concentration in the blood is attained that is never exceeded no matter how long it is inhaled, providing the concentration of the inhaled gas in the air remains constant. For example, 100 parts per million of carbon monoxide inhaled from the air will reach an equilibrium concentration in the blood corresponding to about 13% of carboxyhemoglobin in 4 to 6 hours. No additional amount of breathing the same carbon monoxide concentration will increase the blood carbon monoxide level, but upon raising the concentration of carbon monoxide level in the air a new equilibrium level will eventually be reached.

**INGESTION**

Poisoning by ingestion in the workplace is far less common than by inhalation for the reason that the frequency and degree of contact with toxic agents from material on the hands, food, and cigarettes are far less than by inhalation. Because of this, only the most highly toxic substances are of concern by ingestion.

The ingestion route passively contributes to the intake of toxic substances by inhalation since that portion of the inhaled material that lodges in the upper respiratory tract is swept upward within the tract by ciliary action and is subsequently swallowed, thereby contributing to the body intake.

The absorption of a toxic substance from the gastrointestinal tract into the blood is commonly far from complete, despite the fact that substances in passing through the stomach are subjected to relatively high acidity and on passing through the intestine are subjected to alkaline media.

On the other hand, favoring low absorption are observations such as the following: (1) food and liquid mixed with the toxic substance not only provide dilution but also reduce absorption because of the formation of insoluble material resulting from the combinatory action of substances commonly contained in such food and liquid; (2) there is a certain selectivity in absorption through the intestine that tends to prevent absorption of "unnatural" substances or to limit the amount absorbed; and (3) following absorption into the blood stream the toxic material goes directly to the liver, which metabolically alters, degrades, and detoxifies most substances.

## ACKNOWLEDGMENT

Reprinted from *Occupational Diseases*, Washington, D.C., U.S. Public Health Service Publication No. 1097, pp.7—12, 1964.

### 4.8.4. HEALTH ASSURANCE PROGRAM

Concern for the effect of the work environment on an employee's health should be the primary purpose of any medical monitoring program. The term "health assurance" not only implies that the health of the participant will be scrutinized, but also carries the positive connotation that steps will be taken to eliminate or minimize any negative impact on the employee's health of work-related factors.

### 4.8.4.1. An Overview of Health Assurance Programs*

The laboratory environment, as well as that of many support activities in academic institutions and industrial organizations, has the potential of causing health problems for employees. Since many laboratories are working on the forefront of knowledge, persons working in these environs may be exposed to materials for which significant knowledge of the health effects are not yet known. In other cases, the exposures to known materials may be sufficient to cause an adverse effect on an individual. A very positive component of a health and safety program for any industry, and especially those for which the risks are incompletely known or where stress on an individual's health may be unusually high, is a health assurance program, i.e., a program in which the health of the individual is monitored on a regular basis or as a result of any out-of-the-ordinary work experience. Not only does such a program serve to alert the facility and individuals of the possible health implications of their activities, but employees normally view the program as reflecting an interest in their welfare.

It is incumbent on universities and/or industries to provide well-thought-out and implemented health assurance programs. The size and complexity of the programs should obviously be geared to the size, type, and activity of the facility. Recognition of the elements which contribute to poor environmental conditions or are harmful to employees is mandatory before control or corrective actions can be formulated and effected.

Progressive corporations and academic institutions are recognizing that a pleasant, safe, healthy environment and contented, healthy employees results in:

- Decreased employee turnover — less training for new employees
- Increased productivity — both quantity and quality
- Longevity — employees retire later
- Decreased loss time and insurance costs

The corporate health service can encompass many areas — biomechanics, psychology, ergonomics, health promotion, nutrition, CPR, first aid training, industrial hygiene, statistics, epidemiology, and occupational medicine. Related areas which could be in the program or closely tied to a health assurance program could be data processing, cost analysis, insurance evaluation, and workers' industrial compensation and laws. These areas might be addressed to whatever degree applicable by nurses, hygienists, technicians, or physicians. If the corporation is small or if a large facility is near a specialized clinic, all of these parameters could be handled by that clinic without establishing an in-house health facility, although the organization might have more control over costs with an in-house program.

The interrelated services of a health assurance department must be correlated and administered by a highly motivated, knowledgeable manager. This may vary from a

---

* This section was written by Dr. Richard F. Desjardins.

nurse in a small facility to an experienced health services professional in a large facility, such as a university with a multiplicity of situations. In a large factory, the range of exposures is relatively static, but in a technically oriented organization with multiple research programs covering many disciplines, there are a myriad of factors to be considered such as: (1) radiation isotopes, reactors, X-ray machines, lasers, electron microscopes, etc., (2) diverse chemicals such as solvents, pesticides, herbicides, carcinogens, metals, etc., (3) animal-handling research, veterinarian school, and agriculture, (4) genetic recombination engineering, (5) employees working in power plants, steam tunnels, electric services, plumbing, carpentry, custodial work, etc., (6) air quality in buildings such as gymnasiums, laboratories, laboratory hoods, "energy efficient" buildings, dormitories, and classrooms, and (7) emergencies — crowd control, fire control facilities, injuries, drug or alcohol problems, or whatever.

All of these areas must be dealt with, whether in a university or factory, as they arise in order to maintain a healthy, safe environment.

When environmental problems in the workplace arise, both industry and university personnel are free to consult federal and state agencies for advice, to register complaints, or for help in controlling or solving certain conditions. Consultants who are specialists in nearly any given area can be called upon for help, from such agencies as NIOSH (National Institute for Occupational Safety and Health) and OSHA (Occupational Safety and Health Agency).

Qualifications for health physicists, safety analysts, technicians, and engineers can be obtained from the many texts relating to the organization, function, and running of a good health and safety department. There are many good references with details.[1-5,13]

## HEALTH ASSURANCE MEDICAL DEPARTMENTS

It is desirable that the health assurance program be a division of a health and safety department within an organization. Their missions are strongly interdependent. No nurse or physician has the specific knowledge, experience, or facilities to perform the tests, surveillance, or control as do the other members of the health and safety department, nor do the latter typically have the medical knowledge to operate independently. However, both groups should be able obtain the help required from the other.

The size and complexity of the facility ultimately determines the size and complexity of the medical branch of the health and safety department. It is the responsibility of the employer to know when, where, and how to obtain medical personnel to provide medical services for the employees. The AMA (American Management Association) and American Occupational Medical Association have plans and pamphlets containing much information which can be used to help devise and implement a medical facility.

In a small facility, the number of employees and nature of the work will dictate the extent of health monitoring. Clerical employees might be able to avail themselves of the medical facility, but not need general physicals or specific testing unless, as is becoming more frequent, air quality problems arise stemming from the construction of energy-efficient buildings in recent years. On the other hand, workers exposed to asbestos, toxic chemicals, or radiation would require more complete and frequent examinations, testing, and monitoring. In a small facility without undue usage of toxic chemicals or hazards, a trained nurse, paramedic, or physician's assistant might do all that is necessary and refer more complicated problems to a physician or hospital.

According to Bond,[5,7] a work force of 300 that includes employees who require periodic monitoring and medical examinations can justify an in-plant medical department with a full-time nurse and a part-time physician; between 100 and 300, a part-time nurse, paramedic, or physician's assistant might suffice. For 300 to 800 employees, one full-time nurse is usually sufficient. A part-time physician could be on call or spend a specified number of hours at the plant. A group of employees numbering 800 to 1500, including a substantial number who require physicals and environmental monitoring, will justify a full-time physician and at least two industrial nurses.

This is an opportunity for a family practice physician[6] to participate in a part-time capacity in industry. It must be of some interest to the physician or the proper motivation would be missing for the medical and psychological care of the employees. It is an opportunity for doctors to further involve themselves with patients and the functioning community. Statistics indicate that primary care physicians provide about 80% of employee medical care. Only 20% of plant physicians are occupational physicians. In the U.S., financially secure companies continue to have a full-time physician for approximately each 3000 employees, although some industries are known to have only one for about 15,000 employees. Indeed, there are a few substantial companies which have no full-time physician, regardless of the number of employees. This might be because of low exposure or toxicity in the facility, or because a commercial medical clinic is contracted to care for their employees (HMO, PPO, emergency clinics).[7]

The acceptance of a health program by both management and those actually manufacturing the product, whether it is paperwork, nuts, bolts, chemicals, or research, is by the slow accrual of satisfied patients. There is an inherent suspicion and skepticism by both areas. Management needs to know that they are complying with federal and state mandates, and would like to see a positive productivity gain as a result of health expenditures. The workers find it hard to believe that management would venture into a program other than for financial gain. It is gratifying to observe the progression of acceptance by both vital areas in programs that succeed.[8] It has been amply shown that it is cost effective to have a good health program with caring personnel. The employees are healthier, happier, and more productive; the turnover of workers decreases, lessening training expenditures; loss time for illness or injuries decreases because of both attitude and from instruction and training in "wellness". Workers who feel they are an integral, functional, and productive part of a facility are content to stay on the job to retirement, and environmental stress is diminished and managed.[14-17]

The physician who accepts the challenge of participating in a health program should obviously be able to perform the usual functions of a general physician and, in addition, should be knowledgeable about the psychology of workers, the hazards and conditions in varying work sites, toxicology, and communicable diseases in the workplace.[11,12] Recognition of problems such as drug abuse, alcoholism, and the effects of smoking is mandatory. Although this is a very broad background, we must consider that patients/workers are male/female, young/middle aged/elderly.

The occupational health examination[8,9] may be divided into three areas:

1. Preemployment or preplacement
   a. Medical history
   b. Occupational history
   c. Physical exam
   d. Laboratory and X-ray (if needed)
   e. Multiphasic screening
2. Periodic examination (with interval history)
   a. Annual physical examination or at desirable intervals
   b. Executive physicals[10]
   c. Toxic or hazardous exposure bioassays
3. Special examinations
   a. Food handlers
   b. Job transfers (if markedly different)
   c. Return to work after serious illness or injury
   d. Retirement examinations (document final condition and advise as to future health and wellness)
   e. Fitness classification

These areas of assessment (1) determine the immediate health state, any change since previous examinations, and the suitability to work in any area, (2) will suggest advice and modalities to enhance or improve health, and (3) may indicate conditions of stress or unhealthy situations in the environment needing attention and change.[12]

The results of an examination should be kept confidential. A lay person should not be expected to interpret the results and make employment decisions on the basis of these data. The physician does not hire or recommend that a person should be hired. That is a corporate decision. However, the physician can categorize the preemployee or employee into several levels:

1. Fit for general work — physically and mentally.
2. Fit for work only in specific categories — physically or mentally.
3. Unfit for employment at this time — presence of medical condition requiring attention. When corrected, may be eligible for employment.
4. Incapacitating condition — illness, injury (old or new), or mental illness of a chronic nature. These would prevent employment in either a general or specific category.[8]

The validity of the physician's assessment would depend on his knowledge of the required work conditions of the specific facility and his ability in disability assessment.[11] It is also an introduction to the applicant or employee of a caring medical resource within the company or university. The perception of the physician as a "company doc" is really an uncomplimentary epithet. Fostering the perception of a caring physician really interested in the patient, who also happens to work for the establishment, will contribute to a more accurate assessment as well as help maintain healthiness at work.

Many applicants and employees are educationally deficient, but this must not be interpreted as low intellectual capacity. Very often, appreciative patients will indicate that they had always wondered about a condition, but no doctor had ever taken the time to discuss it in understandable terms. The results of a small amount of expended time are very gratifying.

The initial assessment of an employee should be more comprehensive. The elements of a specific company's products should dictate the specific areas to be assessed beyond the general. Obviously, an individual who is to do heavy lifting of any kind should have musculo-skeletal and neurological systems carefully examined. Whether X-rays are mandated is still a moot point. It is really of marginal value, generally. If there is a suggestion of a problem which would make X-rays of value, either on the physical or from the medical history, X-rays must be taken. On the other hand, exposure to chemicals, gases, heat, cold, radiation, etc. would need more specific scrutiny in other systems. Figure 4.26 is an excerpt from the form used at the author's facility. It is completed by the applicant prior to seeing the physician and carefully reviewed in the applicant's presence. Some areas can be amplified and clarified during this time. Encouraging questions from the patient at this time emphasizes caring. Following this, a careful physical examination is carried out.[8] It is probably a good idea to discuss the findings of the physical examination as it proceeds. When the examination is completed, proceed to other testing. If specific findings during the physical militate against performing certain tests, the routine can be modified. Some tests may be done by the physician, nurse, or a technician. They might include color perception, visual acuity, audiometry, spirometry, glaucoma, and electrocardiography. Some tests, such as a blood panel or urinalysis require laboratory services. All of these are mainly screening tests. Positive findings may, and probably should, be referred back to the patient's physician or to an appropriate specialist. The industrial physician must not be in competition with outside physicians. By the same token, if the patient signs an authorization, a complete copy of the examination can be forwarded to him or to the physician of his choice. Under ordinary conditions, it is not wise or proper to refer an employee to a specific physician unless that physician is the only one able to perform a given function. A list of qualified names can be provided for referral. This also contributes to harmonious relations with the local medical community.

Subsequent physical examinations of employees should be spaced appropriate to the nature of their jobs. The questionnaire may be abbreviated if they indicate the absence of changes. The physical, however, should be as careful and complete as at the beginning. This will naturally reveal any changing status, e.g., needs glasses or hearing aids, dermatitis, tumors, glaucoma, asbestosis, etc. Finding any deviation from normalcy early is a real bonus to treatment. By the time some are obvious to the patient, it may be too late. As an old medical professor once said, "There's a lot of pathology out there. All you have to do is find and recognize it." It is the responsibility of the physician to adhere to three dictums or duties:

1. Prevent disease.
2. Diagnose and treat to the best of your ability.
3. Help the patient's demise to be with as much dignity as possible. Good rapport will allow this.

## 1.0  Health Assurance

Health assurance means knowing what normal health is, what your present condition is, and whether you need to change anything such as life style, diet, or exercise to maintain your best state.

Progress in health maintenance has been so rapid that it is difficult to keep up with doing right by your body, and mind. This pamphlet is compiled to detail some areas that can be used as a basis for your own evaluation of your health condition, or to indicate goals to pursue. It is obvious, but often neglected, that each of us must be responsible for ourselves-eating properly, sleeping adequately, moderating our use of drugs and alcohol, exercizing, protecting ourselves from work exposures and hazards, and dealing with diseases-inherited or developed. If we are responsible for ourselves, learn to both identify and handle problems which exist, and prevent the development of harmful conditions, we can greatly reduce the likelihood of disease, increase the productivity of life, and develop a high state of wellness.

In an effort to achieve and maintain these goals, Virginia Tech has established a formal Health Assurance program. It is primarily for those employees identified as encountering environmental factors during the course of assigned duties that have the potential for affecting the employees' health. Medical or psychological examinations are provided at no cost to the employee and are performed by a University physician at this time. The physician's results, evaluation, and recommendations are the basis for informing the employee and University of his or her health status. Where specific medical conditions are found which may be related to or aggravated by job requirements or environment, the Department of Health and Safety follows a procedure to assure that the employee's health is protected. The procedure is as follows:

1. A conference will be held as soon as possible to discuss the physician's findings and recommendations with the employee, the employee's supervisor or department head, a staff member of the Department of Health and Safety, the Employee Relations Manager, and if necessary the physician.

2. Immediately following the discussion, any action thought necessary to protect the employee's health will be taken. Temporary changes will be made, allowing time to further evaluate the total situation and to review all possible alternatives. Where questions arise concerning the physician's findings or recommendations, consultations will be made.

3. The supervisor, staff member of Health and Safety, and Employee Relations Manager will review the physician's reports and recommendations in depth to determine any necessary job changes related to environment, or job activity requirements. In doing so the options considered to assure employee health are:

   a. Temporary job change to allow treatment of the condition

   b. Job environment change

   c. Use, or better use of protective and/or safety equipment

   d. Change in job activity requirement

   e. Change in job responsibility

   f. Schedule change in work hours

   g. Relocation within the University or related facility

Health Assurance

   h. Changes in an individual situation not already listed that could aid in problem solution

Every effort will be made to assure continued employee health and employment. If the individual's assurance plan requires relocation within the University, the plan will be reviewed by the Director of Employee Relations, and submitted for approval to the Vice-President for Administration and Operations. The completed and approved plan will be discussed with the employee, the supervisor, and the Employee Relations Manager. A copy of the plan will be placed with the employee's personal file, a copy given to the employee, a copy to the supervisor, and a copy retained by the Department of Health and Safety.

As you can see, every effort will definitely be made to involve the employee in the entire decision making procedure, consistent with insuring health and safety of that person.

FIGURE 4.26.    Excerpt from health assurance program booklet.

The industrial physician may want to compose or purchase pamphlets appropriate to the specific facility and/or leaflets or booklets promoting general health: ideas, smoking cessation, cholesterol control, back care, weight control, why and how to exercise, etc. How much good these actually do is not well documented, but employees do pick them up and carry them home. Perhaps the absorption of even small dollops of advice will contribute to the enhancement of healthfulness. Figure 4.27 is part of a pamphlet used in a university setting. As seen, it must be comprehensive to professors, carpenters, plumbers, or family members who may read it. Figure 4.28 is

Safety and Health Programs

TO:        Doctors Administering Health Assurance Tests

SUBJECT:  Tests To Be Given

| TESTS REQUIRED THIS VISIT | | TOTAL TESTS & FREQUENCY |
|---|---|---|
| _____ | A Complete Physical with Blood Series | _____ |
| _____ | A Pulmonary Function Test | _____ |
| _____ | A Cholinesterase Test Only | _____ |
| _____ | A Cholinesterase Test with Blood Series | _____ |
| _____ | E.K.G. | _____ |
| _____ | PPD(TB) | _____ |
| _____ | Hemoccult | _____ |
| _____ | X-Ray | _____ |
| _____ | Other | _____ |

This position works in the _____ . This _____ is _____ is not a pre-employment physical. The position is set for a testing frequency as outlined above. If there should be a change to the frequency of a test or if a test should be added or deleted please indicate the changes in red and return the form with the physical.

The tests of E.K.G., PPD(TB) and Hemoccult or others will be determined on an individual basis and the need for such will be determined by the physician. The physician should mark the frequency as well. Once the test is recommended the frequency will be maintained until changed by the doctor or the position becomes vacant .

FIGURE 4.27.    Sample page from medical exam form (Virginia Polytechnic Institute and State University).

an example of a specific form used to provide ongoing surveillance of those using lasers or ultraviolet light. Figure 4.29 is a form appropriate for testing hearing control and maintenance in loud or higher-decibel areas such as around a printing press, boiler makers, usage of chain saws, etc. Figure 4.30 is one of many available spirometry forms that can be used.

All of the above forms and many more are readily adaptable to computer compilation for easy comparisons of successive testings and can be used for facility statistics to show trends — good or bad. They could also be used as yardsticks to measure the efficacy of any remedial program.

The highly motivated industrial physician should have the facilities to address most environmental, factory, and facility problems. If administrators recognize and adequately fund a health assurance program, tangible evidence of the health improvement (reflected in decreased loss time) should prove to be cost effective. Intangible evidence is hard to accrue, but well employees are likely to feel more content with their work, remain with the company longer, accept healthful ways of work, and pursue similar attitudes at home.

Health assurance is truly a product of a well-thought-out and implemented health and safety program. It is cost effective, humane, and generally good administrative policy to provide such a program.

## REFERENCES (SECTION 4.8.4.1)

1. **LaDou, J., Ed.,** *Introduction to Occupational Health and Safety,* National Safety Council, Chicago, 1986.

OCULAR HISTORY:

- Patient History

- Family History

- Current Complaints About Vision

- General Health Status


VISUAL ACUITY

EXTERNAL OCULAR EXAMINATION:

- Brows
- Lids
- Lashes
- Conjunctiva
- Sclera
- Cornea
- Iris
- Pupillary Size
- Equality, Reactivity & Regularity


LENS OPACITY:

RETINOSCOPY (Manifest Refraction):

INTRAOCULAR PRESSURE:


SKIN SURVEILLANCE (for employees with history of photosensitivity
or working with ultraviolet lasers.)

FIGURE 4.28.    Items covered during eye examination.

2. Council of Occupational Health, AMA: scope, objectives and functions of occupational health programs, *JAMA,* 174, 533, 1960.
3. Council on Occupational Health, AMA: a management guide for occupational health problems, *Arch. Environ. Health,* 9, 408, 1964.
4. *Guide to Developing Small Plant Occupational Programs,* American Management Association, Chicago, 1983.
5. **Bond, M. B.,** Occupational health services for small businesses and other small employee groups, in *Occupational Medicine — Principles and Practical Applications,* Year Book Medical Publ., Chicago, 1988, 89.
6. **Howe, H. F.,** Small industry: an opportunity for the family physician, *Gen. Pract.,* 26, 166, 1962.
7. **Knight, A. L. and Zenz, C.,** Organization and staffing, in *Occupational Medicine — Principles and Practical Applications,* Chicago, 1988.
8. **Collins, T. R.,** The occupational examination: a preventive medicine tool, *Continuing Med. Ed.,* 77, 1982.
9. *Guiding Principles of Medical Examinations in Industry,* American Management Association, Chicago, 1973.
10. **Thompson, C. E.,** The value of executive health examinations, *Occup. Health Safety,* 49, 44, 1980.
11. Disability Evaluation Under Social Security, A Handbook for Physicians, Social Security Administration, Washington, D.C., 1973.
12. Occupational Diseases. A Guide to their Recognition, Publ. No. 1097, Public Health Services, U.S. Department of Health, Washington, D.C., 1966.

**AUDITORY EXAM AND HISTORY**

| | Y or N | Comments: |
|---|---|---|
| Did you have EAR infections during childhood, which ear, how frequently? | | |
| If you have ever had a hearing test, when, where, why? | | |
| Ever had attacks of dizziness? Onset? How frequently? | | |
| Ever had measles, mumps or scarlet fever? | | |
| Ever had a head injury? if unconscious, for how long? | | |
| Anyone in your family have hearing loss before age 50? | | |
| (Give reasons) After age 50? | | |
| If exposed to noise in military service, give dates & types of exposure. | | |
| Do you use firearms now or in past? Type? How often? Which shoulder? | | |
| Do you have noise in your ears? If so, onset & describe (crickets, roar, etc.) | | |
| If ever worked at a noisy job, what and how long? | | |
| Ear surgery? | | |
| Previous Jobs: | | |

| ANNUAL HISTORY UPDATE (N or Y) | 19 | 19 | 19 | 19 | 19 | 19 | 19 | 19 | 19 | 19 | 19 |
|---|---|---|---|---|---|---|---|---|---|---|---|
| Do you now have any hearing problem? | | | | | | | | | | | |
| Do your ears run (infection)? | | | | | | | | | | | |
| Do you have noise in your ears? | | | | | | | | | | | |
| Do you have dizziness? | | | | | | | | | | | |
| Do you have allergies or sinusitis? | | | | | | | | | | | |
| Have you had any injuries, illness or operations? | | | | | | | | | | | |
| Since last audiogram? If yes, explain. | | | | | | | | | | | |
| Do you have a second job? (List under Comments.) | | | | | | | | | | | |
| Any noisy hobbies? (List under Comments.) | | | | | | | | | | | |
| Do you have diabetes, arthritis or other chronic disorders? | | | | | | | | | | | |
| Do you take any medication? | | | | | | | | | | | |
| Do you hunt, skeet or target shoot? | | | | | | | | | | | |
| Do you use tractor, power tools, chain saws, etc.? | | | | | | | | | | | |
| Record BP—History of Hypertension | | | | | | | | | | | |

| COMMENTS & DATE: | |
|---|---|
| | |
| | |
| | |
| | |
| | |

| DATE: | | | | | | | | | | | | | | | | | | | | | | | | |
|---|---|---|---|---|---|---|---|---|---|---|---|---|---|---|---|---|---|---|---|---|---|---|---|---|
| OTOSCOPIC INSPECTION | R | L | R | L | R | L | R | L | R | L | R | L | R | L | R | L | R | L | R | L | R | L | R | L |
| Are drums: visible? | | | | | | | | | | | | | | | | | | | | | | | | |
| normal? | | | | | | | | | | | | | | | | | | | | | | | | |
| Cone of light visible? | | | | | | | | | | | | | | | | | | | | | | | | |
| Malleus prominent? | | | | | | | | | | | | | | | | | | | | | | | | |

FIGURE 4.29. Audiometric test form.

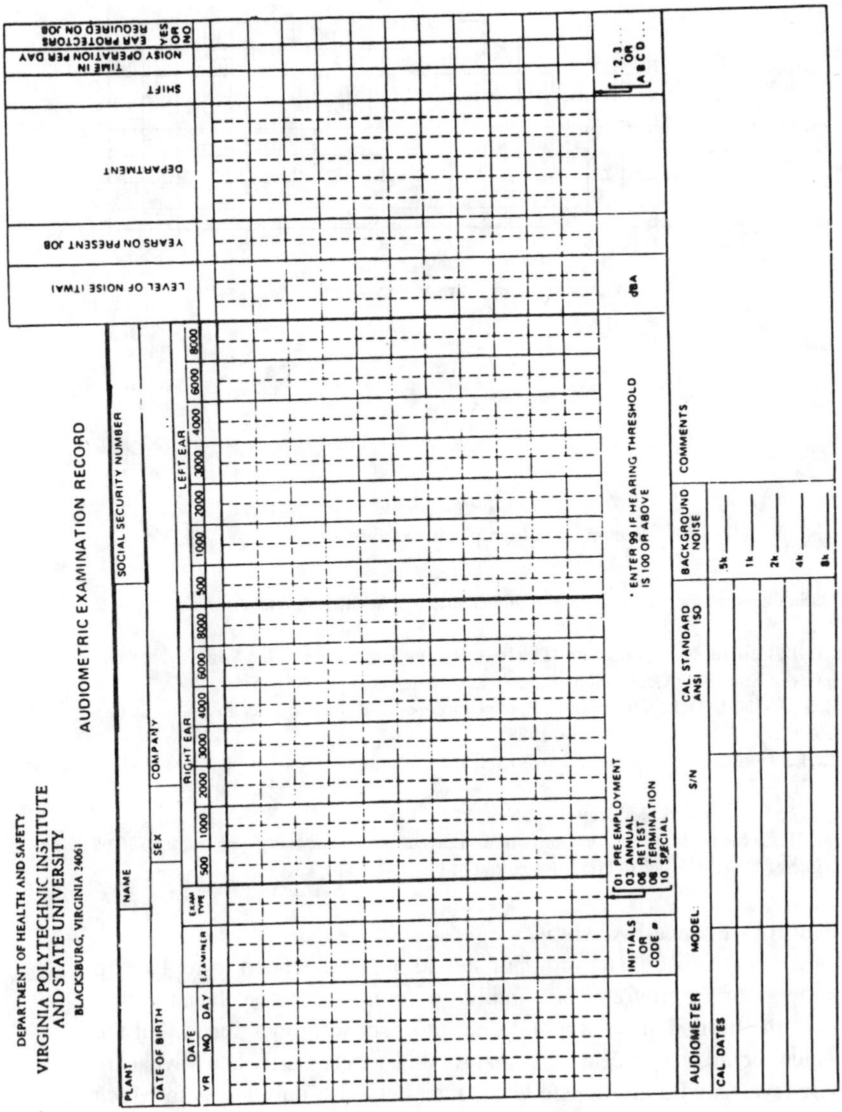


FIGURE 4.29 (continued)

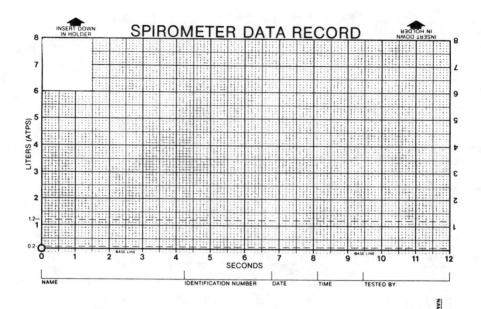

FIGURE 4.30.   Record sheet from pulmonary function test.

13. **Felton, J. S.,** Organization and operation of an occupational health program, *J. Occup. Med.,* 6, 25, 1964.

14. **French, J. R., Caplan, R. D., and Von Harrison, R.,** *The Mechanism of Job Stress and Strain,* Wiley Series of Studies in Occupational Stress, John Wiley & Sons, New York, 1982.

15. **House, J. S., Wells, J. A., Landerman, L. R. et al,** Occupational stress and health among factory workers, *J. Health Soc. Behav.,* 20, 139, 1979.

16. **Kahn, R. L.,** Conflict, ambiguity, and overload: three elements in job stress, *Occup. Ment. Health,* 3, 2, 1973.

17. **Selye, H.,** *The Stress of Life,* McGraw-Hill, New York, 1956.

18. **Stokinger, H. E.,** Means of contact and entry of toxic agents, in *Handbook of Laboratory Safety,* 2nd ed., Steere, N. V., Ed., CRC Press, Boca Raton, FL, 1971, 314.

### 4.8.4.2. A Health Assurance Program

A formal health assurance (HA) program should not be intended to replace an environmental monitoring program, but, rather, to complement one. In an earlier section, some of the limitations of a monitoring program were mentioned, but the main point, which should be appreciated, is that we do not necessarily know what are "safe" exposure levels, and, indeed, safe levels may not exist, only levels in which the negative effects attributable to an exposure disappear into the statistical noise caused by the presence of other parameters. This is especially true for individuals, with their wide range of susceptibilities. In many instances, even this much is not known since the data are not available, simply because the studies have not been performed. For example, the ACGIH tables incorporate several hundred chemicals, but this is very small compared to the 50,000 to 100,000 (according to various estimates) commercial chemicals already in use, to which must be added hundreds of new ones developed yearly. Further, the number of studies establishing safe levels for the synergistic interactions of combinations of materials is virtually nil, as often as not associating the effects of a material with an easily measurable personal habit such as whether the exposed individual smokes. The prospect of conducting synergistic studies, with all the combinations involved, is obviously not bright.

Regulatory requirements for a number of materials now specifically mandate employee access to an employer-supported medical surveillance program designed to monitor problems associated with the specific chemical. Many authorities recommend participation in programs for anyone who works with toxic substances during the normal performance of their duties. However, there are few specific recommendations on what actually constitutes a significant involvement with toxic materials which should trigger participation in an HA program. This will be discussed further later in this section. The content of the examination will depend to a great extent on the duties associated with the job. However, there are some components of an HA program which essentially are universally agreed upon as essential. Among these are

1. A medical history
2. A prior work history
3. A preplacement examination
4. Periodic reexaminations
5. An end-of-employment examination

The entire program should, in addition to a number of standard components, be tailored to the anticipated types of exposures in which the employee might be involved. As these exposures change during alterations in the research program, periodic reexaminations can be modified to reflect the changing conditions.

Even before the HA program is initiated, there are a number of key ethical issues which must be addressed. If participation in an HA program is required as a condition of employment, the advertisements for positions should so state. Further, the examination must clearly be intended to determine if the duties would be such as to make it unsafe for the employee to perform the work or aggravate an existing health problem. If reasonable adjustments can be made in the duties or responsibilities, then the examination cannot be used to discriminate against an otherwise qualified applicant. The employee should have an assurance of confidentiality. There are factors which may be health related, but which have no bearing on the ability of an employee to do the work assigned, and which cannot harm those with whom the employee would come into contact. The employee has the right to expect that any such information remain confidential. Finally, the employee should have access to the results of the examination and any tests which are performed, and should be able to authorize release of the data to others, such as his family physician, if he so desires.

### 4.8.4.2.1. Participation

If the organizational approach is to provide an examination based on need, then the necessity arises to define criteria as to who should be included and who should not. Individuals who do not work with chemical or biologic agents, or whose duties are not unduly stressful, can justifiably be excluded. Since the intent of an HA program, as applied to research personnel, is to monitor the impact of chemicals or pathological organisms on the health of the individual (either directly or indirectly, e.g., wearing a respirator can place a burden on a person with impaired pulmonary function), exposure to chemicals and biologically active (to humans) agents should be a major factor to be considered in the participation of an individual in an HA program. OSHA requires access to a medical program for persons working with regulated carcinogens and certain other materials. However, for other substances, the toxicity of the material, mechanism of exposure, duration and intensity of exposure, safeguards available to prevent exposure, current state of an individual's health, and prior exposures all play a part in the decision.

If an individual is working with an agent which is significantly infectious to humans, there appears to be little question that participation in a medical program is needed. Although the probability of contracting a disease increases with higher exposure rates, once contracted, the characteristics of the disease are not dependent upon continued exposure or the initial level of exposure.

If a major portion of an individual's time is spent working with a regulated carcinogen, other regulated materials such as lead or cotton dust, other materials which meet the criteria for being highly toxic or corrosive, a sensitizer, or an irritant, then again it is usually desirable for the individual to be in a medical program. Even if facilities are available, such as totally enclosed glove boxes in which the work is done, it is arguable that unplanned exposures could occur, and the conservative approach would be to include rather than exclude the person. It also could be argued that if the exposure levels are maintained sufficiently low, then participation in a program is not needed. It is on this basis that OSHA defines exempt levels for meeting some of the regulatory requirements for some of the regulated carcinogens. However, documentation of the low levels would appear to be required to deny access of an individual to a medical surveillance program on this basis. Of course, if a monitoring program provides information which indicates that the individual uses materials of concern or is in an area where others use them, and is actually exposed to airborne concentrations which are typically a significant percentage of acceptable levels, then participation in an HA program is indicated.

The OSHA standard for respiratory protection in 29 CFR 1910.134(b)(10) states: "Persons should not be assigned to tasks requiring the use of respirators unless it has been determined that they are physically able to perform the work and use the equipment. The local physician shall determine what health and physical conditions are pertinent. The respirator user's medical status should be reviewed periodically (for instance annually)."

Persons who do not have a continued exposure to chemicals, but periodically perform tasks requiring intense uses of chemicals for a brief period, such as in agricultural field experimentation, should probably be included in an HA program. Not only are many agricultural chemicals quite toxic, but the working conditions place severe physiological stress on research personnel and their support staff. Respirators should be worn, as should clothing which will not be permeable to the chemical sprays. The first of these places stress on the pulmonary and cardiac system, while protective clothing which is impermeable to fumes and vapors usually does not, but the body temperature will rapidly increase since the clothing prevents heat from being carried away from the body by evaporation, conduction, or convection of perspiration.

Persons who have known health problems which could be aggravated by the exposures involved in their job duties, or who have had prior work histories where they could have had significant exposures to chemicals which could have sensitized them to chemicals in the work place or which could have initiated delayed effects, also would fall in a category which should be considered for participation in an HA program.

It is most difficult to determine whether persons for whom the exposures are marginal — i. e., where the portion of their duties in which they use chemicals is limited, but who do use materials with properties which could cause ill effects a portion of the time — should be included in an HA program. It would be easy to establish a criterion that any use whatsoever should qualify a person for participation. However, many activities of normal, everyday life involve use of such items as gasoline and household products containing toluene, acetone, phenol, isopropyl alcohol, ethyl alcohol, hydrogen peroxide, acid, and caustic materials such as lye, which certainly are toxic materials. The "any use" criterion is undoubtedly too liberal, unless one simply admits that there are no selection criteria and includes every employee. A compromise which appears reasonable, but which has no other scientific justification, is to arbitrarily select a percentage (e.g., 10% of a typical work week) for actual use or exposure to a chemical or combination of chemicals of average health risk as a threshold. An employee approximating an exposure of this level could be asked to fill out a form listing the chemicals which are in use in his vicinity and estimate the average time each is used. This form should be reviewed by a physician (preferably with a background in environmental and occupational medicine, if one is available), and his recommendation should govern the question of participation. However, an individual who wishes to be included, but who might not be recommended, probably should be permitted to do so.

Although every person who has duties which could give rise to health problems should be a participant, it is especially critical to include permanent employees. Many of the tests which are run on the individual have a sufficiently wide "normal" range that, except in extreme cases of acute exposure where the individual will receive medical attention anyway, a single examination may not be particularly informative. However, problems due to environmental work conditions, as will be discussed later, may be shown to reflect trends by comparison of successive examinations.

### 4.8.4.2.2. Medical and Work Histories

The medical history and prior work histories are key components of any health assurance program. There are any number of health-related factors for which a heredity predisposition exists, so the medical history will normally include a segment concerning family members, particularly parents. Obviously, known prior medical conditions will be of importance. Emphysema, for example, would certainly be of concern if the employee were to have to wear a respirator frequently during the course of his duties. Hypertension and heart problems would clearly be of importance if the job involved significant physical stress and, of course, there are chemicals which directly or indirectly affect the heart function. Medical history forms vary substantially in content, but one used for an HA program should be comprehensive. It is part of a record which, along with the prior work history and the actual examination, including tests which may be run, will constitute the baseline against which changes in the employee's health will be compared to determine if occupational exposures are having a negative effect on the employee's health. Some prospective employees are inclined to conceal previous illnesses if it is likely to affect their chances of obtaining a desired position. This is unfortunate, but quite understandable. It is important that these conditions, should they manifest themselves later, are detected, if possible, by appropriate questions or during the preplacement examination. The need to do so stems from a desire (1) to avoid responsibility for the diseases and (2) to explain them on the basis of the occupational exposures experienced by the individual.

The prior work history serves essentially the same purpose as does the medical history. For example, a prospective employee who would be working in agricultural research programs might have had a previous period of employment working with pesticides and herbicides which could have had a depressing effect on his cholinesterase enzyme levels. It would be important to include a test of this parameter in the preplacement examination. Even some nonchemical activities, such as previous work in a heavily dust-laden atmosphere, might have caused a decrease of pulmonary function to the extent that an individual might find it impossible to wear a respirator to provide protection against solvent fumes. Previous exposures to some chemicals or substances may result in effects which are delayed for many years, such as the latency periods generally associated with carcinogens. Among other substances for which any history of prior exposures might be elicited are asbestos, dusts, welding fumes, heavy metals, pesticides, herbicides, acids, alkalis, solvents, dyes, inks, paints, thinners, strippers, gases, radiation, etc. If there are any specific areas of concern because the work regimen will involve materials which are known to have a possible impact on a given physiological function or organ, then the physician should supplement his standard questionnaire for both the medical and work histories with questions designed to elicit as much relevant information as practicable.

There are any number of common diseases which are not necessarily related to an occupational exposure, but which could be a risk in the work environment. Diabetes and hypertension are certainly not necessarily work related, but the individual, unless treated, could be a hazard to himself and, potentially to his co-workers. Similarly, a disease of the eyes such as glaucoma could interfere with a person's ability to see properly, but many persons could be unaware of its onset as it is an insidious disease, primarily a problem to persons over 40. A relatively simple automatic instrument is available to detect pressure increases in the eyes, which is a sign of the disease, and which would permit the physician to refer persons to an ophthalmologist. Loss of hearing could be a problem if persons do not hear warnings and, again, many persons do not realize that this has become a problem or are reluctant to admit it, even to themselves, as a sign of increasing age. Although these problems should have been brought to the attention of the individuals by their family physician, a surprising number of persons either do not have a family physician or do not visit him frequently. At the author's institution, approximately 15 to 20% of the persons participating in the HA program had reasonably serious problems of which they were unaware, and which could have placed them at risk or, at best, reduced their efficiency and productivity. The scope of the examination should be sufficient to detect these conditions, which may not be job related.

### 4.8.4.2.3. Preplacement Examination

It would be highly desirable if a preplacement exam could be given prior to any work exposure to provide a true baseline for the individual. However, unless a medical examination has been an integral part of an organization's employment procedure since the inception of the company or institution, then instituting an HA program will always catch a number of current employees already in the midst of research programs involving exposure to hazardous materials. A medical examination at this time will still have significance in the sense that future examinations can still be compared to the earlier one to detect changes during the interval between examinations. However, the information gained in the exam, including any test results, will not necessarily reflect normal conditions for the employee. If, for example, an individual has been working within an organization using agricultural chemicals and has a very low level of cholinesterase enzyme at the time of the initial examination, it may be suspected that the employee's work has caused the depressed level of the enzyme to occur. However, the individual may have a naturally low level. If a person is tested at the time of initial employment, then the effect of the working environment on the parameters measured in the examination will be much more apparent, although the effect of work exposures on an individual may be confused if similar exposures are likely to occur outside the workplace. In the example just used, if the initial examination revealed a normal enzyme level and a later one showed a depressed value, perhaps after a suspected exposure, then the initial conclusion, barring any alternate exposure mechanism, would be that an exposure had occurred and remedial steps would be taken to prevent further exposures and future incidents of the same kind.

The other major purpose of a preplacement examination would be to avoid placing a person in a position where an existing condition would be aggravated or the individual could be injured by the work environment. A colorblind person, for example, should not be placed in a position where the ability to distinguish colors is essential to safely. An individual with a severely reduced pulmonary function should not be placed in a position that requires wearing a respirator much of the time. These restrictions may make it impossible for an applicant to be offered a position, and it should be clearly stated in the advertised job qualifications, in such a case, that passing a preplacement examination is required as a condition of employment. A byproduct of such a restriction is that the organization may be protected against acquiring a future liability. If, for example, a person with a depressed pulmonary function is hired without an examination and placed in a situation where exposures could cause the same result, it could be difficult to prove that the problem did not arise from a recent exposure. On the other hand, detection of the problem in the preplacement examination might lead to a decision not to hire the individual because of the problem.

There are some pitfalls in using the preplacement examination as an exclusionary device. This has already been alluded to in Section 4.8.3, where a cautionary flag was raised against using the examination as a discriminatory device. This can occur with the best intentions in the world. An organization may decide to exclude women of fertile age from a position in which they may be exposed to a teratogen. Discrimination may be claimed if the installation of engineering controls to reduce the exposures to well below the permissible limits is feasible, but not done. A woman may decide not to work in an area where even low levels of an embryotoxin are present, but the decision should be clearly her own, with no taint of coercion.

A fairly common practice in a preplacement medical examination program, in addition to a thorough physical and a battery of tests, is to take a serum sample to be stored in an ultra-low temperature freezer. These samples take up very little space and are valuable should a question later arise where a comparison between a current serum specimen and a baseline sample would be useful. It is also possible and feasible to lyphilize the serum for storage. This might be cost effective and space saving if a large number of specimens are to be kept.

### 4.8.4.2.4. Reexamination

Periodic reexaminations should be scheduled for all participants of an HA program, whether it is a part of a program mandated by a standard, as is becoming more common in newer regulations, or as a result of an internal decision based upon the level of usage. The frequency of the reexamination need not be any fixed interval, but should be based on the level of exposure. Returning to the use of pesticides, which could cause a depression in the cholinesterase enzyme level, as an example, it might be desirable to test for this one component prior to a period of active use, perhaps at the height of the spraying season and again at the end of the period of activity (assuming the material is significantly dangerous to humans). For pesticides which are less toxic, this amount of testing might be excessive, whereas for the intensive use of an exceptionally dangerous material such as parathion, it might even be desirable to test daily. However, there would normally be no reason to perform a complete examination at an accelerated schedule such as this.

An annual examination is probably the one most often used in HA programs for individuals with typical exposures in a representative laboratory. However, for persons only marginally meeting the requirements for participation, the interval between examinations might be extended to 2, 3, or 5 years. Some programs use a 5-year interval for a complete physical, but recommend special tests more often. The NIH, in their program for animal handlers, recommends taking a new serum sample every 5 years. The medical advisor or occupational physician should evaluate the requirements for each participant to establish the optimum period between examinations.

After each examination, the physician should compare the results of the current examination to the findings of previous examinations. Except in isolated instances, such as the depression of the cholinesterase enzyme which we have been using as an example, or unless there has been a severe exposure where acute effects might be anticipated, the primary means of detecting problems will be the comparison of the results of successive examinations. Changes in various parameters which have been measured, such as pulmonary function, might vary slightly between two successive tests, but a persistent trend toward poorer performance would indicate damage to the respiratory system. Similarly, should a persistent trend develop for the other parameters measured, the examining physician should discuss the work environment and other possible contributing factors, such as leisure-time activities, with the employee. In at least one instance, a spraying program to control insects at a cottage where a person spent weekends was a major threat to the individual's health rather than any personal problem or exposure from any other source. The patient did not mention this factor to the physician for some time because he did not recognize it as a potential problem. As a result, the examining physician had major problems identifying the cause of the individual's illness and was unable to treat it successfully. By the time the problem was recognized, the patient had been highly sensitized to any similar material and had some long-term health problems which affected his capacity to perform many activities.

### 4.8.4.2.5. Utilization of Results

The primary purpose of the examination is to protect the employee, with a secondary purpose being to help protect the institution from the liability associated with unwittingly allowing an individual to become ill due to the work environment.

As noted earlier, a substantial number of persons involved in an HA program may have existing problems which are not job related or, as a normal course of events, develop health problems which are clearly not job related. These may be detected during the HA examination as readily as in any other comparable comprehensive physical examination. Some organizations will assume direct responsibility for treating these illnesses, although most do not, leaving the burden of seeking treatment to the patient. Financially, in many cases there is only a moderate difference to the patient due to the availability of group health insurance plans, but if the patient has the responsibility of seeking out medical treatment, the condition may remain untreated, although the examining physician should certainly encourage the individual to seek assistance. In such a case, the employee should have the right to authorize the release of the medical records to his own physician, and to have this done promptly by the organization for which he works.

Where the physical examination reveals a medical condition which may be job related or is aggravated by the duties of the person's job or the environment in which he works, then steps should be taken to protect the employee's health. One of the first things to do is to confirm that the condition exists or to obtain additional data to better understand the problem by seeking additional tests, obtaining a second opinion, or referring the patient to a specialist. These options should be discussed with the patient. In some cases, the situation is sufficiently straightforward that these followup steps would not be necessary.

Whether one postpones gathering supportive data from additional examinations depends somewhat upon the seriousness of the problem which has been discovered and the work situation. The physician, in consultation with the individual, his supervisor, and usually a representative of the department with overall responsibility for the organization's health and safety program, should determine what temporary steps can be taken to reduce the risk to the employee. In some cases, the head of the department in which the employee works may have to become involved if the supervisor does not have the authority or the flexibility to make changes.

Once all the data are available, the various options to protect the employee should be carefully reviewed. A number of these should be routinely considered:

1. If the condition can be treated, a temporary change in duties may be all that is needed.
2. It may be feasible to make engineering changes to modify the work environment.

3.    Personnel protective equipment or safety devices can be used to reduce an individual's exposure if engineering changes are not practicable.

4.    It may be possible to change the job activity.

5.    If the person has a unique problem, job responsibilities may be distributed differently among personnel in the facility so that the duties causing the difficulties to the individual would not be a problem to the others.

6.    It may be possible to reschedule the activity causing the problem to another time or to reschedule the individual.

7.    If there are no suitable options available within the individual laboratory, then relocation within the organization should be considered.

Some of these options are more easily applied in an industrial environment than in the typical academic laboratory, where each person may be supported by a grant and each laboratory is nearly autonomous. There may be very little flexibility available to the laboratory supervisor or laboratory director. This makes the task of treating the employee fairly much more difficult since the work causing the problem usually must be done and the laboratory supervisor does not have the funds to hire a new person and keep the original employee as well. There may, in fact, be little flexibility of any kind if there are no alternative positions for which the individual is qualified or is willing to accept. However, every avenue must be explored because it is not permissible to maintain the individual in a situation where his health may be endangered, even if the person wishes to do so. A waiver of responsibility for any future problems by the corporation or institution signed by the employee is not an acceptable alternative, nor is it legally defensible.

In an extreme case, where every option has been examined and none are feasible, the person may have to resign or be terminated for his own protection. An employee relations specialist in matching persons to jobs as well as an individual charged with seeing that employees are not discriminated against should be brought into the situation well before this drastic step is considered. In such a case, a financial severance settlement, insurance such as workman's compensation, or disability retirement options may be available to the employee.

### 4.8.4.2.6. Physician Training

Any physician involved in a health assurance program will have had the usual training and exposure to a variety of medical situations. It also would be highly desirable if the individual has had specific training in industrial medicine. However, since the actual conditions of employee exposure to hazardous materials will differ with each organization, the physician should be sufficiently familiar with the types of exposures represented by the job descriptions of the employees to be able to apply his own expertise and experience to these potential exposures. The more complex and diversified the research programs in an organization, the more difficult this task will be. It probably would be desirable for the physician to set aside some time to visit the various research areas as well as the supervisors and individual employees.

Since OSHA requires that a medical surveillance program be made available to employees working with regulated carcinogens and a number of other materials, the physician should be provided with all current information related to these standards as well as appropriate technical information relating to these materials and other hazardous materials used by the employees. The physician should, for example, have access to a set of all current material safety data sheets for the chemicals used by the employees. Subscriptions should be provided to some of the excellent services which are now available to keep track of the rapidly changing regulatory and technical information as well as the usual medical journals. The physician should have an opportunity to attend relevant workshops, seminars, and professional meetings to ensure that his background is maintained at a high standard.

Finally, the way in which the employees perceive the physician is an extremely important component of an organization's health and safety program. He should be perceived as professionally capable. It also is important that employees do not perceive him as a "company" man. They must feel that the their health is important to the physician and that, if they are having a problem on the job, the physician is concerned about it for their sake, not because it will cause a problem for the organization. Certainly, the physician should be concerned about the welfare of the organization, but this can be done by working to make sure that the health and safety of the organization's employees is protected. This is one reason, as noted earlier, why the name "Health Assurance Program" is preferred over "Medical Surveillance Program". The former has a much more positive sound than does the latter. Since a preplacement medical examination is recommended for individuals exposed to hazardous materials, the physician has a superb opportunity to establish from the beginning that the company or institution is concerned about the employee's well being.

### 4.8.4.2.7. Records

Because many materials are now known to have long-term effects and extended latency periods are known to exist for many carcinogens, it would be desirable to maintain all medical examination records as well as those relating to exposures and monitoring for an extended period after the employee had left the organization. Many of the specific OSHA standards describe the records which must be maintained and the period for which the records must be kept. However, 29 CFR 1910.20 covers the topic of health and safety records in general. Some of the more critical portions of this section are given below. Some of these requirements are very detailed and demanding. The reader is referred to the OSHA Standards for General Industry for the complete version of the standard.

**Access to employee exposure and medical records**

(a) *Purpose:* The purpose of this section is to provide employees and their designated representatives a right of access to relevant exposure and medical records, and to provide representatives of the Assistant Secretary a right of access to these records in order to fulfill responsibilities under the Occupational Safety and Health Act. ...

...

(2) This section applies to all employee exposure and medical records, and analyses thereof, of employees exposed to toxic substances or harmful physical agents, whether or not the records are related to specific occupational safety and health standards.

...

(4) "Employee" means a current employee, a former employee, or an employee being assigned or transferred to work where there will be exposure to toxic substances or harmful physical agents. In the case of a deceased or legally incapacitated employee, the employee's legal representative may directly exercise all the employee's rights under this section.

(5) "Employee exposure record" means a record containing any of the following kinds of information concerning employee exposure to toxic substances or harmful physical agents:

(i) environmental (workplace) monitoring or measuring, including personal, area, grab, wipe, or other form of sampling, as well as related collection and analytical methodologies, calculations, and other background data relevant to interpretation of the results obtained;

(ii) biological monitoring results which directly assess the absorption of a substance or agent by body systems (e.g., the level of a chemical in the blood, urine, breath, hair, fingernails, etc.) but not including tests which assess the biological effect of a substance or agent;

(iii) material safety data sheets; or

(iv) in the absence of the above, any other record which reveals the identity (e.g., chemical, common, or trade name) of a toxic substance or harmful physical agent.

(6)(i) "Employee medical record" means a record concerning the health status of an employee which is made or maintained by a physician, nurse, or other health care personnel, or technician, including:

(A) medical and employment questionnaires or histories (including job description and occupational exposures),

(B) the results of medical examinations (pre-employment, pre-assignment, periodic or episodic) and laboratory tests (including X-ray and all biological monitoring),

(C) medical opinions, diagnoses, progress notes, and recommendations,

(D) descriptions of treatments and prescriptions, and

(E) employee medical complaints.

(ii) "Employee medical records" does not include the following:

(A) physical specimens (e.g., blood or urine samples) which are routinely discarded as a part of normal medical practice, and are not required to be maintained by other legal requirements,

(B) records containing health insurance claims if maintained separately from the employer's medical program and its records, and not accessible to the employer by employee name or other direct personal identifier (e.g., social security number, payroll number, etc.), or

(C) records concerning voluntary employee assistance programs (alcohol, drug abuse, or personal counseling programs) if maintained separately from the employer's medical program and its records.

(7) "Employer" means a current employer, a former employer, or a successor employer.

(8) "Exposure" or "exposed" means that an employee is subjected to a toxic substance or harmful physical agent in the course of employment through any route of entry (inhalation, ingestion, skin contact or absorption, etc.), and includes past exposure and potential (e.g., accidental or possible) exposure, but does not include situations where the employer can demonstrate that the toxic substance or harmful

agent is not used, handled, stored, generated, or present in the workplace in any manner different from typical non-occupational situations.

(9) "Record" means any item, collection, or grouping of information regardless of the form or process by which it is maintained (e.g., paper document, microfiche, microfilm, X-ray film, or automated data processing).

...

(d) *Preservation of records.* (1) Unless a specific occupational safety and health standard provides a different period of time, each employer shall assure the preservation and retention of records as follows:

(i) *Employee medical records.* Each employee medical record shall be preserved and maintained for at least the duration of employment plus thirty (30) years, except that health insurance claims records maintained separately from the employer's medical program and its records need not be retained for any specified period;

(ii) *Employee exposure records.* Each employee exposure record shall be preserved and maintained for at least thirty (30) years, except that:

(A) Background data to environmental (workplace) monitoring or measuring, such as laboratory reports and worksheets, need only be retained for one (1) year so long as the sampling results, the collection methodology (sampling plan), a description of the analytical and mathematical methods used, and a summary of other background data relevant to interpretation of the results obtained, are retained for at least thirty (30) years; and

(B) Material safety data sheets and paragraph (c)(5)(iv) records concerning the identity of the substance or agent need not be retained for any specified period as long as some record of the identity (chemical name if known) of the substance or agent, where it was used, and when it was used is retained for at least (30) years; and

(iii) *Analyses using exposure or medical records.* Each analysis using exposure or medical records shall be preserved and maintained for at least thirty (30) years.

(e) *Access to records.*(1) *General.* (i) Whenever an employee or designated representative requests access to a record, the employer shall assure that access is provided in a reasonable time, place, and manner, but in no event later than fifteen (15) days after the request for access is made.

(ii) Whenever an employee or designated representative requests a copy of a record, the employer shall, within the period of time previously specified, assure that either:

(A) a copy of the record is provided without cost to the employee or representative,

(B) the necessary mechanical copying facilities (e.g., photocopying) are made available without cost to the employee or representative for copying the record, or

(C) the record is loaned to the employee or representative for a reasonable time to enable a copy to be made.

...

Employers can charge reasonable direct expenses for additional copies of records, except that a certified collective bargaining agent for the employee can receive a copy without cost, and if new information is added to the record, this information is available to the employee without cost under the same conditions as the original record.

For certain medical records, the privacy of the individual may be protected in making records available:

...

(e)(2)(ii)(E) Nothing in this section precludes a physician, nurse, or other responsible health care personnel maintaining employee medical records from deleting from the requested medical records the identity of a family member, personal friend, or fellow employee who has provided confidential information concerning an employee's health status.

and, under *Analyses using exposure or medical records:*

(e)(2)(iii)(B) Whenever access is requested to an analysis which reports the contents of employee medical records by either direct identifier (name, address, social security number, payroll number, etc.) or by information which could reasonably be used under the circumstances indirectly to identify specific employees (exact age, height, weight, race, sex, dates of initial employment, job title, etc.), the employer shall assure that personal identifiers are removed before access is provided. If the employer can demonstrate that removal of personal identifiers is not feasible, access to the personally identifiable portions of the analysis need not be provided.

New employees have certain rights concerning records from the beginning of their employment.

(g) *Employee information.* (1) Upon an employee's first entering into employment, and at least annually thereafter, each employer shall inform employees exposed to toxic substances or harmful physical agents of the following:
(i) The existence, location and availability of any records covered by this section;
(ii) the person responsible for maintaining and providing access to records; and
(iii) each employee's rights of access to these records.
...

Corporations are often bought out, merge with other firms, or cease to operate, and provision is made in the standard for the retention of records for the required periods by transfer of the records to the successor firm or, under requirements of specific standards, to NIOSH. This is rarely a problem for academic research institutions which seldom cease to operate, although semiautonomous components which retain their own records may cease to exist, in which case their records should be subsumed into those of the parent institution.

### 4.8.4.2.8. CPR and First Aid Training
Subpart K — Medical and First Aid, 29 CHR 1910.151 of the OSHA Standards for General Industry describes the minimal medical care which must be available to employees, although there are references to first aid in several other sections of the standards. This short section is given below in its entirety.

**"Medical services and first aid**
(a) The employer shall ensure the ready availability of medical personnel for advice and consultation on matters of plant health.
(b) In the absence of an infirmary, clinic, or hospital in near proximity to the

workplace which is used for the treatment of all injured employees, a person or persons shall be adequately trained to render first aid. First aid supplies approved by the consulting physician shall be readily available.

(c) Where the eyes or body of any person may be exposed to injurious corrosive materials, suitable facilities for quick drenching of the eyes and body shall be provided within the work area for immediate emergency use."

The need for access to medical services in emergencies has already been discussed at some length in Chapter 1. However, prompt action can frequently save an individual's life or can significantly reduce the seriousness of injuries. Although not intended as instruction manuals, some first aid procedures for accidents involving chemicals and cardiopulmonary resuscitation (CPR) techniques were also discussed in Chapter 1. It would be highly desirable for individuals working in facilities where hazards are present to be trained in both of these subjects. By working carefully so as not to tempt fate too much, and with a great deal of luck, an individual may go through an entire working career without personally experiencing an accident or being present when someone else does, but this cannot be counted upon. Although you cannot perform CPR on yourself, and you may be incapacitated so that even simple first aid is beyond you, if enough personnel in a laboratory do make the effort to become trained, it is likely that someone will be available to start emergency aid while waiting for more skilled personnel to arrive. The training for basic first aid and single-person CPR is not difficult, and everyone should annually devote the few hours neccessary to receive and maintain these skills. Many rescue squads, fire departments, hospitals, and other public service agencies offer the training at a minimal fee, covering only the cost of the manuals and supplies.

### 4.8.4.2.9. Vaccinations

All of us probably received some vaccinations as a child against a number of diseases. A number of common diseases afflicting children born in the first third of this century are now decreasing in frequency as a result of widespread vaccination programs. A recent controversy centered around whether the last smallpox virus in the world, being maintained in a laboratory, should be destroyed. Yet this used to be one of the world's great killers. Relatively recently, vaccinations for other diseases have been developed and diseases such as polio and measles are relatively rare now in the U.S., although, unfortunately, there has been some resurgence in these two illnesses recently. It would appear that with this obvious benefit, vaccination against a disease would be a matter of course, provided a vaccine exists. This is not necessarily the case.

Several factors need to be considered in determining whether vaccination is desirable. The first clearly is, does a safe reliable vaccine exist? At one time, rabies vaccine using duck embryos was the best available. However, it did not always provide a reliable immunization, and a significant fraction of the persons on which it was used had reactions, some of which were neurologically very severe. Now, a much more reliable human diploid rabies vaccine is available which provides a very high percentage of persons with protection, and the incidence of untoward reactions is very low. It is probably desirable to mandate vaccinations for all personnel who face a high risk of exposure to rabies, i.e., persons who work directly with animals that might be rabid, individuals who do necropsies on such animals, and technicians who work with untreated tissue from potentially rabid animals. The second question is, what is the risk-benefit to the individual if the disease is contracted? The disease may not be sufficiently serious to warrant the risk of a possible reaction to a vaccination. On the other hand, if the disease is sufficiently life threatening, then the use of a vaccine would be indicated. Thirdly, is there a satisfactory postexposure treatment? This is really critical for life-threatening diseases. If there is not and the exposure risk is significant, the use of even a less than totally satisfactory vaccine might well be considered. Other considerations would be the state of the individual's health. If the condition of the person is such that the possibility of an adverse reaction could have a strong negative impact on the individual, then one would question the desirability of using a vaccine, but one would also question placing such a person in an environment where vaccination might be considered.

Booster injections are needed for some diseases to ensure an adequate protective level of antibodies. However, some patients may experience reactions to a booster. It is advisable to do a blood titer test prior to repeat injections. If the titer is adequate, no booster should be administered.

Although the laboratory supervisor should have considerable input in deciding whether a vaccine should be used, any decision to institute a mandatory vaccination program should be reviewed by a separate biosafety committee before implementation. Individuals must be fully informed of any possible risks.

## REFERENCES (SECTIONS 4.8.4.2 TO 4.8.4.2.9)

1. **Hogan, J. C. and Bernacki, E. J.,** Developing job-related preplacement medical examinations, *J. Occup. Med.,* 23(7), 469, 1981.
2. Health Monitoring for Laboratory Employees, Research and Development Fact Sheet, National Safety Council, Chicago, 1978.

### 4.8.5. INFECTION FROM WORK WITH HUMAN SPECIMENS

Recently, laboratory personnel working with human blood and other body fluids have become increasingly concerned with contracting diseases from this contact, primarily from fear of becoming infected with AIDS. This is not the only possibility, nor is it the most likely disease which can be contracted in this manner. The following two sections discuss AIDS and hepatitis.

### 4.8.5.1. The Liability of AIDS in the Laboratory*

AIDS is the acronym for acquired immunodeficiency disease.[14] It is considered a fatal disease caused by a retrovirus presently called HIV (human immunodeficiency virus).[4] The disease has previously been called HTLV 111 and LAV. The incubation period is variable, but is considered to be from 6 months to 5 years or more.[7] The statistical acquisition of data will hone it up or down further. It is a relatively new disease, having come on the scene in 1981.

There is, as yet, no cure, nor a vaccine to prevent the infection. Recent reports indicate that AIDS has spread into at least 90+ countries. Over 50,000 cases have been reported to date in the U.S., and at least 28,000 people have died from it. It is

---

* This section was written by Dr. Richard F. Desjardins.

estimated that there probably are as many as 2,000,000 American carriers of the virus who have not evolved the full-blown syndrome. An intermediate condition between being an asymptomatic carrier and having active AIDS is called ARC (AIDS-Related Complex).[11]

The manifestations of AIDS are also variable, but are all related to the compromising of the immune system. The result of destroying the essential cellular manager of the immune protection (the helper thymic lymphocyte) is to lose the capacity to exist healthfully in the presence of viruses, bacteria, fungi, and some forms of cancer.[12] As the disease progresses, the patient loses weight, is very fatigued, develops large, swollen lymph nodes, develops infections such as herpes, moniliasis, pneumocystis carini pneumonia, and severe diarrhea; most also develop progressive neurological signs and symptoms. The usual cancer which develops used to be rare, but it is not now. It is called Kaposi sarcoma and manifests itself as progressive purple to brown skin spots and splotches. Most patients die from massive infections not controlled or cured by any antibiotics. The incidence of suicide by those in whom the diagnosis has been made is not insignificant. It is, after all, an unpleasant death sentence at this time. Hope of developing a preventative or curative vaccine abounds and, indeed, the work is progressing with all possible haste. Government and private funding is burgeoning.

There is, as yet, no readily available test for the presence of the virus. Persons exposed to HIV usually develop detectable antibody against the virus within 6 to 12 weeks of becoming infected. The presence of antibody is considered indicative of current infection. The first test used is called the ELISA (enzyme-linked immunosorbant assay). If this tests positive, it is repeated. Confirmatory testing is then done with the Western blot test. There is a very small incidence of false positive and false negative tests. Elisa has a 99% sensitivity. The Western blot test efficiency rate is <1:100,000 up to 5:100,000 in some laboratories.[4,5] If there is any doubt as to the appropriateness of a positive or negative test when all of the signs and symptoms indicate a likely presence of the disease, the patient must be followed and treated as probably positive. Retesting is mandatory, of course. Making this diagnosis can be devastating, both to the truly afflicted and to the actually negative patient.

It is very important to know how one might contract this dreadful disease.[1] The CDC study reported that the predominant seropositive cases are found in the exclusively homosexual males — 20 to 25% or 500,000 to 625,000. Bisexual males have an incidence of 5% or 125,000 to 375,000. The third-tier incidence is found in intravenous drug users. This group is increasing in proportion since homosexual and bisexual men are using more precautions. The incidence ranges from 5 to 50%, depending on the area tested. It is very much higher in ghettoes and crowded cities. The estimated number infected at this time is 335,000. The fourth tier is found in hemophiliacs and patients having blood diseases requiring administration of human blood extracts. Of the known hemophiliacs, the government believes that 9800 of the 15,000 harbor the virus. Since the blood products used to control these bleeding problems are being tested and treated (since 1985), new infections with AIDS are at a virtual standstill in this group. No one can get AIDS from donating blood. Also, all donated blood is tested for HIV, making it safe to receive transfusions.[6] Heterosexual transmission of AIDS depends on the area surveyed. In this country, it is estimated that 45,000 to 127,000 have become infected from heterosexual activity. In parts of

Africa, where many investigators believe AIDS evolved by mutation from Green monkeys, the incidence is over 50%. A small number of people have been diagnosed as seropositive at birth. Unfortunately, this is increasing and it is evident that the virus passes across the placental barrier. It is possible for antibodies to cross the placental barrier and apparently not infect the child since a few children have become negative a few months after birth. When a child develops AIDS, the progression of the disease is rapid and virulent.[15,18] A newborn child's immune system isn't well developed yet.

In the laboratory, two known cases have occurred from needle sticks.[15] The CDC has followed over 800 health care workers directly exposed to HIV after being stuck with needles previously used on AIDS patients. Also, studies from New York and San Francisco are being followed. Of the total studies of 1750 cases of needle stick, only the two mentioned above resulted in infection.[2,8]

There is no evidence of virus transmission from casual contact with AIDS carriers or active AIDS patients.[10] This includes tears, saliva, respired air, hand shaking, hugging, insect bites or stings from such as mosquitos, or even kissing.[13] The only means of transmission is via body fluids — semen, vaginal secretions, rectally, blood or blood products, and transplacentally. Obviously, a vial of concentrated virus in the laboratory which breaks or is spilled into an open wound could cause infection.

The situation in the workplace is no different from the situation in the classroom.[17] There is no reason to ostracize an AIDS patient. If they are well enough to do normal work, they should be allowed to do so.

Practical suggestions to be used in the laboratory when handling human body fluids are given below, unless the human body fluids are known not to be contaminated by AIDS or other diseases infectious to humans.[19,20] These are conservative. When working with fluids which are likely to be contaminated, and when performing activities which are likely to generate aerosols, e.g., blending, sonicating, and vigorous mixing, the work should be done in a Class 2 biological safety cabinet.

1. Avoid AIDS hysteria — AIDS is not casually contracted.[9]
2. In handling body fluids, extracts of AIDS, or suspected AIDS patients, use gloves (double gloves are used in some labs). Containers of specimens with uncertain histories should be wiped with a 5% solution of sodium hypochlorite.
3. Cover the laboratory bench area with plastic-lined absorbent paper.
4. Use mechanical pipettes for manipulating all possible infectious body fluids. Mouth pipetting must not be done.
5. Use needles cautiously and never resheath used needles. Place used needles and syringes in special boxes to be incinerated.
6. Work carefully so as to avoid breaking glassware.
7. As a matter of principle, wear protective glasses or goggles and a face mask. Blood, body products, semen, etc. can splash, and there are no good data concerning conjunctival exposure.
8. Wear a coat or gown; if it becomes contaminated, remove it at once. Wash yourself with a 5% sodium hypochlorite solution (as a disinfectant).[16] Clothes put in the laundry should be in a properly labeled bag, stating its possible infectious nature.
9. Wash the hands thoroughly and immediately with a sodium hypochlorite solution if they become contaminated.

10. Label all containers with a warning as to blood or body fluid precautions.
11. If a container becomes externally contaminated, clean it immediately with a 5% sodium hypochlorite solution. Place all containers in an impervious bag for transportation.
12. Work safely and deliberately. The life you save might be your own.
13. Autoclave all equipment that might have been contaminated.

There are situations in the workplace where policy has to be formulated both to protect all employees and to help and protect the seriously ill fellow employee.[3] In a case of AIDS, it may be safely stated that healthy individuals are more likely to harm the afflicted, who have a poor immune capacity, than vice versa. It is humane and generally therapeutic for patients to be allowed to work if capable of doing so. The following are a few guidelines which could help in formulating policy, whether in the laboratory, office setting, or classroom:

1. Practice confidentiality; disease states are very personal.
2. Work with the personnel department relative to advice in handling situations involving ill employees. If there are questions about the contagiousness of a disease, inquire — for your sake and that of the patient.
3. If in a supervisory position, ask personnel relations if a statement from the patient's physician is on file, with specified working conditions or helpful suggestions for the welfare of the patient.
4. Consistent with the functioning of the facility, accommodations as to work, place, and timing can be made.
5. Do not give special considerations beyond normal transfer requests for employees who feel threatened by a co-workers's illness.
6. Encourage and advise the afflicted employee to seek care and guidance from support groups. Corporate health facilities and/or employee relations should be able to help.

The immediately preceding information on procedures for working with materials which could possibly cause personnel to become infected with the HIV virus was based on recommendations issued by the CDC in Reference 20. Since that material was prepared, the CDC has issued an update which can be found in Reference 21. the following material is a brief summary of the recommendations found in the latter source. It closely follows the material issued by the CDC and was adapted for use at the author's institution by T. S. Smithwick.

### PROCEDURES FOR HANDLING HUMAN BLOOD
### AND OTHER BODY FLUIDS

The Centers for Disease Control recommends the use of "universal precautions" to prevent transmission of blood-borne diseases to personnel working with human blood and other body fluids. These precautions treat the blood and certain body fluids as being potentially infectious for blood-borne disease regardless of their infectious status. Universal precautions should be applied to blood, semen, vaginal secretions, and body fluids containing visible blood. Blood is the most important source of human immunodeficiency virus (HIV), the virus that causes AIDS, and hepatitis B virus (HBV).

Since the risk of transmission of HIV and HBV from tissues and cerebrospinal, synovial, pleural, peritoneal, pericardial, and amniotic fluids is unknown, universal precautions should be applied to these fluids also.

Universal precautions do not apply to feces, nasal secretions, sputum, sweat, tears, urine, and vomitus unless they contain visible blood. The risk of transmission of HIV and HBV from these materials is extremely low or non-existent.

Universal precautions do not apply to saliva unless contamination with blood is predictable, as in dental procedures. However, the use of gloves and handwashing is recommended after exposure to saliva to further decrease the minute risk.

The following procedures are recommended to ensure the safe handling of potentially infectious materials. These precautions also apply to blood and body fluids from primates other than humans.

### Universal Precautions

1.  Workers should use appropriate barrier precautions, such as gloves, gowns, masks, and protective eyewear to prevent skin and mucous-membrane exposure when working with potentially infectious fluids.
2.  Medical gloves made of vinyl or latex should be worn when drawing blood, performing finger or heel sticks, and for procedures involving contact with mucous membranes or potentially infectious materials. Gloves should be changed between patient contact and disposed of as contaminated waste. Medical gloves should not be washed or disinfected for reuse.
3.  Rubber household utility gloves should be used for housekeeping chores that involve handling items or surfaces contaminated with blood or body fluids to which universal precautions apply. Utility gloves may be washed and disinfected for reuse but should be discarded if punctured or torn.
4.  Masks, gowns, and protective eyewear should be worn during procedures that are likely to generate splashes or droplets of potentially contaminated material.
5.  Workers should wash hands and other skin surfaces contaminated with blood and other potentially infectious body fluids immediately and thoroughly. Hands should be washed immediately after removing gloves.
6.  Needle stick exposure is the main route of transmission of HIV and HBV to workers. Workers should take appropriate precautions to prevent injuries from needles, scalpels, and other sharp instruments. Needles should not be recapped, removed from disposable syringes, bent, broken, or otherwise manipulated by hand. After use, needles, disposable syringes, and other sharp items should be placed in a puncture-resistant container located as close as practical to the user for eventual disposal.

### Additional Precautions for Laboratories

1.  Specimens of blood and body fluids to which universal precautions apply should be placed in a secure secondary container to prevent leaking during transport.
2.  Care should be taken when collecting specimens to ensure that the outside of the container does not become contaminated. Specimen containers should be

wiped with an appropriate germicide, such as a 1:10 dilution of household bleach, prepared daily, before storing.

3.  Biological safety cabinets (Class 1 or Class 2) should be used for procedures that have a high potential for generating aerosols, such as blending, sonicating, and vigorous mixing.

4.  Mouth pipetting must be strictly prohibited. Mechanical pipetting devices must always be used to transfer liquids.

5.  Needles and syringes should be used to transfer liquids only if absolutely necessary.

6.  Laboratory work surfaces should be covered with plastic-lined, absorbant paper. Spills should be decontaminated with an appropriate chemical germicide approved for hospital use. Routine daily cleaning of contaminated surfaces is recommended.

7.  Contaminated materials must be decontaminated or placed in leakproof containers such as plastic bags (preferably double-bagged), labeled as infectious waste, and incinerated or autoclaved before disposal in a sanitary landfill.

8.  Workers must remove protective clothing and wash their hands before leaving the laboratory.

9.  Although the risk of disease transmission is low, soiled clothing should be handled as little as possible, transported in leakproof bags, and washed with detergent for 25 minutes at 71°C (160°F).

The reader might wish to compare these most recent recommendations with the guidelines for organisms requiring Biological Safety Level 2 and 3 practices found elsewhere in this volume.

# REFERENCES (SECTION 4.8.5.1)

1. *Science,* 15, January 1988.
2. *Fam. Pract. News,* 15(23), 63, 1985.
3. **Halcrow, A.,** AIDS: the corporate response, *Personnel J.,* August 1985.
4. *Epidemiol. Bull.,* 87(12), 1987.
5. **Meyer, K. B. and Pauker, S. G.,** Screening for HIV: can we afford the false positive rate? *N. Engl. J. Med.,* 317, 238, 1987.
6. Centers for Disease Control, Human immunodeficiency virus infection in transfusion recipients and their family members, *MMWR,* 36, 137, 1987.
7. **Keeling, R.,** AIDS on the college campus, ACHA Special Report, Am. College Health Assoc., Rockville, MD, 1986.
8. **Harris, A. A., Segreti, J., and Levin, S. J.,** Practical precautions when caring for AIDS patients, *Respir. Dis.,* 26, April 1985.
9. **Frolkis, J. P.,** AIDS anxiety. New faces for old fears, *Postgrad. Med. J.,* 79(6), 1986.
10. AIDS in the Workplace — A Guide for Employees, San Francisco AIDS Foundation, 1986.
11. The Facts About AIDS, Zoe International, Wayne, NJ, 1986.
12. On the front line, *Emerg. Med.,* January 1986.
13. **Booth, W.,** AIDS and insects, *Science,* 237, 24, 1987.
14. **Koop, C. E.,** Surgeon General Report on AIDS, U.S. Public Health Service, Washington, D.C., 1987.
15. *Fam. Pract. News,* 15(23), 63,
16. **Martin, L. S., McDougal, J. S., and Loskoski, S. L.,** Disinfection and inactivation of the human T-lymphotrophic virus type III/lymphadenopathy-associated virus, *J. Infect. Dis.,* 152, 400, 1985.
17. *Epidemiol. Bull.,* 86(2), February 1986.
18. *Epidemiol. Bull.,* 86(3), March 1986.

19. Centers for Disease Control, Recommendations for preventing transmission of infection with human T-lymphotropic virus type III/lymphadenopathy-associated virus in the workplace, *MMWR,* 34, 681, 691, 1985.
20. Centers for Disease Control, Recommendations for prevention of HIV transmission in health-care settings, *MMWR,* 36 (Suppl. 2S), August, 1987.
21. Centers for Disease Control, Update: acquired immunodeficiency virus infection among health-care workers, *MMWR,* 37, 229, 1988.

### 4.8.5.2. Laboratory-Acquired Viral Hepatitis*

Hepatitis refers to inflammation of the liver. Lest we forget, hepatitis may be caused by trauma, toxic shock, hyperthermia, metabolic derangements, chemicals, radiation, bacteria, fungi, parasites, and a number of viruses. For the purposes of this section, we will refer only to the three most common viral etiologies. These are Type A (also called infectious hepatitis), Type B (also called transfusion or serum hepatitis), and Type C (also known as non-A, non-B hepatitis).

The usual onset of hepatitis symptoms occurs 1 to 3 weeks prior to the obvious presence of the disease. These are nonspecific and may be seen with any of the types. There is the gradual development of vague aches and pains, fatigue, and anorexia. As this accelerates, the urine becomes dark — almost a mahogany red — and the patient becomes jaundiced. This becomes evident when the blood bilirubin rises at least two and one half to three times normal. Ten percent of patients have the triad of fever, arthralgia (bone and joint pain), and an urticarial rash. As the jaundice worsens, the patient usually feels better. The usual recovery period is 2 to 4 weeks. Differentiation of the types of hepatitis is helped by epidemiological considerations and by laboratory tests.[3]

### Incubation Period
- A — 15 to 50 d (average, 28 to 30 d)
- B — 45 to 160 d (average, 60 to 120 d)
- C — 18 to 89 d (average, intermediate to A or B)

### Incidence of Viral Hepatitis
- A — 20%
- B — 60%
- C — Variable percent of the rest of cases

### Incidence of Severe or Fulminant Hepatitis
- A — 3%
- B — 60%
- C — 37%

### Source of Viral Etiology
- A — Fecal, possible mucous, saliva, urine
- B — Blood, blood products, saliva, feces, breast milk, semen, mucous (as in the vagina), and probably neonatally

---

* This section was written by Dr. Richard F. Desjardins.

- C — Posttransfusion of whole blood or even blood products or fractions. A variant has recently been described as orally acquired and in epidemic proportions in some areas of Africa.

### Viral Carriers
- A — Neither chronic hepatitis nor the carrier state has ever been demonstrated. Tests reveal that 45 to 50% of Americans have antibodies to this virus, indicating previous infection. A fair proportion of hepatitis A infections are subclinical (few or transient symptoms and no jaundice) and still produce antibodies which render the individual immune to this type of hepatitis for life.
- B — <0.5%
- C — Up to 8% of population

### Chronicity (long-standing active hepatitis)
Many investigators will only consider hepatitis chronic if the disease is active for 9 to 12 months or more.

- A — 0%
- B — 10%
- C — 45%

It must be emphasized that a chronic active hepatitis will most probably result in cirrhosis of the liver or a malignant hepatoma. However, some cases of chronic hepatitis C will gradually subside and heal without sequelae.

### Infectivity
- A — This is usually acquired by eating contaminated foods such as shell fish or vegetables and fruit. Although it might be possible to acquire this disease from blood, it has not been reported.
- B — This is acquired by entry through the skin by cuts or abrasions, blood transfusions or injected blood fractions, or by needle sticks when using fluids containing the virus. It is not uncommon for drug addicts to get this type of hepatitis from shared syringes and needles. It has been shown that a needle stick in AIDS virus research results in less than 1% seroconversion.
  In hepatitis B, however, positive seroconversion results in 19 to 24% of cases. A dilution of 1:1,000,000 of the virus is still infective.[1]
- C — This is 80 to 90% acquired by transfusions. The newer described variant may cause 10 to 20% of this type orally. There is no good test to accurately diagnose this virus as yet.

### Mortality Rates
- A — 0.001%
- B — 1 to 2%
- C — up to 20%

## Diagnosis

Obviously, a detailed history and physical examination must precede any laboratory testing.

- A — IGM-Anti-HAV (no antigen test presently available). By the time this test is positive, the patient is most likely no longer infectious and isolation is of little or no value.[2]
- B — There are five available tests useful in diagnosing the existence of the disease and determining the progression, healing, and possible chronicity of hepatitis B.[4,5]

    $HB_sAG$ — hepatitis B surface antigen
    $HB_eAG$ — hepatitis B e antigen
    Anti-$HB_c$ — hepatitis B anti-core
    Anti-$HB_e$ — hepatitis B anti-e
    Anti-$B_sAG$ — hepatitis B anti-surface

The usefulness of these tests can be seen in the following diagnostic summary:

| Clinical diagnosis | $HB_sAG$ | $HB_eAG$ | anti-$HB_c$ | anti-$HB_e$ | anti-$HB_sAG$ |
|---|---|---|---|---|---|
| Early acute B | + | – | – | – | – |
| Early acute B | + | + | – | – | – |
| Acute B (? chronic) | + | + | + | – | – |
| Chronic carrier | + | + | + | – | + |
| Acute B (good prognosis) | + | – | + | + | – |
| Recent acute B (good prognosis) | – | – | + | + | – |
| Recovery stage | – | – | + | + | + |
| Recovery stage | – | – | + | – | + |
| Recovery stage | – | – | – | + | + |
| Recovery stage | – | – | – | – | + |

This must appear complicated, but is very necessary considering the 10% chronicity with poor prognosis. It has been determined that at least one out of five of laboratory workers, dialysis workers, and physicians will contract hepatitis B at some time.[6,7]

- C — (non-A, non-B). There is no available antigen or antibody test for this type of hepatitis. The diagnosis is made by exclusion. Good progress is being made in isolating etiologic viruses. At present, there are two good candidates known.[8]

    *Treatment* — There is no specific viricide or treatment. The usual rule is to use conservative measures: protect the patient from more damage, use good nutrition, and, in the relatively rare fulminant case, use corticosteroids.

## Prophylaxis[9]

1a. Hepatitis A (HAV) Preexposure

   Indications — travelers to foreign countries where hygienic conditions may be poor.

Agent — Immune serum globulin (ISG)

Dosage — single dose, 0.02 ml/kg. For prolonged travel, 0.06 ml/kg every 6 months.

Effect — A 0.02-ml/kg dose is effective for up to 3 months.

1b.  Hepatitis A (HAV) Postexposure

Indications — household contacts, institutions with poor hygiene such as nursery schools and homes for the retarded or insane.

Agent — Immune serum globulin (ISG)

Dosage — 0.01 to 0.02 ml/kg single dose

Effect — Attenuates or aborts clinical signs or symptoms. It is best given immediately after the exposure. Has some good effect if given up to 6 weeks post exposure. It would be of no help if given later.

2.  Hepatitis B (HBV)

Preexposure — Here we have good news: there is a good hepatitis B vaccine which provides a good level of protection with little or no untoward reactions. This is called HEPTAVAX-B (made by Merck, Sharp, and Dohme) and is rated as at least 96% effective after three doses. It works in vaccine responders against acute hepatitis B, asymptomatic infection, and prevents the chronic carrier state. If Anti-HB$_s$ develops, there is virtually 100% protection. There is now available a Pasteur vaccine and a Dutch vaccine. Although they are prepared differently, the effects are similar.

**Vaccination guidelines**

1.  Normal subjects (HBV contacts) — if at risk, parenteral transmission, or work/travel in high-risk countries; depends on the type of exposure.

2.  Health care personnel — if at any risk of parenteral transmission from needles or from getting the virus on the hands where there might be scratches, etc.

3.  High-risk patients — hemophiliacs, hemodialysis patients, immunosuppressed patients, and members of high-risk social groups such as male homosexuals.

*A. Postexposure to hepatitis B virus*

| Hepatitis B | Immune globulin | | Vaccine | |
|---|---|---|---|---|
| Exposure | Dose | Timing | Dose | Timing |
| Perinatal | 0.5 ml i.m. | Within 12 h | 0.5 ml i.m. | Within 12 h, repeat vaccine in 1 and 6 months |
| Sexual | 0.06 ml/kg single dose i.m. | Within 14 d of contact | — | — |

*B. Recommendations for hepatitis B prophylaxis following a percutaneous exposure*

| Source | Unvaccinated | Vaccinated |
|---|---|---|
| HB$_s$AG+ | HBIG × 1 stat<br>Initiate HB vaccine | Test exposed person for anti-HBsAG<br>If inadequate titer, give HBIG+ vaccine booster |

| Source | Unvaccinated | Vaccinated |
|---|---|---|
| Known source | —high risk, $HB_sAG$ positive | |
| | Initiate HB vaccine | Test for $HB_sAG$ only if exposed is |
| | Test source for $HB_sAG$; | nonresponder; if positive, give HBIG+ |
| | if positive, administer HBIG × 1 | booster vaccine. |
| Low risk, $HB_sAG+$ | Initiate vaccine | Nothing required |
| Unknown source | Initiate vaccine | Nothing required |

## C. Hepatitis C (non-A, non-B)

There is no presently available vaccine for protection. The use of hyperimmune globulin (HBIG) may have some beneficial effect, but it is certainly not specific.

There is a lifelong immunity secondary to having had any of the viral hepatitides. There also is no cross-protection, one with the other.

The recommended methods of sterilizing or inactivating the viruses which cause hepatitis are

| Method | Contact time |
|---|---|
| Heat | |
| Boiling water | 10 to 20 min |
| Autoclaving | 30 min |
| Dry | 60 min |
| Chemical disinfectants | |
| Hypochlorite, 0.5—1% | 30 min |
| Formalin | 12 h |
| Alkalinized glutaraldehyde, 2% | 10 h |
| Iodine solution 1% | 30 min |
| Ethylene oxide gas | According to recommendations of manufacturers |

The liver is a beautifully productive factory within our bodies. It processes over 1500 known chemical reactions. It is estimated that we have about 80% more liver mass than we need, but when it is affected by such as a viral hepatitis, 100% of it is involved. Fortunately, recovery is possible and probable in most cases. If recovery is not complete, the result could very well be a digestive cripple. Then, too, the development of cirrhosis is devastating and ultimately fatal. A hepatoma might presumably be removed if discovered early, but often it is not.

Working in a laboratory with human body products constitutes a risk, and even more so if the specific hepatitis viruses are being tested or researched. Preventative measures, precautions, and vaccination whenever possible are, indeed, mandated, and should be policy.

## REFERENCES (SECTION 4.8.5.2)

1. *Respir. Times,* 3(1), January 1988.
2. **Miller, D. J.,** Seroepidemiology of viral hepatitis, *Postgrad. Med.,* 68(3), 1980.
3. **Hoofnagle, J.,** The challenge of acute viral hepatitis, *Patient Care,* October 1979.
4. **Pribor and Duello,** Hepatitis, diagnostic profiles, *Lab. Manage.,* 19(7, 8).
5. **Fife, K. H. and Corey,** Recent advances in the diagnosis of hepatitis, *Lab. Med.,* 11(10), 650, 1980.
6. **Denes, A. E. et al.,** Hepatitis B infection in physicians, results of a nationwide survey, *JAMA,* 239(3), 210, 1978.

7. **Dienstag, J. L., Wands, J. R., and Koff, R.S.,** Acute hepatitis, in *Harrison's Principles of Internal Medicine,* 9th ed., Isselbacher, K. L. et al., Eds., McGraw-Hill, New York, 1980, 1459.
8. *Epidemiol. Bull.,* 85(8), August 1985.
9. **Waldvogel, F.,** Viral hepatitis: diagnose it fast and prevent its spread, *Mod. Med.,* October 1983.
10. ACIP, Recommendations for protection against viral hepatitis, *MMWR,* 34, 313, 329, 1985.

### 4.8.5.3. Other Disease-Causing Agents*
#### *4.8.5.3.1. Zoonotic Diseases*

Zoonotic diseases are those transmitted from animals to humans, with a wide range of manifestations in humans from simple illness to death.

An understanding of the modes of transmission of these diseases and the clinical signs observed in affected animals will help facility managers establish preventative measures to protect individuals who come into contact with the animals or their tissues.

#### *4.8.5.3.1.1. Modes of Transmission*

Disease agents can be transmitted either directly or indirectly. Bacterial, viral, fungal, and parasitic disease agents can be transmitted through direct contact with animal saliva, feces, urine, other body secretions, bites, scratches, aerosols, or excised body tissues. Humans can be protected through use of gloves, masks, gowns and other protective clothing, and through use of restraint techniques which minimize the possibility of bites and scratches.

One indirect means of disease agent transmission involves fomites, inanimate objects (boots, brooms, cages, instruments, etc.) which can transport the agent following contact with animals, secretions, or wastes. Disease agents may be short-lived when outside the body or persist for years on fomites if the object is not cleaned or disinfected. The Orf virus from sheep and goats has remained viable for 15 years in dried scabs. Rooms, cages, and equipment should be adequately disinfected with virucidal disinfectant agents.

A second indirect means of transmission involves vectors, living organisms (insects) which can extract or carry the disease agent from one animal to other animals or humans. A mechanical vector extracts and carries the disease agent without any change occurring in the agent. In a biological vector, the disease agent undergoes changes in one or more stages of its life cycle before becoming an infective form.

An effective vermin and insect control program is needed to eliminate these indirect means of transmission

#### *4.8.5.3.1.2. Routes of Exposure*

Barkley and Richardson[1] listed the four primary routes of exposure or entry of a disease agent:

1. Ingestion (i.e., placing contaminated fomites in one's mouth, or contaminated hand contact with food)
2. Inhalation (i.e., aerosolized material — urine, feces, saliva, or other bodily secretions; these materials may also become aerosolized when using high-pressure water hoses to clean rooms or cages)

---

* This section was written by Dr. David M. Moore.

3.   Contact with mucous membranes (i.e., contact with nose, mouth, or eyes through spills, contaminated hands, or aerosolized material)
4.   Direct parenteral injection (i.e., bites, cuts, scratches, or accidental needle sticks)

Each should be handled accordingly by prohibiting food consumption in animal holding areas, requiring the practice of good hygiene, altering sanitation procedures to lessen aerosol production, providing protective garments and safety items (safety glasses, respirators, gloves, and masks), and establishing safety awareness training programs to advise employees of risks and preventative measures.

## REFERENCES (SECTION 4.8.5.3.1)

1.   **Barkley, E. W. and Richardson, J. H.,** Control of biohazards associated with the use of experimental animals, in *Laboratory Animal Medicine,* Fox, J. G. et al., Eds., Academic Press, Orlando, 1984, 595.
2.   **Fox, J. G., Newcomer, C. E., and Rozmiarek, H.,** Selected zoonoses and other health hazards, in *Laboratory Animal Medicine,* Fox, J. G. et al, Eds., Academic Press, Orlando, 1984, 613.
3.   Biological Hazards in the Nonhuman Primate Laboratory, Office of Biohazard Safety, NCI, Bethesda, MD, 1979.
4.   **Richardson, J. H. and Barkley, E. W.,** Biosafety in Microbiological and Biomedical Laboratories, Public Health Service/National Institutes of Health, HHS Publ. No. (CDC) 84-8395, Bethesda, MD, 1984, 37.
5.   **Hellman, A., Oxman, M. N., and Pollack, R.,** *Biohazards in Biological Research,* Cold Spring Harbor Laboratory, Cold Spring Harbor, New York, 1973.

### *4.8.5.3.2. Allergies*

Some investigators and animal care technicians who have prolonged contact with laboratory animals may develop allergies to animal dander, hair, urine, tissues, or secretions. Reactions to skin contact or inhalation of these materials vary from a wheal and flare phenomenon (a firm, red raised area at the site of skin contact which develops within several minutes) to life-threatening anaphylactic shock.

Olfert[2] lists the species most commonly associated with allergic reactions in a laboratory setting: rat, rabbit, guinea pig, and mouse. When transfer of personnel to a nonanimal area is not a viable option, other measures should be taken to avoid exposure to specific allergens. Lutsky et al.[1] suggest the use of gloves, masks, protective outer garments, and filtered cages as methods to reduce exposure. Additionally, eliminating recirculation of room air will decrease the levels of allergens, as will more frequent cage cleaning.

## REFERENCES (SECTION 4.8.5.3.2)

1.   **Lutsky, I. T., Kalbfleisch, J. H., and Fink, J. N.,** Occupational allergy to laboratory animals: employer practices, *J. Occup. Med.,* 25(5), 272, 1983.
2.   **Olfert, E. D.,** Allergy to laboratory animals — an occupational disease, *Lab Anim.,* 15(5), 24, 1986.
3.   **Krueger, B.,** Lab animal allergies: a manager's perspective, *LAMA Lines,* 2(6), 16, 1987

### *4.8.5.3.3. Waste Collection and Storage*

Shearing of hypodermic needles following injection of infectious or toxic agents, or following routine clinical use in animals, should be avoided. Aerosolization of the contents of the needle can occur during shearing, posing a hazard to humans or other animals in the room.

### 4.8.5.3.4. Bedding

Bedding from cages housing animals treated with biohazardous microbial or chemical agents should be considered contaminated and disposed of appropriately.[2] If an incinerator is not on site or available for direct dumping of bedding, then bedding should be double bagged and tagged as hazardous material prior to transport to the incinerator to avoid contamination of personnel or the work environment. The incineration of carcinogen-contaminated bedding requires an incinerator capable of operating at a temperature range of 1800 to 1900°F, with a retention time of 2 sec.[3]

## REFERENCES (SECTIONS 4.8.5.3.3 TO 4.8.5.3.4)

1.  **Barkley, W. E. and Richardson, J. H.,** Control of biohazards associated with the use of experimental animals, in *Laboratory Animal Medicine,* Fox, J. G. et al., Eds., Academic Press, Orlando, 595, 1984.
2.  **Wedum, A. G.,** Biohazard control, in *Handbook of Laboratory Animal Science,* Melby, E. C. and Altman, N. H., Eds., CRC Press, Cleveland, 1974, 196.
3.  Chemical Carcinogen Hazards in Animal Research Facilities, Office of Biohazard Safety, National Cancer Institute, Bethesda, MD, 1979, 15.
4.  **Dimmick, R. L., Vogl, W. F., and Chatigny, M. A.,** Potential for accidental microbial aerosol transmission in the biological laboratory, in *Biohazards in Biological Research,* Hellman, A. et al., Eds., Cold Spring Harbor Laboratory, Cold Spring Harbor, New York, 1973, 246.

## 4.8.6. PREGNANCY IN THE LABORATORY*

Pregnancy is not a disease. It is the ultimate expression of a couple's heritage, the desirable as well as the less-than-desirable elements. During the nine-month development period of the fetus, there is a dynamically changing physical and chemical status of both the mother and the child. It might be stated that the pregnant state imbues the fetus with a metabolic priority to obtain oxygen, chemicals, and nutrition from the only source available — the mother. It would therefore behoove the mother to sustain her physicochemical status at an optimum, both to provide the requirements for development of the fetus and to prevent depletion of her own requirements for a healthful state.

By the same token, chemical exposure by the mother will result in fetal absorption of sometimes deleterious concentrations of either the parent elements or the compounds and their metabolites. The noxious potential may be more or less in the parent element, compound, or metabolites. Estrogens are less stimulatory to the liver, where most of the elements or compounds are metabolized, than the male testosterone. Therefore, if a parent element or compound is more toxic than the metabolite, it would have a more injurious effect on a female than it would on a male. This is at least theoretically valid, but there have been so few relevant studies that it must be held as unsubstantiated at this time. In 1946, Baetjer[1] made a critical review of the literature and data which indicated that women were more susceptible than men to chemical exposures. She found that the conclusions were based on an inadequate

---

* This section was written by Dr. Richard F. Desjardins.

number of poorly constructed studies as well as failures to include both the chemical quantity and duration of exposures. She found no significant basis for the conclusion that nonpregnant women were more liable to injury by chemicals than were males.

As a result of legislation passed in the last two decades by OSHA (1970), Title VII of the Civil Rights Act of 1964, and guidelines adopted by The Equal Employment Opportunity Council, more women are entering the work force than ever (presently, the work force is over 40% women), and there can be no sex discrimination in the area of work. Basic to the provisions of the federal regulations is the inherent assumption that tolerance of industrial stresses, susceptibilities to toxic substances, and performance capabilities are similar in all workers. Furthermore, the "preservation of our human resources" implies not only the protection of the health of our working men and women, but also the preservation of their reproductive capacities and products of such capacity.

With a few exceptions, current standards for occupational exposures to chemicals and physical agents in the work environment do not consider the reproductive effects, including mutagenesis, teratogenesis, transplacental carcinogenesis, or infections. The latest available data[4] indicate that of 3,034,000 women who had a live birth during a 12-month period, an estimated 1,260,000 women (41.5%) had worked during their pregnancy. This, however, represents only about 8.8% of the estimated 14,357,000 married women of reproductive age in the workforce at that time. A significant number of the women work in laboratories and in the health services.

It should be noted that consideration of the effects of toxicants on the paternal elements is mandatory, if seldom addressed. The long-term effects of vinyl chloride and dibromochloropropane are well documented in this respect, but other potentially toxic substances may emerge as studies are effected.

The American Management Association[5] has recognized reproductive hazardous materials in the workplace and prepared a report in 1985, reviewing 120 chemicals. They pointed out general principles, clinical applications, and aids to recognizing human teratogens. Also presented are reviews and opinions for three representative chemicals: acrylonitrile, inorganic arsenic, and carbon disulfide. Table 4.19 considers the dynamics of reproductive toxicology.

Chemicals or physical agents can act directly on germ cells so that fertilization or implantation may not occur. The result could be spontaneous abortion. Anomalies are about 60 to 100 times more frequent in spontaneous abortions than in live births. Birth defects may occur in up to 8% of births, being revealed up to the age of 10, and 6% of these are due to known environmental agents. The causes of the rest of the birth defects are not known at this time. It is believed some will be of occupational environmental etiology.

Specific data and tables can be found in the section titled "Reproductive Toxicology and Occupational Exposure", in Chapter 54 of the excellent book by Zenz.[3]

Rarely recorded or discussed are the acquired infections of the mother and developing fetus. AIDS and hepatitis are known to cross the placental barrier, and other viruses and infectious agents might also do so. Laboratory precautions are particularly essential during pregnancy due to the increased vulnerability at that time.

In general, it is prudent for any pregnant woman to be assiduous in her precautions at work. Considering the dearth of relevant data on the effects of environmental toxicants while pregnant, the best advice should be to minimize exposures to any

TABLE 4.19
Dynamics of Reproductive Toxicology

| Stage of development | Stage of pregnancy | | |
| --- | --- | --- | --- |
| | Preconception | Intrauterine | Perinatal |
| | Gametes (sperm, ova) organogenesis; | 1st trimester organogenesis; 2nd, 3rd trimester fetus | Infant |
| Vulnerable areas | Spermatogenesis Oogenesis Fertilization | | Lactation |
| Major developmental effects | Mutagenesis | Teratogenesis | CNS—late transplacental carcinogenesis |
| Adverse manifestations | Sterility, decreasing fertility, chromosomal aberrations | Implantation defects, spontaneous abortions | Stillbirth; structural, behavioral, or functional alternations |
| Parental source of problem | Maternal and paternal | Maternal (3rd trimester) | Maternal (lactation) |

chemical which can be inhaled, swallowed, or absorbed cutaneously. We must consider the chemical type, quantity, means and time of exposure (both the length of exposure and at what stage of pregnancy), and individual susceptibility. What is known is that chemicals can cause maternal and fetal problems, including death. The liability of these problems is real and must be addressed by both knowledge of the possible effects of the specific chemical being used and exerting proper laboratory precautions.

A beautiful newborn is happiness. A defective child evokes guilt and unhappiness. It is then too late to fantasize as to the etiology, if any can be known, which was causative. A few precautions taken appropriately will at least reassure the parents that they did nothing wrong in the formative period of their child.

## REFERENCES (SECTION 4.8.6)

1. **Baetjer, A.,***Women in Industry, Their Health and Efficiency,* W. B. Saunders, Philadelphia, 1946, 145.
2. **Henschel, A.,** Women in industry — the difference, in Trans. 33rd Meet. Am. Conf., Governmental Industrial Hygienists, Toronto, 1971, 73.
3. **Messite, J. and Bond, M. B.,** Reproductive toxicology and occupational exposure, in *Occupational Medicine,* Zenz, C., Ed., Year Book Medical Publishers, Chicago, 1988.
4. National Center for Health Statistics, Statistics needed for determining the effects of the environment on health, U.S. Department of Health, Education, and Welfare, *Health Resources,* 77, 1457, 1977.
5. Council on Scientific Affairs: effects of toxic chemicals on the reproductive system, *JAMA,* No. (23), 253, 1985.
6. Assessment of reproductive and teratogenic hazards, in *Advances in Modern Environmental Toxicology,* Vol. 3, Christian, M. S., Galbraith, W. W., Voytek, P., and Mehlman, M. A., Eds., Princeton Scientific Publ., Princeton, NJ, 1983.

## 4.8.7. REGULATED AND POTENTIAL CARCINOGENS
An individual planning to work with carcinogenic material must be prepared to

meet the stringent standards imposed by OSHA for regulated material and inform the laboratory employees of the risks associated with the research program, as required by the Hazardous Communication Standard, for materials considered to be carcinogenic under that standard. These considerations should arise at the time of purchase of the material or before. Therefore, the list of chemicals considered as probable carcinogens was placed in the section on purchasing earlier in this chapter.

### 4.8.7.1. Carcinogens (Ethylene Oxide)

Ethylene oxide is used here as an example of a carcinogenic compound, one for which a specific standard was fairly recently (August 24, 1984) adopted by OSHA. The appendices of the standard provide an unusually complete guide to the safe use of this material.

Ethylene oxide ($C_2H_4O$; CAS NO. 75-21-8) is a gas at normal temperatures (boiling point = 10.7°C (51.3°F). The specific gravity of the gas with respect to the density of air is 1.49. It dissolves readily in water. It has an ether-like odor when concentrations are well above the OSHA PEL, so its odor cannot be considered to warn adequately of its presence. It is a significant fire hazard in addition to being a health hazard. The lower and upper explosion limits are, 3 and 100%, respectively. It will burn without the presence of air or other oxidizers, with a flash point below 0°F (−18°C) and may decompose violently at temperatures above 800°F (444°C). It will polymerize violently when contaminated with aqueous alkalis, amines, mineral acids, and metal chlorides and oxides. It would be classified as a Class B fire hazard for purposes of compliance with 29 CFR 1910.155. Locations defined as hazardous due to its use would be Class 1 locations for purposes of compliance with 29 CFR 1910.307.

Although dangerous because of its physical properties, the primary reasons for regulating the material by a separate standard were related to health effects — specifically, its identification as a human carcinogen, adverse reproductive effects, and ability to cause chromosome damage. Because of the latter two problems, women who suspect or know that they are pregnant should take special care to avoid exposures above the acceptable limits. There are a number of other adverse health effects in addition to these relatively newly identified problems.

Acute effects from inhalation include respiratory irritation and lung damage, headache, nausea, vomiting, diarrhea, shortness of breath, and cyanosis. Ingestion can cause gastric irritation and liver damage. It is irritating on contact with the eye and skin and can cause injury to the cornea and skin blistering on extended contact. Contact with pressurized, expanding vapor can cause frostbite. Individuals using this material should not wear contact lenses. It has also been associated with mutagenic, neurotoxicity, and sensitization effects.

Safe work practices for ethylene oxide fall into two areas: (1) normal practices associated with the use of a flammable gas and (2) health practices needed to reduce exposure to the vapors. For the former set of problems and the more common health problems, the procedures are relatively straightforward, the same as for other chemicals with similar properties: keep ignition sources and reactive materials away from the material; do not smoke, eat, or drink in the area; and wear personal protective equipment (goggles, gloves, respiratory protection, and protective clothing as needed to prevent exposure).

The means of preventing exposure to the gas to protect against the carcinogenic and reproductive hazards are spelled out in considerable detail in the appendices to the OSHA standard (29 CFR 1910.1047) as well as the specific measures needed to comply with the standard. A major use of the material is as a sterilizing agent in medical care operations, so much of the material in the standard is concerned with means of avoiding release of the gas or to capture gas that has been released. Other portions of the appendices deal with achieving compliance with other parts of the standard. The basic features of the standard are briefly given below as representative of those of the other regulated carcinogens. Many details are omitted, for which the reader is referred to the complete current standard, available from any local OSHA office.

The use of ethylene oxide does not automatically invoke all of the regulatory requirements if it can be shown by careful, objective measurements and documentation that the procedures to be used are such as to make it unlikely that any individual on a worst-case basis will exceed a 0.5 ppm action level for an 8-h time-weighted average exposure. These levels are measured without taking into account any respiratory protection provided by personal protective equipment. When any circumstance changes in such a way that the levels of exposure may increase, it will probably be necessary to demonstrate anew that the levels are lower than the action levels. Records of these data must be kept and available for examination. If the action levels are exceeded, in general, compliance with the terms of the standard is required.

Probably the most critical requirement of the standard is to ensure that no employee is allowed to be exposed to an airborne concentration level of ethylene oxide in excess of 1 ppm as an 8-h, time-weighted average. This must be demonstrated for each employee although, where the exposure conditions are sufficiently similar, measurements need only be made for representative employees. The method used to make the measurements must be capable of providing an accuracy of plus or minus 25% in the range of the PEL of 1 ppm, with 95% confidence limits, and plus or minus 35% in the range of the action limit of 0.5 ppm. It should be noted that a STEL of 5 ppm has recently been adopted.

If the measured levels are between 0.5 and 1 ppm, monitoring must be repeated at least every 6 months; for levels in excess of 1 ppm, monitoring must be done at least every 3 months. The results of the monitoring program as well as the outcome of any corrective actions taken to reduce the levels must be made available to the employees.

A regulated area must be established wherever the airborne concentrations may exceed 1 ppm. Access to this area must be limited to authorized personnel and the number of these persons must be kept at a minimum. The entrances to this area must be clearly marked with the following sign:

**DANGER
ETHYLENE OXIDE
CANCER HAZARD AND REPRODUCTIVE HAZARD
AUTHORIZED PERSONNEL ONLY
RESPIRATORS AND PROTECTIVE CLOTHING MAY BE REQUIRED
TO BE WORN IN THIS AREA**

Any containers of ethylene oxide with the potential for causing an exposure at or above the action level must be labeled with the legend:

**DANGER**
**CONTAINS ETHYLENE OXIDE**
**CANCER HAZARD AND REPRODUCTIVE HAZARD**

The label also must warn against breathing ethylene oxide. If it is to be used as a pesticide, the container labeling requirements of the Federal Insecticide, Fungicide, and Rodenticide Act (FIFRA) preempt the OSHA requirements.

If the PEL of 1 ppm is exceeded, the employer must establish and implement a written program to reduce actual employee exposures to below this level. Preferably, this protective program should be based on engineering controls and work practices, but also may include the use of approved respiratory protection where alternate measures are not feasible. Approved respiratory protection means those respiratory devices specifically approved for protection from ethylene oxide exposure by either the Mine Safety and Health Administration (MSHA) or the National Institutes for Occupational Safety and Health (NIOSH), under the provisions of 30 CFR, Part 11. Employee rotation is not an acceptable means of achieving compliance. The plan must also include means of leak detection and an emergency plan. The compliance plan must be reviewed and revised as needed at least annually. The written emergency plan must provide for equipping employees with respiratory protection. It must include those elements required under 29 CFR 1910.38. Provision must be made for alerting the employees of an emergency and for evacuation of employees from the danger area.

A medical surveillance and consultation program must be available for any employee who may be exposed to ethylene oxide at or above the action level of 0.5 ppm for 30 or more days a year, or in an emergency situation. There must be a preemployment examination, a medical and work history, an annual examination for each year the 30-d criterion is met, and a postemployment examination. In addition to these requirements, examinations may be indicated for employees exposed in an emergency, and as soon as possible after any employees believe that they are exhibiting symptoms of exposure to ethylene oxide. Employees may also request and be given medical advice about the effects of their exposures to ethylene oxide on their ability to produce a child. The physician may recommend other examinations. For example, employees may wish to obtain fertility and pregnancy tests, and they are to be given these tests if the physician considers the tests appropriate under the circumstances.

The surveillance program must include a medical and work history, with emphasis on the pulmonary, hematologic, neurologic, and reproductive systems, the eyes, and the skin. The physical examination must emphasize the same areas. A complete blood count is to be part of the examination as well as any other appropriate tests designated by the physician.

The results of the tests must be made available to the employer, employee, and others (such as the employee's physician) upon the employee's written authorization. The physician must provide a written opinion, including the results of the examination, whether the examination revealed any conditions that the employee's occupational exposure would aggravate, and whether there should be any restrictions placed on the employee or modification to the employee's duties to reduce exposure. The physician must also state that he has discussed the results of the examinations with the employee and any followup actions that should ensue. If there are any extraneous medical factors not pertinent to the work-related activities of the employee discovered in the course of the examination, the employee has the right to expect complete confidentiality of this information. The medical records must be maintained, according to the provisions of 29 CFR 1910.20, by the employer for the duration of the employee's employment plus 30 years. Much of the supporting data records, such as exposure information, must be kept for a similar period.

In addition to the labeling and signs that have already been discussed, a hazard communication and training program must be established by the employer. Before an employee is assigned duties which could result in an exposure above the action level, he must be provided training and information about ethylene oxide. This communication program must be repeated at least annually and must be made current as needed during the course of the year. The program must include information about the requirements of the OSHA standard, including where a copy of it can be obtained, operational procedures, the medical surveillance program, methods available to detect ethylene oxide, measures taken by the employer to ensure compliance with the standard, measures the employees can take to protect themselves, the emergency plan, the hazards of ethylene oxide, where the current material safety data sheet can be found and how to interpret the information on it, and the details of the employer's hazard communication plan.

## REFERENCE (SECTION 4.8.7.1)

1. Occupational Safety and Health Administration, General Industry Standards, Subpart Z — Occupational Health and Environmental Controls, Ethylene Oxide, 1910.1047, Washington, D.C.

### 4.8.8. NEUROLOGICAL HAZARDS OF SOLVENTS*

Many solvents have been recognized or suspected of being carcinogenic or of posing reproductive problems (see Section 4.2.4.2.4). However, many of the recognized health effects due to solvents are neurotoxic. In commonly available references such as Sax or Merck, much of the descriptive material on health symptoms due to exposure is based on neurotoxic actions. As with most other health effects, the neurotoxic problems may be divided into acute or immediate effects and those due to chronic exposures which lead to delayed and possibly persistent health changes.

Much of the epidemiological data on health effects have been due to exposures in an industrial setting rather than the laboratory since the exposures are liable to be at relatively stable levels and involve a relatively small number of solvents, in contrast to the extremely complicated and rapidly changing laboratory environment.

The primary modes of uptake of solvents by the body are inhalation and absorption through the skin. The rate of absorption by inhalation is affected by a number

---

* The information in this section is derived primarily from Current Intelligence Bulletin 48, Organic Solvent Neurotoxicity, DHHS (NIOSH) Publ. No. 87-104, National Institute for Occupational Safety and Health, U.S. Department of Health, Education and Welfare, Washington, D.C., 1987.

of factors, some due to the properties of the material and the interaction of the solvent vapors with the lungs, and some due to other factors such as the concentration of solvent fumes in the air, the duration of the exposure, and the level of exertion at the time of exposure by the exposed individual. The rate of intake is significantly increased by elevated levels of physical activity.

The rate of intake through the skin is dependent upon the duration of the contact, skin thickness, degree of hydration of the skin, and possible breaks in the integrity of the skin, i.e., injuries or skin disorders. An example of the comparative rates of intake by the two routes is that an immersion of both hands in xylene for 15 min gives about the same levels in the blood as an exposure to an airborne concentration of 100 ppm for the same period. However, this cannot be assumed to be an accurate reflection of the comparative rates for other solvents.

After exposure, the material is often transformed by the liver into less toxic water-soluble compounds or, in some cases, into more toxic intermediate metabolites. In other cases, the solvents are lipophilic, i.e., may be taken up and accumulate in lipid-rich tissues such as the nervous system.

The acute, short-term neurotoxic effects result from action of the solvent on the central nervous system (CNS). The effects may range from symptoms resembling intoxication to CNS depression, psychomotor impairment, narcosis, and death from respiratory failure. At intermediate levels of exposure, common effects are drowsiness, headache, dizziness, dyspepsia, and nausea. Short-term effects may cause mood changes as well, as reflected by increasing feelings of physical and mental tiredness during an exposure period corresponding to a typical workshift, for example.

The effects of extended or chronic exposures have been divided into three categories of varying severity at two relatively recent international workshops (minimal, moderate, and pronounced), although the nomenclature differed slightly at the two conferences. The last category, which so far has not been observed in an occupational situation, but has in persons deliberately exposing themselves to solvent fumes, would be reflected in serious deterioration of the nervous system, including mental capacity and function. The effects of such extreme exposures would be only partially reversible at best.

The least severe category would be characterized by deterioration in memory function and ability to concentrate, physical fatigue, and irritability, while the second level would involve sustained mood and personality changes as well as further deterioration of intellectual functions, including learning capacity. At least some of the effects of chronic exposures appear to persist well beyond the termination of the exposure and may be permanent.

Acute effects appear to be caused by the solvent itself, while the chronic effects may be associated with the intermediate metabolic reaction products. The effects of short-term exposures appear to be reversible, while the effects of prolonged exposures leading to changes in nerve tissue may be irreversible. Although the data on chronic exposure toxicity are not as abundant or definitive as could be desired, the data that are available definitely appear to support a conservative approach to personnel exposure. Levels to which an individual is exposed, especially over an extended period, should emulate the NRC As Low As Reasonably Achievable (ALARA) philosophy. The various levels established as regulatory by OSHA (permissible exposure levels, PEL) and NIOSH (recommended exposure levels, REL) or the threshold limit values (TLVs) recommended by the ACGIH should be adopted as a maximum permitted occupational level, using the lowest of the three values as a guide. An action level of 50% of the appropriate level is recommended as a trip point for initiating remedial steps.

Adequate general ventilation, provision of appropriate containment equipment (fume hoods, safety cabinets, or localized exhaust systems), procedural controls, or use of personal protective equipment, respiratory protection, and protective gloves, clothing, and eye protection are all part of a possible program to reduce exposures.

It is difficult in the laboratory to monitor the exposure levels as recommended above, but monitoring should be performed wherever possible. Even partial data or data taken under nonstandard conditions are better than none at all. As a supplement to a monitoring and exposure limitation program, those individuals who actively use solvents as a routine part of their job a significant portion of the time should participate in a medical surveillance program. Individuals should receive a preemployment examination, which should include a prior work history and medical history. The examination should emphasize the nervous, cardiovascular, respiratory, and reproductive systems as well as the liver, kidneys, blood, gastrointestinal tract, eyes, and skin. A comprehensive blood panel should be run as well as a complete blood count and urinary test. Some special tests might be suggested by the physician, such as a cholinesterase enzyme test for information on the CNS. A blood serum sample might be stored for later comparison. Periodic reexaminations should be given annually if the exposure is heavy, but no less often than every 5 years.

Table 4.20 provides, for a number of common solvents, current recommended PELs, RELs, and TLVs for 8-h time-weighted averages.

As can be seen from the Table 4.20, the values from the three sources frequently disagree. Usually, the ACGIH values are the most current, while the OSHA levels often go back to the original adoption of the OSHA act. The OSHA values are legally binding, however, while those of the other two are guidelines only. The NIOSH values, where available, are often the most conservative. Currently, OSHA is proposing to change its values in favor of, primarily, the ACGIH values or the NIOSH values.

### TABLE 4.20
### OSHA PELs, NIOSH RELs, and ACGIH TLVs® for Some Organic Solvents
### All Values in Parts Per Million Except Where Noted

| Compound (CAS No.) | OSHA PEL (ceiling) | NIOSH REL (ceiling) | | ACGIH TLV® (STEL, 15 min) | |
|---|---|---|---|---|---|
| Alkanes (C5—C8) | | | | | |
| Pentane (109-66-0) | 1000 | 120 | (610/15 min) | 600 | (750) |
| Hexane (110-54-3) | 500 | 100 | (510/15 min) | 50 | (1000) |
| Heptane (142-82-5) | 500 | 85 | (440/15 min) | 400 | (500) |
| Octane (111-65-9) | 500 | 75 | (385/15min) | 300 | (375) |
| Allyl chloride (107-5-1) | 1 | 1 | (3/15 min) | 1 | (2) |
| Benzene (71-43-2) (carcinogen) | 1 | 0.1 | (1/15 min) | 10 | |
| Benzyl chloride (100-44-7) | 1 | | (1/15 min ceiling) | 1 | |

**TABLE 4.20 (continued)**
**OSHA PELs, NIOSH RELs, and ACGIH TLVs® for Some Organic Solvents**
**All Values in Parts Per Million Except Where Noted**

| Compound (CAS No.) | OSHA PEL (ceiling) | | NIOSH REL (ceiling) | | ACGIH TLV® (STEL, 15 min) | |
|---|---|---|---|---|---|---|
| Carbon disulfide (75-15-0) | 20 | (30 ceiling; 100/30 min) | 1 | (10/15 min) | 10 | |
| Carbon tetrachloride (56-23-5) (carcinogen) | 10 | (25 ceiling; 200/5 min in 4 h) | 2 | (1 h, 45-¹ sample) | 5 | |
| Chloroethane (75-00-3) | 1000 | | Handle with care | | 1000 | |
| Chloroform (67-66-3) (carcinogen) | | (50 ceiling) | 2 | (1 h, 45-i sample) | 10 | |
| Chloroprene (126-99-8) (carcinogen) | 25 | | (1/15 min ceiling) | | 10 | |
| Cresol (1319-77-3) | 5 | | 2.3 | | 5 | |
| Di-2-ethyl-hexyl-phthalate (117-81-7) (carcinogen) | 5 mg/M³ | | Reduce exposure to lowest feasible level | | 5 | (10)mg/M³ |
| Dioxane (123-91-1) (carcinogen) | 100 | | (1/30 min ceiling) | | 25 | |
| Epichlorohydrin (106-89-8) (carcinogen) | 5 | | Minimize occupational exposure | | 2 | |
| Ethylene dibromide (106-93-4) (carcinogen) | 20 | (50/5 min) | 0.045 (0.13/15 min) | | Suspect skin carcinogen | |
| Ethylene dichloride (107-06-2) (carcinogen) | 50 | (200/5 min in 3 h) | 1 | (2/15 min) | 10 | |
| Furfuryl alcohol (98-00-0) | 50 | | 50 | | 10 | (15) |
| Glycol ethers | | | | | | |
| 2-Methoxyethanol (110-86-4) | 25 | | Reduce exposure to lowest feasible level | | 5 | (skin) |
| 2-Ethoxyethanol (110-80-5) | 200 | | Reduce exposure to lowest feasible level | | 5 | (skin) |
| Isopropyl alcohol (67-63-0) | 400 | | 400 | (800/15 min) | 400 | (500) |
| Ketones | | | | | | |
| Acetone (67-64-1) | 1000 | | 250 | | 750 | |
| Methyl ethyl ketone (78-93-3) | 200 | | 200 | | 200 | |
| Methyl *n*-propyl ketone (107-87-9) | 200 | | 150 | | 200 | |
| Methyl *n*-butyl ketone (591-78-6) | 100 | | 1 | | 5 | |
| Methyl *n*-amyl ketone (110-43-0) | 100 | | 100 | | 50 | |
| Methyl isobutyl ketone (108-10-1) | 100 | | 50 | | 50 | |
| Disobutyl ketone (108-83-8) | 50 | | 25 | | 50 | |
| Cyclohexanone (108-94-1) | 50 | | 25 | | 25 | (skin) |
| Mesityl oxide (141-79-7) | 25 | | 10 | | 15 | |
| Diacetone alcohol (123-42-2) | 50 | | 50 | | 50 | |

**TABLE 4.20 (continued)**
**OSHA PELs, NIOSH RELs, and ACGIH TLVs® for Some Organic Solvents**
**All Values in Parts Per Million Except Where Noted**

| Compound (CAS No.) | OSHA PEL (ceiling) | | NIOSH REL (ceiling) | ACGIH TLV® (STEL, 15 min) | |
|---|---|---|---|---|---|
| Isophorone (78-59-1) | 25 | | 4 | 5 | (maximum value, not to be exceeded) |
| Methyl alcohol (67-56-1) | 200 | | 200   (800/15 min) | 200 | (250) |
| Methyl bromide (74-83-9) (carcinogen) | | (20 ceiling; skin) | Reduce exposure to lowest feasible level | 5 | (skin) |
| Methyl chloride (74-87-3) (carcinogen) | 100 | (200 ceiling; 300/5 min in any 3-h period in an 8-h shift) | Reduce exposure to lowest feasible level | 50 | (100) |
| Methylene chloride (75-09-2) (carcinogen) | 500 | (1000 ceiling; 2000/5 min in any 2-h period in an 8-h shift) | Reduce exposure to lowest feasible level | 100 | (50 pending) |
| Methyl iodide (74-88-4) (carcinogen) | 5 | (skin) | Reduce exposure to lowest feasible level | 2 | (skin) |
| Nitriles | | | | | |
| Acetonitrile (75-05-8) | 40 | | 20 | 40 | (60)(skin) |
| Tetramethyl succino-nitrile (3333-52-6) | 0.5 | | 1 | 0.5 | |
| 2-Nitropropane (79-46-9) (carcinogen) | 25 | | Reduce exposure to lowest feasible level | 10 | |
| Styrene, monomer (100-42-5) | 100 | (800 ceiling; 600/5 min in 3-h period) | 50   (100 ceiling) | 50 | (100) |
| 1,1,2.2-Tetrachloro-ethane (79-34-5) (carcinogen) | 5 | | Reduce exposure to lowest feasible level | 1 | (skin) |
| Tetrachloroethylene (127-18-4) (carcinogen) | 200 | (200 ceiling; 300/5 min in 3 h) | Minimize workplace exposure levels; limit number of workers | 50 | (200) |
| Butanethiol (butyl mercaptan) (109-79-5) | 10 | | (0.5/15 min) | 0.5 | |
| Ethanethiol (Ethyl mercaptan) (75-08-1) | | (10 ceiling) | (0.5/15 min) | 0.5 | |
| Toluene (108-88-3) | 200 | (300 ceiling; 500/10 min) | 100   (200/10 min) | 100 | (150) |
| 1,1,1-Trichloro-ethane (71-55-6) | 350 | | (350 ceiling/ 15 min) | 350 | (450) |
| 1,1,2-Trichloro-ethane (79-00-5) (carcinogen) | 10 | | Reduce exposure to lowest feasible level | 10 | (skin) |
| Trichloroethylene (79-01-6) (carcinogen) | 100 | (200 ceiling; 300/5 min in 2 h | 25 | 50 | (200) |
| Vinyl acetate (108-05-4) | None | | (4/15 min) | 10 | (20) |

**TABLE 4.20 (continued)**
**OSHA PELs, NIOSH RELs, and ACGIH TLVs® for Some Organic Solvents**
**All Values in Parts Per Million Except Where Noted**

| Compound (CAS No.) | OSHA PEL (ceiling) | NIOSH REL (ceiling) | ACGIH TLV® (STEL, 15 min) | |
|---|---|---|---|---|
| Vinylidene chloride (75-35-4) | None | Control as vinyl chloride; OSHA regulated carcinogen | 5 | (20) |
| Xylene (1330-20-7) | 100 | 100    (200/10 min) | 100 | (150) |

# REFERENCES (SECTION 4.8.8)*

1. *TLVs© Threshold Limit Values and Biological Exposure Indices for 1986-87,* American Conference of Governmental Industrial Hygienists, Cincinnati, 1986.
2. **Astrand, I.,** Uptake of solvents in the blood and tissues of man, *Scand. J. Work Environ. Health,* 199, 1975.
3. **Baker, E. L. and Seppalainen, A. M.,** Session 3, human aspects of solvent neurological effects, report on the workshop session on clinical and epidemiological medicine, *Arch. Ind. Health,* 13, 581, 1986.
4. **Baker, E. L. and Vyskocil, J.,** The neurotoxicity of industrial solvents: a review of the literature, *Am. J. Ind. Med.,* 8, 207, 1956.
5. **Bird, M.,** Industrial solvents: some factors affecting their passage into and through the skin, *Ann. Occup. Hyg.,* 24, 235, 1981.
6. **Browning, E.,** *Toxicity and Metabolism of Industrial Solvents,* Elsevier, Amsterdam, 1965.
7. **Cherry, N., Venables, H., and Waldron, H.,** The acute behavioral effects of solvent exposure, *J. Soc. Occup. Med.,* 33, 13, 1983.
8. **Gamberale, F.,** Behavioral effects of exposure to solvents, experimental and field studies, in *Adverse Effects of Environmental Chemicals and Psychotropic Drugs,* Vol. 2, Horvath, M., Ed., Elsevier, Amsterdam, 1976, 111.
9. **King, M., Day, R., Oliver, J., Lush, M., and Watson, J.,** Solvent encephalopathy, *Br. Med. J.,* 283, 663, 1981.
10. **Knave, B., Anselm-Olson, B., Elofsson, S., Gamberale, F., Isaksson, A., Mindus, P., Persson, H. E., Struwe, G., Mennberg, A., and Westerholm, P.,** Long-term exposure to jet fuel. II. A cross-sectional epidemiologic investigation on occupationally exposed industrial workers with special reference to the nervous system, *Scand. J. Work. Environ. Health,* 4, 19, 1978.
11. **Lindstrom, K.,** Changes in psychological performances of solvent-poisoned and solvent-exposed workers, *Am. J. Ind. Med.,* 1, 69, 1980.
12. Criteria for a Recommended Standard: Occupational Exposure to Refined Petroleum Solvents, DHEW (NIOSH) Publ. No. 77-192, National Institute for Occupational Safety and Health, U.S. Department of Health, Education and Welfare, Cincinnati, 1977.
13. **Orbaek, P., Risberg, J., Rosen, I., Haeger-Aronson, B., Hagstadius, S., Hjortsberg, U., Regnell, G., Rehnstrom, S., Svensson, K., and Welinder, H.,** Effects of long-term exposure to solvents in the paint industry, *Scand. J. Work Environ. Health,* 11 (Suppl. 2), 1, 1985.
14. **Politis, M., Schaumburg, H., and Spencer, P.,** Neurotoxicity of selected chemicals, in *Experimental and Clinical Neurotoxicology,* Spencer, P. and Schaumburg, H., Eds., Williams & Wilkins, Baltimore, 1980, 613.
15. **Seppalainen, A. M.,** Neurophysiological aspects of the toxicity of organic solvents, *Scand. J. Work. Environ. Health,* 11 (Suppl. 1), 61, 1985.
16. **Seppalainen, A. M. and Antti-Poika M.,** Time course of electrophysiological findings for patients with solvent poisoning, *Scand. J. Work. Environ. Health,* 9, 15, 1983.

* These references are the more general and comprehensive ones from the list of references in the original article.

17. **Seppalainen, A. M., Husman, K., and Martenson, C.,** Neurophysiological effects of long-term exposure to a mixture of organic solvents, *Scand. J. Work. Environ. Health,* 4, 304, 1978.

18. **Seppalainen, A. M., Lindstrom, K., and Martelin, T.,** Neurophysiological and psychological picture of solvent poisoning, *Am. J. Ind. Med.,* 1, 37, 1980.

19. **Spencer, P. and Schaumburg, H.,** Organic solvent neurotoxicity, facts and research needs, *Scand. J. Work. Environ. Health,* 11 (Suppl. 1), 53, 1980.

20. **Toftgard, R. and Gustafsson, J.,** Biotransformation of organic solvents, a review, *Scand. J. Work. Environ. Health,* 6, 1, 1980.

21. **Valciukas, J., Lilis, R., Singer, H., Glickman, L., and Nicholson, W.,** Neurobehavioral changes among shipyard painters exposed to solvents, *Arch. Environ. Health,* 40, 47, 1985.

22. **Waldon, H. A.,** Solvents and the brain, *Br. J. Ind. Med.,* 43, 73, 1986.

23. Organic Solvents and the Central Nervous System, EH5, World Health Organization and Nordic Council of Ministers, Copenhagen, 1985, 1.

# 4.9. SPILLS AND EMERGENCIES

A chemical spill is probably the most common type of laboratory accident and potentially one of the most serious if the material gives rise to hazardous vapors, interacts with the laboratory environment in a violent physical fashion, or is toxic or corrosive upon contact with a person's body. Most accidents involving chemical spills do not have such dramatic consequences, but they must all be handled correctly.

### 4.9.1. SMALL- TO MODERATE-SCALE SPILLS

Chemical spills generally involve only small quantities of materials, such as would be contained in a single reagent container or reaction flask. Unless the chemical has unusually hazardous properties, procedures to correct such minor spills are relatively straightforward. If the material is a solid, it is simply swept up into a container with which it will not react for disposal. However, relatively few materials may then be disposed of as ordinary trash. Custodians should not, under normal circumstances, be expected to remove trash containing chemical materials from laboratories. Chemicals which are known to be harmless if placed in a municipal landfill (which are to be disposed of in this way) should be placed in a separate container, labeled with the identity of the material, and marked "safe for disposal in regular trash". Some materials which are soluble in water or, if liquid, miscible in water, may also be disposed of in the sanitary system, while others may be rendered harmless by reaction with other chemicals in the laboratory. However, before any of these things are done, it should be confirmed that the material does not fall under the provisions of the Resource Conservation and Recovery Act, which prohibits such disposal procedures. If restrictions exist, the materials must be treated as hazardous waste.

Spilled liquids may often be diluted with water and simply mopped up or, in some cases, eliminated by spreading an absorbent material such as vermiculite or a clay absorbent (such as calcium bentonite) on the spilled material or, a bit more neatly but much more expensively, by placing pillows containing an absorbent material on the liquid, after which the absorbent material or pillows are collected into containers for later disposal.

Spills of larger quantities of solids are usually no different from smaller spills, except in scale. The spilled solids simply remain as they fell. As long as the solids do not react with materials with which they may have come into contact, or the dust is not a breathing hazard, the cleanup is simply mechanically larger, with more containers being set aside for evaluation as to the mode of disposal, i.e., to determine whether they are to be treated as hazardous waste.

Spills of liquids, even of relatively innocuous materials, are messier and more likely to come into contact with other materials with which they may interact. If absorbent materials such as loose absorbent, spill control pillows, and absorbent pads are immediately available within the laboratory to quickly confine the area of the spill, the immediate problems can be minimized and the cleanup made considerably easier.

Relatively few laboratory spills involve totally innocuous materials. However, there are degrees of hazards associated with spilled materials, from the mildly troublesome to IDLH materials. The risk also depends upon the exposure mode or the character of an ensuing physical hazard. A material which emits deadly fumes is infinitely more dangerous to most laboratory personnel than one which is highly corrosive to tissue. Unless an individual is directly injured by contact with the material, the latter hazard may be avoided by removing oneself from the immediate area, and protective clothing is often enough to protect those engaged in the ensuing cleanup. However, generation of deadly vapors or gases will usually mandate an evacuation of at least the laboratory and perhaps the entire building. Individuals correcting the situation will require air-supplied breathing apparatus and often wear garments protecting their entire body from contact with the airborne materials. Similar, and often more cumbersome, protective gear will be required if a fire occurs after a spill of flammable materials. Remedying the situation involving a corrosive spill is very likely to be within the capacity of laboratory employees, but incidents involving toxic gases or fire would almost certainly mandate the participation of trained emergency personnel.

The following material will address handling spills and emergencies which are confined to the laboratory. Those emergencies which are not confined to at least the building in which they originate will be treated separately.

Some basic emergency equipment and supplies should be readily available to every laboratory using hazardous chemicals. It is essential that some items be available within the laboratory itself. For example, in any laboratory in which flammables are stored, OSHA standards (1910. 106[d][7][b]) require that at least one 12-B portable fire extinguisher be located not less than 10 ft nor more than 25 ft from a flammable material storage area within a building. If other types of chemicals are in use, other types of fire extinguishers should be available, such as Class-D units if reactive metals are stored or in use, pressurized water units if there are substantial amounts of ordinary combustible materials present, and perhaps even some portable Halon™ units if there is a concern for damaging delicate electronic gear.

Spill control materials to absorb spilled materials are available commercially. Some materials are universal, and some are intended to not only absorb materials, but also to neutralize specific materials such as acids, hydrofluoric acid, caustics, and mercury. Some of these products for coping with acids and caustics contain color indicators to determine when the spill is neutralized. If bought in convenient small packages suitable for cleaning up modest spills of about half a liter, the cost per unit of absorbent is high for these commercial products. However, they can usually be bought in small drums at a substantial saving. If bought in this way, the materials can be repackaged into convenient smaller sizes for use in individual laboratories. For many purposes, calcium bentonite, bought in bulk, serves very nearly as well and is much less costly.

Whether commercial kits of absorbent materials or cheaper substitutes are used, some capability to quickly soak up spilled materials needs to be readily available. Bulk quantities can be kept in a central location for replenishing supplies in individual laboratories or for use in atypical larger spills, but a sufficient supply should be kept within an individual laboratory to use on a spill of up to a gallon or two.

In addition to absorbent material, the following items should be maintained in an individual laboratory (Chapter 1 provides a more comprehensive list of items and equipment which should be available within the building, although not necessarily within the laboratory).

- Bags, large, 6-mil polyethylene
- Brooms
- Brush, hand
- Bucket, plastic (polyethylene)
- Containers, plastic (5 gal)
- Coveralls, lightweight, chemical resistant, treated
- Dust pan

- Gloves, chemical resistant
- Goggles, chemical splash protective
- Mops
- Paper, plastic-backed absorbent roll
- Respirators, organic, acid, dust, caustic
- Scoops
- Shoe covers, high topped, chemical resistant
- Tape, duct

For typical laboratory spills, many of these items will not be needed, but all of them could have a use, depending on the type of spill. For example, an acid spill could quickly destroy a person's shoes during the cleanup, and a drop of acid on a person's clothes would ruin them. Since the persons doing this cleanup work are likely to be laboratory personnel, not only should they be protected against injury, but they also should not be expected to incur any economic loss.

Large quantities of each item should not be needed in a typical spill kit. Unless it is an unusually large incident, one or two persons actually working on the clean up, with one person bringing supplies and taking waste away, is probably about optimum. Aisle widths in the typical laboratory would preclude the easy access of more than a few persons to the spill at a time, so that even if relatively complete protective clothing and equipment were needed, no more than three sets would be needed at a time.

None of the equipment listed above requires an extraordinary level of training to use properly. It should be possible for most technically trained personnel to clean up fairly substantial spills safely. Some straightforward guidelines will aid the user in preventing any material from getting on his person.

Chemically resistant clothing generally either snaps, buttons, or zips up the front, and although the two edges overlap, material can still enter the front seam. Overlapping layers of duct tape will seal this opening. The sleeves usually fit loosely and should be folded over around the outside of the glove and duct taped, with the tape in contact with both the sleeve and the glove. Trouser cuffs should be brought down over the top of the shoe covers and folded tightly around them. Duct tape is then used to seal this opening by wrapping it around the ankle so that the adhesive is in contact with both the shoe cover and the coverall. With this done, the entire body below the neck is protected from incidental contact with materials. The head can be covered with a hood of the same material as the coverall and the face protected with a full-face respirator, if this is required. Otherwise, a half-face respirator and goggles will provide almost as complete protection.

There are two notes of caution about the apparel described above. Light-weight coated garments are usually not rated for chemical protection, although for the level of contact which should be experienced if one works carefully, the wearer should be reasonably well protected against light exposures. If substantial contact with the chemical or its vapors is anticipated, a heavier coverall designed to provide protection against chemicals should be substituted. Another problem with the clothing is that it does not "breathe" and after a relatively short period — 30 min to 1 h, depending upon the level of exertion — it will be necessary to cease work for a period of time, open the coveralls, and "cool off". However, it should be possible to clean up most small to moderate laboratory spills in less than an hour.

If the outside of the coverall is contaminated, there is a definite procedure which can be used to avoid the outside from coming into contact with your skin and clothes. If a hood and face mask are worn, they should be removed first. The duct tape sealing the hood to the coverall is peeled off and then the hood is grasped at about the level of the cheek bones and peeled backwards, so that it tends to turn inside out. The hood is put in a plastic bag. The respirator (and goggles if worn separately) is then removed and, since it is likely to be worn again whereas the coveralls may not be reused, placed in a separate plastic bag. The duct tape down the front of the coverall is then removed, and the front unbuttoned. The two sides are then separated and pulled back off the shoulders so that the inside is exposed, the outside being doubled back over itself. The duct tape is removed from one sleeve, and the arm of the coverall on that side is pulled down over the hand so that the sleeve is turned inside out. The operation is then repeated on the other side, to free both arms. The duct tape is removed from the cuffs of the pants and the leg of the coverall pulled down over the foot so that the leg is turned inside out. The operation is then repeated on the other leg. The coverall is now off, with the clean interior surface exposed. The coverall is put in the plastic bag with the hood. The shoe covers can then be removed carefully (so that the outside does not come into contact with the skin and clothing) and placed in the plastic bag with the clothing. Finally, the gloves are removed in such a way that they also are turned inside out and placed into the contaminated clothing bag. If carefully done, there should be no opportunity for the contaminated outer surface of any of the protective gear to come into contact with the wearer.

The cleanup itself also has an optimum procedure. If two persons are working directly on the spill, then a third person should be available to bring fresh supplies and take waste away. Individuals working directly on the cleanup should not leave the immediate vicinity in order to avoid contaminating a wider area. For spills of substantial sizes and involving materials of significant hazard, it may be desirable to establish a formal entrance and exit control point, one side of which is to be considered dirty and the other maintained clean. The cleanup should start at the perimeter of the contaminated area and move inward so as to steadily reduce the area involved. In principle, the workers should clean in front of themselves so that they create a clean area in which to work. A previously cleaned area should not be readily recontaminated if this procedure is followed.

Any waste from the dirty area should be placed in plastic pails and passed to the person outside the dirty area. If the pails are contaminated, they should be placed in clean, heavy-duty (preferably 6 mil or better) double plastic bags before being set on a floor area covered with plastic-backed absorbent paper.

Once the spilled chemical has been removed from the surface, it is always good practice to thoroughly scrub or mop the previously contaminated area.

These procedures seem very formal on first reading if one is not familiar with decontamination work. However, it takes very little additional time to do the job properly. The procedures ensure that no one is likely to be injured or that no one's clothes are damaged in the process, and it is unlikely that the cleanup will have to be repeated.

For large spills, especially as the hazards associated with the material become more serious, it would be highly desirable for the work to be done under the guidance and assistance of an experienced safety professional. At some higher level of risk, the danger to inexperienced personnel, as most laboratory personnel usually are in the context of a serious chemical emergency, should cause one to consider not using laboratory personnel at all, other than as an information resource. These are judgment calls to be made by the organization's emergency coordinator, who should be called to the scene of any major chemical emergency, especially ones in which a building has been evacuated. Really large emergencies may be beyond even the most experienced local personnel, and hazard material response teams should be called in from outside agencies as quickly as possible. One of the more important factors in reducing the scope of most emergencies is prompt, effective action. If local capacity to deal with an emergency is questionable, outside aid should be called for immediately, while local efforts continue to protect human life and confine the scope of the incident. Most emergency groups would rather be called for unnecessarily than to arrive at a scene where the situation has deteriorated to the point of being out of control.

### 4.9.2. LARGE-SCALE RELEASES OF CHEMICALS (SARA)

In the last few years, there have been some noteworthy releases of chemicals from facilities using chemicals which have caused a substantial number of injuries and deaths in nearby communities. There had already been a substantial movement in many states to require operators of facilities using hazardous chemicals to provide information on these chemicals to nearby communities and to work with the communities on emergency planning. Many responsible chemical firms had individually begun to provide such data. The accidents which occurred in 1984 in Bhopal, India and in 1985 in Institute, West Virginia undoubtedly provided additional motivation for federal action in this area. On October 17, 1986, the Emergency Preparedness and Community Right-To-Know Act was signed into law. This law is more commonly known as Title III of the Superfund Amendments and Reauthorization Act (SARA, Title III), and this appellation will be used henceforth. This act extended and revised the authority established under the original Superfund Act (the Comprehensive Environmental Response, Compensation, and Liability Act of 1980, [CERCLA]). The extended act specifically establishes new authority for emergency planning and preparedness, community right-to-know reporting, and toxic chemical release reporting.

The SARA provisions apply to those industries covered under the OSHA Hazard Communication Act, which has been extended to all chemical users in addition to the originally covered industry groups in Standard Industrial Codes 20 to 39 and which handle, use, or store certain extremely hazardous chemicals or quantities of hazardous chemicals in excess of the limits established by the EPA under the act. The limits are such that many facilities do not have quantities in excess of the limits. These levels are likely to be lowered substantially in 1989 so that far more facilities will be covered.

At the moment, there are a number of important exemptions under SARA, one of which is specifically applicable to the research facility. In 40 CFR 370.2(5), a "Hazardous Chemical" is defined to mean any hazardous chemical as defined under 29 CFR 1910.1200(c), "except that such term does not include the following substances:"

1.    Any food, food additive, color additive, drug, or cosmetic regulated by the Food and Drug Administration.
2.    Any substance present as a solid in any manufactured item to the extent exposure to the substance does not occur under normal conditions of use.
3.    Any substance to the extent it is used for personal, family or household purposes, or is present in the same form and concentration as a product packaged for distribution and use by the general public.
4.    Any substance to the extent that it is used in a research laboratory or a hospital or other medical facility under the direct supervision of a technically qualified individual.
5.    Any substance to the extent it is used in routine agricultural operations or is a fertilizer held for sale by a retailer to the ultimate customer.

The fourth exemption is obviously the one of importance in the context of this book. Note that it does not exempt research facilities, it simply redefines the definition of a hazardous chemical to exclude those used in small quantities in laboratories under the direct supervision of a technically knowledgeable person. It does not exclude the same chemicals stored in a warehouse or stores area, nor does it exclude chemicals used in maintenance or support operations, except as they fall under the other exemptions.

There are a number of critical requirements under SARA for a facility which falls under the provisions of the act. A facility is subject to the provisions of the act if it has any (nonexempt quantities) of the 366 extremely hazardous chemicals currently on the list which EPA originally issued (with 406 chemicals) on April 22, 1987, in excess of the threshold planning quantity established by the EPA for that substance.

Individual facilities were sent the list of 406 extremely hazardous chemicals by their state commissions in the spring of 1987 and required to inventory their holdings and make a report of those which met the threshold planning quantity by May 17, 1987. It was also necessary to list the chemicals which the facility had in excess of higher thresholds for emergency planning. Eligible facilities were then required to notify the local planning committee established under the act of their facility emergency planning coordinator by September 17, 1987. The local committee was required to complete its local emergency plan by October 17, 1988. Large facilities may be represented directly on the planning committee since membership is required to be drawn from elected state and local officials, law enforcement, civil defense, fire fighting, first aid, health, local environmental, hospital, and transportation personnel, broadcast and print media, community groups, and facility owners and operators subject to SARA.

The first request for information was originally designed to identify agencies which would be likely to fall under the provisions of SARA. Organizations subject to SARA are required to prepare certain inventory forms and make them available to (1) the appropriate local emergency planning committee, (2) the state emergency response commission, and (3) the fire department with jurisdiction over the facility.

There are basically two inventory forms containing certain levels of information, defined as Tier 1 and Tier 2. The inventory form with Tier 1 information was required to be submitted by March 1, 1988 and annually thereafter. If Tier 2 forms are requested by the groups listed in the preceding paragraphs for a given year, then Tier 1 forms are not required for that year.

Tier I information required on the inventory form is

1.    An estimate (in ranges) of the maximum amount of hazardous chemicals in each category of health and physical hazards, as set forth by the Occupational Health and Safety Act and regulations published under that act, at the facility during the preceding calendar year.
2.    An estimate (in ranges) of the average daily amount of hazardous chemicals in each category present at the facility during the preceding calendar year.
3.    The general location of hazardous chemicals in each category.

Tier 2 information is required to be provided only upon request to the same three groups. They, in turn, can make this information available to others such as other state and local officials or the general public under certain conditions.

Tier 2 information applies to each hazardous chemical at the facility. The following information is required:

1.    The chemical name or the common name of the chemical, as provided on the material safety data sheet for the chemical.
2.    An estimate (in ranges) of the maximum amount of the hazardous chemical present at the facility at any time during the preceding calendar year.
3.    A brief description of the manner of storage of the hazardous chemical.
4.    The location at the facility of the hazardous chemical.
5.    An indication of whether the owner elects to withhold location information of a specific hazardous chemical from disclosure to the public under Section 324 (which allows the owner to do so upon request).

Leaks, spills, and other releases of specified chemicals into the environment require emergency notification under both SARA Title III and Section 103 of CERCLA. Under CERCLA, those in charge of a facility must report any spill or release of a chemical on a list of hazardous chemicals included in that act in excess of a reportable quantity (RQ), which is substance specific. The report must be made immediately to the National Response Center (1988 telephone numbers: 800-424-8802 or 202-267-2675). Under SARA Tile III, the notification process now includes all 366 extremely hazardous chemicals. In addition to the National Response Center, the state commission and the local committee must be notified. Note that if the release is confined totally within a facility and does not enter the "environment", so that it affects only the employees within the building, it does not have to be reported unless it falls under an OSHA regulation that would require it to be reported.

Generally, the following information must be reported:

1.  The name of the chemical(s) (trade name protection of the chemical is not permitted)
2.  Identification of whether the chemical is on the extremely hazardous chemical list
3.  The quantity released or an estimate of the quantity released
4.  The location, time, and duration of the release
5.  The medium (air, water, soil) into which the chemical(s) was released
6.  Known acute or chronic risks and any available helpful medical data
7.  Precautions to take, including evacuation if necessary
8.  The names and telephone numbers of persons to be contacted for further information

Obviously, if a facility has an emergency contingency plan covering releases of materials from its facilities to the environment, these should be coordinated with emergency plans for the local region developed by the district emergency planning committee. Although, under the current standard, it appears that most laboratories are, in effect, exempt from the act due to the provision that chemicals in research laboratories shall not be considered hazardous if they are used under the direct supervision of a qualified individual, it should be remembered that the facilities themselves are not exempt and there may be chemicals elsewhere on site which trigger coverage by SARA Title III. Also, these provisions may well be modified at a later date. Finally, if an emergency does occur within a laboratory such that a substantial portion of the stock becomes involved in an incident, resulting in a significant release to the environment, a responsible position may be to have actively participated in the emergency planning process in the local district.

## REFERENCES (SECTION 4.9.2)

1.  Resource Conservation and Recovery Act (RCRA) of 1986, SARA, Title III, Sections 300 to 330, October 17, 1986.
2.  Extremely Hazardous Substance List, Sections 302 to 304, *Fed. Reg.*, 13378, 1987.
3.  Emergency Planning and Hazardous Chemical Forms and Community Right to Know Reporting R; Final Rule, Title III, Sections 311 to 312, *Fed. Reg.*, 52, 38344, 1986.
4.  Toxic Chemical Release Reporting; Community Right-to-Know; Final Rule Title III, Section 313, *Fed. Reg.*, 53, 4500, 1986.

## 4.10. CHEMICAL WASTES

Until a relatively few years ago, disposal of chemical wastes from laboratories was mostly down the drain. However, this practice is no longer acceptable as a general procedure, although to a limited degree it is still permissible for certain wastes. The change has been due to the growing concern about the impact of chemicals on the environment. Virtually all the chemical wastes in this country are due to industrial sources, with less than 1% being due to laboratory wastes. However, regulations established to govern the disposition of chє     l wastes include the wastes generated by laboratories, so organizations of which ι     aboratories are a part must have a hazardous waste program. Locally, a major research university or corporate research laboratory may be one of the largest, if not the largest, generators of hazardous waste.

### 4.10.1. RESOURCE CONSERVATION AND RECOVERY ACT

Many individuals who are now responsible for the management of laboratory facilities received their initial training when there were no regulations governing the disposal of laboratory wastes, and may not fully accept the need for the newer procedures. However, especially in the academic area, the training of students should be in the context of compliance with regulations so that when these students embark on their own careers, they will be trained to comply with legally applicable standards. Senior personnel need to adapt, if they have not already done so, and manage their operations in accordance with current standards.

Organizations involved with excess hazardous chemical materials must conform to the provisions of the Environmental Protection Agency Resource Conservation and Recovery Act (RCRA), 40 CFR Parts 260 to 265, as most recently amended. In 1984, the Hazardous and Solid Waste Amendments (HSWA) of 1984 required that the EPA ban the land disposal of over 400 waste streams unless the wastes are treated or unless it can be demonstrated that there will be no migration of the waste while the waste remains hazardous. The portion of these rules covering dioxin and solvents went into effect on November 8, 1986. Although there was a partial extension for some of the wastes, the extension did not include solvents. The rationale for the extension was the lack of alternate capacity for the large amounts of soil contaminated by dioxin or solvents. The second portion covering the California list, including a number of heavy metals, went into effect on July 8, 1987. The remainder of the restrictions go into effect on three different dates: August 8, 1988, June 8, 1989, and July 8, 1990. These amendments have already begun to significantly change the options available to laboratories for disposing of their waste chemicals.

Laboratory facilities or their parent organizations have important decisions to make in regard to the scope of their activities which will be subject to the provisions of the RCRA regulations. Unless they produce less than 100 kg per month, at which level they are defined as a small generator, they are subject to some portions of 40 CFR Part 262 which apply to generators. If they produce more than 1000 kg per month, they become "large" generators and must comply with all of the requirements applicable to generators.

Other portions of the act cover the operations of treatment, storage, disposal, and recycling (TSDR) facilities, which require additional permitting. Many large industrial organizations, in which laboratory operations represent only a small part of their activities involving hazardous chemicals, may choose or have chosen to go through the permitting process. Some also transport their own hazardous waste and are subject to the regulations pertaining to transporters. However, only a few academic institutions have the capacity or interest required to meet the stringent regulations covering operations other than those of generators of the waste chemicals, and the same is true of many small industrial operations and their affiliated laboratories. Those organizations which intend to treat, store, and dispose of their hazardous waste normally will be sufficiently sophisticated and knowledgeable about the procedures to not need additional information, so this article will concern itself with waste management programs for generators only.

Before a program of waste management appropriate to a research organization is discussed, some additional background information will be helpful. What constitutes a hazardous waste needs to be defined and the essential portions of 40 CFR Part 262 need to be addressed. It is virtually a full-time job to keep up with the changing requirements in the context of laboratory operations, even for relatively small organizations. In many cases, it is desirable to transfer as much of the responsibilities as possible to an outside contractor, but there are essential functions that must be performed by laboratory workers and support personnel.

### 4.10.1.1. Definition of a Hazardous Waste

There are relatively frequent changes in detail as to what constitutes a hazardous waste according to the EPA, but the main features of the definitions have been consistent. Frequently, there are evaluations and rulings pertaining to the hazardous nature of various waste streams, for example, reported in the *Federal Register,* but these are based on certain guidelines in 40 CFR 261, Subpart D. There are indications that there may be some changes in some of the definitions and additional restrictions in the future, which will make waste management programs more demanding. The announced requirements for treatment previous to land disposal are examples of the more stringent requirements that appear to be forthcoming.

The criteria are (1) it is a listed waste (currently, there are about 400 materials listed in 40 CFR 261.31 to 261.33) and (2) it meets certain criteria of reactivity, ignitability, corrosiveness, and toxicity.

Wastes which meet certain hazardous criteria may be placed on the lists under three different categories:

**Hazardous wastes from nonspecific sources** — Examples of laboratory wastes that would fall in this category are spent solvents, the residue resulting from distillation recovery of used solvents, materials left over from silk screening and electroplating procedures in electronic laboratories, and other sources of used chemicals.

**Hazardous wastes from specific sources** — Unless the laboratory is a pilot operation simulating an industrial process, it is unlikely that most research laboratories would fall within this category. Note, however, that hazardous chemicals used in a pilot plant operation would normally not be exempt from the regulatory provisions of SARA Title III, discussed in Section 4.9.2.

**Discarded commercial chemical products, off-specification species, containers, and spill residues thereof** — There are two levels in this category: (1) materials which are acutely hazardous and (2) those which are less so. There are more stringent limitations on the former.

The following excerpts are from 40 CFR 261.11: *Criteria for listing hazardous waste.*

(a) The Administrator shall list a solid* waste as a hazardous waste only upon determining that the solid waste meets one of the following criteria:

---

* "Solid" waste need not be physically solid, as will be noted from some of the definitions in this section.

(1) It exhibits any of the characteristics of hazardous waste identified in Subpart C. [ignitability, corrosivity, reactivity, and EP toxicity.]

(2) It has been found to be fatal to humans in low doses or, in the absence of data on human toxicity, it has been shown in studies to have an oral $LD_{50}$ toxicity (rat) of less than 50 mg per kilogram, an inhalation $LC_{50}$ toxicity (rat) of less than 2 mg per liter, or a dermal $LD_{50}$ toxicity (rabbit) of less than 200 mg per kilogram or is otherwise capable of causing or significantly contributing to an increase in serious irreversible, or incapacitating reversible, illness. (Waste listed in accordance with these criteria will be designated Acute Hazardous Waste.)

(3) It contains any of the toxic constituents listed in Appendix VIII [in 40 CFR 261], unless after considering any of the following factors, the Administrator concludes that the waste is not capable of posing a substantial present or potential hazard to human health or the environment when improperly treated, stored, transported, or disposed of, or otherwise managed:

(i)     The nature of the toxicity presented by the constituent.

(ii)    The concentration of the constituent in the waste.

(iii)   The potential of the constituent or any toxic degradation product of the constituent to migrate from the waste into the environment under the types of improper management considered in paragraph (a)(3)(vii) of this section.

(iv)    The persistence of the constituent or any toxic degradation product of the constituent.

(v)     The potential for the constituent or any toxic degradation product of the constituent to degrade into non-harmful constituents and the rate of degradation.

(vi)    The degree to which the constituent or any degradation product bioaccumulates in ecosystems.

(vii)   The plausible types of improper management to which the waste could be subjected.

(viii)  The quantities of the waste generated at individual generation sites or on a regional or national basis.

(ix)    The nature and severity of the human health and environmental damage that has occurred as a result of the improper management of wastes containing the constituent.

(x)     Action taken by other governmental agencies or regulatory programs based on the health or environmental hazard posed by the waste or waste constituent.

(xi)    Such other factors as may be appropriate.

Substances will be listed in Appendix VIII only if they have been shown in scientific studies to have toxic, carcinogenic, mutagenic or teratogenic effects on humans or other life forms.

...

If the material is not a listed waste, it is a hazardous waste if it meets any of the criteria in Subpart C — Characteristics of Hazardous Waste. Listed below are the essential sections of this part of the regulations.

261.21 Characteristic of Ignitability

(a) A solid waste exhibits the characteristics of ignitability if a representative sample has any of the following properties:

(1) It is a liquid, other than an aqueous solution containing less than 24% alcohol by volume and has a flash point less than 60°C (140°F)...

(2) It is not a liquid and is capable, under standard temperature and pressure, of causing fire through friction, absorption of moisture or spontaneous chemical changes and, when ignited, burns so vigorously and persistently that it creates a hazard.

(3) It is an ignitable compressed gas as defined in 49 CFR 173.300 and as determined by the test methods described in that regulation or equivalent test methods approved by the Administrator under paragraphs 260.20 and 260.21.

...

261.22 Characteristic of Corrosivity*

(a) A solid waste exhibits the characteristics of corrosivity if a representative sample of the waste has either of the following properties:

(1) It is aqueous and has a pH less than or equal to 2 or greater than or equal to 12.5.

...

(2) It is a liquid and corrodes steel (SAE 1020) at a rate greater than 6.35 mm (0.250 inch) per year at a test temperature of 55°C (130°F)...

261.23 Characteristic of Reactivity

(a) A solid waste exhibits the characteristic of reactivity if a representative sample of the waste has *any* of the following properties:

(1) It is normally unstable and readily undergoes violent change without detonating.

(2) It reacts violently with water.

(3) It forms potentially explosive mixtures with water.

(4) When mixed with water, it generates toxic gases, vapors, or fumes in a quantity sufficient to present a danger to human health or the environment.

(5) It is a cyanide or sulfide bearing waste which, when exposed to pH conditions between 2 and 12.5, can generate toxic gases, vapors or fumes in a quantity sufficient to present a danger to human health or the environment.

(6) It is capable of detonation or an explosive reaction if it is subjected to a strong initiating source or if heated under confinement.

(7) It is readily capable of detonation or explosive decomposition or reaction at standard temperature and pressure.

(8) It is a forbidden explosive as defined in 49 CFR 173.53 or a Class B explosive as defined in 49 CFR 173.88.

...

261.24 Characteristic of EP Toxicity

(a) A solid waste exhibits the characteristic of Extraction Process (EP) toxicity if, using the test methods described in Appendix II [of the regulation] or equivalent methods approved by the Administrator under the procedures set forth in paragraphs 260.20 and 260.21, the extract from a representative sample of the waste contains any

---

* Note that the OSHA definition of a corrosive material in terms of a health hazard was explicitly limited to the action of a material on tissue.

of the contaminants listed in Table 1 [of the regulation] at a concentration equal to or greater than the respective value given in that table. Where the waste contains less than 0.5 percent filterable solids, the waste itself, after filtering, is considered to be the extract for the purposes of this section.

...

Since the appendices and lists referred to in the above material change as materials are added to and deleted from the list, laboratory personnel should subscribe to an information service which will provide information to permit maintenance of a current valid list.

A generator must evaluate his waste to determine if it may be exempt from being considered a hazardous waste or, if not, if it is a listed waste, as defined above, or meets any of the criteria for a hazardous waste. Most laboratory wastes are likely to fall under one of these provisions and must be treated as a hazardous waste. A major problem with many chemicals produced by laboratory activities, as opposed to chemicals that are purchased, is that they are insufficiently identified. The identity of the chemical in a bottle, for example, labeled "solution A" may have been known perfectly well by the individual who labeled it at the time the label was affixed to the bottle, but even this person may not recall the contents several months later. As may often be the case, the chemical is an "orphan" left behind by a departed graduate student, in which case the identity of the contents may be even more uncertain. Commercial waste disposal firms will probably not accept these unknown containers. The contents must be identified. Other examples of unknown chemicals which also must be identified are older containers which have lost their labels, or the labels may have become damaged so that the contents cannot be determined. In all cases of unknowns, there is a certain amount of risk associated with opening an unlabeled container for any purpose, including that of identifying the contents. The container may contain, for example, degraded chemicals that have become explosive, such as ether with a high content of peroxides or dry picric acid. Procedures should be adopted that require all personnel to properly identify the contents of any container which will not be used in its entirety by the maker or the person transferring a quantity from an original container to a secondary container. The latter requirement is written into the OSHA Hazard Communication Standard, and the former should be adopted as a matter of good laboratory practice.

Most generators of laboratory waste use commercial firms for removing hazardous waste from their facility rather than doing it themselves. In many cases, the commercial firms do virtually a turnkey job, i.e., they go to the individual laboratories and perform all the tasks necessary to prepare and dispose of the waste properly. This is especially appropriate for smaller facilities which do not have the staff to do many of these tasks. Larger operations often find it economical to perform virtually all of the work themselves and use the commercial firm only to transport the material to an ultimate disposal site. Single laboratories rarely have the resources to use the latter method, and in most comprehensive universities and research organizations, the process of handling hazardous waste has been assigned to specialists within the organization, usually as an adjunct of the safety and health organization.

### 4.10.1.2. Requirements for Generators of Hazardous Waste

Title 40 CFR Part 262 defines the requirements for generators of hazardous waste. As in other parts of this section, it will be assumed that the generators are not owners or operators of treatment, storage, or disposal facilities, but, except for the modest amount of processing permitted under the law, dispose of most of their waste by shipping it away from the generating facility. As will be noted, the generator assumes substantial responsibilities in initiating a shipment of hazardous waste from his facility. This information is taken directly from the standard. In some cases, the language has been simplified or made more concise for brevity and clarity, while in other cases, the exact wording of the standard has been followed where it did not lend itself to modification.

1.    The generator must first determine if the waste meets the criteria for hazardous waste. The previous section describes the basic criteria which define a hazardous waste, and the generator should evaluate his waste in terms of these criteria. In addition, it may be decided that a given material is a hazardous waste based on the materials or processes used. This last interpretation might apply to much of the materials synthesized or generated in a research laboratory, where the detailed characteristics of the waste might be too much trouble to determine, but the procedures make it likely that it is a hazardous material.

2.    If the generator intends to initiate a shipment of hazardous waste, he must first secure an EPA identification number from the EPA administrator.

3.    The generator must be sure that anyone he engages to handle or receive his hazardous waste has a valid EPA identification number. This step is required, but is by no means sufficient to assure that waste generated at a facility will, in fact, be handled legally and environmentally soundly. Although it may consume additional time and resources, visits to a waste broker's storage facility and/or to the eventual site where the material will be placed in a landfill, incinerated, or otherwise processed is highly recommended. All organizations willing to bid on laboratory waste disposal services do not offer the same degree or quality of service. Many apparently reputable firms have had to close facilities, pay fines, or substantially modify their procedures because they did not comply with acceptable operating practices. In the worst possible case, where the operator of a facility did not have the resources available to correct a problem which had been identified by the EPA, the waste generators have had to share the cost of cleaning up a substandard facility.

4.    A manifest describing the contents of the shipment must be prepared. Ostensibly, this must be done by the generator, but it may be done with the aid of the waste disposal firm to whom the waste is being transferred, especially if the latter is collecting and packing the waste for the generator, as is the case for many smaller generators.

5.    The manifest must include the following information according to §262.21 (modified slightly):

     a.    A manifest document number;

     b.    The generator's name, mailing address, telephone number, and EPA identification number;

     c.    The name and EPA identification number of each transporter;

     d.    The name, address, and EPA identification number of the designated facility and an alternate facility, if any;

     e.    The description of the wastes, required by the regulations of the U.S.

Department of Transportation (DOT) in 49 CFR 172.101, 172.202, 172.203 (including proper shipping name, Hazardous Class and ID number (UN or NA);

f. The total quantity of each hazardous waste by units of weight or volume, and the type and number of containers as loaded into or onto the transport vehicle.

g. The following certification must appear on the manifest: "This is to certify that the above named materials are properly classified, described, packaged, marked, and labeled and are in proper condition for transportation according to the applicable regulations of the DOT and the EPA.

A representative of the generator must sign the manifest by hand. The initial transporter must also sign and date the manifest. The generator is to retain one copy of the manifest and give the transporter the remaining copies.

The generator must mark each package of hazardous waste as required by DOT regulations under 49 CFR Part 172. Each container of 110 gal or less must be labeled with a label, bearing the following legend (many commercial firms sell pads of these labels):

HAZARDOUS WASTE — Federal law prohibits improper disposal. If found, contact the nearest police or public safety authority or the U.S. Environmental Protection Agency.

Generator's Name and Address _____

Manifest Document Number _____

The vehicle carrying the hazardous waste must be properly placarded as required by the DOT under 49 CFR 172, Subpart F. It is the responsibility of the generator to see that this is done, if necessary by doing it himself or by providing placards to the transporter. Most transporters routinely provide the correct placards.

Within 35 days after the generator has transferred his hazardous waste to the initial transporter, he must receive a copy of the manifest with a handwritten signature of the owner or operator of the waste facility. Otherwise, he must contact the transporter or the facility to determine the status of his waste. If he has not received this document within 45 days after the initial transporter accepted his waste, the generator must submit an exception report to the EPA regional administrator for the region in which his facility is located. The report must contain a legible copy of the manifest and a letter outlining the efforts made to locate the waste and the results obtained.

A generator, unless he has been granted an extension by the EPA, must either have a permit as a storage facility or must not collect and store hazardous waste for more than 90 d. During this storage interval, the materials must be kept in appropriate containers, the beginning of the accumulation period must be shown on each container, and the containers must be labeled as hazardous waste. Generators who do not keep hazardous materials more than 90 d must comply with 40 CFR 265, Subparts C and D and §265.16.

It may be appropriate to store the material within the individual laboratories until shortly before a waste disposal firm is scheduled to pick up an organization's waste. Under appropriate conditions, as much as 55 gal of hazardous waste or 1 qt of acutely hazardous waste may be accumulated at or near the point of generation, i.e., the laboratory, beyond 90 days without a permit as a storage facility. The containers used must be in good condition so that they do not leak and they must be constructed so that the hazardous waste does nor react with the container to impair the ability of the container to store the material. Except for periods during which material is to be added to the container, the container must remain closed. The container must be treated so as to keep it from leaking. As a minimum, the containers must be labeled with the words "Hazardous Waste" or a more explicit description of the contents.

### 4.10.1.3. Recordkeeping Required of the Generator

Copies of the manifests signed by the transporter must be maintained for a period of at least 3 years. Retention substantially beyond this date would probably be advisable. Similarly, copies of required biennial reports (see following paragraph) and exception reports must be kept for at least 3 years beyond the date on which they were due. Any test data relevant to the shipment must also be kept for 3 years beyond the date of the shipment.

Biennial reports must be prepared by each generator initiating shipments of hazardous waste away from his facility. These must be prepared on EPA Form 8700-13A and submitted by March 1 of each even-numbered year, and must cover the previous calendar year. The report must provide (1) the EPA identification number, name, and address of each facility to which waste was shipped during the year, (2) the name and EPA identification number of each transporter used during the year, (3) a description, EPA waste number, and amount of each hazardous waste shipped off-site. This information must be identified further with the identification number for each site to which it was shipped. The report must be signed by the generator or an authorized representative to certify that the report is correct.

### 4.10.1.4. Personnel Training

A record of each position related to management of the waste must be maintained at the facility engaged in hazardous waste activities. The name of the current person filling each of these positions must be recorded. A written job description must be available for each position, describing the skills, education, and other qualifications needed for each job and the actual duties involved for each position.

The owners and operators of a facility engaged in a hazardous waste management program must ensure that the personnel identified as working in the facility are properly trained to ensure that the program will be conducted safely and with minimal risk to the environment and the general public. This can be done by providing either classroom instruction or effective on-the-job training. The person(s) responsible for the training must be knowledgeable about hazardous waste management procedures. The training should be tailored to the specific duties of each person working in the facility. However, as a minimum, individuals must be trained to respond effectively in an emergency. They must be familiar with emergency procedures, including initiation of an alarm, evacuation procedures, and appropriate responses to a fire, explosion, or environmental contamination incident. They should know how to safely secure the facility in an emergency.

A written description of the initial and maintenance training programs for each person or position must be available. Persons assigned to duties in a hazardous waste management program must successfully complete the training programs before they are allowed to work unsupervised. Each employee shall take part in an annual review of the initial training appropriate to his duties. Records of the training for each continuing employee must be kept until closure of the facility, and for employees who leave the facility, records must be kept for at least 3 years after the employee has left.

### 4.10.1.5. Preparedness and Prevention

Every facility involved with hazardous waste must be operated in such a manner as to minimize the chances of any incident such as a fire, explosion, or release of a toxic substance into the environment or which could endanger human life. However, planning for such emergencies must take place even though it is intended that no emergency response will ever be needed.

Communications are a vital part of any emergency system. Provision must be made for an emergency alarm in the facility. Ready access must be available to either a telephone or two-way radio system to summon emergency help from appropriate agencies, e.g., fire department, police, emergency response teams. Whenever hazardous material is being handled, access to communication equipment is especially important and must be available. Another occasion when communication capability is especially critical is when an employee is alone in a facility while operations are being conducted. Such a situation is undesirable and should be avoided.

Other emergency equipment that must be present are portable fire extinguishers (of different fire suppression classes, if needed), spill control equipment, and decontamination equipment. Other desirable emergency equipment would be positive air-supplied breathing apparatus, fire and chemical protective clothing, and turnout clothing. Access to any area of the facility must be available in an emergency, so adequate aisle space must be provided.

Participation in a local area emergency plan, such as required under Title III SARA, is essential. Arrangements should be made with local fire departments and police departments to familiarize them with the facility, its operations, and access to the facilities for emergency vehicles. Agreements should be established with state and local emergency response agencies as well as commercial firms that might need to assist in an emergency. Other internal groups, such as physical plant or building maintenance groups as well as internal security personnel, should be similarly involved. Working agreements with nearby medical facilities should be established, and they should be made aware of the type of exposures or medical injuries which they might be asked to handle as a result of the operations within the facility. Periodic drills would be highly desirable.

It is the responsibility of the owners and operators of an organization engaging in hazardous waste operations to make these emergency arrangements. If an agency refuses to accommodate the facility in establishing an appropriate arrangement, this refusal must be documented.

### 4.10.1.6. Contingency Plan

The previous section described some of the emergency provisions that must be made. However, a requirement of every hazardous waste facility is that a written contingency plan be adopted. Copies of this plan must be maintained at the facility and at any group that may be called upon to assist in an emergency, including police, fire department, emergency response teams, and hospitals. This plan must be designed to minimize the adverse effects of any emergency incident due to the operations of the facility. The plan must be immediately activated in the event of any such emergency. There are a number of essential components of the plan.

1.  The contingency plan must describe the actions facility personnel will take in the event of emergencies such as fires, explosions, or releases of hazardous substances into the environment.
2.  It must describe the agreements or arrangements made with the emergency response groups to coordinate the activities of these agencies in the event of an emergency.
3.  A person must be identified as the primary emergency coordinator. Others who are qualified to act as emergency coordinator are to be listed in the order they would be expected to act as alternates. The addresses and home and business phone numbers of these individuals must be included on the list. At all times, at least one of these persons must be at the facility or on call in case of an emergency.
4.  A current list of all the emergency equipment at the facility and the specific location of each item must be included. Each item must be identified by a brief physical description and its capabilities.
5.  The plan must include an evacuation plan for facility personnel, identifying the signal(s) used to initiate the evacuation and the primary and alternate escape routes.

### 4.10.1.7. Emergency Procedures

If an emergency appears likely or actually occurs, the individual acting as the emergency coordinator must immediately initiate an emergency response, including the following actions:

1.  Activate the alarm system to alert facility personnel.
2.  If necessary, notify emergency response groups to secure their aid.
3.  If there has been a release of hazardous substances, a fire, or an explosion, the emergency coordinator must identify the nature of the material released, the exact source of the release, and the actual extent of the release.
4.  The emergency coordinator must also immediately assess the possible impact of the release on human health or the environment away from the facility, from both direct causes and secondary effects, and include in the latter any adverse effects such as water run-off from efforts to control the incident (e.g., putting out a fire).
5.  If it is determined that adverse effects on humans or the environment could occur and evacuation of areas near the facility is needed, the emergency coordinator must immediately notify appropriate local authorities and remain available to assist in making decisions as to which areas require evacuation.
6.  In the event of an emergency of possible harm to humans and the environment away from the facility, the emergency coordinator must also report the incident to the government coordinator at the scene or directly to the National Response

Center (24-hour telephone number at the time of writing, 1-(800)-424-8802). The report must include:

a.   Name and telephone number of person making the report

b.   Name and address of facility

c.   Time and character of incident

d.   Name and quantity of materials involved, to the extent known

e.   Extent of injuries, if any

f.   Potential hazards to humans or the environment outside the facility

7.   The emergency coordinator must take all reasonable steps to contain the scope of the incident. If operations have not already ceased, shutting down operations must be considered. Waste containers and other hazardous materials not already involved in the incident should be moved to safer locations. The shutdown of operations must be done in such a way as to ensure there will not be any untoward equipment failures which could cause additional problems.

8.   After the emergency, all recovered waste as well as contaminated soil, runoff water, etc. must be properly handled in accordance with the provisions applicable to generators of hazardous waste.

9.   All emergency equipment listed in the contingency plan must be cleaned and fit for its intended use before operations are resumed.

10.   The operator or owner must notify the regional administrator and appropriate state and local authorities that the facility is again in compliance with the required standards before operations are resumed.

11.   The operating records of the facility must reflect the time, date, and any cogent details of any incident that require implementation of the contingency plan. A report must be made to the regional administrator within 15 days after the incident. In addition to the information listed under item 6 above, the report must include:

a.   An assessment of the actual or potential damages to human health or the environment as a result of the incident, where applicable

b.   The estimated quantity and disposition of recovered material that resulted from the incident

## 4.10.2. PRACTICAL HAZARDOUS WASTE MANAGEMENT PROGRAM

Sections 4.10.1.1 to 4.10.1.7 briefly covered basic portions of the Resource Conservation and Recovery Act (RCRA). Hazardous waste management programs for generators must comply with applicable portions of this act, including requirements which were not provided directly or paraphrased in those sections. It is incumbent on generator waste management personnel to become familiar with all of the requirements, including changes as they occur.

### 4.10.2.1. Internal Waste Management Organization

In most organizations, it would be impractical for each individual laboratory or even department to establish a separate program, even if the regulatory organization in the state in which the facility was located were willing to deal separately with a dozen different generator facilities at a single institution. Thus, in most academic institutions and many corporate research facilities, the responsibility for the overall management of the hazardous waste program has been assigned to a centralized unit.

This does not mean that the individual laboratory and laboratory personnel do not play a key role in an effective hazardous waste management program. Since they are the generators of the wastes, they can control, to a major degree, the amount of hazardous waste generated and are the primary source for identifying waste generated in their facility, especially that which is not in the original containers.

The hazardous waste organization thus starts with the laboratory itself. It is at this level that the waste is generated, collected, identified, and, in some cases, processed for reuse or rendered harmless. Each individual in the laboratory need not know all of the requirements under RCRA, but he must be trained in the procedures by which the waste generated in the laboratory is to be handled prior to removal. He must know what precautions to take to ensure that the waste will be acceptable for disposal, what information is required for each waste generated, and what records must accompany waste offered for disposal.

Unless the facility is so small that outside waste disposal firms come directly to the individual laboratories to package and remove the material, a waste management agency will normally be established within the organization to pick up the waste, classify it (according to the RCRA criteria), store it temporarily (up to 90 d), prepare appropriate documentation for each container, package it (in some cases), and make arrangements for the material classified as excess hazardous waste to be taken from the facility, again according to the RCRA requirements.

It will be the responsibility of this internal agency to keep up with the current regulations, make all of the required reports, and maintain all of the required records as well as handle the waste itself. It will also probably be called upon to help develop and publish procedures for the institution's researchers to use in their part of the waste management program.

It is also likely that the same internal agency would be assigned responsibility for preparing and implementing the required emergency contingency plan. This program must be integrated with the disaster emergency planning for the entire organization. As a trained group, some emergency responsibilities outside the waste program may also be assigned to waste management personnel.

### 4.10.2.2. Reduction of Hazardous Waste Volume

An effective hazardous waste management program cannot concern itself solely with disposal of unneeded materials. This is far too expensive and contradicts the RCRA concept. Chemicals are a resource which should be conserved, and, where practical and legal, it is desirable for surplus chemicals to be put to a beneficial use rather than buried in a landfill, burned, or otherwise rendered unusable. However, as will be discussed later, there are limitations to how much effort an organization can devote to salvaging or treating a chemical waste without becoming a treatment and disposal facility.

One of the current requirements on generators is that they must document that they are making an effort to reduce the amount of waste which they generate. There are a number of effective ways in which laboratory operators can aid in this effort. Among the more successful are

1.  Planning of experiments. Anticipation of waste generation should be a part of each operation. The research should be designed to reduce or eliminate the

amount of dangerous substances generated as a byproduct of the research to the extent possible, either as an ultimate or intermediary waste. Alternative reagents should be evaluated which would result in waste that is less dangerous or which would minimize the volume of hazardous waste. The experiment should be planned to include rendering harmless any excess material generated as a part and logical continuation of the work.

2.  Purchasing in smaller quantities. It was noted earlier that a substantial premium per unit chemical purchased is paid by buying chemicals in smaller containers rather than in somewhat larger ones. The cost of eight 1-pt bottles of a given material is virtually certain to be much greater than that of a 1-gal container of the same material, especially if one adds in the surcharge for buying the material in glass bottles covered with a protective plastic safety film. However, the total cost to the organization, if the cost of disposal of unused material is included, might actually favor the smaller container. If individual containers are packed properly in a drum filled completely, except for the containers, with an absorbent material, one typically can get only the equivalent of about 15 gal of waste in the drum. If the original cost of the material is paid for by the laboratory facility, while waste costs are paid for out of the organization's total budget, it is difficult to persuade the laboratory to pay the additional costs. It may require an explicit organizational policy to mandate the purchasing of all chemicals in smaller sizes, unless the laboratory can document that its rate of use is sufficient to ensure the use of all the material in a short period of time, so that it will not become degraded or suspected of possible contamination and, hence, unusable. If all materials in an institution or organization are purchased and distributed centrally, this becomes an easier policy to enforce. Proposals to charge for waste are usually self-defeating. Either the cost, in time and money, of collecting the charges exceeds the funds collected or some individuals attempt to evade the charges by improperly pouring their waste down the sink.

3.  Using smaller quantities in experiments. New technology has made it possible to perform much research involving chemicals with much smaller amounts of materials than was possible only a few years ago. More sensitive analytical instruments, balances, and instruments operating on new principles, or more versatile and reconfigured instruments based on older principles, often make it possible to perform an experiment with very small amounts of the reagents involved. For example, interfacing a small computer to an experiment may permit the simultaneous recording of many variables with automatic control of the experiment to continuously modify the experiment, so that one run may take the place of a dozen or more. Not only will this reduce the volume of waste generated, but the experiments should also become inherently safer.

4.  Redistribution of used chemicals. Excess chemicals are not necessarily waste chemicals. A specific part of any chemical waste program should be a classification and evaluation step, prior to which a chemical is only "surplus", not "waste". Clearly, many materials are obviously unusable and must be classified as waste, and if they meet any of the RCRA hazardous waste criteria, they must be classified as hazardous waste or acutely hazardous waste if they fall in this category. However, many chemicals are still usable for research or, if not, often

are adequate for instructional laboratories. Unopened containers are the most likely to be attractive to potential users, but even partially used containers of reasonably fresh materials which do not normally degrade by contact with air or moisture are often desirable to someone. A list of these materials should be maintained and circulated frequently to potential "customers" within the same organization. In some instances, these materials are offered at a lower price than is fresh material, but in most cases charging will reduce the amount of surplus redistributed and may wind up increasing the cost to the organization because of the cost of disposing of the surplus material. There are "waste exchanges" to which lists of unwanted materials may be submitted and which circulate the information to organizations which might be interested.

Another category of materials that can often be reused is contaminated solvents. A number of chemicals such as xylene, methanol, and ethanol are heavily used materials and frequently are available in the waste in sufficient volume to make it practical to reclaim them by distillation. An automatic spinning band still can easily handle 20 to 30 l/d. There will still be residues which will have to be treated as waste, but the volume will be far less than the original amount. The wastes must be treated carefully to ensure that two materials which form azeotropic mixtures are not mixed together; otherwise, simple distillation will not produce pure solvents. Recent restrictions on the disposal of liquids in landfills make this method of waste reduction even more attractive.

Other materials such as mercury and silver from photographic processes are also readily recovered. Moreover, it is possible for value of the reclaimed silver to more than pay for the effort involved.

5.    Management by tracking materials which degrade and become dangerous with time. There are a number of chemicals which deteriorate over time, usually as they react with air or moisture. Ethers are a well-known example of this class of materials. The contents of partially empty containers of ethers kept in storage are likely to form peroxides which are heat, shock, and friction sensitive, so an attempt to use them could result in a fatal explosion. Chemical waste vendors will not collect these potentially dangerous containers and remove them from the facility as part of a normal waste shipment. There are firms which specialize in disposing of such dangerous materials, but their services are extremely expensive. An effective program should be established for tracking such materials that dates each container when it is acquired, sets a target date for beginning to check the containers for dangerous degradation products, and establishes a date when they must be removed from the facility. Other materials which should be included in a tracking program are pyrophoric materials, highly water-reactive materials, or the converse, materials which become dangerous as they dry out such as perchloric acid and picric acid.

6.    Elimination of unknowns. There are two ways in which unknowns are generated in a laboratory. One is when information on the original labels on commercial containers is defaced or lost. The other is when inadequate information is placed on labels to begin with, such as when a graduate student labels a container "Solution A" and then leaves the facility, so that no useful records are available. Unknown chemicals are among the most troublesome of all

materials to handle as waste. There is a certain amount of risk in opening an unknown container, although the type of operations conducted in the laboratory in which it is found may provide enough information to reduce or eliminate a concern about shock or friction causing an explosion, or the possibility of a reaction with air or moisture in the air when the container is opened. Even after opening the container, the characteristics of the material may be sufficiently ambiguous to make identification very difficult, especially if it is not originally a commercial product, but has been generated in the laboratory.

The solution to the degraded or destroyed label problem is relatively straightforward. If the label on a container is damaged, it should be laboratory policy for a new label to be affixed immediately by the first individual to be aware of the problem. With the advent of the new OSHA Hazard Communications Standard, with its secondary container requirements, every laboratory should have an ample supply of generic labels available for use. These labels should have space on them for the common name of the chemical, its CAS number, and basic hazard data such as the NFPA 704 color-coded numbering system or diamond.

The failure to properly label containers in the first place is a more difficult management problem since there typically is only one person who knows what the contents of the container actually are. In academic laboratories, the turnover in graduate students and post-doctoral research associates (and, to some extent, faculty as well) is high. A policy should be adopted, and enforced, that requires every person leaving the facility to identify the contents of each container for which he is responsible. Withholding graduation or requiring a financial deposit have been suggested as ways of enforcing such a policy. In most cases, however, there would be few difficulties if the laboratory director or supervisor simply met with the individual shortly before he was scheduled to depart and went over his materials with him. Unfortunately, some laboratory directors do not wish to take the time. There also have been cases where the individual in charge has compounded the problem by having the unknowns surreptitiously carried to an ordinary trash container outside the laboratory. When this occurs and the responsible person can be identified, he should be subject to severe punishment by the organization's authorities because, in addition to the inappropriateness of such behavior, the organization can be held responsible for permitting such actions under current environmental standards.

7.  Placing severe restrictions on "free" research materials or chemicals accepted by research personnel. It is frequently simpler for suppliers of agricultural chemicals to provide a researcher with a drum of one of their products (which may be a "numbered" experimental material) than it is to weigh out the amount the person actually needs. A policy should be introduced that a supplier of a "free" material must either take back or guarantee to pay for the disposal cost of any excess material. Implementation of such a policy at the author's institution substantially reduced the amount of hazardous agricultural waste chemicals which required disposal.

8.  Miscellaneous. A number of waste reduction techniques which are more appropriate to a laboratory than to an internal waste disposal agency could, if employed on a useful scale, be considered a treatment and disposal facility. For

example, a container holding only a small quantity of ether should be placed in a suitable fume hood and the ether allowed to evaporate. This eliminates the material as a solvent waste as well as any chance of dangerous peroxides forming. Small quantities of other solvents can be dealt with in a similar fashion. Small quantities of acids and bases can be used in elementary neutralization processes. Other purification, reclamation, and neutralization techniques are suggested in the literature for specific materials.

Since a nearly empty bottle requires as much space as a full one when packed in a drum for disposal, containers should be filled in the laboratory with either the same or compatible waste materials before being transferred from the laboratory to the waste collection group. In the latter case, a complete, accurate record of the contents of the container must be maintained and placed on the waste label affixed to the container when the transfer takes place.

A modest amount of material can be safely disposed of into the sanitary system. However, local restrictions may preclude this type of disposal and, if encouraged, some materials may be illegally disposed of in this fashion. Unless done carefully under the direct supervision of a qualified laboratory employee, preferably a senior person who normally exercises managerial authority, disposal of materials into the sanitary systems should be discouraged. The amount of money saved over a period of several years would probably be minor compared to the cost of a single incident which subsequently causes a problem.

An excellent reference for ways to render chemicals in the laboratory less hazardous is *Prudent Practices for Disposal of Chemicals from Laboratories*, published by the National Academy Press in 1983. Although the regulatory material included in the book is outdated, the chemistry remains valid.

### 4.10.2.3. Waste Collection

The RCRA regulations permit temporary storage of hazardous waste at or near the point of generation, which in the context of the present document means in or near a laboratory. The amount is limited to 55 gal of hazardous waste material and 1 qt of acutely hazardous material. If the procedure is to pick up the waste directly from the laboratory, then the only requirement will be to provide the proper documentation to the collection firm. In most cases, storage in the laboratory will only be a stage in the internal handling of the waste, with the waste typically being picked up and transferred to a central location where most of the waste management activities take place.

There are advantages in delaying the determination of whether the surplus chemicals meet the RCRA criteria for hazardous waste until after the material is taken to the central facility. If the excess chemical is defined at the laboratory as hazardous waste, transportation of the material may be a problem since one of the restrictions on transporting hazardous waste without meeting the requirements of a hazardous waste transporter is that it cannot be transported *along* a public street, although the waste can be moved *across* a public street. Internal roads at many corporate sites meet this condition since public access to their property is often restricted. Most academic campuses, however, are open to the public and the use of few streets in the vicinity of academic research buildings is restricted to the public. In some instances, the strict interpretation of this requirement may be circumvented by using small vehicles that can move on sidewalks and across limited-access parking lots and other open spaces.

The delay of classification until after the material is taken to a central location is a dubious circumvention of the standard if there is no real effort to classify the material brought to the facility as usable material, harmless waste, or hazardous waste. Only modest amounts of material should be transported at any one time across a crowded campus to limit the consequences of an accident. Except in unusual circumstances, it would be desirable to limit the size of individual containers to 5-gal pails for liquids and the equivalent for solids, which would be approximately 50 to 75 lb, again as a safety measure. Containers of these sizes can be handled by hand, while 55-gal drums would require lifting equipment if they were spilled in an accident. No chemicals should be moved in containers that are damaged and not sealed tightly. The containers should be securely loaded on the vehicle being used for transportation so that they will not shift while the vehicle is in motion.

It may be desirable to make separate trips if two or more highly reactive or incompatible materials are to be transported.

It would be a good idea to have emergency supplies on the vehicle used for transportation of the chemicals, including: respiratory protection, gloves, chemical splash goggles, chemically protective coveralls, two Class ABC fire extinguishers, a bag or two of an absorbent material, a package or two of Plug-n-Dike™ for plugging holes in a leaking container, a coil of rope and at least three light-weight standards for establishing a temporary barrier, and a two-way radio. All of this emergency gear should be located where it is likely to be accessible in the event of an accident. The individuals who pick up the material must be trained in how to respond in the event of an emergency.

### 4.10.2.4. Packaging

An upper limit to the size of the container packages was recommended in the previous section. Most surplus chemicals are not in 5-gal metal cans, but are typically in smaller glass bottles of various English or metric sizes. The caps on containers should be screw type and neither cracked nor corroded. Bottles with other types of stoppers such as rubber, cork, or glass may come open during transportation, allowing the contents to spill and cause a dangerous incident. The outside of any container offered for transportation should be CLEAN. Small vials containing liquids may be placed in a larger glass container for transportation.

Container sizes for flammable solvents, if the material has been transferred from the original container, cannot exceed the permissible sizes allowed by OSHA for flammable liquids of the appropriate class. The quality of the containers must be comparable to those used for the commercial versions of the chemicals. Chemical wastes should not be placed in polyethylene milk jugs or the equivalent. These containers are not sufficiently sturdy.

Some solid materials normally come in bags. The original bags should be placed within one or two additional bags. One may suffice if the original bag is intact, while two should be used if the original bag is damaged and leaking. The external bag(s) preferably should be clear so the contents can be seen, especially if the original bag bears a legend identifying the contents. As noted earlier, compatible chemicals can be combined in a single container.

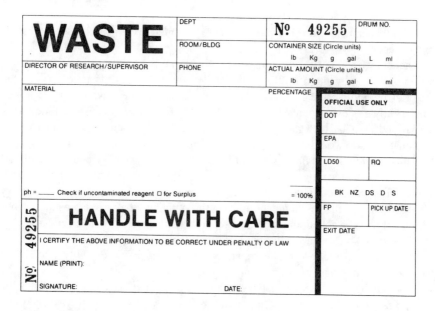

FIGURE 4.31.    Chemical waste label.

### 4.10.2.5. Characterization of the Waste

A typical waste label which might be used at any facility to identify and document waste internally is shown in Figure 4.31. A form such as this should be attached to each container of surplus chemical before it is removed from the laboratory and should remain with the container until final disposition of the material. A copy should be retained by the waste facility as part of their records. There are two sections to this form, one identifying the source of the material and the other characterizing the material itself. The latter information is to be used by the personnel responsible for classifying the material and, if necessary, to complete the preparations for its disposal as a hazardous waste.

As can be seen, the information pertinent to the source is sufficiently complete to trace the material to its origin. A very important piece of information is the name of the person originating the waste, both as a printed name and as a full signature. The latter is important since it represents an acknowledgment by the individual signing the form that the remainder of the information is correct. If problems develop with the material, the information can be used to trace it to its source.

Providing information identifying the contents is the responsibility of the generator, not waste management personnel. They can provide a reference or telephone numbers, but it is not their responsibility to perform this task for the generator. They are justified in refusing to take containers for which the contents have not been properly identified. All constituents of the material in the container must be listed by their proper chemical names. Abbreviations and formulas may cause confusion and should not be used as the sole identification. Similarly, trade names may be helpful, but again should not be used as a substitute for the correct chemical name. One useful identifier would be a CAS number which, if not conveniently available to the researcher elsewhere, can be found for most common chemicals in virtually any chemical catalog. The term "inert ingredients" should not be used. All materials should be completely identified on the label.

Waste liquids containing acidic or basic materials must show the approximate pH of the liquid. If the liquid is not homogeneous, but is stratified into layers, the pH of each layer must be provided.

Any material which is equivalent to a commercial product and is uncontaminated should be specifically marked as such, so as to be a candidate for the redistribution program.

The approximate percentage of each constituent should be provided, and the sum of these must add up to 100%. If the container is one to which materials have been added from time to time, the log of these additions can be used to estimate the percentages of the various materials in the container.

It is advisable for the waste management personnel to at least spot check materials which they have collected from the various laboratories. A simple gas chromatograph analysis can be performed on most organic solvents in about 15 min if the machine has the proper column and has been calibrated with appropriate standards. The pH of aqueous solutions can also be easily checked. If there is any doubt about the identity of most solid materials, the physical characteristics may be enough to resolve the question and at least a rough qualitative identification can be made for most commonly used materials. Unknowns, of course, are a different story. The analysis of these materials should be the responsibility of the originating laboratory, if at all possible. There are times when this is impossible, but these should be rare.

### 4.10.2.6. Packing of Waste for Shipment

The manner of packing the waste in preparation for transportation to a disposal facility depends somewhat upon the eventual means of disposal. For materials intended to be placed in landfills, the normal means of packaging is in 55-gal steel drums, either DOT 17C closed-head, which are used for bulk liquids, or DOT 17H open-head steel drums, which are used for packing smaller containers. These are intended to be used only once, i.e., single-trip containers (STC), and are identified as such by stamping them with this legend. Normally, containers are placed in the drum with enough filler material to completely fill the empty space. If the waste is a liquid, the filler material should be sufficient to absorb the liquid should the containers break. The space occupied by the awkward shape of most chemical containers, plus the need for enough inexpensive filler to essentially absorb completely the liquid, limits the total volume of the bottles which can be placed in a 55-gal drum to between 12 and 17 gal. Smaller containers of incompatible chemicals, i.e., ones which can react vigorously together, cannot be placed in the same container. Smaller drums or pails also can be used if they meet DOT specifications for the materials placed in them. Occasionally there is a need to place a drum inside a larger one if there is a risk of a leak. This is called an overpack and is relatively expensive.

Some firms which accept materials for incineration will accept waste in metal drums if they remove the waste containers from the drums at the disposal site. However, most firms accepting waste for incineration prefer the waste to be placed in combustible fiber drums which can be placed directly in the incinerator. These combustible fiber drums cannot be placed in landfills.

### 4.10.2.7. Restrictions on Wastes

DOT restrictions forbid some materials to be transported as waste. Among these are

1.  Reactive wastes that can explode or release toxic vapors, gases, or fumes at standard temperature or pressure
2.  Reactive materials that, when mixed with water or exposed to moisture, can explode or generate toxic vapors, fumes, or gases
3.  Materials which are shock sensitive
4.  Materials which will explode, etc., if heated while confined
5.  Class A or Class B explosives

These restrictions are imposed to ensure that the containers will not explode or evolve toxic gases, vapors, or fumes during transportation on a waste carrier and endanger the public. If a container of material were to explode on a truck loaded with other wastes, a major catastrophe could ensue.

The hazardous and solid waste amendments prohibit land disposal of some wastes unless the wastes are treated or it can be shown that there will be "no migration as long as the waste remains hazardous." Dioxin and solvent-contaminated soils as well as dilute waste waters contaminated with solvents were originally to be excluded, but there was not sufficient capacity to handle these otherwise and so an extension was granted to continue placing them in landfills.

A more important restriction affecting laboratories went into effect on November 8, 1986, when it became illegal to place in a landfill wastes containing more than 1% of a number of solvents. These are

F001 — The following spent halogenated solvents used in degreasing: tetrachloroethylene, trichloroethylene, methylene chloride, 1,1,1-trichloromethane, carbon tetrachloride, and chlorinated fluorocarbons, all spent solvent mixtures/blends containing, before use, a total of 10% or more (by volume) of one or more of the above halogenated solvents or those solvents listed in F002, F004, and F005; and still bottoms from the recovery of these spent solvent mixtures.

F002 — The following spent halogenated solvents: tetrachloroethylene, methylene chloride, trichloroethylene, 1,1,1-trichloromethane, chlorobenzene, 1,1,2-trichloro-1,2,2-trifluoroethane, ortho-dichlorobenzene, and trichlorofluoromethane; ortho-dichlorobenzene, and trichlorofluoromethane; all spent solvent mixtures/blends containing, before use, a total of 10% or more (by volume) of one or more of the above halogenated solvents or those solvents listed in F001, F004, and F005; and still bottoms from the recovery of these spent solvents and spent solvent mixtures.

F003 — The following spent non-halogenated solvents: xylene, acetone, ethyl acetate, ethyl benzene, ethyl ether, methyl isobutyl ketone, *n*-butyl alcohol, cyclohexanone, and methanol; all spent solvent mixtures/blends containing solely the above spent non-halogenated solvents; and all spent solvent mixture/blends containing before use, one or more of the above nonhalogenated solvents, and a total of 10% or more (by volume) of one or more of those solvents listed in F001, F002, F004, and F005; and still bottoms from the recovery of these spent solvents and spent solvent mixtures.

F004 — The following spent non-halogenated solvents; cresols and cresylic acids and nitrobenzene: all spent solvent mixtures/blends containing, before use, a total of 10% or more (by volume) of one or more of the above non-halogenated solvents or those solvents listed in F001, F002, and F005; and still bottoms from the recovery of these spent solvents and spent solvent mixtures.

F005 — The following spent non-halogenated solvents; toluene, methyl ethyl ketone, carbon disulfide, isobutanol, and pyridine; all spent solvent mixture/blends containing, before use, a total of 10% or more (by volume) of one or more of the above non-halogenated solvents or those solvents listed in F001, F002, and F004; and still bottoms from the recovery of spent solvents and solvent mixtures.

F020 — Wastes (except waste water and spent carbon from hydrogen chloride purification) from the production or manufacturing use (as a reactant, chemical intermediate, or component in a formulating process) of tri- or tetrachlorophenol, or of intermediates used to produce their pesticide derivatives. (This listing does not include wastes from the production of hexachlorophene from highly purified 2,4,5-trichlorophenol.)

F021 — Wastes (except waste water and spent carbon from hydrogen chloride purification) from the manufacturing use (as a reactant, chemical intermediate, or component in a formulating process) of pentachlorophenol, or of intermediates used to produce its derivatives.

F022 — Wastes (except waste water and spent carbon from hydrogen chloride purification) from the manufacturing use (as a reactant, chemical intermediate, or component in a formulating process) of tetra-, penta-, or hexachlorobenzenes under alkaline conditions.

F023 — Wastes (except waste water and spent carbon from hydrogen chloride purification) from the production of materials on equipment previously used for the production or manufacturing use (as a reactant, chemical intermediate, or component in a formulating process) of tri- and tetrachlorophenols. (This listing does not include wastes from equipment used only for the production or use of hexachlorophene from highly purified 2,4,5-trichlorophenol.)

F026 — Wastes (except waste water and spent carbon from hydrogen chloride purification) from the production of materials on equipment previously used for the manufacturing use (as a reactant, chemical intermediate, or component in a formulating process) of tetra-, penta-, or hexachlorophene under alkaline conditions.

F027 — Discarded unused formulations containing tri-, tetra-, or pentachlorophenol or discarded unused formulations containing compounds derived from these chlorophenols. (This listing does not include formulations containing hexachlorophene synthesized from prepurified 2,4,5 trichlorophenol as the sole component.)

On December 11, 1986, the EPA proposed lowering the acceptable limits in waste

## TABLE 4.21
### Statutory Limits for Specific Filtrates

| Constituent | Suggested filtrate prohibition level (mg/l) |
|---|---|
| Arsenic | 5.0 |
| Cadmium | 1.0 |
| Chromium | 5.0 |
| Lead | 5.0 |
| Mercury | 0.2 |
| Nickel | 50.0 |
| Selenium | 1.0 |
| Thallium | 0.9 |
| Cyanide | 20.0 |

of certain "California list" metals and cyanide. On August 12, 1987, the EPA announced it was considering the following statutory limits (Table 4.21) for these materials in wastes. The numbers are the levels, in milligrams per liter, in the filtrate from a waste, "as derived using the assumptions discussed in the May 19, 1980 FR notice (45 FR 33119) which promulgated the Extraction Procedure Toxicity Characteristic."

As time passes, it is becoming clearer that there will be more limitations on the materials to be placed in landfills. It is also clear that there will be additional restrictions on the construction and management of landfills which will make them more expensive to use. These costs will necessarily be passed on to the generator. Obviously, it will become increasingly important for the academic institution or corporation to restrain the amount of waste being produced. Disposal costs already are not a negligible item for a facility that is actively pursuing a waste program that fully complies with the standards.

### 4.10.2.8. Shipping Waste

Once waste has been collected and classified as hazardous, it must be disposed of by shipping it off-site for all those who are only classified as generators and do not have a permit for treatment, storage, disposal, or recycling facilities. A substantial amount of money can be saved if the packing and preparation of the manifest is done by internal waste management personnel rather than the disposal firm. This requires the management personnel to be thoroughly familiar with the current requirements for generators initiating shipments of hazardous waste equivalent to those given in Section 4.10.1.2. Some hazardous waste disposal firms will not permit this, feeling that they cannot be sure that the waste is packaged properly. Others will accept waste packaged by the generator, but will spot check individual drums or check the pH, for example, of some of the individual containers to confirm that the materials are packed properly, that the drums contain materials which agree with the accompanying lists, and that the chemicals are identified properly. It is essential that the generators subscribe to a reference service which will provide them with a current set of regulations for both the EPA and DOT requirements since the latter change too frequently for the preparers of the shipments to otherwise be sure that they are in compliance and, in any event, they need the references for the chemical identification numbers.

They will also need to conform to the specific requirements of the firm taking the waste. Some firms are essentially pure brokers, having no facilities of their own for the ultimate disposal of the waste. These firms sometimes collect wastes from several small generators at a central facility, group the materials to be landfilled separately from those to be incinerated, and then take them to firms which provide these services. If the size of the shipment is large enough, they may take it directly from the generator to the appropriate disposal site. In a case such as this, the generator will, in effect, have to conform to both the broker's requirements and those of the second firm, although it may make little difference to the generator.

Some firms offer a complete range of services for disposing of prepared waste. They offer transportation and either landfill disposal or incineration facilities. In principal at least, these firms should provide disposal at a lower cost per unit volume since they do not have to share the profits with an intermediary firm.

As noted earlier, transporters are forbidden to carry some materials because of the risk of an accident. However, there are firms which specialize in on-site destruction of materials such as shock-sensitive chemicals, explosives, highly reactive chemicals, and gas cylinders which cannot be returned to the manufacturer or which pose other special hazards. Some firms bring special equipment to a facility which permits them to take the materials such as very old ethers containing large quantities of peroxides and other comparably dangerous materials, directly from the laboratory in cases of extreme risk. Local destruction will normally require permits and an isolated area in which to perform the task. Usually, these will be the responsibility of the generator, and in some instances both may be difficult to obtain. Disposal of these items will be expensive.

The generator must accept one last responsibility which is not explicitly stated in the standards, and that is that the disposal firm must be capable of handling the waste properly. It would be assumed that the possession of an EPA permit and an Identification Number, as well as being financially secure, and being able to provide a certificate of insurance would suffice to guarantee proper handling of the waste. However, there have been numerous fines and closures of facilities operated by even major firms. It could be cost effective for a knowledgeable representative of the waste management group, or a colleague in the area whose judgement is known to be reliable, to visit a facility before selecting a firm. Again, at the author's institution, such a practice has resulted in negative decisions in a number of cases. In cases where this was a critical factor in the selection process, the facility eventually had to take corrective actions and, in some cases, ceased operations. There are many cases where the cost of remedying problems at disposal sites have eventually had to be borne by the generators who shipped waste to the sites.

### 4.10.2.9. Landfill Disposal

Until recent years, landfill disposal has been the favored method of disposing of laboratory wastes. Because of potential problems such as damage to the incinerators and control of toxic emissions and hazardous constituents in the residual ash due to the variety of materials in mixed laboratory wastes, most incineration facilities did not wish to accept this category of materials, although some would accept properly segregated materials. In addition, or partially because of the operational problems, incineration was somewhat more expensive. However, because of the growing number of restrictions on materials which can be put into landfills (see Section 4.10.2.7) as well as the increasing costs of constructing and managing landfills, incineration is becoming more popular and more cost competitive.

Landfills are essentially elongated rectangular trenches cut into the earth in which the drums of hazardous waste are placed. Landfills are preferably sited in areas in which the water table is low, compared to the depth of the trench, and where the underlying earth consists of relatively deep beds of clay, which has very low permeation properties. Clay layers underlying some of the more heavily used hazardous waste landfills are several hundred feet thick. Any material leaking into these landfills from the waste containers should be contained much as water in a bathtub. Although steel drums would not deteriorate as rapidly as a combustible fiber drum, they still have a short lifetime compared to that of many of the materials contained within the drums. Orderly placement of the drums as well as having them completely filled is essential to keep the surface from sagging as the exterior of the drums deteriorates. In order to prevent leakage from these sites, they should be located in areas where surface flooding is highly unlikely. All other things being equal, it would be desirable to have them in areas of low precipitation, but if the surface is properly configured and maintained, reasonable amounts of precipitation should not be a problem.

Low population density, the absence of commercial deposits of valuable minerals, and no historic or scenic attractions in the immediate vicinity also are desirable. The availability of good roads for bringing waste to the site is essential.

The typical size of these facilities is not overly large, ranging from a few tens to a few hundreds of acres, and if managed properly, does not constitute an environmental problem to the surrounding acres. A significant number of existing landfills were not designed or managed properly and are among the Superfund sites which must be cleaned up. Permitted landfills today must be lined with synthetic materials and have a leachate collection system and a groundwater monitoring system.

Because relatively few landfills meet all of the essential criteria, waste disposal firms often transport waste for extremely long distances to these sites. Presently, there are only about 33 operating commercial landfills.

A fundamental concern about the landfilling of hazardous waste is the fact that many of the materials retain their hazardous properties for very long periods of time. Although currently these sites can be managed properly, it is possible to devise circumstances that could arise in which good management would no longer be possible, and even where records identifying the material buried at the site would be lost. Individuals raising these issues admit that these scenarios are remote, but the possibility of a negative impact on future generations has been used to argue against the use of landfills for disposal of hazardous chemicals. The past history of mismanagement by some collectors of hazardous waste has also contributed to an emotional reaction against landfills. Hence, siting additional landfills (or other hazardous waste operations) has become increasingly difficult because of public concerns.

### 4.10.2.10. Incineration of Hazardous Waste

The two primary options of off-site waste disposal available to generators of hazardous chemical waste have been landfills and incineration. It is relatively unusual for a laboratory facility or even a facility with a large number of laboratories at the same site to generate enough of a waste stream which is consistent enough to be usefully reprocessed or recycled by a waste processor, although, occasionally, some operations may generate enough clean nonhalogenated organic solvents to be of interest as fuel. Landfilling has been the predominant disposal mechanism in the past, but incineration is being used more often due to regulatory restrictions on landfills and a narrowing price differential between incineration and burying the waste.

Incineration has a number of major advantages. Many of the materials which are currently prohibited from being placed in landfills can be destroyed by incineration. Many of the incinerated materials can be totally destroyed and the total volume of hazardous waste is substantially reduced whene there is a hazardous residue. There-fore, the size of the long-term management problem, if not totally eliminated, is at least greatly diminished. As a result, the residual liability of the generator for waste that has been incinerated is eliminated in most instances. There may be sufficient energy generated to be useful.

There are some disadvantages. The emission of toxic materials such as $NO_x$, $SO_x$, HCl, and some particulates containing metals from the stack must be controlled within stringent standards established by the EPA. This can be expensive for some waste streams. The ash residue may contain concentrated quantities of toxic sub-stances, which will have to be disposed of as hazardous waste in a landfill. The combustion chamber itself may be damaged by physical damage or corrosion and maintenance may be a problem. On the whole, however, especially as EPA standards on alternatives become more restrictive, the advantages outweigh the disadvantages, as long as the facilities are operated properly.

The following article[5] provides a review of the factors which must be considered in establishing a local incineration program for an organization's hazardous waste.

## STATE-OF-THE-ART HOSPITAL & INSTITUTIONAL WASTE INCINERATION, SELECTION, PROCUREMENT, AND OPERATIONS*

### INTRODUCTION

On-site incineration is becoming an increasingly important alternative for the treatment and disposal of institutional waste. Incineration reduces the weight and volume of most institutional solid waste by upwards of 90 to 95%, sterilizes patho-genic waste, detoxifies chemical waste, converts obnoxious waste, such as animal carcasses, into innocuous ash and also provides heat recovery benefits. At most institutions, these factors provide a substantial reduction in off-site disposal costs such that on-site incineration is highly cost-effective. Many systems have payback periods of less than 1 year. In addition, on-site incineration reduces potential expo-sures and liabilities associated with illegal or improper waste disposal activities.

Clearly, the most important factor currently affecting the importance of on-site incineration for health-care organizations and research institutions across the country relates to infectious waste management and disposal. First of all, recent legislation and guidelines have dramatically increased the quantities of institutional waste to be disposed of as "potentially infectious". For many institutions, particularly hospitals,

---

* This article was written by Lawrence G. Doucet, P.E.

incineration is the only viable technology for processing the increased, voluminous quantities of waste. Secondly, about half of the states and several major cities currently mandate that infectious waste be treated on-site, restrict its off-site transport and/or prohibit it from being landfilled. Many additional states are planning similar, restrictive legislative measures within the next few years.

Off-site disposal difficulties and limitations probably contribute the greatest incentives for many health-care and other institutions to consider or select on-site incineration as the preferred infectious waste treatment method. It has become increasingly difficult, if not impossible, to locate reliable, dependable infectious waste disposal service contractors. Many institutions able to obtain such services are literally required to transport their infectious waste across the country to disposal facilities. Furthermore, such services are typically very costly, if not prohibitive. Off-site disposal contractors are typically charging from about $0.30 to about $0.80 per pound of infectious waste, and some are charging as much as $1.50 per pound. For many hospitals and other institutions, this equates to hundreds of thousands of dollars per year, and several are paying in excess of a million dollars per year.

## INCINERATION TECHNOLOGIES

Before the early 1960s, institutional incineration systems were almost exclusively multiple-chamber types, designed and constructed according to Incinerator Institute of America (IIA) *Incinerator Standards*. Since these systems operated with high excess air levels, most required scrubbers in order to comply with air pollution control standards. Multiple-chamber type systems are occasionally installed at modern facilities, because they represent proven technology. However, the most widely and extensively used incineration technology over the past 20 years is "controlled-air" incineration. This has also been called "starved-air" incineration, "two-stage" incineration, "modular" combustion and "pyrolytic" combustion. More than 7,000 controlled air type systems have been installed by approximately two dozen manufacturers over the past 2 decades.

Controlled-air incineration is generally the least costly solid waste incineration technology — a factor that has undoubtedly influenced its popularity. Most systems are offered as low cost, "pre-engineered" and prefabricated units. Costly air pollution control equipment is seldom required, except for compliance with some of the more current, highly stringent emission regulations, and overall operating and maintenance costs are usually less than for other comparable incineration technologies.

The first controlled air incinerators were installed in the late 1950s, and the first U.S. controlled air incinerator company was formed in 1964. The controlled air incineration industry grew very slowly at first. The technology received little recognition because it was considered unproven and radically different from the established and widely accepted IIA *Incinerator Standards*.

Approximately every five years the controlled air incineration industry has gone through periods of rapid growth. In the late 1960s, this was attributable to the Clean Air Acts, in the early and late 1970s to the Arab oil embargoes, in the early 1980s to the enactment of hazardous waste regulations, and, recently, to the enactment of infectious waste disposal regulations. Dozens of "new" vendors and equipment suppliers appeared on the scene during each of these growth periods. However, increased competition and rapid changes in the technology and market structure forced most of the smaller and less progressive companies to close. Generally, the controlled air incineration industry has been in a state of almost constant development and change, with frequent turnovers, mergers and company failures.

Today there are approximately a dozen listed "manufacturers" of controlled air incinerators. However, only about half of these have established successful track records with demonstrated capabilities and qualifications for providing first-quality installations. In fact, some of the "manufacturers" listed in the catalogs have yet to install their first system, and a few are no more than brokers who buy and install incinerator equipment manufactured by other firms.

Controlled-air incineration is basically a two-stage combustion process. Waste is fed into the first stage, or primary chamber, and burned with less than stoichiometric air. Primary chamber combustion reactions and turbulent velocities are maintained at very low levels to minimize particulate entrainment and carryover. This starved air burning condition destroys most of the volatiles in the waste materials through partial pyrolysis. Resultant smoke and pyrolytic products, along with products of combustion, pass to the second stage, or secondary chamber. Here, additional air is injected to complete combustion, which can occur either spontaneously or through the addition of auxiliary fuel. Primary and secondary combustion air systems are usually automatically regulated, or controlled, to maintain optimum burning conditions despite varying waste loading rates, composition, and characteristics.

Rotary kiln type incineration systems have been widely promoted within the past few years. A rotary kiln basically features a cylindrical, refractory-lined combustion chamber which rotates slowly on a lightly inclined, horizontal axis. Kiln rotation provides excellent mixing, or turbulence, of the solid waste fed at one end — with high quality ashes discharged at the opposite end. However, in general, rotary kiln systems have relative high costs and maintenance requirements, and they usually require size reduction, or shredding, in most institutional waste applications. There are only a handful of rotary kiln applications in hospitals and other institutions in the U.S. and Canada.

"Innovative" incineration technologies also frequently appear on the scene. Some such systems are no more than reincarnations of older "failures", and others feature unusual applications and combinations of ideas and equipment. Probably the best advice when evaluating or considering an innovative system is to first investigate whether or not any similar successful installations have been operating for a reasonable period of time. Remember, so-called innovative systems should still be designed and constructed consistent with sound, proven principles and criteria.

## SIZING AND RATING

Classifications systems have been developed for commonly encountered waste compositions. These systems identify "average" characteristics of waste mixtures, including such properties as ash content, moisture, and heating value. The classification system published in the IIA *Incinerator Standards* is the most widely recognized and is almost always used by the incinerator manufacturers to rate their equipment. In this system, shown in Table A, wastes are classified into seven types. Types 0 through 4 are mixtures of typical, general waste materials, and Types 5 and 6 are industrial wastes requiring special analysis.

## TABLE A
## Classification of Wastes

| Type | Description | Principal components | Approximate composition (% by weight) | Moisture content (%) | Incombustible solids(%) | Value per lb of refuse as fired |
|------|-------------|----------------------|----------------------------------------|----------------------|--------------------------|----------------------------------|
| 0 | Trash | Highly combustible waste, paper, and wood cardboard cartons, including up to 10% treated papers, plastic or rubber scraps; commercial and industrial sources | Trash (100) | 10 | 5 | 8500 |
| 1 | Rubbish | Combustible waste, paper, cartons, rags, wood scraps, combustible floor sweepings; domestic, commercial, and industrial sources | Rubbish (80) Garbage (20) | 25 | 10 | 6500 |
| 2 | Refuse | Rubbish and garbage; residential sources | Rubbish (50) Garbage (50) | 50 | 7 | 4300 |
| 3 | Garbage | Animal and vegetable wastes, restaurants, hotels, markets; institutional, commercial, and club sources | Garbage (65) Rubbish (35) | 70 | 5 | 2500 |
| 4 | Animal solids and organic wastes | Carcasses, organs, solid organic wastes; hospital, laboratory, abattoirs, animal pounds, and similar sources | Animal and human tissue (100) | 85 | 5 | 1000 |

## Primary Combustion Chambers
*Heat Release Rates*

Incinerator capacities are commonly rated as pounds of specific waste types, usually Types 0 through 4, that can be burned per hour. Incinerators usually have a different capacity rating for each type. For example, an incinerator rated for 1000 pounds per hour of Type 1 waste may only be rated for about 750 pounds per hour of Type 0 waste or about 500 pounds per hour of Type 4 waste. Such rating variations exist because primary chamber volumes are sized on the basis of internal heat release rates, or heat concentrations. Typical design heat release values range from about 15,000 to 25,000 Btu per cubic foot of volume per hour (Btu/cu-ft/hr). In order to maintain design heat release rates, waste burning capacities vary inversely with the waste heating values (Btu/lb). As heating values increase, less waste can be loaded.

Since Type 3 waste, food scraps, and Type 4 (pathological) waste have heat contents of only 3500 and 1000 Btu per pound respectively, it might be assumed that

## TABLE B
### Maximum Burning Rate (lb/ft²/h) of Various Type Wastes

| Capacity (lb/h) | Logarithm | Type of waste | | | |
|---|---|---|---|---|---|
| | | 1 (factor 13) | 2 (factor 10) | 3 (factor 8) | 5 (no factor) |
| 100 | 2.00 | 26 | 20 | 16 | 10 |
| 200 | 2.30 | 30 | 23 | 18 | 12ᵃ |
| 300 | 2.48 | 32 | 25 | 20 | 14ᵃ |
| 400 | 2.60 | 34 | 26 | 21 | 15ᵃ |
| 500 | 2.70 | 35 | 27 | 22 | 16ᵃ |
| 600 | 2.78 | 36 | 28 | 22 | 17ᵃ |
| 700 | 2.85 | 37 | 28 | 23 | 18ᵃ |
| 800 | 2.90 | 38 | 29 | 23 | 18ᵃ |
| 900 | 2.95 | 38 | 30 | 24 | 18ᵃ |
| 1000 | 3.00 | 39 | 30 | 24 | 18ᵃ |

[a]  The maximum burning rate in lb/ft²/h for Type 4 Waste depends to a great extent on the size of the largest animal to be incinerated. Therefore, whenever the largest animal to be incinerated exceeds one third the hourly capacity of the incinerator, use a rating of 10 lb/ft²/h for the design of the incinerator.

even higher capacity ratings could be obtained for these waste types. However, this is not the case. The auxiliary fuel inputs required to vaporize and superheat the high moisture contents of Types 3 and 4 wastes limit effective incineration capacities.

In essence, primary chamber heat release criterion establishes primary chamber volume for a specific waste type and charging rate. Heat release values are simply determined by multiplying burning rate (lb/hr) by heating value (Btu/lb) and dividing by primary volume (cu-ft).

### Burning Rates

The primary chamber burning rate generally establishes burning surface, or hearth area, in the primary chamber. It indicates the maximum pounds of waste that should be burned per square foot of projected surface area per hour (lb/cu-ft/hr). Recommended maximum burning rates for various waste types are based upon empirical data, and are published in the IIA *Incinerator Standards*. Table B tabulates these criteria.

The figures in Table B are calculated as follows: Maximum Burning Rate Lbs Per Sq. Ft. Per Hr for Types #1, #2, & #3 Wastes Using Factors as Noted in the Formula:

$$B_R = \text{Factor for type waste} \times \text{log of capacity/hr}$$

$$B_R = \text{Max. burning rate lb/ft}^2\text{/hr}$$

For example, for an incinerator capacity of 100 lb/hr, for Type #1 waste,

$$B_R = 13 \text{ (factor for \#1 waste)} \times \log 100 \text{ (capacity/hr)}$$

$$= 13 \times 2 = 26 \text{ lb/ft}^2\text{/hr}$$

*Secondary Combustion Chambers*

Secondary chambers are generally sized and designed to provide sufficient time, temperature, and turbulence for complete destruction of combustibles in the flue gases from the primary chamber. Unless specified otherwise, secondary chamber design parameters are usually manufacturer specific. Typical parameters include:

- Flue gas retention times ranging from 0.25 s to at least 2.0 s.
- Combustion temperatures ranging from 1400°F to as high as 2200°F.
- Turbulent mixing of flue gases and secondary combustion air through the use of high velocity, tangential air injectors, internal air injectors, abrupt changes in gas flow directions, or refractory orifices, baffles, internal injectors and checkerwork in the gas flow passages.

Retention times, temperatures, and turbulence are interdependent. For example, secondary chambers that are specially designed for maximum turbulence but that have relatively short retention times may perform as well as other designs with longer retention times but less effective turbulence. On many applications, increased operating temperatures may allow for decreased retention times, or vice versa, without significantly affecting performance. Regulatory standards and guidelines often dictate secondary chamber retention time and temperature requirements.

Flue gas retention time (s) is determined by dividing secondary chamber volume (cu-ft) by the volumetric flue gas flow rate (ft³/s) quantities and operating temperatures. They can be calculated or measured. However, during normal incinerator operations, flue gas flow rates vary widely and frequently.

*Shapes and Configurations*

Primary and secondary chamber shapes and configurations are generally not critical as long as heat release rates, retention time, and air distribution requirements are satisfactory. Chamber geometry is most affected by the fabrication and transport considerations of the equipment manufacturers. Although some primary and secondary chambers are rectangular or box-like, most are cylindrical.

Controlled air incinerators with a capacity of less than about 500 lb/hr are usually vertically oriented, with primary and secondary chambers integral, or combined, within a single casing. Larger capacity controlled air incinerators are usually horizontally oriented and have non-integral, or separated, primary and secondary chambers. A few controlled air incinerator manufacturers offer systems with a third stage, or tertiary chamber, following the second stage. One manufacturer offers a fourth stage, termed a "reburn tunnel", which is primarily used to condition flue gases upstream of a heat recovery boiler.

Most manufacturer "variations" are attempts to improve efficiency and performance. However, some of these may be no more than "gimmicks" that offer no advantages or improvements over standard, conventional systems. Adherence to proper design fundamentals, coupled with good operations, are the overall keys to the success of any system. Acceptance of unproven variations or design deviations is usually risky.

## SELECTION AND DESIGN FACTORS

Highly accurate waste characterization and quantification data are not always required for selecting and designing incineration systems. However, vague or incomplete data can be very misleading and result in serious problems.

Waste characterization involves identification of individual waste constituents, relevant physical and chemical properties, and presence of any hazardous materials. A number of terms commonly used to characterize waste can be very misleading when used in specifications. As examples, vague terms such as "general waste", "trash", "biological waste", "infectious waste", and "solid waste" provide little information about the waste materials. An incineration system designed for waste simply specified as "general" waste would probably be inadequate if waste contained high concentrations of plastics. Likewise, the term "pathological" waste is frequently, but incorrectly, used to include an assortment of materials, including not only animal carcasses but also cage waste, laboratory vials and biomedical waste items of all types. "Pathological" incinerators are usually specifically designed for burning animal carcasses, tissues, and similar types of organic wastes. Unless the presence of other materials is clearly specified, resultant burning capacities may be inadequate for waste streams to be incinerated.

Waste characterization can range from simple approximations to complex and costly sampling and analytical programs. As discussed, the most frequently used approximation method is to categorize "average" waste mixtures into the five IIA classes, Types 0 through 4. The popularity of this waste classification system is enhanced by the fact that most of the incinerator manufacturers rate their equipment in terms of these waste types. However, it should be noted that actual "average" waste mixtures rarely have the exact characteristics delineated for any of these indicated waste "types".

The other end of the characterization spectrum involves sampling and analysis of specific "representative" waste samples or items in order to determine "exact" heating values, moisture content, ash content and the like. This approach is not generally recommended because it is too costly and provides no significant benefit over other acceptable approximation methods.

Virtually all components found in typical institutional-type solid waste have been sufficiently well characterized in various engineering textbooks, handbooks, and other technical publications. An example is presented in Table C. In many cases, a reasonable accurate compositional analysis of the waste stream, in conjunction with such published data and information, could provide reliable and useful characterization data.

Table C shows the various Btu values of materials commonly encountered in incinerator designs. The values given are approximate and may vary based on their exact characteristic or moisture contents.

A key factor is that incineration systems must be designed to handle the entire range of the waste stream properties and characteristics, not just the "averages". System capacity and performance may be inadequate if the waste data does not indicate such ranges.

*Capacity Determination*

One of the primary criteria for selecting incineration system capacity is the quantity of waste to be incinerated. Such data should include not only average waste generation rates, but also peak rates and fluctuation cycles. The most accurate

## TABLE C
## Waste Data Chart

| Material | Btu value per lb as fired | Wt in lb/ft³ (loose) | Wt in lb/ft³ | Content by weight (%) | |
| --- | --- | --- | --- | --- | --- |
| | | | | Ash | Moisture |
| Type 0 waste | 8,500 | 8—10 | | 5 | 10 |
| Type 1 waste | 6,500 | 8—10 | | 10 | 25 |
| Type 2 waste | 4,300 | 15—20 | | 7 | 50 |
| Type 3 waste | 2,500 | 30—35 | | 5 | 70 |
| Type 4 waste | 1,000 | 45—55 | | 5 | 85 |
| Acetic acid | 6,280 | | 65.8 | 0.5 | 0 |
| Animal fats | 17,000 | 50—60 | | 0 | 0 |
| Benzene | 18,210 | | 55 | 0.5 | 0 |
| Brown paper | 7,250 | 7 | | 1 | 6 |
| Butyl sole composition | 10,900 | 25 | | 30 | 1 |
| Carbon | 14,093 | | 138 | 0 | 0 |
| Citrus rinds | 1,700 | 40 | | 0.75 | 75 |
| Coated milk cartons | 11,330 | 5 | | 1 | 3.5 |
| Coffee grounds | 10,000 | 25—30 | | 2 | 20 |
| Corn cobs | 8,000 | 10—15 | | 3 | 5 |
| Corrugated paper | 7,040 | 7 | | 5 | 5 |
| Cotton seed hulls | 8,600 | 25—30 | | 2 | 20 |
| Ethyl alcohol | 13,325 | | 49.3 | 0 | 0 |
| Hydrogen | 61,000 | | 0.0053 | 0 | 0 |
| Kerosene | 18,900 | | 50 | 0.5 | 0 |
| Latex | 10,000 | 45 | 45 | 0 | 0 |
| Linoleum | 11,000 | 70—100 | | 20—30 | 1 |
| Magazines | 5,250 | 35—50 | | 22.5 | 5 |
| Methyl alcohol | 10,250 | | 49.6 | 0 | 0 |
| Naphtha | 15,000 | | 41.6 | 0 | 0 |
| Newspaper | 7,975 | 7 | | 1.5 | 6 |
| Plastic-coated paper | 7,340 | 7 | | 2.6 | 5 |
| Polyethylene | 20,000 | 40—60 | 60 | 0 | 0 |
| Polyurethane (foamed) | 13,000 | 2 | 2 | 0 | 0 |
| Rags (linen or cotton) | 7,200 | 10—15 | | 2 | 5 |
| Rags (silk or wool) | 8,400—8,900 | 10—15 | | 2 | 5 |
| Rubber waste | 9,000—11,000 | 62—125 | | 20—30 | 0 |
| Shoe leather | 7,240 | 20 | | 21 | 7.5 |
| Tar or asphalt | 17,000 | 60 | | 1 | 0 |
| Tar paper ($^1/_3$ tar, $^2/_3$ paper) | 11,000 | 10—20 | | 2 | 1 |
| Toluene | 18,440 | | 52 | 0.5 | 0 |
| Turpentine | 17,000 | | 53.6 | 0 | 0 |
| $^1/_3$ wax-$^2/_3$ paper | 11,500 | 7—10 | | 3 | 1 |
| Wax paraffin | 18,621 | | 54—57 | 0 | 0 |
| Wood bark | 8,000—9,000 | 12—20 | | 3 | 10 |
| Wood bark (fir) | 9,500 | 12—20 | | 3 | 10 |
| Wood sawdust | 7,800—8,500 | 10—12 | | 3 | 10 |
| Wood sawdust (pine) | 9,600 | 10—12 | | 3 | 10 |

method of determining such data is a comprehensive weighing program over a period of about 2 weeks. However, the most common procedure has been to estimate waste quantities from the number and volume of waste containers hauled off-site to land disposal. Large errors have resulted from such estimates because of failures to account for container compaction densities or from faulty assumptions that the waste containers were always fully loaded.

Three major variables affect the selection of incineration system capacity, or hourly burning rate: waste generation rates; waste types, forms, and sizes; and operating hours.

The quantity of waste to be incinerated is usually the primary basis for selecting system capacities. When waste generation rates are grossly underestimated, incineration capacity may be inadequate for the planned, or available, periods of operation. In such cases, the tendency is to overload the system, and operational problems ensue. On the other hand, incineration systems must be operated near their rated, or design, capacities for good performance, an oversized system must be operated less hours per day than may have been anticipated. Such reduced operating hours could cause difficult problems with waste handling operations, particularly if waste storage areas are marginal. Furthermore, if waste heat recovery is necessary to justify system economics, insufficient waste quantities could be a serious problem.

Since incinerators are primarily sized according to heat release rates, waste heating value is a fundamental determinant of capacity. However, the physical form, or consistency, of waste may have a more significant impact on burning capacities. For example, densely packed papers, books, catalogs and the like may have an effective incinerability factor of only about 20% compared to burning loosely packed paper. Likewise, animal bedding, or cage waste, typically has high ash formation tendencies that may reduce burning rates by as much as 50%. Furthermore, highly volatile wastes, such as plastics and containers of flammable solvents, may require burning rate reductions of as much as 65% to prevent smoking problems.

The physical size of individual waste items and containers is also an important factor in the selection of incineration capacity. One rule-of-thumb is that an average incinerator waste load, or largest item, should weigh approximately 10% of rated hourly system capacity. On this basis, a minimum 300 lb/h incinerator would be required for, say, Type 1 waste packaged in up to about 30-lb containers or bags. This capacity would be required regardless of the total daily quantity of Type 1 waste requiring incineration.

A typical daily operating cycle for a controlled air incinerator without automatic ash removal is as follows:

| Operating steps | Typical durations |
| --- | --- |
| • Clean-out | 15—30 min |
| • Preheat | 15—60 min |
| • Waste loading | 12—14 h |
| • Burn-down | 2—4 h |
| • Cool-down | 5—8 h |

It is important to note that waste *loading* for systems with manual clean-out is typically limited to a maximum of 12 to 14 h per operating day.

*Burning Rate vs. Charging Rate*

When evaluating incineration equipment, it is important to distinguish between the terms "burning or combustion rate" and "charging or loading rate." Manufacturers may rate their equipment or submit proposals using either term. "Burning rate" refers to the amount of waste that can be burned or consumed per hour, while "charging rate" is the amount of waste that can be loaded into the incinerator per hour. For systems operating less than 24 h per day, "charging rates" typically exceed "burning rates" by as much as 20% Obviously, failure to recognize this difference could lead to selecting a system of inadequate capacity.

## INCINERATOR SYSTEM AUXILIARIES

The incinerator proper is only a single component in a typical incineration "system." Other components, or sub-systems, which require equal attention in the design and procurement process, include:

- Waste handling and loading systems
- Burners and blowers
- Residue, or ash, removal and handling systems
- Waste heat recovery boiler systems
- Emission control systems
- Breeching, stacks and dampers
- Controls and instrumentation

Features of the major sub-systems are as follows:

*Waste Handling and Loading Systems*

Incinerators with capacities less than about 200 lb/h are usually available only with manual loading capabilities. Manual loading entails charging waste directly into the primary chamber without the aid of a mechanical system. Units with capacities in the 200 to 500 lb/h range are usually available with mechanical loaders as a special option. Mechanical loaders are standardly available for most incinerators with capacities of more than about 500 lb/h.

The primary advantage of mechanical loaders is that they provide personnel and fire safety by preventing heat, flames and combustion products from escaping the incinerator. In addition, mechanical loading systems prevent, or limit ambient air infiltration into the incinerator. In most cases, air infiltration affects combustion conditions and, if excessive, substantially lowers furnace temperatures and causes smoking at the stack and into charging room areas. Infiltration also increases auxiliary fuel usage and usually accelerates refractory deterioration. Several states have recently enacted regulations requiring mechanical loading on *all* institutional waste incinerators.

Mechanical loaders enable incinerators to be charged with relatively small batches of waste at regulated time intervals. Such charging is desirable because it provides relatively stabilized combustion conditions and approximates steady-state operations. Limiting waste batch sizes and loading cycles also helps protect against overcharging and resultant operating problems.

The development of safe, reliable mechanical loaders has been a major step toward modernizing institutional waste incineration technology. The earliest incinerators were restricted to manual charging, which limited their capacities and applications. Of the loader designs currently available, most manufacturers use the hopper/ram system. With this system, waste is loaded into a charging hopper, a hopper cover closes, a primary chamber fire-door opens and a charging ram then pushes the waste into the primary chamber. Hopper/ram systems are available with charging hopper volumes ranging from several cubic-feet to nearly 10 cubic yards. The selection of proper hopper volume is a function of waste type, waste container size, method of loading the hopper and incinerator capacity. An undersized hopper could result in spillage during waste loadings, an inability to handle bulky waste items, such as empty boxes, or the inability to charge the incinerator at rated capacity. On the other hand, an oversized hopper could result in frequent incinerator overcharging and associated operational problems.

A few manufacturers have recently developed mechanical loader systems which are capable of accepting as much as an hour's worth of waste loading at one time. These systems use internal rams to charge the primary chamber at intervals, as well as to prevent hopper bridging. Although these systems have had reportedly good success, they are still generally in the developmental stage.

One particular rotary kiln manufacturer uses an integral shredder at the bottom of the waste feed hopper. This system is termed an "auger feeder". It basically serves to process waste into a size that is compatible with the kiln dimensions.

With small capacity incinerators, less than about 500 lb/h waste is usually loaded manually, bag by bag, into the charging hopper. Larger capacity systems frequently employ waste handling devices such as conveyors, cart dumpers and, sometimes, skid-steer tractors to charge waste into the hopper. Pneumatic waste transport systems have been used to feed incinerator loading hoppers at a few institutions, but these have had limited success.

A cart-dumper loader basically combines a standard hopper/ram system with a device for lifting and dumping waste carts into the loading hopper. Several manufacturers offer these as integrated units. Cart dumpers can also be procured separately from several suppliers and adapted or retrofitted to almost any hopper/ram system. Cart-dumper loader systems have become increasingly popular because using standard, conventional waste carts for incinerator loading reduces extra waste handling efforts and often eliminates the need for intermediate storage containers and additional waste handling equipment.

Most modern hopper/ram assemblies are equipped with a water system to quench the face of the charging ram face after each loading cycle. This prevents the ram face from overheating due to constant, direct exposures to high furnace temperatures during waste injection. Without such cooling, plastic waste bags or similar materials could melt and adhere to the hot ram face. If these items did not drop from the ram during its stoking cycle, they could ignite and be carried back into the charging hopper, where they could ignite other waste remaining in the hopper or new waste loaded into the hopper. For additional protection against such possible occurrences, loading systems can also be equipped with hopper flame scanners and alarms, hopper fire spray systems and/or an emergency switch to override the normal charging cycle timers and cause immediate injection of hopper contents into the incinerator.

*Residue Removal and Handling Systems*

Residue, or ash, removal has always been a particular problem for institutional type incineration systems. Most small capacity incinerators (less than about 500 lb/h) and most controlled air units designed and installed before the mid-1970s, must be cleaned manually. Operators must rake and shovel ashes from the primary chamber into outside containers. Small capacity units can be cleaned from the outside, but large capacity units often require operators to enter the primary chamber to clean ashes. The practice of manual clean-out is especially objectionable from many aspects, including:

- Difficult labor requirements.
- Hazards to operating personnel because of exposures to hot furnace walls, pockets of glowing ashes, flaming materials, airborne dusts and noxious gases.
- Daily cool-down and start-up cycling requirements which substantially increase auxiliary fuel usage and reduce available charging time.
- Detrimental effects of thermal cycling on furnace refractories.
- Severe aesthetic, environmental and fire safety problems when handling hot, unquenched ashes outside the incinerator.
- Possible regulatory restrictions.

In multiple-chamber incinerators, automatic ash removal systems usually feature mechanical grates, or stokers. In rotary kiln systems, ash removal is accomplished via the kiln rotation. However, automatic, continuous ash removal has historically been difficult to achieve in controlled air systems which have conventionally featured stationary, or fixed, hearths.

Early attempts at automatic ash removal in controlled air incinerators employed a "bomb-bay" door concept. With these systems, the bottom of the primary chamber would swing open to drop ashes into a container or vehicle located below. Serious operating problems led to the discontinuance of these systems. More recent automatic ash removal systems use rams or plungers to "push" a mass of residue through the primary chamber and out a discharge door on a batch basis. Most of these systems have had only limited success.

Controlled air incinerator automatic ash removal systems that have shown the most promise use the waste charging ram of the hopper/ram system to force waste and ash residues through the primary chamber to an internal discharge, or drop chute for removal. Although charging rams usually extend no more than about 12 to 18 in. into the furnace during loading, this is sufficient to move materials across the primary chamber via the repetitive, positive-displacement actions of the rams. With proper design and operations, the waste should be fully reduced to ash by the time it reaches the drop chute. For incinerators with capacities greater than about 800 to 1000 lb/h internal transfer rams are usually provided to help convey ashes through the furnace to the drop chute. Transfer rams are necessary because the ash displacement capabilities of charging rams are typically limited to a maximum length of about 8 ft Primary chambers longer than about 16 ft usually have two or more sets of internal transfer rams.

The most innovative residue removal system uses a "pulse hearth" to transfer ashes through the incinerator. The entire floor of the primary chamber is suspended on cables and pulses intermittently via sets of end-mounted air cushions. The pulsations cause ash movement across the chamber and toward the drop chute.

After the ashes drop from the primary chamber through the discharge chute, there are two basic methods, other than manual, for collecting and transporting them from the incinerator. The first is a semi-automatic system using ash collection carts positioned within an air-sealed enclosure beneath the drop chute. A door or seal gate at the bottom of the chute opens cyclically to drop ashes into ash carts. Falling ashes are sprayed with water for dust suppression and a minor quenching. Because of weight considerations, ash cart volumes are usually limited to about 1 cubic yard.

Loaded ash carts are manually removed from the ash drop enclosure and replaced with empty carts. After the removed carts are stored on-site long enough for the hot ashes to cool, they are either emptied into a larger container for off-site disposal or are brought directly to the landfill and dumped. Adequate design and proper care are needed when dumping ashes into larger on-site containers to avoid severe dusting problems. In addition, some ashes could still be hot and may tend to ignite when exposed to ambient air during the dumping operations.

The second method of ash removal is a fully automatic system using a water quench trough and ash conveyor that continuously and automatically transports wet ashes from the quench trough to a container or vehicle. With these systems, the discharge chute terminates below water level in a quench trough in order to maintain a constant air seal on the primary chamber. Most manufacturers use drag, or flight, type conveyors, but a few offer "backhoe" or "scoop" type designs to batch grab ashes from the quench trough. The important factor is that the selected ash conveyor system be of proven design and of heavy-duty construction for the severe services of ash handling.

*Waste Heat Recovery*

In most incineration systems, heat recovery is accomplished by drawing the flue gases through a waste heat boiler to generate steam or hot water. Most manufacturers use conventional firetube type boilers for reasons of simplicity and low costs. Both single and multi-pass firetube boilers have been used successfully at many installations. Several facilities incorporate supplemental fuel-fired waste heat boilers so that steam can be generated when the incinerator is not operating. Also, automatic soot blowing systems are being installed on an increasing number of firetube boilers, in order to increase on-line time and recovery efficiencies.

One manufacturer uses single-drum, watertube type waste heat boilers on incineration systems. Watertube boilers are also used by other manufacturers on installations where high steam pressures and flow rates are required. Another manufacturer offers heat recovery systems with waterwall, or radiant, sections in the primary chamber. These waterwall sections, which are usually installed in series with a convective type waste heat boiler, can increase overall heat recovery efficiencies by as much as 10 to 15%.

Many incinerator manufacturers typically "claim" system heat recovery efficiencies for their equipment ranging from 60 to as high as 80%. However, studies and EPA-sponsored testing programs have shown that realistic heat recovery efficiencies are typically on the order of 50 to 60%. The amount of energy, or steam, that can be recovered is basically a function of flue gas mass flow rates and inlet and outlet temperatures. Depending on boiler type and design, gas inlet temperatures are usually limited to a maximum of 2200°F. Outlet temperatures are limited to the dewpoint temperature of the flue gases in order to prevent condensation and corrosion of heat exchanger surfaces. Depending upon flue gas constituents, incinerator dewpoint temperatures are usually on the order of 400°F.

For estimating purposes, about 3 to 4 lb of steam can be recovered for each pound of typical institutional type solid waste incinerated. However, the economic feasibilities of providing a waste heat recovery system usually depend upon the ability to use the recovered energy. If only half of recovered steam can be used because of low seasonal steam demands, heat recovery may not be cost-effective.

Some controlled air incinerator manufacturers offer air preheating, or "economizer packages", with their units. These primarily consist of metal jacketing, or shrouds, around sections of the primary or secondary chambers. Combustion air is heated by as much as several hundred degrees when pulled through the shrouds by combustion air blowers. This preheating can reduce auxiliary fuel usage by as much as 10 to 15%. In addition, the shrouding on some systems also helps limit incinerator skin temperatures to within OSHA limits.

For safety and normal plant shutdown, waste heat boilers are equipped with systems to divert flue gases away from the boiler and directly to a stack. One such system comprises an abort, or dump, stack upstream of the incinerator and stack. Another system includes a bypass breeching connection between the incinerator and stack. Modern, well-designed bypass systems are equipped with isolation dampers either in the dump stack or in the bypass breeching section. In systems without isolation dampers, either hot flue gases can bypass the boiler or ambient air can dilute gases to the boiler. Because of these factors, boiler isolation dampers may improve overall heat recovery efficiencies by at least 5%.

### Chemical Waste Incineration

An increasing number of institutions are disposing of chemical waste in their incineration systems. Incinerated chemicals are usually flammable waste solvents that are burned as fuels with solid waste. A simple method of firing solvents has been to inject them through an atomizer nozzle into the flame of an auxiliary fuel burner. Larger capacity and better designed systems use special, packaged burners to fire waste solvents. Such burners are either dedicated exclusively for waste solvent firing or have capabilities for switching to fuel oil firing when waste solvents are not available. Waste solvent firing is usually limited to the primary chamber in order to assist in the burning of solid wastes and to maximize retention time by fully utilizing secondary chamber volumes. Injectors and burners must be located and positioned so as not to impinge on furnace walls or other burners. Such impingement results in poor combustion and often causes emission problems.

Chemical waste incineration systems must also include properly designed chemical waste handling systems. These include a receiving and unloading station, a storage tank, a pump set to feed the injector or burner, appropriate diking and spill protection, monitoring, and safety protection devices. Most of these components must be enclosed within a separate, fire-rated room that is specially ventilated and equipped with explosion-proof electrical fixtures.

When transporting, storing and burning chemical waste, local, state and federal hazardous waste regulations must be followed. If the incinerated waste is regulated as a "hazardous waste", very costly trial burn testing, (Part B) permitting and monitoring equipment are required. In addition, obtaining the permits could delay starting a new facility by as much as 12 to 18 months. Incinerators burning chemical solvents which are only hazardous due to "ignitability" are not likely to be considered "hazardous waste incinerators", and the costly and lengthy hazardous waste incinerator permitting process is avoided. However, the storage and handling of these solvents will likely require a hazardous waste (Part B) permit.

At many institutions, bottles and vials of chemical wastes are often mixed with solid waste for incineration. If the quantities, or concentrations, of such containers and chemicals are very small with respect to the solid waste, incinerator operations may be unaffected. However, whenever solid waste loads are mixed with excessive concentrations of chemical containers, serious operating problems are likely, including rapid, uncontrolled combustion and volatilization resulting in heavy smoke emissions and potentially damaging temperature excursions. In addition, glass vials and containers tend to melt and form slag that can damage refractory materials and plug air supply ports.

*Emission Control Systems*

In general, only controlled air incinerators are capable of meeting the stringent emission standard of 0.08 grains of particulate per dry standard cubic foot of flue gas (gr/DSCF), corrected to 12% carbon dioxide, without emission control equipment. However, no incineration systems can meet the emission limits recently enacted by many states which require compliance with the best available control technology (BACT) levels. The BACT particulate level identified by many of the states is 0.015 gr/DSCF, corrected to 12% carbon dioxide. However, this is a controversial level which is being challenged by some in that it is only applicable to municipal waste incineration technology. Compliance with a 0.015 level will likely require a very high pressure drop, energy intensive, venturi scrubber system. Although "dry scrubbers," which comprise alkaline injection into the flue gas stream upstream of a baghouse filter, may also achieve a 0.015 level, as of this writing, this technology has yet to be demonstrated on an institutional waste incineration system.

Most institutional solid waste streams, particularly hospitals, include significant concentrations of polyvinyl chloride (PVC) plastics. Upon combustion, PVC plastics break down and form hydrogen chloride (HCl) gas. The condensation of HCl gases results in the formation of highly corrosive hydrochloric acid. Therefore, flue gas handling systems, and particularly waste heat boilers, must be designed and operated above the dewpoint of the flue gases. Protection of scrubbing systems typically includes the provision of an acid neutralization system on the scrubber water circuitry and the use of acid resistant components and materials.

Some states have identified BACT for HCl emissions as either 90% removal efficiency or 30 to 50 PPM, by volume, in the exhaust gases. For most well-designed wet scrubbers, 99% removal efficiencies are readily achievable. With respect to minimizing emissions of products of incomplete combustion (PICs), such as carbon monoxide and even dioxins and furans, the keys are proper furnace sizing, good combustion controls designed to accommodate varying waste compositions and charging rates, good operations and proper care and adjustment of system components. Inadequacies in any of these could result in objectionable emissions.

## TABLE D
### Incineration System Performance Problems

| Major performance difficulties | Examples |
|---|---|
| Objectionable stack emissions | Out of compliance with air pollution regulations |
| | Visible emissions |
| | Odors |
| | Hydrochloric acid gas (HCl) deposition and deterioration |
| | Entrapment of stack emissions into building air intakes |
| Inadequate capacity | Cannot accept "standard" size waste containers |
| | Low hourly charging rates |
| | Low daily burning rates (throughput) |
| Poor burnout | Low waste volume reduction |
| | Recognizable waste items in ash residue |
| | High ash residue carbon content (combustibles) |
| Excessive repairs and downtime | Frequent breakdowns and component failures |
| | High maintenance and repair costs |
| | Low system reliability |
| Unacceptable working environment | High dusting conditions and fugitive emissions |
| | Excessive waste spillage |
| | Excessive heat radiation and exposed hot surfaces |
| | Blowback of smoke and combustion products from the incinerator |
| System inefficiencies | Excessive auxiliary fuel usage |
| | Low steam recovery rates |
| | Excessive operating labor costs |

## INCINERATION PERFORMANCE AND PROCUREMENT

*Success Rates*

Incineration is considered proven technology in that a great many systems readily comply with stringent environmental regulations and performance requirements. Properly designed and operated incineration systems provide "good" performance if they satisfy specific user objectives in terms of burning capacity, or throughput, burnout, or destruction, environmental integrity and on-line reliability. However, many incineration stems of both newer and older designs perform poorly. Performance problems range from minor nuisances to major disabilities, and needed corrective measures range from simple adjustments to major modifications or even total abandonment. Furthermore, performance problems occur as frequently and as extensively in small, dedicated systems as in large, complex facilities. The most common incineration system performance problems are shown in Table D.

It has been estimated that roughly 25% of incineration systems installed within the last 10 years either do not operate properly or do not satisfy user performance objectives. A 1981 University of Maryland survey of medical and academic institutions incinerating low-level radioactive wastes indicated that only about 50% of the institutions surveyed (23 total) "reported no problems," and about 47% of the institutions (20 total) reported problems ranging from mechanical difficulties to combustion difficulties. A survey conducted by the U.S. Army Corps of Engineers Research Laboratory in 1985 at 52 incineration facilities reported that 17% of the users were "very pleased with their systems," 71% were "generally satisfied with the

## TABLE E
## 20 Common Problems Found in Small
## Waste-To-Energy Plants[a]

| Problems | Installations reporting (%) |
|---|---|
| Castable refractory | 71 |
| Underfire air ports | 35 |
| Tipping floor | 29 |
| Warping | 29 |
| Charging ram | 25 |
| Fire tubes | 25 |
| Air pollution | 23 |
| Ash conveyor | 23 |
| Not on-line | 21 |
| Controls | 19 |
| Inadequate waste supply | 19 |
| Water tubes | 17 |
| Internal ram | 15 |
| Low steam demand | 13 |
| Induced draft fans | 12 |
| Feed hopper | 10 |
| High pH quench water | 8 |
| Stack damper | 4 |
| Charging grates | 2 |
| Front-end loaders | 2 |

### Consensus

17% very pleased
71% generally satisfied — minor improvements needed
12% not happy

[a]  Results of 1983 survey of 52 heat recovery incineration
systems (5-50 TPD) conducted by U.S. Army Construction Engineering Research Laboratory.

performance of their systems" (but indicated that minor changes were needed to
reduce maintenance and improve efficiency) and 12% were "not happy with their
systems" (reporting severe problems). Results of this Army survey are summarized
in Table E.

### Fundamental Reasons for Poor Performance

Underlying causes or reasons for poor incineration system performance are not
always obvious. When performance difficulties are encountered, a typical reaction is
often to "blame" the incinerator contractor for furnishing "inferior" equipment.
While this may be the case on some installations, there are other possible reasons
which are more common and sometimes more serious. Generally, incineration system performance problems can be related to deficiencies or inadequacies in any or
all of three areas:

1.    Selection and/or design — *before* procurement
2.    Fabrication and/or installation — *during* installation
3.    Operation and/or maintenance — *after* acceptance

Examples of deficiencies in these three areas are as follows:

**1. System Selection and/or Design Deficiencies**

Deficiencies in this area are usually the result of basing incineration system selection and design decisions on incorrect or inadequate waste data, as well as failures to address specific, unique facility requirements. The resultant consequences are that system performance objectives and design criteria are also inadequate. An example of this is the procurement of an incineration system of inadequate capacity because of underestimated waste generation rates. Not so obvious examples include the relationships between operating problems and inadequate waste characterization data. Since incinerators are designed and controlled to process specific average waste compositions, vague identification of waste types or wide variances between actual waste parameters and "selected" design parameters often result in poor system performance. Significant deviations in parameters such as heating values, moisture, volatility, density and physical form could necessitate a capacity reduction of as much as two-thirds in order to avoid objectionable stack emissions, unacceptable ash quality and other related problems. Table F indicates examples of improper waste characterization affecting incineration capacity.

The establishment of good performance objectives based upon sound data and evaluations is only the initial step toward procuring a successful installation. The next step would be to ensure that system design criteria and associated contract documents are adequate to satisfy the performance objectives. A prime example of design inadequacies is the failure to relate incinerator furnace volumes to any specific criteria such as acceptable heat release rates. Another example is the specification of auxiliary components, such as waste loaders and ash removal systems, that are not suitable for the required operating schedules or rigors.

**2. Fabrication and/or Installation Deficiencies**

Deficiencies in this area relate to inferior workmanship and/or materials in either the fabrication or installation of the system. The extent and severity of such deficiencies are largely dependent upon the qualifications and experience of the incinerator contractor. Unqualified incineration system contractors may be incapable of or disinterested in providing a system in compliance with specified criteria. This could be either because of general inexperience in the field of incineration or because of a disregard of criteria that are different from their "standard way of doing business or furnishing equipment."

It is typical for even the most experienced and qualified incineration system contractors to deviate to some extent from design documents or criteria. This is largely because there are no such things as "standard" or "universal" incineration systems or "typical" applications or facilities. Unless design documents are exclusively and entirely based upon and awarded to a specific, preselected incinerator manufacturer, different manufacturers usually propose various substitutions and alternate methodologies when bidding a project. The key to evaluating such proposed

**TABLE F**
**Waste Characterization Data Deficiencies Necessitating**
**System Capacity Reductions**[a]

| Actual waste characterizations (deviations from selected "design" values) | Typical examples | Basic reasons for reduced capacities |
|---|---|---|
| Heating values (Btu/lb) excessive | Greater concentrations of paper and plastic components (or less moisture) than originally identified and specified | Incinerator volumetric heat release rates (Btu/ft³/h) exceed design limits[b] |
| Moisture concentrations excessive | Greater concentrations of high water content wastes, such as animal carcasses or food scraps (garbage), than originally identified and specified | Increased auxiliary fuel firing rates and additional time required for water evaporation and superheating |
| Volatiles excessive | Greater concentrations of plastic (such as polyethylene and polystyrene) or flammable solvents than originally identified and specified | Rapid (nearly instantaneous) releases of combustibles (volatiles) in large quantities along with excessively high temperature surges |
| Densities excessive | Computer printout, compacted waste, books, pamphlets, and blocks of paper | Difficulties in heat and flames penetrating and burning through dense layers of waste |
| High ash formation tendencies | Animal bedding or cage wastes — wood chips, shavings or sawdust | Ash layer formation on surface of waste pile insulates bulk of waste from heat, flames, and combustion air |

[a] Failure to reduce capacities, or hourly waste loading rates, to accommodate indicated deviations would likely result in other, more serious operational problems.
[b] Based upon accepted empirical values, primary chamber heat release rates should be in the range of 15,000 to 20,000 Btu/ft³/h.

variations is to assess whether they comply with fundamental design and construction criteria and whether they reflect proven design and application. On the other hand, allowing such variations without proper assessment could have unfortunate consequences.

The number and severity of fabrication and installation deficiencies are also directly related to quality control efforts during construction phases of a project. For example, a review of contractor submittals, or shop drawings, usually helps assure compliance with contract documents *before* equipment is delivered to the job site. Site inspections during installation work may detect deficiencies in design or workmanship before they lead to operational problems and performance difficulties. In addition, specific operating and performance testing as a prerequisite to final acceptance is a key element in assuring that a system is installed properly.

Table G lists some of the most common reasons for deficiencies in the fabrication and installation of incineration systems.

## TABLE G
## Common Reasons for Fabrication and Installation Deficiencies

Incineration equipment vendor (manufacturer) unqualified
Equipment installation contractor (GC) unqualified
Inadequate instructions (and supervision) from the manufacturer for system installation by the GC
No clear lines of system performance responsibility between the manufacturer and the GC
Failure to review manufacturer's shop drawings, catalog cuts, and materials and construction data to
   assure compliance with contract (design) documents
Inadequate quality control during and following construction to assure compliance with design (contract)
   documents
Payment schedules inadequately related to system performance milestones
Final acceptance testing not required for demonstrating system performance in accordance with contract
   requirements

## TABLE H
## Common Reasons for Operational and
## Maintenance Deficiencies

Unqualified operators
Negligent, irresponsible, and/or uncaring operators
Inadequate operator training programs
Inadequate operating and maintenance manuals
No record keeping or operating logs to monitor and verify performance
Inadequate operator supervision
Lack of periodic inspections, adjustments, and preventative maintenance
Extending equipment usage when repairs and maintenance work are needed

### 3. Operational and/or Maintenance Deficiencies

Deficiencies in this area are basically "self-inflicted" in that they usually result from owner, or user, omissions or negligence, and related problems occur *after* a system has been successfully tested and officially accepted.

Successful performance of even the best designed, most sophisticated and highest quality incineration systems is ultimately contingent upon the abilities, training, and dedication of the operators. The employment of unqualified, uncaring, poorly trained and unsupervised operators is one of the most positive ways of debilitating system performance in the shortest time.

Incineration systems are normally subject to severe operating conditions, and they require frequent adjustments and routine preventive maintenance in order to maintain good performance. Failures to budget for and provide such adjustments and maintenance on a regular basis leads to increasingly bad performance and accelerated equipment deterioration. Also, operating incineration equipment until it "breaks down" usually results in extensive, costly repair work and substantially reduced reliability.

Table H lists some of the most common operational and maintenance deficiencies which could result in poor incineration system performance.

The above problems are usually inter-related, and they usually occur in combination. They occur as frequently and as extensively in small, dedicated facilities as in large, complex facilities. They may range in severity from objectionable nuisances to major disabilities. Also, required corrective measures may range from minor adjustments to major modifications or even total abandonment.

Selection and design deficiencies are probably the most common as well as the most serious causes of problem incineration systems. Reputable incinerator contractors usually make every effort to satisfy specified design and construction criteria and meet their contractual obligations. Operating and maintenance deficiencies can usually be corrected. However, once a system has been installed and started, very little can be done to compensate for fundamental design inadequacies. Major, costly modifications and revisions to performance objectives are usually required.

The relatively frequent occurrence of design deficient systems may largely be attributable to a general misconception of the incineration industry as a whole. Incinerators are often promoted as standard, off-the-shelf equipment that can be ordered directly from catalogs, shipped to almost any job-site and, literally, "plugged in". This impression has been enhanced by many of the incinerator vendors in a highly competitive market. Exaggerations, half-truths and, sometimes, false claims are widespread relative to equipment performance capabilities. In addition, attractive, impressive brochures often suggest that implementation of an incineration system is simpler than it really is.

Incineration systems are normally subject to extremely severe operating conditions. These include very high and widely fluctuating temperatures, thermal shock from wet materials, slagging residues which clinker and spall furnace materials, explosions from items such as aerosol cans, corrosive attacks from acid gases and chemicals and mechanical abrasion from the movement of waste materials and from operating tools. These conditions are compounded by the complexity of the incineration process. Combustion processes are complicated in themselves, but in incineration this complexity is magnified by frequent, unpredictable and often tremendous variations in waste composition and feed rates. To properly manage such severe and complex operating conditions, incineration systems require well-trained, dedicated operating personnel, frequent and thorough inspections, maintenance and repair, and administrative and supervisory personnel attuned to these requirements.

At many facilities, the practice is to operate the incineration system continuously until it breaks down because of equipment failures. This type of operation accelerates both bad performance and equipment deterioration rates. Repairs done after such breakdowns are usually far more extensive and costly than those performed during routine, preventive maintenance procedures. Also, items which are typically capable of lasting many years can fail in a fraction of that time if interrelated components are permitted to fail completely.

## KEY STEP

A first step in procuring a good incineration system is to view the incineration "industry" in a proper perspective. There are four basic principles to bear in mind:

1.  Incineration technology is not an "exact" science — it is still more of an art than a science, and there are no shortcuts, simplistic methods or textbook formulas for success.
2.  There is no "universal" incinerator — no design is universally suited for all applications. Incinerators must be specifically selected, designed and built to

## TABLE I
### Recommended Incineration System Implementation Steps

Evaluation and selections
    Collect and consolidate waste, facility, cost, and regulatory data
    Identify and evaluate options and alternatives
    Select system and components
Design (contract) documents
    Define wastes to be incinerated; avoid generalities and ambiguous terms
    Specify performance requirements
    Specify *full* work scope
    Specify minimum design and construction criteria
Contractor selection
    Solicit bids on quality and completeness, not strictly least cost
    Evaluate and negotiate proposed substitutions and deviations
    Negotiate payment terms
    Consider performance bonding
Construction and equipment installation
    Establish lines of responsibility
    Require shop drawing approvals
    Provide inspections during construction and installation
Startup and final acceptance
    Require "punch-out" system for contract compliance
    Require comprehensive testing: system operation, compliance with performance requirements, and
      emissions
    Obtain operator training
After final acceptance
    Employ qualified and trained operators
    Maintain operator supervision
    Monitor and record system operations
    Provide regular inspections and adjustments
    Implement preventive maintenance and prompt repairs; consider service contract

meet the needs of each facility on an individual basis. Manufacturers' catalogs identify typical models and sizes, but these are rarely adequate for most facilities without special provisions or modifications.

3.     There is no "typical" incinerator application — even institutions of similar type, size and activities have wide differences in waste types and quantities, waste management practices, disposal costs, space availability and regulatory requirements. Each application has unique incineration system requirements that must be identified and accommodated on an individual basis.

4.     Incinerator manufacturers are not "equal" — there are wide differences in the capabilities and qualifications of the incinerator equipment manufacturers. Likewise, there are wide differences in the various systems and equipment which are offered by different manufacturers.

## RECOMMENDED PROCUREMENT STEPS

Table I outlines six steps recommended for implementing an incineration system project. Each is considered equally important toward minimizing or eliminating the deficiencies discussed above and for increasing the likelihood of obtaining a successful installation.

Performance difficulties on *most* problem incineration systems can usually be traced to a disregard or lack of attention to details in the first two steps; namely (1) evaluations and selections and (2) design documents. For example, many facilities have been procured strictly on the basis of "purchase orders" containing generalized requirements such as:

> "Furnish an incineration system to
> burn _____ lb/h of institutional
> waste in compliance with applicable
> regulations."

Obviously, the chances for success are marginal for any incineration system procured on the basis of such specifications.

On many projects, incinerator contractor evaluation and selection, under Step 3, involve no more than a solicitation of prices from a random listing of vendors with the award of a contract to that firm proposing a system for the "least cost." There are two basic problems with this approach. First, the selected incinerator contractors are assumed to have equivalent capabilities and qualifications. Second, "least cost" acceptance assumes that the equipment offered by each of the contractors is equivalent, or identical. A comparative "value" assessment of proposals usually results in the procurement of a superior quality system for a negligible price difference. It is not uncommon to see cost proposals "low" by no more than 10% but the equipment offered only half of the quality of the competition.

Again, although incineration is considered a proven technology, in many ways it is still more of an art than a science. There are no shortcuts, textbook formulas or shortcut methods for selecting and implementing a successful system, and there are no guarantees that a system will not have difficulties and problems. However, the probabilities of procuring a successful, cost-effective system increase proportionally with attention to details and utilization of proven techniques, methodologies and experience.

# REFERENCES

1. **Bleckman, J., O'Reilly, L., and Welty, C.,** *Incineration for Heat Recovery and Hazardous Waste Management,* American Hospital Association, Chicago, 1983.
2. **Boegly, W. J., Jr.,** *Solid Waste Utilization — Incineration with Heat Recovery,* Publ. No. ANL/ CES/TE 78-3, prepared for the Argonne National Laboratory under Contract W-31-109-Eng-38 with the U.S. Department of Education, Washington, D.C., 1978.
3. **Cooley, L. R., McCampbell, M. R., and Thompson, J. D.,** *Current Practice of Incineration of Low-Level Institutional Radioactive Waste,* Publ. No. EGG-2076, prepared for the U.S. Department of Education, Washington, D.C., 1981.
4. **Doucet, L. G.,** Waste handling systems and equipment, *NFPA Fire Protection Handbook,* chap. 14, Section 12, NFPA, Quincy, MA, 1985.
5. **Doucet, L. G.,** *Incineration: State-of-the-Art Design, Procurement and Operational Considerations,* Tech. Doc. No. 055872, American Society for Hospital Engineering — Environmental Management File, Chicago, 1988.
6. **Doucet, L. G. and Knoll, W. G., Jr.,** The craft of specifying solid waste systems, *Actual Specif. Eng.,* 107, May 1974.
7. **Doucet, L. G.,** Institutional waste incineration problems and solutions, in *Proc. Incineration of Low Level & Mixed Wastes Conf.,* St. Charles, IL, April 1987.

8. **Ducey, R. A., Joncich, D. M., Griggs, K. L., and Sias, S. R.,** Heat Recovery Incineration: A Summary of Operational Experience, Tech. Rep. No. CERL SRE-85/06, prepared for the U.S. Army Construction Engineering Research Laboratory, Champaign, IL, 1985.
9. **English, J. A., II,** Design aspects of a low emission, two-stage incinerator, in Proc. 1974 National Incinerator Conf., ASME, New York, 1974.
10. Small Modular Incinerator Systems with Heat Recovery: A Technical, Environmental and Economic Evaluation, EPA Publ. SW-177c, Environmental Protection Agency, Washington, D.C., 1979.
11. **Hathaway, S. A.,** Application of the Packaged Controlled Air-Heat Recovery Incinerator of Army Fixed Facilities and Installations, Tech. Rep. No. CERL-TR-E-151, prepared for the U.S. Army Construction Engineering Research Laboratory, Champaign, IL, 1979.
12. *Incinerator Standards,* Incinerator Institute of America, New York, 1968.
13. **Martin, A. E.,** Small-Scale Resource Recovery Systems, Noyes Data Corporation, Park Ridge, MN, 1982.
14. **McColgan, I. J.,** Air Pollution Emissions and Control Technology: Packaged Incinerators, Economic and Technical Review, EPS-3-AP-77-3, Canadian Environmental Protection Service, Burlington, Ont., 1977.
15. **McRee, R. E.,** Controlled-air incinerators for hazardous waste application theory and practice, in Proc. APCD Int. Workshop Ser. Hazardous Waste, New York, April 1, 1985.
16. **McRee, R. E.,** Waste heat recovery from packaged incinerators, in Proc. ASME Incinerator Division Conf., Arlington, VA, January 25, 1985.
17. **Theoclitus, G., Liu, H., and Dervay, J. R., II,** Concepts and behavior of the controlled air incinerator, in Proc. 1972 National Incinerator Conf., ASME, New York, 1972.

## GENERAL REFERENCES (SECTIONS 4.10 TO 4.10.2.10*)

1. The Hazardous Waste System, Office of Solid Waste and Emergency Response, U.S. Environmental Protection Agency, Washington, D.C., 1987.
2. *Prudent Practices for Disposal of Chemicals from Laboratories,* National Academy Press, Washington, D.C., 1983.
3. Hazardous Materials Emergency Planning Guide, NRT-1, National Response Team, Washington, D.C., 1987.
4. **Diehl, J.,** *How to Comply with Hazardous Waste Regulations,* Bureau of Hazardous Waste Management, Commonwealth of Virginia Department of Health, Richmond, VA, 1986.

## 4.11. LABORATORY CLOSEOUT PROCEDURES

Occasionally, laboratories cease operations and totally close down due to the laboratory director retiring, changing jobs or locations, or for other reasons. In these circumstances, there are almost always large quantities of surplus chemicals for which disposal must be arranged. Prior planning and a cooperative effort involving laboratory personnel and the waste management group will ensure that substantial quantities of chemicals will not be left behind which cannot be identified. It should be mandatory, wherever possible, that the persons responsible for the laboratory give the waste group at least 30 days notice, and preferably more, to allow the identification procedure to be done carefully and thoroughly.

Most of the chemicals in the laboratory should be in their original containers and should pose no difficulty in identification. Many of these will still be useful and laboratory personnel should either distribute them directly to others who might use them or transfer them to the organization's redistribution program. Unfortunately,

Other references and references to specific sections of the RCRA regulations were included in the text.

some will not be useful and, hence, will require disposal. The worst situation for materials which are readily identified is that they have become unstable and require special measures to move and handle safely. Often, stuck at the back of cabinets and long forgotten, will be ancient bottles of ether, perchloric acid, picric acid, etc. However, even in this difficult situation, at least the material is known and is subject to known procedures.

It will require the assistance of laboratory personnel to identify those materials which do not have labels or whose labels have no significance to those who have not been working in the laboratory. However, the laboratory personnel at least should be able to help identify the contents of the containers for which they are directly responsible, and if the labels bear the initials of the preparer, then from laboratory notebooks, reports, theses, dissertations, or even some familiarity with what the individual had been working on, it may be possible to determine what the contents are likely to be, which will greatly facilitate identification if tests are required. In some cases, it may be possible to make a reasonable estimate from other containers in the vicinity if, for example, they are obviously grouped by category.

There will almost always be some containers left over which cannot be identified. Some may be dangerous, while others may not. If materials are routinely used in the laboratory which degrade to dangerous materials, then these unknown containers must be treated with care. In some cases, it may be necessary to call in a firm which specializes in handling dangerous materials, rather than take the risk of handling the material at all. Where the materials normally generated in the laboratory are not shock, heat, or friction sensitive and procedures have not changed in character for several years, it is probably permissible to handle these unknown materials with only normal care. If the laboratory personnel are no longer available, it might not be possible to make this distinction.

Even if unknowns are still left over after the best efforts of the laboratory personnel and waste group working together, there will certainly be far fewer than if the latter group had to do all the work themselves, or if the evaluation of the containers had to be done hastily because of inadequate warning. Most individuals will be surprised at the effort and time required to close out an average laboratory. It will rarely be done in a single day. Providing adequate warning will allow waste personnel to schedule their own duties to take maximum advantage of their own time and that of the laboratory personnel. The reduction in the amount of waste generated and number of containers requiring analysis will amply repay the effort.

# Nonchemical Laboratories 5

## 5.0. INTRODUCTION

The emphasis in the previous chapters has been on laboratories in which the primary concerns were due to the use of chemicals, although, in order to not completely avoid a topic unnecessarily, some of the problems arising in other types of operations were covered briefly. For example, some aspects of health implications involving biological hazards were briefly discussed in the previous chapter; hence, key issues involving these problems were mentioned as part of the larger topic of health effects. In this chapter, laboratory operations which involve special problems will be discussed in greater detail. However, in responding to these special problems, one should be careful not to neglect the safety measures associated with more common hazards.

## 5.1. RADIOISOTOPE LABORATORIES

Exposure of individuals to ionizing radiation is a major concern in laboratories using radiation as a research tool or in which radiation is a byproduct of the research. Although there are many types of research facilities in which ionizing radiation is generated by equipment, e.g., accelerator laboratories, X-ray facilities, and laboratories using electron microscopes, the most common research application in which ionizing radiation is a matter of concern is the use of unstable forms of the common elements which emit radiation. A given element must have a fixed number of protons in the nucleus, but can differ in the number of neutrons, the different forms being called isotopes. It is the property of the unstable forms, or radioisotopes, to emit radiation which makes them useful, since their chemical properties are essentially identical to the stable form of the element (where a stable form exists; for some of the heavier elements, there are no completely stable forms). The radiation which the radioisotopes emit allows them to be distinguished from the stable forms of the element in an experiment.

The radiation which makes radioisotopes useful also makes their use a matter of concern to the users and the general public. Exposure to high levels of radiation is known to cause health problems; at very high levels, immediate death can occur. At lower, but still substantial levels, other health effects are known to occur, some of which, including cancer, can be delayed for many years. At very low levels, knowledge of the potential health effects is much more uncertain. The generally accepted practice at this point is to extrapolate the known effects on individuals exposed to higher levels to large groups of persons exposed to low levels of radiation in a statistically linear fashion. The concept is similar to the use of higher concentrations of chemicals on a limited number of animals in health studies of chemical effects, instead of using more normal concentrations on a very large number of test animals. There are some who question the validity of this assumption in both cases, but it is a conservative assumption and, in the absence of data, is a generally safe course of action to follow. However, the practice may have led to a misleading impression of the risks of many materials. When a scientist states that he doesn't know if a given material is harmful, he is often simply indicating in a very honest way that the data do not clearly show whether, at low levels of use or exposure, a harmful effect will result. It does not necessarily imply, as many assume, that there is a lack of research in discovering possible harmful effects. In many cases, major efforts have been made to unambiguously resolve the issue, as in the case of radiation, but the data do not support a definite answer. In the case of radiation, there is even a substantial body of experimental data, to which proponents of a concept called "hormesis" call attention, that support possible positive effects of radiation at very low levels. This position is, of course, very controversial. However, in chemical areas, there are many examples of chemicals essential to health in our diets in trace amounts which are poisonous at higher levels. It is not the intent of this section to attempt to resolve the issue of the effects of low-level radiation, but, rather, to emphasize that many employees and the general public are concerned. It may well be that, by being very careful to not go beyond known information, scientists have actually contributed to these concerns. Radiation levels which normally accompany the use of radioisotopes are deliberately kept low, and the perception of risk by untrained individuals may be overstated. However, a linear dose-effect relation is the accepted basis for regulatory requirements at this time, and until better data are available, scientists using radioactive substances must conform to the standards. Users owe it to themselves and the public to use the materials in ways known to be safe. However, as a general concept, it would be well for scientists, when speaking to persons not trained in their field, to make sure that the lack of knowledge of possible harmful effects of a given material is understood to be an informed uncertainty, where this is the case, as opposed to being based on a lack of effort.

It is unfortunate that there is so much concern about radiation since there are many beneficial effects which, because of the dramatization of the concerns, have caused many individuals to fear radiation out of all proportion to any known risks. In a recent opinion poll where members of the general public were asked to rank the relative risks of each of a number of hazards, nuclear radiation was ranked highest. However, based on known data, the least dangerous of all the other risks was much more likely to cause death or injury than was radiation. Individuals who have badly needed X-rays, the application of diagnostic uses of radioactive materials, or radiation therapy have declined to have them because of this heightened fear. Used properly, radiation is an extremely valuable research tool and has many beneficial aspects. Used improperly, it can be dangerous, but so can many other things.

## 5.1.1. NATURAL RADIOACTIVITY

A common misconception is that radiation is an artificial phenomenon. Some forms have been created artificially, but there are abundant sources of natural radiation. Elements such as potassium and carbon, which are major constituents of our body, have radioactive isotopes. Many other elements, such as the rare earths, have radioactive versions. Every isotope of elements with atomic numbers (i.e., the number of protons in the nucleus of the element or the number of electrons around the nucleus in a neutral atom) above 83 is unstable, and these elements are common in the soils and rocks which make up the outer crust of the earth. There are areas in the world where the natural levels of radiation substantially exceed that permitted for the general public which result from the operation of any licensed facility using radioisotopes. Radiation constantly bombards us from space due to cosmic rays. Persons who frequently take long airplane flights receive a significantly increased amount of radiation over a period of time, compared to persons who fly rarely or not at all. Arguments that these natural forms of radiation are acceptable because they are natural has absolutely no basis in fact. As will be discussed later, there are only a modest number of varieties of radiation, and these are produced by both natural and artificially produced radioactive materials. Similarly, there are only a few ways in which radiation may interact with matter, and they also are the same for all sources of radiation.

The argument that there are natural sources of radiation is not intended to belittle concerns about radiation, even the natural forms, but, rather, to point out that if there are concerns about low levels of radiation, then these natural levels must be considered as well as the artificial sources. At the time of this writing, one of the naturally occurring radioactive materials, radon, has been receiving much attention and may be a significant hazard, perhaps contributing to a 1 to 5% increase in the number of lung cancer deaths each year. This estimate, as in most cases, deals with the attribution of specific effects of low levels of radiation, supported by some and disputed by others. Note, however, that even in this case, 95 to 99% of lung cancer deaths are attributable to other causes. Radon as an issue will be discussed in a separate section later in this chapter.

There are various estimates of the average source of radiation exposure for most individuals. An article by Komarov,[1] who is associated with the World Health Organization, provides the following data on radiation exposure: 37% from cosmic rays and the terrestrial environment, 28% from building materials in the home, 16% from food and water, 12% from medical usage (primarily X-rays), perhaps 4% from the daily watching of color television, 2% from long-distance airplane flights, and 0.6% (under normal operating conditions) from living near a nuclear power plant. This article was written before the Chernobyl incident, but even this outstanding example of poor management is not sufficient to change the general picture. Unlike the Chernobyl reactor, commercial nuclear power plants in the U.S. are protected by very strong confinement enclosures to prevent unscheduled releases. In the case of Three-Mile Island, the confinement enclosure performed as designed and minimal amounts of radioactive material were released. As the news media reported some time after the initial furor, "the biggest danger from Three-Mile Island was psychological fear".

In summary, radiation is a valuable research tool. In order to prevent raising public concerns and perhaps leading to further restrictions on its use, scientists need to scrupulously conform to accepted standards governing releases or overexposures.

## 5.1.2. BASIC CONCEPTS

Each scientific discipline has its own special terms and basic concepts on which it is founded. This section is, of course, not necessary for most scientists who work with radiation, but it may be useful for establishing a framework within which to define some needed terms. As scientists work with accelerators of higher and higher energies, the concept of matter is at once growing more complex and simpler — more complex in that more entities are known to make up matter, but simpler in that theorists working with the data generated by these gigantic machines are constructing a coherent concept unifying all of the information. For the purposes of this discussion, a relatively simple picture of the atom will suffice.

### 5.1.2.1. The Atom and Types of Decay

In the simple model of the atom which will be employed here, the atom consists of a very small, dense nucleus containing positively charged particles called protons and neutrally charged particles called neutrons, surrounded by a cloud of negatively charged electrons. The number of protons and electrons balance for a neutral atom, but the number of neutrons can vary substantially, resulting in different forms, or, as already noted, isotopes of an element. Some elements have only one stable isotope, while tin has ten. There are unstable isotopes, logically called radioisotopes, in which, over a statistically consistent time, a transition of some type occurs within the nucleus. Different types of transitions lead to different types of emitted radiation. Hydrogen, for example has two stable forms and one unstable form in which a transition occurs that allows an electron to be generated and emitted from the nucleus, producing a stable isotope of helium. Prior to the transition, the electron did not exist independently in the nucleus. A neutron is converted to a proton in the process, and the electron is created by a transformation of energy into matter. This process is called beta decay. No element with more than 83 protons in the nucleus has a completely stable nucleus, although some undergo transitions (including those by processes other than beta decay) extremely slowly.

In some cases, the energy of the nucleus favors emission of a positive electron (positron) instead of a normal electron with a negative charge. This is called positive beta decay or positron decay. In these cases, a proton is converted into a neutron. A competitive process to positive beta decay is electron capture ($\varepsilon$), in which an electron from the electron cloud around the nucleus is captured by the nucleus, a proton being converted into a neutron in the process. In the latter process, X-rays are emitted as the electrons rearrange themselves to fill the vacancy. However, following positron emission, the positive electron eventually interacts with a normal electron in the surrounding medium, and the two vanish or annihilate each other in a flash of energy. The amount of energy is equal to the energy of conversion of the two electron masses, according to $E = mc^2$. This amounts to 1.02 MeV. In order to conserve momentum, two photons or gamma rays of 0.511 MeV each are emitted 180° apart in the process.

In many cases, the internal transitions accompanying adjustments in the nucleus result in emission of electromagnetic energy only, or gamma rays. These can be in

## TABLE 5.1
### Properties of Radioactive Emissions

| Type | Mass (in amu) | Charge (electron units) | Range of energy |
|------|---------------|-------------------------|-----------------|
| Alpha (α) | 4 | +2 | ≈4—6 MeV |
| Beta (β) | $1/_{1840}$ | ±1 | ≈eVs—4 MeV |
| Gamma | 0 | 0 | ≈eVs—4 MeV |
| X-rays | 0 | 0 | ≈eVs—100 KeV |

the original or parent nucleus, in which case they are called internal transitions and the states are called metastable states. More often, the gamma emitting transitions occur in the daughter nucleus after another type of decay, such as beta decay (metastable states can exist in the daughter nucleus, too). The gamma emission distribution can be very complex. In some instances, the internal transition energy is directly transferred to one of the electrons close to the nucleus in a process called internal conversion, and the electron is emitted from the atom. In this last case, energy from transitions in the orbital electron cloud is also emitted as X-rays.

Finally, the most massive entity normally emitted as radiation is the alpha (α) particle, which consists of a bare (no electrons) small nucleus consisting of two protons and two neutrons. The nucleons making up an alpha particle are very strongly bound together and, unlike electrons, the alpha particle appears to exist in the parent nucleus as a unit prior to the decay.

The processes briefly described above are the key ones in terms of safety in the use of radioisotopes. There is another aspect of the decay processes which is also very important, and that is the energy of the emitted radiation. The electrons emitted in beta decay can have energies ranging from a few eVs to between 3 and 4 MeV. An unusual feature of the beta decay process is that the betas are not emitted monoenergetically from the nucleus as might be expected, and as does occur for alpha and gamma decay. The reason is that, in addition to a beta, another particle called a neutrino, of either zero mass or very close to it, is emitted simultaneously which shares the transitional energy. This particle reacts minimally with matter and, except for modifying the energy of the electrons, does not play a role in radiation safety, although its existence is very important for many other reasons. Gammas can have a range of energies similar to that of electrons, but the energies of the gammas are discrete instead of a distribution.

Alpha particles have a relatively high energy, normally ranging between 4 to 6 MeV. The decay of alphas with lower energies is so slow that it occurs very rarely, whereas a nucleus, which emits alphas with higher energies, decays very rapidly. The high energy accompanying the high mass and the double positive charge make the alpha particle a particularly dangerous type of radiation if it is emitted in the proximity of tissue, which can be injured. This is an important qualification, as will be seen later.

Table 5.1 summarizes the properties of the types of radiation.

Graphically, the decay process can be depicted as shown in Figure 5.1 below, where A = the atomic mass number and Z = the nuclear charge. The box with A, Z is the parent nucleus and the others are the possible daughters for the processes indicated.

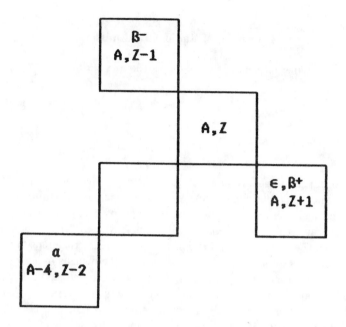

FIGURE 5.1. Relationship of daughter nuclei to parent nucleus for various emissions.

### 5.1.2.2. The Fission Process

A major omission from the above discussion is the process of fission, which will not normally be of direct concern in most laboratories using radioisotopes. However, without this process, many of the commonly used radioisotopes would not be available since they are obtained from reprocessing spent fuel and recovery of the remnants left over after the fission process. Fission describes the process by which some very heavy atoms decay by splitting into two major components and usually a few neutrons, accompanied by the release of large amounts of energy, $\approx$200 MeV. The process can be spontaneous, but also can be initiated by exposing specific heavy nuclei to neutrons. There are no radioisotopes which normally emit neutrons, but there are several interactions in which a neutron is generated. Among these are several reactions in which a gamma ray interacts with beryllium to yield neutrons, so that a portable source of neutrons can be created. There are many other ways to generate neutrons, but there is no need to describe these in this book. However, if a source of neutrons, n, is available and used to bombard an isotope of uranium, $^{235}$U, the following reaction can occur.

$$n + {}^{235}U \rightarrow X + Y + \approx 2.5n + energy \qquad (1)$$

Here, X and Y are two major atomic fragments or isotopes resulting from the fission process. The process is enhanced if the neutrons are slowed down first until they are in thermal equilibrium with their surroundings. X and Y themselves will typically decay after the original fission event, a few by emitting additional neutrons as well as betas and gammas. As noted earlier, about 200 MeV of energy are released in the process, much of it as kinetic energy shared by the particles. Some of these fission fragments are long lived and can be chemically separated to provide radioisotopes of use in the laboratory. These are the major source of the byproduct radioisotopes regulated by the NRC. The fission reaction can, under appropriate circumstances, be self-sustaining in a chain reaction. In some configurations, the chain reaction is extremely rapid, and an atomic bomb is the result. However, by using the neutrons emitted by the fission fragments (called delayed neutrons), the process can be controlled safely in a reactor. Over a period of time, the fission products build up in the uranium fuel and eventually can be recovered when the fuel element is reprocessed.

Other byproduct materials or radioisotopes are made by the following reaction:

$$n +{}^{A}X \rightarrow {}^{(A+1)}Y^* + a \tag{2}$$

The asterisk indicates that the product nucleus, Y, may be unstable and will undergo one (or more) of the modes of decay discussed previously. The "a" indicates that there may be a particle resulting directly from the reaction. In most cases, the source of neutrons for radioisotopes created by this reaction is a nuclear reactor, so these also are "byproduct materials". Plutonium is made in nuclear reactors by the above reaction, where ${}^{238}U$ is the target nucleus. Although there are other reactions, using different combinations of particles in the reaction, in most cases, these require energetic bombarding particles generated in accelerators. Also, since there are no common radioisotopes that generate neutrons, there is essentially no probability that other materials in laboratories will be made radioactive by exposure to radiation from byproduct materials.

Materials which will undergo fission and can be used to sustain a chain reaction are, in the nomenclature of the NRC, "special" nuclear materials. These include the isotopes of uranium with mass numbers 233 and 235 and materials which have been enriched in these isotopes or the artificially made element plutonium. Materials which have 0.05% or more uranium or thorium in them are called source materials.

### 5.1.2.3. Radioactive Decay

An important relationship concerning the actual decay of a given nucleus is that it is purely statistical, dependent only upon the decay constant for a given material, i.e., the activity, A, is directly proportional to the number, N, of unstable atoms present.

$$\text{Activity} = A = dN/dt \equiv C\,N \tag{3}$$

This can be reformulated to give the number of radioactive atoms, N, at time t in terms of the number originally present.

$$N(t) = N0e^{-Ct} \tag{4}$$

where $C = \ln2/\tau$.

Equation 4 shows that during any interval, $\tau$, approximately one half of the unstable nuclei at the beginning of the interval will decay. This is illustrated in Table 5.2.

**TABLE 5.2**
**Typical Decay of A Group of Radioactive Atoms**

| Number | Time ($\tau$) | Number | Time ($\tau$) |
|--------|------|--------|------|
| 1000 | 0 | 14 | 6 |
| 502 | 1 | 7 | 7 |
| 249 | 2 | 4 | 8 |
| 125 | 3 | 1 | 9 |
| 63 | 4 | 1 | 10 |
| 31 | 5 | 0 | 11 |

The data in this table illustrate clearly that when small numbers are involved, the statistical variations cause the decrease to fluctuate around a decay of about one half of the remaining atoms during each successive half-life, but obviously between three and four half-lives in this table, it would have been impossible to go down by precisely half. The table also illustrates a fairly often used rule of thumb that after radioactive waste has been allowed to decay by ten half-lives, the activity has often decayed sufficiently to allow safe disposal. This, of course, depends upon the initial activity.

The daughter nucleus can also decay, as can the second daughter, and so forth. However, eventually, a nucleus will be reached which will be stable. This is, in fact, what occurs, starting with the most massive natural elements, uranium and thorium. All of their isotopes are unstable, and each of their daughters decay until stable isotopes of lead are reached. All of the elements above atomic number 83 owe their existence to the very long half-lives of the most massive members of these chains.

### 5.1.2.4. Units of Activity

Dimensionally, the units of activity are the number of decays or nuclear disintegrations per unit time. Until recently, the standard unit to measure practical amounts of activity was the curie (Ci), which was defined to be $3.7 \times 10^{10}$ dis/sec (dps). Other units derived from this were the millicurie (mCi), or $3.7 \times 10^7$ dps, the microcurie ($\mu$Ci), or $3.7 \times 10^4$ dps, the nanocurie (nCi), or 37 dps, and the picocurie (pCi), or 0.037 dps. The curie was originally supposed to equal the amount of activity of 1 g of radium. This unit, and the derivative units, are still the ones most widely used in this country. However, an international system of units or SI system has been established, and in this system, 1 dis/sec is defined as a becquerel (Bq). Larger units, which are multiples of $10^3$, $10^6$, $10^9$, and $10^{12}$, are indicated by the prefixes kilo, mega, giga, and tera, respectively. In most laboratories which use radioisotopes as tracers, the quantities used are typically of the order of $10^4$ to $10^8$ dps. There are other uses of radioisotopes (e.g., therapeutic) which use substantially larger amounts.

### 5.1.2.5. Interaction of Radiation with Matter
### 1. Alphas

As an alpha particle passes through matter, its electric field interacts primarily with the electrons surrounding the atoms. Because it is a relatively massive particle, it moves comparatively slowly and spends a significant amount of time passing each atom. Hence, the alpha particle has a good opportunity to transfer energy to the electrons by either removing them from the atom (ionizing them) or raising them to higher energy states. Because it is so much more massive than electrons, it moves in relatively short, straight tracks through matter and causes a substantial amount of ionization per unit distance. It is said to have a high linear energy transfer (LET). A typical alpha particle has a range of only about 0.04 mm in tissue or about 3 cm in air. Since the thickness of the skin is about 0.7 mm, a typical alpha particle will not penetrate the skin. However, if a material which emits alphas is ingested, inhaled, or, in an accident, becomes embedded in an open wound such that it lodges in a sensitive area or organ, the alpha radiation can cause severe local damage. Since many of the heavier radioactive materials emit alpha radiation, they often are more dangerous than materials which emit other types of radiation, especially if they are chemically likely to simulate an element which is retained by the body in a sensitive organ. If they are not near a sensitive area, they may cause local damage to nearby tissue, but this may not cause appreciable damage to the organism as a whole.

### 2. Betas

Since beta particles are electrons, they have a single negative or positive charge and are the same mass as the electrons around the atoms in the material in which they are moving. Normally, they also are considerably less energetic than an alpha particle. They typically move about two orders of magnitude more rapidly than alpha particles. They still interact with matter by ionization and excitation of the electrons in matter, but the rate of interaction per unit distance traveled in matter is much less. Typically, beta radiation ($\approx 1$ MeV) can travel perhaps 0.5 cm in tissue or about 4 m in air, although this is strongly dependent upon the energy of the beta. Low-energy betas, such as from $^{14}$C, would penetrate only about 0.02 cm in tissue or about 16 cm in air. Therefore, only those organs lying close to the surface of the body can be injured by external beta irradiation and then only by the more energetic beta emitters. Radioactive materials emitting betas which are taken into the body can affect tissues further away than those which emit alphas.

There is a secondary source of radiation from beta emitters. As the electrons pass through matter, they cause electromagnetic radiation called "bremsstrahlung" or braking radiation to be emitted as their paths bend while passing through matter. The energy which appears as bremsstrahlung is approximately ZE/3000 (where Z is the atomic charge number of the absorbing medium and E is the $\beta$ energy in MeV). This is not a problem for alpha radiation. Bremsstrahlung radiation can have important implications for certain energetic beta emitters such as $^{32}$P. Protective shielding for energetic beta emitters should be made of plastic or other low-Z material, instead of a high-Z material such as lead. Because of the silicon, even keeping $^{32}$P in a glass container can substantially increase the radiation dose to the hands while handling the material in the container.

### 3. Gammas

Since gamma rays are not charged and do not have any mass, they interact differently with matter than do alpha and beta particles. They can, however, interact with the electrons in matter by three different mechanisms.

**Photoelectric effect** — The gamma ray interacts with electrons around an atom which normally are completely removed from the atom, i.e., the atom is ionized. In

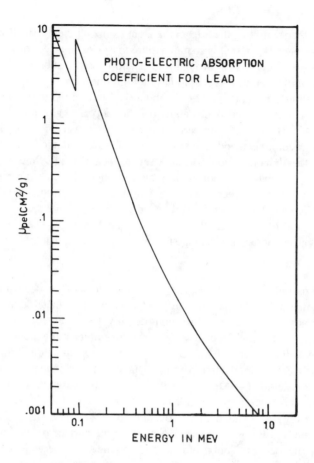

FIGURE 5.2.   Photoelectric absorption coefficient for lead as a
function of energy.

this case, all the energy of the gamma ray is transferred to the electron and the gamma
ray no longer exists. The photoelectric effect mechanism is dependent upon the
gamma energy, as shown in Figure 5.2. For low-energy photons, the photoelectric
effect depends upon the charge of the absorber, approximately as $Z^4$. As energies
increase, the importance of the atomic number of the absorber decreases.

**Compton effect** — The gamma ray can also scatter from an electron, transferring
part of its energy to the electron, and thus become scattered as a lower-energy
gamma. There is an upper limit to the amount of energy which can be transferred to
the electron by this mechanism, so in every scattering event, a lower energy gamma
ray remains after the interaction. Dependent upon the energy transferred, the residual
gamma can be scattered in any direction, relative to the original direction, up to 180°.
This has important implications on shielding since gammas can be scattered into
areas that are shielded from a direct beam by the shielding itself or by other nearby
materials. Equation 5 gives the energy of the scattered gamma as a function of the
angle of scattering. The interaction with matter is considerably less strongly dependent upon the energy of the gamma. This is also shown in Figure 5.3. Compton

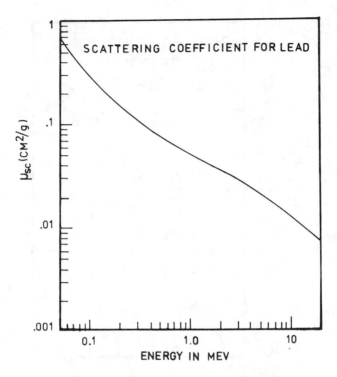

FIGURE 5.3.   Scattering coefficient for lead as a function of energy.

scattering is the primary mechanism of interaction for low atomic number elements and decreases as the atomic number increases.

$$E_s = \frac{E}{1 + (E/m_e c^2)(1 - \cos \theta)} \tag{5}$$

**Pair production** — If the energy of the gamma is greater than the energy needed to create an electron-positron pair, 1.02 MeV, then the gamma can interact with the absorbing medium to create a pair of electrons, an electron and a positron. The probability of this process increases as the energy increases. The energy in excess of 1.02 MeV is shared by the two particles. The energy dependence of this process is also shown in Figure 5.4. The probability of the reaction increases with the atomic number of the absorber, approximately proportionally to $Z^2 + Z$.

Gamma rays can penetrate deeply into matter, in theory, infinitely since, unless the gamma interacts with an electron, it will go on unimpeded, just as, in theory, a rifle bullet fired into a forest can continue indefinitely unless it hits a tree (assuming no loss of energy for the bullet due to air friction). The intensity, I, of the original radiation at a depth, x, in an absorbing medium, compared to the intensity of the radiation at the surface, $I_o$, is

$$I = I_o e^{-\mu x} \tag{6}$$

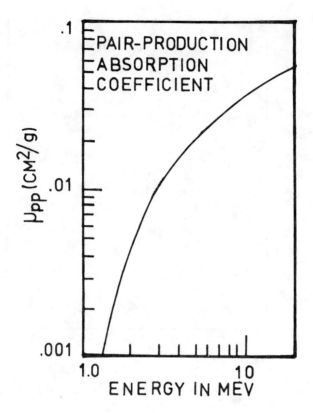

FIGURE 5.4.    Pair-production coefficient for lead as a function of
energy.

This equation is literally true only if gammas of the original energy are considered. If Compton scattering and the pair-production process are included, the decrease in the total number of gammas is less than that given by Equation 6 because of the scattered gammas from the Compton process and the contribution of the annihilation gammas as the positron eventually is destroyed by interacting with a normal electron.

If the total effect of all three mechanisms is considered, at low and high energies, higher Z absorbers interact with gammas more strongly. However, between about 1 and 3 MeV, there is little dependence in the total absorption coefficient on atomic number.

As can be noted, all of the mechanisms by which a gamma interacts with matter (except the small number of instances in which the gamma ray interacts with a nucleus) result in the energy being transferred to an electron, so a gamma is considered to have the same low LET characteristics as betas (at very low energies, the LET characteristics of electrons increase somewhat). However, unless a beta emitter is taken into the body, most internal organs will not be affected by beta radiation, while gammas can penetrate deeply into the body and injure very sensitive organs such as the blood-forming tissues.

### D. Neutrons

As noted earlier, neutron radiation is rarely encountered in most laboratories which use radioisotopes in research programs. However, it is useful to understand the difference in the mechanisms by which a neutron interacts with matter, compared to those involving other types of radiation, since neutron radiation *may* make the matter it interacts with radioactive. The neutron has no charge, but it does have about one fourth of the mass of an alpha particle, so it does have an appreciable mass compared to the atoms with which it interacts.

Equation 7, similar to Equation 6, gives the number of neutrons, N, with an initial energy, E, of an original number, $N_o$, penetrating to a depth, x, in matter. Note that in both Equations 6 and 7, the units of x are usually converted into $mg/cm^2$ for the commonly tabulated values of $\mu$ and $\sigma$.

$$N = N_o e^{-\sigma x} \tag{7}$$

A neutron does not, as do electrons, alpha particles, and gammas, interact with the orbital electrons, but, instead, interacts directly with the nucleus. The neutron is not repelled by the positive charge on the nucleus, as is the alpha, because it has no charge. It is either scattered (elastically or inelastically) or captured by the nucleus. In a typical capture process, several "capture" gammas with a total energy of about 8 MeV are emitted (it is somewhat less for some lighter nuclei). Thus, by this mechanism alone, the neutron could be considered more harmful than other radiations. Further, the nucleus in which it is captured may have been made radioactive and the charge on the nucleus could change so that the atom would no longer be chemically equivalent to its original form. In any event, the energy transferred to the participants in the interaction normally would be sufficient to break the chemical bonds.

Scattering events also typically transfer enough energy to break the chemical bonds, as long as the initial energy of the neutron is sufficiently high. As with Compton-scattered gammas, the scattered neutrons can be scattered in virtually any direction, so the equivalent of Equation 7 for neutrons of all energies would, as for gammas, have to be modified to include a buildup factor.

No figure showing the systematics of the reaction mechanisms will be given here because the relationships are extremely complex, varying widely not only between elements, but also between isotopes of the same element. In addition, the interaction probabilities can vary extremely rapidly as a function of energy, becoming very high at certain "resonant" energies and far less only a few electronvolts above or below the resonances. However, a few generalizations are possible. The probability of the capture process, excluding resonance effects, typically increases as the energy of the neutrons becomes lower, and for specific isotopes of certain elements, such as cadmium, gadolinium, samarium, and xenon, is extremely high at energies equivalent to thermal equilibrium (about 0.025 eV for room-temperature matter). Energy can be lost rapidly by neutrons in scattering with low-Z materials, such as hydrogen, deuterium ($^{21}H$), helium, and carbon. Interposing a layer of water, paraffin, or graphite only a few inches thick, backed up by a thin layer ($\approx 1/_{32}$ in.) of cadmium, in a beam of fast neutrons makes an effective shield for a beam of neutrons. Paraffin wax, in which boric aid ($^{10}B$ has quite a respectable capture cross section at thermal neutron energies) has been mixed also makes an effective and cheap neutron shield.

Overall, the danger of neutrons interacting with matter is estimated to be about ten times that of a gamma or electron, although this varies with the energy of the neutrons. Thermal neutrons are about 2 times as effective in causing atoms in tissue to be ionized, for example, as are betas and gammas, while neutrons of 1 to 2 MeV neutrons are about 11 times more damaging.

### 5.1.2.6. Units of Exposure and Dose

There are two important concepts in measuring the relative impact of radiation on matter: the intensity of the radiation field, which represents a potential exposure problem, and the actual energy deposited in matter, or the dose. Further, as far as human safety is concerned, the amount of energy absorbed in human tissue is more important than that absorbed in other types of matter. Each of these quantities has been assigned specific units in which they are measured.

The original unit of measuring radiation intensity was the roentgen, defined as the amount of X-ray radiation which would cause an ionization of $2.58 \times 10^{-4}$ C/kg of dry air at standard temperature and pressure. As noted, the dose or energy deposited in matter is more important, so another unit was defined, the rad, which was defined as the deposition of 0.01 J/kg gram of matter. An exposure to one roentgen would result in an absorbed dose of 0.87 rads in air. A third unit, the rem, was subsequently defined which measured the equivalent dose, taking into account the relative effectiveness of the various types of radiation in causing biological damage. This originally was taken into account by multiplying the absorbed dose in rads by a relative biological effectiveness (RBE) factor to obtain a dose equivalent for tissue for the different varieties of radiation. Later, it was decided to restrict the term RBE to research applications and an equivalent multiplier, called the quality factor, Q, was substituted.

The terms rad and rem are still used by most American health physicists, and the current regulations use these terms, as they do the curie and its derivative units. However, there is an international system (SI) of units for dose as well as for activity. These equivalent units are given below.

> 1 Gray (Gy) $\equiv$ 1 J/kg = 100 rad = absorbed dose
> 1 Sievert (Sv) $\equiv$ 1 Gy $\times$ Q $\times$ N = dose equivalent (N is a possible modifying
>    factor, assigned a value of 1 at this time)
> 1 Sv = 100 rem = dose equivalent

The quality factors for the various types of radiation are shown in Table 5.3.

This concludes this very brief discussion of some of the basic terms and concepts in radiation physics which will be employed in the next few sections. Many important points and significant features have been omitted which will be of importance primarily to professional health physicists, but which are of less importance to those individuals who use radiation as a research tool to serve their more direct interests.

### 5.1.3. LICENSING

This section will be restricted to a discussion of licensing of radioisotopes or byproduct materials, rather than other types of applications such as a research reactor. It has been some time since any new application for construction of a nuclear power plant in the U.S. has been submitted, and the number of operating nongovernmental

**TABLE 5.3**
**Quality Factors**

| Type of radiation | Q |
|---|---|
| Betas, gammas, X-rays | 1 |
| Alphas | 20 |
| Thermal neutrons | 2 |
| Fast neutrons ($\approx 1$ MeV) | 20 |
| Neutrons (unspecified energy) | 10 |

research reactors has been diminishing. A number of these research units are either in the process of terminating their license or going into an inactive status. At least some of the research reactors have closed rather than renew their license, as they must do periodically, because of excessive costs needed to meet the concerns of the public. The other major type of facility involved with radiation, laboratories using X-ray units, are usually regulated by state agencies, although the federal Food and Drug Administration sets standards for the construction of the machines and their applications. X-ray facilities will be discussed in a separate section.

Radioactive materials fall into two classes as far as regulation is concerned. Radioactive materials "yielded in or made radioactive by exposure to the radiation incident to the process of producing or utilizing special nuclear material" are regulated by the Nuclear Regulatory Commission (NRC) or by equivalent regulations in states with whom the NRC has entered into an agreement for the states to act as the regulatory agency within their borders. Radioactive materials which are naturally radioactive or produced by means such as a cyclotron are regulated by the states in most cases.

The licensing of byproduct material is regulated under Title 10, CFR, Part 30 or 33 (10, CFR, 30 or 33). Licenses are issued to "persons", which may be an individual, but may also mean organizations and groups of persons, associations, etc. It is possible for individuals with an organization to have separate licenses, although it is more likely that instead of several individuals having separate licenses, an institution will apply for and be granted a license covering the entire organization if it can show that the individual users will conform to the terms of the license and regulations governing the use of radioactive materials.

For most types of licenses of interest to research laboratories, the NRC has delegated licensing authority to five regional offices in Pennsylvania, Georgia, Illinois, Texas, and California. The current addresses of these regional offices can be obtained by writing to:

Director
Office of Nuclear Material Safety and Safeguards
U. S. Nuclear Regulatory Commission
Washington, D.C. 20555

Not all uses of radioisotopes require securing a license. There are a number of commercial products, such as watch dials, other self-luminous applications, and some types of smoke detectors which contain very small quantities of radioactive materials, which the owner obviously does not require a license to own. However, the most significant class of exemptions are the "exempt quantities" listed in Schedule B, 10 CFR, §30.71. Schedule A lists exempt concentrations in gas, liquids, and solids. In Table 5.4, the units are in microcuries. To convert to Becquerals, multiply the number given in microcuries by 37,000.

Any byproduct material which is not listed in Table 5.4, other than alpha-emitting byproduct material, has an exempt quantity of 0.1 µCi or 3,700 Bq.

Most users of radioisotopes would find it necessary to use more than the exempt quantities in Table 5.4 and should apply for a license. This is done through NRC form 113, which can be obtained from the NRC office in the local region. If the activity planned has the potential for affecting the quality of the environment, the NRC will weigh the benefits against the potential environmental effects in deciding whether to issue the license. For most research-related uses of radioisotopes, environmental considerations will not usually apply, although where isotopes will be used in the field, outside of a typical laboratory, the conditions and restrictions on their use to ensure that there will be no meaningful release into the environment will need to be fully included in the application.

There are three basic conditions which the NRC expects the applicant to meet in their application. In this context, "applicant" is used in the same sense as "person," which can be an individual or an organization, as noted earlier.

1.    The purpose of the application is for a use which is authorized by the act. Legitimate basic and applied research programs in the physical and life sciences, medicine, and engineering are acceptable programs.
2.    The applicant's proposed equipment and facilities are satisfactory in terms of protecting the health of the employees and the general public and being able to minimize the risk of danger to persons and property. The laboratories in which the radioisotopes are to be used need to be in good repair and contain equipment suitable for use with radioisotopes. Depending upon the level of radioactivity to be used and the scale of the work program, this may mandate the availability of hoods designed for radioisotope use. It could require specific areas designated and restricted for isotope use only or the level of use and the amounts of activity may make it feasible to perform the research on an open bench in a laboratory. In any event, it must be shown in the application that the level of facilities and equipment is adequate for the proposed uses of radiation.
3.    The applicant must be suitably trained and experienced so as to be qualified to use the material for the purpose requested in a manner which will protect the health of individuals and minimize danger to life and property. The experience and training must be documented in the application.

Prior to granting the license, the NRC may require additional information or may require the application to be amended. The license is issued to a specific licensee and cannot be transferred without specific written approval of the NRC. The radioisotopes identified in the license can be used only for the purposes authorized under the license and at the locations specified in the license. If the licensee wishes to change the isotopes permitted to be used, to modify significantly the program in which they

## TABLE 5.4
## Exempt Quantities

| Byproduct material | μCi | Byproduct material | μCi | Byproduct material | μCi |
|---|---|---|---|---|---|
| Antimony 122 | 100 | Holmium 166 | 100 | Potassium 42 | 10 |
| Antimony 124 | 10 | Hydrogen 3 | 1000 | Praseodymium 142 | 100 |
| Antimony 125 | 10 | Indium 113m | 100 | Praseodymium 143 | 100 |
| Arsenic 73 | 100 | Indium 114m | 10 | Promethium 147 | 10 |
| Arsenic 74 | 10 | Indium 115m | 100 | Promethium 149 | 10 |
| Arsenic 76 | 10 | Indium 115 | 10 | Rhenium 186 | 100 |
| Arsenic 77 | 100 | Iodine 125 | 1 | Rhenium 188 | 100 |
| Barium 131 | 10 | Iodine 126 | 1 | Rhodium 103m | 100 |
| Barium 133 | 10 | Iodine 129 | 0.1 | Rhodium 105 | 100 |
| Barium 140 | 10 | Iodine 131 | 1 | Rubidium 86 | 10 |
| Bismuth 210 | 1 | Iodine 132 | 10 | Rubidium 87 | 10 |
| Bromine 82 | 10 | Iodine 133 | 1 | Ruthenium 97 | 100 |
| Cadmium 109 | 10 | Iodine 134 | 10 | Ruthenium 103 | 10 |
| Cadmium 115m | 10 | Iodine 135 | 10 | Ruthenium 105 | 10 |
| Cadmium 115 | 100 | Iridium 192 | 10 | Ruthenium 106 | 1 |
| Calcium 45 | 10 | Iridium 194 | 100 | Samarium 151 | 10 |
| Calcium 47 | 10 | Iron 55 | 100 | Samarium 153 | 100 |
| Carbon 14 | 100 | Iron 59 | 10 | Scandium 46 | 10 |
| Cerium 141 | 100 | Krypton 85 | 100 | Scandium 47 | 100 |
| Cerium 143 | 100 | Krypton 87 | 10 | Scandium 48 | 10 |
| Cerium 144 | 1 | Lanthanum 140 | 10 | Selenium 75 | 10 |
| Cesium 131 | 1000 | Lutecium 177 | 100 | Silicon 31 | 100 |
| Cesium 134m | 100 | Manganese 52 | 10 | Silver 105 | 10 |
| Cesium 134 | 1 | Manganese 54 | 10 | Silver 110m | 1 |
| Cesium 135 | 10 | Manganese 56 | 10 | Silver 111 | 100 |
| Cesium 136 | 10 | Mercury 197m | 100 | Sodium 24 | 10 |
| Cesium 137 | 10 | Mercury 197 | 100 | Strontium 85 | 10 |
| Chlorine 36 | 10 | Mercury 203 | 10 | Strontium 89 | 1 |
| Chlorine 38 | 10 | Molybdenum 99 | 100 | Strontium 90 | 0.1 |
| Chromium 51 | 1000 | Neodynium 149 | 100 | Strontium 91 | 10 |
| Cobalt 58m | 10 | Neodymium 149 | 100 | Strontium 92 | 10 |
| Cobalt 58 | 10 | Nickel 59 | 100 | Sulfur 35 | 100 |
| Cobalt 60 | 1 | Nickel 63 | 10 | Tantalum 182 | 10 |
| Copper 64 | 100 | Nickel 65 | 100 | Technetium 96 | 10 |
| Dysprosium 165 | 10 | Niobium 93m | 10 | Technetium 97m | 100 |
| Dysprosium 166 | 100 | Niobium 95 | 10 | Technetium 97 | 100 |
| Erbium 169 | 100 | Niobium 97 | 10 | Technetium 99m | 100 |
| Erbium 171 | 100 | Osmium 185 | 10 | Technetium 99 | 10 |
| Europium 152 (9.2 h) | 100 | Osmium 191m | 100 | Tellurium 125m | 10 |
| Europium 152 (13 years) | 1 | Osmium 191 | 100 | Tellurium 127m | 10 |
| Europium 154 | 1 | Osmium 193 | 100 | Tellurium 127 | 100 |
| Europium 155 | 10 | Palladium 103 | 100 | Tellurium 129m | 10 |
| Fluorine 18 | 1000 | Palladium 109 | 100 | Tellurium 129 | 100 |
| Gadolinium 153 | 10 | Phosphorous 32 | 10 | Tellurium 131m | 10 |
| Gadolinium 159 | 100 | Platinum 191 | 100 | Tellurium 132 | 10 |
| Gallium 72 | 10 | Platinum 193m | 100 | Terbium 160 | 10 |
| Germanium 71 | 100 | Platinum 193 | 100 | Thallium 200 | 100 |
| Gold 198 | 100 | Platinum 197m | 100 | Thallium 201 | 100 |
| Gold 199 | 100 | Platinum 197 | 100 | Thallium 202 | 100 |
| Hafnium 181 | 10 | Polonium 210 | 0.1 | Thallium 204 | 10 |

## TABLE 5.4 (continued)
## Exempt Quantities

| Byproduct material | μCi | Byproduct material | μCi | Byproduct material | μCi |
|---|---|---|---|---|---|
| Thulium 170 | 10 | Xenon 131m | 1000 | Yttrium 93 | 100 |
| Thulium 171 | 10 | Xenon 133 | 100 | Zinc 65 | 10 |
| Tin 113 | 10 | Xenon 135 | 100 | Zinc 69m | 100 |
| Tin 125 | 10 | Ytterbium 175 | 100 | Zinc 69 | 1000 |
| Tungsten 181 | 10 | Yttrium 90 | 10 | Zirconium 93 | 10 |
| Tungsten 185 | 10 | Yttrium 91 | 10 | Zirconium 95 | 10 |
| Tungsten 187 | 100 | Yttrium 92 | 100 | Zirconium 97 | 10 |
| Vanadium 48 | 10 | | | | |

are used, or to change the locations where they are to be used, the license will have to be amended. This typically takes a substantial length of time, 1 to 3 months or even more not being unusual. For this reason, most substantial users of radioisotopes often apply for a "broad" license under 10 CFR 33.

Under the terms of a broad license, the application usually covers a request to use radioisotopes with atomic numbers from 3 to 83, with individual limits on the quantities of specific isotopes held and an overall limit of the total quantity of all isotopes held at one time. In addition, there should be specific identification of sealed sources held separately by the applicant on the license. The purposes still must be spelled out and the locations still must be identified. However, under the terms of a broad license, an internal procedure can be established to review individual uses, personnel, and spaces in which the materials can be used and to monitor the operations. This will necessitate the formation of a radiation safety committee and the naming of a radiation safety officer. This infrastructure must be described in the license application and the credentials of the individual members of the committee and the safety officer included as part of the application. In order to obtain a license of broad scope, applicants must prove that they have the personnel and administrative structure to make sure that the radioisotopes will be used safely and in compliance with regulations.

The license will be granted for a specific period and the ending date will be written into the license. If the licensee wishes to renew the license as the end of the license period approaches, the applicant must submit a renewal request at least 30 d prior to the expiration date of the license. If this deadline is met, the original license will remain in force until such time as the NRC acts on the request.

While the license is in effect, the NRC has the right to make inspections of the facility, the byproduct material, and the areas where the byproduct material is in use or stored. These inspections have to be at reasonable hours, but they normally are unannounced. The inspector also will normally ask to see records of such items as surveys, personnel exposure records, transfers and receipts of radioactive materials, waste disposal records, instrument calibrations, and any other records relevant to compliance with the terms of the license and with other parts of 10 CFR, such as 19 and 20. Failure to be in compliance can result in citations of various levels or financial penalties. Enforcement will be discussed further later. The NRC can require tests to be performed to show that the facility is being operated properly, such as asking for tests of the instruments used in monitoring the radiation levels.

Under §30.51, records of all transfers, receipts, and disposal of radioactive materials normally must be kept for at least 2 years after transfer or disposal of a radioactive material, or in some cases, until the NRC authorizes the termination of the need to keep the records. There are other recordkeeping requirements in other parts of Title 10.

If, for any reason, there is a desire to terminate the license on or prior to the expiration date, under 10 CFR 30.36 there are procedures which must be followed.

1. Terminate the use of byproduct material.
2. Remove radioactive contamination to the extent practicable.
3. Properly dispose of byproduct material.
4. Submit a completed form NRC-314.
5. Submit a radiation survey documenting the absence of radioactive contamination or the levels of residual contamination. In the latter case, an effort will be required to eliminate the contamination.
   A. The survey instruments used for the survey must be specified and certified to be properly calibrated and tested.
   B. The radiation levels in the survey must be reported as follows:*
      (1) Beta and gamma levels in rads per hour at 1 cm from the surface and gamma levels at 1 m from the surface
      (2) Levels of activity in microcuries per 100 cm$^2$, of fixed and removable surface contamination
      (3) Microcuries per milliliter in any water
      (4) Picocuries per gram in contaminated solids and soils
6. If the facility is found to be uncontaminated, the licensee shall certify that no detectable radioactive contamination has been found. If the information provided is found to be sufficient, the NRC will notify the licensee that the license is terminated.
7. If the facility is found to be contaminated, the license will continue after the normal termination date. However, the use of byproduct materials will be restricted to the decontamination program and related activities. The licensee must submit a decontamination plan for the facility and must continue to control entry into restricted areas until they are suitable for unrestricted use and the licensee is notified in writing that the license is terminated.

In principle, the NRC has the right to modify, suspend, or revoke a license for a facility which is being operated improperly or if the facility were to submit false information to the NRC. If the failure to comply with the requirements of the license and other requirements for safely operating a facility can be shown to be willful or if the public interest, health, or safety can be shown to demand it, the modification, suspension, or revocation can be done without institution of proceedings allowing the licensee an opportunity to demonstrate or achieve compliance.

Normally, an inspection will be followed up with a written report by the inspector

---

* In SI units, a microcurie is equal to 37,000 Bq and a microrad is equal to $10^{-8}$ Gy.

in which any compliance problems will be identified. These may be minimal, serious (which would require immediate abatement) or between these two extremes. The facility can (1) appeal the findings and attempt to show that it was, in fact, complying with the regulations or that the violation was less serious than the citation described or (2) accept the findings. Unless the facility can show compliance, it must show how it will bring itself into compliance within a reasonable period.

In recent years, there have been increasing numbers of occasions when the NRC has imposed substantial financial penalties on research facilities, including academic institutions, as it is entitled to do under §30.63 for violations which are sufficiently severe. Further, one city has filed 179 criminal charges against a major university and several of its faculty members for failure to comply with radiation safety standards. Many of the individual violations were relatively minor, but apparently the city attorney thought he had a substantial case of a pattern of failure to comply with the terms of the license and the regulations.

The use of radioisotopes in research is continuing to increase, while the public concern about the safety of radiation continues unabated. It behooves all licensees to follow all regulations scrupulously to avoid aggravating the concerns of the public unnecessarily.

### 5.1.3.1. Radiation Safety Committees

The primary function of the radiation safety committee (RSC), which is required under 10 CFR 33, is to monitor the performance of the users of ionizing radiation in a facility. It is, in effect, a local surrogate of the NRC or the equivalent state agency in an agreement state. Usually, it is the ultimate local authority in radiation matters. In this one area at least, it is assigned more responsibility than the usual senior administrative officials. It is an operational committee charged with an important managerial role in the use of ionizing radiation within the organization.

In addition to the responsibility of the RSC to ensure compliance with the provisions of the byproduct license and the other regulatory requirements of Title 10 CFR, it also must establish internal policies and procedures to guide those wishing to use radiation and to provide the internal operational structure in which this is done. The committee has other duties as well, which will be discussed after the makeup of the committee is considered.

The members of an RSC should be carefully selected. It would be highly desirable to select much of the membership from among the active users of radiation within the organization and across the major areas or disciplines represented by the users. Each prospective member should be scrutinized very carefully. An RSC has to enforce regulations set by one of the stronger regulatory agencies, and it must be willing to accept the delegated authority. Individuals on the committee must be willing, if necessary, to establish policies which many users may feel are too restrictive. As active users themselves, they have a better chance of achieving compliance if the other users realize that the members of the RSC have accepted imposition of these same policies on their own activities. The members of the committee should have a reputation for objectivity, fairness, and professional credibility. A prima donna has no place on such a committee.

As professional scientists in their own right, they also will understand the impact of a given procedure or policy on laboratory operations and can often find legitimate ways to develop effective policies and procedures which are less burdensome on the users to implement than would otherwise be the case.

The radiation safety officer (RSO) of the organization must be part of the RSC, as a person who must maintain a current awareness of the rules and regulations required by the NRC and of radiation safety principles. This individual will serve to carry out the policies of the committee and also should be the individual to perform the direct day-to-day monitoring of the operations of the laboratories using radioactive material. The decisions of the committee cannot only be burdensome to the users, but, without the input of the RSO, can be equally burdensome for this individual to implement.

The working relationship between the RSO and the RSC is extremely important. No committee can effectively administer a program of any size on a daily basis. It must delegate some of its authority to a person, such as the RSO, or to an alternate agency, such as a safety and health department, which is charged with the daily administration of the area of responsibility assigned to an operational committee. However, especially where the RSO is a dynamic, effective person, there is a tendency to defer to this person and to abrogate some of the committee's oversight responsibility. Both the RSC and the RSO should guard against this possibility. The RSO should have a voice, and an influential one, in the committee's deliberations, but should not be allowed to dictate policies independently.

The membership need not be limited to the persons already defined. The head of the health and safety department, if different from the radiation safety office, might well be a member because this individual would bring a wider perspective than would the RSO alone to the implications of some of the issues brought before the committee. Some large organizations may wish to have a representative of the organization's legal department as a member. Some may wish to have a representative of the public relations area as a member, especially if the facility is in an area where there has been vigorous public opposition to the use of ionizing radiation. Some may wish to include a lay person, if not as a voting member, perhaps as an observer. The membership should not become too large, however, so that it will be practical to set up meetings without too much concern for having a quorum. Committees which are too large also tend to be less efficient because of the time required for all the members to participate in discussions. On the other hand, each major scientific discipline using radiation should be represented. A reasonable size might be between 9 to 15 members, with a quorum established at between 5 to 8 members.

It would be desirable for the chairman of the committee to have had prior experience with radiation, but it is also desirable for the chairman to have administrative credentials. Such a person will normally be able to ensure that committee meetings will be conducted efficiently, but if the administrative experience is at a level carrying budgetary and personnel responsibilities, the chairman will bring still another dimension to the committee. Some actions of the committee may carry cost or manpower implications which an individual with managerial experience will recognize and perhaps have a feel for the feasibility of accommodating these requirements.

In addition to monitoring existing programs, establishing policies, and providing guidance to radiation safety personnel, there are at least three other important functions that the committee must perform. The first of these is to perform the same function as the NRC in authorizing new participants to use radiation or radioactive materials. Basically, the same information that the NRC requires of new applicants for a license should be required when a new facility is involved. The adequacy of the facility, the purpose of the program for which the use of radiation is involved, and the qualifications of the users should all be reviewed. At academic institutions especially, there is a considerable turnover in users, represented by graduate students, postdoctoral research associates, and even faculty. Often, individuals come from other facilities where internal practices may differ from local practices. In order to ensure that all users are familiar with not only the basic principles of radiation safety, but also with local internal procedures, a simple written test, administered as part of the authorization procedure, is an effective and efficient means of documenting that the prospective users have familiarized themselves with the information. In order to avoid setting standards on who should take the test, it should be administered universally. Some faculty may object, but it serves an important legal point. A passed quiz demonstrates unequivocally that the individual is familiar with the risks and requirements associated with the use of radiation at the facility.

An internal authorization should be issued to an individual. Others may be added to the authorization, but one person should be designated as ultimately responsible for compliance with applicable safety and legal standards related to the use of radiation under the authorization.

The second additional function is to carefully review research or "new experiments" substantially different in the application of radioactive materials or radiation envisioned from work previously performed under the license. This role is relatively easy to perform when it is part of a new request for an authorization, but when an ongoing operation initiates a new direction in its program, it will be necessary for the committee to make it clear that the user must address the question, "is this application covered under the scope of work previously reviewed by the committee in my application?" If the answer is "no", or "possibly not", then the responsible individual should ask for a review by the committee. The need to do this must be explicitly included in the internal policies administered by the Committee. The RSC then must consider the proposed program in the same context as the institution's application to the NRC. Is the purpose of the work an approved purpose? This question must be answered in the affirmative in the context of the NRC facility license. Are the facilities adequate to allow the work to be done safely? Are the persons qualified by reason of training or experience to carry out the proposed research program safely? Incidentally, it is NOT within the purview of the committee's responsibility to judge the validity or worth of the research program, but only if the proposed research can be performed safely according to radiation safety and health standards. Every experiment need not be reviewed, only those which are sufficiently different in an important aspect of radiation usage. A completely new research program may still use tried-and-proven radiological procedures.

The third function not previously discussed is the role of the RSC as a disciplinary body. Occasions will arise when individual users will be found to not be in full compliance with acceptable standards. Often, this will be done by the RSO in his periodic inspections, but many will be reported by the users themselves. The NRC will expect these situations to be evaluated and appropriate actions taken, which can include disciplinary measures. Not all violations are equally serious. Categories of violations should be established by the RSC to guide the RSO and the users. Faulty record keeping is not as serious as poor control over byproduct material usage. Allowing material to be lost or radioactive material to escape into the environment is a serious violation. If the loss or discharge is due to an unforeseen accident, it is less serious than if the cause is negligence. However, a continuing pattern of minor violations may be indicative of carelessness or lack of concern about compliance with the standards, which could eventually lead to a more serious incident. Possible penalties should be spelled out in the internal policies and guidelines issued by the RSC.

Every case where noncompliance is discovered should be carefully investigated by the RSO and a report made to the RSC. The person responsible for the noncompliance and the responsible individual from the facility (if not the same person) should be invited to meet with the committee and present their sides of the issue. After hearing both sides, the committee should take the appropriate measures. Issues are rarely black or white, and the penalties or corrective actions should be adapted to the circumstances. An initial minor violation, for example, might elicit no more than a cautionary letter from the committee. However, a series of minor violations within a relatively short interval probably should result in a mandatory cessation of usage of radioisotopes until such time as the user can demonstrate a willingness and capability to comply with acceptable practice. A proven serious accidental violation should result in an immediate cessation of operations until procedures can be adopted which will prevent future reccurrence of the problem. A serious violation due to willful noncompliance should result in a mandatory cessation of the use of radioisotopes for a substantial length of time or even permanently. The elimination of the right to use radioisotopes is a very serious penalty since the user's research program may depend upon this capability. In an academic institution, even a relatively short hiatus in a research program could result in the loss of a research grant or failure to get tenure. As a result, a permanent or extended loss of the right to use radioactive material should not be imposed lightly, but if the user shows, by action and attitude, that future violations are likely to occur, the committee may have no practical alternative except to do so in order to protect the rest of the organization's users from the loss of the institution's license or the imposition of a substantial fine by the NRC. They must be willing to accept the responsibility, unpleasant though it may be. If the users believe that the RSC is willing to be firm as well as fair, they will be more likely to comply with the required procedures.

The role of the RSO could be construed as the enforcement arm of the RSC and the radiation safety office (which may be part of a larger organization). If this were the case, the RSO would be the equivalent of an NRC inspector. However, as with many other persons working in safety and health programs who have enforcement duties, their primary function is service to the users. In later sections, the other duties will make this clear.

### 5.1.4. RADIATION PROTECTION, DISCUSSION, AND DEFINITIONS

Many of the basic terms have been defined in some detail in prior subsections of Section 5.1. However, several additional concepts will be introduced in the next subsections, and some additional terms need to be defined.

The original definitions of dose units primarily were employed by users of radioactive materials and other applications of ionizing radiation for external exposure. Concerns relating to internal exposure were generally covered by establishing maximum permissible concentrations of radionuclides in air and water, in terms of the workplace and the general public, in Appendix B to 10 CFR 20, Tables 1 and 2 respectively. Protection was provided by considering the amount of radiation given to the most critical organ by the intake of specific radioisotopes and their physical or chemical form. The exposure limits to the whole body were established by the organs which had been assigned the lowest dose limits, the bone marrow, gonads, and lens of the eye. The reason for establishing these organs as the most sensitive to radiation were concerns about leukemia, hereditary effects, and cataracts, respectively.

The amounts of a radionuclide in organs, or the "burden", were calculated based on a constant exposure rate, maintained for a period of time long enough for an equilibrium to be established between the intake of the material and the effective elimination rate. The effective elimination rate, or effective half-life, is a combination of the radioactive half-life and the biological half-life based on the rate at which the material would be eliminated from the body. The relationship is given by the following equation:

$$\frac{1}{T_{eff}} = \frac{1}{T_r} + \frac{1}{T_b} \tag{8}$$

where $T_{eff}$ = effective half-life, $T_r$ = radiological half-life, and $T_b$ = biological half-life.

The maximum permissible concentrations (MPCs) are those which correspond to an organ burden which would cause the annual dose limits to be attained. Control measures, therefore, are designed to maintain the concentrations below the MPCs.

The current 10 CFR 20 which sets standards for protection for users of radioisotopes does not require combining internal and external exposures. However, the NRC has proposed a revision of 10 CFR 20 which does combine the two classes of exposures. The proposed revision contains a number of other changes in Part 20 which are based on many of the recommendations contained in International Commission on Radiological Protection (ICRP) publications 26, 30, and 32. In anticipation of the enforcement of the proposed revision, the remainder of this section and some of the succeeding sections will discuss both the current and proposed standards.

### 5.1.4.1. Selected Definitions

In order to compare the current and proposed Part 20, some terms introduced into the latter from ICRP 26 and 30 are needed. The information below is adapted from the proposed Part 20.[15]

The first six major definitions are related to dose terms:

1. Dose Equivalent means the product of absorbed dose, quality factor, and all other necessary modifying factors at the location of interest in tissue.
2. External dose refers to that portion of the dose received from radiation sources outside the body.
   a.   Deep dose equivalent ($H_d$) applies to the external whole-body exposure and is taken as the dose equivalent at a tissue depth of 1 cm.

TABLE 5.5
Definitions

| Organ or tissue | Weighting factor (WT) | Risk coefficient per rem | Probability per rem |
|---|---|---|---|
| Gonads | 0.25 | $4 \times 10^{-5}$ | 1 in 25,000 |
| Breast | 0.15 | $2.5 \times 10^{-5}$ | 1 in 40,000 |
| Red bone marrow | 0.12 | $2 \times 10^{-5}$ | 1 in 50,000 |
| Lung | 0.12 | $2 \times 10^{-5}$ | 1 in 50,000 |
| Thyroid | 0.03 | $5 \times 10^{-6}$ | 1 in 200,000 |
| Bone surfaces | 0.03 | $5 \times 10^{-6}$ | 1 in 200,000 |
| Any remaining organs or tissues receiving the highest dose at a relative sensitivity of 0.06 each | 0.30 | $5 \times 10^{-5}$ | 1 in 20,000 |
| Total | 1.0 | $1.65 \times 10^{-4}$ | 1 in 6,000 |

    b.    Eye dose equivalent ($H_e$) applies to the external exposure of the lens of the eye and is taken as the dose equivalent at a tissue depth of 0.3 cm.

    c.    Shallow dose equivalent ($H_s$) applies to the external exposure of the skin or an extremity and is taken as the dose equivalent at a tissue depth of 0.007 cm.

3. "Internal dose" is that portion of the dose equivalent received from radioactive material taken into the body.

    a.    Committed dose equivalent ($H_{c,T}$) means the dose equivalent to organs or tissues of reference (T) that will be received from an intake of radioactive material by an individual during the 50-year period following the intake ($H_{50,T}$).

    b.    Effective dose equivalent ($H_E$) is the sum of the products of the dose equivalent ($H_T$) to the organ or tissue (T) and the weighting factors ($W_T$) applicable to each of the body organs or tissues which are irradiated.

$$H_E = \Sigma W_T H_T \tag{9}$$

    c.    Committed effective dose equivalent is the sum of the products of the weighting factors applicable to each of the body organs or tissues which are irradiated and the committed dose equivalent.

    d.    Collective effective dose equivalent is the sum of the products of the individual weighting dose equivalents received by a specified population from exposure to the given source of radiation.

4. $W_T$ is the weighting factor assigned to an organ, which is proportional to the estimate of risk to that organ relative to the estimate of risk per unit dose for a uniform whole-body exposure. In Table 5.5, except for the gonads, for which the risk is based on serious hereditary effects in the first two generations cf the exposed person, the weighting factors are based on the risk of inducing cancer in the organ.

5. Occupational dose means the dose received by an individual in a restricted area or in the course of employment in which the individual's assigned duties involve exposure to radiation and to radioactive material from licensed and

unlicensed sources of radiation, whether in the possession of the licensee or other person. This dose does not include exposure from natural background, medical exposure as a patient or due to participation in a research program, or as a member of the public.

6. Public dose is an exposure of a member of the public to radiation or to the release of radioactive material, or to another source, either in a licensee's controlled area or in unrestricted areas. This does not include medically related exposures.

The next six definitions relate to dose control factors.

7. ALARA is an acronym which stands for "as low as reasonably achievable". It is a policy that involves making every reasonable effort to reduce the exposures of personnel below the dose limits and to find the purpose of the utilization of radiation, technological feasibility, economics, benefits, etc. It does not mean taking extreme measures, where the benefit would not be cost-effective.

8. Annual limit of intake (ALI) means the derived limit for the amount of radioactive material taken into the body of an adult worker by inhalation or ingestion in a year. It is the smaller of (1) the value of the intake of a given radionuclide in a year by a reference man which would result in a committed dose equivalent of 5 rems (0.05 Sv), (2) a committed dose equivalent of 50 rem (0.5 Sv) to an organ or tissue, i.e., the quantity of radioactive material such that:

$$0.05 \text{ Sv} = 5 \text{ rem} = \Sigma H_T W_T \tag{10}$$

or

$$0.5 \text{ Sv} = H_{c,T} \tag{10a}$$

9. Derived air concentration (DAC) means the concentration of a given radionuclide in air which, if breathed by a reference man for a working year of 2000 h under conditions of light activity (corresponding to an inhalation rate of 1.2 $m^3$ air per h), results in an inhalation of one ALI. These are comparable to the MPCs in the current Part 20.

10. Dose limits means the permissible upper bounds of radiation doses. These are usually set for a calendar year. They apply to the dose equivalent received during the set interval, the committed effective dose equivalent resulting from the intake of radioactive material during the interval, or the effective dose equivalent received in a year.

    The external and internal doses must be combined so as not to exceed the permissible limits. Equation 11 can be used to compute the relative amounts of each for the annual intake, $I_j$ of nuclide j:

$$\frac{H_D}{5} + \frac{\Sigma_j I_j}{(ALI)_j} \leq 1 \tag{11}$$

Two terms are used to describe two different classes of effects of radiation.

11. Stochastic effects refers to health effects which occur randomly, so that the probability (generally assumed to be linear, without a threshold) of an effect such as the induction of cancer occurring is a function of the dose rather than the severity of the effect.
12. Nonstochastic effects are health effects for which the severity depends upon the dose and for which there is probably a threshold.

With these definitions in mind, Table 5.6 presents the current radiation standards in 10 CFR 20, as they currently exist, on the left-hand side of the page, and as they will be in the proposed Part 20 (assuming no revisions), on the right. Most of the material is taken directly from Reference 15, but some additional comments are interspersed within the table.

The proposed revision to Part 20 was published in the January 9, 1986 *Federal Register*. It has not been adopted at the time of this writing, and the expressed intention is to phase in the requirements over a period of time after the rule becomes final. Therefore, the need to consider measures to meet the new procedures is not immediate, but should become so in the not-too-distant future. For most research laboratories, the inclusion of the internal contributions to the total dose should not be too significant for laboratory personnel. However, those individuals working in areas where there could be significant airborne radioactive materials would likely receive a significant contribution from the inhalation of this material, and the new standard will mandate documentation of this contribution. This will require more effort on behalf of the licensee and may require or cause to be initiated additional measures to protect the workers.

### 5.1.5. RADIATION WORKING AREAS

Areas in which radiation is used should meet good laboratory standards for design, construction, equipment, and ventilation, as described in Chapter 3. The International Atomic Energy Agency (IAEA) has defined three classes of laboratories suitable for working with radionuclides. Their Class 2 facility is essentially equivalent to a good quality Level 2 facility described in Chapter 3, while their Class 1 facility would be similar to the Level 3 or 4 laboratory depending upon the degree of risk, especially equipped to handle even high levels of radioactive materials safely.

One additional feature which should be included in the design of a laboratory using radioactive materials, is provision for using a HEPA filter or an activated charcoal filter for nonparticulate materials on the exhaust of any fume hood in which substantial levels of radioactive materials are used.* A problem with this requirement is the possibility of the filter rapidly becoming "loaded up" by the chemicals used in the hood. A velocity monitor should be mandatory on any fume hood equipped with a HEPA filter to provide a warning should the face velocity fall below 100 fpm. The definition of "substantial" will depend upon the type of radioactive material used.

---

\* If appropriate filters are not available or are too costly, the maximum levels of activity exhausted by the hood must be restricted to those which will not exceed levels permitted by Part 20 for unrestricted areas, unless the area to which the hood discharges is made a restricted area.

## TABLE 5.6
## Selected Radiation Protection Standards

| **Current Part 20** | **Proposed Part 20** |
|---|---|

### Occupational Limits for Employees — External Limits

| Current Part 20 | Proposed Part 20 |
|---|---|
| Whole body,[a] head, trunk, active blood-forming organs, lens of eye, or gonads 1.25 rems/quarter or 3 rems/quarter with lifetime occupational exposure history and within 5(N-18) dose-averaging formula | Whole body, head, trunk, arm above elbow, and leg above knee 5 rems/year (0.05 Sv/year) — includes summation of (external) deep-dose equivalent and (internal) committed[b] effective dose equivalent 3 rems (0.03 Sv)(external) maximum deep-dose equivalent in any quarter 15 rems/yeatr (0.15 Sv/year) to lens of eye |
| Hand and forearms, feet and ankles | Hand, elbow, arm below elbow, foot, knee, and leg below knee[c] |
| 18.75 rems/quarter (75 rems/year) | 50 rems/year (0.5 Sv/year) |
| Skin of whole body, 7.5 rems/quarter (30 rems/year); no summation of internal (organ) doses | Skin (10 cm²), 50 rems/year (0.5 Sv/year); weighted organ doses for all organs are summed |
| Planned special; 5(N-18)dose averaging provided, with quarterly limits; ALARA program recommended | Planned special exposures are allowed in addition to the annual limits from routine exposures. Limits are set at 1 × annual limits per year from all events in a year and 5 × annual limits per lifetime from all events. The 5(N-18) dose-averaging provision is eliminated. The dose to an embryo/fetus is not addressed; 0.5 rem (0.05 Sv) during the entire pregnancy due to occupational exposure of the "declared" pregnancy No reference level; licensee sets investigation level below annual limit ALARA program required. |

### Occupational Limits for Employees — Internal Limits

| Current Part 20 | Proposed Part 20 |
|---|---|
| Intake equivalent to 520 MPC — h/quarter; calculated to result in a 50-year committed dose of: | Annual limit of intake (ALI) is equivalent to 2000 DAC—h/year. Calculated DACs are based on the following: organs are assigned weighting factors, based on the estimate of risk to that organ per unit of dose for uniform whole-body exposure. A "capping" dose limit of 50 rems/year (0.5 Sv/year) is used to avoid nonstochastic effects For body parts other than those listed above: |

Whole body    1.25 rems (5 rems/year)
Bone, thyroid,  7.5 rems (30 rems/year)
and skin
Other organs   3.75 rems (15 rems/year)

| Tissue | $w_T$ | Inferred dose limits (rems/year) | Actual dose limits (rems/year) |
|---|---|---|---|
| Gonads | 0.25 | 20 | 20 |
| Breast | 0.15 | 33 | 33 |
| Red bone marrow | 0.12 | 42 | 42 |
| Lung | 0.12 | 42 | 42 |
| Thyroid | 0.03 | 167 | 50 |
| Bone surfaces | 0.03 | 167 | 50 |
| Each of five remaining organs with the largest dose | 0.06 | 83 | 50 |

## TABLE 5.6 (continued)
## Selected Radiation Protection Standards

| Current Part 20 | Proposed Part 20 |
|---|---|

### Limits to the Public

| | |
|---|---|
| Implied limit for individuals of 0.5 rem/year to whole body, blood-forming organs, and gonads; 3 rems/year to bone and thyroid; and 1.5 rems/year to other organs; no summation of external and internal dose or consideration of food pathways | Explicit limit of 0.5 rem/year (0.005 Sv/year) for individuals from all sources; includes summation of external and internal doses and food pathways |
| No reference level | 0.1 rem/year (0.001 Sv/year) to member of the public as action level for licensee |
| No collective dose cutoff level | 0.001 rem per year (0.01 mSv/year) per person cutoff level for evaluating collective doses to general population |

### Occupational Monitoring Requirements

| | |
|---|---|
| Adults: required at 25% of the basic quarterly limit (0.312 rem) | Adult: required at 10% of the annual limit for deep-dose equivalent (0.5 rem or 0.005 Sv); required at 10% of the annual limit for eyes, skin, or extremities |
| Required for intakes greater than 25% of 520 MPC—h/quarter | Required at 30% of the ALIs |
| Minors: required at 5% of the basic quarterly limit (0.0625 rem) for adults | Minors: required at 5% of the external annual limits and 5% of the ALIs for adults |

### Miscellaneous Requirements Related to Doses to Employees or General Public

| | |
|---|---|
| The NRC must be notified immediately of incidents in which persons are exposed to 20 times the basic quarterly dose limits, a release of 5000 times (over a 24-h period) the concentration of a material listed in Appendix B, Table 2, a loss of 1 week or more of facility operations, or property damage in excess of $200,000 | The NRC must be notified immediately of incidents in which persons are exposed to five times the annual dose limits. The loss of facility use and property damage criteria are deleted. |
| A report also must be made if the limits for short-term radiation levels to the public or annual releases to unrestricted areas are exceeded | A report is required if any individual in an unrestricted area exceeds 0.5 rem (0.05Sv) in one year. |
| Radiation dose history reports to individuals required under §19.13(d) for any information reported to the NRC, but applies only to overexposures and termination reports. Other reports are available to individuals on request. | Same as the current Part 20, except that doses will be effective dose equivalents. In addition, licensees would report any planned special exposures to individuals and licensees operating under §20.205 (the exception for certain uranium and transuranic nuclides having very long, effective half-lives) would report estimates of both annual effective dose equivalent and 50-year committed dose equivalent to their employees. |

[a] The definition of "body" in the proposed standards has been limited by deleting the extremities.

[b] Excludes selected uranium and transuramic radionuclides for which the derived air concentrations (DACs) and annual limits of intake (ALIs) are hard to measure at levels found in the workplace. For these nuclides, the regulation may be based upon the effective dose-equivalent received in the year rather than the committed effective dose equivalent. The previous standard did not require the addition of external and internal doses. Also, the basic interval over which radiation is measured has been extended from a calendar quarter to a year. The 5(N-18) formula has also been deleted. There is a higher limit to the eye exposure, reflecting more information on the sensitivity of this organ.

[c] The extremities are more specifically defined in the new rule, with the calf of the leg now being considered equivalent to the forearm. Also, the new limit conforms to the restriction, based on nonstochastic effects, of no more than 50 rems to any organ.

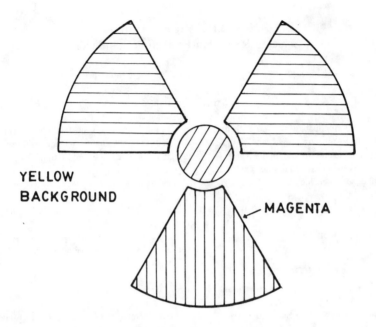

YELLOW
BACKGROUND

MAGENTA

FIGURE 5.5.    Standard radiation symbol.

The levels of activity from an unfiltered hood exhaust should not exceed the levels permitted in Appendix B, Table 2 (10 CFR 20), under the worst possible circumstances, such as a spill of an entire container of a radioisotope in the hood. The volume of material, the physical properties of the material, and the rate at which air is pulled through the system should permit the maximum concentration in the exhausted air to be computed.

Most laboratories using radionuclides are not required to be limited-access facilities. However, any area in which radioactive material is used or kept should be identified with a standard radiation sign, such as shown in Figure 5.5 below. Unless the amount of material or exposure levels in the facility trigger a more explicit warning, the legend should say only "Caution, Radioactive Material". Individuals using the radioactive materials must be trained in safe procedures for working with radiation, i.e., in the rules and regulations pertaining to the use of radiation and measures to be taken in the event of a spill or other emergency. Individuals working in the same area who do not use radioactive materials need not be as thoroughly trained, but should be sufficiently informed so that they understand the reasons for the care others take with radioactive materials and will not inadvertently become exposed to radiation levels in excess of those permitted for members of the general public or, by their actions or inactions, cause an incident involving radiation.

Licensed materials which are stored in an unrestricted area must be securely locked to prevent their removal from the area. If radioactive materials in an unrestricted area are not in storage, they must be under the constant surveillance and immediate control of the licensee.

Some laboratory facilities, or portions of them, may need to be made into "restricted areas" because of the type of activities conducted within them or in order to ensure that members of the general public will not be exposed to radiation in excess of that permitted by Part 20. The dose levels in the current Part 20, given in Table 5.6, are intended to apply to individuals in restricted areas. Access to a restricted area is not prohibited to a member of the general public, but it must be controlled access in order to provide the proper assurance of protection. A sign should be posted at the entrance to a restricted area, stating: "RESTRICTED AREA, ACCESS LIMITED TO AUTHORIZED PERSONNEL ONLY". The entrance should be locked at all times when the area is unoccupied and at such times when the occupants would be unaware of or unable to control the entrance.

Within a restricted area, there may be specifically defined areas where the levels of radiation may be significantly above those which would be acceptable for personnel to work on a normal 40 hour per week schedule.

### 5.1.5.1. External Radiation Exposure Areas

A "radiation area" within a restricted area is one accessible to personnel where there exists radiation, according to the legal definition, arising in whole or in part within licensed material, at such levels that a major portion of the body could receive a dose of 5 mrem (0.05 mSv)/h or in excess of 100 mrem (1 mSv) in five (5) consecutive days. A radiation area must be conspicuously posted with one or more signs bearing the radiation symbol and the words:

**CAUTION**
**(or DANGER)**
**RADIATION AREA**

Similarly, a "high radiation area" is one within a restricted area accessible to personnel where there exists radiation, arising in whole or in part within licensed material, at such levels that a major portion of the body could receive a dose of 100 mrem (1 mSv)/h. A high radiation area must be posted with one or more signs carrying the radiation symbol and the legend:

**CAUTION**
**(or DANGER)**
**HIGH RADIATION AREA**

In addition to the warning signs, additional measures must be taken to prevent individuals from accidentally entering a high radiation area. In the current Part 20, if the high radiation area does not continue for more than 30 d, direct surveillance of the area can be used to meet this requirement. In the proposed Part 20, this 30-d limitation is eliminated. If the high radiation area is to continue for 30 d or longer, one of the following precautions must be taken:

1.  An automatic device must reduce the level to the 100 mrem/h level upon entry of a person into the area.
2.  The area must be equipped with an automatic visual or audible alarm to warn the individual and the licensee, or a supervisor, of the entry into the area.
3.  The area must be kept locked, except at such times when entry into the area is required and positive control is maintained over entry to the area at these times.

None of the control measures which might be adopted can be configured so as to restrict individuals from leaving the high radiation area.

In the current Part 20, there is a third area, which has not been given a specific name, where levels in excess of 500 rems/h (5 Sv/h) could be received at 1 m from a sealed source which is used for irradiation of materials. In the new Part 20, this type of area is defined as a "very high radiation area", with the dose being specified as 500 rads (5 Gy) at 1 m.

To employ the designation from the proposed Part 20, which seems to be very appropriate, a very high radiation area* requires very stringent control measures to either prevent entry into the area or automatically reduce the radiation levels to where an individual would receive less than 100 mrem/h. The requirements address the possibility that the control measures may fail and mandate additional control devices to effect the same reduction in radiation levels and to provide a visual and audible alarm of the failure of the safety barriers to the person trying to enter the area, the licensee, and at least one other knowledgeable person who can render or secure assistance. Similar protective devices must be provided to reduce the exposure level and provide alarms should the shielding (other than that used in storing the irradiation device) fail or be removed.

The irradiation facility must be equipped with devices to automatically provide an adequate and timely warning to allow persons in the area to leave the irradiation area prior to such an irradiation device being operated. A clearly identified backup control must be provided in the irradiation area so that anyone in the area can prevent activation of the irradiation source. Administrative controls must also be in place to ensure that no one who might have entered the area since the previous use of the irradiation source is still in the area prior to the irradiation source device being put into operation. No access control measures may be established which would prevent a person from leaving the area. The entry control devices must be checked for proper functioning each time before the initial operation on any given day or after any occasion when operations were unintentionally interrupted. A periodic test schedule of the entry control devices is required.

Direct radiation measurements must be made prior to the entry of the first person into an area after a use of the source to determine that the level is below the 100 mrem/h level. A fixed radiation monitoring system or a portable instrument with suitable response characteristics (will not saturate in high radiation fields) and ranges may be used.

Any means of access to the area other than the ones normally used for facility personnel entering or leaving the irradiation area must be controlled so as to prevent the inadvertent entry of anyone (e.g., for the delivery of equipment or removal of processed materials) through these additional entrances.

### 5.1.5.2. Areas With Possible Internal Exposures

The previous areas within a restricted area were based on external exposures to a major portion of the whole body. The most probable means of radioactive materials

---

* This material does not apply to radioactive sources used in teletherapy, in radiography, to completely self-shielded irradiators so designed that high radiation fields cannot be created in an area accessible to individuals, or to certain other sources of radiation described in §20.203(c)(6).

entering the body to cause an internal exposure is through inhalation.* Therefore, §20.203(d) establishes requirements for spaces in which airborne radioactivity is present. Any area in which airborne radioactive material exceeds the concentrations listed in Appendix B, Table 1, (Column 1), to Part 20 or which, when averaged over the number of hours in any week in which persons are in the area, exceeds 25% of the same amounts is defined as an "airborne radioactivity area". Each area meeting or exceeding these limits must be conspicuously posted with one or more signs with the radiation symbol and with the legend:

**CAUTION**
**(or DANGER)**
**AIRBORNE RADIOACTIVITY AREA**

Normally, individuals without respiratory protection should not work in such areas. If at all possible, engineering practices should be used to eliminate the need for individual respiratory protective devices. In most research laboratory situations, it would be unusual to find an airborne radioactivity area on other than a short-term basis. However, a discussion of the usage of respiratory protection will be found in Section 6.1.

### 5.1.6. MATERIAL CONTROL PROCEDURES

Licensees may order, store, use, and transfer radioactive materials to other licensees and dispose of waste amounts, following regulated procedures. For relatively short-lived isotopes, there will be a continuing reduction of the amount on hand due to decay alone. Written records should be maintained of the amounts involved in each of these processes, and it should be possible to account for virtually all of the material from the time it is received until it is disposed of as waste. Some uncertainty will almost inevitably be introduced during actual experimentation, especially if there is a gaseous metabolic or combustion product which is exhausted through a hood. Small quantities retained on the interior of vessels containing radioactive liquids will escape into the sanitary system when the container is washed. However, it should be possible to estimate these types of losses with reasonable accuracy.

Facility records of receipt, transfer, use, and disposal should be maintained at each laboratory authorized to use radioisotopes. This is the responsibility of the individual in charge of the facility who has been assigned the internal equivalence of a radioisotope license. A technician may perform the actual record keeping, but the ultimate responsibility for the radioactive material in the facility belongs to the principal authorized user. Except for the removal of material from a storage container for use within the facility, the organization's RSO should be involved in each of these transactions and should be able to detect any anomalies or disparities. It is the RSO's responsibility to maintain overall inventory records of the radioisotopes within the organization, and these records can be used to audit the records for each facility.

It has already been noted that materials in unrestricted areas must be kept in secure

---

* Radiation also can enter the body by ingestion, by absorption through the skin, or through a break in the skin such as a cut. However, the latter is more likely to be due to unusual circumstances rather than to the presence of a continuing source.

storage when the user is not present. This security requirement is applicable whenever the facility is left vacant and open, even for short periods such as when going to the restroom, the stockroom, checking the mail, etc. In restricted areas, which should be locked at all times when they are not occupied or when access by unauthorized persons cannot be controlled, radioactive material can, in principal, be kept in storage areas which are not necessarily locked. However, if substantial amounts of radioactive materials are to be kept on hand, an effective key-control program should be in place or other security measures taken to protect against loss of the material.

Whenever radioactive material is removed from or returned to storage, the type of material transferred, the amount, the time, and the person performing the transfer should be entered into a log.

Perhaps the most common type of loss is the disposal of waste material as ordinary trash. Up to this stage, the radioactive material is usually maintained in containers labeled as containing radioactive materials, but contaminated trash often appears the same as ordinary waste. This can be due to laboratory workers inadvertently putting the material in with or near ordinary trash. Under such circumstances, a custodian is liable to take the material away due to failure to recognize the material as different from other waste, even if the radioactive waste containers are labeled "radioactive waste", marked with the radiation symbol, or a distinctive color. Unfortunately, an increasing percentage of the population is becoming functionally illiterate (a recent estimate is 30%), individuals may be color blind and if the symbol were concealed, the custodian could remove the material as ordinary trash unless additional precautions were taken. These errors can be reduced if the internal locations of radioactive waste (as well as broken glass, waste, or surplus chemicals) and ordinary solid waste are well separated and radioactive waste is placed in distinctively shaped containers used for no other purpose. These additional measures have been found to be helpful, provided all personnel have been fully informed and cooperate. It would be desirable for new custodial persons to receive special training, not only to reduce the possibility of loss of material, but also to ensure that they are not unduly concerned about servicing a laboratory in which radioactive materials are used.

If it is suspected or known that radioactive material has been lost from a laboratory, either by direct knowledge of the disappearance or inferred by an examination of the records, the organization's RSO must be notified immediately. If, after investigation, the loss or possible theft is confirmed, it probably is desirable in all cases to notify the NRC. This must be done, according to §20.402, if the amount unaccounted for and the circumstances of its disappearance are such that the licensee believes a substantial hazard could result to persons outside the facility. Since this requires a judgment decision on the part of the licensee, §20.1201(a)(ii) of the proposed Part 20 requires a report if the amount lost is ten times the quantity listed, for the material involved. The situation should be thoroughly investigated to attempt to recover the material or to determine its actual fate. If the loss appears to be deliberate, it may be necessary to solicit police assistance. Although the actual monetary loss may be small, individuals who were not knowledgeable about the potential hazards have suffered severe injuries from the possession of radioactive materials. If the loss is a relatively strong sealed source used for irradiation, it is very important, as will be discussed in a later section, that the material be recovered quickly.

It is rare for radioactive material to be taken deliberately, one of the most common loss mechanisms being loss as trash, as already discussed. If it is, in fact, a reportable incident, the NRC must be informed by a written report, filed within 30 days, specifying what was lost and the circumstances concerning the disappearance of the material. They will wish to know, to the best of the licensee's knowledge, what happened to the material and the possible risks to individuals in unrestricted areas. The steps taken to recover the material and, perhaps most importantly, the steps the facility intends to take to ensure that a similar incident will not reoccur will need to be in the report.

Even if the amounts lost are minimal and pose no danger to the general public, and often the amounts in waste are in this category, the RSO and the person in charge of the laboratory in which the loss occurred should be required to make a full report to the organization's RSC. The RSC should consider whether disciplinary measures are needed. Whether the loss was due to a failure on the part of the laboratory to adhere to established policies or whether the internal policies are sufficient to prevent future incidents of the same type must be determined. If the policies need to be modified, the committee should determine the new requirements and inform all users within the organization.

In a well-run organization with sound policies guiding the use of radiation, incidents involving loss of radioactive material should occur very rarely. In any active research institution, especially academic ones where graduate students and even faculty change frequently, mistakes will occur. A fairly common defense in instances where radioactive material is lost is that the amounts lost were small and no one will be harmed. The RSC, however, needs to determine whether procedures or the enforcement of procedures is the basic problem, not the quantity that is lost. Individuals who have made a mistake should not be treated harshly if they recognize that a mistake has been made and learn from it. On the other hand, the RSC should take very firm steps to correct the situation if a facility shows a continuing pattern of laxity in following sound procedures or, worse, deliberate neglect of safety and regulatory policies.

### 5.1.6.1. Ordering and Receipt of Materials

A very sound procedure to follow in ordering and receiving radioactive materials is for all orders and receipts of radioactive materials to be handled by the radiation safety office. This has a number of advantages for the overall organization in terms of record maintenance, as discussed in the previous section, and for the individual facility since it provides a parallel set of records. Some suppliers of radioactive materials will provide a discount for volume purchases if all the orders go through a common ordering center, even if the materials are for different users. Virtually the only problems involve the occasional user who wishes to have a custom compound prepared with the radioisotope in a specific location within the compound. Such users may wish to discuss their requirements by telephone with the vendor. Even telephone orders can be handled through the radiation safety office satisfactorily if the vendor, the user, and the radiation safety office work together. Since many biological materials deteriorate with time and at normal temperatures, they are sent packed in dry ice to keep them cool. If the radiation safety office is aware of the anticipated delivery date, it can ensure that the shipment be delivered promptly.

A major safety function of the radiation safety office is to receive all packages of radioisotopes and process them according to the requirements of §20.205. Unfortunately, there are occasional errors in shipping and packages can be damaged. On at least two occasions at the author's institution, packages have been received which contained substantially larger amounts of material, with consequently higher radiation levels, than should have been the case.

Most packages of radioactive materials are shipped in what are called Type A packages, which are only required to withstand normal transportation conditions without a loss of their contents. The amounts of the various nuclides which can be shipped in a Type A package are given in 10 CFR 71, Table A. It is the responsibility of the shipper to conform to these limits. (The NRC has currently proposed that Part 71 be modified to upgrade it to international shipping requirements).

The receiver of packages of radioactive materials must be prepared to receive them when delivered, pick them up at the carrier's location at the time of arrival, or promptly upon notification of the arrival of the material. Most packages are probably delivered to the purchaser's location by the numerous freight and parcel delivery services that are available. Some of these deliver 24 h/d while, except in the largest organizations that conduct research on a 24-h basis, most radiation safety groups work a normal daytime schedule. At a university, for example, the only operation likely to be functioning on a 24-h basis is the security or police unit, which would not normally know how to perform the required checks on packages that have been received. It also would be desirable for packages to be placed in a secure area until such time as they can be inspected. One alternative would be to place a few permanently mounted, lockable boxes in the security area in which packages could be placed until the following morning.

Other than exceptions for packages containing small quantities of radioactive materials spelled out in §20.205(b)(1)(i-v), packages are to be checked for external radiation levels, including loose surface contamination as well as direct radiation, within 3 h of receipt during the day or within 18 h of receipt if received after normal working hours. Levels which would require immediate notification of the NRC would be 200 mrem (2 mSv)/h hour on the surface or 10 mrem (0.1 mSv)/h 3 ft from the surface of the package. A limit of 22,000 dpm is set as well for loose contamination on the surface. There are a few problems with this requirement. The 18-h limit is feasible for Sunday through Thursday nights, but would require a person coming in during the weekend for packages received from Friday evening until mid-afternoon on Sunday. Holidays also would be a problem. In the proposed Part 20, the requirement would change to 3 h after the beginning of the next working day after receipt of the package.

The RSO normally should not open individual packages (although some do) unless there is reason to suspect that the material contained within the package is not consistent with the packing list, but, rather should either deliver the material to the users or have them pick up their packages. Users should exercise due care in opening packages to minimize exposures to themselves. The precautions which should be taken will depend upon the nature of the material, the anticipated radiation levels, or the potential for loose material becoming airborne should the package not be tightly sealed. If there is any risk of personnel becoming contaminated while opening a package, it should be done in a radioisotope hood, with the package placed in a pan large enough to retain any spilled material and the pan sitting on a layer of plastic-backed absorbent paper. The employee should be gloved. The need for shielding would depend upon the radiation characteristics of the material.

A package known to be damaged should be opened in the equivalent of a Class 3, biological safety cabinet, i.e., a glove box, with the interior surfaces of the box previously prepared by lining it with plastic-backed absorbent paper to reduce the need for decontamination. The individual handling the package prior to its being placed in the glove box should wear a complete set of protective equipment, including gloves, a coverall with head covering, respiratory protection, and goggles (unless the latter are part of the respirator). Any package damaged such that contamination is released in excess of the allowable limits must be reported at once to the NRC by telephone and telegraph mailgram (or facsimile) and to the final delivering carrier. It would probably be desirable to report any significant problems with the package, regardless of the amounts.

### 5.1.7. OPERATIONS

Operations is an all-inclusive term encompassing the program in which radioactive materials are used as well as the support programs necessary to allow the materials to be used safely and in compliance with all regulations. Many of the latter requirements have been discussed in some detail — including, in some instances, the means by which compliance or safety can be enhanced — in preceding sections.

Although a direct relationship between low-level exposures and adverse biological effects has not been demonstrated conclusively, it is assumed, as a conservative premise, that many effects which are known to be caused by high doses will occur on a statistical basis to a portion of a large population exposed to lower levels. Stochastic effects which are based on probabilities, such as induction of cancer or genetic damage, are assumed to have no threshold, while nonstochastic effects, such as cataracts caused by radiation, are assumed to be related to the intensity of the exposure and to have a threshold. In most cases, the stochastic effects are the ones which limit the permissible exposures in using radioisotopes.

There is a wide variation in the susceptibility of individuals to radiation, as with most other agents which can cause harm to humans. Some individuals may be much more resistant to radiation than others, while other persons may be hypersensitive to radiation. There are numerous factors which can affect the damage done to an individual. Some of these may be sex linked and some dependent upon the age of the individual. Some may depend on exposure to other agents (synergistic effects) and others could depend upon the health of the individual at the time of the exposure. This section will not attempt to pursue these modifying factors, but will assume, as did the sections in which chemical agents were discussed, that the workers are normal, healthy individuals in their normal working years (unless otherwise specified).

The following sections will be concerned with means to reduce exposures while using radioisotopes.

### 5.1.7.1. Reduction of Exposures, ALARA Program

The goal of an ALARA program is to make personnel exposures to radiation as low as reasonably achievable in the context of reasonable cost, technical practicality, and cost-benefit considerations. In such a program, each aspect of an operation is evaluated to determine if there are alternatives by which the procedure can be modified to reduce exposures.

One of the first considerations is to select the radionuclide to be used in the research. The half-life, the type of radiation emitted, the energies of the emissions, and the chemical properties of the radioisotopes are factors to be evaluated.

Control of external hazards is based on manipulation of three primary variables: time, distance, and shielding.

Control of internal hazards is more complex. The properties of the materials must be considered, but the experimental techniques to be used, the design and construction of the facility, the way it is equipped, and the use of personal protective equipment are all factors which must be considered in establishing an effective ALARA program.

### 5.1.7.1.1. Selection of Radioisotopes

In order to select the appropriate radioisotope for use in a given research program, a number of factors must be considered. If the choice is strictly based on nuclear properties, a beta emitter would generally be preferable to a gamma emitter since gammas typically are much more penetrating than betas. Although alphas are much less penetrating than betas, an alpha emitter would normally not be selected over a beta emitter because, among other reasons, if ingested, the damage caused by an alpha as it moves through matter is much greater than that caused by a beta. For the same type of emitter, the isotope with the lower energy emission should normally be chosen. When half-lives are considered, the shorter half-life is preferable because the problem of disposal of the nuclear waste can be solved simply by allowing the radiation in the waste products to decay, rather than having to ship it away for burial. Moreover, if the materials may be ingested by humans or used as a diagnostic aid, the total dose to an individual could be much less if the isotope used has a short half-life.

The desired chemical properties and costs may dictate the isotope chosen. Some isotopes are much more expensive than others, so for economic reasons, it might be preferable to use the less expensive choice. A custom-prepared, labeled compound is usually much more expensive than a commercially available, labeled compound, so unless the safer isotope is available in a preprepared compound, economic reasons would probably dictate using what is readily available.

Chemical properties may dictate the choice of materials because of the risk to personnel ingesting them. An examination of Appendix B, Table 1 (column 1) in 10 CFR 20 shows wide variation in the permissible air concentrations in a work area. Soluble $^{239}$Pu, for example, which behaves much like calcium chemically, has an MPC of $2 \times 10^{-12}$ µCi/ml for the airborne concentration in a restricted area, while for $^{14}$C, the MPC is 500,000 times greater, partially due to the type of radiation emitted and its different chemical role in the body.

Carbon 14 and tritium are two of the most frequently used radioisotopes since they can readily be incorporated into organic compounds. Unfortunately, $^{14}$C has a very long half-life, 5730 years, while tritium has a much shorter one, 12.3 years. Even the latter is sufficiently long to represent a disposal problem. Otherwise, these two isotopes have excellent radiological properties. The beta emitted by carbon has an energy of only 0.156 MeV, so its range in tissue is less than 0.3 mm and about 10

**TABLE 5.7**
**Properties of Some Selected Nuclides**

| Nuclide | Half-life | Beta(s) (MeV) | Gamma(s) (MeV) |
|---------|-----------|---------------|----------------|
| $^{45}$Ca | 163 d | 0.257 | 0.0124 |
| $^{109}$Cd | 453 d | ε | 0.088 |
| $^{36}$Cl | 3.0 x 10$^5$ years | 0.709 | — |
| $^{60}$Co | 5.27 years | 0.318 | 1.173, 1.332 |
| $^{51}$Cr | 27.7 d | ε | 0.320 |
| $^{137}$Cs | 30.2 years | 0.512, 1.173 | 0.6616 |
| $^{59}$Fe | 44.6 d | 0.467, 0.273 | 1.099, 1.292 |
| $^{203}$Hg | 46.6 d | 0.212 | 0.279 |
| $^{125}$I | 59.7 d | ε | 0.0355 |
| $^{131}$I | 8.04 d | 0.606, others | 0.364, others |
| $^{54}$Mn | 312.5 d | ε | 0.835 |
| $^{22}$Na | 2.6 years | ε, β$^+$ 0.545 | 1.27, 0.511 (annual) |
| $^{63}$Ni | 100 years | 0.0659 | — |
| $^{65}$Zn | 243.8 d | ε, β$^+$ 0.325 | 1.116, 0.511 (annual) |

in. in air. The beta emitted by tritium is much weaker (0.0186 MeV), with a range in tissue of only about 0.006 mm and about 0.2 in. in air. Clearly, both of these isotopes would pose little risk as far as external exposures are concerned, although both can represent an internal problem. Tritium, for example, can easily become part of a water molecule which would be treated by the body virtually the same as would any other water molecule. Other popular beta-emitting isotopes are $^{35}$S, which emits a 0.167 MeV beta and has an 87.2-d half-life, and $^{32}$P, which emits a 1.71 MeV beta and has a 14.3-d half-life. As noted in an earlier section, the high-energy beta from $^{32}$P can cause a substantial radiation exposure problem due to emission of bremsstrahlung radiation if the betas are allowed to interact with materials of significant atomic numbers. Another radioisotope of phosphorous, $^{33}$P, can be used which has a longer half-life, 25.2 d, but a much lower-energy beta, 0.248 MeV.

Some of the other more commonly used radioactive isotopes used in research are listed in Table 5.7.

### 5.1.7.1.2. Shielding

Shielding is usually the first protective measure which comes to mind when considering radiation protection. In discussing properties of radioactive particles, the ranges or penetrating capacity of the various types were briefly discussed to provide some measure of understanding of their characteristics. These properties will now be considered in order to determine how best to provide shielding. The discussion will be limited to shielding against beta and gamma radiation.

**Betas**

The range of betas in any material can be calculated by the following two equations:

$$\text{for energies} \quad 0.01 \leq \text{energy (MeV)} \leq 2.5$$
$$R = 412E^{(1.265 - 0.0954\ln E)} \tag{12}$$

$$\text{and for energies} \quad E \geq 2.5 \text{ MeV}$$
$$R = 530 \, E - 106 \tag{13}$$

The range, R, in these equations is given in mg/cm². This can be converted into centimeters in a given material by dividing the ranges by the density of the material in mg/cm³. The range, R, can be considered as the area density of a material, in contrast to the usual volume density.

The range of the 1.71-MeV beta from ³²P calculated from Equation 12 is 785 mg/cm². Tissue and many plastic materials have a density near 1, so this beta would have a range of about 0.8 cm, or 0.3 in., in such materials. In lead, the range would be 11.34 times less, or 0.7 mm (0.027 in.). However, it will be recalled that bremsstrahlung would be a major problem for such an energetic beta emitter if a high atomic number absorber were used. Therefore, a low-Z material should be used for shielding pure beta emitters. A piece of plastic ³/₈ in. thick would shield against virtually all common betas used as radioisotopes since there are few commercially available beta emitters which emit more energetic betas (quite a few neutron-activated nuclides emit higher energy betas, but these are not often used as radioisotope materials).

Research personnel who work with ³²P can minimize hand exposures from bremsstrahlung by slipping thick-walled plastic tubing over test tubes and other containers.

## Gammas

Equation 6, repeated below as Equation 14, can be used to calculate the attenuation of a monoenergetic narrow beam of gamma rays by a shielding material.

$$I = I_o e^{-\mu x} \tag{14}$$

This equation, as noted earlier, will not take into account gammas scattered into the shielded area by Compton interactions by the shield and the walls (and the floor and ceiling if the gammas are not collimated so as to avoid these surfaces). Figure 5.6 shows an optimal shielding geometry where a shield just subtends a solid angle sufficient to completely block the direct rays from the source. This geometry minimizes in-scatter. In Figure 5.7, a more typical geometry is shown where a worker is using a source in a fume hood. In this geometry, the shielding would not be quite as effective.

For work with gamma emitters and low-energy betas, lead is a good shield. It is dense, 11.4 g/cm³, and has a high atomic number, 82. It is relatively cheap and is more effective than any other material that is comparably priced. The mass attenuation coefficient for lead is shown in Figure 5.8. The coefficients read from the graph can be converted to linear coefficients by multiplying them by the density, expressed in g/cm³. If the absorption coefficients for lead at 0.5, 1.0, and 1.5 MeV are used to compute the thickness of lead required to attenuate a beam of lead by factors of two and ten, the half-value thicknesses would be 0.4, 1.1, and 1.5 cm, respectively, and for the tenth-value thicknesses, 1.25, 3.5, and 5.0 cm for the three energies. Thus, a lead brick 2 in. thick would reduce the intensity of the gammas from most of the common radioisotopes used in research by a factor of ten or more, which is often enough for the average source strengths used in research. Shields should, however, be built of at least two layers of overlapping bricks to prevent radiation streaming in a direct path through a seam.

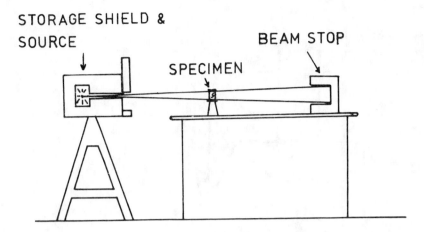

FIGURE 5.6.   Optimum shielding to minimize scattered radiation from shield.

Occasionally, individuals will set up a shield to protect themselves and forget that radiation from an open source extends over $4\pi$ steradians and levels through the top, bottom, sides, and rear of the work area, perhaps in the next room, may be excessive.

Extremely high-level gamma radiation operations are conducted in a special shielded enclosure called a "hot cell". Usually, the shielding in such units is provided by making the enclosure of high-density cast concrete. The thickness needed to reduce the radiation to levels acceptable for occupational exposures would depend upon the strength of the source. Often, the top of a hot cell as well as the sides is covered in order to avoid "sky-shine" radiation, i.e., radiation scattered from the air and ceiling due to radiation from the source within the cell. The researcher would perform the operations required in the cell using mechanically or electrically coupled manipulators. Vision typically would be through a thick leaded glass window, often doped with a material such as cerium to reduce the tendency of glass to discolor upon exposure to radiation, or closed-circuit television.

### 5.1.7.1.3. Distance

Distance is an effective means of reducing the exposure to radiation. The radiation level from a point source decreases proportionally to $1/R^2$, while from a point near an extended source such as a wall, the level falls off in a more complicated manner since the distance from each point on the wall to the measuring location will vary and the contribution of each point must be considered. There are a number of excellent references where specific geometrical sources are discussed, but, unfortunately, most real world situations often do not lend themselves to simple mathematical treatment. If the distance to the source is large compared to the size of the source, then the approximation of a point source is reasonably accurate.

Many radionuclides emit more than one gamma and for many, the gamma spectrum is extremely complex. For each original nuclear disintegration, which defines the number of curies or becquerals represented by the source, a specific gamma will occur a certain fraction, $f_i$, of the time. If a gamma is emitted for every original nuclear decay, then $f_i$ for that gamma would equal 1. With this concept in

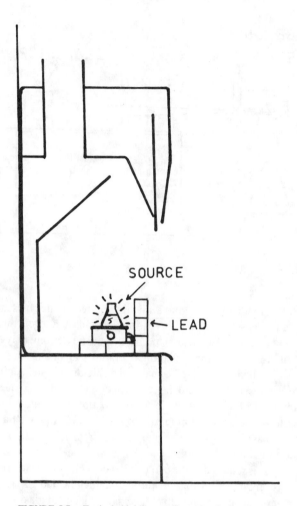

FIGURE 5.7.   Typical shielding configuration for hood applications of radiation.

mind, there are two simple expressions which can be used to calculate the dose at a
distance, R, from a point source. The difference in the two expressions is simply a
matter of units.

To compute the dose, D, in μSv/h, the expression is

$$D = \frac{\Sigma_i M f_i E_i}{6R^2} \tag{15}$$

Here, M is the source strength in MBq, $E_i$ is the energy of the ith gamma in MeV,
and R is the distance in meters.

The same expression, expressed in the more traditional units, is

$$D = \frac{6\Sigma_i C_i f_i E_i}{R^2} \tag{16}$$

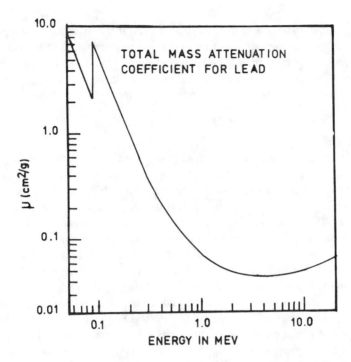

FIGURE 5.8. Total mass absorption coefficient for lead as a function of energy.

Here, the source strength is in $C_i$, $E_i$ is the energy in MeV, R is in feet, and the dose is given in rads per hour.

It should always be kept in mind that, even from a point source, both of these expressions give only the dose due to *direct* radiation from the source. They do not include the radiation scattered from objects in the room, the walls, floors, ceiling, or the workbench on which the source is sitting. It also should be kept in mind that the radiation dose can be reduced further by placing a shield in the path of the radiation. If this is done, the levels calculated by either Equation 12 or 13 would be reduced by the factor $e^{-\mu x}$, where $\mu$ would be the linear attenuation coefficient for the shield material at the energy, E, of the gamma. The equation would have to be calculated for each of the gammas and the contribution of all the gammas would have to be summed. The result would be the minimum dose that would be received from the source. The actual dose would be higher.

Because the exposure rate does go down rapidly with distance, sources should always be handled with tongs rather than directly. Whenever a source is to be transferred, only an individual trained to handle the equipment and the source should perform the transfer, perhaps with an assistant nearby, and the transfer should be done as expeditiously as possible.

When thinking of distance as a protective measure, the mental image is of a source with the user standing back from it. However, one way of thinking of contamination on the skin is of there being virtually a zero distance separating the irradiating material and the tissue, so that even small amounts of radioactive material could eventually cause a significant local exposure, even if the material did not penetrate or permeate through the skin. In order to prevent contamination of the skin, gloves and protective covering such as a laboratory coat should always be worn when working with material which is not a single solid piece, such as a liquid or a powder which could be spilled or otherwise contaminate the worker and the experimental area.

### 5.1.7.1.4. Time

Time is perhaps the most easily achieved means of reducing the dose received. If the radiation level remains constant, reducing the exposure time reduces the dose received by exactly the same proportion that the time is shortened. If the radioactivity of the material in use decays during the interval, the reduction in exposure would be even greater. This should be kept in mind when developing experimental procedures. Personnel should leave sources within their containers as much of the time as possible while working.

Another way in which time can be used effectively to at least reduce the level of exposure to a single individual is to share the exposure time among more individuals. Statistically, for those events which depend on the probability of an event being caused by the exposure of individuals, i.e., stochastic events, little has been gained if the exposure of the entire group of persons is considered, but it does reduce the probability of a problem developing for a specific individual and reduces the nonstochastic effects to the individual which depend upon the dose received.

### 5.1.7.1.5. Quantity

Modern techniques often allow work with much smaller amounts of material than was possible only a few years ago. When working with radioactive materials, full advantage should be taken of any efficient and effective procedure for using less of the radioactive substance, not only at any given time, but also over the course of the research. A carefully constructed, written research plan in which the amounts of materials needed for each procedure are calculated will assist the minimization process.

### 5.1.7.1.6. Example of Time and Distance

Small sources are frequently used in laboratories. If a source is dropped from a storage container, an employee may pick it up in his hands to return it to the storage container, handling it in direct contact with the hands for perhaps 1 min. Even this short interval with relatively weak sources can give rise to significant doses to the hands. Published data have shown that, at or very near the surface of a steel-clad, sealed gamma source, electrons from the source cladding, caused by interactions of the gammas with the material, can contribute between 25 to 45% of the dose to the hands. The primary exposure to the hands still is due to direct radiation from the source, which is high because of the minimal distance between the source and the tissue. The tissue of the hands is estimated to absorb about 5 to 10% of the gamma energy per centimeter. With these facts in mind, if the source is $1/_4$ in. diameter and is clad with $1/_{32}$ in. of 304 stainless steel, the approximate surface dose rate per curie per minute for a $^{60}$Co source would be 2075 rads/min. In less than 2 $1/_4$ sec, the hand of a person picking up a 1-Ci source would receive, locally, the entire year's permissible exposure of 75 rads. The rate falls off rapidly with depth in the tissue; at 1 cm deep, the rate is estimated to be 114 rads/min and at 3 cm, 16 rads/min. A reduction of any one of the three contributing factors — the quantity of radioactive material in the source, the time it is handled, or increasing the distance from the source — will diminish the harm done to the individual.

Although this may be dismissed as unlikely, there have been a number of cases where persons unfamiliar with what they were handling have picked up sources and placed them in their pockets. Often, the persons have been custodians or lower-level support personnel. In some cases, many persons outside of laboratory facilities have been exposed to high levels of radiation because pellets of material from irradiation sources which were no longer being used were disposed of improperly and wound up in trash. Deaths have resulted from such instances. Not all high-exposure incidents have been among the untrained and poorly educated. There have been instances of laboratory personnel who have had substantial exposures because of failure to exercise sufficient care in working with radioactive materials. In several cases, the workers were exposed to substantial radiation levels due to exposed sources, of which they were unaware, because an interlock had failed. Safety features which depend upon a single microswitch do not provide adequate protection. A number of persons have died and a number have been seriously injured. It should be noted that the exposures in these cases have been in the several hundreds of rads (or rems) range or greater and in some cases, localized exposures have been in the many tens of thousands of rad range, with the person surviving, but with extremely severe damage over the parts of the body exposed to these extraordinarily high levels.

In summary, to minimize external exposure:

1.  Work with the safest isotope appropriate to achieve the desired results.
2.  Use the smallest amount of radioactive material possible, consistent with the requirements of the experimental program.
3.  Minimize the time exposed to the radiation.
4.  Use shielding wherever possible.
5.  Take advantage of distance as much as possible. Use extension tools to avoid direct handling of the material.
6.  Make sure that any protective devices are fully functional. A checklist which must be followed prior to each use is a good and inexpensive safety device.

### 5.1.7.1.7. Internal Dose Limiting

There are, as in the case of chemicals, only four basic means for radiation to enter the body: inhalation, ingestion, absorption through the skin, or through a break in the skin. The basic means of controlling internal exposure is to restrict entry into the body by any of these routes. However, other factors will affect the consequences of radioactive material which has succeeded in bypassing the defenses to prevent entry into the body. The form of the material is important. Is it in a soluble or insoluble form? Insoluble particles carried into the lungs following inhalation may remain there for long intervals. Soluble materials may be absorbed into body fluids. If the material is incorporated in a form in which it is likely to be metabolized, it is more likely to reach a critical organ, if ingested, than otherwise. For airborne contaminants, the size of the particle is critical in determining in what part of the respiratory system the material will be deposited: a portion of the material which is not exhaled may be swallowed.

The chemical properties of the material are important in determining what organs are likely to be involved. In some cases, such as tritium, the tissue of the entire body is likely to be involved since the chemical form of released tritium often will be as HTO, and the behavior within the body will be the same as that of ordinary water. Other materials, such as strontium, are bone seekers in soluble form, while soluble iodine will most likely go to the thyroid.

### 5.1.7.1.7.1. Entry Through the Skin

Entry through the skin, either percutaneously or through a break, normally is the least likely to occur and the most easily prevented. In most cases, entry would follow an inadvertent spill in which material reached the skin by the worker handling contaminated equipment without proper equipment or by failure of protective items. However, in the case of exposure of an otherwise unprotected worker to water vapor containing an appreciable amount of HTO, a significant fraction of the total exposure could be through skin absorption (normally assumed to be one third of the total, the remainder being due to inhalation).

Prevention of contamination of items likely to be handled is the first step in prevention of skin exposure. For example, the work surface should be covered with a layer of plastic-backed absorbent paper to ensure that the permanent work surface is not contaminated by spilled materials or from vapors that might diffuse from a container since the vapors of most chemicals are heavier than air. Setting up the apparatus in a shallow container, such as an inexpensive plastic or aluminum pan which can be obtained from a department store, that is large enough to contain all the spilled liquid in use at the time is another way to confine the consequences of an accident to items which are disposable rather than ones which would have to be decontaminated. Covering some items of equipment with plastic-backed absorbent paper, or aluminum foil for warm surfaces, may prove profitable for some levels of work using materials of high toxicity, amounts of radioactive materials in excess of the typical laboratory application, or in procedures in which the possibility of contamination is higher than normal. Frequent checks of the apparatus in use and the immediate work area near the apparatus with survey instruments, or smear or wipe tests of work surfaces and equipment being handled, is another means of limiting contamination before it becomes widespread.

Protective gloves and laboratory coats or coveralls are the first level of protective gear which should be adopted in laboratory operations involving the use of radioisotopes. Gloves should be chosen for their resistance to the chemicals of the materials in use. The dexterity permitted by the gloves also is a factor. Where the possibility of damage to the gloves is significant or the properties of the radioactive materials are especially dangerous if contact occurs, wearing two pair of gloves should be considered. Gloves used for protection while working with radioactive material should be discarded as radioactive waste after use. The cuffs on gloves should be long enough to allow sealing (with duct tape or the equivalent) to the sleeves of a garment worn to provide body protection, if needed.

Any cuts or abrasions on the hands or other exposed skin areas on the forearms should be covered with a waterproof bandage while actively working with radioactive materials.

Cotton laboratory coats provide reasonable protection for the body of the worker in most laboratory uses of radioisotopes. However, if they do become contaminated, they must be washed to decontaminate them. If laundering facilities appropriate for cleaning radioactive materials may not be available, consideration of disposable protective clothing made of materials such as Tyvek™ may be desirable. These garments are not launderable, but if a thorough survey shows that they are free of contamination and they are handled with reasonable care, they can be worn several times before they need to be discarded. However, they are inexpensive enough to allow disposal whenever they become contaminated. Disposable garments should have both welded and sewn seams to provide assurance that the seams will not split during use (wearing a garment one or even two sizes larger than is actually needed also helps avoid this).

Hands should be carefully washed at the conclusion of any operation in which radioactive materials are handled in procedures which offer any opportunity for contamination. If the skin has become contaminated, the affected area should be washed with tepid water and soap. The use of a soft brush aids in the removal of material on the surface. Harsh or abrasive soap should not be used. After washing for a few minutes, the contaminated area should be dried and checked for contamination. If the area is not free of radioactivity, rewashing can be tried. A mild detergent also can be tried, but repeated or prolonged application of detergents to an area may damage the skin and increase the likelihood that surface contamination will penetrate the skin. Organic solvents and acid or alkaline solutions should not be used since they will increase the chances of skin penetration. Difficult-to-clean areas should be checked with extra care.

Any contamination which cannot be removed by the above procedures should be reviewed by a radiation safety specialist. Further measures which could cause abrasion or injury to the skin should only be taken under the advice and supervision of a physician.

### 5.1.7.1.7.2. Ingestion

The inadvertent ingestion of radioactive material is often the consequence of transferral of radioactive material from the hands to the mouth while eating or drinking. No food or drink should be permitted in the active work area of a laboratory in which radioactive materials are used. Prior to leaving the laboratory after working with such materials, as noted in the previous section, the hands should always be carefully washed and surveyed to ensure that no material is on the hands. If gloves and protective outer wear have been worn, the gloves should be discarded, and the outer wear left at the entrance to the work area after ensuring that it is not contaminated. Hands should be washed after handling the clothing. It should be possible to effectively prevent ingestion of radioactive materials by using reasonable care to prevent the transfer of such materials to food and drink or to any other item which might be put in the mouth such as a cigarette or gum.

### 5.1.7.1.7.3. Inhalation of Radioactive Materials

Airborne radioactivity is the most likely means by which radioactive materials can enter the body, resulting in an internal exposure. Most mechanisms by which skin contamination or ingestion occur involve contact of the user with the active material in a relatively fixed location. However, airborne materials are not constrained to a given space, and unless the work is confined to a glove box, they can fill the entire volume of the laboratory, surround the occupants of the room, and contaminate the air they breathe. The radioactive material also can be discharged from the laboratory by the building exhaust system or through the fume hoods.

The size of particles which are deposited in the deep respiratory tract reaches a maximum at 1 to 2 μ. Unfortunately, particles in this size range tend to remain suspended in air for relatively long periods of time. The gravitational settling rate for a 1-μ particle with a density of one is about 0.0035 cm/sec. It would require about 12 h for such a particle to settle about 5 ft under the influence of gravity, if allowed to do so without being disturbed. Thus, the particles of airborne radioactive contamination, which have the greatest potential for deposit in the deep respiratory tract are among those which would tend to remain airborne long enough for a worker to have an opportunity to inhale them.

It has been estimated that about 25% of all soluble particles which have been inhaled are exhaled; about the same fraction are dissolved and absorbed into the body fluids. The remaining 50% are estimated to be deposited in the upper respiratory tract and swallowed within 24 h after intake. For insoluble particles, about seven eighths are exhaled or deposited in either the upper or deep respiratory tract, but swallowed within 24 h after intake. However, the remaining one eighth is assumed (in the absence of specific biological data appropriate for the material in question) to be deposited and retained in the deep respiratory tract for 120 d. After being swallowed, the radioactive particles may either descend through the gastrointestinal system and be excreted or pass into the body. Varying portions may be excreted through the kidneys.

Engineering controls are the preferred means of keeping the MPCs below the limits in 10 CFR 20, Appendix B, Table 1 (column 1). Areas in which radioactive materials are used should be adequately ventilated. If the workplace is designed properly so that air movement is away from the worker's breathing zone, work in which airborne releases are possible are performed in an effective fume hood, and the quantities of activity are typical of most research programs which do not require access control, six to ten air changes per hour should be sufficient to provide the needed environmental control for normal operations. The preferred hood type would be a good quality standard hood rather than an add-air type since the latter, with two air streams and a more complicated design, occasionally is more prone to spillage if not used and maintained properly. An airflow velocity through the face of the hood of approximately 125 fpm (38 Mpm) would be a good design target. The hood should be equipped with an alarm to warn if the airflow through the face falls below 100 fpm (30.5 MpM). Making sure that any source of radioactive gaseous or vapor effluent is at least 8 in. (20 cm) from the entrance to the hood and keeping the sash down approximately half way will further significantly improve the ability of a hood to capture and confine airborne hazardous materials.

If, for a valid reason, full control of the levels of airborne contaminants cannot be maintained within the legal limits, it is permissible to use respiratory protection under some conditions. These are found in §20.103 and Appendix A of the current Part 20. No credit can be taken for the use of sorbants to provide for respiratory protection against radioactive gases or vapors. Air-supplied respiratory protection is required in such cases.

There are a number of different types of respirators which provide various levels of protection. The types and their characteristics will be discussed briefly below and more extensively in Chapter 6. Their function is to reduce the level of radioactive material in the air being breathed by the user as far below the acceptable levels as is reasonably achievable.

Respirators fall into two primary classes: those in which air is supplied and those in which the air is purified by being passed through filters. For chemicals, the latter type can be used for particulates and for vapors and gases for which suitable sorbants are available. As noted above, the use of sorbants is not permitted where the risk is due to radiation. Air-supplying respirators fall into several different classes. Units in which air is supplied on demand and which do not provide a positive pressure inside the face mask with respect to the ambient air are the least effective because contaminants can enter should the seal between the mask and the face fail. These are rated to provide a protection factor of five. Self-contained units in which air is supplied continuously or in which the air is always at positive pressure are rated highest and provide protection factors up to 10,000. There are two other types, one in which the air is recirculated internally and purified chemically, but is always at a positive pressure with respect to the outside, and another in which air is supplied to a hood which does not fit snugly to the face, but which has a skirt which comes down over the neck and shoulders. The latter types provide protection factors of 5000 and up to 1000, respectively. Self-contained units have a severe limitation in that they provide air for only a limited period of time. The most popular SCUBA type with a 30-min air tank may provide air for only 15 min if the wearer is under heavy physical stress. Larger versions of the air-recirculating and -purifying type can provide air for up to 4 h, but even these are not suitable for continuous wearing.

Respirators which purify the air by filtering require a snug fit to the face to ensure that they are effective. Unfortunately, as a person works, the contact may be temporarily broken and the protection provided by the mask diminished. Since facial hair will prevent a good seal, bearded workers cannot be allowed to use respirators requiring a good facial seal for protection. Some facial types are very hard to fit successfully, although there are increasingly more models available which allow different types of features to be fitted. However, NIOSH has recently proposed new standards which virtually all of the current models on the market cannot meet.

A major requirement is that the individual be physically able to wear a respirator. It takes some effort to breathe through a cartridge respirator, and a person with emphysema or some other breathing impairment would not be able to wear such a unit for extended periods. Passing a physical examination, including a pulmonary function test, is required of employees who would be expected to wear a respirator in performing some of the tasks assigned them. The employer is required to have a comprehensive respirator program if this method of achieving compliance with the permissible air limits of airborne concentrations of radioactive materials is used for the employees.

Although respiratory protection does offer an alternative to engineering controls for providing suitable breathing air quality, this option is more commonly used to provide suitable air for emergency situations.

In summary, to minimize internal exposure:

1. Use the isotope that would be the least dangerous if taken into the body.
2. Use the smallest amount of radioactive material possible, consistent with the requirements of the experimental program.
3. Minimize the time in which an airborne exposure could occur.
4. Do the hands-on work in a radiological fume hood for most radioisotope research or in a glove box if the possibility of airborne exposure mandates it because of the type, amount, or form of the material.
5. Take measures to minimize contamination with which it would be possible to come into contact, e.g., by using absorbent plastic-backed paper to line the work surface and a tray underneath the experimental apparatus to catch any spills.
6. Wear gloves (two pair when working with material in solvents that could cause a glove material to soften or weaken).
7. Use protective outer wear that can be discarded if contaminated.
8. Do not eat, drink, or smoke in the area where radioactive materials are used.
9. Wash your hands after any use of radioactive material, before leaving the work area, and before eating, drinking, smoking, or any other activity in which radioactive material could be ingested.
10. Supplement the protection provided by the hood with the use of respiratory protection, if needed to achieve compliance with airborne concentration limits.
11. Cover any cut with a waterproof bandage. If a minor cut occurs while working, wash the wound and allow it to bleed freely for a while. Then check for contamination. If uncontaminated, cover the wound with a waterproof bandage. If still contaminated, immediately contact radiation safety personnel for assistance.

### 5.1.7.1.7.4. Limitation of Dose to the Fetus

There is nothing in the current 10 CFR 20 to explicitly restrict a radiation dose to an unborn child. However, Part 19, §19.12, requires that workers "...shall be instructed in the health protection problems associated with exposure to such radioactive materials...." This requirement, in conjunction with Regulatory Guide 8.13, *Instruction Concerning Prenatal Radiation Exposure,*[20] and the radiation protection standards in Part 20 for all workers, provides such guidance as currently exists in the NRC standards. (Note that a revised version of 8.13 will be issued shortly which should incorporate the more explicit restrictions of the proposed Part 20.) Responsibility for reducing the radiation exposure to a fetus is currently placed largely upon the woman carrying the child. The guidelines recommend that exposure to the fetus be limited to 0.5 rem (0.005 Sv) for the entire gestation period.

It is generally accepted that rapidly dividing cells are unusually radiosensitive, which, of course, is the situation which obtains as the embryo develops. The early stages of fetal development, especially between the 10th to the 17th weeks, appears to be a critical time when the fetus is unusually susceptible to radiation damage. Unfortunately, during these first few critical months of pregnancy, the woman might not realize that she is pregnant and, hence, begin to take precautions to limit the exposure to the developing fetus. As noted above, Regulatory Guide 8.13 indicates that it is desirable for the woman to limit the total exposure of the fetus to 0.5 rem (0.005 Sv) during the entire gestation period. In order to avoid any unusually high exposures at potentially critical stages of development, the recommendation is that the exposure be spread uniformly throughout the prenatal period, or an average of 0.054 rem (0.54 mSv) per month.

Because of the uncertainty in the early, critical stages of pregnancy of whether an individual is pregnant, a restriction on all fertile women to an annual dose of 0.5 rem (0.005 Sv) might be considered desirable. Since only a limited number of women of fertile age are pregnant at any given time, this could give rise to discrimination against hiring women and unnecessarily affect their employment opportunities. Since radiation workers, on average, receive a dose equivalent of less than 0.5 rem/year, it does not appear justifiable to impose such a limit on fertile women as a class. It would, however, appear reasonable for women to continue to exercise reasonable care in minimizing the radiation exposure of any unborn child which they might be unknowingly carrying.

The proposed Part 20, §20.208 does set limits on the occupational exposure of a woman who has voluntarily *declared* or informed her employer that she is pregnant. These limits include both external and internal exposures. Many materials pass from the woman to the unborn child through the placenta. In some instances, the fetus receives a disproportionate share. In addition, the fetus may be exposed to radiation from materials in the organs of the woman's body. The proposed rule requires that the licensee ensure that the embryo/fetus in such a case not receive an effective dose in excess of 0.5 rem (0.005 Sv) during the entire pregnancy. The effective dose is the sum of the deep-dose equivalent to the declared pregnant woman and twice the committed effective-dose equivalent that would otherwise be assessed due to the intake of radioactive materials by the woman. The standard provides an alternative way to compute the latter component of the dose in some cases. Because of the uncertainty in establishing the actuality of the pregnancy, some women may not inform the licensee that they are pregnant until the fetus has already received a dose of 0.5 rem (0.005 Sv) or more. In such a case, the licensee is enjoined from allowing the fetus to receive more than 0.05 rem (0.5 mSv) during the remainder of the pregnancy.

This procedure should not involve an economic penalty on the part of the woman, nor should it jeopardize the woman's job security or employment opportunities.

Although the proposed standard is not in effect, it does provide a significant degree of protection to the embryo/fetus and might well be considered as a useful goal or supplement to current standards.

### 5.1.7.1.7.5. *Personnel Monitoring*

Determination of whether individuals receive doses in excess of the permissible levels requires that a personnel monitoring program which measures the doses to individuals be in effect. Under the current Part 20, most programs generally measure and maintain records for external doses only since the exposures to airborne concentrations are limited to a relatively small number of specialized facilities. The latter types of operations must measure the airborne concentrations of contaminants in areas which exceed 25% of the limits in Appendix B, Table 1 (column 1), of Part 20, averaged over the number of hours per week that individuals work in the laboratory. Bioassays may be required for some individuals under §20.108.

Personnel monitoring requirements are given in §20.202. The licensee must provide personal monitoring devices and require their use by employees who may receive an occupational dose in excess of 25%, in any calendar quarter, of the limits specified in §20.101, or of any individual who may enter a high radiation area. The percentage threshold is reduced to 5% for individuals under 18 years of age. According to §20.401, personnel monitoring records "…shall be preserved until the Commission authorizes disposition."

Part 19, §19.13, provides that workers or former workers can have access to their radiation monitoring records. Individual workers may request an annual report of their radiation exposure. Former workers may request their exposure history from an employer, and this must be provided within 30 d, or, if the records have not been completed for the current quarter by the time the request is made, within 30 d after the data are available. An employee who is terminating employment can also request and receive his monitoring records. These data must be organized by calendar quarter and must provide the dates and locations at which the exposures occurred. There also are additional technical requirements for the content and form of the data provided.

Maintenance of these records by each licensee is important since each employer must be aware of a worker's prior exposure history before the employer can allow entrance to a restricted area where an additional exposure in excess of 25% of the permissible quarterly limits may occur.

### 5.1.7.1.8. Methods of Monitoring Personnel Exposures

There are a number of devices, or dosimeters, which can be used to implement a personnel radiation monitoring program. These devices record the actual radiation received by the wearer at the location on the body where they are worn. Since the most radiologically sensitive organs are the blood-forming tissues, genitals, and eyes, the most logical place to wear a dosimeter is on a shirt pocket. For persons working at a laboratory bench or in front of a hood in which the radioactive material is located, the distance to the source will probably be least at this location and so the reading will most likely be higher than the average over the body. This is a conservative procedure. The most common devices are film badges, thermoluminescent dosimeter badges, and direct or indirect reading pocket ionization chambers (the latter are becoming muuch less common). Except for the latter type of dosimeter and those used to measure the dose to the extremities, §20.202(c) requires that personnel dosimeters utilized by licensees to comply with the monitoring requirements "…must be processed and evaluated by a dosimetry processor:

1.    Holding current personnel dosimetry accreditation from the National Voluntary Laboratory Accreditation Program (NVLAP) of the National Bureau of Standards, and
2.    Approved in this accreditation process for the type of radiation or radiations included in the NVLAP program that most closely approximates the type of radiations for which the individual wearing the dosimeter is monitored."

In order to accurately reflect the dose received by the body, a dosimeter ideally should have an energy response which would be the same as the tissue of the body. None of the dosimeters mentioned above fully meet this requirement, although the use of various filters in badges allows correction factors to be applied to the results.

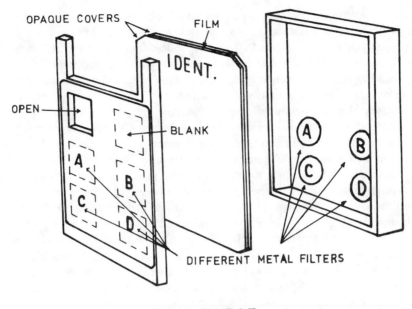

## FILM BADGE

FIGURE 5.9.  Film badge for personal radiation monitoring.

Each of the three types of units mentioned above will be discussed in the following three sections.

Other dosimetry materials are used in certain circumstances, such as very high exposure levels. There also are devices which electronically integrate the output of an electronic radiation detector and provide a digital reading of the accumulated dose, serving the same purpose as a simpler pocket ionization chamber, but at a considerably higher cost.

Versions of each of the following devices can be used to measure exposure to neutrons. However, since the use of neutrons outside of reactor facilities or some types of accelerator facilities is relatively uncommon, these will not be discussed here.

### Film Badges

The sensitivity of film to radiation was the means by which Roentgen discovered the existence of penetrating radiation, and it still serves as one of the more commonly used materials to measure radiation exposures. The radiation level to which the film has been exposed is proportional to the blackening of the film. For gammas below about 200 KeV, however, the energy dependence of the film depends very strongly on the energy, increasing rapidly as the gamma (or X-ray) energy decreases. In a typical case, the sensitivity of the film peaks at about 40 KeV, falling off below this level because of the attenuation of the soft gammas by the light-protective cover on the film. The sensitivity curve can be modified by using an appropriate filter, as shown in Figure 5.9. With filters, the sensitivity of a dosimeter using film can be made to be uniform to within about ±20% over an energy range of about 0.12 to 10

MeV. For intermediate-energy gammas, film is usable (with care) as a dosimeter for exposures of about 10 mrad (0.1 mSv) to 1800 rad (18 Sv). Film covered only by a light-tight cover also can be used to measure the dose from a beta emitter if the energy of the beta is above about 0.4 MeV, for exposure levels of about 50 mrad to 1000 rad (0.5 mSv to 10 Sv). There is no point in providing badges to individuals who work with low-energy pure beta emitters such as tritium, $^{14}$C and $^{35}$S since their betas would not pass through the light-tight cover of the film.

Film badges can be fairly sophisticated in their design. Even the simplest will have at least one film packet (or a portion of a larger piece of film) covered only by its light cover to permit detection of betas and soft gammas or X-rays.

Another area of the film is covered only by the plastic of which the badge holder is typically made, which will help distinguish between the betas and less energetic gammas. Usually one or more metal filters of different atomic numbers will be used which will change the response characteristic of the film as well as help define the energies of the incident radiation. Usually the film has two emulsion layers, one of which is "faster" (i.e., more sensitive than the other, in order to provide a wider response to different levels of radiation. Additional filters can be placed within the badge to allow it to detect and distinguish between thermal and fast neutrons. A diffuse source of radiation (such as scattered radiation) can be distinguished from a point source since the former will give diffuse edges to the images, while the latter will cause the edges of the filter images to be more sharply defined.

There are factors other than the varying energy response of the film which cause the reproducibility of their readings to vary. Each batch of film is a little different, the processing is subject to variation, and temperature will affect the results. The average accuracy of film badge dosimeters to doses of 100 mrad (1 mSv) or more is about ±25%. Values below about 50 mrad (0.5 mSv) are often relatively meaningless. The requirement that dosimeter evaluation be done by processors belonging to NVLAP is a measure taken to control as many of the variable factors as possible. Additional variations are introduced by the way they are handled and worn. They should be worn on the outside of the worker's clothes since clothes would filter the radiation received. The wearer should be careful to orient the badge properly since the various sections of the badge depend upon radiation being incident to the front of the badge. Obviously, the badges should never be placed in the employee's trouser pocket. Not only will the orientation be incorrect, but the badge will be shielded by coins, keys, and any other objects which might be present.

Although film badges have some disadvantages — specifically, relatively poor accuracy and a nonuniform energy response for low-energy gammas — the major advantage is that they represent a long-lasting record. If the original developed film is stored properly, it can be reevaluated at a later time if there is reason to suspect the original record is incorrect.

**Thermoluminescent Dosimeters (TLDs)**

The two most serious problems with a film badge — nonuniform energy response and reproducibility — are eliminated by a TLD dosimeter. Since a lithium fluoride (LiF) TLD with an effective atomic number of 8.1 is similar to tissue, which has an effective atomic number of 7.4, the LiF energy response is about the same as tissue. The response of an LiF TLD is almost energy independent over an energy range of

0.1 to about 3 MeV. A TLD can be better than 10% more accurate than a film badge over an exposure range of 10 mrad to 1000 rad (0.1 mSv to 10 Sv). The upper limit for a quantitative response is on the order of 100 times as great.

TLD materials are crystals in which radiation excites the atoms of the material, leaving them in long-lived metastable energy levels, locking the excitation energy into the crystal. Heating the crystals releases the energy in the form of light. If the heating process takes place at a uniform, controlled rate, the crystal will release the trapped energy in a reproducible "glow" or light curve as the temperature of the crystal rises. The area under the light curve generated in this manner is proportional to the quantity of the original radiation. TLD material is often in the form of a small chip, although one vendor of TLD badges uses Teflon™ impregnated with LiF instead. Obviously, the light seen by the reader from the detector will be affected by the condition of the surfaces, but if handled with care, a TLD dosimeter can be recycled many times.

As noted above, the energy response and reproducibility of a LiF TLD dosimeter is much better than that of a film badge. The former can be configured, as with film badges, to provide dose information on betas, X-rays, soft gammas, and more energetic gammas. However, unlike film, the process of reading the dosimeter to obtain the exposure information will destroy the information stored in the crystal. They also cannot readily distinguish between a point and a diffuse source. TLD dosimeters have gained a large share of the market in recent years because of their overall advantages, despite the deficiencies noted.

**Ionization Dosimeters**

Both of the two previous types of dosimeters require processing in order to be read. The two most common types of ionization dosimeters can be read immediately, although the indirect type requires an auxiliary reading device to do so. Ionization dosimeters used as personnel dosimeters are typically about the size of a pen and are usually called pocket dosimeters.

The indirect reading type is simply a high-quality cylindrical capacitor in which the outer cylinder, made of an electrically conducting plastic or having a conducting surface, forms one side or plate of the capacitor, and a wire in the center forms the other side. The capacitor is charged to a predetermined voltage, and radiation impinging on the walls and air in the cavity acts to discharge it, lowering the potential difference between the two electrodes. This difference is then read by an electrostatic voltmeter, calibrated so that the scale reads the amount of radiation to which the dosimeter and the wearer were exposed. The process of reading this type of ionization chamber discharges them so they must be recharged to be useful again.

The direct reading type also is a simple device. A fine, gold-plated quartz fiber, acting as a gold-leaf electroscope, is charged to a predetermined potential. As the air surrounding the fiber is exposed to radiation, the "electroscope" discharges and the quartz fiber moves toward the discharged position. The motion of the fiber, superimposed on a scale, is viewed through a simple microscope incorporated in the device. The scale is calibrated so that the change in position can be read as the amount of radiation to which the dosimeter has been exposed.

Both of these types of dosimeters are primarily intended to read gammas and provide a response which is within about 15% of the actual exposure over gamma energy range of about 0.04 to 2 MeV. Both can be configured to read over a wide range of exposures. Most of those used in laboratories are usually set up to read from 0 to 200 mrem. The reading of the units are conservative since charge leakage, or discharging due to factors other than radiation, will cause the readings to be too high.

### 5.1.7.1.9. Bioassays

The dosimeters discussed in the previous section measure the external fields of radiation to which the wearer is exposed. They do not measure the intake of any materials which will contribute to internal exposures. Methods of measuring airborne concentrations of radionuclides will be discussed in the following section, as part of the general subject of area radiation surveying. Measurement of the actual uptake of radioactive materials by the body is done by performing bioassays on individuals. There are routine laboratory uses of radioactive materials which are biologically active, such as the use of $^{125}I$ in certain laboratory analytical tests, which call for tests of specific organs to see if any of the material has been taken into the body. On other occasions, bioassays are used to determine if there has been any intake of material after a spill or other types of accidents in which there has been an unplanned release.

There are two distinct aspects to performing a bioassay: (1) measurement of an activity which is related to the original intake and (2) inferring the exposure from the measured activity. Of the two, measurement of an activity associated with the intake is the easier.

The basic assumption on which most bioassay procedures are based is that activity in the excreta, i.e., the urine or feces, will be related to the amount of activity taken into the body. Urine is the usual choice for monitoring, especially if the original material is soluble, while for particulates which are less soluble, both the feces and urine might be used. The radioactivity from $^{40}K$ (0.012% isotopic abundance, $1.28 \times 10^9$ year half-life, 1.31 MeV beta, 1.461 MeV gamma) in the potassium in urine typically results in about 45 dps/l as an interfering activity when trying to measure the activity due to an intake of some other radioactive material. Chemical removal of the potassium activity is possible and usually desirable to facilitate the measurement of the intruding isotope. Where the activity to be measured is due to a pure beta emitter, determining the activity in the urine and feces is almost the only way to obtain a measurement, unless bremsstrahlung radiation accompanies the beta emission.

Interpretation of the data obtained from urine and fecal measurements in terms of the activity in the body and the original amount taken in is complicated. Soluble materials containing radioisotopes which are distributed throughout the body lend themselves to the easiest analysis. Soluble materials which concentrate themselves in an organ represent a more complicated situation since they first must build up in the organ from the body fluids and then leave the organ to enter the excreta. Those which are absorbed in the bone have very long clearance times. Unfortunately, there are many factors which affect the metabolic processing of the materials and affect the interpretation of the results in terms of the actual internal exposure.

Urine analysis can be performed for gamma emitters as well, but in cases where the isotope is strongly concentrated in a single organ, a direct measurement of the activity in the organ using a scintillation counter (or perhaps a germanium detector) may be feasible. It also is possible to obtain a relatively accurate measurement of the amount in the organ by comparing the data with data obtained when using a known amount of the isotope (or an isotope with a comparable gamma spectrum) in a model (or phantom) which simulates the physical characteristics of a person as far as absorption and scattering of radiation from the organ are concerned. This is commonly done for iodine uptake measurements in the thyroid.

An extension of the concept of measuring the activity in a single organ is to use a "whole-body counter" which provides a measurement of the entire body burden of gamma-emitting radioisotopes. These are large, expensive devices and are not often found in a typical radiation safety operation, although there are a number of locations where access to one is available in emergencies.

If it is suspected that a radioactive material has been inhaled, measuring the activity from nose swipes or in tissue after the nose is blown can be used to confirm the inhalation. Measurement of radon concentrations in the breath is a technique which can be used if materials are inhaled which have radon as a decay product. Carbon 14 can be detected in exhaled breath as well.

### *5.1.7.1.10. Radiation Surveys*

Measurements of the radiation levels in an area are a vital aspect of radiation safety. Every laboratory using radioisotopes should have the capability of measuring the ambient activity in the facility at any time. In areas where there are fixed sources of substantial radiation, such as a $^{60}$Co irradiator, some types of accelerators, or a nuclear reactor, fixed radiation monitoring systems should be installed. These permanently installed systems should be supplemented by portable instruments use in determining radiation levels in localized areas which might not be "seen" by the fixed system or which arise from movable sources.

A number of types of instruments can be used for radiation monitoring, each with its own characteristics, advantages, and limitations. There have been serious radiation exposures when individuals used an instrument which was not appropriate for the radiation field involved. This section will be limited to the type of radiation detection instruments which are commonly used for radiation surveys, rather than those used for laboratory research.

A steady exposure to radiation which would result in a dose of approximately 100 mrem (1 mSv)/week or 2.5 mrem/h (25 μSv/h) would permit a worker to meet the current limits of occupational exposure for the whole body of 1.25 rem per calender quarter. Hands are often placed in radiation fields which would be at much higher levels than this, although they would rarely remain exposed to higher levels for an entire work week. The current limit of 18.75 rem per quarter for extremities would translate to about 1.5 rem (15 mSv)/week or approximately 37 mrem/h (0.37 mSv/h). A survey instrument for a laboratory using byproduct materials should be able to read levels which are a small percentage of the average whole-body exposure rate. Title 10, CFR (§35.72) stipulates that a survey instrument should be able to read a minimum of 0.1 mrem/h and should, as a minimum, be able to read levels at or above the equivalent rate for extremities. The high end of the instrument's range probably should be selected to be somewhat greater than this guide would indicate, especially in the case of spills in which an amount of material larger than that typically used in a single procedure may be involved. The maximum amounts of radioactive material that could be in the laboratory at a given time could be used to determine the high range of a survey instrument that might be selected (§35.51 requires survey instruments to be calibrated on all scales, up to 1000 mrem/h).

Most laboratory survey instruments use gas-filled chambers as their detectors. In this type of instrument, as discussed earlier, the passage of radioactive emissions (alpha, beta, or gamma) through matter creates a large number of ion pairs which can be translated into an electrical signal to detect particles. There are a number of parameters which affect the ability of an instrument to determine whether an interaction with the gas in the chamber is a discrete event. The combination of these factors is called the resolving time, with a short resolving time permitting more events per unit time to be detected individually than a longer resolving time. Other factors would affect the ability of the detectors to distinguish between emissions of different energies or types. In most cases, survey instruments for daily use in a laboratory are needed to merely detect the presence or absence of radiation, without regard to the energy or type. Most, however, have the ability, by mechanical means, to distinguish between types of radiation (taking readings with and without a cap over the detector will permit the instrument to distinguish between betas and gammas). Since the radioisotope in use is usually known and laboratories often use only one type of material or at least a limited number, the person doing the survey rarely will need an instrument which can measure energies or varieties of radiation for simple surveys. When an instrument with more sophisticated capabilities is needed, the radiation safety office can normally be counted upon to provide one. For more information on the properties of gas-filled counters, the reader is referred to any number of references on nuclear instrumentation.

Although a large number of commercial varieties of instruments are available, many with advanced and useful features, they often are elaborations, with more versatile and complicated circuitry, of the simple Geiger counters and ionization chambers which have been available for decades. The basic instruments are adequate to illustrate the basic safety features which are most critical.

All survey instruments that are kept in the laboratory should be calibrated at least once per year on all scales. If facilities are not available to do this locally, there are a number of commercial firms which will provide the service.

### Geiger-Müller Counter

The G-M counter, or Geiger counter as it is usually called, is the least sophisticated type of instrument which might be employed as a laboratory survey instrument. In general, its simplicity allows it to be built inexpensively as a rugged and, if used properly, relatively foolproof device.

In its basic form, a simple Geiger counter usually has a cylindrical probe containing a Geiger counter tube approximately 1 in. in diameter and about 6 in. long. The outer cylinder forms the cathode of the detector (the electrode toward which the positive ions from the ion pairs created by the passage of a particle or ray move ). A center wire forms the anode toward which the negative ions from the ion pairs move. The center wire is connected to the external circuitry. One end usually has a thin "window" with an aerial density of about 1 to 2 mg/cm$^2$ through which betas of approximately 30 KeV and greater can pass. The fragile window usually is covered with a removable cap to protect it. Alpha particles also can be detected with a thin window counter such as this if the counter is held very close to the source of the alphas (a 4.5 MeV alpha has a range in air of about 3 cm, but loses energy as it passes through the air so that at a distance of a little less than 2 cm, the alpha particle would not retain sufficient energy to penetrate even a 1 mg/cm$^2$ window). The geometry just described is not ideal for surveying a surface contaminated with alpha and beta emitters where the detector has to be held very close to the contamination, but many thousands of such instruments have been constructed and used, primarily for gamma surveys.

When a charged particle, gamma, or X-ray enters the active volume of a Geiger counter, not only does it create ion pairs, but the potential between the cathode and anode is set sufficiently high so that the ion pairs create an "avalanche" of charge along the entire length of the central anode, thereby developing a substantial charge on the center wire. The negative ions are electrons and move very swiftly to the anode, while the positive ions, being atoms, move far more slowly toward the cathode, represented by the outer wall of the tube. While the positive ions are moving, the detector tube is not able to initiate another pulse, or at least not one as large as the first. The minimum time which must elapse before another event can be detected is the resolving time. For most geiger counters, the resolving time is on the order of 100 to 200 μsec. This seems short, but it places a severe limitation on the number of events per second which can be detected. For example, let us calculate what a counter with a 200 μsec resolving time would measure were it operated such that it had a true count rate of 10,000 cps, assuming no problems with resolving time.

The equation which relates the true count rate, C, the observed count rate, c, and the resolving time, t, is given by Equation 17.

$$C = \frac{c}{1 - ct} \tag{17}$$

If the assumed numbers are inserted in this equation, the observed count rate is 3,333 cps, or a discrepancy of two thirds. Obviously, a Geiger counter cannot be used at a high interaction rate. In fact, if the number of particles or rays becomes too high, the counter will saturate, giving the impression that no pulses are occurring, and the apparent count rate will become zero. Under these conditions, the instrument would be worse than useless since a zero reading might be construed by the user as the true one and, on that basis, enter a dangerously high radiation field. This has happened, and serious radiation injuries have occurred as a result.

Because of the inability to function properly at high count rates, simple Geiger counters are not usable in high fields. A Geiger counter can be calibrated against a known source of a known specific activity in terms of millirads (mGy) or millirems (mSv). A basic Geiger counter usually has a maximum scale of 20 to 50 mrem/h (0.2 to 0.5 mSv/h). If the activity being used is small enough, this may be adequate, although instruments which can read levels 10 to 20 times as high, but still have a scale on which the maximum reading is 1 to 2 mrem/r (10 to 20 μSv), would be a more versatile general purpose instrument. An ionization chamber can be purchased which will meet this criterion, as discussed below.

**Ionization Chambers**

The basic geometry is similar to that of the Geiger tube, except that the diameter of the chamber usually is substantially larger, so the volume of the gas in which the radiation may interact is much larger. An ion chamber is operated at a high enough potential between the electrodes so that the ion pairs created by the impinging radiation do not recombine, but not so high that the moving ions create any significant number of secondary ions. The ions are collected on the anode and cathode and can, in principal, be measured as a current. A more intense (stronger) source will create more ion pairs and give rise to a stronger current. Actually, the current is not measured, but a large resistor ($10^{10}$ $\Omega$ or larger) is connected from the anode to the ground so that a voltage is developed across the resistor. A sensitive and stable electrometer circuit is used to measure the voltage. The larger the resistor, the more sensitive the counter. However, using larger resistors results in a penalty due to the time required for the current through the resistor to reach equilibrium (the time constant of the response is determined by the load resistor and the capacitance of the chamber). On a very sensitive scale, it could take 1 min or more for a steady-state voltage to be attained. An ion chamber requires more care than does a Geiger counter. Accuracy requires that the resistance values not change, and dirt and moisture can cause the resistance to decrease.

The advantage of the ion chamber is not that it can read very low fields, but that it can be designed, if the necessity arises or exists routinely, to read higher levels which represent a danger to the workers. At the higher levels, the slow response time observed at lower levels should pose no problems since the resistor and, hence the time constant, can be substantially lower and still provide an adequate voltage which can be measured.

As with the simple Geiger counter, the end window of the counter chamber can be made thin so that the unit is capable of counting alphas, betas, and gammas. If only one survey instrument can be afforded by a facility, an ion chamber with an upper range of 10 rem/h (0.1 Sv/h) would be a good choice.

**Other Concepts**

There are units which are designed to operate in what is called the proportional region of the confined gas, in which there is some multiplication of the charge deposited by the impinging radiation, but which permit much shorter resolving times so that count rate losses are much less severe. There are other arrangements of the electrodes such that the detector presents a large, sensitive area to the surface being surveyed and a shallow depth for the sensitive volume, instead of a long cylindrical geometry. This usually will result in increased sensitivity, especially for alphas and betas which have short ranges in matter. It also permits "seeing" a larger surface area and facilitates a survey of an extended area such as a work surface. If energy discrimination and sensitivity are needed, portable scintillation counters are available which incorporate NaI(Tl) crystals. If there is a possibility of personnel contamination, counters can be set up at the points of egress to automatically check the hands and feet of employees as they leave the facility. Friskers, which use a pancake-shaped detector, are available for rapidly scanning a person's body to detect contamination on areas other than the hands and feet. There are specialized counters for virtually any application.

**General Surveys**

Maps of the laboratory area should be prepared and, periodically, a general contamination survey of the area should be made using a general-purpose survey meter, especially if the isotope in use emits gammas or strong beta emitters such as $^{32}P$. The frequency of these general surveys should depend upon the level of use of material in the facility. If the use is very heavy, a daily survey of the immediate work area and a weekly general survey of the entire laboratory might be necessary, while, for very light use, once a quarter might be adequate. The records of these surveys should be maintained in a permanent log and audited periodically by the institution's RSO. They should also be retained for NRC use, should the need arise.

More detailed surveys of the immediate work area and certain critical areas should be carried out as part of contamination control, as discussed in the following sections.

### 5.1.7.1.11. Measurement of Airborne Activities

The collection of samples for testing airborne radioactive materials follows essentially the same procedures as those described in Section 4.8.2 for airborne monitoring of chemical contaminants. The basic procedure is to use a pump to circulate a known quantity of air through a collection device with known characteristics for capturing contaminants. If individuals are working in an area where airborne radioactivity may be a problem, it is recommended that not only routine periodic sampling of the air for activity be performed, but that the activities to which individuals may be exposed also be directly and specifically monitored. This can be done for an entire work day or, in some cases, during times of potential peak exposures by using small, light-weight pumps and collectors worn near the breathing zone.

If the suspected contaminants are particulates, membrane filters are frequently used which have collection efficiencies of 100% for particles which would be most likely to enter the deep respiratory tract (>0.3 to around 2 $\mu$m). A membrane filter traps particles on the surface, which is desirable for beta and alpha emitters since energy losses in the filter material could be important if the particles were carried into the filter. The concentration of a specific airborne contaminant in air is derived by measuring the activity of the collected sample, correcting for counting efficiencies, and dividing by the volume of air circulated through the filter.

Natural airborne activity will contribute to most airborne measurements of activity. This arises primarily from the decay products of the various radon isotopes. Radon is a gaseous radioactive element arising from the decay of thorium and radium. The two principal radon isotopes of concern are $^{220}Rn$ (55-sec half-life) and $^{222}Rn$ (3.82-d half-life). The first of these has such a short half-life that it will tend to decay before the gas diffuses very widely and hence the daughter products, which are not gaseous, will tend to remain in the same vicinity as the $^{224}Rn$, the parent nuclide for $^{220}Rn$. With its much longer half-life, $^{222}Rn$ will tend to diffuse away from the original location of its parent. The short-lived daughter products of $^{222}Rn$, with half-lives ranging from 164 $\mu$sec for $^{214}Po$ to 26.8 min for $^{214}Pb$, tend to become ionized and attach themselves to dust particles.

The activity of the particulates in the air is typically at an approximate state of equilibrium since additional radon gas is diffusing into the air constantly. However, the activity due to the daughter products of $^{222}Rn$ on the particulates that have been collected on the filter is not being replenished and will decay until a much longer half-life daughter is formed, $^{210}Pb$ (half-life, 22.3 years), which will have a much lower specific activity. A 4- or 5-h wait will usually see most of the activity from the daughter products of $^{222}$Rn gone. Although not as significant originally, the activity from the daughter products of $^{220}$Rn, such as that of $^{212}$Pb and its daughter products, may dominate the natural activity contribution after 4 or 5 h. Since the half-life of $^{212}$Pb is longer (10.6 h) than that of any of the succeeding (or preceding) daughters until a completely stable isotope is reached, the activity in the collected sample due to the daughters of $^{220}$Rn will have an equilibrium half-life of 10.6 h. Since the activity of most radioisotopes used in the laboratory is typically much longer than this, the activity of the airborne research isotope can readily be mathematically separated from the background natural activity by counting the collected sample several times over the next 2 to 3 d. If $t_c$ is the half-life of the contaminant and $t_n$ the half-life of the natural background radiation in the air sample, the activity due to the contaminant in the collected sample can be obtained from any two of the delayed counts by using the expression:

$$C_{c1} = \frac{C_2 - C_1 e^{-(0.693/t_n)t}}{e^{-(0.693/t_c)t} - e^{-(0.693/t_n)t}} \tag{18}$$

Here, t is the difference in time interval between the two counts employed in the calculation.

If several counts were taken, Equation 18 can be used for several pairs to make sure that only two half-lives are contributing to the activity. Within statistical counting variations, the results should be the same for each pair. If they change systematically as the time interval from the beginning of the initial count changes, then additional components are contributing to the activity.

If the airborne contaminant is contained in a solvent aerosol or is gaseous instead of being a particulate, the basic sampling procedures for nonparticulate materials in Section 4.8.2 can be used. The counting procedures may differ. Instead of using a standard counter such as a proportional counter for betas and gammas or a solid scintillation detector for gammas, a liquid scintillation detector might be used. The latter type is particularly effective for betas since its efficiency can approach 100%. If the contaminant is a gamma emitter, a more sophisticated counting system using either an NaI(Tl) or germanium detector and a dedicated multichannel analyzer or computer can differentiate the energies of the gammas. The latter types of counting systems, with either a hard-wired or software-based analytical program, cannot only positively identify the isotopes contributing to the energy but can also automatically determine the activity in the sample. These sophisticated analytical instruments are expensive compared to the type of instruments previously discussed, and normally are found only in laboratories heavily engaged in work with gamma emitters or in the radiation safety facility.

### 5.1.7.1.12. Fixed and Loose Surface Contamination

Contamination is the presence of undesirable radioactivity. It may be either fixed or loose. Usually, the acceptable limits for these in laboratories are set by the organization, except for certain items such as packages received in the laboratory for which Part 20 establishes acceptable standards at levels which should be sufficient to assure the occupational safety of personnel and that of members of the general public. Although acceptable levels are generally set by the licensee, 10 CFR 35.70 stipulates that in medical facilities involving humans, the licensee must have instruments capable of measuring radiation levels as low as 0.1 mrem/h or detect contamination of 2000 dpm with a wipe test. Although not explicitly stated, it could be inferred that these would be acceptable limits for surface contamination by the NRC in these facilities.

**Fixed Contamination**

Fixed contamination, by definition, is unlikely to either become airborne so that it may be inhaled or adhere to the hands when handled, so that it is also not likely to become ingested. Material which has soaked into a porous surface would be a typical fixed source of radioactive contamination. The hazard is an external one. Under the circumstances, an acceptable level could be defined by the acceptable levels for external exposures. In an unrestricted area, a level of 0.25 mrem/h (2.5 µSv/h) would barely meet this criterion. A level substantially below this would be preferable, such as 50 µrem/h (0.5 µSv/h). NRC Regulatory Guide No. 1.86, for purposes of releasing an area to unrestricted use, uses a value of 5 µrem/h (0.05 µSv/h) above background at a distance of 1 m from the source. This is roughly one third of the average background in most parts of the country. Such a level would be for a permanent cessation of operations and is probably lower than is necessary while the facility is licensed. While normal operations are being conducted and radiation safety survey and monitoring programs are actively performed, a level five times as high would correspond to a conservative value of 10% of the permissible level for unrestricted areas. For betas, a value of 50 µrem/h (0.5 µSv/h) for a reading close to the surface would not appear unreasonable for an unrestricted area. A detectable alpha level in an unrestricted area would indicate a possible dispersal of some of the more toxic radionuclides in the area, which should not be permitted to occur as a result of a research program. A nondetectable level of alpha contamination for fixed material in an unrestricted area would be desirable. However, levels equivalent to about 0.3 nCi or 10 Bq of loose contamination per 100 cm$^2$ are considered acceptable by some agencies.

For a restricted area, a level of fixed contamination well below the occupational level of 1.25 rem per calendar quarter (12.5 mSv per calendar quarter) or about 2.5 mrem/h (25 µSv/h) would be appropriate. Some organizations have set the level at 1 mrem/h (10 µSv/h). Another option might be to set it below a level which would require personnel monitoring, currently the equivalent of about 0.6 mrem/h (6 µSv/h), or, under the proposed Part 20, at about 0.25 mrem/h (2.5 µSv/h).

**Loose Surface Contamination**

The limits for loose surface contamination should be related to the amount of airborne contamination which would result if the material were to be disturbed in such a way that it could enter the body by inhalation or indirectly by ingestion through handling objects on which the material rested. The limiting factor is internal exposure. An examination of either the current or proposed Part 20, readily reveals that the permissible MPCs or DACs for the various isotopes vary by more than seven orders of magnitude, although the majority cluster around intermediate levels. For the isotopes most commonly used in the laboratory, the maximum permissible concentrations still differ by almost three orders of magnitude. This disparity makes it difficult to establish a single limit for loose surface contamination.

As noted above, a level of removable alpha contamination between not detectable and about 0.3 nCi (10 Bq)/100 cm$^2$ is considered acceptable by many different groups for an unrestricted area. In restricted areas, some organizations still do not accept any stray alpha contamination, except in an immediate work area such as the inside of a hood. However, a level of about 0.8 nCi or 30 Bq/100 cm$^2$ is considered acceptable by others.

Loose surface contamination is usually measured by rubbing or wiping an area of about 100 cm$^2$ with a filter paper and then counting the activity which adheres to the paper. An estimate must be made of the fraction of the loose material which is removed from the surface and remains on the paper to make a judgment on the amount of surface contamination.

For beta and gamma contamination, otherwise unspecified, levels of 0.1 to 1 nCi/100 cm$^2$ are considered acceptable by various organizations for unrestricted areas. For restricted areas, these organizations usually make the acceptable levels ten times higher, i.e., 1 to 10 nCi/100 cm$^2$. The United Kingdom and some others attempt to roughly take into account the differences in the MPCs and DACs by grouping the nuclides into toxicity classifications and recommending acceptable levels for each group. For the most toxic radionuclides (which includes the alpha emitters), an acceptable level of loose contamination in restricted areas is set at 0.8 nCi (30 Bq)/100 cm$^2$. The levels for medium toxicity radioisotopes are increased by a factor of 10, while the levels for lesser toxicity materials are increased by another factor of 10 to 100. This range of four orders of magnitude does not span the entire range of MPCs, but it is more reasonable than a single number for all species of radioactive materials.

All the numbers are for general area contamination. For personal clothing, they should be at least ten times less and for the skin of the body, the levels should be either zero or as close as can be obtained by decontamination efforts that do not require damage to the tissue.

### Frequency of Surveys for Contamination

In laboratories where unsealed radioactivity is in use, a contamination survey of the work areas directly involved in the operations should be made at the end of each workday or at the conclusion of that day's operations involving radioactive materials. Surveys of the workers hands and clothing also should be made at the end of each day's operations or more frequently if they leave the work area or take a break to get something to eat or drink. In any week that radioactive materials have been used, a contamination survey of the entire facility should be made. Particular attention should be paid to the work areas, radioactive material and waste storage areas, and the points of egress from the facility. The latter data are needed to demonstrate that no radioactive materials have been transported from the room on the soles of the employee's shoes.

### Accidents and Decontamination

Contamination may arise from poor laboratory practices, in which case the organization's radiation safety program must be prepared to take whatever steps are necessary to ensure that these practices are corrected. No such situation should be permitted to persist for long in a well-managed organization. In a properly run laboratory, accidents should be the major source of contamination of the facility. Even accidents should be infrequent in a well-run facility.

The scale of the accident will determine the response to the incident. If, for example, the accident is a spill in a hood, the base of which has been lined with absorbent paper and the apparatus set up in a pan or tray which is sufficient to contain all the spilled fluids, there may be little or no contamination or dispersal of airborne activity within the room because the vapors will be discharged in the hood exhaust. The individual working at the hood, if not personally contaminated, should notify the laboratory supervisor and the material should be removed from the hood and placed in a suitable container. The container can be sealed and placed in the area reserved for radioactive waste. Any additional protection required for cleaning up, such as the use of two pairs of gloves, respirators, and coveralls, should be decided upon prior to beginning the work. Every accidental spill should be reported to the radiation safety office, which may wish to assist, even in the case of a minor spill, in order to ensure that all personnel are protected. After the cleanup is completed, the entire area involved in the incident should be surveyed as well as all personnel involved in the original incident and the cleanup. The controls of instruments that may have been handled should not be overlooked. The problem of personal contamination is covered below.

Larger accidents pose more risk that individuals may have become directly contaminated or may have inhaled or ingested radioactive materials. Individuals in the laboratory other than those directly using the material may have been exposed as well. A contingency plan should have been developed which should be put into effect when a radioactive emergency occurs. The following two scenarios illustrate some of the more common problems that should be considered in preplanning for an emergency

### 1. A Spill Directly Upon A Person

The individual should stay where he is or, at most, move a few feet away to avoid standing in spilled material and remove all garments which are contaminated by the spilled material. The spread of contamination should be minimized. The garments should be placed in a pail or other liquid-tight container, preferably one that can be sealed. The pail can be placed in a garbage bag and the bag closed with a twist if a covered container is not available. After removing all contaminated clothing, the individual involved in the spill should wash all contaminated areas of the body. Be especially careful to wash areas where tissue creases or folds. If the spilled material is volatile, the contaminated individual and those assisting him should don appropriate respirators as soon as possible. Individuals who assist the contaminated person also should thoroughly wash their hands. Potentially contaminated persons should be checked for contamination. Persons not immediately involved in the incident and not needed should immediately leave the area until the potential for aerosol dispersal is determined. However, those leaving the area should be checked as they leave to confirm that they are not carrying contamination with them.

Radiation safety personnel and laboratory employees should jointly decide how to clean up the spill after the immediate contamination to personnel has been resolved, and determine a schedule which would allow operations to resume. An investigation should take place for any incident of this magnitude or greater, and the RSC should review the circumstances, immediate actions taken, and decontamination program to ensure that all actions taken were consistent with maximum protection of personnel and compliance with regulatory standards.

## 2. A Large Incident in which Significant Airborne Activity Is Generated

In any incident in which there is a likelihood of airborne contamination being generated, the most conservative approach is to immediately evacuate all personnel. This is based on the assumption that the situation will not rapidly worsen if it is left alone. An example of a situation which could worsen would be a small fire involving small quantities of radioactive material, but which could spread to involve larger quantities of radioactive materials and other dangerous materials as well. If the fire appears controllable with portable fire extinguishers (which should be conveniently available), one or two persons might choose to remain to attempt to rapidly put out the fire, but all other occupants should immediately evacuate. The individuals staying behind should leave as soon as possible. Note that this approach will depend upon such factors as the toxicity of the materials, the possibility of inhalation of airborne material, and the amounts involved as well as the possibility of controlling the fire. If staying to fight the fire only to minimize loss of property would endanger a person's life, the recommendation is to evacuate all personnel immediately.

Where the only problem is the generation of airborne radioactivity, individuals should leave the space in which the airborne activity is present, closing the doors behind them and turning off any ventilation in the area if the controls are in the room. If the ventilation controls are in a switch room or breaker room elsewhere, they can be turned off from this location. If the accident occurs in or near a fume hood which would vent the material from the laboratory, it may be appropriate to allow the hood to remain on to reduce the airborne concentration of activity within the room. However, this would depend upon the radioactive toxicity of the material involved. If it is one of the more toxic materials, it would usually be preferable to avoid dispersal into a public area, even if the levels are low, because of the possible exposure to persons in these areas to the material. The laboratory represents a controlled space which can be decontaminated and, in general, the best approach is to restrict the contaminated area as much as possible.

Individuals within the laboratory should, as in any other emergency, evacuate the area via routes which would minimize their risks. Any location within the facility should have two evacuation routes. The storage location of portable radiation survey instruments should be near the entrance to the facility, along the evacuation route deemed most likely to be safe in a radiation-related emergency.

Once the occupants have left the immediate area of danger and it has been isolated, consideration of the next steps to take in the emergency should take place. The first order of business is for an individual (preferably one with the least potential of having experienced personal exposure in the incident) to request assistance from the radiation safety department and other responsible authorities. Concurrently, others should begin to evaluate any personal radiation exposure which may have taken place by using the laboratory's portable instrumentation to check each individual. No one should leave the immediate area, assuming it is safe to remain, until they have been thoroughly checked for contamination, for their own safety and in order to prevent the spread of contamination. In the case of airborne contamination, it would be essential to check the accessible portions of the respiratory tract for traces of inhalation. All individuals who are knowledgeable about the incident should remain available to provide information to those persons arriving to take charge of the emergency response.

Once radiation safety personnel and other persons such as the building authority have arrived and been made thoroughly aware of the circumstances of the accident, a plan can be developed to respond to the situation. Other temporary steps to isolate the problem can be taken, such as taping around the edges of the doors and defining a secure area with barricades to prevent persons from entering or approaching the area and interfering with the response program.

Groups involved in planning the response should include the participation of laboratory personnel, radiation safety specialists, managerial personnel, media relations staff, and other support groups, as dictated by the nature of the incident.

An operations center should be established outside the immediate area of the incident and all operations and information releases should be managed from this center in order to insure that all actions to control and remedy the situation are properly coordinated. Since every incident is different, the correctional plans also would be different. However, in general, the approach is to establish a control point at the boundaries of the affected area through which all personnel, supplies, and waste would enter and leave. All persons and material leaving this area must be checked for contamination and all individuals entering should be checked to see that they are suitably protected. The normal procedure is to start at the edge of the contaminated area and work to reduce the area involved until all contamination has been removed or reduced to an acceptable level.

Assistance should be sought immediately for individuals who might have received internal and external contamination. Contact with a regional radioactive incident response center is advisable since few local medical centers have sufficient training, experience, or facilities to respond to an internal radiation exposure incident. If the telephone number of the appropriate center is not immediately available, the regional NRC office, which should have been notified as part of the response plan, will be able to provide it. The regional center may recommend transfer of the affected personnel to a facility in their area. Nasal swipes should have been taken as soon as possible to see if any material had been inhaled. Any such information which could possibly aid medical personnel in their evaluation and treatment is invaluable. The response center may be able to suggest short-term measures to reduce the amount of radioactive material taken up by the body.

## Decontamination Techniques

Regardless of whether contamination has been caused by routine operations or by a major or minor spill, the problem of decontamination arises. The cost and feasibility of decontamination of individual pieces of equipment and the added exposure risk to personnel must be weighed against the cost of replacement equipment and the cost of disposal of any material considered as radioactive waste. The latter costs are not negligible.

The facility itself, including major items of equipment such as hoods, work benches, and valuable specialized items of laboratory apparatus, is generally worth substantial decontamination efforts since it and the equipment are expensive to replace. At current average building costs of $150 to $200 per square foot for laboratory space, a facility is not lightly abandoned or left idle for extended periods of time. If the facility had been built originally to facilitate decontamination of all types of spills and the major items of equipment specified for ease of maintenance, the decontamination process is relatively straightforward. Personnel, dressed in coated Tyvek™ coveralls, shoe covers, and head covers, and wearing gloves and respirators, can clean most such areas by washing the surfaces with detergent and water since the surfaces should have been selected to resist absorption of water as well as chemicals. The equipment in the room should have few seams or cracks in which contamination can become lodged.

Unfortunately, many laboratory facilities using radiation have been located in older buildings with tile, unsealed concrete, or even wooden floors in which radio-active contamination can become fixed, and the equipment has many seams and cracks which lend themselves to becoming contaminated. If surfaces are porous or if water would collect in the cracks, the use of detergent and water may not be desirable. In addition, since materials tend to adhere more strongly to surfaces as time passes due to moisture, oils, etc. in the air, if it is not desirable to use a cleansing solution to remove the material, it would be best to begin decontamination as soon as practicable while the material is still relatively loose. If a very limited area is involved, loose particulates often may be picked up by pressing the sticky side of masking or duct tape against the contaminated surface. If a larger area is contaminated, much of the loose material can be picked up by using a HEPA filter-equipped vacuum cleaner. The filter in a vacuum cleaner of this type is capable of removing 99.97% of all particles of 0.3 µm or greater from the air moving through it. Brushing lightly with a soft brush can break loose additional material which can be vacuumed with the special vacuum. In order to keep airborne material from spreading, the entrance to the area being cleaned can be isolated with a temporary "airlock" made of 6-mil polyethylene plastic which will block any air moving to and from the isolated area, but will permit passage of workmen in and out of the room. Airborne material also can be continuously removed from the area by placing a movable, HEPA filter-equipped air circulator within the space being cleaned. The circulator should be sized to pass the air in the space through the filter frequently. A 2000 cfm unit would pass the air (if thoroughly mixed) in one of the standard laboratory modules described in Chapter 3 through the filter about every 2 min.

If contamination has permeated deeply into the work and floor surfaces, as well as the seams and cracks in the floor, it may be simpler to remove some of the work surface and replace the floor tile than to attempt cleaning. If the floor tile contains asbestos, the removal would have to be done according to OSHA restrictions on asbestos removal.

Every decontaminated area must be thoroughly surveyed by the organization's Radiation Safety Office and certified as meeting the limits for surface contamination established by the organization prior to its release for use.

Pieces of electronic equipment are perhaps the most difficult items to clean since they frequently are equipped with fans which draw air through them, and over a period of time, they all accumulate at least some dust and grime to which the contamination is likely to adhere, even in rooms in which the air is carefully filtered. Normally, washing interior components is impractical because of the potential damage to the components. It may be possible to use a very fine nozzle on a HEPA vacuum to clean the bulk of the removable dust from the inside of the instrument, and the remainder may be loosened and removed with the careful use of small swabs and solvent, followed by another vacuuming. The decision to make the effort will depend on the cost of the equipment, and the difficulty of the decontamination effort. One factor that might influence such a decision would be the possibility of a successful damage claim on the organization's insurer. The disposal cost of the equipment should be considered as well as the replacement cost.

Small tools and glassware with hard, nonporous surfaces may be cleaned by standard techniques. Washing with a detergent and water may be sufficient. More resistant contaminating material in glassware can be cleaned with any of the standard solutions used in laboratories for cleaning glassware, such as a chromic acid solution. Contaminated metal tools that resist cleaning with detergent and water, accompanied by brushing, can be washed with a dilute solution of nitric acid. Sulfuric acid can be used on stainless steel tools. Again, the choice must be made between the relative cost of the tool, the cost of disposal, and the effort required to decontaminate it. Tools which have grease in and on them, and hidden crevices where radioactivity may become embedded, may be too difficult to decontaminate to make the effort cost effective.

If disposable protective clothing is used throughout the decontamination process, the problem of laundering contaminated garments can be minimized. However, any items that need to be cleaned should be checked by the radiation safety office prior to sending them to a commercial laundry. If the quantity is not large, manual washing of some items might be possible, taking care not to contaminate skin. Where more clothes are involved, purchase of a washing machine might be cost effective. A simple, inexpensive model would normally be sufficient and, unless the clothes are heavily contaminated in a form physically difficult to remove, probably would be able to eliminate or reduce the contamination to an acceptable level. Any wastewater should be captured and retained until approved by the RSO for disposal into the sanitary system.

### 5.1.7.1.13. Radioactive Waste Disposal

In recent years, almost all off-site, low-level* radioactive waste has been sent to three commercially operated waste disposal burial grounds located in South Carolina, Nevada, and Washington. Other commercial facilities were closed because of various environmental problems. The choice of the location to which generators of radioactive waste sent their waste was dictated by a number of operating restrictions for the three facilities that remained operating. As a result, many institutions found themselves sending much of their laboratory-type radioactive waste all the way from the east coast and elsewhere to the Washington facility.

Because Congress deemed it inappropriate for one region to have to handle another region's wastes, a system of regional compacts was to be established and, when the system became fully operational, each region would be required to take care

---

* Low-level waste, as opposed to the high-level waste engendered by the use of uranium as a fuel.

of its own radioactive waste. Some of the regional compacts are fully operational, while others are not. At this time, it is still possible for generators in some regions to send waste to another area, but at some point in the future, this will no longer be allowed. The date on which this will occur is not wholly certain because it may be modified by Congress to allow additional time for regions who have not set up facilities to do so.

Although the nation's research and medical facilities do not produce the volume of waste generated by the electric utilities, they are much more vulnerable to the lack of a disposal facility for their waste because they do not have adequate facilities in which to store low-level waste on a long-term basis. It is essential for many research applications to use certain isotopes which, because of their radiation characteristics, must go to a long-term disposal facility. The inability to use these isotopes would seriously hamper many of the research programs. These research applications often are in the life sciences and involve such critical areas as cancer, the search for cures to other diseases, recombinant DNA, and other comparably important areas of basic and applied research.

The actual amount of waste shipped off-site at many research facilities has decreased or stabilized as more facilities use waste-reduction techniques to minimize the amount of radioactive waste which must be shipped away.

Decay-in-storage is the simplest method available to reduce the amount of waste which must be disposed of in a burial facility. Many of the most commonly used radioisotopes have half-lives so short that the radioactive content will decay to a "safe" level within a reasonably short time. For example, all of the following have half-lives of less than 65 d, so that any material in the waste will go through at least ten half-lives in less than 2 years, which is generally accepted as a reasonable storage period: $^{131}$I, $^{32}$P, $^{47}$Ca, $^{33}$P, $^{51}$Cr, $^{59}$Fe, $^{203}$Hg, and $^{125}$I. Also, the combination $^{99}$Mo $\rightarrow$ $^{99m}$Tc, which has important nuclear medicine applications, has a short effective half-life, so any waste from its use also can be allowed to decay in storage. The initial activity of any radioactive material when placed in storage will determine if ten half-lives is sufficient to reduce the level to a point where an unshielded reading with a survey meter on its most sensitive scale shows no reading in excess of background. If this condition is not met, then additional decay time must be allowed before the material is disposed of as ordinary waste.

Another significant means of reducing the amount of waste is to take use the provisions of §20.303 to dispose some of the radioactive waste into the sanitary system. However, §20.303(a) prohibits the disposal of licensed material into the sanitary system unless it is readily soluble or dispersible in water. Many of the fluids used in liquid scintillation fluids do not meet this criteria, but there are now commercially available alternative scintillation liquids which do. The use of these newer fluids permits licensed or unlicensed material to be put into the sanitary system as long as the amount in any one day does not exceed the larger of (1) the quantity, diluted by the average daily quantity of sewage released by the licensee, provided the former does not exceed the concentrations permitted in Appendix B, Table 1 (column 2), (2) the same restriction, except averaged over a month, or (3) ten times the amount listed in Appendix C of Part 20. Another restriction limits the yearly amounts to 5 Ci of $^3$H or 1 Ci of $^{14}$C and a total of 1 Ci of all other radioisotopes combined. $^{14}$C and $^3$H, two of the most commonly used materials, have long half-lives so advantage cannot be taken of the decay-in-storage technique; the capability of using the sanitary system is of significant help in reducing the waste volume requiring commercial burial. Records of the amounts disposed of by this method must be maintained, as well as the amounts disposed of by other means.

Very low specific activity (0.05 μCi or less per gram) $^3$H or $^{14}$C in scintillation fluid or in animal tissue (averaged over the weight of the entire animal) can be disposed of without consideration of the radioactivity, but is normally incinerated. Some generators have licensed incinerators for the disposal of radioactive waste.

Some dry solid wastes can be compacted to reduce their volume. Some generators have developed centralized local facilities to do this, while others place these materials into separate containers and have a waste disposal firm perform the waste volume reduction.

These methods, plus encouraging users to choose a material which can be allowed to decay in storage, may be employed to substantially reduce the amount and expense of radioactive waste disposal.

In order to take advantage of these waste- and cost-reduction measures, users must segregate the waste into appropriate categories: (1) aqueous, (2) nonaqueous, (3) aqueous containing $^3$H and $^{14}$C from other aqueous, (4) vials of nonmiscible scintillation fluids containing less than 0.05 μCi/g $^3$H and $^{14}$C from other scintillation fluids, (5) dry solid wastes, and (6) animal carcasses, separated into those containing more than 0.05 μCi/g/d of $^3$H and $^{14}$C and those which contain less than that amount. The generators of the waste must label each container of waste with the amount of each radioactive isotope contained in it. This information is used by the radiation safety office to prepare the waste for shipping and to prepare the required manifests and forms as specified in §20.311 or by a waste disposal firm which may perform this service as well as transport the waste to a burial ground. Each container or package of waste which is shipped must be classified and labeled according to 10 CFR 60, §60.55 to 60.57.

Almost all waste is originally packed into 55-gal steel drums which meet DOT specifications, segregated by type. Generally, landfills will not take liquid wastes, even if they have been absorbed in an appropriate absorber. They must be solidified before offered for burial. Some organizations have a commercial service perform this procedure. In most cases, the decision to perform a given operation locally or have it done by an outside contractor is usually based on economic factors. Nonaqueous waste is usually treated as a Resource Conservation and Recovery Act (RCRA) waste as well.

Radioactive waste which is allowed to decay in storage may be disposed of according to the following schedule:

- Dry solids are taken to an ordinary landfill. Any labels on any container indicating that the contents are radioactive must be removed or defaced.
- Nonhazardous water-miscible waste is put into the sanitary system.
- Chemically hazardous waste is disposed of via the hazardous waste disposal program. Labels on a container indicating that the contents are radioactive must be removed or defaced.
- Carcasses are normally incinerated.

Proper disposal of all radioactive waste is a major responsibility of the organization's radiation safety office. Proper records must be maintained of all transfers of radioactivity from the laboratory into the waste stream.

### 5.1.7.1.14. Individual Rights and Responsibilities

Individuals who are employed in licensed programs in which they work with byproduct materials (or in a number of other activities regulated under Title 10, Chapter I, of the Code of Federal Regulations) have a number of legal rights and responsibilities spelled out in the regulations, primarily in Parts 19 and 21. Part 21 requires the responsible parties in any licensed activity to report to the NRC any safety deficiency, improper operations, or defective equipment which could pose a "substantial safety hazard…to the extent that there could be a major reduction in the degree of protection provided to public health and safety…." However, any individual can make such a report. If a responsible party fails to do so, substantial civil penalties can be imposed.

Part 19 defines the rights of the employees and those of the NRC, as well as the responsibilities of the licensee to make these rights available. The major provisions of Part 19 are summarized below.

### 5.1.7.1.14.1. Information Requirements

The licensee must post a number of items of information relating to the operations under the license or post information describing the documents and where they may be examined. The latter option is the one most often exercised and involves the following documents:

- The regulations in 10 CFR, Part 19
- The regulations in 10 CFR, Part 20
- The license, license conditions, and documents incorporated into the license by reference
- The operating procedures applicable to the licensed activities
- Form NRC-3, Notice to Employees

The following items must be posted, without the option of simply informing the employees of their existence and where they may be examined.

- Any violations of radiological working conditions
- Proposed civil penalties or orders
- Any response of the licensee

A critical section of Part 19 is "§19.12, Instructions to Workers", given in its entirety below. It is brief, but functionally equivalent to the OSHA Hazard Communication Standard for employees engaged in activities regulated by the NRC.

### §19.12 Instructions to Workers

All individuals working in or frequenting any portion of a restricted area shall be kept informed of the storage, transfer, or use of radioactive materials or of radiation in such portions of the restricted areas; shall be instructed in the health protection problems associated with exposure to such radioactive materials or radiation, in precautions or procedures to minimize exposure, and in the purposes and functions of protective devices employed; shall be instructed in, and instructed to observe, to the extent within the worker's control, the applicable provisions of Commission regulations and licenses for the protection of personnel from exposure to radiation or radioactive materials occurring in such areas; shall be instructed of their responsibility to report promptly to the licensee any condition which may lead to or cause a violation of Commission regulations and licenses or unnecessary exposure to radiation or to radioactive material; shall be instructed in the appropriate response to warnings made in the event of any unusual occurrence or malfunction that may involve exposure to radiation or radioactive material; and shall be advised as to the radioactive exposure reports which workers may request pursuant to §19.13. The extent of these instructions shall be commensurate with potential radiological health protection problems in the restricted areas.

Implicit in the above instructions is that the employees must comply with the regulations and procedures adopted to protect themselves and others, including the general public outside the areas in which the licensed activities take place. However, it is not sufficient for an employer to place the responsibility on the employee. The employer must see that the employees comply with all applicable standards. There have been several instances where heavy fines have been imposed on corporations and academic institutions which have failed to enforce compliance.

### 5.1.7.1.14.2. Monitoring Data
Under a number of provisions of §19.13, individual workers have the right to information concerning their radiation exposures and directly related supporting data.

- The worker can request an annual report.
- A former worker can request a report of his exposure records. The report must be provided within a 30-d interval or within 30 d after the exposure of the individual has been determined, whichever is later.
- An employee terminating work involving exposure to radiation, either for the licensee or while working for another person, can request the information. An estimate, clearly labeled as such, may be provided if the final data are not available at the time of the request.
- If the licensee is required to make a report to the NRC of an individual's exposure data, a report must be made to the individual no later than the report made to the NRC.

### 5.1.7.1.14.3. Inspection Rights
The NRC has the right to make unannounced inspections of a licensed facility, including the "...materials, activities, premises, and records...." During an inspection, the inspectors have the right to consult privately with the workers. However, at other times, the licensee or a representative of the licensee can accompany the inspector.

The workers may have an official representative who may accompany the inspector during the inspection of physical working conditions. Normally, this is an individual who is, himself, a worker engaged in licensed activities under the control of the licensee and, hence, could be expected to be familiar with the instructions provided employees by the employer. However, if agreeable to the employer and to the employees, an outside person can be permitted to accompany the inspector during the inspection of the physical working conditions. Different employee representatives can accompany the inspector during different parts of the inspection, but no more than one at any given time. If a person deliberately interferes with an inspector while conducting a reasonable inspection, he can refuse to let that person continue to accompany him. Certain areas, such as classified areas or those where proprietary information is involved, would require that only individuals normally having access to these areas or the information could accompany the inspector.

During an inspection, an employee can bring privately to the attention of the inspector any matter relating to radiological safety pertaining to the licensee's operations that he wishes, either orally or in writing.

An employee or employee representative can report a violation of the terms of the license or regulations covering the radiological activities engaged in by the employee. This should be done in writing to the Commission or to the director of the Commission's regional office and should contain pertinent details of the alleged violations. The identity of the employee making the report will not appear in any public report, except under demonstrated good cause.

If the director of the regional office feels that there is valid cause to believe that a problem exists as described by the complainant, he will initiate an inspection as soon as practical. If an inspection does take place, the inspector will not necessarily confine himself to the original problem. The director may decide that an inspection is not justified. However, a complainant may request an informal conference to press his case. A licensee also may request one, although in this case the complainant has the right to not allow his identity to be made known.

The licensee is specifically prohibited against any act of discrimination against an employee who makes a complaint or asks for an inspection.

### 5.1.7.1.14.4. Penalties

Violations of Part 19, as with other regulatory sections, can be classed as *de minimis* or higher. If the violations are sufficiently serious, the Commission can obtain a court order prohibiting a violation or imposing a fine, or a license may be revoked. If a violation is willful, the person responsible can, in principle, be considered guilty of a crime and be subject to a fine and/or imprisonment. Of course, none of the penalties can be invoked without the licensee having ample opportunity to respond to the charges and to propose a plan of correction.

## REFERENCES (SECTIONS 5.1 TO 5.1.7.1.14.4)

1. **Komarov, E. I.,** Radiation in daily life, *Int. Civ. Defense,* No. 295, 3.
2. New a-bomb studies alter radiation estimates, *Science,* Vol. 212, May 22, 1981.
3. Ionizing Radiation: Sources and Biological Effects, United Nations Scientific Committee on the Effects of Atomic Radiation (UNSCEAR), 1982 Report to the General Assembly, with Annexes, United Nations, New York, 1982.

4. **Fabrikant, J. I.,** Carcinogenic effects of low-level radiation, *Health Phys. Soc. Newsl.,* X (No. 10), October 1982.

5. Radiation Exposure of the Population of the United States, NCRP Report No. 93, National Council on Radiation Protection, Bethesda, MD, 1987.

6. **Luckey, T. D.,** Physiological benefits from low levels of ionizing radiation, *Health Phys.,* 43(6), 771, 1982.

7. Special issue on radiation hormesis, *Health Phys.,* 52(5), May 1987.

8. **Cohen, B. L.,** Alternatives to the BEIR relative risk model for explaining atomic-bomb survivor cancer mortality, *Health Phys.,* 52(1), 55, 1987.

9. *Location and Design Criteria for Area Radiation Monitoring Systems for Light Water Nuclear Reactors,* ANSI/ANS-HPSSC-6.8.1, American National Standards Institute, New York, 1981.

10. *Performance Requirements for Pocket-Sized Alarm Dosimeters and Alarm Rate Meters,* ANSI N13.27, American National Standards Institute, New York, 1981.

11. **Hodges, H. D., Gibbs, W. D., Morris A. C., Jr., and Coffey, W. C., II,** An improved high-level whole-body counter, *J. Nucl. Med.,* 15(7), 610, 1974.

12. **Graham, C. L.,** A survey-instrument design for accurate β dosimetry, *Health Phys.,* 52(4), 485, 1987.

13. **Kathren, R. L.,** *Radiation Protection, Medical Physics Handbook 16,* Adam Hilger, Accord, MA, 1985.

14. Standards for Protection Against Radiation, 10 CFR, Part 20, *Fed. Reg.,* May 31, 1983.

15. Standards for Protection Against Radiation, 10 CFR, Part 20, proposed rules, *Fed. Regist.,* 51(6), 1092, 1986.

16. Rules of General Applicability to Domestic Licensing of Byproduct Material, 10 CFR, Part 30, *Fed. Regist.,* March 31, 1987.

17. 10 CFR Part 61, Licensing Requirements for Land Disposal of Radioactive Waste, *Fed. Regist.,* 47(248), 57446, 1982.

18. Decontamination Limits, Regulatory Guide 1.86, Nuclear Regulatory Commission, Washington, D.C., 1974.

19. Decontamination Limits, ANSI/ANS-N13.12, American National Standard Institute, New York.

20. Instruction Concerning Prenatal Radiation Exposure, Regulatory Guide 8.13, Nuclear Regulatory Commission, Washington, D.C., 1975.

21. Information Relevant to Ensuring that Occupational Exposures at Medical Institutions Will Be As Low As Reasonably Achievable, Regulatory Guide 8.18, Nuclear Regulatory Commission, Washington, D.C., 1982.

22. Instructions Concerning Risks from Occupational Radiation Exposure, Regulatory Guide 8.29, Nuclear Regulatory Commission, Washington, D.C., 1981.

23. **Steere, N. V., Ed.,** Safe handling of radioisotopes, in *Handbook of Laboratory Safety,* CRC Press, Boca Raton, FL, 1971, 42.

24. **Strom, D. J.,** The four principles of external radiation protection: time, distance, shielding, and decay, *Health Phys.,* 54(3), 353, 1988.

25. **Saenger, E. L.,** Acute local irradiation injury, in *Proc. REAC/TS Int. Conf. Medical Basis for Radiation Accident Preparedness,* Hübner, K. F. and Fry, S. A., Eds., Elsevier/North-Holland, Amsterdam, 1980.

26. **Martin, A. and Harbison, S. A.,** *An Introduction to Radiation Protection,* 3rd ed., Chapman and Hall, New York, 1986.

27. **Morgan, K. Z. and Turner, J. E.,** *Principles of Radiation Protection,* John Wiley & Sons, New York, 1967.

28. **Pelá, C. A., Ghilardi, A. J. P., and Netto, G. T.,** Long-term stability of electret dosimeters, *Health Phys.,* 54(6), 669, 1988.

29. **Cember, H.,** *Introduction to Health Physics,* Pergamon Press, Oxford, 1969.

30. Radiological Health Handbook, Public Health Service, U.S. Department of Health. Education and Welfare, Washington, D.C., 1970.

## 5.1.8. RADON

The problem of radon is not necessarily connected with laboratory operations involving radiation, other than being a contributor to the background for counters

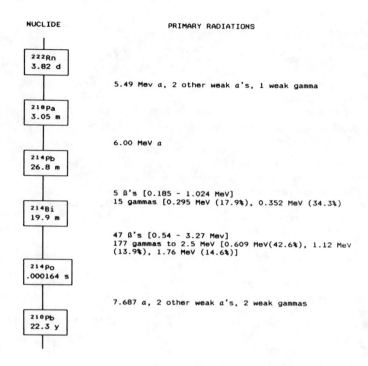

NUCLIDE                                    PRIMARY RADIATIONS

222Rn
3.82 d

5.49 Mev α, 2 other weak α's, 1 weak gamma

218Pa
3.05 m

6.00 MeV α

214Pb
26.8 m

5 β's [0.185 - 1.024 MeV]
15 gammas [0.295 MeV (17.9%), 0.352 MeV (34.3%)

214Bi
19.9 m

47 β's [0.54 - 3.27 MeV]
177 gammas to 2.5 MeV [0.609 MeV(42.6%), 1.12 MeV
(13.9%), 1.76 MeV (14.6%)]

214Po
.000164 s

7.687 α, 2 other weak α's, 2 weak gammas

210Pb
22.3 y

FIGURE 5.10.    Primary decay path for radon[222].

used in these facilities and representing a problem when taking an air sample. However, many laboratories are built of concrete blocks and have concrete floors, drains which open to storm sewers, and other means by which radon can enter the facilities. The sand and aggregate used for the concrete in both the blocks and cement slabs may contain substantial amounts of natural radioactivity. Currently, a considerable amount of concern is being raised about the possibility of radon-induced lung cancer from this source. Since a typical research worker may spend an average of 8 or more h/d in a research facility, it may be useful to provide a brief section on radon. A discussion of the physics of radon activity was provided in Section 5.1.7.1.10 Figure 5.10 provides some additional information on the decay products of $^{222}$Rn, the major environmental problem

A recent study by the National Research Council[1] has resulted in a new estimate of the risk of lung cancer which is intermediate between those of earlier estimates. The report also suggests that the risk to smokers is 10 or more times greater than that to nonsmokers. Although compiled by an eminent group of scientists using a more sophisticated set of assumptions than in the past, other well-known scientists have questioned some of the assumptions made and some features of the model. The results are given below, but before presenting them, a brief discussion of some of the terms is needed.

Risk estimates are made in terms of working-level months (WLM). A working-level month is defined as "an exposure to a concentration of any combination of the short-lived radon daughters in 1 L of air that results in the ultimate release of 1.3 × 10⁵ MeV of potential alpha energy for a working month of 170 hours." This amount of radiation is approximately equivalent to the radiation of a radon daughter in equilibrium with 100 pCi of radon. A level of 4 pCi/l of radon is equal to about 0.02 working levels. Over an entire working year, equal to 2000 h, the equivalent exposure would be equal to about 0.25 WLM.

The BEIR IV report[1] estimates that the excess lifetime risk of death from lung cancer from this source is 350 deaths per million person WLM. Based on this model and the assumed levels of exposures throughout the U.S., this is equivalent to an estimated 5,000 to 20,000 lung cancer deaths due to radon per year in this country.

At the time of this writing, there are reputable individuals who both support and object to these findings. However, because there is some concern about the potential of radon as a cause of lung cancer, a number of commercial services have been established which will provide devices which can be placed in the home for a period of time, and then can be sent to their laboratories for analysis of the ambient radon levels. These firms place advertisements in the media so that anyone wishing the analysis can readily have it done. In some states, state agencies will make a survey upon request.

## REFERENCES (SECTION 5.1.8)

1. *Health Risks of Radon and Other Internally Deposited Alpha-Emitters; BEIR IV,* National Academy Press, Washington, D.C., 1988.
2. **Cohen, B. L. and Gromicko, N.,** Adequacy of time averaging with diffusion barrier charcoal adsorption collectors for $^{222}$Rn measurements in homes, *Health Phys.,* 54, 195, 1988.
3. Radon in Buildings, NBS Special Publ. 581, Collé, R. and McNall, P. E., Jr., Eds., National Bureau of Standards, Gaithersburg, MD, 1980.
4. **Puskin, J. S. and Yang, Y. A.,** A retrospective look at Rn-induced lung cancer mortality from the viewpoint of a relative risk model, *Health Phys.,* 54(6), 635, 1988.

### 5.1.9. ACUTE RADIATION SYNDROME

It is highly improbable that an individual working with byproduct materials will ever be exposed to radiation levels which would cause him to be concerned about the immediate acute effects of radiation, leading to serious injury or death. However, there have been a relatively small number of incidents in which massive exposures have occurred. These have normally involved large sources used for radiography, radiation therapy, sterilization of insects, machines producing intense beams of radiation, X-ray units, and critical reactor assemblies. Individuals have survived very high localized radiation doses, in the range of 10,000 rems or more, if the areas affected were relatively small and did not expose the more sensitive organs to a substantial dose. Following is a brief discussion of the immediate effects of a short-term exposure to high levels of radiation. It should be noted that the levels which cause immediate harm to individuals are hundreds of times greater than that permitted individuals over a period of months.

There is significant variation in the responses of individual persons to exposures of the whole body. The least amount of radiation for which clinical evidence is discernible in a "typical" person is approximately 25 rad (0.25 Gy) The effect will be noted as changes in the blood system. Above about 50 rad (0.5 Gy), most persons will show some effect on their blood system. There may be temporary sterility in men whose gonads have been exposed to these levels. However, it is unlikely that an individual will personally feel any immediate discomfort at these relatively low levels.

As the exposure levels increase to approximately 100 rad (1 Gy), a small percentage of sensitive persons will begin to experience some of the symptoms of discomfort associated with acute exposures to substantial levels of radiation — nausea, fatigue, loss of appetite, sweating, and a general feeling of malaise. As the exposure level increases, the percentage of individuals exhibiting symptoms increases, At an exposure of 200 rad (2 Gy), most persons will be affected to some degree and at about this same level, some individuals who do not receive treatment may not survive. At an exposure of 300 to 450 rad (3 to 4.5 Gy), the survival rate without aid decreases to 50% by 30 d after the exposure. With proper treatment, survival is a distinct possibility, up to about 500 rad (5 Gy). Even with optimal treatment, survival chances decrease rapidly above 500 rad; at about 800 rad (8 Gy), survival is highly unlikely.

Many of the individuals who have a finite chance of survival will begin to feel better after a day or two, but after a few weeks, their situation appears to worsen. They may experience fevers, infections, loss of hair, severe lethargy, hemorrhaging, and other problems with their cardiovascular systems. During this 2- to 3-week interval, blood tests will reveal that significant changes are occurring in the blood system which reflect the damage to the blood-producing bone marrow. At about 200 rad (2 Gy), depression of the bone marrow function is apparent. At higher doses, complete ablation of the bone marrow occurs. Up to some exposure level, if the patient can be kept alive, the bone marrow will regenerate, but at a level in the area of 400 to 600 rad (4 to 6 Gy), this will become unlikely and bone marrow transplants will be necessary for survival. There are problems with this technique, including the need for a close genetic match with the donor material and the possibility of infection. However, if the blood-forming function can be restored and death does not result from secondary infections because of the lack of white blood cells, survival for 60 d or more is a promising sign of eventual survival.

As the level of radiation increases, the survival period decreases. At about 1000 rad (10 Gy), there will be serious damage to the gastrointestinal tract and survival may be a few weeks. At exposures of several thousand rad, the central nervous system will be strongly affected and survival will be measured in terms of hours.

# REFERENCES (SECTION 5.1.9)

1. **Upton, A. C. and Kimball, R. F.,** Radiation biology, acute radiation syndrome, in *Principles of Radiation Protection,* Morgan, K. Z. and Turner, J. E., Eds., John Wiley & Sons, New York, 1968, 427.
2. **Wald, N.,** Effects of radiation over exposure in man, in *Principles of Radiation Protection,* Morgan, K. Z. and Turner, J. E., Eds., John Wiley & Sons, New York, 1968, 457.
3. **Andrews, G. A., Sitterson, B. W., Kretchner, A. L., and Brucer, M.,** Criticality accident at the Y-12 Plant, in Proc. Scientific Meeting on the Diagnosis and Treatment of Acute Radiation Injury, World Health Organization, Geneva, October 17, 1960.
4. **Andrews, G. A., Hübner, K. F., and Fry, S. A.,** Report of 21-year medical followup of survivors of the Oak Ridge Y-12 accident, in *The Medical Basis for Radiation Accident Preparedness,* Hübner, K. F. and Fry, S. A., Eds., New York, Elsevier/North-Holland, 1980, 59.
5. Radiation accident grips Goiania, *Science,* 238, 1028, 1987.

# 5.2. X-RAY FACILITIES

There are a number of radiation safety issues associated with the use of X-rays in the laboratory. Much of the health physics information in standard texts, as applied to X-ray systems, address radiation safety in terms of a determination of the amount of shielding necessary in medical X-ray installations to protect persons in adjacent areas from the effects of radiation. The radiation sources are the primary beam, scattering from the patient or target, and leakage from the radiation source. Except in specialized texts, there is little discussion of operational personnel protection. Although there are many research applications in which X-ray machines are used in a manner equivalent to medical usage, even here there are many applications in which there are significant differences in safety considerations. For example, in veterinary medicine, X-rays of large animals are not only likely to cover a larger area and involve more scattering mass, but it is far more likely that a holder (or more than one) will be needed than in the case of most human patients. In addition to briefly covering shielding requirements, the sections which follow will emphasize the exposures which the researcher receives, the doses to the patient, and the means by which these exposures can be reduced. Shielding design cannot be neglected, but will often be decided by the architect and his consultant prior to the research scientist being involved. In most cases, the research scientist will have relatively little control over his physical environment, especially those who begin working in existing facilities. Further, most will seek out a consultant to assist them should the need arise to create an X-ray facility in an existing structure. Most X-ray users' expertise and their interests are primarily in applications, not facility design, other than those features which enhance its usefulness to them. However, the scientist will exercise essentially complete control over the operations of the X-ray facility and should be held responsible for establishing operating procedures which maximize radiation protection for everyone.

Many individuals associate anything to do with radiation with the NRC, but it neither regulates the use of X-ray machines nor establishes exposure levels for individuals working in these facilities. The Food and Drug Administration, in 21 CFR, provides standards for X-ray devices, but for the most part, the responsibility for regulating the use of equipment which emits X-rays is left up the states. The degree of regulation provided varies among the various states. The National Committee on Radiation Protection and Measurements (NCRP) has recommended limits of exposure which are similar to those required by the NRC, basically, 100 mrem/week (1 mSv/week) as an occupational limit and 10 mrem/week (0.1 mSv/week) for the general public.

## 5.2.1. GENERATION OF X-RAYS

The general form of a device in which X-rays are generated is one in which electrons emitted from a cathode are accelerated through a voltage, V, and focused on a target which has a high atomic number. The primary mechanism by which the X-rays are produced is bremstrahlung, as the electrons slow down within the target. A substantial amount of heat is generated in the process since only about 1% or less of the energy of the electron beam is converted to X-rays. The target, in order to have a high atomic number for the maximally efficient conversion of energy into X-rays, is usually made of tungsten, which also has a high melting point. The target is usually embedded in an efficient heat conductor, usually copper, serving as the anode. Some designs add fins to conduct heat away from the anode, some cause the anode to rotate so the beam will not be continuously focused on the same spot, and other designs circulate cooling liquids through the anode.

The maximum value of the energy of the X-rays generated in an X-ray tube is determined by the accelerating potential, usually denoted by kVp. However, the energy spectrum of the X-rays is a continuum extending downward from this maximum energy. Thus, in any X-ray spectrum there are many "soft" X-rays which often are not of any significant value, although if the normal operating voltage is low, a portion of the wall of the tube may be made of a thin piece of beryllium to permit more of the lower energy component to escape from the tube. In most cases, these unwanted X-rays are eliminated by interposing aluminum filters (in the case of higher energy machines, aluminum and copper are used to extend the energy of the filtered component to higher levels) in the beam. These energy-modifying filters are typically 0.5 to 3 mm thick, depending upon the kVp of the X-ray machine.

### 5.2.2. TYPES OF MACHINES

As noted above, X-ray machines for medical applications represent only one type of X-ray-emitting device used in the laboratory. There are open- and closed-beam analytical instruments used for X-ray diffraction work, cabinet systems, electron microscopes, and microprobes, among others. Each of these devices can, under appropriate circumstances, pose radiation exposure problems. In addition to installing these in properly designed facilities, operating procedures must be established to limit occupational exposures as well as those of patients (for medical machines) and members of the general public. Users of these machines should have access to a copy of *A Guide to Radiation Protection in the Use of X-Ray Optics Equipment*,[2] which provides much practical guidance to assist X-ray equipment users to do so safely.

#### 5.2.2.1. Diagnostic Machines

There are three sources of radiation for a diagnostic X-ray machine (and for an open-beam analytical machine): (1) direct radiation from the primary beam, (2) scattered radiation from the target, and (3) leakage from the protective tube housing. If the orientation of the machine can change, it may be necessary to design all the walls of the structure to provide adequate shielding for the primary beam. If the machine has a permanent fixed orientation, one wall can be designed to protect against the primary beam, while the other walls, and possibly the floors and ceiling, will only have to be designed to reduce the intensity of the scattered and leakage radiation to an acceptable value. An acceptable value for the radiation level in a given space will depend upon whether the area is a controlled-access area and whether the exposures of the personnel working within the space are monitored. Maximum permissible exposure levels, P, are 100 mrem/week (1 mSv/week) in a controlled area and 10 mrem/week (0.1 mSv/week) in areas outside the controlled area. The primary shielding barrier thickness required can be obtained from the characteristics of the machine, the manner in which it is used, and the occupancy level of the spaces on the other side of the barrier.

### Primary Beam Shielding

The information required to design an adequate shield for the primary beam is (1) the peak voltage (kVp) at which the machine may be operated, (2) the maximum current, I, of the electron beam used to excite the X-rays, usually expressed in milliamperes (mA), (3) the amount of time per week the unit is in use, usually expressed in minutes, and the fraction of time the machine is in use that it is aimed at the wall for which shielding is being calculated, U, and (4) a factor, T, which gives the occupancy level of the adjacent space. The required thicknesses of shielding can be computed by using this information and experimental data for the effectiveness of shielding of different materials for broad-beam X-ray irradiation at different kVps. The compiled data is given in terms of the factor, K.

For the primary beam, K can be calculated from:

$$K = \frac{Pd^2}{ItUT}$$

Here, d is the distance from the beam target in the X-ray tube to the desired point on the other side of the protective barrier. Usually, the product, It, is replaced by the single factor, W. All the factors on the right-hand side of the equation are normally known, so K can be calculated. All that remains to obtain the required thickness is to find K, select the given kVp curve from a family of published curves for a given shielding material, and read the shielding thickness from the abscissa.

The factor, T, ranges from 1 for full occupancy (space occupied by full-time employees, children play areas, living quarters, employee restrooms, occupied space in immediately adjacent buildings, etc.) to 0.25 for areas with pedestrian traffic, but unlikely to have long-term occupancy (corridors too narrow for desks, utility rooms, public parking lots, public restrooms) and $1/16$ for areas occupied occasionally (stairways, automatic elevators, closets, outside areas used by pedestrians or vehicular traffic).

### Scattered Radiation

The corresponding equation for radiation scattered from the object being X-rayed is

$$K_{ux} = \frac{P \times (d_{sca})^2 \times (d_{sec})^2 \times 400}{a \times W \times T \times F \times f}$$

Here: $K_{ux}$ = scattered radiation rate/workload in rads per milliampere per minute @ 1 m

$a$ = scattered-to-incident ratio

$d_{sec}$ = distance from a point to the scattering object in the X-ray beam in meters

$F$ = area of beam impinging on scatterer (cm$^2$)

$d_{sca}$ = distance from X-ray source to scatterer

$f$ = a factor which depends on the kVp and adjusts the equation for the enhanced production of X-rays as the energy increases. Below 500 kVp, f is taken to be 1.

### TABLE 5.8
### Half-Value and Tenth-Value Layers

| Voltage (kVp) | Lead | | | Concrete | | |
|---|---|---|---|---|---|---|
| | HVL | (mm) | TVL | HV | (cm) | TVL |
| 50 | 0.06 | | 0.17 | 0.43 | | 1.5 |
| 70 | 0.17 | | 0.52 | 0.84 | | 2.8 |
| 100 | 0.27 | | 0.88 | 1.6 | | 5.3 |
| 125 | 0.28 | | 0.93 | 2.0 | | 6.6 |
| 150 | 0.30 | | 0.99 | 2.24 | | 7.4 |
| 200 | 0.52 | | 1.7 | 2.5 | | 8.4 |
| 250 | 0.88 | | 2.9 | 2.8 | | 9.4 |
| 300 | 1.47 | | 4.8 | 3.1 | | 10.4 |

### Leakage Radiation

A properly shielded diagnostic X-ray tube is limited to a maximum of 100 mrem/h (1 mSv/h) at a distance of 1 m. In order not to exceed the maximum weekly exposure at any distance, an attenuation of the leakage radiation is required, as given by Equation 19.

$$B_{Lx} = \frac{P \times d^2 \times 600I}{WT} \tag{19}$$

The thickness of shielding required to provide this degree of attenuation for a broad beam can be found by computing the number of half-value layers (thickness required to reduce the intensity of an incident beam by one half) which would be needed. In order to attenuate both the scattered and leakage radiation, the shielding required for each component is computed using the equation for $K_{ux}$ and Equation 19. If one of the answers is more than a tenth-value layer larger than the other, the thicker shield can be considered adequate. If not, then the thicker shield should be increased by one half-value layer.

Equation 19 can also be used for a therapeutic X-ray machine if the factor 600 in the numerator is changed to 60, reflecting the ten-factor higher leakage allowed for this type of machine. Tables 5.8 and 5.9, adapted from NCRP 49, provide half- and tenth-value layers for lead and concrete for a number of peak voltages and the scattered-to-incident ratio, a.

It will be noted that the minimum for the scattered ratio, a, in the energy range covered by Table 5.9 comes at an angle ranging from 45 to 90° with respect to the incident beam. At lower energies, the minimum is closer to 45°, while at the upper end of the voltage range, the minimum shifts toward 90°. Since the range of energies in Table 5.9 covers a large portion of the diagnostic voltages used, persons acting as holders should position themselves accordingly and at as great a distance as feasible from the scatterer and the X-ray tube.

A recent article concludes that the procedure described gives a value which will be too thick and, consequently, more expensive than necessary and provides an

**TABLE 5.9**
**Scattered to Incident Exposure Ratio, a**

| Voltage (kV) | Scattering angle (relative to incident beam) | | | | | |
|---|---|---|---|---|---|---|
|  | 30 | 45 | 60 | 90 | 120 | 135 |
| 50 | 0.0005 | 0.0002 | 0.00025 | 0.00035 | 0.0008 | 0.0010 |
| 70 | 0.00065 | 0.00035 | 0.00035 | 0.0005 | 0.0010 | 0.0013 |
| 100 | 0.0015 | 0.0012 | 0.0012 | 0.0013 | 0.0020 | 0.0022 |
| 125 | 0.0018 | 0.0015 | 0.0015 | 0.0015 | 0.0023 | 0.0025 |
| 150 | 0.0020 | 0.0016 | 0.0016 | 0.0016 | 0.0024 | 0.0026 |
| 200 | 0.0024 | 0.0020 | 0.0019 | 0.0019 | 0.0027 | 0.0028 |
| 250 | 0.0025 | 0.0021 | 0.0019 | 0.0019 | 0.0027 | 0.0028 |
| 300 | 0.0026 | 0.0022 | 0.0020 | 0.0019 | 0.0026 | 0.0028 |

alternative procedure which is somewhat more complicated. Since the method in NCRP 49, and briefly outlined here, will result in a conservative shield, the additional thickness and cost may be justified since it does provide some margin for error and provides for lower exposures.

### Exposure to Users

Many people are abnormally afraid of radiation because of a lack of knowledge or because of misinformation which they may have. A recent questionnaire given to second year medical students before they had taken a course in biophysics or radiation applications in medical school, but after they had completed an undergraduate, presumably scientifically orientated, program, revealed a striking and troubling level of confusion about X-rays. If this is the situation which obtains for a well-educated group (the entrance requirements for medical schools are notoriously high), the lack of knowledge on the part of the general public is likely to be far worse. Four of the questions and the results are given below.

1.  Following the completion of an X-ray radiographic examination, objects within the room:
    a.  emit a large amount of radiation   29.3%
    b.  emit a small amount of radiation   43.9%
    c.  do not emit radiation              26.8%
    d.  don't know                         0.0
2.  Intravenous contrast materials used in angiograms and intravenous pyelograms are radioactive.
    a.  true        37.0%
    b.  false       61.0%
    c.  don't know  2.0%
3.  Gamma rays are more hazardous than X-rays.
    a.  true        58.5%
    b.  false       22.0%
    c.  don't know  19.5%
4.  Nuclear materials used in nuclear medicine are potentially explosive.

a. true          24.4%
b. false         68.3%
c. don't know    7.3%

The normal practice in taking medical X-rays is for the patient to be situated with respect to the source and film cassette by the technician and, where possible, the patient is asked to "hold still" while the technician retreats to a shielded control room and activates the X-ray unit. The technician, under such circumstances, does not receive a dose other than that permitted by the scattered and leakage radiation levels behind the shield. Since the permissible limits are 100 mrem/week (1 mSv/week), the X-ray technician must be provided with a dosimeter capable of reading X-rays accurately. Both TLD and film badges are commercially available which are satisfactory for the purpose. The reproducibility and inherent radiological similarity to tissue of commonly used TLD dosimeters are desirable features for this type of dosimeter. The ability of film to distinguish between a diffuse, extended source, as would be the case for the scattered radiation, and that of a localized source is an advantage of film dosimeters. The patient is exposed to not only the direct beam, but also to the scattered and leakage radiation. However, the exposure of the patient will usually be a single episode, or at least a limited number, while the occupational exposures of the technician will continue. When X-ray personnel are not wearing the dosimeter badges, they should be stored or kept in an area at background radiation levels so that the dosimeter readings will reflect true occupational exposures.

The control room booth should have walls and a viewing window with sufficient shielding capability to ensure that the operator's exposure should not exceed the permissible limits for a controlled area. The booth must be of adequate height and width to ensure complete shielding while the beam is on, and the distance between the activation switch for the beam and the entrance to the door to the X-ray area should be sufficient to guarantee that an X-ray technician cannot operate the machine in an unshielded location. The entrances to an X-ray area should be posted with a sign bearing the words (or equivalent) *"CAUTION: X-RAY EQUIPMENT"*. It would also be desirable if a visual or audible warning were provided should anyone attempt to enter the beam area while the beam is on. Another precaution which can be taken is to have the doors leading into the X-ray area, other than those from the control area, interlocked so that they can only be opened from the outside when the beam is off or so that the X-ray machine will be turned off automatically when the doors are opened.

Not all X-rays are taken under ideal circumstances. The patient may be in pain and find it difficult to stay in position. Children may be frightened or simply too young to follow instructions. In certain fluoroscopic examinations, even healthy, cooperative patients may need assistance. Passive restraints should be used wherever possible, but, on occasion, patients must be held. This must not always be the responsibility of a single individual or a limited group of persons. Excessive personal doses are likely unless this duty is shared among a number of persons. Since the X-ray beam is not on except for well-defined and determinable intervals, there should be no increase in the collective dose over that of the single individual by following this practice. No one not essential to the procedure should remain in the room while the X-ray is being taken.

Holders should remain as far from the primary beam as possible. As noted earlier, the intensity of the scattered radiation is a minimum of 45 to 90° with respect to the incident beam for most commonly used diagnostic X-ray voltages. A lead apron or movable shield can substantially reduce the whole-body exposure. As noted in Table 5.8, less than 1 mm of lead will reduce the intensity of a broad beam of 150-kVp X-rays by a factor of ten. In an apron, this amount of lead might be too heavy for comfort for some individuals, but even 0.5 mm of lead will substantially reduce the level of radiation to the torso.

If a lead apron is not worn, the dosimeter badge should be worn at the waist or above so that the measured dose will be representative of a whole-body exposure. If worn on the upper part of the body, the dosimeter will not be partially shielded by the X-ray table, in a vertical beam exposure, when holding a patient lying down.

When a lead apron is worn, the dosimeter should be worn above the apron, on the shirt collar or lapel of a laboratory coat, so that it provides a reading of the dose to the exposed area above the apron. To avoid exposure to the fetus, a pregnant woman should not act as a holder, especially in the early stages of the pregnancy. Should it be necessary for a pregnant woman to act as a holder, she should wear a second dosimeter under the lead apron in the vicinity of the fetus. If an employee is aware that she is pregnant, she should "declare" (as under the proposed Part 20) or inform her supervisor of her pregnancy. The supervisor, with the cooperation of the pregnant employee, should see that she limits the fetal exposure to 0.5 rem (5 mSv) over the entire pregnancy. In order to confirm that this is so, a more frequent monitoring schedule should be established for this individual, unless work assignments can be made which preclude her exceeding the recommended levels.

Persons who assist with fluoroscopic procedures should wear ring badges as well as body dosimeters. The beam stays on for longer periods of time than for a normal X-ray film exposure (sometimes longer than necessary when the radiologist is not conscientious about turning off the beam as promptly as possible at intervals during the procedure) and, consequently, the potential for exposure of the assistant is higher. Additional shielding to protect the viewer and any aides from scattered radiation should be provided as needed. If a direct viewing screen is employed, the entire primary beam must be intercepted by the screen or the equipment should be inter-locked so that it will not work. Units are now available that permit the image to be viewed on a high-resolution television screen instead of directly, which substantially reduces the amount of radiation exposure to the radiological physician.

Film and TLD dosimeters are normally read on a monthly or quarterly basis instead of immediately, although the latter can be read in a few minutes if a calibrated and certified TLD dosimeter reader is available. If there is concern about potential exposures and an immediate reading is needed or desired, pocket dosimeters designed to be properly responsive to X-ray fields should be used to supplement the standard dosimeters and provide an immediate exposure reading.

Many of the devices which decrease the dose to the patient, such as faster film or screen intensifiers, also reduce the exposure to persons required to be in the room. Beam filters, usually of aluminum, are used to eliminate the soft X-rays from the X-ray beam. This reduces the skin exposure to the patient and the exposure to scattered radiation from this source to the worker. The recommended filter thicknesses are 0.5 mm aluminum (or equivalent) for machines operating at up to 50 kVp, 1.5 mm for operations between 50 and 70 kVp, 2 mm for 70 to 125 kVp, and 3 mm for 125 to 300 kVp. Collimators, which are used to define and limit the incident beam to only the area of clinical interest, also reduce the amount of scattering mass within the beam. The irradiated area is required by current standards to be defined by a light which is accurately coincident with the area exposed to the X-ray beam. The visible target will allow any required assistant to take a position which will be as far from the primary beam as possible and which will minimize the exposure to scattered radiation.

The genital areas of patients of reproductive age must be protected with a shield unless the shield would interfere with the diagnostic procedure. It would be desirable to know if a woman is pregnant prior to an X-ray of the pelvic area, and it is recommended that the physician suggest a pregnancy test to fertile women if they are sexually active. This is obviously a very sensitive topic and must be done extremely tactfully. However, failure to discuss the risk and warn the patient of possible problems with the development of the fetus could result in allegations of liability on the part of the physician. A sign prominently posted in the X-ray facility such as *"IF YOU ARE PREGNANT, IT IS IMPORTANT THAT YOU TELL THE TECHNICIAN"*, could be helpful. In an emergency situation, this concern should not delay taking an X-ray. Since the patient's body usually is exposed to scattered and leakage radiation from the areas being X-rayed, if a substantial series of shots are to be taken, shields for other sensitive parts of the patients body should be considered.

No more X-rays than necessary should be taken. No X-rays of humans should be taken for training, demonstration, or other purposes not directly related to the treatment of the patient.

One of the more disturbing problems in diagnostic X-rays is the frequent failure of either the physician or the radiologist to respond to the patient's question of "how much radiation will be received". Too often, the attitude appears to be that "it has been decided that the patient needs the X-ray and that is all that matters" There is substantial concern about the effects of radiation as well as a substantial number of misconceptions. This fear and confusion may inhibit the patient from having a needed X-ray. Not only should a reasonable answer be given, but the significance of the answer should be explained since it is unlikely that most persons will be knowledgeable about radiation safety terminology. The output from a typical X-ray machine is generally quite substantial. At 50 kvP, with no filtration and a thin beryllium window, the output of an X-ray tube is about 10 rad/min/mA at 1 m or 100 mGy/min/mA at 1 m. For a 100-kVp machine with a 3mm-external aluminum filter, a typical exposure rate is 3 rad/min/mA at 1 m or 30 mGy/min/mA at 1 m. These rates or, preferably, the actual rates for the machine can be used to provide an estimate of the doses to the patient. Techniques to minimize the exposures to the patient should be used. A normal "good" level of exposure to the patient for a chest X-ray is 10 mrad (1 mGy) or less.

The performance of a diagnostic X-ray machine should be checked at least annually with appropriate instruments by a qualified person. If this capability is not available in-house, a qualified consultant should be hired to perform the task. Survey instruments for X-ray systems require an ability to accurately read lower-energy radiation than most commonly used survey meters for radioisotope applications. If any maintenance is performed or if the machine is relocated, a survey should be performed for leakage radiation from the source. If the unit is moved to another facility, the exposure levels in the adjacent areas should be tested to ensure that the exposure levels are within the permissible limits for controlled and uncontrolled areas. Records of all maintenance, surveys, leakage checks, calibration, personnel monitoring, etc. should be maintained at the facility and at the radiation safety office. Because of the long latency period for cancer resulting from radiation exposures, it would not be unreasonable to maintain personnel exposure records for up to 40 years.

### 5.2.2.2. Open-Beam Analytical Machines

Open-beam analytical X-ray machines have many research applications which involve frequent manipulations of the equipment and samples. There are many opportunities for excessive occupational exposures. In crystallography studies, for example, a very-high-intensity beam, although confined to small areas, is involved. If an individual were to insert his hands (or any other portion of his body) into the beam, dangerously high localized exposures could occur in a few seconds. The occupational radiation levels considered acceptable for general whole-body exposure are the same as for diagnostic machines, 100 mrem/week (1 mSv/week) in controlled areas. The maximum hourly levels are those which would be equivalent to this rate for a 40-h week, 2.5 mrem/h (25 $\mu$Sv/h). Acceptable levels for hand exposures are 15 times greater than those for whole body exposures, 37.5 mrem/h (37.5 $\mu$Sv/h). For exposures in uncontrolled areas, the levels should not exceed 2 mrem/h (20 $\mu$Sv/h) or 100 mrem/week (1 mSv/week) on a short-term basis or 500 mrem (5 mSv) averaged over a year.

Laboratories which use open-beam analytical machines should possess a survey instrument capable of measuring the scattered radiation from the machine and associated apparatus around the experimental area. The scattered radiation from a crystal is nonuniform, which can cause narrowly defined "hot" spots if the radiation is permitted to escape from the apparatus into the laboratory area. A careful survey needs to be made whenever maintenance is performed on any part of the system which could permit radiation to escape from the apparatus or when the system is reconfigured. If an alignment or maintenance procedure must be done with the beam on, surveys must be performed while this work is in progress, preferably by two persons, one of whom does the actual work while the other person monitors the radiation levels. Periodic surveys should also be made to ensure that no changes have been made or occurred, without everyone's knowledge, which could affect the radiation field near the apparatus. A quarterly schedule is recommended while work is actively in progress. A survey should also be made at any time anyone suspects that an abnormal condition could exist which might increase occupational exposures.

Personal dosimeters should be worn by individuals in the laboratory whenever the machine is operated and finger badges should be worn whenever working directly on the apparatus, making adjustments, or performing any action that could cause the fingers or hand to receive an abnormal exposure. All badges, when not being worn by the individual, should be in a background-level radiation area so that they will reflect the actual occupational exposure of the person to whom they have been assigned.

A procedure manual must be prepared for systems as potentially dangerous as open-beam analytical systems and all persons must be made familiar with the procedures which are applicable to safe use of the equipment. The person in charge of the facility is responsible for ensuring that all personnel are informed of safe procedures and that it is required that all personnel comply with the written safety policies. This should be done not only by instruction, but also by the laboratory director or supervisor setting an appropriate example as well. Every individual working in the facility should be trained in the basic principles of radiation safety, the permissible exposure levels for controlled and uncontrolled areas, what dosimeters to wear and when, what surveys are required, and what records need to be maintained. All employees should also be informed of their rights to limit their exposures and their rights to their exposure records.

Laboratories in which analytical X-ray machines are located should be posted with signs at all entrances, bearing the legend **Caution — Apparatus Capable of Producing High-Intensity X-Rays** or the equivalent. At any time the machine is on, it should be attended or the doors to the experimental area should be locked. Keys should be restricted to those persons who work within the facility. It is especially critical to prevent access or not to leave the apparatus unattended if the equipment is on and partially opened for alignment.

Measures must be taken to warn persons unfamiliar with the risks associated with the equipment, should they enter the facility while no one is present, even if the equipment is not on. The X-ray source housing of analytical X-ray equipment should have a conspicuous sign attached to it bearing the words, **Caution — High-Intensity X-Rays**, and a similar sign placed near any switch which could energize the beam, stating **Caution — This Apparatus Produces High-Intensity X-Rays When the Switch is Energized.** In addition, a lighted sign stating **X-Ray System On** or the equivalent should be near the switch. The light should be incorporated in the activation circuit so that if the light fails, it will be impossible to activate the system. A key-operated switch would be an added safety feature.

It is highly desirable that an interlock mechanism be incorporated into the system which will prevent any part of the body from being directly exposed to the primary beam, either by making it physically impossible to do so or by entry causing the beam to be shut off. Where, for some compelling reason, this is not feasible, alternatives need to be in place to minimize the chance of an accidental overexposure. Conspicuous warning devices to alert persons working on or near the machine should be in place if the status of the apparatus is such that a danger could exist. These may take the form of lighted signs, for example, stating *"Caution — System Is On"* or *"Danger — Shutter Open"* if a port is open. These warning devices should be fail-safe. If they are not functional, the system should not be able to operate. Shutters on any unused ports should be firmly secured in the closed position. It would be desirable if the system could not operate with a port open unless deliberately bypassed, in which case a fail-safe warning sign would have to be used. Similarly, it should be impossible to open a shutter unless a collimator or an experimental device were connected to the port. This feature must be incorporated into commercial systems constructed after January 1, 1980. Beam catchers should be in place to intercept any beam beyond the point at which it has served its purpose.

Interlock systems should not depend solely upon the operation of a device such as a microswitch. These devices can operate for years without failing and then fail without warning. A more basic device, such as incorporating a male and female plug assembly into the shutter mechanism to complete a circuit, is virtually foolproof. A light in the circuit confirms that the circuit is complete. The completed circuit should always provide a warning and preferably be a condition of a safe system, i.e., if the light is on, the system is safe. If it is off, the activation circuit is interrupted by an unsafe condition or the light is burned out, so the system will not operate in either case. In some cases, as in the case of the open shutter above, a warning system needs to operate to warn of an unsafe condition. In this case, the failure of the system to operate if the light is not on simply guarantees that the warning sign is on during the less safe condition.

Repair and alignment activities provide some of the best opportunities for accidental overexposures. Every precaution should be taken to prevent these incidents. Several measures are listed below. Most of these should be made mandatory by the organization's written radiation safety policies.

1. Alignment procedures recommended by the manufacturer should be used. Departures from these procedures should require prior approval by radiation safety personnel.
2. Alignment procedures must be in writing and the personnel performing them must have specific training on performing the procedures.
3. The radiation from an X-ray tube shall be confined by a suitable housing during alignment and maintenance activities.
4. The main switch, instead of safety interlocks, must be used to turn the equipment off.
5. If the interlocks are bypassed to permit alignment or other work to be done, a sign must be placed on the equipment stating **Danger — Safety Interlocks Bypassed**.
6. Personal dosimeters and finger badges must be worn at all times the beam is on, by both the person doing the work and the assistant (see item 11 below).
7. The smallest practical voltage and beam current which permit the alignment to be done shall be used.
8. Temporary shielding should be used to reduce the exposure to scattered radiation.
9. Long-handled tools and extension devices should be used to reduce the risk of direct exposure to the hands.
10. If the alignment and maintenance operations are such as to cause the radiation in an uncontrolled area to exceed the permissible levels for such an area, the area should be secured against entry by locks, surveillance, or both.
11. Alignment and repair operations with the beam on should not be done while working alone. A second person should be present as a safety observer, equipped with a survey meter to check radiation and prepared to immediately shut off the beam in case of a direct exposure of any part of the person working on the equipment.
12. The system should be checked to ensure that all interlocks have been reconnected and that leakage radiation is at an acceptable level.
13. A record of all repairs and the results of surveys taken subsequent to the work performed should be recorded in the permanent log.

### 5.2.2.3. Closed-Beam Analytical Systems

Closed-beam diffraction cameras are used for purposes similar to those of open-beam systems. However, they are designed so that the X-ray tube, sample, detector, and diffracting crystal are enclosed in a chamber, which should prevent entry into the system by any part of the body. However, some older systems permitted the hands to be put into the sample chamber without turning off the beam, even though this was not supposed to be done. Instances of exposures of 50,000 to 100,000 rems (500 to 1000 Sv) to the hands are known to have occurred. The ports on such units must now be equipped with safety interlocks incorporating a warning light in series with the interlock, which guarantees that if the light is off, the beam is off. Because there should be little or no leakage requirements with a closed-beam analytical system, personnel are not required to wear dosimeters.

### 5.2.2.4. Cabinet X-Ray Systems

Cabinet X-ray systems are enclosed systems which not only enclose the X-ray system, but also the object to be irradiated. They are normally designed with a key-activated control so that X-rays cannot be generated when the key is removed. a conspicuous sign with the legend, **CAUTION: X-RAYS PRODUCED WHEN ENERGIZED** must be placed near the controls used to generate the X-rays.

The doors to the cabinet are required to have two or more safety interlocks and each access panel is required to have at least one. If X-ray generation is interrupted through the functioning of a safety interlock, resetting the interlock shall not be sufficient to resume generation of X-rays. A separate control must be provided to reinitiate X-ray production.

The generation of X-rays must be indicated by two different indicators, one being a lighted warning light labeled **X-RAYS ON** the other can be the meter which reads the X-ray tube current.

The radiation at 5 cm from the surface of the cabinet must be no more than 0.5 mrem/h (5 $\mu$Sv/h). The radiation limits in uncontrolled areas due to the operation of the machine are 2 mrem/h (20 $\mu$Sv/h) or 100 mrem (1 mSv) in a 7-d period or 500 mrem/year (5 mSv/year). At least an initial radiation survey and an annual survey thereafter should be made to ensure conformity with these limits. The interlock systems should be tested periodically if the mode of operation does not automatically cause their function to be tested.

Personnel using a cabinet X-ray unit should be provided with a dosimeter badge which they should wear whenever the machine is in operation.

### 5.2.2.5. Miscellaneous Systems

There are any number of systems which generate X-rays, some as dangerous as or more so than the ones described above. Precautions appropriate to the system, including personnel dosimetry, surveys, safety interlocks, etc., must be taken to ensure the safety of all workers as well as those persons in uncontrolled areas nearby. An appropriate safety program must be worked out with the radiation safety department within the organization to ensure that the program not only meets all safety requirements recommended by the manufacturer, but also conforms with regulatory requirements, including those policies established within the organization.

# REFERENCES (SECTIONS 5.2 TO 5.2.2.5)

1. **Marlin, E. B. M.,** *Guide to Safe Use of X-Ray Diffraction and Spectrometry Equipment,* Science Reviews, Leeds, England, 1983.
2. *A Guide to Radiation Protection in the Use of X-Ray Optics Equipment,* Science Reviews, Leeds, England, 1986.
3. **Simpkin, D. J.,** A general solution to the shielding of medical X and gamma rays by the NRCP report no. 49 methods, *Health Phys.,* 52, 431, 1987.
4. *Structural Shielding Design and Evaluation for Medical Use of X-Rays and Gamma Rays of Energies Up to 10 MeV,* NCRP Report No. 49, National Council on Radiation Protection and Measurements, Washington, D.C., 1976.
5. *X-Ray Equipment,* UL-187, Underwriters Laboratories, Chicago, 1974.
6. Radiological Safety Standard for the Design of Radiographic and Fluoroscopic Industrial X-Ray Equipment, NBS Handbook 123, National Bureau of Standards, Washington, D.C., 1976.
7. *Radiation Safety for Diffraction and Fluorescence Analysis* Equipment, ANSI N43.2, American National Standards Institute, New York, 1971.
8. Performance Standards for Ionizing Radiation Emitting Products, 21 CFR CHAP I, 1020.30, 1988.
9. **Kaczmarek, R., Bednarek, D., and Wong, R.,** Misconceptions of medical students about radiological physics, *Health Phys.,* 52, 106, 1987.
10. **Fleming, M. F. and Archer, V. E.,** Ionizing radiation, health hazards of medical use, *Consultant,* p. 167, January 1984.
11. **Lubenau, J. O., David, J. S., McDonald, D. J., and Gerusky, T. M.,** Analytical X-ray hazards: a continuing problem, *Health Phys.,* 16, 739, 1969.
12. **Weigensberg, I. J., Asbury, C. W., and Feldman, A.,** Injury due to accidental exposure to X-rays from an X-ray fluorescence spectrometer, *Health Phys.,* 39, 237, 1980.
13. **Cember, H.,** *Introduction to Health Physics,* Pergamon Press, New York, 1969.

# 5.3 NONIONIZING RADIATION

The International Non-Ionizing Radiation Committee of the International Radiation Protection Association (IRPA/INIRC) recently developed a revised version of the 1984 Interim Guidelines on Limits of Exposure to RF Electromagnetic Fields in the Frequency Range From 100 kHz to 300 GHz. The revised guidelines were approved by the IRPA Executive Council on June 2, 1987. The following material is an abbreviated version of the guidelines prepared under the auspices of a committee chaired by Jammet.[1] Only the basic material and the recommended exposure numbers are presented below. The interested reader should refer to the complete document as well as a special section on nonionizing radiation found in the *Health Physics Journal.*[2]

The current federal standards are given in 29 CFR 1910.97. From 10 MHz to 100 GHz, the radiation protection guide provides that the incident radiation not exceed 10 mW/cm$^2$, as averaged over any possible 0.1-h period. This refers to the power density. The guide provides that the energy density be no more than 1 mW/cm$^2$ for any 6-min period.

## 5.3.1. GUIDELINES ON LIMITS OF EXPOSURE TO RADIO FREQUENCY ELECTROMAGNETIC FIELDS IN THE FREQUENCY RANGE FROM 100 KHZ TO 300 GHZ

The exposure limits can be expressed in terms of either the energy absorbed by the presence of a body in the area or the power density in an area in the absence of a physical body to absorb the energy. The basic limits of exposure are given by the

## TABLE 5.10
## Occupational Exposure Limits to Radio Frequency Electromagnetic Fields

| Frequency (MHz) | Unperturbed RMS field strength | | Equivalent plane wave power density $P_{eq}$ (W/m$^2$) |
|---|---|---|---|
| | Electric $E$(V/m) | Magnetic $H$(A/m) | |
| 0.1—1 | 614 | 1.6/f | — |
| >1—10 | 614/f | 1.6/f | — |
| >10—400 | 61 | 0.16 | 10 |
| >400—2,000 | $3f^{1/2}$ | $0.008f^{1/2}$ | f/40 |
| >2,000—300,000 | 137 | 0.36 | 50 |

specific absorption rate (SAR) and expressed in units of power absorbed per unit mass or W/kg). Obviously, the SAR represents the body-present situation. The absorption can be averaged over the entire body mass of a person present and a defined time interval a modulation period, or a single pulse of the radiation.

In the absence of a body being present, the electromagnetic power density can be calculated if the electric and magnetic field strengths are known. In the far-field region, well away from the source of radiation, the relationship between the electric and magnetic field strengths is straightforward since the field is essentially a plane wave. In this case, the electromagnetic power density is given by:

$$S = E^2/120\pi = 120\pi/H^2 \text{ W/m}^2$$

In the near-field region (and where there are multiple paths), the relationship between the electric and magnetic field components is not straightforward and both must be measured in order to compute the equivalent plane-wave power density. This is generally the situation which obtains in cases below 100 MHz.

Below 10 MHz, the occupational exposure to RF radiation "should not exceed the levels of the unperturbed RMS electric and magnetic field strengths given in"[1] Table 5.10 below "when the squares of the electric and magnetic field strengths are averaged over any 6 minute period during the working day, provided that the body to ground current does not exceed 200 mA and that any hazards of RF burns are eliminated." Currents at the point of contact above 50 mA pose the risk of burns.

Above 10 MHz, the occupational exposures should not exceed 0.4 W/kg averaged over the whole body for a 6-min period, provided that 2 W/0.1 kg will not be exceeded in the hands, wrists, arms, and legs and that 1W/0.1kg will not be exceeded in any part of the body. The values for the field strengths in Table 5.10 are the working limits derived from the 0.4 W/kg SAR limit. The guidelines also recommend that, as long as the occupational limits over a 6-min interval are not exceeded, the equivalent plane wave power density averaged over the pulse width for a pulsed field not exceed 1000 times the $P_{eq}$ limits for the frequency involved.

The limits of exposure to RF radiation for the general public are given in Table 5.11. Below 10 MHz, the limits are defined in the same way as for an occupational exposure, but are lower. Above 10 MHz, the limits for the various frequencies are derived from an SAR of 0.08 W/kg averaged over the entire body over any 6-min period.

**TABLE 5.11**
**General Public Exposure Limits to Radio Frequency Electromagnetic Fields**

| Frequency (MHz) | Unperturbed RMS field strength | | Equivalent plane wave power density $P_{eq}$ (W/m²) |
|---|---|---|---|
| | Electric $E$(V/m) | Magnetic $H$(A/m) | |
| 0.1—1 | 87 | $0.23/f^{1/2}$ | — |
| >1—10 | $87/f^{1/2}$ | $0.23/f^{1/2}$ | — |
| >10—400 | 27.5 | 0.073 | 2 |
| >400—2,000 | $1.375f^{1/2}$ | $0.0037f^{1/2}$ | $f/200$ |
| >2,000—300,000 | 61 | 0.16 | 10 |

# REFERENCES (SECTION 5.3)

1. **Jammet, H.P., et al.,** Guidelines on limits to radio frequency electromagnetic fields in the frequency range from 100 khz to 300 GHz, *Health Phys.*, 54, 115, 1988.
2. Special section: non-ionizing radiation, *Health Phys.*, 53, 567, 1987.
3. Nonionizing Radiation, 29 CFR 1910.97, Nuclear Regulatory Commission, Washington, D.C., 1988.
4. *Safety Level of Electromagnetic Radiation with Respect to Personnel,* ANSI C-95.1, American National Standards Institute, New York, 1982.

# 5.4. LASER LABORATORIES*

The risk to personnel from the use of a laser in the laboratory depends upon a number of factors. The first consideration is the classification of the laser itself, which is based on several parameters. Among these are (1) the frequency or frequencies of radiation emitted, (2) for a pulsed system, the pulse repetition frequency (PRF), the duration of each pulse, the maximum or peak power, P, in watts or maximum energy, Q, in joules per pulse, the average power output, and the emergent beam radiant exposure, (3) for a continuous wave (CW) machine, the average power output, and (4) for an extended-source laser, the radiance of the laser and the maximum viewing angle subtended by the laser. Since relatively few lasers used in the laboratory fall in the latter class, the following discussion will be limited to lasers which are not considered extended-source units.

All commercial units currently being built must have the classification identified on the unit. The classification must be according to that given in ANSI-Z 136.1 and 21 CFR 1040. However, facilities which work with experimental units should determine the class in which their lasers fall. The specifications of noncommercial units should be compared with Table 5.12 and Table 5.13 to determine the classification.

The least-powerful class of laser is Class 1. In this class, the power and energy of the unit are such that the TLV©** for direct viewing of the laser beam, if the entire beam passes through the limiting aperture of the eye, cannot be exceeded for the

---

* The material in this section complies with 21 CFR 1040.1 for laser products as well as ANSI Z-136-1 and the Threshold Limit Values© of the American Conference of Governmental Industrial Hygienists.[2]

** Copyrighted trademark of the American Conference of Governmental Industrial Hygienists.

## TABLE 5.12A
### Class 1 Accessible Emission Limits for Laser Radiation

| Wavelength (nanometers) | Emission duration (sec) | Class I accessible emission limits (power [W] or energy [J]) |
|---|---|---|
| 180—400 | $t \geq 3 \times 10^4$ | $2.4 \times 10^{-5} k_1 k_2$ J |
| 180—400 | $t \geq 3 \times 10^4$ | $8.0 \times 10^{-10} k_1 k_2$ W |
| >400—1400 | $t > 1 \times 10^{-9} — 2 \times 10^{-5}$ | $2.0 \times 10^{-7} k_1 k_2$ J |
| >400—1400 | $t > 2 \times 10^{-5} — 1 \times 10^1$ | $7.0 \times 10^{-4} k_1 k_2 t^{0.75}$ J |
| >400—1400 | $t > 1 \times 10^1 — 1 \times 10^4$ | $3.9 \times 10^{-3} k_1 k_2$ J |
| >400—1400 | $t > 1 \times 10^4$ | $3.9 \times 10^{-7} k_1 k_2$ W |
| >1400—2500 | $t > 1 \times 10^{-9} — 1 \times 10^{-7}$ | $7.9 \times 10^{-5} k_1 k_2$ J |
| >1400—2500 | $t > 1 \times 10^{-7} — 1 \times 10^1$ | $4.4 \times 10^{-3} k_1 k_2 t^{1/4}$ J |
| >1400—2500 | $t > 1 \times 10^1$ | $7.9 \times 10^{-4} k_1 k_2$ W |
| >2500—$1 \times 10^6$ | $t > 1 \times 10^{-9} — 1 \times 10^{-7}$ | $1.0 \times 10^{-2} k_1 k_2$ J/cm² |
| >2500—$1 \times 10^6$ | $t > 1 \times 10^{-7} — 1 \times 10^1$ | $5.6 \times 10^{-1} k_1 k_2 t^{1/4}$ J/cm² |
| >2500—$1 \times 10^6$ | $t > 1 \times 10^1$ | $1 \times 10^{-1} k_1 k_2 t$ J/cm² |

[a]  Class 1 accessible emission limits for wavelengths $\geq$180—400 nm shall not exceed the Class 1 accessible limits for wavelengths $\geq$1400—$1 \times 10^6$ nm with a $k_1$ and $k_2$ of 1.0 for comparable sampling intervals.

## TABLE 5.12B
### Class 2 and Class 3 Accessible Emission Limits for Laser Radiation
### (Class 2a, 2, 3 and 3b limits are the same as for Class 1 except within the following wavelength and emission durations)

| Wavelength (nm) | Emission duration (sec) | Accessible emission limits (power [W] or energy [J]) |
|---|---|---|
| **Class 2a Accessible Emission Limits for Laser Radiation** | | |
| 400—710 | $t > 1 \times 10^3$ | $3.9 \times 10^{-6}$ W |
| **Class 2 Accessible Emission Limits for Laser Radiation** | | |
| 400—710 | $t > 2.5 \times 10^{-1}$ | $1 \times 10^{-3}$ |
| **Class 3a Accessible Emission Limits for Laser Radiation** | | |
| 400—710 | $t > 3.8 \times 10^{-4}$ | $5 \times 10^{-3}$ W |
| **Class 3b Accessible Emission Limits for Laser Radiation** | | |
| 180—400 | $t \leq 2.5 \times 10^{-1}$ | $3.8 \times 10^{-4} k_1 k_2$ J |
| 180—400 | $t > 2.5 \times 10^{-1}$ | $1.5 \times 10^{-3} k_1 k_2$ W |
| >400—1400 | $t > 1 \times 10^{-9} — 2.5 \times 10^{-1}$ | $10 k_1 k_2 t^{1/3}$ J/cm² to a maximum value of 10 J/cm² |
| >400—1400 | $t > 2.5 \times 10^{-1}$ | $5.0 \times 10^{-1}$ W |
| >1400—$1 \times 10^6$ | $t > 1 \times 10^{-9} — 1 \times 10^1$ | 10 J/cm² |
| >1400—$1 \times 10^6$ | $t > 1 \times 10^1$ | $5.0 \times 10^{-1}$ W |

**TABLE 5.13**
**Values of Wavelength-Dependent Correction Factors $k_1$ and $k_2$**

| Wavelength (nm) | $k_1$ | $k_2$ |
|---|---|---|
| 180—302.4 | 1 | 1 |
| 302.4—315 | $10^{\left[\frac{\lambda-302.4}{5}\right]}$ | |
| >315—400 | 330 | 1 |
| >400—700 | 1 | 1 |
| >700—800 | $10^{\left[\frac{\lambda-700}{515}\right]}$ | if $t \le \dfrac{10,100}{\lambda-699}$, then $k_2 = 1$; if $\dfrac{10,100}{\lambda-699} < t \le 10^4$, then $k_2 = \dfrac{t(\lambda-699)}{10,100}$; if $t > 10^4$, then $k_2 = \dfrac{\lambda-699}{1.01}$ |
| >800—1060 | $10^{\left[\frac{\lambda-700}{515}\right]}$ | |
| >1060—1400 | 5 | if $t \le 100$, then $k_2 = 1$; if $100 < t \le 10^4$, then $k_2 = \dfrac{t}{100}$; if $t > 10^4$, then $k_2 = 100$ |
| >1400—1535 | | 1 |
| 1535—1545 | if $t \le 10^{-7}$ then $k_1 = 100$; if $t > 10^{-7}$ then $k_1 = 1$ | |
| >1545—$1 \times 10^6$ | 1 | |

classification duration (the maximum duration of the exposure inherent in the design of the laser). In the spectral region of 400 to 1400 nm (1 nm = $10^{-9}$ m), the limiting aperture of the eye is taken to be 7 mm. Class 1 lasers must emit levels below the accessible exposure limit (AEL) which are given in the first part of Table 5.12. Many of the operating procedures are based on not exceeding the accessible exposure limits for Class 1 lasers.

These low-power lasers require no control measures to prevent eye damage. Some Class 1 lasers incorporate more powerful lasers, but are designated Class 1 because they are in an enclosure. If the more powerful enclosed laser is accessible, control measures appropriate to the higher class are required. Class 2 lasers are not considered hazardous for momentary (0.25 sec) and unintentional direct viewing of the laser beam, but should not be deliberately aimed at anyone's eyes. A Class 2a laser must have a label affixed to the unit which bears the words **Class 2a Laser Product — Avoid Long-Term Viewing of Direct Laser Radiation.** A Class 2 laser must have

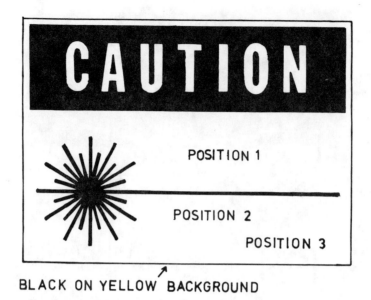

BLACK ON YELLOW BACKGROUND

FIGURE 5.11.    Caution sign for low- to moderate-power lasers.

a label, as shown in Figure 5.11, affixed to the unit in a conspicuous location, bearing the words **LASER RADIATION — DO NOT STARE INTO BEAM** in position 1 on the label. The words **CLASS 2 LASER PRODUCT** must appear in position 3. The labels of all Class 2, 3, and 4 units must give in position 2, in appropriate units, the maximum output of laser radiation, the pulse duration (when appropriate), and the laser medium or emitted wavelength(s).

Class 3 laser devices are hazardous to the eyes if one looks directly at the direct beam or at specular reflections of the beam. Class 3a units must have a label, as in Figure 5.11, affixed to the unit with the words **LASER RADIATION — DO NOT STARE INTO BEAM OR VIEW DIRECTLY WITH OPTICAL INSTRU- MENTS** in position 1 and **CLASS 3a LASER PRODUCT** in position 3. A Class 3b unit would use a label, as in Figure 5.12, and have the words **LASER RADIA- TION — AVOID DIRECT EXPOSURE TO THE BEAM** in position 1 and the words **CLASS 3b LASER PRODUCT** in position 3. Mirrorlike, smooth surfaces of any material that would reflect the beam, as a beam, should be eliminated as much as possible from the area in which the laser is situated. As a minimum, the user should take every precaution to avoid aiming the laser at such surfaces. It would be desirable to have the beam terminate on a surface which would only provide a diffuse reflection. It should, however, reflect well enough for a well-defined beam spot to be observable so that it is possible to visibly determine the point of contact. The laser should be set up in an area to which access can be controlled. When no responsible person is in the room, the room should be locked in order to prevent others from entering the room and changing the physical configuration or accidentally exposing their eyes to the laser beam.

If there is any way in which it would be possible for either the direct or specularly reflected beam to enter the eye, appropriate eye protection must be worn in an area

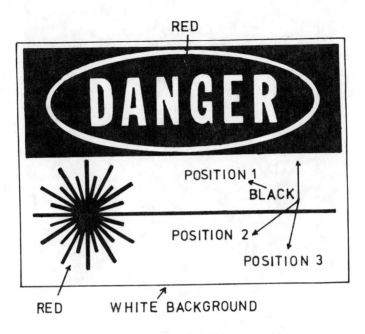

FIGURE 5.12.    Danger sign for higher-powered lasers.

where a Class 3a laser is operated. Operation of Class 3b lasers generally should follow the practices for Class 4 lasers.

Class 4 lasers must use a label as in Figure 5.12, and bear the words: **LASER RADIATION — AVOID EYE OR SKIN EXPOSURE TO DIRECT OR SCATTERED RADIATION** in position 1 and the words **CLASS 4 LASER PRODUCT** in position 3. Class 4 lasers require many precautions to be used safely. The radiation from pulsed or CW units which lie in the visible and near-infrared regions can be focused by the eye and can cause damage to the retina from either direct and specular beams or diffuse reflections. The skin can also be injured. Pulsed units and CW units operating in the infrared and ultraviolet regions are a danger to the skin and external portions of the eye and can provide enough energy to cause a fire in combustible materials.

Lasers of any class, except medical units and Class 2 lasers that do not exceed the accessible exposure limits of Class 1 lasers for any exposure duration of 1000 sec or less, must have a label affixed near any aperture through which laser radiation is emitted in excess of the limits for Class 1 lasers bearing the words **AVOID EXPOSURE — LASER RADIATION EMITTED THROUGH THIS APERTURE**.

Table 5.13 provides the means to calculate the values of the wave dependent factors $k_1$ and $k_2$. Table 5.14 provides threshold limit values for direct ocular exposure from a laser beam.

### Protective Procedures for Class 3b and Class 4 Lasers

The primary means of protection is to physically prevent exposure. For laboratory workers, baffles may be used to physically intercept or terminate the primary beam and any reflected or secondary beams. Any windows in the facility should be covered

## TABLE 5.14
### Threshold Limit Values for Direct Ocular Exposures (Intrabeam Viewing) from A Laser Beam

| Spectral region | Wavelength (nm) | Exposure time (t) (sec) | TLV | |
|---|---|---|---|---|
| UVC | 200—280 | $10^{-9}$—$3 \times 10^4$ | 3 mJ/cm$^2$ | |
| UVB | 280—302 | $10^{-9}$—$3 \times 10^4$ | 3 mJ/cm$^2$ | |
| | 303 | | 4 mJ/cm$^2$ | |
| | 304 | | 6 mJ/cm$^2$ | |
| | 305 | | 10 mJ/cm$^2$ | |
| | 306 | | 16 mJ/cm$^2$ | |
| | 307 | | 25 mJ/cm$^2$ | Not to |
| | 308 | | 40 mJ/cm$^2$ | exceed |
| | 309 | | 63 mJ/cm$^2$ | $0.56t^{1/4}$ J/cm$^2$ |
| | 310 | | 100 mJ/cm$^2$ | t ≤ 10 sec |
| | 311 | | 160 mJ/cm$^2$ | |
| | 312 | | 250 mJ/cm$^2$ | |
| | 313 | | 400 mJ/cm$^2$ | |
| | 314 | | 630 mJ/cm$^2$ | |
| UVA | 315—400 | $10^{-9}$—10 | 0.56 $t^{1/4}$ J/cm$^2$ | |
| | | 10—$10^3$ | 1.0 J/cm$^2$ | |
| | | $10^3$—$3 \times 10^4$ | 1.0 mW/cm$^2$ | |
| Light | 400—700 | $10^{-9}$—$1.8 \times 10^{-5}$ | $5 \times 10^{-7}$ J/cm$^2$ | |
| | 400—700 | $1.8 \times 10^{-5}$—10 | 1.8 $(1/t^{0.75})$ mJ/cm$^2$ | |
| | 400—549 | 10—$10^4$ | 10 mJ/cm$^2$ | |
| | 550—700 | 10—$T_1$ | 1.8 $(1/t^{0.75})$ mJ/cm$^2$ | |
| | 550—700 | $T_1$—$10^4$ | 10 $C_B$ mJ/cm$^2$ | |
| | 400—700 | $10^4$—$3 \times 10^4$ | $C_B$ μW/cm$^2$ | |
| IR-A | 700—1049 | $10^{-9}$—$1.8 \times 10^{-5}$ | 5 $C_A \times 10^{-7}$ J/cm$^2$ | |
| | 700—1049 | $1.8 \times 10^{-5}$—$10^3$ | 1.8 $C_A$ $(1/t^{0.75})$ mJ/cm$^2$ | |
| | 1050—1400 | $10^{-9}$—$10^{-4}$ | $5 \times 10^{-6}$ J/cm$^2$ | |
| | 1050—1400 | $10^3$—$3 \times 10^4$ | 320 $C_A$ μW/cm$^2$ | |
| IR-B and C | 1.4—$10^3$ μm | $10^{-9}$—$10^{-7}$ | $10^{-2}$ J/cm$^2$ | |
| | | $10^{-7}$—10 | 0.56 $t^{0.75}$ J/cm$^2$ | |
| | | 10—$3 \times 10^4$ | 0.1 W/cm$^2$ | |

*Note:*  For $C_A$, see Figure 5.13; $C_B$ =1 for λ = 400—549 nm; $C_B = 10^{[0.015 (\lambda - 550)]}$ for λ = 550—700 nm; $T_1$ = 10 sec for λ = 400—549 nm; $T_1 = 10 \times 10^{[0.02 (\lambda - 550)]}$ for λ = 550—700 nm.

during operations. Safety glasses also are to be worn by workers while within the facility during operations. Interlocks of various sorts are another avenue of protection. Entrance to the facility by unauthorized personnel or unexpected entry by laboratory employees should be prevented by safety interlocks while the laser is in an operating condition, i.e., when it is on and in a condition to emit radiation. A warning light should be placed at the entrance. The interlocks should be capable of being bypassed to allow individuals to pass in and out of the controlled area in an emergency or to allow controlled access as needed to allow operations to be conducted. In the latter case, which might be considered as routine bypassing of the interlocks, the activation of the bypass should be limited to the person in charge of the operations at the time and who is aware of the condition of the system. In an

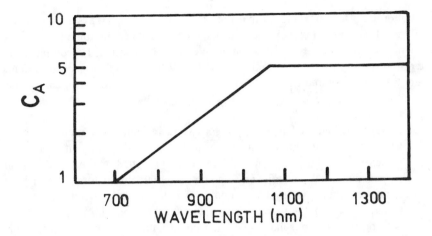

FIGURE 5.13.  Correction factors for TVL values.

emergency situation, there should be one or more readily available rapid shutoff switches to immediately disconnect power to the laser. One of these can be the required remote control unit stipulated in 21 CFR 1040.1. The access points to the controlled area should be so situated that there is little or no likelihood of dangerous levels of emitted radiation to persons entering or leaving through the portals or in areas beyond the entrances. Guests or visitors should be allowed in the controlled area during operations only under carefully controlled conditions and with everyone in the area aware of their presence. Special care must be taken to provide such persons with protective eyewear or other protective gear as needed to ensure their safety. The applicable TLV© levels should not be exceeded for either employees or visitors.

It should be made impossible, by designing the firing circuit with sufficient fail-safe safeguards, for a Class 4 pulsed laser to be discharged accidentally. A warning mechanism should be incorporated into the design to insure that all persons in the room are aware that the discharge cycle has been initiated. If the laser unit and its power supply are more than 2 m apart, both the laser and the power supply must have separate emission warning devices. Both the fail-safe system and the warning system should be designed so that no single component failure or a shorted or open circuit can disable the protective and warning features of the system. The system should not be capable of being operated if the redundancy of these circuits or of the emergency shutoff system has been compromised.

Very-high-power infrared CW lasers, such as $CO_2$ units, not only represent a danger to operating personnel, they also constitute a fire hazard if their beams come into contact with combustible material. It would be desirable to have these units operated remotely or totally within an enclosure which affords good fire-stopping capability. Asbestos should be avoided because it is a regulated carcinogen under current OSHA regulations and, under intense radiation, could become friable.

Commercial lasers incorporate many safety features. Below is a brief list of system safety features, in addition to those already mentioned, that should either be available in the laser as manufactured or should be incorporated in the system when set up for use. The items refer primarily to Class 3b and Class 4 systems; they also apply to Class 2 units, except those in which the acceptable emission limits of Class 1 are not exceeded for any emission duration up to 1000 sec. This list complies with the provisions of 21 CFR 1040 and ANSI Z-136. When possible, systems should be upgraded to the requirements of the latest editions of these two documents as they are revised.

1.    All lasers are required to be in a protective housing, but safety interlocks should be provided on any portion that could be removed when the unit is operating if the exposure limits for the class would be exceeded. Normally, the enclosures should limit the radiation to no more than the limits for a Class 1 laser. Some powerful laser systems should be required to be within enclosures, including the target or irradiation area, to protect personnel. Any portion of the housing which is not interlocked and could emit radiation in excess of the accessible emission limits for each class must have a label attached to it bearing the words previously mentioned for position 1 of the basic identifying label for each class.

2.    Interlocks which are designed to prevent firing of a pulsed laser by turning off the power supply or interrupting the beam must not automatically permit the power supply to be reactivated when they have been reset after serving their protective function.

3.    An audible or visible warning device is needed if a required interlock is bypassed or defeated. The warning system should be fail-safe, i.e., if it became inoperative, the unit should be inoperable.

4.    Class 4 lasers must be key interlocked or activated, and it is recommended that Class 3b lasers have the same feature. The key must be removable and the unit must not be operable without the key in place and in the on position.

5.    A portal, viewing window or an attached optical device must be designed so as to prevent any exposure above the permissible TLV©.

6.    Any Class 2 (except units emitting less than Class 1 limits for no more than 1000 sec), 3, or 4 laser should be equipped with an appropriate safety device, such as a beam stop or optical attenuator, which will prevent emission of radiation in excess of those for a Class 1 unit in the controlled area.

7.    The control units for laser units should be located in areas where the accessible emission limits for a Class 1 device are not exceeded.

8.    Signs bearing essentially the same information as that on the labels of lasers in use within a controlled area should be posted at each entrance to the laser laboratory.

9.    There is a possibility of radiation being emitted by other parts of the laser system, such as the power supply. This radiation, associated with the operation of the laser system, is called collateral radiation. The limits on this collateral radiation are (from Table 4, 21 CFR 1040.1):

   1.    *Accessible emission limits* for collateral radiation having wavelengths greater than 180 nanometers but less than or equal to $1.0 \times 10^6$ nanometers are identical to the accessible emission limits of Class I radiation. ...

      i.    In the wavelength range of less than or equal to 400 nm, for all emission durations.

ii. In the wavelength range of greater than 400 nm, for all emission durations less than or equal to $1 \times 10^3$ seconds and, when applicable ...[see f(8) of 21 CFR 1040.1], for all emission durations.

2. *Accessible emission limits* for collateral radiation within the X-ray range of wavelengths are 0.5 milliroentgen in an hour, averaged over a cross-section parallel to the external surface of the product, having an area of 10 square centimeters with no dimension greater than 5 centimeters.

**Eye Protection**

The shorter wave-length ultraviolet radiations, UV-C (100 to 280 nm) and UV-B (280 to 315 nm) are primarily absorbed within the conjunctiva and corneal portions of the eyes, causing corneal inflammation. The ultraviolet frequencies just below the visible range, UV-A (315 to 400 nm), are absorbed largely within the lens of the eye. Although little UV-B radiation reaches the lens, UV-B is much more effective in causing cataracts to form than is UV-A. Radiation in the range of 400 to 1400 nm, which includes visible light and the near infrared, are focused by the lens of the eye on the retina. The focusing properties of the lens may increase the energy per unit area for a point source by as much as 100,000 times. The energy is primarily absorbed by the retinal pigmented epithelium and the choroid. At wavelengths longer than 1400 nm, the water in the eye tissue becomes opaque, so that most of the energy is absorbed in the corneal region. The major mechanism for damage is thermal absorption, although there is also some due to photochemical reactions.

Protective eyewear appropriate to the laser system in use should be worn if there is any eye hazard. The filters in the protective goggles should be matched to the wavelength of the laser's emissions. Since some lasers emit radiation at more than one wavelength, it may be necessary to have filters covering each range of frequencies. It would be desirable to have the filters in the protective eyewear attenuate only narrow wavelength regions spanning those emitted by the laser, thereby allowing as much visible light through as possible to facilitate seeing by the wearer.

The term optical density (OD) is a convenient way to define the attenuation of incident radiation by a filter. A difference of one unit in the OD of two filters corresponds to a transmission difference of a factor of ten. Thus, a filter which attenuates the incident radiation by a factor of ten would have an optical density of 1; attenuation by a factor of 100 would mean an optical density of 2, and so forth. Since this is a logarithmic scale (to the base 10), the optical density of two filters stacked together is the sum of the optical densities of the two filters. Note, however, that this is true only if the two filters are for the same wavelength. If they are for different wavelengths and transmit essentially all of the radiation at the other's wavelength, the filtration of the stacked filter would be the same as that of the two filters considered individually.

When working with lasers of very high power or beam intensities, the absorption of energy in the filter can damage the filters. For pulsed lasers, the threshold for damage to glass filters is approximately 10 to 100 J/cm$^2$ and for plastic and dielectric coatings, between 10 and 100 J/cm$^2$. A continuous-wave laser operating at 10 W or more can cause glass filters to fracture and can burn through plastic filters. If these numbers appear small, it should be recalled that the power per square centimeter of

## TABLE 5.15
### Selection Guide for Eye Protection for Direct Viewing of Laser Beams Emitting Radiation Between 400- and 1400-Nanometer Wavelengths

| Q-switched (1 nsec—0.1 msec) | | Non-Q-switched (0.4—10 msec) | | Continuous | | | | | |
| --- | --- | --- | --- | --- | --- | --- | --- | --- | --- |
| | | | | Momentary (0.25—10 sec) | | Long-term (staring ≥3 h) | | Attenuation | |
| Maximum output energy (J) | Maximum beam radiant (J/cm²) | Maximum output energy (J) | Maximum beam radiant (J/cm²) | Maximum power output (W) | Maximum beam irradiance (W/cm²) | Maximum power output (W) | Maximum beam irradiance (W/cm²) | Attenuation factor | OD |
| 10 | 20 | 100 | 200 | NR | NR | NR | NR | $10^8$ | 8 |
| 1 | 2 | 10 | 20 | NR | NR | NR | NR | $10^7$ | 7 |
| $10^{-1}$ | $2 \times 10^{-1}$ | 1 | 2 | NR | NR | 1 | 2 | $10^6$ | 6 |
| $10^{-2}$ | $2 \times 10^{-2}$ | $10^{-1}$ | $2 \times 10^{-1}$ | NR | NR | $10^{-1}$ | $2 \times 10^{-1}$ | $10^5$ | 5 |
| $10^{-3}$ | $2 \times 10^{-3}$ | $10^{-2}$ | $2 \times 10^{-2}$ | 10 | 20 | $10^{-2}$ | $2 \times 10^{-2}$ | $10^4$ | 4 |
| $10^{-4}$ | $2 \times 10^{-4}$ | $10^{-3}$ | $2 \times 10^{-3}$ | 1 | 2 | $10^{-3}$ | $2 \times 10^{-3}$ | $10^3$ | 3 |
| $10^{-5}$ | $2 \times 10^{-5}$ | $10^{-4}$ | $2 \times 10^{-4}$ | $10^{-1}$ | $2 \times 10^{-1}$ | $10^{-4}$ | $2 \times 10^{-4}$ | 100 | 2 |
| $10^{-6}$ | $2 \times 10^{-6}$ | $10^{-5}$ | $2 \times 10^{-5}$ | $10^{-2}$ | $2 \times 10^{-2}$ | $10^{-5}$ | $2 \times 10^{-5}$ | 10 | 1 |

*Note:*  NR = not recommended.

a heating element on a range is approximately 15 to 20 W. The filters should be inspected routinely to be sure that they are not damaged.

In Table 5.15, there are two columns for each type of laser, one labeled "maximum output power or energy", which should be used for a focused beam, and the other labeled "beam radiant exposure", which should be used for an unfocused beam larger than the pupil of the eye. Note that the pulsed laser outputs are defined in terms of energy (joules and joules per square centimeter), while the continuous-wave laser outputs are defined in terms of power (watts or watts per square centimeter).

### Medical Surveillance

All persons who work routinely with Class 3 or Class 4 lasers should have a preemployment medical examination. Others who occasionally work with lasers or at tasks where their eyes could be exposed to the laser radiation should at least have an eye examination. The examination for those that normally work with lasers should include a medical history, stressing the visual and dermatologic systems. The examination should measure visual acuity and special attention should be paid to those tissues most likely to be affected by the wavelengths emitted by the lasers the employee will use. If employees suffer or suspect that they have had a potential injury to the eye or skin, an examination of the potentially affected tissues should be performed by a qualified physician and the employee provided the treatment determined to be appropriate by the physician.

Although periodic examinations are not required by current guidelines, it would not be unreasonable to have a thorough eye examination on a 3- to 5-year schedule to determine if subtle changes are occurring in the various systems of the eye.

# REFERENCES (SECTION 5.4)

1. Performance Standards for Light-Emitting Products — Laser Products, 21 CFR, Chapter 1, Part 1040.1, FDA, Washington, D.C., 1987.
2. *A Guide for the Control of Laser Hazards,* American Conference of Governmental Industrial Hygienists, Cincinnati, OH, 1981.
3. ANSI Z-136-1, American National Standards Institute, New York, 1976.
4. **Goldman, L., Rockwell, J. R., Jr., and Hornby, P.,** Laser design and personnel protection from high energy lasers, in *Handbook of Laboratory Safety,* 2nd ed., Steere, N. V., Ed., CRC Press, Boca Raton, FL, 1971.

# 5.5. MICROBIOLOGICAL AND BIOMEDICAL LABORATORIES*

## 5.5.1. INTRODUCTION

Individuals who work in microbiological laboratories face a number of special problems when they work with organisms that are infectious to humans. There is evidence to show that biological laboratory personnel in such circumstances do have higher rates of incidence for a number of diseases associated with selected types of organisms than do comparable personnel working elsewhere, but there is no comprehensive system of data collection that defines the extent of the problem. However, virtually none of the primary laboratory infections which have occurred appear to have led to secondary infections for families, friends, or members of the general public. Only about 20% of the laboratory infections that have been reported have been attributed to specific incidents. The remainder have been assumed to be related to work practices within the laboratories, especially those which generate aerosols. In recent years, the fear which many laboratory workers feel about contracting the HIV virus, with the consequences which that implies, has emphasized the need for providing guidelines for safety in laboratories working with all types of infectious organisms.

Persons who work in microbiological laboratories in which the research involves organisms which are infectious to humans are usually well aware of the risks to themselves posed by the infectious agents. However, not everyone recognizes the many different operations and procedures which may result in significant quantities of aerosols being generated. Some of this awareness can be taught and some can be dictated by firm rules of laboratory practice, but some must be gained through experience. On the other hand, familiarity with procedures can occasionally lead to a casual attitude toward the risks associated with research activities. Whether due to inexperience, too relaxed an attitude, or poor work performance, it only requires one mistake to cause a problem for everyone. Unlike radiation, there is no simple way to detect contamination, should it occur. There are methods of surveying for biological contamination, but they are more complicated than those for radiation, so it is likely that an unsafe condition can persist for some time without the workers being aware of the situation. Since many diseases can be transmitted with only a minor exposure, it does not require a major incident to create a problem. In addition, regardless of whether a disease has been contracted by exposure to a few or a multitude, of organisms, one can become equally as ill.

* This section is adapted from Biosafety in Microbiological and Biomedical Laboratories.[15]

## 5.5.2. HAZARD COMMUNICATION STANDARD

As of May 23, 1988, virtually all employees working with hazardous chemicals became covered by the OSHA Hazard Communication Standard. This standard requires that all employees be made aware, by documented training programs, of the hazards associated with the toxic and dangerous chemicals with which they work. This standard does not cover the hazards associated with infectious organisms, but many of the procedures in the microbiological laboratory involve chemicals which are covered by the standard. Individuals already working in the laboratory and all new employees, at the time of employment, must be instructed about the dangers associated with the specific chemical agents involved in their work and the safety procedures which have been adopted. Instruction on new materials brought into the laboratory must take place before employees begin using the material. As noted, the legal standard only applies to chemicals, but the concept is entirely appropriate to any microbiological hazards to which the workers may be exposed.

## 5.5.3. LABORATORY DIRECTOR

The key person in the microbiological laboratory is the person in charge of the facility, the laboratory director. Although many agencies have guidelines and recommendations, there are relatively few mandatory governmental requirements for this class of laboratory. In the absence of a substantial body of formal regulations other than the internal policies which may have been set within the organization, the laboratory director usually is the individual assigned the responsibility and discretionary authority to set work practices. The attitude of this person will be reflected by others working in the facility.

## 5.5.4. MISCELLANEOUS SAFETY PRACTICES

The microbiological laboratory shares many of the same safety needs as a standard chemical laboratory. Some of these are briefly summarized below. For a more detailed discussion of good general laboratory safety practices, refer to the comparable sections in the first four chapters.

### 5.5.4.1. Laboratory Line of Authority

The telephone numbers of the laboratory authorities and emergency phone numbers should be posted near a telephone in the laboratory and near the primary entrance outside the laboratory. In addition, a list of the significant hazards in the laboratory should be posted at the entrances to the area in the event that emergency personnel cannot reach knowledgeable persons or if the time scale of the emergency requires immediate entrance. Emergency workers should be alerted to any risk to which they might be exposed.

### 5.5.4.2. Spills and Emergencies Involving Chemically Dangerous Materials

Minor spills should be cleaned up immediately by laboratory personnel, provided the material is not immediately dangerous to life and health (IDLH) and the equipment and supplies to do so are available.

Moderate spills of ordinarily dangerous materials may require technical guidance, supplies in excess of those normally kept on hand, or the assistance of the organizations safety or emergency response personnel.

For moderate to large spills of IDLH-level materials or for large spills of ordinarily dangerous materials, e.g., acids, etc.:

1.   Evacuate the area and initiate an evacuation of the building either personally or with the assistance of the building authority. Persons leaving the building should gather at a point upwind from the building.
2.   Call the local emergency number (911, if available) and report the incident. The type of emergency must be clearly described so the dispatcher can send the appropriate emergency responders. Do this from a location away from the affected area.
3.   Those individuals involved in the incident remain available outside the building to assist the emergency groups. After the initial notification of emergency personnel, the laboratory authority and the department head should be notified.

The building should be evacuated and the fire department called for any fire other than a very small one, where there is confidence that it can be put out without risk of spreading or danger to anyone.

If there is an emergency involving an injury, be sure to alert the dispatcher of the circumstances of the injuries. If chemicals are involved, specific members of the local rescue squad having special training may need to respond. If there is exposure to the eye or skin, assist the injured person to an eyewash station, a deluge shower, or a combination unit.

### 5.5.4.3. Emergency Equipment for Nonbiohazardous Spills

Everyone should be familiar with the location and use of all equipment in their laboratory area. This includes means to initiate an evacuation (fire alarm pull stations, etc.), fire extinguishers, fire blankets, eyewash stations, deluge showers, first aid kits, spill kit materials, respiratory protective devices, and any other materials normally kept in the area for emergency response.

### 5.5.4.4. Evacuation

Everyone should be familiar with the primary and secondary evacuation routes from their area to the nearest exit or an alternative one if the primary exit is blocked. Everyone should be told what method is used to signal a building evacuation, where to go, to check in with a responsible person, and not to reenter the building until an OFFICIAL clearance to do so is given.

### 5.5.5. ACCIDENTS AND SPILLS OF BIOHAZARDOUS MATERIALS

Most public emergency agencies are not equipped to handle biohazardous incidents, so laboratories handling organisms which are infectious to humans or which could harm the environment should plan for emergencies involving these materials.

A minor spill would be one that remains contained within a biological safety cabinet which provides personnel protection. It is assumed that no one is contaminated by the spill. Essentially, the procedures to be used in such a case should take advantage of the protection afforded by the cabinet, i.e., the cabinet should continue to operate. Decontamination of the cabinet would be the measure to be taken to remedy the situation. In some minor cases, this can be done by using the routine procedures normally used for surface contamination. However, this will not decontaminate the fan, filters, and airflow plenums, which are not normally accessible. When there has been a substantial spill within the cabinet, the cabinet should be decontaminated by formaldehyde gas. Standard procedures for doing this are available from a number of government agencies. Individuals should be trained in how to perform this procedure safely and should use it on a number of occasions, such as before filter changes, maintenance work, relocating a cabinet, and upon instituting a different program in a biosafety cabinet.

A major spill would be one that is not contained within a biological safety cabinet. The response would depend upon the nature of the organism involved and the size of the spill, i.e., the probability of individuals being infected. In a later section, the design, special eqipment, and practices will be described for laboratories of biosafety levels 1 through 4. At the upper end of this scale, biosafety levels 3 and 4, there is real concern about the possibility of infections being carried out of the laboratory, so the response would be different for these facilities than for areas designed to work with less dangerous organisms.

Following a spill of an infectious agent, one should immediately evacuate the room, breathing as little as possible of any aerosols. The door to the contaminated area should be closed to avoid airborne transfer of the material from the area in which the spill occurred. Any outer garments that were contaminated, including laboratory coats, shoes, trousers, gloves, etc., should be removed and placed in sealed containers. These can be pails with covers, autoclavable bags, or, if no other alternatives are available, double-plastic garbage bags. Everyone in the room should at least thoroughly wash their face and hands with a disinfectant soap and, if possible, shower thoroughly. Once the area is isolated and the possibility of retaining infectious material on individuals is minimized, the situation should be reviewed and a plan of correction determined by the responsible parties. Because of the possibility of liability, institutional and corporate management personnel, as well as safety specialists, should participate in the discussion. Any persons potentially infected should be referred to a physician as soon as possible. They should not participate in the cleanup efforts until it is determined whether they were infected.

The potentially contaminated materials should be autoclaved as soon as possible or, depending on the value of the materials and the risk of infections to someone trying to sort the material, it may be decided that disposal as a hazardous waste is the proper action. Most organizations carry insurance to cover losses due to accidents.

If the material is not highly infectious and the quantity is not great, it may be sufficient to have someone reenter the area, wearing a coverall covering the entire body, gloves, shoe covers, and a respirator, to decontaminate the immediate area of the spill with a suitable disinfectant and to wipe down nearby surfaces onto which some material may have splashed or areas where the ventilation system in the laboratory may have carried aerosols. Once the preliminary cleanup is completed, it may be sufficiently clean to warrant others entering the room wearing protective outer garments, gloves, and felt protective masks to proceed with a more thorough cleanup. All disposable materials used in the cleanup should go into hazardous waste containers and materials which are to be kept, into autoclavable containers. It will probably be desirable to test the air to determine if organisms are still present. All of these actions are not necessarily steps that should be taken in every case, but represent a conservative approach which might be considered at the time of an incident, based on the circumstances.

For an organism which is highly infectious to humans or a massive spill, it may be necessary to seal off the area thoroughly, including shutting off ventilation ducts to and from the area, to ensure that the organisms remain isolated. A specially designed decontamination program may be necessary. Consultation with specialists at the Centers for Disease Control and elsewhere may be desirable for advice and aid on the care of exposed individuals and correction of the situation. Two things should be kept in mind: equipment can be replaced, people cannot. If the situation is not deteriorating, it is best to leave it alone while a plan of correction is developed which will provide maximum safety for everyone involved.

Custodial personnel should not be asked to assist in cleaning up a spill involving materials that pose biohazards. They are not qualified to perform the work safely. Decontamination is a technical problem, not a custodial one.

### 5.5.6. GENERATION OF AEROSOLS

It is generally conceded that aerosols are the primary means by which infectious diseases are contracted or spread in the microbiological laboratory, although some cases are known to have occurred due to animal bites, sticking with needles, and similar situations where direct contact can happen.

There are many opportunities for aerosols to be generated through normal laboratory procedures. Studies have been conducted of the averge number of droplets created by many typical operations, and some procedures are prolific generators of aerosols. Each droplet often contains several organisms. There are far more of these daily releases than there are accidents, where the potential for many thousands of infectious organisms being released exists.

Only about 20% of all laboratory infections are traced to specific incidents. The majority of the remainder probably come from ordinary, routine activities. In one comprehensive study, it was found that over 70% of the infections occurring in a laboratory were to scientific personnel and that 98% of all laboratory-acquired infections were in institutions doing research or diagnostic work.

Some of the laboratory operations which release a substantial number of droplets are almost trivial in nature, such as breaking bubbles on the surface of a culture as it is stirred, streaking a rough agar plate with a loop, a drop falling off the end of a pipette, inserting a hot loop into a culture, pulling a stopper or a cotton plug from a bottle or flask, taking a culture sample from a vaccine bottle, opening and closing a petri dish in some applications, or opening a lyophilized culture, among many others. Most of these only take a second or so and are often repeated many times daily. Other, more complicated procedures might be considered more likely to release organisms into the air, such as grinding tissue with a morter and pestle, conducting an autopsy on a small animal, harvesting infected tissue from animals or eggs, intranasal innoculation of small animals, opening a blender too quickly, etc. Some incidents have occurred by failing to take into account the possibility that accidents can happen, such as a tube breaking in a centrifuge. The possibility of aerosol production should always be considered while working with infectious organisms.

### 5.5.7. INFECTIOUS WASTE

Any item that has been in contact with infectious organisms or with materials such as blood, serums, excreta, tissue, etc. that may be infected must be considered infectious unless it has been treated. There are individuals who have excellent laboratory technique and take every precaution to avoid infection who do not consider the possibility that someone who handles the materials which they discard may become infected from these materials. Some of the items used to carry out the procedures discussed in the previous paragraph are potentially infectious, but are sometimes discarded as ordinary trash, in which case custodial workers can come into contact with them. Infected tissue has been found in lab clothing sent to the local laundry. Carcasses of animals that had rabies have been sent for disposal with no warnings about careful handling.

Waste from areas where the potential exists for coming into contact with infectious diseases should be treated as if it were hazardous and prepared so that it can be handled safely. It should be bagged or put in a container which is not likely to break or rupture, then incinerated, steam sterilized, or perhaps chemically treated before disposal. Every organization which generates infectious waste should establish procedures to make sure that the waste is treated to render it harmless and that all personnel follow these procedures. One method which is suitable for a small generator of biological waste to safely collect waste is to double-bag it in heavy-duty plastic bags and freeze it until enough has been collected to run an incinerator economically or to justify the cost of a pickup by a biological waste-disposal firm. A number of commercial firms are available for disposing of dangerous biological waste, just as there are firms for disposing of hazardous chemical wastes.

### 5.5.8. LABORATORY FACILITIES — DESIGN AND EQUIPMENT

A microbiological laboratory must be designed properly to allow for safe working conditions. It will share many of the same basic features of a good chemistry laboratory. The ventilation should be a 100% fresh air system in most cases, with perhaps more stringent temperature and humidity controls than other types of laboratories. The temperature should be controlled over a relatively narrow range around 72°F (22°C) and the humidity should be maintained between 45 and 60%. There should be more than one means of egress in an emergency, although this is typically not a code requirement in most cases. The layout should be conducive to free movement of personnel. Aisles should be sufficiently wide so that stools, chairs, or equipment placed temporarily in the aisles will not block them. The floors, walls, and surfaces of equipment should be of easily decontaminated material. The junctures of the floors with the walls, and the equipment, should be as seamless as possible to avoid cracks in which organic materials could collect and microorganisms thrive. There should be areas, either near the entrance to the laboratory or adjacent to it, where paperwork and records can be kept and processed, and where social conversation, studying, eating, etc. can be done safely outside the work area. Adequate utilities should be provided. Proper experimental equipment which will allow laboratory operations to be done safely should be provided and properly maintained. An eyewash fountain and deluge shower should be provided and emergency equipment liable to be needed in the laboratory should be kept in a readily accesible place. The laboratory should be at a negative pressure with respect to the corridor servicing it and the air flow within the laboratory should be away from the "social" area toward the work area.

### 5.5.9. BIOSAFETY LEVELS*

This section is adapted from Biosafety in Microbiological and Biomedical Laboratories published by the U.S. Department of Health, Education and Welfare.[15] It is somewhat shorter than the original guide, but contains virtually all of the guide's recommendations. A few additional recommendations have been added which have been found useful. The supportive material in the original publication has been deleted.

Microbiological laboratories are divided into four different classifications or levels, with Level 1 intended for work with the lowest risk and Level 4 designated for work with the highest risk. An ordinary laboratory would approximate a Level 2 facility. Each level features a combination of design, standard, and special laboratory practices and procedures and standard and special equipment needed to allow the work appropriate for each level to be done safely.

The following four sections describe each of these biosafety levels, after which several lists of organisms will be provided which are appropriate for each level.

### 5.5.9.1. Biosafety Level 1

The work at this level normally involves well-defined and characterized strains of viable microorganisms which are not known to cause disease in healthy adult humans. However, individuals who may be immunodeficient or immunosuppressed for any reason (e.g., poor health) may be at a higher risk and should inform the person responsible for the laboratory operation to assist him in reducing the possibility of their acquiring an infection. The facilities and equipment are appropriate for instruction of undergraduates and secondary-level students in good laboratory practices using microorganisms. Work is generally conducted on open bench tops. No special containment equipment is required.

The guidelines in this area are perhaps slightly more restrictive than necessary, but, as facilities used in many cases for teaching, it is desirable to stress basic laboratory safety practices from the beginning for students entering the field.

### 1. Standard Microbiological Practices

- Access to the laboratory may be limited or restricted, at the discretion of the laboratory director, when experiments are in progress.
- Work surfaces are to be decontaminated after each day in which operations are performed and after any spill of viable material.
- All contaminated liquid or solid wastes must be decontaminated before disposal.
- Mechanical or automatic pipetting devices must be used; mouth pipetting is prohibited.
- Eating, drinking, smoking, and applying cosmetics are not permitted in the work area. Food can only be stored in cabinets or refrigerators specifically designated for that purpose and which may not be used for storage of any laboratory materials. They must be located outside the work area, although they can be in the area set aside for paperwork and social activities as long as no laboratory research materials are brought into these areas. It would be preferable if food storage units were outside the laboratory entirely.

---

* An earlier version of this and succeeding sections was prepared by the author for use at his institution as part of its biosafety manual.

- Persons must wash their hands with a disinfectant soap or detergent after they handle viable materials and before leaving the laboratory.
- All procedures must be performed so as to minimize the creation of aerosols.
- Laboratory coats, gowns, or uniforms should be worn over street clothes while working in the laboratory. These should not be worn away from the laboratory work area.

## 2. Special Practices

- Contaminated materials that are to be decontaminated at a site away from the laboratory are to be placed in a durable, leakproof container which is closed before being removed from the laboratory.
- Freezers should be provided for storage of biological waste materials that will not be collected promptly so that they will not putrefy. No solvents are to be placed in these freezers.
- An effective insect and rodent control program must be in place.

## 3. Containment Equipment

- No special containment equipment is usually required for the work practices and agents assigned to Biosafety Level 1.

## 4. Laboratory Facilities

- The laboratory should be designed so that it is easily cleaned. Seamless or poured floor coverings are recommended. Epoxy paint is recommended for the walls. Junctures with the wall, floor, ceiling, and on equipment should be rounded, with no corner seams. Laboratory furniture should be well built and incorporate as few seams and cracks as possible.
- Spaces between benches, cabinets, and equipment should be readily accessible for cleaning.
- Bench tops must be impervious to water and resistant to acids, alkalies, organic solvents, and moderate heat.
- Each laboratory is to contain a sink for handwashing.
- Laboratory windows that can be opened shall be equipped with fly screens.

### 5.5.9.2. Biosafety Level 2

This class of laboratory is suitable for work involving agents of moderate potential hazard to personnel and the environment. In addition to the characteristics of a Biosafety Level 1 facility, (1) laboratory personnel have specific training in handling pathogenic agents and are directed by competent scientists, (2) access to the laboratory is limited whenever work is being conducted, and (3) some procedures in which infectious aerosols are created will be performed in biological safety cabinets or other physical containment equipment.

## 1. Standard Microbiological Practices

- Access to the laboratory may be limited or restricted, at the discretion of the laboratory director, when work involving infectious agents are in progress.
- Work surfaces are to be decontaminated at least once each day work is in progress and after any spill of viable material.

- All contaminated liquid or solid wastes must be decontaminated before disposal.
- Mechanical pipetting devices must be used; mouth pipetting is prohibited.
- Eating, drinking, smoking, and applying cosmetics are not permitted in the work area. Food can only be stored in cabinets or refrigerators specifically designated for that purpose, and these must be located outside the work area.
- Persons must wash their hands after they handle viable infectious materials and animals and before leaving the laboratory.
- All procedures must be performed so as to minimize the creation of aerosols.

**2. Special Practices**
- Contaminated materials that are to be decontaminated at a site away from the laboratory are to be placed in a durable leakproof container which is closed before being removed from the laboratory.
- The laboratory director shall limit access to the laboratory. In general, persons who are at increased risk of acquiring infection or for whom infection might be unusually hazardous are excluded from the facility. The laboratory director has final responsibility for determining who may enter or work in the laboratory and under what circumstances.
- The laboratory director shall establish written policies and procedures whereby only persons who have been advised of the hazards and meet any specific entry requirements, e.g., special training or immunization, enter the facility.
- When the infectious agents in use in the laboratory require special provisions for entry, e.g., vaccination, a hazard warning sign incorporating the universal biohazard symbol are to be posted on the door to the laboratory work area. The warning sign identifies the infectious agent, lists the names of the laboratory director, laboratory supervisor, and department head and identifies the special requirements for entering the laboratory.
- An effective insect and rodent control program must be in effect.
- Laboratory coats, gowns, smocks, or uniforms are to be worn while working in the laboratory. Before leaving the laboratory for nonlaboratory areas, these garments are to be removed and left behind in the laboratory or, alternatively, a clean coat may be worn over the potentially contaminated garments, but the outer garment must be considered contaminated upon removal.
- Animals not involved in the research being performed are not permitted in the laboratory.
- Special care is to taken to avoid skin contamination with infectious materials. Gloves are to be worn when handling infected animals and when skin contact with infectious materials is unavoidable.
- All wastes from laboratories and animal rooms are to be decontaminated before disposal.
- Hypodermic needles and syringes are to be used only for parenteral injection and aspiration of fluids from laboratory animals and diaphragm bottles. Only needle-locking syringes or disposable syringe-needle units are to be used for the injection or aspiration of infectious fluids. Extreme caution should be used when handling needles and syringes to avoid autoinoculation and the generation of aerosols during use and disposal. Needles are not to be bent, sheared, replaced in the sheath or guard, or removed from the syringe following use. The entire

needle and syringe unit should be promptly placed in a puncture-proof container and decontaminated, preferably by autoclaving, before being discarded or reused.

- Spills and accidents which result in overt exposures to infectious agents are to be reported immediately to the laboratory director. Medical evaluation, surveillance, and treatment are to be provided as needed at no cost to the employee. Written records are to be maintained of the incident, including a determination of the causes, any personnel exposed, and any actions which were taken. Medical records are to be maintained for at least 40 years.
- For those individuals working with infectious materials (including possibly infectious animals, animal tissues that have not been rendered safe, or derivative materials), baseline serum samples are to collected and stored. Additional serum samples shall be taken at 5-year intervals or after an overt exposure. In addition, as a minimum, a medical history is required. It is recommended also that the employees participate in a medical surveillance program if one is available through the organization.
- A biosafety manual should be prepared and adopted. Personnel are to be advised of special hazards and are required to become familiar with and follow experimental practices and procedures, including all safety precautions.

### 3. Containment Equipment

Class 1 or 2 biological safety cabinets or other appropriate personal protective or physical containment devices are used in the following situations.

- When procedures with a high potential for creating infectious aerosols are conducted. These include, but are not limited to, centrifuging, grinding, blending, vigorous shaking or mixing, sonic disruption, pipetting, opening pressurized containers, inoculating animals intranasally, and harvesting infected tissues from animals or eggs.
- When high concentrations or large volumes of infectious agents are used, such materials may be centrifuged in the open laboratory if sealed heads or centrifuge safety cups are used and if the materials are opened only in a biological safety cabinet.

### 4. Laboratory Facilities

- The laboratory should be designed so that it is easily cleaned. Seamless or poured floor coverings are recommended. Epoxy paint is recommended for the walls. Junctures with the wall, floor, ceiling, and on equipment should be rounded, with no corner seams. Laboratory furniture should be well built and incorporate as few seams and cracks as possible.
- Spaces between benches, cabinets, and equipment should be readily accessible for cleaning.
- Bench tops must be impervious to water and resistant to acids, alkalies, organic solvents, and moderate heat.
- Each laboratory is to contain a sink for handwashing.
- Laboratory windows that can be opened shall be equipped with fly screens.
- An autoclave for decontaminating infectious laboratory waste is to be available.

### 5.5.9.3. Biosafety Level 3

This level is applicable to clinical, diagnostic, teaching, and research facilities in which work is done with indigenous or exotic agents which may cause serious or potential lethal disease as a result of exposure by inhalation. All laboratory personnel are to be specifically trained in handling pathogenic and potentially lethal agents and should be supervised by competent scientists who are experienced in working with these agents. All procedures involving manipulation of infectious agents are conducted within biological safety cabinets or other physical containment devices or by personnel wearing appropriate protective clothing or devices. If the last option is used, others in the laboratory other than the person directly involved will probably also require the same protective clothing or devices. Some routine operations with a very low probability for exposure may, upon the explicit review and decision of the laboratory director, be performed at Biosafety Level 2.

### 1. Standard Microbiological Practices

- Work surfaces are to be decontaminated at least once a day and after any spill of viable material.
- All contaminated liquid or solid wastes must be decontaminated before disposal.
- Mechanical pipetting devices must be used; mouth pipetting is prohibited.
- Eating, drinking, smoking, and applying cosmetics are not permitted in the work area.
- Persons must wash their hands after they handle infectious materials and animals and before leaving the laboratory.
- All procedures must be performed so as to minimize the creation of aerosols.

### 2. Special Practices

- Laboratory doors must be kept closed when experiments are in progress.
- Contaminated materials that are to be decontaminated at a site away from the laboratory are to be placed in a durable, leakproof container which is closed before being removed from the laboratory.
- The laboratory director shall control access to the laboratory and restrict access to persons whose presence is required for program or support purposes. Persons who are at increased risk of acquiring infections or for whom infection may be unusually hazardous are not allowed in the laboratory or animal rooms. The laboratory director has final responsibility for assessing each circumstance and determining who may enter or work in the laboratory.
- The laboratory director shall establish written policies and procedures whereby only persons who have been advised of the hazards and meet any specific entry requirements, e.g., special training or immunization, and who comply with all entry and exit procedures may enter the facility.
- When infectious agents or infected animals are being used in the laboratory, a hazard warning sign incorporating the universal biohazard symbol is to be posted on all laboratory and animal room access doors. The warning sign identifies the infectious agent, lists the names and telephone numbers of the laboratory director, laboratory supervisor, and department head and identifies the special requirements for entering the laboratory.
- All activities involving infectious materials are to be conducted in biological

safety cabinets or other physical containment devices within the containment module. No work in open vessels is permitted on the open bench.

- The work surfaces of biological safety cabinets and other containment equipment must be decontaminated when work with infectious materials is finished. Plastic-backed paper toweling used on nonperforated work surfaces within biological safety cabinets is to be used to facilitate cleanup.
- An effective insect and rodent control program must be in effect.
- Laboratory clothing that protects street clothing, such as solid-front or wrap-around gowns, scrub suits, or coveralls is to be worn in the laboratory. Laboratory clothing is not to be worn outside the laboratory and it is to be decontaminated before being laundered.
- Special care should be taken to avoid skin contamination with infectious materials. Gloves are to be worn when handling infected animals and when skin contact with infectious materials is unavoidable.
- Molded surgical masks or respirators are to be worn in rooms containing infected animals. Individuals who are to wear respirators must be fitted and trained in the care of the respirators, as required by OSHA standards, and must be given a pulmonary function test by or under the supervision of a physician.
- Animals and plants not involved in the research being performed are not permitted in the laboratory.
- All wastes from laboratories and animal rooms are to be decontaminated before disposal.
- Vacuum lines are to be protected with HEPA filters and liquid disinfectant traps.
- Hypodermic needles and syringes are to be used only for parenteral injection and aspiration of fluids from laboratory animals and diaphragm bottles. Only needle-locking syringes or disposable syringe-needle units used for the injection or aspiration of infectious fluids. Extreme caution should be used when handling needles and syringes to avoid autoinoculation and the generation of aerosols during use and disposal. Needles are not to be bent, sheared, replaced in the sheath or guard, or removed from the syringe following use. The entire needle and syringe unit should be promptly placed in a puncture-proof container and decontaminated, preferably by autoclaving, before being discarded or reused.
- Spills and accidents which result in overt or potential exposures to infectious agents are to be immediately reported to the laboratory director. Medical evaluation, surveillance, and treatment are to be provided as needed at no cost to the employee. Written records are to be maintained of the incident, including a determination of the causes, any personnel exposed, and any actions which were taken. Medical records are to be maintained for at least 40 years.
- For those individuals working with infectious materials (including possibly infectious animals, animal tissues which have not been rendered safe, or derivative materials), baseline serum samples are to be collected and stored. Additional serum samples shall be taken at 5-year intervals or after an overt exposure. In addition, as a minimum, a medical history is required and it is recommended that employees participate in a medical suveillance program if one is available through the organization.
- A biosafety manual is to be prepared and adopted. Personnel are to be advised of special hazards and are required to become familiar with and follow experimental practices and procedures, including all safety precautions.

### 3. Containment Equipment

Biological safety cabinets (Class 1, 2, or 3) or other combinations of personal protective or physical containment devices, such as special protective clothing, masks, gloves, respirators, centrifuge safety cups, sealed centrifuge rotors, and containment caging for animals, are to be used for all activities with infectious materials that pose a threat of aerosol exposure. These include, but are not necessarily limited to, manipulation of cultures and those clinical or environmental materials which may be a source of infectious aerosols, the aerosol challenge of infected animals, and harvesting of tissues or fluids from infected animals and embryonated eggs and necropsy of infected animals. Special procedures shall be developed for necropsies of animals that are too large for biological safety cabinets, which will provide sufficient personnel protection.

### 4. Laboratory Facilities

- The laboratory should be separated from areas which are open to unrestricted traffic flow within the building. Passage through two sets of doors for entry into the laboratory from access corridors or other contiguous areas is required. Physical separation of the containment laboratory may also be achieved by a double-doored clothes changing room (which is normally provided with showers), airlock, or other access facility which requires passage through two sets of doors before entering the laboratory. The two sets of doors are to be far enough apart so that they cannot be opened simultaneously.
- The laboratory should be designed so that it is easily cleaned. Seamless or poured floor coverings are recommended. Epoxy paint is recommended for the walls. Junctures with the wall, floor, ceiling, and on equipment should be rounded. Laboatory furniture should incorporate as few seams and cracks as possible. Penetrations through all six sides of the room are to be sealed.
- Spaces between benches, cabinets, and equipment should be readily accessible for cleaning.
- Bench tops must be impervious to water and resistant to acids, alkalies, organic solvents, and moderate heat.
- Each laboratory shall contain a sink for handwashing. The sink is to be foot, elbow, or automatically operated and located near the laboratory exit door. Disinfectant soap or detergent is to be provided.
- Windows in the laboratory are to be closed and sealed.
- Access doors to the laboratory or confinement module are to be self-closing.
- An autoclave for decontaminating laboratory wastes is to be available, preferably within the laboratory.
- A ducted exhaust system is to be provided. The system should be designed so as to create directional air flow through the entry area into the laboratory. The direction of flow must be directly confirmed by laboratory personnel. The system is to provide 100% makeup air and is to be discharged so as to minimize the possibility of it reentering the building air intakes or being discharged into an occupied area. The exhaust air is not required to be filtered or otherwise treated.
- The HEPA-filtered exhaust air from Class 1 or 2 biological safety cabinets is to be exhausted directly to the outside or through the building exhaust system if this

can be done in a manner that avoids any interference with the air balance of the cabinets or the building exhaust system and there is no possibility of the exhaust being discharged into occupied spaces. It may be recirculated within the laboratory if the cabinet is certified initially and recertified after no more than a 12-month interval or after moving the cabinet or performing maintenance which could break the seals.

### 5.5.9.4. Biosafety Level 4

Biosafety Level 4 is required for work with dangerous and exotic agents which pose a high individual risk of life-threatening disease. Members of the laboratory staff are to receive specific and thorough training in handling infectious agents. They are to understand the primary and secondary containment functions of the standard and special practices, the containment equipment, and the laboratory design characteristics. They should be supervised by competent scientists who are trained and experienced in working with the specific agents involved in the research. Access to the laboratory is strictly controlled by the laboratory director. The facility is to be in either a separate building or, if in a multiuse building, in a controlled area within the building that is completely isolated from all other areas of the building.

Normally, activities conducted within a Class 4 biosafety facility are performed in a Class 3 biosafety cabinet. If work is performed in a Class 1 or 2 biosafety cabinet, isolation of the workers is to be achieved by use of self-contained, positive-pressure containment suits. The facility shall be designed to prevent discharge of microorganisms into the environment.

### 1. Standard Microbiological Practices

- Work surfaces are to be decontaminated at least once a day and after any spill of viable material.
- Mechanical pipetting devices must be used; mouth pipetting is prohibited.
- Eating, drinking, smoking, and applying cosmetics are not permitted in the work area.
- All procedures must be performed so as to minimize the creation of aerosols.

### 2. Special Practices

- Biological materials to be removed from the Class 3 cabinet or from the maximum containment laboratory in a viable or intact state are to be transferred to a nonbreakable, sealed primary container and then enclosed in a nonbreakable, sealed secondary container, which is removed from the facility through a disinfectant dunk tank, fumigation chamber, or an airlock designed for this purpose.
- No materials, except for biological materials that are to remain in a viable or intact state, are to be removed from the maximum containment laboratory unless they have been autoclaved or decontaminated before they leave the facility. Equipment or materials which may be damaged by high temperatures or steam are to be decontaminated by gaseous or vapor methods in an airlock or chamber designed for the purpose.
- Only persons whose presence in the facility or individual laboratory rooms is required for program or support purposes are authorized to enter. Persons who may be at increased risk of acquiring infection or for whom infection may be

unusually hazardous are not allowed in the laboratory or animal rooms. The laboratory director shall determine who has authorized access. Access to the facility is by means of secure and normally locked doors. Accessibility is managed by the laboratory director, biohazards control officer, or other person responsible for the physical security of the facility. All persons entering the facility are to be informed in advance of the potential biohazards and instructed in the appropriate safeguards for ensuring their safety. Authorized persons must comply with the instructions and all other applicable entry and exit procedures. A logbook is to be maintained for all personnel entering the facility, with the date and time of each entry and the person's signature. Practical and effective emergency procedures for rapid evacuation are to be established.

Personnel are to enter and leave the facility only through the clothing change and shower rooms. Personnel are to shower each time they leave the facility. Personnel are to use the airlocks as entrances or exits only in an emergency.

Street clothing is to be removed in the outer clothing change room and kept there. Complete laboratory clothing, including underclothes, pants, shirts or jumpsuits, shoes, and gloves, is to be provided for and used by all personnel entering the facility. Head covers are to be provided for personnel who do not plan to wash their hair during the exit shower. When leaving the laboratory and before proceeding into the shower area, personnel are to remove their laboratory clothing and store it in a locker or hamper in the inner change room.

When infectious material or infected animals are present in the laboratory or animal rooms, a hazard warning sign incorporating the universal biohazard symbol is to be prominently posted on all access doors. The sign is to identify the agent, list the name of the laboratory director or other responsible, knowledgeable persons, and indicate any special requirements for entering the area, such as immunizations, respiratory protection, or complete isolation gear.

Supplies and materials needed in the facility are to be brought in by way of the double-doored autoclave, fumigation chamber or airlock which is to be appropriately decontaminated between each use. After the outer doors are secured, personnel within the facility are to retrieve the materials by opening the interior doors of the autoclave, fumigation chamber, or airlock. The interior doors are to be secured after the materials are brought in.

An effective insect and rodent control program must be in place.

Materials such as plants, animals, and clothing not related to the experiment being conducted shall not be permitted in the facility.

Hypodermic needles and syringes are to be used only for parenteral injection and aspiration of fluids from laboratory animals and diaphragm bottles. Only needle-locking syringes or disposable syringe-needle units are to be used for the injection or aspiration of infectious fluids. Needles are not to be bent, sheared, replaced in the sheath or guard, or removed from the syringe following use. The needle and syringe should be promptly placed in a puncture-proof container and decontaminated, preferably by autoclaving, before being discarded or reused. Whenever possible, cannulas are to be used instead of sharp needles (e.g., for gavage).

A system is to be established for proper reporting of laboratory accidents, exposures, and employee absenteeism, and for medical surveillance of potential laboratory-associated illnesses. Written records are to be maintained of the absentee records, exposures, and incidents. Reports of incidents and exposures shall include a determination of the causes, the personnel exposed, and any actions which were taken. Medical records are to be maintained for at least 40 years. An essential adjunct to such a monitoring system is to have available a facility for the quarantine, isolation, and medical care of persons with potential or identified laboratory-associated illnesses.

### 3. Containment Equipment
All procedures within the facility assigned to Biosafety Level 4 are to be conducted in Class 3 biological safety cabinet or in Class 1 or 2 biological safety cabinets used in conjunction with one-piece, positive-pressure personnel suits ventilated by a life support system. Exception: work with viral agents that require Biosafety Level 4 secondary containment capabilities for which highly effective vaccines are available and used can be conducted in Class 1 and 2 cabinets without the suit if (1) the facility has been decontaminated, (2) no other work is being conducted in the facility with other agents assigned to Biosafety Level 4, and (3) all other standard and special practices are followed.

### 4. Laboratory Facility
- The maximum containment facility required for work at Biosafety Level 4 is to consist of either a separate building or a clearly demarcated and isolated zone within a building. Outer and inner change rooms separated by a shower are to be provided for personnel entering or leaving the facility. A double-doored autoclave, fumigation chamber, or ventilated airlock is to be provided for passage of those materials, supplies, or equipment which are not brought into the facility through the change room.
- Walls, floors, and ceilings of the facility are be constructed to form a sealed internal shell which facilitates fumigation and is animal and insect proof. The internal surfaces of the shell are to be resistant to liquids and chemicals to facilitate cleaning and decontamination. All penetrations of the shell are to be sealed. Any drains in the floor are to contain traps filled with a chemical disinfectant with demonstrated efficacy against the target agent. The drains are to be connected directly to the liquid waste decontamination system. Sewer "vents" and other ventilation lines are to contain HEPA filters.
- Internal facility appurtenances such as light fixtures and utility pipes are to be installed in such a way as to minimize the horizontal surface area on which dust can settle.
- Bench tops are to have seamless surfaces which are impervious to water and resistant to acids, alkalies, organic solvents, and moderate heat.
- Laboratory furniture is to be of simple and sturdy construction and spaces between benches, cabinets, and equipment are to be accessible for cleaning.
- A foot, elbow, or automatically operated handwashing sink is to be provided near the door of each laboratory room in the facility. A disinfectant soap or detergent is to be available from a similarly operated dispenser.
- Any installed central vacuum system is not to serve any areas outside the facility. In-line HEPA filters are to be placed as near as practicable to each use point or

service dock. Filters are to be installed to permit inplace decontamination and replacement. Other liquid and gas services to the facility are to be protected by backflow prevention devices.

- If water fountains are provided within the facility, they are to be foot operated and are to be located outside the laboratory area in the facility corridors. The water service to the fountains is not to be connected to the backflow-protected system supplying water to the laboratory areas.

- Access doors to the laboratory are to be self-closing and lockable. They are never to be blocked open.

- Any windows are to be sealed, breakage resistant, and nonopenable.

- A double-doored autoclave is to be provided for decontaminating materials passing out of the facility. The autoclave door that opens to the area external to the facility is to be sealed to the outer wall and automatically controlled so that the outside door cannot be opened until after the "sterilization" cycle has been completed.

- A pass-through dump tank, fumigation chamber, or an equivalent decontamination method is to be provided so that materials and equipment that cannot be decontaminated in the autoclave can be safely removed from the facility.

- Liquid effluents from laboratory sinks, biological safety cabinets, floors, and autoclave chambers are to be decontaminated by heat treatment before being released from the maximum containment facility. Liquid wastes from shower rooms and toilets are to be decontaminated with chemical disinfectants or by heat in the liquid waste decontamination system. The performance of the heat decontamination system is to be monitored by a recording thermometer and by using an indicator microorganism with a defined heat susceptability pattern. If liquid wastes from the shower rooms are decontaminated with chemical disinfectants, the chemical used is to be of demonstrated efficacy against the target or indicator microorganisms.

- A dedicated supply and exhaust air ventilation system is to be provided. The system is to maintain pressure differentials and directional air flow to assure that air flow is into the facility from outside and toward the areas of highest potential risk within the facility. Manometers are to be used to sense pressure differentials between adjacent areas maintained at different pressures. The manometers are to trigger an alarm if a system malfunctions. Systems are to be interlocked to assure inward or, at worst, zero air flow at all times. It would be desirable if dual sensors were used to provide redundancy.

- The exhaust air from the facility is to be filtered through HEPA filters and discharged away from the air intakes and occupied areas. The HEPA filters are be located within the facility as near as practicable to the laboratories in order to reduce the length of potentially contaminated air ducts. The filter chambers are designed to allow *in situ* decontamination before filters are removed and to facilitate certification testing after they are replaced. Coarse filters and HEPA filters are be used on the supply air to extend the life of the exhaust HEPA filters and to protect the supply air mechanical system should the air pressures become unbalanced in the facility.

- If Class 1 or 2 biosafety cabinets are used in the facility, the exhaust air from them can be discharged into the room or into the building exhaust system if the cabinets are tested and recertified at 6-month intervals. The treated exhaust air from Class 3 biological safety cabinets is to be discharged, without recirculation, through two sets of HEPA filters in series via the facility exhaust air system. If the treated exhaust air from any of these cabinets is discharged through the facility exhaust air system, it is to be connected to this system in such a manner as to avoid any interference with the air balance of the cabinets or the facility exhaust air system.

• A specially designed suit area may be provided in the facility. Personnel who enter this area are to wear a one-piece, positive-pressure suit that is ventilated by a life support system. The life support system is to include alarms and emergency backup breathing air tanks. Entry into this area is to be through an airlock fitted with airtight doors. A chemical shower is to be provided to decontaminate the surface of the suit before the worker leaves the area. The exhaust air from the suit area is to be filtered by two sets of HEPA filters installed in series. A duplicate filtration unit, exhaust fan, and an automatically starting emergency power source are to be provided. The air pressure within the suit area is to be lower than that of any adjacent area. Emergency lighting and communication systems are to be provided. All penetrations into the internal shell of the suit area are to be sealed. A double-doored autoclave is to be provided for decontaminating waste materials to be removed from the suit area.

### 5.5.10. RECOMMENDED BIOSAFETY LEVELS

The subsections of this part will contain, for a large number of potentially infectious agents, the biosafety level appropriate for typical laboratory-scale operations involving these agents. However, selection of an appropriate safety level for work with a specific agent or animal study depends upon a large number of factors. Some of the most important are (1) the virulence, pathogenicity, biological stability, route of spread, and communicability of the agent, (2) the nature or function of the laboratory, the procedures and manipulations involving the agent and the endemicity of the agent, and (3) the availability of effective vaccines or therapeutic measures. The following sections will not present the rationale for the recommendations. For additional information, the reader should consult Reference 15, the base document from which the majority of the material in this section is taken.

The risk assessments and biosafety levels reccommended presuppose a population of healthy immunocompetent individuals. There are a number of parameters which influence this factor, including age, heredity, race, sex, pregnancy, predisposing diseases, surgery, and prior exposure to immunosuppressing agents.

Recommendations for the use of vaccines and toxoids are included where effective and safe versions are available. Appropriate precautions should be taken in the administration of live attenuated virus vaccines to individuals with altered immunocompetence. However, these specific recommendations should in no way preclude the routine use of such products as diphtheria-tentanus toxoids, poliovirus vaccine, or influenza vaccine.

The basic biosafety level assigned to an agent is based on the activities typically associated with the growth and manipulation of quantities and concentrations of infectious agents required to accomplish identification or typing. If activities with clinical materials pose a lower risk to personnel than those associated with the manipulation of cultures, a lower biosafety level is recommended. On the other hand, if the activities involve large volumes or highly concentrated preparations ("production quantities") or manipulations which are likely to produce aerosols or which are otherwise intrinsically hazardous, additional personnel precautions and increased levels of primary containment may be indicated. It may be possible to adapt biosafety levels up or down to compensate for the appropriate level of safety.

It is the responsibility of the laboratory director to make these decisions. Risk assessment is ultimately a subjective process, but it is recommended that decisions should be biased toward more safety rather than less.

### 5.5.10.1. Agent Summaries

All of the following recommendations exclude working with production quantities. In such cases, a higher level of protection is recommended.

### 1. Parasitic Agents
- Cestode parasites of humans
  | | |
  |---|---|
  | *Echinoccus granulosus* | Level 2 |
  | *Taenia solium* (*Cysticercus cellulosae*) | Level 2 |
- Nematode parasites of humans
  | | |
  |---|---|
  | *Strongyloides* spp. | Level 2 |
  | *Ascaris* spp. | Level 2 |
- Protozoal parasites of humans
  | | |
  |---|---|
  | *Toxoplasma* spp. | Level 2 |
  | *Plasmodium* spp. (including *P. cynomologi*) | Level 2 |
  | *Trypanosoma* spp. | Level 2 |
  | *Leishmania* spp. | Level 2 |
  | *Entamoeba histolytica* | Level 2 |
  | *Giardia* spp. | Level 2 |
  | *Naegleria fowleri* | Level 2 |
  | *Toxoplasma cruzi* | Level 2 |
- Trematode parasites of humans
  | | |
  |---|---|
  | *Fasciola* spp. | Level 2 |
  | *Schistosoma* spp. | Level 2 |

### 2. Fungal Agents
- *Blastomyces dermititis*
  | | |
  |---|---|
  | Clinical materials, animal tissues, infected animals | Level 2 |
  | Mold cultures, soil, etc. likely to contain infectious conidia | Level 3 |
- *Coccidioides immitis*
  | | |
  |---|---|
  | Clinical specimens and animal tissue | Level 2 |
  | Animal studies when route of challenge is parenteral | Level 2 |
  | Sporulating mold form cultures, samples likely to contain infectious arthrospores | Level 3 |
- *Cryptococcus neoformans*                                          Level 2
- *Histoplasma capsulatum*
  | | |
  |---|---|
  | Clinical specimens, animal tissues, and animal tissues when route of challenge is parenteral | Level 2 |

| | |
|---|---|
| Processing of mold cultures, soil, etc. when likely to contain infectious conidia | Level 3 |
| •   *Sporothrix schenckii* | Level 2 |
| •   Pathogenic members of the genera *Epidermophyton, Microsporum,* and *Trichophyton* | Level 2 |

### 3. Bacterial Agents

•  *Bacillus anthracis*                                                          Level 2

    A licensed vaccine is available, but is not normally recommended for ordinary use except for workers having frequent contact with clinical specimens or diagnostic specimens

•  *Brucella (B. abortus, B. canis, B. melitensis, B. suis)*

| | |
|---|---|
|     Activities with clinical materials of human or animal origin | Level 2 |
|     Manipulation of cultures of pathogenic *Brucella* spp. | Level 3 |

•  *Chlamydia psittaci, C. trachomatis*

| | |
|---|---|
|     Diagnostic examination of tissues or cultures, necropsies, contact with clinical materials | Level 2 |
|     Activities with concentrations of infectious materials or high potential for aerosol production | Level 3 |

•  *Clostridium botulinum*

| | |
|---|---|
|     All activities except the following: | Level 2 |
|     Activities involving purified toxins or high potential for aerosol production | Level 3 |

    An Investigational New Drug (IND) toxoid is available through the Centers for Disease Control and is recommended for personnel working with cultures of *Cl. botulinum* or its toxins

•  *Cl. tetani*                                                       Level 2

    Administration of an adult diphtheria-tetanus toxoid at 10-year intervals is recommended

•  *Corynebacterium diphtheriae*                             Level 2

    Administration of an adult diphtheria-tetanus toxoid at 10-year intervals may be desirable

•  *Francisella tularensis*

| | |
|---|---|
|     Activities with clinical materials of human or animal origin | Level 2 |
|     Manipulations of cultures and for experimental animal studies | Level 3 |

    An investigational live attenuated virus is available through the Centers for Disease Control and is recommended for those working with the agent or having potential contact in the laboratory or with infected animals

•  *Leptospira interrogans* — all serovars

    All activities                                                     Level 2

•  *Legionella pneumophila;* other legionella-like agents

    All activities                                                     Level 2

•  *Mycobacterium leprae*

    All activities (special care with syringes)                        Level 2

•  *Mycobacterium* spp. other than *M. tuberculosis, M. bovis,* and *M. leprae*

All activities — Level 2
- *M. tuberculosis, M. bovis*
  Working with acid-fast smears or culturing sputa; other — Level 2
  clinical specimens provided aerosol generating manipulations
  done in Cass 1 or 2 biosafety cabinet; a few other restricted
  exceptions
  Propagation and manipulation of cultures and studies using — Level 3
  nonhuman primates
- *Neisseria gonorrhoeae*
  All activities except: — Level 2
  Activities generating aerosols or droplets — Level 3
- *N. meningitidis*
  All activities except: — Level 2
  Activities generating aerosol or droplets — Level 3
  Use of a licensed vaccine should be considered for work with
  high concentrations of infectious materials
- *Pseudomonas pseudomallei*
  All activities except: — Level 2
  Activities with high potential for aerosol or droplet — Level 3
  production
- *Salmonella cholerasuis, S. enteritidis* — all serotypes
  All activities — Level 2
- *S. typhi*
  All activities — Level 2
  A reasonably effective licensed vaccine is available and should
  be considered
- *Shigella* spp.
  All activities — Level 2
- *Treponema pallidum*
  All activities — Level 2
  Periodic serological monitoring should be considered for
  personnel
- *Yersinia pestis*
  All activities except: — Level 2
  Activities with a high potential for aerosol or droplet — Level 3
  production or work with antibiotic resistant strains
  Licensed inactivated vaccines are available and recommended

## 4. Rickettsial Agents
- *Coxiella burnetii*
  Serological examinations, staining of impression smears — Level 2
  Innoculation, incubation, etc. of eggs or tissue cultures, — Level 3
  necropsies, manipulation of tissue cultures
  New Q fever vaccine(IND) available from Fort Detrick; use
  should be limited to those at high risk; no demonstrated
  sensitivity to Q fever antigen
- *Rickettsia akari, Rochalimaea quintana,* and *R. vinsonii*
  All activities — Level 2

- *Rickettsia prowazekii, R. typhi (R. mooseri),*
  *R. tsutsugamushi, R. canada,* and Spotted Fever group agents
  of human disease other than *R. rickettsii* and *R. akari*

  Nonpropagative laboratory procedures, including serological and        Level 2
     fluorescent antibody procedures, staining of impression smears

  All other manipulations of known or potentially infected               Level 3
     materials

- *R. rickettsii*

  Nonpropagative laboratory procedures, including serological            Level 2
     and fluorescent antibody procedures, staining of impression
     smears

  All other manipulations of known or potentially infected               Level 3
     materials

  Close monitoring of febrile illnesses is recommended because
     of the proven value of antibiotic therapy in the early stage
     of infection

## 5. Viral Agents

- Hepatitis A virus

  All activities                                                          Level 2

- Hepatitis B, Hepatitis nonA and nonB

  All activities except:                                                  Level 2

  Activities with high potential for aerosol or droplet production        Level 3

  A licensed inactivated vaccine is available and recommended
     for laboratory personnel

- *Herpes virus simiae* (B-virus)

  All activities involving the use or manipulation of tissues,            Level 2
     body fluids, and primary tissue culture materials from macacques

  Activities involving the use or manipulation of any material            Level 3
     known to contain *H. virus simiae*

  Activities involving the propagation of *H. simiae,* housing of         Level 4
     vertebrate animals with proven infection with the agent

- *Herpes* viruses

  All activities                                                          Level 2

- Influenza

  All activities                                                          Level 2

- Lymphocytic choriomeningitis (LCM) virus

  All activities utilizing possibly infectious body fluids or             Level 2
     tissues and for tissue culture passage of mouse-brain
     passaged strains; manipulation of possibly infectious
     passage and clinical materials should be done in biosafety
     cabinet

  Activities with high potential for aerosol or droplet production        Level 3

- Poliovirus

  All activities                                                          Level 2

  All laboratory personnel working with the agent must have
     documented polio vaccinations or demonstrated evidence of
     immunity to all three types of polio virus types

- Poxviruses
  All activities                                                                          Level 2
  Persons working in or entering facilities where activities with
  vaccinia, monkey pox, or cow pox should have documented
  evidence of vaccination within the preceding 3 years
- Rabies virus
  All activities                                                                          Level 2
  Preexposure immunization is required for personnel working in
  facilities involved with diagnostic activities or research with
  rabies infected materials. Care or level precautions should be
  used for activities with a high potential for dropletor aerosol
  production
- Transmissible spongiform encephalopathies (Creutzfeidt-Jakob
  and kura agents)
  All activities                                                                          Level 2
- Vesicular stomatitis virus (VSV)
  Activities utilizing laboratory-adapted strains of demonstrated                         Level 2
  low virulence
  Activities involving the use or manipulation of infected                                Level 3
  tissues and virulent isolates from infected livestock

### 6. Arboviruses Assigned to Biosafety Level 2

The classification of two of the following, marked with an asterisk, depend upon personnel being immunized. When performing some operations with highly infectious materials, some adaptation of Biosafety Level 3 conditions may be needed.

| | | |
|---|---|---|
| Abu Hammad | Bakau | Bunyamwera |
| Acado | Baku | Burg el Arab |
| Acara | Bandia | Bushbush |
| Aguacate | Bangoran | Bussuquara |
| Alfuy | Bangui | Buttonwillow |
| Almpiwar | Banzi | Bwamba |
| Amapari | Barur | Cacao |
| Anhanga | Batai | Cache Valley |
| Anhembi | Batu | Caimito |
| Anopheles B | Bauline | California Encephalitus |
| Anopheles A | Bebaru | Calovo |
| Apeu | Belmont | Candiru |
| Apoi | Bertioga | Cape Wrath |
| Aride | Bimiti | Capim |
| Arkonam | Birao | Caraparu |
| Aruac | Bluetongue (indigenous) | Carey Island |
| Arumowot | Boraceia | Catu |
| Aura | Botambi | Chaco |
| Avalon | Boteke | Chagres |
| Bagaza | Bouboui | Chandipura |
| Bahig | Bujaru | Changuinola |

Charleville
Chenuda
Chilibre
Chobar Gorge
Clo Mor
Colorado Tick Fever
Corriparta
Cotia
Cowbone Ridge
D'Aguilar
Dakar Bat
Dendue-3
Dengue-1
Dengue-2
Dengue-3
Dengue-4
Dera Ghazi Khan
Eastern Equine
Encephalomyelitis
Edge Hill
Entebbe Bat
Epizootic Hemorrhagic
 Disease
Eubenangee
Eyach
Flanders
Fort Morgan
Frijoles
Gamboa
Gomoka
Gossas
Grand Arbaud
Great Island
Guajara
Guama
Guaroa
Gumbo Limbo
Hart Park
Hazara
Huacho
Hughes
Icoaraci
Ieri
Ilesha
Ilheus
Ingwavuma
Inkoo

Ippy
Irituia
Isfahan
Itaporanga
Itaqui
Jamestown Canyon
Japanaut
Jerry Slough
Johnston Atoll
Joinjakaka
Juan Diaz
Jugra
Jurona
Jutiapa
Kadam
Kaeng Khoi
Kaikalur
Kaisodi
Kamese
Kammavanpettai
Kannamangalam
Kao Shuan
Karimabad
Karshi
Kasba
Kemerovo
Kern Canyon
Ketapang
Keterah
Keuraliba
Keystone
Klamath
Kokabera
Kolongo
Koongol
Kowanyama
Kunjin
Kununnurra
Kwatta
La Crosse
Lagos Bat
La Joya
Landjia
Langat
Lanjan
Latino
Lebombo

Le Dantec
Lipovnik
Lokern
Lone Star
Lukuni
M'Poko
Madrid
Maguari
Mahogany Hammock
Main Drain
Malakal
Manawa
Manzanilla
Mapputta
Maprik
Marco
Marituba
Matariya
Matruh
Matucare
Melao
Mermet
Minatitlan
Minnal
Mirim
Mitchell River
Modoc
Moju
Mono Lake
Mont. Myotis Leukemia
Moriche
Mossuril
Mount Elgon Bat
Murutucu
Navarro
Nepuyo
Ngaingan
Nique
Nkolbisson
Nola
Ntaya
Nugget
Nyamanini
Nyando
O'nyong-nyong
Okhotskiy
Okola

Olifantsvlei
Oriboca
Ossa
Pacora
Pacui
Pahayokee
Palyam
Parana
Pata
Pathum Thani
Patois
Phom-Penh Bat
Pichinde
Pixuna
Pongola
Pretoria
Puchong
Punta Salinas
Punta Toro
Qalyub
Quaranfil
Restan
Rio Bravo
Rio Grande
Ross River
Royal Farm
Sabo
Saboya
Saint Floris
Sakhalin
Salehabad
San Angelo
Sandfly F. (Naples
Sandfly F. (Sicilian)
Sandjimba
Sathuperi
Sawgrass

Sebokele
Seletar
Sembalam
Shamonda
Shark River
Shuni
Silverwater
Simbu
Simian Hemorrhagic Fever
Sindbis
Sixgun City
Snowshoe Hare
Sokuluk
Soldado
Sororoca
Stratford
Sunday Canyon
Tacaiuma
Tacaribe
Taggert
Tahyna
Tamiami
Tanga
Tanjong Rabok
Tataguine
Tembe
Tembusu
Tensaw
Tete
Tettnang
Thimiri
Thottapalayam
Timbo
Toure
Tribec
Triniti
Trivittatus

Trubanaman
Tsuruse
Turlock
Tyuleniy
Uganda S
Umatilla
Umbre
Una
Upola
Urucuri
Usutu
Uukuniemi
Vailore
Venezuelan Equine
  Encephalomyelitis (TC-83)*
Vankatapuram
Vesicular Stomatitis
Wad Medani
Wallal
Wanowrie
Warrego
Western Equine
  Encephalomyelitis
Whataroa
Witwatersrand
Wongal
Wongorr
Wyeomyia
Yaquina Head
Yata
Yellow Fever (17D)*
Yogue
Zaliv Terpeniya
Zegla
Zika
Zingilamo
Zirqa

## 7. Arboviruses and Arenaviruses Assigned to Biosafety Level 3

Superscripts: a — importation, possession, use restricted; b — HEPA filtration of all laboratory exhaust air; c — immunization recommended if vaccine available.

Aino
Akabane
Araguari
Batama

Batken
Bhanja
Bimbo
Gluetongue (exotic)[a]
Bobaya

Bobia
Buenaventura
Cabassou[b,c]
Chikungunya[b,c]
Chim

Cocal
Dhori
Dugbe
Everglades[b,c]
Garba
Germiston[b,c]
Getah
Gordil
Guaratuba
Ibaraki
Ihangapi
Inini
Israel Turkey Meningo
Issyk-Kul
Itaituba
Japanese Encephalitis
Kairi
Khasan
Korean Hemorrhagic
 Fever (Hantaan)
Koutango
Kyzylagach
Louping III[a]
Lymphocytic
 choriomeningitis

Mayaro
Middelburg
Mosqueiro
Mucambo[b,c]
Murray Valley Encephalitis
Nariva
Ndumu
Negishi
New Minto
Nodamura
Northway
Oropouche[b,c]
Orungo
Ouango
Oubangui
Paramushir
Piry
Ponteves
Powassan
Razdan
Rift Valley Fever[a,b,c]
 same as Zinga
Rochambeau
Rocio[b,c]

Sagiyama
Sakpa
Salanga
Santa Rosa
Saumarez Reef
Semlika Forest
Sepik
Serra do Navio
Slovakia
Spondweni
St. Louis Encephalitis
Tamdy
Telok Forest
Thogoto
Tlacotalpan
Tonate[b,c]
VSV-Alagoas
Venezuelan Equine
 Encephalomyelitis[b,c]
Wesselsbron[a,b,c]
West Nile
Yellow Fever[b,c]
Zinga[a,b,c]
 (same as Rift Valley Fever)

## 8. Arboviruses, Arenaviruses, and Filoviruses Assigned to Biosafety Level 4

Congo-Crimean Hemorrhagic Fever   Lassa
Ebola                              Machupo
Junin                             Marburg
Tick-borne encephalitis virus complex (Absettarov, Hanzalova, Hypr, Kumlinge, Kyasanur Forest Disease, Omsk hemorrhagic fever, and Russian Spring-Summer Encephalitis)

## 5.5.10.2. Importation and Interstate Shipment of Human Pathogens and Related Materials

The importation or subsequent receipt of etiologic agents and vectors of human disease is subject to the Public Health Service Foreign Quarantine Regulations (42 CFR 71.156). Permits authorizing the importation or receipt of regulated materials and specifying conditions under which the agent or vector is shipped, handled, and used are issued by the Centers for Disease Control.

The interstate shipment of indigenous etiologic agents, diagnostic specimens, and biological products is subject to the applicable packaging, labeling, and shipping requirements of the Interstate Shipment of Etiologic Agents (42 CFR 72). Additional information on the importation and interstate shipment of etiologic agents of human disease and other related materials may be obtained by writing to:

Centers for Disease Control
Attention: Office of Biosafety
1600 Clifton Road, N.E.
Atlanta, GA 30333
Telephone: (404)-329-3883
FTS: 236-3883

## 8. Restricted Animal Pathogens

Nonindigenous pathogens of domestic livestock and poultry may require special laboratory design, operation, and containment features not generally addressed in this handbook. The importation, possession, or use of the following agents is prohibited or restricted by law or by U.S. Department of Agriculture regulations or administrative policies.

| | |
|---|---|
| African horse sickness virus | *M. mycoides* |
| African swine fever virus | Nairobi sheep disease virus |
| *Besnoitia besnoiti* | (Ganjam virus) |
| Borna disease virus | Newcastle disease virus |
| Bovine ephemeral fever | *Pseudomonas mallei* |
| Bovine infectious petechial fever agent | *Rickettsia ruminantium* |
| Camelpox virus | Rift Valley fever virus |
| Foot and mouth disease virus | Rinderpest virus |
| Fowl plague virus | Swine vesicular disease virus |
| *Histoplasma (Zymonema) farciminosum* | Teschen disease virus |
| Hog cholera virus | *Theileria annulata* |
| Louping III virus | *T. bovis* |
| Lumpy skin disease virus | *T. hirci* |
| *Trypanosoma evansi* | *T. llawerencia* |
| *T. vivax* | Vesicular exanthema virus |
| *Mycoplasma agalactiae* | Wesselsbron disease virus |

The importation, possession, use, or interstate shipment of animal pathogens other than those listed above may also be subject to regulations of the U.S. Department of Agriculture.

Additional information may be obtained by writing to:

Chief Staff Veterinarian
Organisms and Vectors
Veterinary Services
Animal and Plant Health Inspection Service
U.S. Department of Agriculture
Hyattsville, MD 20782
Telephone: (301)436-8017
FTS: 436-8017

# REFERENCES (SECTIONS 5.5 TO 5.5.10.2)

1. **Favero, M. S.,** Biological hazards in the laboratory, in *Proceedings of Institute on Critical Issues in Health Laboratory Practice,* Richardson, J. W., Schoenfeld, E., Tullis, J. W., and Wagner, W. W., Eds., Du Pont, Wilmington, DE, 1986, 1.
2. **Pike, R. M.,** Laboratory-associated infections: incidence, fatalities, cases and prevention, *Annu. Rev. Microbiol.,* 33, 41, 1979.
3. **Pike, R. M.,** Past and present hazards of working with infectious agents, *Arch. Pathol. Lab. Med.,* 102, 333, 1978.
4. **Pike, R. M.,** Laboratory-associated infections: summary and analysis of 3,921 cases, *Health Lab. Sci.,* 13, 105, 1976.
5. **Litsky, B. Y.,** Microbiology of sterilization, *AORN J.,* 26, 340, 1977.
6. Laboratory Safety at the Centers for Disease Control, Publ. No. CDC 75-8118, U.S. Department of Health, Education and Welfare, Washington, D.C., 1974.
7. National Cancer Institute Safety Standards for Research Involving Oncogenic Viruses, Publ. No. (NIH)75-790, U.S. Department of Health, Education and Welfare, Washington, D.C., 1974.
8. National Institutes of Health Biohazards Safety Guide, Stock No. 1740-00383, Public Health Service, National Institutes of Health, U.S. Department of Health, Education and Welfare, 1974.
9. **Hellman, A., Oxman, M. N., and Pollack, R., Eds.,** *Biohazards in Biological Research,* Cold Spring Harbor Laboratory, Cold Spring Harbor, New York, 1974.
10. **Steere, N. V., Ed.,** *Handbook of Laboratory Safety,* 2nd ed., CRC Press, Boca Raton, FL, 1971.
11. **Bodily, J. L.,** General administration of the laboratory, in *Diagnostic Procedures for Bacterial, Mycotic and Parasitic Infections,* Bodily, H. L., Updyke, E. L., and Mason, J. O., Eds., American Public Health Association, New York, 1970, 11.
12. **Darlow, H. M.,** Safety in the microbiological laboratory, in *Methods in Microbiology,* Norris, J. R. and Robbins, D. W., Eds., Academic Press, New York, 1969, 169.
13. **Collins, C. H., Hartley, E. G., and Pilsworth, R.,** *The Prevention of Laboratory Acquired Infection,* Monogr. Ser. No. 6, Public Health Laboratory Service, 1974.
14. **Chatigny, M. A.,** Protection against infection in the microbiological laboratory: devices and procedures, in *Advances in Applied Microbiology,* Vol. 3, Umbreit, W. W., Ed., Academic Press, New York, 1961, 131.
15. Biosafety in Microbiological and Biomedical Laboratories, Publ No. (CDC) 84-8395, Centers for Disease Control and National Institutes of Health, U.S. Department of Health and Human Services, Washington, D.C., 1984.

# 5.6. RECOMBINANT DNA LABORATORIES

The basic guidelines now in effect for recombinant DNA research appeared in the *Federal Register* on May 7, 1986 (51 FR 16958). The material in Section 5.6 represents selected portions taken directly from the guidelines. There will be a few additional comments and occasional bridging comments where intermediate material is deleted. Only those portions which apply to research organization personnel are included, although the deletions are not extensive. These are evolving regulations and any organization working in recombinant DNA research or planning to do so should conform to the latest revisions of the guidelines. A number of significant changes were published in the August 1987, *Federal Register* as proposed actions (Vol. 52, No. 154, pp. 29800—29813).

## 5.6.1. DEFINITION OF RECOMBINANT DNA MOLECULES

In the context of these Guidelines, recombinant DNA molecules are defined as either (i) molecules which are constructed outside living cells by joining natural or synthetic DNA segments to DNA molecules that can replicate in a living cell, or (ii) DNA molecules that result from the replication of those described in (i) above.

Synthetic DNA segments likely to yield a potentially harmful polynucleotide or polypeptide (e.g., a toxin or a pharmacologically active agent) shall be considered as equivalent to their natural DNA counterpart. If the synthetic DNA segment is not expressed *in vivo* as a biologically active polynucleotide or polypeptide product, it is exempt from these guidelines.

### 5.6.2. GENERAL APPLICABILITY

The Guidelines are applicable to all recombinant DNA research within the United States or its territories which is conducted at or sponsored by an institution that receives any support for recombinant DNA research from the National Institutes of Health (NIH). This includes research performed by NIH directly.

An individual receiving support for research involving recombinant DNA must be associated with or sponsored by an institution that can and does assume the responsibilities assigned in these Guidelines. The Guidelines are also applicable to projects done abroad if they are supported by NIH funds. If the host country, however, has established rules for the conduct of recombinant DNA projects, then a certificate of compliance with these rules may be submitted to NIH in lieu of compliance with the NIH Guidelines. The NIH reserves the right to withhold funding if the safety practices to be employed abroad are not reasonably consistent with the NIH Guidelines.

### 5.6.3. CONTAINMENT

Effective biological safety programs have been operative in a variety of laboratories for many years. Considerable information, therefore, exists for the design of physical containment facilities and the selection of laboratory procedures applicable to organisms carrying recombinant DNAs. The existing programs rely upon mechanisms that, for convenience, can be divided into two categories: (i) a set of standard practices that are generally used in microbiological laboratories [see Section 5.5 and subsections thereof] and (ii) special procedures, equipment and laboratory installations that provide physical barriers which are applied in varying degrees according to the estimated biohazard [again, see Section 5.5 and subsections thereof].

### 5.6.4. GUIDELINES IN COVERED EXPERIMENTS

This part addresses experiments involving recombinant DNA. The experiments are divided into four classes:

* III-A. Experiments which require specific RAC* review and NIH and IBC** approval before initiation of the experiment
* III-B. Experiments which require IBC approval before initiation of the experiment
* III-C. Experiments which require IBC notification at the time of initiation of the experiment
* III-D. Experiments which are exempt from the procedures of the Guidelines

* The Recombinant DNA Advisory Committee (RAC) advises the Secretary and Assistant Secretary of Health and the Director of NIH concerning recombinant DNA research.
** The Institutional Biosafety Committee reviews, approves, and oversees projects involving recombinant DNA at the institutional level.

IF AN EXPERIMENT FALLS INTO BOTH CLASSES III-A AND ONE OF THE OTHER CLASSES, THE RULES PERTAINING TO CLASS III-A MUST BE FOLLOWED. If an experiment falls into Class III-D and into either of Class III-B or III-C as well, it can be considered exempt from the requirements of the Guidelines.

Changes in containment levels from those specified in the Guidelines cannot be made without the express approval of the Director, NIH.

### 5.6.4.1. III.A. Experiments That Require RAC Review, NIH Review, and NIH and IBC Approval Before Initiation

Experiments in this category cannot be initiated without submission of relevant information on the proposed experiment to NIH, the publication of the proposal in the Federal Register for thirty days of comment, review by the RAC, and specific approval by NIH. The containment conditions for such experiments will be recommended by RAC and set by NIH at the time of approval. Such experiments also require the approval of the IBC before initiation. Specific experiments already approved in this section and the appropriate containment conditions are listed in Appendices D and F [All appendices referred to are those in the Guidelines]. If an experiment is similar to those in Appendices D and F, the NIH Office of Recombinant DNA Activities (ORDA) may determine appropriate containment conditions according to case precedents.

**III-A-1.** Deliberate formation of recombinant DNAs containing genes for the biosynthesis of toxic molecules lethal for vertebrates at an $LD_{50}$ of less than 100 ng per kilogram body weight (e.g., microbial toxins such as the botulinum toxins, tetanus toxin, diphtheria toxin, *Shigella dysenteriae* neurotoxin). Specific approval has been given for the cloning in *E. coli* K-12 of DNAs containing genes coding for the biosynthesis of toxic molecules which are lethal to vertebrates at 100 ng to 100 μg/kg body weight. Containment levels for these experiments are given in Appendix F.

**III-A-2.** Deliberate release into the environment of any organism except certain plants, as described in Appendix L.

**III-A-3.** Deliberate transfer of a drug resistance trait to microorganisms that are not known to acquire it naturally, if such acquisition could compromise the use of the drug to control disease agents in human or veterinary medicine or agriculture.

**III-A-4.** Deliberate transfer of recombinant DNA or DNA or RNA derived from recombinant DNA into human subjects. The requirement for RAC review should not be considered to preempt any other required review of experiments with human subjects. Institutional Review Board (IRB) review of the proposal should be completed before submission to NIH.

### 5.6.4.2. III.B. Experiments That Require IBC Approval Before Initiation

Investigators performing experiments in this category must submit to their IBC, prior to initiation of the experiments, a registration document that contains a description of: (i) The source(s) of DNA; (ii) the nature of the inserted DNA sequences; (iii) the hosts and vectors to be used; (iv) whether a deliberate attempt will be made to obtain an expression of the foreign gene, and if so, what protein will be produced; and (v) the containment conditions specified in these Guidelines. This registration document must be dated and signed by the investigator and filed only with the local

IBC. The IBC shall review all such proposals prior to initiation of the experiments. Requests for lowering of containment for experiments in this category will be considered by NIH.

### 5.6.4.2.1. III.B.1. Experiments Using Human or Animal Pathogens (Class 2, Class 3, Class 4, or Class 5 Agents) as Host-Vector Systems

**III-B-1-a.** Experiments involving the introduction of recombinant DNA into Class 2 agents can be carried out in a BL2 containment.

**III-B-1-b.** Experiments involving the introduction of recombinant DNA into Class 3 agents can be carried out in a BL3 containment.

**III-B-1-c.** Experiments involving the introduction of recombinant DNA into Class 4 agents can be carried out in a BL4 containment.

**III-B-1-d.** Experiments involving the introduction of recombinant DNA into Class 5 agents will be set on a case-by-case basis following ORDA review. A U.S. Department of Agriculture (USDA) permit is required for work with Class 5 agents.

### 5.6.4.2.2. Experiments in Which DNA From Human or Animal Pathogens (Class 2, Class 3, Class 4, or Class 5 Agents) Is Cloned in Nonpathogenic Prokaryotic or Lower Eukaryotic Host-Vector Systems

**III-B-2-a.** Recombinant DNA experiments in which DNA from Class 2 or Class 3 agents is transferred into nonpathogenic prokaryotes or lower eukaryotes may be performed under BL2 containment. Recombinant DNA experiments in which DNA from Class 4 agents is transferred into nonpathogenic prokaryotes or lower eukaryotes can be performed at BL2 containment after demonstration that only a totally and irreversibly defective fraction of the agent's genome is present in a given recombinant. In the absence of such a demonstration, BL4 must be used. Specific lowering of containment BL1 for particular experiments can be approved by the IBC. Many experiments in this category will be exempt from the Guidelines.

Experiments involving the formation of recombinant DNAs for certain genes coding for molecules toxic for vertebrates require RAC review and NIH approval or must be carried out under conditions as described in Appendix F.

**III-B-2-b.** Containment conditions for experiments in which DNA from class 5 agents is transferred into nonpathogenic prokaryotes or lower eukaryotes will be determined by ORDA following a case-by-case review. A USDA permit is required for work with Class 5 agents.

### 5.6.4.2.3. III.B.3. Experiments Involving the Use of Infectious Animal or Plant DNA or RNA Viruses or Defective Animal or Plant DNA or Viruses in the Presence of Helper Virus in Tissue Culture Systems

**Caution:** Special care should be used in the evaluation of containment levels for experiments which are likely to either enhance the pathogenicity (e.g., the insertion of a host oncogene) or to extend the host range (e.g., introduction of novel control elements) of viral vectors under conditions which permit a productive infection. In such cases, serious consideration should be given to raising the physical containment by at least one level.

**Note** — Recombinant DNA molecules or RNA molecules derived therefrom, which contain less than two-thirds of the genome of any eukaryotic virus (all virus from a single family being considered identical) may be considered defective and can be used in the absence of a helper virus under the conditions specified in Section III-C.

**III-B-3-a.** Experiments involving the use of infectious Class 2 animal viruses or defective Class 2 animal viruses in the presence of a helper virus can be performed at BL2 containment.

**III-B-3-b.** Experiments involving the use of infectious Class 3 animal viruses or defective Class 3 animal viruses in the presence of a helper virus can be carried out at BL3 containment.

**III-B-3-c.** Experiments involving the use of infectious Class 4 animal viruses or defective Class 4 animal viruses in the presence of a helper virus may be carried out under BL4 containment.

**III-B-3-d.** Experiments involving the use of infectious Class 5 animal viruses or defective Class 5 animal viruses in the presence of a helper virus will be determined on a case-by-case basis following ORDA review. A USDA permit is required for work with Class 5 pathogens.

**III-B-3-e.** Experiments involving the use of infectious animal or plant viruses in the presence of helper viruses not covered by [the four preceding sections] may be carried out under BL1 containment.

### 5.6.4.2.4. III.B.4. Recombinant DNA Experiments Involving Whole Animals or Plants

**III-B-4-a.** Recombinant DNA, or RNA molecules derived therefrom, from any source except for greater than two-thirds of an eukaryotic viral genome may be transferred to any non-human vertebrate organism and propagated under conditions of physical containment comparable to BL1 and appropriate to the organism under study. It is important that the investigator demonstrate that the fraction of the viral genome being utilized does not lead to productive infection. A USDA permit is required for work with Class 5 agents.

**III-B-4-b.** For all experiments not covered by Section II-B-4-a, the appropriate containment will be determined by the IBC.

### 5.6.4.2.5. Experiments Involving More Than 10 Liters of Culture

The appropriate containment will be decided by the IBC. Where appropriate, Appendix K of the Guidelines should be used by the IBC.

### 5.6.4.3. III-C. Experiments That Require IBC Notice Simultaneously With Initiation of Experiments

Experiments not included in III-A to C and subsections of those sections are included in this section. "All such experiments can be carried out at BL1 containment. ... A registration document [as previously described] must be dated and signed by the investigator and filed with the local IBC at the time of initiation of the experiment. The IBC shall review all such proposals, but IBC review prior to initiation of the experiment is not required.

**CAUTION: Experiments Involving Formation of Recombinant DNA Molecules containing no more than Two-Thirds of the Genome of any Eukaryotic Virus.** Recombinant DNA molecules containing no more than two-thirds of the genome of any eukaryotic virus (all viruses from a single family being considered identical) may be propagated and maintained in cells in tissue culture using BL1 containment. For such experiments, it must be shown that the cells lack helper virus for the specific Families of defective viruses being used. If helper virus is present, procedures specified under Section III-B-3 should be used. The DNA may contain fragments of the genome of viruses from more than one Family, but each fragment must be less than two-thirds of a genome."

### 5.6.4.4. III.D. Exempt Experiments

The following recombinant DNA molecules are exempt from these Guidelines and no registration with the IBC is necessary:

**III-D-1.** Those that are not in organisms or viruses.

**III-D-2.** Those that consist entirely of DNA segments from a single nonchromosomal or viral DNA source, though one or more of the segments may be a synthetic equivalent.

**III-D-3.** Those that consist entirely of DNA from a prokaryotic host, including its indigenous plasmids or viruses, when propagated only in that host (or a closely related strain of the same species) or when transferred to another host by well-established physiological means: also, those that consist entirely of DNA from an eukaryotic host, including its chloroplasts, mitochondria or plasmids (but excluding viruses) when propagated only in that host (or a closely related strain of the same species).

**III-D-4.** Certain specified recombinant DNA molecules that consist entirely of DNA segments from different species that exchange DNA by known physiological processes, though one or more of the segments may be a synthetic equivalent. A list of such exchangers will be prepared and periodically revised by the Director, NIH, with advice of the RAC, after appropriate notice and opportunity for public comment. Certain classes are exempt as of publication of these revised Guidelines [May 7, 1986]. This list is in Appendix A. An updated list may be obtained from the Office of Recombinant DNA Activities, National Institutes of Health, Building 31, Room 3B10, Bethesda, Maryland 20892.

**III-D-5.** Other classes of recombinant DNA molecules — if the Director, NIH with advice of the RAC, after appropriate notice and opportunity for public comment, finds that they do not present a significant risk to health or the environment. Certain classes are exempt as of publication of these revised Guidelines [May 7, 1986]. This list is in Appendix C. An updated list may be obtained from the Office of Recombinant DNA Activities, National Institutes of Health, Building 31, Room 3B10, Bethesda, Maryland 20892.

### 5.6.5. ROLES AND RESPONSIBILITIES
### 5.6.5.1. IV.A. Policy

Safety in activities involving recombinant DNA depends on the individual conducting them. The Guidelines cannot anticipate every possible situation. Motivation and good judgment are the key essentials to protection of health and the environment.

The Guidelines are intended to help the Institution, Institutional Biosafety Committee (IBC), Biological Safety Officer (BSO) and Principal Investigator (PI) determine the safeguards that should be implemented. These Guidelines will never be complete or final, since all conceivable experiments involving recombinant DNA cannot be foreseen. Therefore, *it is the responsibility of the Institution and those associated with it to adhere to the intent of the Guidelines as well as to their specifics.*

Each Institution (and the IBC acting on its behalf) is responsible for ensuring that recombinant DNA activities comply with the Guidelines. General recognition of institutional authority and responsibility properly establishes accountability for safe conduct of the research at the local level.

### 5.6.5.2. IV.B. Responsibility of the Institution
#### 5.6.5.2.1. IV.B.1. General Information

Each Institution conducting or sponsoring recombinant DNA research covered by these Guidelines is responsible for ensuring that the research is carried out in full conformity with the provisions of the Guidelines. In order to fulfill this responsibility, the Institution shall:

**IV-B-1-a.** Establish and implement policies that provide for the safe conduct of recombinant DNA research and that ensure compliance with the Guidelines. The Institution, as part of its general responsibilities for implementing the Guidelines, may establish additional procedures as deemed necessary to govern the Institution and its components in the discharge of its responsibilities under the Guidelines.

**IV-B-1-b.** Establish an IBC that meets the requirements set forth in Section IV-B-2 and carries out the functions detailed in Section IV-B-3.

**IV-B-1-c.** If the Institution is engaged in recombinant DNA research at the BL3 or BL4 containment level, appoint a BSO, who shall be a member of the IBC and carry out the duties specified in Section IV-B-4.

**IV-B-1-d.** Require that investigators responsible for research covered by these Guidelines comply with the provisions of Section IV-B-5. and assist investigators to do so.

**IV-B-1-e.** Ensure that appropriate training for the IBC chairperson and members, the BSO, PIs and laboratory staff regarding the Guidelines, their implementation, and laboratory safety. Responsibility for training IBC members may be carried out through the IBC chairperson. Responsibility for training laboratory staff may be carried out through the PI. The Institution is responsible for seeing that the PI has sufficient training but may delegate this responsibility to the IBC.

**IV-B-1-f.** Determine the necessity in connection with each project for health surveillance of recombinant DNA research personnel, and conduct, if found appropriate, a health surveillance program for the project. [If it appears that a medical surveillance program might be appropriate, a copy of the Laboratory Safety Monograph should be obtained from ORDA for guidance.]

**IV-B-1-g.** Report within 30 days to ORDA any significant problems with and violations of the Guidelines and significant research-related accidents and illnesses, unless the Institution determines that the PI or IBC has done so.

#### 5.6.5.2.2. IV.B.2. Membership and Procedures of the IBC

The Institution shall establish an IBC whose responsibilities need not be restricted to recombinant DNA. The committee shall meet the following requirements:

**IV-B-2-a.** The IBC shall comprise no fewer than 5 members so selected that they collectively have experience and expertise in recombinant DNA technology and the capability to assess the safety of recombinant DNA research experiments and any potential risk to public health or the environment. At least two members shall not be affiliated with the Institution (apart from their membership on the IBC) and shall represent the interest of the surrounding community with respect to health and protection of the environment. Members meet this requirement if, for example, they are officials of state or local public health or environmental protection agencies, members of other local governmental bodies, or persons active in medical, occupational health, or environmental concerns in the community. The Biological Safety Officer (BSO), mandatory when research is being conducted at the BL3 and BL4 levels, shall be a member.

**IV-B-2-b.** In order to ensure the competence necessary to review recombinant DNA activities, it is recommended that (i) the IBC include persons with expertise in recombinant DNA technology, biological safety, and physical containment; (ii) the IBC include or have available as consultants, persons knowledgeable in Institutional commitments and policies, applicable law, standards of professional conduct and practice, community attitudes, and at least one member be from the laboratory technical staff.

**IV-B-2-c.** The Institution shall identify the committee by name in a report to ORDA and shall include relevant background information on each member in such form and at such times as ORDA may require.

**IV-B-2-d.** No member of an IBC may be involved (except to provide information requested by the IBC) in the review or approval of a project in which he or she has been or expects to be engaged or has a direct financial interest.

**IV-B-2-e.** The Institution, who is ultimately for the effectiveness of the IBC, may establish procedures that the IBC will follow in its initial and continuing review of applications, proposals, and activities.

**IV-B-2-f.** Institutions are encouraged to open IBC meetings to the public whenever possible, consistent with protection of privacy and proprietary interests.

**IV-B-2-g.** Upon request, the Institution shall make available to the public all members of IBC meetings and any documents submitted to or received from funding agencies which the latter are required to make available to the public. If comments are made by members of the public on IBC actions, the Institution shall forward to NIH both the comments and the IBC's response.

### 5.6.5.2.3. IV.B.3. Functions of the IBC

On behalf of the Institution, the IBC is responsible for:

**IV-B-3-a.** Reviewing for compliance with the NIH Guidelines recombinant DNA research as specified in Part III conducted at or sponsored by the Institution, and approving those research projects that it finds in conformity with the Guidelines. This review shall include:

**IV-B-3-a-(1).** An independent assessment of the containment levels required by these Guidelines for the proposed research, and

**IV-B-3-a-(2).** An assessment of the facilities, procedures, and practices, and of the training and expertise of recombinant DNA personnel.

**IV-B-3-b.** Notifying the PI of the results of their review.

**IV-B-3-c.** Lowering containment levels for certain experiments as specified in Sections III-B-2.

**IV-B-3-d.** Setting containment levels as specified in Section III-B-4-b and III-B-5.

**IV-B-3-e.** Reviewing periodically recombinant DNA research being conducted at the Institution to ensure that the requirements of the Guidelines are being fulfilled.

**IV-B-3-f.** Adopting emergency plans covering accidental spills and personnel contamination resulting from such research. [Information on this topic is also available in the Laboratory Safety Monograph.]

**IV-B-3-g.** Reporting within 30 days to the appropriate Institutional official and to ORDA any significant problems with or violations of the Guidelines and any significant research-related accidents or illnesses unless the IBC determines that the PI has done so.

**IV-B-3-h.** The IBC may not authorize initiation of any experiments not explicitly covered by the Guidelines until NIH (with the advice of the RAC when required) establishes the containment requirement.

**IV-B-3-i.** Performing such other functions as may be delegated to the IBC under Section IV-B-1.

### 5.6.5.2.4. IV.B.4. Biological Safety Officer

The Institution shall appoint a BSO if it engages in recombinant DNA research at the BL3 or BL4 containment level. The officer shall be a member of the IBC and his duties shall include (but need not be limited to):

**IV-B-4-a.** Ensuring through periodic inspections that laboratory standards are rigorously followed;

**IV-B-4-b.** Reporting to the IBC and the Institution all significant problems with and violations of the Guidelines and all significant research-related accidents and illnesses of which the BSO becomes aware unless the BSO determines that the PI has done so;

**IV-B-4-c.** Developing emergency plans for dealing with accidental spills and personnel contamination and investigating recombinant DNA research laboratory accidents;

**IV-B-4-d.** Providing advice on laboratory security;

**IV-B-4-e.** Providing technical advice to the PI and the IBC on research safety procedures.

**Note** — See the LSM for additional information on the duties of the BSO.

### 5.6.5.2.5. IV.B.5. Principal Investigator (PI)

On behalf of the Institution, the PI is responsible for complying fully with the Guidelines in conducting any recombinant DNA research.

**IV-B-5.PI — General.** As part of this general responsibility, the PI shall:

**IV-B-5-a-(1).** Initiate or modify no recombinant DNA research requiring approval by the IBC prior to initiation (see Sections III-A and III-B) until that research or the proposed modification thereof has been approved by the IBC and has met all other requirements of the Guidelines;

**IV-B-5-a-(2).** Determine whether experiments are covered by Section III-C and follow the appropriate procedures;

**IV-B-5-a-(3).** Report within 30 days to the IBC and NIH (ORDA) all significant problems with and violations of the Guidelines and all research-related accidents and illnesses;

**IV-B-5-a-(4).** Report to the IBC and NIH (ORDA) new information bearing on the Guidelines;

**IV-B-5-a-(5).** Be adequately trained in good microbiological techniques;

**IV-B-5-a-(6).** Adhere to IBC-approved emergency plans for dealing with accidental spills and personnel contamination; and

**IV-B-5-a-(7).** Comply with shipping requirements for recombinant DNA molecules. (See appendix H for shipping requirements and the LSM for technical recommendations.)

**IV-B-5-b. Submissions by the PI to NIH.** The PI shall:

**IV-B-5-b-(1).** Submit information to NIH (ORDA) in order to have new host-vector systems certified;

**IV-B-5-b-(2).** Petition NIH with notice to the IBC for exemptions to these Guidelines;

**IV-B-5-b-(3).** Petition NIH with concurrence of the IBC for approval to conduct experiments specified in Section III-A of the Guidelines;

**IV-B-5-b-(4).** Petition NIH for determination of containment for experiments requiring case-by-case review;

**IV-B-5-b-(5).** Petition NIH for determination of containment for experiments not covered by the Guidelines.

**IV-B-5-c. Submissions by the PI to the IBC.** The PI shall:

**IV-B-5-c-(1).** Make the initial determination of the required levels of physical and biological containment in accordance with the Guidelines;

**IV-B-5-c-(2).** Select appropriate microbiological practices and laboratory techniques to be used in the research;

**IV-B-5-c-(3).** Submit the initial research protocol if covered under Guideline Sections III-A to C (and also subsequent changes — e.g., changes in the source of DNA or host-vector system) to the IBC for review and approval or disapproval; and

**IV-B-5-c-(4).** Remain in communication with the IBC throughout the conduct of the project.

**IV-B-5-d. PI Responsibilities Prior to Initiating Research.** The PI is responsible for:

**IV-B-5-d-(1).** Making available to the laboratory staff copies of the protocols that describe the potential biohazards and the precautions to be taken;

**IV-B-5-d-(2).** Instructing and training staff in the practices and techniques required to ensure safety and in the procedures for dealing with accidents; and

**IV-B-5-d-(3).** Informing the staff of the reasons and provisions for any precautionary medical practices advised or requested, such as vaccinations or serum collection.

**IV-B-5-e. PI Responsibilities During the Conduct of the Research.** The PI is responsible for:

**IV-B-5-e-(1).** Supervising the safety performance of the staff to ensure that the required safety practices and techniques are employed;

**IV-B-5-e-(2).** Investigating and reporting in writing to ORDA, the BSO (where applicable), and the IBC any significant problems pertaining to the operation and implementation of containment practices and procedures;

**IV-B-5-e-(3).** Correcting work errors and conditions that may result in the release of recombinant DNA materials;

**IV-B-5-e-(4).** Ensuring the integrity of the physical containment (e.g., biological safety cabinets) and the biological containment (e.g., purity and genotypic and phenotypic characteristics).

The sections of the Guidelines dealing with the responsibilities of the NIH, RAC, and ORDA are omitted here. The areas that directly affect the research facility are generally incorporated in the instructions to the institutions, IBC, PI, and BSO.

### 5.6.5.3. IV.D. Compliance

As a condition for NIH funding of recombinant DNA research, Institutions must ensure that such research conducted at or sponsored by the Institution, irrespective of the source of funding, shall comply with these Guidelines. The policies on noncompliance are as follows.

**IV-D-1.** All NIH-funded projects involving recombinant DNA techniques must comply with the NIH Guidelines. Noncompliance may result in (i) suspension, limitation, or termination of financial assistance for such projects and of NIH funds for other recombinant DNA research at the Institution, or (ii) a requirement for prior NIH approval of any or all recombinant DNA projects at the Institution.

**IV-D-2.** All non-NIH funded projects involving recombinant DNA techniques conducted at or sponsored by an Institution that receives NIH funds for projects involving such techniques must comply with the NIH Guidelines.

Noncompliance may result in (i) a requirement for prior approval of any or all recombinant DNA projects at the Institution.

**IV-D-3.** Information concerning noncompliance with the Guidelines may be brought forward by any person. It should be delivered to both NIH (ORDA) and the relevant Institution. The Institution, generally through the IBC, shall take appropriate action. The Institution shall forward a complete report of the incident to ORDA, recommending any further action.

**IV-D-4.** In cases where NIH proposes to suspend, limit, or terminate financial assistance because of noncompliance with the Guidelines, applicable DHHS and Public Health Service procedures shall govern.

**IV-D-5. Voluntary Compliance.** Any individual, corporation, or Institution that is not otherwise covered by the Guidelines is encouraged to conduct recombinant DNA research activities in accordance with the Guidelines through the procedures set forth in Part VI of the Guidelines.

The most significant components of Part VI of the Guidelines are the sections dealing with protection of proprietary data, in addressing the concerns of Institutions or individuals as to whether their interests will be protected. These sections are given immediately below. Section V, which consists of references and some notes to Sections I to IV, will be combined with the references to the other sections and moved to the end of the DNA material.

### 5.6.5.4. Selected Portions of Section VI. Voluntary Compliance
### VI-C. Certification of Host-Vector Systems

A host-vector system may be proposed for certification by the Director, NIH, in accordance with the procedures set forth in Appendix I-II-A.

In order to ensure protection for proprietary data, any public notice regarding a host-vector system which is designated by the Institution as proprietary under Section-VI-E-1 will be issued only after consultation with the Institution as to the content of the notice.

### VI-D. Requests for Exemptions and Approvals

Requests for exemptions or other approvals required by the Guidelines should be requested by following the procedures set forth in the appropriate sections in Parts I-IV of the Guidelines.

In order to ensure protection for proprietary data, any public notice regarding a request for an exemption or other approval which is designated by the Institution as proprietary under Section VI-E-1 will be issued only after consultation with the Institution as to the content of the notice.

### VI-E. Protection of Proprietary Data

In general, the Freedom of Information Act requires Federal agencies to make their records available to the public upon request. However, this requirement does not apply to, among other things, "trade secrets and commercial and financial information obtained from a person and privileged or confidential." 18 U.S.C. 1905, in turn makes it a crime for an officer or employee of the United States or any Federal department or agency to publish, divulge, disclose, or make known "in any manner or to any extent not authorized by law any information coming to him in the course of his employment or official duties or by reason of any examination or investigation made by, or return, report or record made to be filed with, such department or agency or officer or employee thereof, which information concerns or relates to the trade secrets, (or) processes... of any person, firm, partnership, corporation or association." This provision applies to all employees of the Federal Government, including special Government employees. Members of the Recombinant DNA Advisory Committee are "special Government employees."

**VI-E-1.** In submitting to NIH for purposes of complying voluntarily with the Guidelines, an Institution may designate those items of information which the Institution believes constitute trade secrets, privileged, confidential commercial, or financial information.

**VI-E-2.** If NIH receives a request under the Freedom of Information Act for information so designated, NIH will promptly contact the Institution to secure its views as to whether the information (or some portion) should be released.

**VI-E-3.** If the NIH decides to release this information (or some portion) in response to a Freedom of Information request or otherwise, the Institution will be advised; and the actual release will not be made until the expiration of 15 days after the Institution is so advised except to the extent that earlier release in the judgment of the Director, NIH is necessary to protect against an imminent hazard to the public or the environment.

### VI-E-4. Presubmission Review

**VI-E-4-a.** Any Institution not otherwise covered by the Guidelines, which is considering submission of data or information voluntarily to NIH, may request presubmission review of the records involved to determine whether if the records are submitted NIH will or will not make part or all of the records available upon request under the Freedom of Information Act.

**VI-E-4-b.** A request for presubmission review should be submitted to ORDA along with the records involved. These records must be clearly marked as being the property of the Institution on loan to NIH solely for the purpose of making a determination under the Freedom of Information Act. The ORDA will then seek a determination from the HHS Freedom of Information Officer, the responsible official under HHS regulations (45 CFR Part 5) as to whether the records involved (or some portion) are or are not available to members of the Public under the Freedom of Information Act. Pending such a determination, the records will be kept separate from ORDA files, will be considered records of the Institution and not ORDA, and will not be received as part of ORDA files. No copies will be made of the records.

**VI-E-4-c.** The ORDA will inform the Institution of the HHS Freedom of Information Officer's determination and follow the Institution's instruction as to whether some or all of the records involved are to be returned to the Institution or to become a part of ORDA files. If the Institution instructs ORDA to return the records, no copies or summaries of the records will be made or retained by the HHS, NIH, or ORDA.

**VI-E-4-d.** The HHS Freedom of Information Officer's determination will represent that official's judgement at the time of the determination as to whether the records involved (or some portion) would be exempt under the Freedom of Information Act if at the time of the determination the records were in ORDA files when a request was received for them under the Act.

### 5.6.6. APPENDICES
### 5.6.6.1. Appendix A. Exemptions Under Section III.D.4

Under Section II-D-4 of these Guidelines are recombinant DNA molecules that are (1) composed entirely of DNA segments from one or more of the organisms within a sublist and (2) to be propagated in any of the organisms within a sublist. (Classification of *Bergey's Manual of Determinative Bacteriology,* 8th edition, R.E. Buchanan and N.E. Gibbons, editors, Williams and Wilkins Co., Baltimore, 1974.)

Although these experiments are exempt, it is recommended that they be performed at the appropriate biosafety level for the host or recombinant organism.

**Sublist A**
1. Genus *Escherichia*
2. Genus *Shigella*
3. Genus *Salmonella* (including *Arizona*)
4. Genus *Enterobacter*
5. Genus *Citrobacter* (including *Levinea*)
6. Genus *Klebsiella*
7. Genus *Erwinia*
8. *Pseudomonas aeruginosa, P. putida* and *P. fluorescens*
9. *Serratia marcescens*
10. *Yersinia enterocolitica*

**Sublist B**
1. *Bacillus subtilis*
2. *Bacillus licheniformis*

3.  *Bacillus pumilus*
4.  *Bacillus globigii*
5.  *Bacillus niger*
6.  *Bacillus nato*
7.  *Bacillus amyloliquefaciens*
8.  *Bacillus aterrimus*

**Sublist C**
1.  *Streptomyces aureofaciens*
2.  *Streptomyces rimosus*
3.  *Streptomyces coelicolor*

**Sublist D**
1.  *Streptomyces griseus*
2.  *Streptomyces cyaneus*
3.  *Streptomyces venezuelae*

**Sublist E**
1.  One way transfer of *Streptococcus mutans* or *Streptococcus lactis* DNA into *Streptococcus sanguis*

**Sublist F**
1.  *Streptococcus sanguis*
2.  *Streptococcus pneumoniae*
3.  *Streptococcus faecalis*
4.  *Streptococcus pyogenes*
5.  *Streptococcus mutans*

**5.6.6.2. Appendix B. Classification of Microorganisms on the Basis of Hazard**
**Appendix B-I. Classification of Etiologic Agents.**
The original reference for this classification was the publication, "*Classification of Etiological Agents on the Basis of Hazard,* 4th edition, July 1974, U.S. Department of Health, Education and Welfare, Public Health Service, Centers for Disease Control, Office of Biosafety, Atlanta, GA 30333." For the purposes of these Guidelines, this list has been revised by the NIH.

**Appendix B-I-A. Class 1 Agents.** All bacterial, parasitic, fungal, viral, rickettsial, and chlamydial agents not included in higher classes.

**Appendix B-I-B. Class 2 Agents**
**Appendix B-I-B-1. Bacterial Agents.**
*Acinetobacter calcoaceticus*
*Actinobacillus* — all species
*Aeromonas hydrophila*
*Arizona hinshawii* — all serotypes
*Bacillus anthracis*
*Bordetella* — all species
*Borrelia recurrentis, B. vincenti*

*Campylobacter fetus*
*Campylobacter jejuni*
*Chlamydia psittaci*
*Chlamydia trachomatis*
*Clostridium botulinum, Cl. chauvoei, Cl. haemolyticum, Cl. histolyticum, Cl. novyi, Cl. septicum, Cl. tetani*
*Corynebacterium diphtheriae, C. equi, C. haemolyticum, C. pseudotuberculosis, C. pyogenes, C. renale*
*Edwardsiella tarda*
*Erysipelothrix insidiosa*
*Escherichia coli* — all enteropathogenic, enterotoxigenic, enteroinvasive, and strains bearing K1 antigen
*Haemophilus ducreyi, H. influenzae*
*Klebsiella* — all species and all serotypes
*Legionella pneumophila*
*Leptospira interrogens* — all serotypes
*Listeria* — all species
*Moraxella* — all species
*Mycobacteria* —all species except those listed in Class 3
*Mycoplasma* — all species except *Mycoplasma mycoides* and *Mycoplasma* agalactiae which are in Class 5
*Neisseria gonorrhoeae, N. meningitidis*
*Pasteurella* — all species except those listed in Class 3
*Salmonella* — all species and all serotypes
*Shigella* — all species and all serotypes
*Sphaerophorus necrophorus*
*Staphylococcus aureus*
*Streptobacillus moniliformis*
*Streptococcus pneumoniae*
*Streptococcus pyogenes*
*Treponema carateum, T. pallidum, and T. pertenue*
*Vibrio cholerae*
*Vibrio parahemolyticus*
*Yersinia enterocolitica*

## Appendix B-I-B-2. Fungal Agents.
*Actinomycetes* (including *Nocardia* species, *Actinomyces* species, and *Arachnia propionica*)
*Blastomyces dermatitidis*
*Cryptococcus neoformans*
*Paracoccidioides braziliensis*

## Appendix B-I-B-3. Parasitic Agents.
*Endamoeba histolytica*
*Leishmania* sp.
*Naegleria gruberi*
*Schistosoma mansoni*

*Toxoplasma gondii*
*Toxocara canis*
*Trichinella spiralis*
*Trypanosoma cruzi*

**Appendix B-I-B-4. Viral, Rickettsial and Chlamydial Agents.**
Adenoviruses — human — all types
Cache Valley virus
Coxsackie A and B viruses
Cytomegaloviruses
Echoviruses — all types
Encephalomyocarditis virus (EMC)
Flanders virus
Hart Park virus
Hepatitus — associated antigen material
Herpes virus — except *Herpesvirus simiae* (Monkey B virus) which is in Class 4
Corona viruses
Influenza viruses- all types except A/PR8/34 which is in Class I
Langat virus
*Lymphogranuloma venereum* agent
Measles virus
Mumps virus
Parainfluenza virus — all types except Parainfluenza virus 3, SF4 strain, which is in
  Class 1
Polioviruses — all types, wild and attenuated
Poxviruses — all types except Alastrim, Smallpox, and Whitepox which are Class 5
  and Monkey pox which, depending on experiments, is in Class 3 or Class 4.
Rabies virus — all strains except Rabies street virus, which should be classified in
  Class 3
Reoviruses — all types
Respiratory syncytical virus
Rhinoviruses — all types
Rubella virus
Simian viruses — all types except *Herpesvirus simiae* (Monkey B virus) and Mar-
  burg virus which are in Class 4
Sindbis virus
Tensaw virus
Turlock virus
Vaccinia virus
Varicella virus
Vesicular stomata virus
Vole rickettsia
Yellow fever virus, 17D vaccine strain

**Appendix B-I-C. Class 3 Agents.**
**Appendix B-1-C-1. Bacterial Agents.**
*Bartonella* — all species

*Brucella* — all species
*Francisella tularensis*
*Mycobacterium avium, M. bovis, M. tuberculosis*
*Pasteurella multocide* type B ("buffalo" and other foreign virulent strains)
*Pseudomonas pseudomallei*
*Yersinia pestis*

### Appendix B-I-C-2. Fungal Agents.
*Coccidioides immitis*
*Histoplasma capsulatum*
*Histoplasma capsulatum* var. *duboisii*

### Appendix B-I-C-3. Parasitic Agents.
None

### Appendix B-I-C-4. Viral, Rickettsial, and Chlamydial Agents.
Monkey pox virus, when used *in vitro*
Arboviruses — all strains except those used in Class 2 and 4 (Arboviruses indigenous
   to the United States are in Class 3 except those listed in Class 2. West Nile and
   Semliki Forest viruses may be classified up or down depending on the condition of
   use and geographical location of the laboratory.)
Dengue virus, when used for transmission or animal inoculation experiments
Lymphocytic choriomeningitis (LCM)
*Rickettsia* — all species except *Vole rickettsia* when used for transmission or animal
   inoculation experiments
Yellow fever virus — wild, when used *in vitro*

### Appendix B-I-D. Class 4 Agents.

### Appendix B-I-D-1. Bacterial Agents.
None

### Appendix B-I-D-2. Fungal Agents.
None

### Appendix B-I-D-3. Parasitic Agents.
None

### Appendix B-I-D-4. Viral, Rickettsial, and Chlamydial Agents.
Ebola fever virus
Monkey pox, when used for transmission or animal inoculation experiments
Hemorrhagic fever agents, including Crimean hemorrhagic fever, (Congo), Junin,
   and machupo viruses, and others as yet unidentified.
*Herpesvirus simiae* (*Monkey B virus*)
Lassa virus
Marburg virus
Tick-borne encephalitis virus complex, including Russian spring-summer encepha-

litis, Kyasanur forest disease, Omsk hemorrhagic fever, and Central European encephalitis viruses

Venezuelan equine encephalitis virus, epidemic strains, when used for transmission or animal inoculation experiments

Yellow fever virus — wild, when used for transmission or animal inoculation experiments

### Appendix B-II. Classification of Oncogenic Viruses on the Basis of Potential Hazard
### Appendix B-II-A. Low-Risk Oncogenic Viruses.

| | |
|---|---|
| Rous sarcoma | Rat leukemia |
| SV-40 | Hamster leukemia |
| CELO | Bovine leukemia |
| Ad7-SV40 | Dog sarcoma |
| Polyoma | Mason-Pfizer monkey virus |
| Bovine papilloma | Marek's |
| Rat mammary tumor | Guinea pig herpes |
| Avian leukosis | Lucke (Frog) |
| Murine leukemia | Adenovirus |
| Murine sarcoma | Shope fibroma |
| Mouse mammary tumor | Shope papilloma |

### Appendix B-II-B. Moderate-Risk Oncogenic Viruses.

| | | |
|---|---|---|
| Ad2-SV40 | EBV | HV ateles |
| FeLV | SSV-1 | Yaba |
| HV Saimiri | GaLV | FeSV |

### Appendix B-III. Class 5 Agents.

### Appendix B-III-A. Animal Disease Organisms Which Are Forbidden Entry into the United States by Law.
Foot and mouth disease virus

### Appendix B-III-B. Animal Disease Organisms Which Are Forbidden Entry into the United States by USDA Policy

| | |
|---|---|
| African Horse sickness virus | Newcastle disease virus (Asiatic strains) |
| African swine fever virus | *Mycoplasma mycoides* (contagious bovine |
| *Besnoitia besnoiti* | pleuropneumonia) |
| Borna disease virus | *Mycoplasma agalactiae* (contagious agalactia |
| Bovine infectious petechial fever | of sheep) |
| Camel pox virus | *Rickettsia ruminatium* (heart water) |
| Ephemeral fever virus | Rift Valley fever virus |
| Fowl plague virus | Rhinderpest virus |
| Goat pox virus | Sheep pox virus |
| Hog cholera virus | Swine vesicular disease virus |
| Louping ill virus | *Trypanosoma vivax* (Nagana) |
| Nairobi sheep disease virus | *Trypanosoma evansi* |

*Theileria parva* (East Coast fever)
*Theileria annulata*
*Theileria lawrencei*
*Theileria bovis*

*Theileria hirci*
Vesicular exanthema virus
Wesselsbron disease virus
Zyonema

### Appendix B-III-C. Organisms Which May Not Be Studied in the United States Except at Specified Facilities.

Small pox            Alastrim            White pox

### 5.6.6.3. Appendix C. Exemptions Under Section III-D-5

The following classes of experiments are exempt under Section III-D-5 of the Guidelines:

### Appendix C-I. Recombinant DNAs in Tissue Culture.

Recombinant DNA molecules containing less than one-half of any eukaryotic genome (all viruses from a single Family being considered identical) that are propagated and maintained in cells in tissue culture are exempt from these Guidelines with the exceptions listed below.

**Exceptions.** Experiments described in Section III-A which require specific RAC review and NIH approval before initiation of the experiment.

Experiments involving DNA from Class 3, 4, or 5 organisms or cells known to be infected with these agents.

Experiments involving the deliberate introduction of genes coding for the biosynthesis of molecules toxic for vertebrates (see Appendix F).

### Appendix C-II. Experiments Involving *E. coli* K-12 Host-Vector Systems.

Experiments which use *E. coli* K-12 host-vector systems, with the exception of those experiments listed below, are exempt from these Guidelines provided that (i) the *E. coli* host shall not contain conjugation proficient plasmids or generalized transducing phages; and (ii) lambda or lambdoid or Ff bacteriophages or nonconjugative plasmids shall be used as vectors. However, experiments involving the insertion into *E. coli* K-12 of DNA from prokaryotes that exchange genetic information with *E. coli* may be performed with any *E. coli* K-12 vector (e.g., conjugative plasmid). When a nonconjugative vector is used, the *E. coli* K-12 host may contain conjugative-proficient plasmids, either autonomous or integrated, or generalized transducing phages.

For these exempt laboratory experiments, BL1 physical containment conditions are recommended.

For large scale (LS) fermentation experiments, BL1-LS physical containment conditions are recommended. However, following review by the IBC of appropriate data for a particular host-vector system, some latitude in the application of BL1-LS requirements as outlined in Appendix K-II-A through K-II-F is permitted.

**Exceptions.** Experiments described in Section III-A which require specific RAC review and NIH approval before initiation of the experiment.

Experiments involving DNA from Class 3, 4, or 5 organisms or from cells known to be infected with these agents may be conducted under containment conditions specified in Section III-B-2 with prior IBC review and approval.

Large scale experiments (e.g., more than 10 liters of culture) require prior IBC review and approval (see Section III-B-5).

Experiments involving the deliberate cloning of genes coding for the biosynthesis of molecules toxic for vertebrates (see Appendix F).

### Appendix C-III. Experiments Involving *Saccharomyces* Host-Vector Systems.

Experiments which use *Saccharomyces cerevisiae* host-vector systems, with the exception of experiments listed below, are exempt from these Guidelines.

Experiments which use *Saccharomyces uvarum* host-vector systems, with the exception of experiments listed below, are exempt from these Guidelines.

For these exempt laboratory experiments, BL1 physical containment conditions are recommended.

For large scale (LS) fermentation experiments, BL1-LS physical containment conditions are recommended. However, following review by the IBC of appropriate data for a particular host-vector system, some latitude in the application of BL1-LS requirements as outlined in Appendix K-II-A through K-II-F is permitted.

**Exceptions.** Same as for Appendix C-II.

### Appendix C-IV. Experiments Involving *Bacillus subtilis* Host-Vector Systems.

*Any asporogenic Bacillus subtilis* strain which does not revert to a spore former with a frequency greater than $10^{-7}$ can be used for cloning DNA, with the exception of those experiments listed below.

For these exempt laboratory experiments, BL1 physical containment conditions are recommended.

For large scale (LS) fermentation experiments, BL1-LS physical containment conditions are recommended. However, following review by the IBC of appropriate data for a particular host-vector system, some latitude in the application of BL1-LS requirements as outlined in Appendix K-II-A through K-II-F is permitted.

**Exceptions.** Same as for Appendix C-II.

### Appendix C-V. Extrachromosomal Elements of Gram Positive Organisms.

Recombinant DNA molecules derived entirely from extrachromosomal elements of the organisms listed below (including shuttle vectors constructed from vectors described in Appendix C), propagated and maintained in organisms listed below are exempt from these Guidelines.

| | |
|---|---|
| *Bacillus subtilis* | *Bacillus natto* |
| *Bacillus pumilus* | *Bacillus niger* |
| *Bacillus licheniformis* | *Bacillus aterrimus* |
| *Bacillus thuringiensis* | *Bacillus amylosacchariticus* |
| *Bacillus cereus* | *Bacillus anthracis* |
| *Bacillus amyloliquefaciens* | *Bacillus globigii* |
| *Bacillus brevis* | *Bacillus megaterium* |
| | |
| *Staphylococcus aureus* | *Staphylococcus carnosus* |
| *Staphylococcus epidermidis* | *Clostridium acetobutylicum* |

| | | |
|---|---|---|
| *Pediococcus damnosus* | *Listeria grayi* | *Lactobacillus casei* |
| *Pediococcus pentosaceus* | *Listeria murrayi* | |
| *Pediococcus acidilactici* | *Listeria monocytogenes* | |

| | | |
|---|---|---|
| *Streptococcus pyogenes* | *Streptococcus avium* | *Streptococcus equisimilis* |
| *Streptococcus agalactiae* | *Streptococcus faecalis* | *Streptococcus thermophylus* |
| *Streptococcus sanguis* | *Streptococcus anginosus* | *Streptococcus milleri* |
| *Streptococcus salivarious* | *Streptococcus sobrinus* | *Streptococcus durans* |
| *Streptococcus cremoris* | *Streptococcus lactis* | *Streptococcus mitior* |
| *Streptococcus pneumoniae* | *Streptococcus mutans* | *Streptococcus ferus* |

**Exceptions.** Experiments described in Section III-A which require specific RAC review and NIH approval before initiation of the experiment.

Large scale experiments (e.g., more than 10 liters of culture) require prior IBC review and approval (see Section III-B-5).

Experiments involving the deliberate cloning of genes coding for the biosynthesis of molecules toxic for vertebrates (see Appendix F).

### 5.6.6.4. Appendix D. Actions Taken Under the Guidelines

These are usually specific to a given institution for specific procedures and are deleted for that reason.

### 5.6.6.5. Appendix E. Certified Host-Vector Systems

While many experiments using *E. coli* K-12, *Saccharomyces cerevisiae* and *Bacillus subtilis* are currently exempt from the Guidelines under section III-D-5, some derivatives of these host vector-systems were previously classified as HV1 or HV2. A listing of those systems follows:

#### Appendix E-I. *Bacillus subtilis.*

**HV1.** The following plasmids are accepted as the vector components of certified *B. subtilis* HV1 systems: pUB110, pC194, pS194, pSA2100, pE194, pT127, pUB112, pC221, pC223, and pAB124. *B. subtilis* strains RUB 331 and BGSC 1S53 have been certified as the host component of HV1 systems based on these plasmids.

**HV2.** The asporogenic mutant derivative of *Bacillus subtilis,* ASB 298 with the following plasmids as the vector component: pUB110, pC194, pS194, pSA2100, pE194, pT127, pUB112, pC221, pC223, and pAB124.

#### Appendix E-II. *Saccharomyces cerevisiae.*

**HV2.** The following sterile strains of *Saccharomyces cerevisiae,* all of which have the ste-VC9 mutation, SHY1, SHY2, SHY3, and SHY4. The following plasmids are certified for use: YIp1, YEp2, YEp4, YIp5, YEp6, YRp7, YEp20, YEp21, YEp24, YIp25, YIp26, YIp27, YIp28, YIp29, YIp30, YIp31, YIp32, and YIp33.

#### Appendix E-III. *Escherichia coli.*

**EK2 Plasmid Systems.** The *E. coli* K-12 strain chi-1776. The following plasmids are certified for use: pSC101, pMB9, pBR313, pBR322, pDH24, pBR325, pBR327, pGL101, and pHB1. The following *E. coli/S. cerevisiae* hybrid plasmids are certified as EK2 vectors when used in *E. coli* chi-1776 or in the sterile yeast strains, SHY1, SHY2, SHY3, and SHY4: YIp1, YEp2, YEp4, YIp5, YEp6, YRp7, YEp20, YEp21, YEp24, YIp25, YIp26, YIp27, YIp28, YIp29, YIp30, YIp31, YIp32, and YIp33.

**EK2 Bacteriophage Systems.** The following are certified EK2 systems based on bacteriophage lambda:

| Vector | Host |
|--------|------|
| λgtWES.λB′ | DP50*sup*F |
| λgtWES.λB* | DP50*sup*F |
| λgtZJvir.λB′ | *E. coli* K-12 |
| λgtALO.λB | DP50*sup*F |
| Charon 3A | DP*sup*50 or DP50 F |
| Charon 4A | DP*sup*50 or DP50 F |
| Charon 3A | DP*sup*50 or DP50 F |
| Charon 16A | DP*sup*50 or DP50 F |
| Charon 21A | DP50 F |
| Charon 23A | DP*sup*50 or DP50 F |
| Charon 24A | DP*sup*50 or DP50 F |

*E. coli* K-12 strains chi-2447 and chi-2281 are certified for use with lambda vectors that are certified for use with strain DP*sup*50 or DP50 F provided that the *sup* strain not be used as a propagation host.

### Appendix E-IV. *Neurospora crassa.*

**HV1.** The following specified strains of *Neurospora crassa* which have been modified to prevent aerial dispersion:

Inl (inositolless) strains 37102, 37401 46316, 64001, and 89601.

Csp-1 strain UCLA37 and csp-2 strains FS 590, UCLA101 (these are conidial separation mutants).

Eas strain UCLA191 (an "easily wettable" mutant).

### Appendix E-V. *Streptomyces.*

**HV1.** The following *Streptomyces* species: *Streptomyces coelicolor, S. lividans, S. parvulus,* and *S. grieseus.* The following are accepted as vector components of certified *Streptomyces* HV1 systems: *Streptomyces* plasmids SCP2, SLP1.2, pJ101, actinophage phi C31, and their derivatives.

### Appendix E-VI. *Pseudomonas putida.*

**HV1.** *Pseudomonas putida* strains KT2440 with plasmid vectors pKT262, pKT263, pKT264.

### 5.6.6.6. Appendix F. Containment Conditions for Cloning of Genes Coding for the Biosynthesis of Molecules Toxic for Vertebrates

(see Section III-A) ... No specific restrictions shall apply to the cloning of genes if the protein specified by the gene has an $LD_{50}$ of 100 micrograms or more per kilogram of body weight. Experiments involving genes coding for toxic molecules with an $LD_{50}$ of 100 micrograms or less per kilogram shall be registered with ORDA prior to initiating the experiments. A list of toxic molecules classified as to $LD_{50}$ is available from ORDA. Testing procedures for determining toxicity of toxic molecules not on the list are available from ORDA. The results of such tests shall be forwarded to ORDA which will consult with the RAC Working Group on Toxins to inclusion of the molecules on the list (see Section IV-C-1-b-(2)-(e)).

**Appendix F-II. Containment Conditions for Cloning of Toxic Molecule Genes in *E. coli* K-12.**

**Appendix F-II-A.** Cloning of genes coding for molecules toxic for vertebrates that have an $LD_{50}$ in the range of 100 nanograms to 1000 nanograms per kilogram of body weight (e.g., *Clostridium perfringens* epsilon toxin) may proceed under BL2 + EK2 or BL3 + EK1 containment conditions.

**Appendix F-II-B.** Cloning of genes for the biosynthesis of molecules toxic for vertebrates with an $LD_{50}$ in the range of 1 microgram to 100 micrograms per kilogram body weight may proceed under BL1 + EK1 containment conditions (e.g., *Staphylococcus aureus* alpha toxin, *Staphylococcus aureus* beta toxin, ricin, *Pseudomonas aeruginosa* exotoxin A, *Bordatella pertussis* toxin, the lethal factor of *Bacillus anthracis*, the oxygen-labile hemolysins such as streptolysin O, and certain neurotoxins present in snake venoms and other venoms).

**Appendix F-II-C.** Some enterotoxins are substantially more toxic when administered enterally than parenterally. The following enterotoxins shall be subject to BL1 + EK1 containment conditions: cholera toxin, the heat-labile toxins of *E. coli* and of *Yersinia enterocolitica*.

**Appendix-F-III. Containment Conditions for Cloning of Toxic Molecule Genes in Organisms Other Than *E. coli* K-12.**

Requests involving the cloning of genes coding for molecules toxic for vertebrates in host-vector systems other than *E. coli* will be evaluated by ORDA, which will consult with the Working Group on Toxins (see Section IV-C-1-b-(3)-(f)).

**Appendix F-IV-A. Specific Approvals.**

This section is deleted because the material is generally specific to a given user application.

**5.6.6.7. Physical Containment**

This material has been presented for the most part in Section 5.5.11.

In most cases, the material for recombinant DNA facilities is virtually identical, often requiring only the substitution of the words "containing recombinant DNA material" for "containing viable infectious material".

As stated in Appendix G-II: The purpose of physical containment is to confine organisms containing recombinant DNA molecules and thus to reduce the potential for exposure of the laboratory worker, persons outside of the laboratory, and the environment to organisms containing recombinant DNA molecules. ...The selection of alternative methods of primary containment is dependent, however, on the level of biological containment provided by the host-vector system used in the experiment.

### 5.6.6.8. Appendix H. Shipment

Recombinant DNA molecules contained in an organism or virus shall be shipped only as an etiologic agent under requirements of the U.S. Public Health Service and the Department of Transportation (Section 72.3, Part 72, Title 42, and Sections 173.386-388, Part 173, Title 49, United States Code of Federal Regulations). [Reference should be made to the shipping regulations current at the time of shipping.]

### 5.6.6.9. Appendix I. Biological Containment

**Appendix I-I. Levels of Biological Containment.** In consideration of biological containment, the vector (plasmid, organelle, or virus) for the recombinant DNA and the host (bacterial, plant, or animal cell) in which the vector is propagated in the laboratory will be propagated together. Any combination of vector and host which is to provide biological containment must be chosen or constructed so that the following types of "escape" are minimized: (i) survival of the vector in its host outside in the laboratory and (ii) transmission of the vector from the propagation host to other non-laboratory hosts.

The following levels of biological containment (HV, or Host-Vector Systems) for prokaryotes will be established; specific criteria will depend on the organisms to be used.

**Appendix I-I-A. HVI.** A host-vector system which provides a moderate level of containment.

**Specific Systems**

**Appendix I-I-A-1. EK1.** The host is always *E. coli* K-12 or a derivative thereof, and the vectors include nonconjugative plasmids (e.g., pSC101, ColE1, or derivatives thereof and variants of bacteriophages, such as lambda. The *E. coli* K-12 hosts shall not contain conjugation-proficient plasmids, whether autonomous or integrated, or generalized transducing phages.

**Appendix I-I-A-2. Other HV1.** Hosts and vectors shall be, at a minimum, comparable in containment to *E. coli* K-12 with a nonconjugative plasmid or bacteriophage vector. The data to be considered and a mechanism for approval of such HV1 systems are described below (**Appendix I-II**).

**Appendix I-I-B. HV2.** These are host-vector systems shown to provide a high level of biological containment as demonstrated by data from suitable tests performed in the laboratory. Escape of the recombinant DNA either via survival of the organisms or via transmission of recombinant DNA to other organisms should be less than $10^{-8}$ under specified conditions.

**Specific Systems**

**Appendix I-I-B-1.** For EK2 host-vector systems in which the vector is a plasmid, no more than $10^8$ host cells should be able to perpetuate a cloned DNA fragment under the specified nonpermissive laboratory conditions designed to represent the natural environment, either by survival of the original host or as a consequence of transmission of the cloned DNA fragment.

**Appendix I-I-B-2.** For EK2 host-vector systems in which the vector is a phage, no more than $10^8$ phage particles should be able to perpetuate a cloned DNA fragment under the specified nonpermissive laboratory conditions designed to represent the natural environment either (i) as a prophage (in the inserted or plasmid form) in the laboratory host used for phage propagation or (ii) by surviving in natural environments and transferring a cloned DNA fragment to other hosts (or their resident prophages).

### Appendix I-II. Certification of Host-Vector Systems

**Appendix I-II-A. Responsibility.** HV1 systems other than *E. coli* and HV2 host-vector systems, may not be designated as such until they have been certified by the Director, NIH.

Application for certification of a host-vector system is made by written application to ORDA, NIH, Building 31, Room 3B10, Bethesda, Maryland 20892.

Host-vector systems that are proposed for certification will be reviewed by the RAC. (See Section IV-C-I-b-(i)-(e)). This will first involve review of the data on construction, properties and testing of the proposed host-vector system by a Working Group composed of one or more members of the RAC and other persons chosen because of their expertise in evaluating such data.

When new host-vector systems are certified, notice of the certification will be sent by ORDA to the applicant and to all IBCs and will be published in the *Recombinant DNA Technical Bulletin*. Copies of a list of all currently certified host-vector systems may be obtained from ORDA at any time.

The Director, NIH, may at any time rescind the certification of any host-vector systems. (See Section IV-C-I-b-(3)-(d)). If certification of a host-vector system is rescinded, NIH will instruct investigators to transfer cloned DNA into a different system, or use the clones at a higher physical containment level unless NIH determines that the already constructed clones incorporate adequate biological containment.

Certification of a given system does not extend to modifications of either the host or vector component of that system. Such modified systems must be independently certified by the Director, NIH. If modifications are minor, it may only be necessary for the investigator to submit data showing that the modifications have either improved or not impaired the major phenotypic traits on which the containment of the system depends. Substantial modifications of a certified system require the submission of complete testing data.

### Appendix I-II-B. Data to be Submitted for Certification.

**Appendix I-II-B-1. HV1 Systems Other than *E. coli* K-12.** The following types of data shall be submitted, modified as appropriate for the particular system under consideration: (i) A description of the organism and vector; the strain's natural habitat and growth requirements; its physiological properties, particularly those related to its reproduction and survival and the mechanisms by which it exchanges genetic information; the range of organisms with which this organism exchanges genetic information and what sort of information is exchanged; and any relevant information on its pathogenicity or toxicity; (ii) A description of the history of the particular strains and vectors to be used, including data on any mutations which render this organism less able to survive or transmit genetic information; (iii) A general description of the range of experiments contemplated, with emphasis on the need for developing such an HV1 system.

**Appendix I-II-B-2. HV2 Systems.**

Investigators planning to request HV2 certifications for host-vector systems can obtain instructions from ORDA concerning data to be submitted. In general, the following types of data are required: (i) Description of construction steps with indication of source, properties, and manner of introduction of genetic traits; (ii) Quantitative data on the stability of genetic traits that contribute to the containment of the system; (iii) Data on the survival of the system of the host-vector system under nonpermissive laboratory conditions designed to represent the relevant natural environment; (iv) Data on transmissibility of the vector and/or a cloned DNA fragment under both permissive and nonpermissive conditions; (v) Data on all other properties of the system which affect containment and utility, including information on yields of phage or plasmid molecules, ease of DNA isolation, and ease of transfection or transformation; (vi) In some cases, the investigator may be asked to submit data on survival and vector transmissibility from experiments in which the host-vector is fed to laboratory animals and human subjects. Such *in vivo* data may be required to confirm the validity of predicting *in vivo* survival on the basis of *in vitro* experiments.

Data must be submitted in writing to ORDA. Ten to twelve weeks are normally required for review and circulation of the data prior to the meeting at which such data can be considered by the RAC. Investigators are encouraged to publish their data on the construction, properties, and testing of proposed HV2 systems prior to consideration of the system by the RAC and its subcommittee. More specific instructions concerning the type of data to be submitted to NIH for proposed EK2 systems involving either plasmids or bacteriophages in *E. coli* K-12 are available from ORDA.

**Appendix J. Federal Interagency Advisory Committee on Recombinant DNA Research**

[Omitted]

**5.6.6.10. Appendix K. Physical Containment for Large-Scale Uses of Organisms Containing Recombinant DNA Molecules**

This part of the NIH Guidelines specifies physical containment guidelines for large-scale (greater than 10 liters of culture) research or production involving viable organisms containing recombinant DNA molecules. It shall apply to large-scale research or production activities as specified in Section III-B-5 of the Guidelines.

All provisions of the NIH Guidelines shall apply to large-scale research or production activities with the following modifications;

- Appendix K shall replace appendix G when quantities in excess of 10 liters of culture are involved in research or production.
- The Institution shall appoint a biological safety officer if it engages in large-scale research or production activities involving viable organisms containing recombinant DNA molecules.
- The Institution shall establish and maintain a health and surveillance program for personnel engaged in large-scale research or production activities involving viable organisms containing recombinant DNA molecules which require BL3 containment at the laboratory scale. The program shall include preassignment

and periodic physical and medical examinations; collection, maintenance and analysis of serum specimens for monitoring serologic changes that may result from the employee's work experience and provisions for the investigation of any serious, unusual or extended illnesses of employees to determine possible occupational origin.

Since most laboratory-scale research will not involve this scale of operation, the remainder of this appendix is omitted. Operations which involve this level of use or are considering large-scale applications will need to refer to the published guidelines.

## 5.6.6.11. Appendix L. Release Into the Environment of Certain Plants
### Appendix L-I. General Information.

Appendix L specifies conditions under which certain plants, as specified below, may be approved for release into the environment. Experiments in this category cannot be initiated without submission of relevant information on the proposed experiment to NIH, review by the RAC Plant Working Group, and specific approval by NIH. Such experiments also require the approval of the IBC before initiation. Information on specific experiments which have been approved will be available in ORDA and will be listed in Appendix L-III when the Guidelines are republished. [Experiments which do not meet the specifications of Appendix L-II fall under Section III-A and require RAC review and NIH and IBC approval before initiation.]

### Appendix L-II. Criteria Allowing Review by the RAC Plant Working Group Without the Requirement for Full RAC Review.

Approval may be granted by ORDA in consultation with the RAC Plant Working Group without the requirement for full RAC review (IBC review is also necessary) for growing plants containing recombinant DNA under the following conditions:

**Appendix L-II-A.** The plant species is a cultivated crop of a genus that has no species known to be a noxious weed.

**Appendix L-II-B.** The introduced DNA consists of well-characterized genes containing no sequences harmful to humans, animals or plants.

**Appendix L-II-C.** The vector consists of DNA: (i) from exempt host-vector systems (Appendix C); (ii) from plants of the same or closely related species; (iii) from nonpathogenic prokaryotes or nonpathogenic lower eukaryotic plants; (iv) from plant pathogens only if sequences causing disease have been deleted; or (v) chimeric vectors constructed from sequences defined in (i) to (iv) above. The DNA may be introduced by any suitable method.

**Appendix L-II-D.** Plants are grown in controlled access fields under specified conditions appropriate for the plant under study and the geographical location. Such conditions should include provisions for using good cultural and pest control practices, for physical isolation from plants of the same species outside of the experimental plot in accordance with pollination characteristics of the species, and for further preventing plants from becoming established in the environment. Review by the IBC should include an appraisal by scientists knowledgeable of the crop, its production practices, and the local geographical conditions. Procedures for accessing alterations in and the spread of organisms containing recombinant DNA must be developed. The results of the outlined tests must be submitted for review by the IBC. Copies must also be submitted to the Plant Working Group of the RAC.

**Appendix L-III. Specific Approvals**

As of publication of the revised Guidelines, no specific proposals have been approved. An updated list may be obtained from the Office of Recombinant DNA Activities, National Institutes of Health, Building 31, Room 3B10, Bethesda, Maryland 20892.

# REFERENCES (SECTIONS 5.6 TO 5.6.6.11)

1. National Institutes of Health guidelines for research involving recombinant DNA molecules, *Fed. Reg.*, 51, 16958, 1968.
2. Classification of Etiologic Agents on the Basis of Hazard, 4th ed., Centers for Disease Control, Office of Biosafety, U.S. Department of Health, Education and Welfare, Atlanta, GA, 1974.
3. Laboratory Safety at the Center for Disease Control, Publ. No. CDC 75-8118, U.S. Department of Health, Education and Welfare, 1974.
4. National Cancer Institute Safety Standards for Research Involving Oncogenic Viruses, Publ. No. (NIH) 75-790, U.S. Department of Health, Education and Welfare, Washington, D.C., 1974.
5. National Institutes of Health Biohazards Safety Guide, Stock No. 1740-00383, Public Health Service, National Institutes of Health, U.S. Department of Health, Education and Welfare, Washington, D.C., 1974.
6. **Hellman, A., Oxman, M. N., and Pollack, R., Eds.,** *Biohazards in Biological Research,* Cold Spring Harbor Laboratory, Cold Spring Harbor, New York, 1974.
7. **Steere, N. V., Ed.,** *Handbook of Laboratory Safety,* 2nd ed., CRC Press, Boca Raton, FL, 1971.
8. **Bodily, J. L.,** General administration of the laboratory, in *Diagnostic Procedures for Bacterial, Mycotic and Parasitic Infections,* Bodily, H. L.,Updyke, E. L., and Mason, J. O., Eds., American Public Health Association, New York, 1970, 11.
9. **Darlow, H. M.,** Safety in the microbiological laboratory, in *Methods in Microbiology,* Norris, J. R. and Robbins, D. W., Eds., Academic Press, New York, 1969, 169.
10. **Collins, C. H., Hartley, E. G., and Pilsworth, R.,** The Prevention of Laboratory Acquired Infection, Monogr. Ser. No. 6, Public Health Laboratory Service, 1974.
11. **Chatigny, M. A.,** Protection against infection in the microbiological laboratory: devices and procedures, in *Advances in Applied Microbiology,* Vol. 3, Umbreit, W. W., Ed., Academic Press, New York, 1961, 131.
12. Design Criteria for Viral Oncology Research Facilities, Publ. No. (NIH) 75-891, U.S. Department of Health, Education and Welfare, Washington, D.C., 1975.
13. **Kuehne, R. W.,** Biological containment facility for studying infectious disease, *Appl. Microbiol.,* 26, 239, 1973.
14. **Runkle, R. S. and Phillips, G. B.,** *Microbial Containment Control Facilities,* Van Nostrand Reinhold, New York, 1969.
15. **Chatigny, M. A. and Clinger, D. I.,** Contamination control in aerobiology, in *An Introduction to Experimental Aerobiology,* Dimmick, R. L. and Akers, A. B., Eds., John Wiley & Sons, New York, 1969, 194.
16. **Matthews, R. E. F., Ed.,** Third report of the international committee on taxonomy of viruses: classification and nomenclature of viruses, *Intervirology,* 12, 129, 1979.
17. **Buchanan, R. E. and Gibbons, N. E., Eds.,** *Bergey's Manual of Determinative Bacteriology,* 8th ed., Williams and Wilkins, Baltimore, 1974.
18. Biosafety in Microbiological and Biomedical Laboratories, HHS Publ. No. (CDC) 84-8395, 1st ed., Centers for Disease Control and National Institutes of Health, U.S. Department of Health and Human Services, 1984.
19. Laboratory Safety Monograph — A Supplement to the NIH Guidelines for Recombinant DNA Research, Office of Recombinant DNA Activities, National Institutes of Health, Bethesda, MD.
20. **Hershfield, V., Boyer, H. W., Yanofsky, C., Lovett, M. A., and Helinski, D. R.,** Plasmid Col E1 as a molecular vehicle for cloning and amplification of DNA, *Proc. Natl. Acad. Sci. U.S.A.,* 71, 3455, 1974.
21. **Wensink, P. E., Finnegan, D. J., Donelson, J. E., and Hogness, D. S.,** A system for mapping DNA sequences in the chromosomes of *Drosophila melanogaster, Cell,* 3, 315, 1974.

22. **Tanaka, T. and Weisblum, B.,** Construction of a colicin E1-R factor composite plasmid in vitro: means for amplification of deoxyribonucleic acid, *J. Bacteriol.,* 121, 354, 1975.
23. **Armstrong, K. A., Hershfield, V., and Helsinki, D. R.,** Gene cloning and containment properties of plasmid Col E1 and its derivatives, *Science,* 196, 172, 1977.
24. **Bolivar, F., Rodriguez, R. L., Batlach, M. C., and Boyer, H. W.,** Contruction and characterization of new cloning vehicles I. Ampicillin-resistant derivative of pMB9, *Gene,* 2, 75, 1977.
25. **Cohen, S. N., Chang, A. C. W., Boyer, H., and Helling, R.,** Construction of biologically functional plasmids in vitro, *Proc. Natl. Acad. Sci. U.S.A.,* 70, 3240, 1973.
26. **Bolivar, F., Rodriguez, R. L., Greene, R. J., Batlach, M. C., Reyneker, H. L., Boyer, W., Crosa, J. H., and Falkow, S.,** Construction and characterization of new cloning vehicles. II. A multi-purpose cloning system, *Gene,* 2, 95, 1977.
27. **Thomas, M., Cameron, I. R., and Davis, R. W.,** Viable molecular hybrids of bacteriophage lambda and eukaryotic DNA, *Proc. Natl. Acad. Sci. U.S.A.,* 71, 4579, 1974.
28. **Murray, N. E. and Murray, K.,** Manipulation of restriction targets in phage lambda to form receptor chromosomes for DNA fragments, *Nature,* 251, 476, 1974.
29. **Rambach, A. and Tiolais, P.,** Bacteriophage having ecor1 endonuclease sites only in the non-essential region of the genome, *Proc. Natl. Acad. Sci. U.S.A.,* 71, 3927, 1974.
30. **Blattner, F. R., Williams, G. G., Bleche, A. E., Denniston-Thompson, K., Faber, H. E., Furlong, L. A., Gunwald, D. J., Kiefer, D. O., Moore, D. D., Shumm, J. W., Sheldon, E. L., and Smithies, O.,** Charon phages: safer derivatives of bacteriophage lambda for DNA cloning, *Science,* 196, 163, 1977.
31. **Donoghue, D. J. and Sharp, P. A.,** An improved lambda vector: construction of model recombinants coding for kanamycin resistance, *Gene,* 1, 209, 1977.
32. **Leder, P., Tiemeier, D., and Enquist, L.,** EK2 derivatives of bacteriophage lambda useful in the cloning of DNA from higher organisms: the gt WES system, *Science,* 196, 175, 1977.
33. **Skalka, A.,** Current status of coliphage EK2 vectors, *Gene,* 3, 29, 1978.
34. **Szybalski, W., Skalka, A., Gottesman, S., Campbell, A., and Botstein, D.,** Standardized laboratory tests for EK2 certification, *Gene,* 3, 36, 1978.

# 5.7. RESEARCH ANIMAL CARE AND HANDLING*

## 5.7.1. INTRODUCTION

Animal use in research, teaching, and testing has provided advances in health care and preventative medicine for both animals and humans. Experimental results are greatly dependent upon the humane care and treatment of animals used in research. There is a large body of laws, regulations, and guidelines governing the use of animals in research to assure humane animal care and use. These regulations should not be feared as inhibitory to scientific freedom. Rather, as Aristotle said, "shall we not like the archer who has a mark to aim at, be more likely to hit upon that which is right?" Compliance with these laws and guidelines assures healthy, high quality animal models for use in research, assuring consistency from laboratory to laboratory throughout the nation, thus enhancing experimental reliability. The use of high quality, healthy animals and experimental methodologies which seek to minimize or eliminate pain or discomfort should be incorporated not only because it's the law, but also because it makes scientific sense and is the most humane thing to do. The following sections briefly describe the laws and regulations governing animal care and use, and programs for health maintenance of animals and research personnel who come into contact with those animals.

---

* This section was written by Dr. David M. Moore, D.V.M.

### 5.7.2. LAWS AND REGULATIONS RELATING TO ANIMAL CARE AND USE

Two major types of regulatory activities impacting the use of animals at a research facility involve voluntary and involuntary regulations. Involuntary regulations are statutory in nature, uncompromising, and include federal and state laws which dictate minimum standards for the acquisition of animals, provision of veterinary and husbandry care, and the use and disposition of laboratory animals. Personal and institutional compliance is mandatory. Voluntary regulations are those which a research facility imposes on itself, above and beyond the minimum standards set forth by the government. Knowledge of and compliance with applicable institutional, state, and federal policies, regulations, and laws will assure humane care of animals, improve scientific reliability, and deflect criticism from the small segment of society which questions whether animals used in research are humanely treated.

#### 5.5.2.1. Animal Welfare Act

The major federal law affecting and regulating use of animals in research is the Federal Animal Welfare Act (PL 89-544 and its amendments — PL 91-579, PL 94-279, and PL 99-198). Full text copies of the Act are published in the Code of Federal Regulations (CFR), Title 9 — Animals and Animal Products, Subchapter A — Animal Welfare, Parts 1, 2, and 3. Copies of these laws and regulations can be obtained from the Animal and Plant Health Inspection Service, Veterinary Services, Room 700, U.S. Department of Agriculture, Hyattsville, MD 20782.

The Act is administered and enforced by the U.S. Department of Agriculture Animal, Plant Health Inspection Service (USDA/APHIS). All research institutions using animals defined by the Act (dog, cat, nonhuman primate, guinea pig, hamster, or rabbit) must complete and submit VS Form 18-11, "Applications for Registration of a Research Facility", to the USDA/APHIS-VS veterinarian in charge for the state in which the facility is located (contact the national office listed above for the local address). The form asks for the location of animal holding facilities and the species and number of animals to be used. Failure to register as a research facility using covered animals may result in fines or sanctions prohibiting future use of animals at that institution. Currently, rodent species, birds, farm animals, and exotic species are not considered animals under the definition contained in the Act. These may be included in subsequent amendments, so it would be advisable to contact the veterinarian in charge for your state to determine if you must register. The Act addresses and sets minimum standards for the care and use of animals in research in the areas of facilities, construction, caging and operations of the facility; animal health and husbandry standards covering feeding, watering, sanitation, employee qualifications, separation of species, record keeping, and provision of adequate veterinary care; and transportation standards of animals to and from the facility. Please refer to a copy of the Act for detailed specifications. The size of a facility and its research program may determine whether a full-time veterinarian is required on the staff or if the animal health care needs can be met with a part-time or consulting veterinarian with laboratory animal training or experience. The institution must file a "Program of Veterinary Care" with the USDA/APHIS-VS veterinarian in charge for that state, detailing programs of disease control and prevention, euthanasia methods, and use of appropriate anesthetic, analgesic, or tranquilizing drugs, when necessary, as determined by the institutional veterinarian.

Each registered research facility must submit VA 18-23, "Annual Report of Research Facility", with the USDA/APHIS each year, listing the number of animals by species and by category of potential pain or discomfort used during that year, with the signatures of the attending institutional veterinarian and a designated senior administrative official at the institution. Federal penalties may be invoked against the signators for falsification of information in the body of the report.

The Improved Standards for Laboratory Animals Act (PL 99-198), the most recent amendment to the Animal Welfare Act, sets even more important requirements. Each research facility must have an institutional animal care committee of not fewer than three members, to be appointed by the chief executive officer of the facility, including a doctor of veterinary medicine (usually the institutional veterinarian), another facility employee, and a nonemployee, community member who has no family member affiliated with the facility. The committee must inspect all animal study areas and facilities twice annually, keeping all inspection reports on file for 3 years and notifying the administrative representative of the facility of any deficiencies or deviations from the Act. Additionally, each research facility must establish a program for the training of scientists, animal technicians, and other personnel involved with animal care and treatment, including instruction on:

1. Humane practice of animal maintenance and experimentation
2. Research or testing methods that minimize or eliminate animal pain or distress
3. Utilization of the information service at the National Agricultural Library
4. Methods whereby deficiencies in animal care and treatment should be reported

This amendment also calls for establishing institutional standards for the exercise of dogs and provision of environmental enrichment for nonhuman primates.

### 5.7.2.2. The Good Laboratory Practices Act

The Good Laboratory Practices Act (December 22, 1978 issue of the *Federal Register*, 43FR59986-60025) regulates "nonclinical laboratory studies that support applications for research or marketing permits for products regulated by the Food and Drug Administration, including food and color additives, animal food additives, human and animal drugs, medical devices for humane use, biological products, and electronic products."

Standards addressed by GLP regulations include (1) compliance with the NIH Guide for the Care and Use of Laboratory Animals (to be discussed later), (2) establishment of standard operating procedures (SOPs) for animal husbandry and experimental treatment, (3) meticulous record keeping and documentation of activities and the establishment of a functional quality assurance unit reporting to the highest administrative levels of the facility. Please consult this document for further details and applicability to your studies.

### 5.7.2.3. Guide for the Care and Use of Laboratory Animals

The "Guide" (NIH Publication 85-23) was prepared by the Committee on Care and Use of Laboratory Animals of the Institute of Laboratory Animal Resources, National Research Council. The Guide presents recommendations and basic guidelines for (1) appropriate cage and enclosure sizes for a variety of commonly used laboratory species, (2) social environmental enrichment, (3) appropriate environmental temperature and humidity ranges, (4) ventilation of animal facilities, levels of illumination, and noise, (5) separation of species and sanitation of caging and facilities, and (6) provision for quarantine and provision of adequate veterinary care.

Whereas the standards in the Guide are more stringent than those in the Animal Welfare Act, they are simply recommendations and do not by themselves carry legal penalties. However, other government agencies use the Guide as a measure for animal care and will terminate funding support for an institution for noncompliance with the Guide. Single copies of the Guide can be obtained from the Animal Resources Program, Division of Research Resources, National Institutes of Health, Bethesda, MD 20205.

### 5.7.2.4. Public Health Service Policies

Institutions receiving federal grant support for animal research activities from the Public Health Service (including the NIH) must file an Animal Welfare Assurance statement with the Office for Protection from Research Risks (OPRR), National Institutes of Health, 9000 Rockville Pike, Building 31, Room 4B09, Bethesda, MD 20892. The components of the Assurance are listed in the publication, Public Health Service Policy on Humane Care and Use of Laboratory Animals, which can be obtained from OPRR. The intent of this policy is to "require Institutions to establish and maintain proper measures to ensure the appropriate care and use of all animals involved in research, research training, and biological lasting activities."

The PHS "requires that institutions, in their Assurance Statement, use the Guide for the Care and Use of Laboratory Animals as the basis for developing and implementing an institutional program for activities involving animals." In contrast to the Animal Welfare Act, PHS mandates an Institutional Animal Care and Use committee with a minimum of five members (the membership credentials mirroring those specified in the Act). Institutions with approved Assurance statements must file an annual report with OPRR detailing any changes in the animal program and listing the dates of the twice yearly facility inspections by the committee. Failure to comply with the provision of this policy will result in nonfunding or withdrawal of funding for ongoing activities.

As with the Animal Welfare Act, investigators submitting proposals to PHS agencies must submit a research protocol detailing animal use for review of and approval by the Institutional Animal Care and Use committee. This committee must approve the project before funding is released and may require modification of the project if it is not in compliance with Animal Welfare Act standards or PHS policies. The Institutional Animal Care and Use committee may halt or terminate an ongoing project for noncompliance with federal laws and policies.

### 5.7.2.5. Voluntary Regulations

Institutions may develop their own internal policies regarding animal care and use, provided they are equal to or more stringent than those contained in the Guide or the Animal Welfare Act. Internal policies might include SOPs for animal care and use, mechanisms for selection of and purchase from commercial animal vendors, quarantine policies, and human health monitoring programs.

### 5.7.3. PERSONNEL

The Guide for the Care and Use of Laboratory Animals promotes institutional personnel policies requiring the use of technicians qualified to provide proper, humane animal care and husbandry and recommends that these individuals apply for and receive certification from the American Association for Laboratory Animal Science (70 Timbercreek Drive, Suite 5, Cordova, TN 38018). There are three levels of certification, based upon educational background and training and on-the-job experience dealing with laboratory animals: Assistant Technician, Technician, and Laboratory Animal Technologist. Most facilities require facility supervisors or managers to have the AALAS technologist certification. In-house training of technicians using AALAS course materials will satisfy the training requirement set forth in the Improved Standards for Laboratory Animals Act. Training of the scientific staff in humane animal care would be best accomplished by the lab animal veterinarian or an AALAS technologist. The qualifications for full-time or consulting veterinarians to the research facility should include either specialty board certification by the American College of Laboratory Animal Medicine (ACLAM) or indications of postdoctoral training or experience with laboratory animals. The role of the veterinarian in assuring the provision of "adequate, veterinary care" (as referred to in the laws and policies section of this chapter) is described in a report by ACLAM on "Adequate Veterinary Care" issued in October 1966.

### 5.7.4. ANIMAL HOLDING FACILITIES

Animal holding facilities should be designed and constructed following the recommendations of the Guide for the Care and Use of Laboratory Animals, which also assures compliance with the Animal Welfare Act. Consult the references at the end of this section for further information on animal facility design and management.

Facilities should be designed and operated for the comfort of the animals and the convenience of the investigator. Another critical factor in facility design operations involves the prevention of transmission of latent diseases from animal to animal or animals to humans. The first step involves the purchase of animals from "clean" sources who have a documented animal health quality assurance program. Newly arrived animals should be held in a quarantine area in the facility to prevent potential contamination of existing research animal populations. The quarantine facility should be located in an area adjacent to the main colony, but with separate access to prevent cross-contamination of the colony, as would be the case with common traffic flows.

The second step for maintaining clean animals involves control of the microenvironment through the use of appropriate housing units (i.e., micro-isolator caging), laminar air flow housing racks, or mass air displacement "clean" rooms. The items suggested above may be cost prohibitive for some facilities, and adequate care could involve simply following sanitation and hygiene recommendations in the Guide. Air pressure in the room can be changed to protect either the animals or humans working in the facility. Making the room air pressure slightly positive with respect to the hallways will minimize the chances of entry of airborne disease agents into an animal room. However, if the research involves animals infected with animal or human pathogens or if they are treated with toxic or carcinogenic agents, then the room air pressure should be made slightly negative, compared to the adjacent hallways, to prevent contamination of animals in other nearby rooms or humans who use that hallway. Hessler and Moreland[5] discuss the use of HEPA filtration in rooms using nonvolatile carcinogens.

### 5.7.5. ANIMAL CARE AND HANDLING

The Improved Standards for Laboratory Animals Act requires that experimental procedures "ensure that animal pain and distress are minimized". A stressed animal is not a good experimental model since its biochemical and physiological attributes are altered during stress. Minimizing animal stress can be easily accomplished by:

1.  Purchasing animals free of latent and overt clinical diseases, and providing adequate veterinary care to maintain their health
2.  Familiarizing animals with experimental devices or rooms prior to the start of the experiment
3.  Limiting restraint to that which is necessary to accomplish the experimental goals, preconditioning the animals to the restraint apparatus, or using other nonrestraint alternatives
4.  Controlling or eliminating environmental stressors (i.e., inappropriate temperature, humidity, light, noise, aggressive cage mates, cage size)
5.  Providing environmental enrichment programs and/or exercise for animals, especially dogs, cats, and nonhuman primates
6.  Selecting and using appropriate anesthetics, tranquilizers, sedatives, and analgesic drugs for procedures where pain or distress are likely
7.  Allowing only skilled, trained individuals to perform surgery
8.  Providing training for technicians and investigators in humane animal care and use techniques

### 5.7.6. HUMAN HEALTH MONITORING

It is imperative that a human health monitoring program be established for those individuals having limited or full-time contact with research animals. Caretakers and investigators can be exposed to hazardous aerosols, bites, scratches, bodily wastes and discharges, and fomites contaminated with zoonotic agents. A preemployment physical should be conducted to obtain baseline physical and historical data, and a serum sample should be drawn and frozen for future reference. Additional examinations should be scheduled periodically, depending on the nature and risks in the work environment.

Training programs should be established to acquaint personnel with biologic (zoonotic), chemical, and physical hazards within the animal facility. Appropriate hygiene should be stressed, and protective clothing and equipment (gloves, protective outer garments, masks, respirators, face shields, or eye protectors) should be made available, and their use made mandatory where appropriate, in SOPs. Technicians should be aware of clinical signs of disease, notifying the facility veterinarian for confirmation, treatment, isolation of the animal(s), or euthanasia of the affected animal.

# REFERENCES (SECTIONS 5.7 TO 5.76)

1. **Clark, J. D.,** Regulation of animal use: voluntary and involuntary, *J. Vet. Med. Educ.,* 6(2), 86, 1979.
2. **McPherson, C. W.,** Legislation regulations pertaining to laboratory animals — United States, in *Handbook of Laboratory Animal Science,* Vol. 1, Melby, E. C., and Altman, N. H., Eds., CRC Press, Cleveland, OH, 1974, 3.
3. **McPherson, C. W.,** Laws, regulations, and policies affecting the use of laboratory animals, in *Laboratory Animal Medicine,* Fox, J. G., Eds., Academic Press, Orlando, FL, 1984, 19.
4. **Poiley, S. M.,** Housing requirements — general considerations, in *Handbook of Laboratory Animal Science,* Vol. 1, Melby, E. C. and Altman, N. H., Eds., CRC Press, Cleveland, 1974, 21.
5. **Hessler, J. R. and Moreland, A. F.,** Design and management of animal facilities, in *Laboratory Animal Medicine,* Fox, J. G., Eds., Academic Press, Orlando, 1984, 505.
6. **Simmonds, R. C.,** The design of laboratory animal homes, *Aeromedical Review,* Vol. 2, U.S. Air Force School of Aerospace Medicine, Brooks Air Force Base, San Antonio, TX, 1973.
7. **Runkle, R. S.,** Laboratory animal housing — part II, *Am. Inst. Architect. J.,* 41, 77, 1964.
8. Comfortable Quarters for Laboratory Animals, Animal Welfare Institute, Washington, D.C., 1979.

# Personal Protective Equipment 6

## 6.0. INTRODUCTION

This section is intended to be a reasonably extensive, but not exhaustive, treatment of personal protective equipment. It is designed to provide an overview of the topic in the context of normal laboratory usage. However, the concept of laboratories will be extended to include field studies, where workers are often exposed to hazardous materials such as agricultural chemicals.

## 6.1. RESPIRATORY PROTECTION

Respiratory protective devices range all the way from the simple soft felt mask, which is often used to provide protection against nuisance levels of dusts and particulates, to the self-contained, positive-pressure, fully enclosing suit, which, if properly matched to the anticipated exposure, offers total body and respiratory protection from the toxic substance involved. As will be noted in a later section on gloves and protective clothing, there is no single material that will protect against all possible chemicals in the work place.

If a laboratory is designed and operated properly, with adequate ventilation and efficient fume hoods or other types of safety cabinets, additional respiratory protection normally will not be needed for most procedures. However, in the event of an accident or an unusual operation which cannot be performed in a hood, laboratory persons working with toxic materials should be included in a respiratory protection program managed by the institution or corporation.

The selection of the proper respiratory protective device will depend upon a number of factors, the most important of which relate to the properties of the chemical or material for which protection is needed. What is the permissible exposure-limit (PEL) or threshold level value (TLV) of the material? Does it pose a skin absorption problem? Is it immediately dangerous to life and health (IDLH)? What type of material is it, i.e., acid, solvent, dust, radioactive, carcinogen, asbestos, ammonia, etc.? Will air or oxygen need to be supplied? What levels of the material are expected to be present? Does the material provide an adequate warning of its presence by an odor or by irritation of the respiratory system or the eyes? The units to be used must comply with the specifications of ANSI Z88.2, bear an appropriate

NIOSH or MSA approval number, and provide the degree of protection needed under the existing working conditions.

The OSHA respirator standard, 29 CFR, Part 1910.134(b)(10), requires that the individual who is to wear the respirator must be capable of wearing it under working conditions, i.e., he must be physically able to perform his work and use the equipment. A physician must determine if the respirator user's medical condition permits wearing a respirator. Normally, this will include a pulmonary function or spirometer test. An individual with poor respiratory function should not be asked or permitted to wear many types of respirators which require breathing through a protective filter or cartridge. The medical status of the respirator user should be checked on a regular basis. A number of medical conditions preclude wearing of a respirator. Among these are emphysema, asthma, reduced pulmonary function (variety of causes other than the preceding two), severe hypertension, coronary artery disease, cerebral blood vessel disease, epilepsy, claustrophobia (brought on by wearing the unit), or other relevant conditions as determined by the examining physician. Since the worker is being asked to wear a respirator to protect his health by reducing or eliminating inhalation of noxious vapors, it would appear logical that the employee should participate in a comprehensive medical surveillance program to check on the status of his health.

The respirator must be properly fitted to the user. It is not possible to obtain a proper seal of the respirator to the user's face if there is facial hair where the respirator comes into contact with the face. Facial hair also may interfere with the operation of the exhalation valve. There are skin conditions which also make it very difficult to obtain a proper seal. Facial structure also will have a bearing on the quality of fit. For example, a small or petite face is often difficult to fit, as is a narrow face with a prominent nasal bridge. In recent years, significant advances have been made in the design of half-face cartridge respirators, which are the most common type used, to improve the seal to the face. Different size units are now available. There are units which provide a second seal to the skin to aid in keeping toxic fumes from entering. The means of holding respirator units more securely in place have been improved. Hypoallergenic materials can be used in the construction of respirators to prevent the skin from becoming irritated when wearing a unit for an extended period is required.

Each person wearing a respirator must be individually fitted to ensure that the respirator is providing the needed protection to the wearer. The worker should be trained in the proper use of the respirator to enable it to fulfil its function and to maintain it in good working condition. Respirators kept for common use in a facility are not acceptable. An individual should be issued a personal respirator and should be held responsible for maintaining the unit in a clean, good operating condition.

The fitting program, as a minimum, must include a qualitative fit test administered by a knowledgeable person (usually someone from the safety or medical departments), using a particulate irritant, such as a nontoxic smoke generated by a smoke tube, or an organic solvent generating a distinctive odor, such as isoamyl acetate, to challenge the respirator. It would be preferable to perform a quantitative test, using known concentrations of an appropriate test material while the wearer performs simulated work movements. It has been found that even in qualitative tests, simulated work movements are helpful in detecting poor fits.

The training program should include, as a minimum, (1) how to care for the unit, including how to inspect it for proper functioning as well as normal care, (2) how to put the respirator on and check to see that it is performing its function, (3) the function and limitations of the respirator, and (4) the health risks associated with either not using the protection or failing to use it properly. Refresher training should be given on an annual basis.

### 6.1.1. "DUST" MASKS

The use of the term "dust" mask for the nonrigid soft wool, felt mask is somewhat of a misnomer since, in modified form, it can be used for a number of other applications such as protection against paint fumes, moderate levels of organics, acid fumes, mercury, etc., although the biggest use is for protection against dust. These units are the simplest form of the air-purifying respirator. These respirators normally should not be employed for hazardous dusts, but are helpful for exposures to inert or nuisance dust levels below 15 mg/M$^3$. More elaborate versions of these felt masks include such features as exhalation valves, molded bridge pieces, and small sections of metal over the bridge of the nose which can be bent to help the mask remain in contact with the nose and cheeks. Some have chemical absorbent incorporated in the mask to absorb some fumes and gases. In most cases, these inexpensive masks are meant to be worn for relatively low levels of air pollutants (although the better versions which provide good facial contact are considered to provide a protection factor of 10) and disposed of after a limited period of wear). For particulates, the felt mask tends to become more effective as it is worn in a contaminated atmosphere since the space for air to pass through the filter becomes more limited as loading of the filter increases. This increases the difficulty in breathing as the filter offers more resistance to the flow of air. The use of an exhalation valve eases the outflow of air, but does not decrease the effort required during the intake of air.

Filters used to absorb chemical gases or vapors do not use mechanical action to trap the material, but, rather, use an absorbent material (or, in some cases, a chemical reactant) to keep the material from passing through. When the absorbent is saturated or the reactant exhausted, the filter will no longer be effective. The relatively small amount of absorbent material incorporated in simple felt masks limits their lifetime. Since they are usually discarded after use, they are typically intended to be used for about 8 h.

Unless the contaminant in the air has an effective warning property such as a distinctive odor or acts as an irritant, respirators which only purify the air should not be worn for protection. The sensations experienced by an individual due to lack of oxygen do not constitute a sufficient warning signal for this hazard. At oxygen concentrations of 10 to 16%, it is possible to continue to function for short intervals, but with significantly impaired judgment as the oxygen supply to the brain is decreased. At levels below 6%, death occurs in only a few minutes. It is quite likely that more persons have died wearing air-purifying respirators because of the failure to recognize the lack of oxygen in the air than have died from the direct effects of toxic materials. As noted earlier, air-purifying respirators are not approved for use in atmospheres which are immediately dangerous to life and health.

### 6.1.2. HALF-FACE CARTRIDGE RESPIRATORS

The half-face cartridge respirator is the type most frequently used, especially in atmospheres where there is little or no problem of irritation or absorbtion of material through the skin. The face piece of most of these units is molded of a flexible plastic or silicone rubber, which provides a seal to the face when adjusted properly. As noted earlier, facial hair between the mask and the face will prevent the seal from being effective, and it is not permitted for a person with a beard or extended sideburns in the area of the seal to be fitted with a respirator. The face pieces of most of these units are provided with receptacles for two sets of cartridges and/or filters. The respirators are certified as complete units, i.e., the face piece equipped with specific filters. Cartridges from one vendor cannot be used on another manufacturer's face piece. The major advantage of this type of unit is that by interchanging cartridges and filters or by putting one or more filters and cartridges in series, a single face piece can be adapted to provide protection against a large variety of contaminants. However, some cartridge respirators are now being built with nonremovable and noninterchangeable cartridges. These are disposable units since the protective devices cannot be replaced. However, the capacity of the cartridges is considerably larger than the felt mask type and often can be worn for several shifts if the levels of contaminants challenging the filter are not excessive. The models of these disposable cartridge respirators currently available are shaped somewhat differently than the usual half-face respirator and some individuals prefer them for this reason. The price is usually competitive with the replacement costs of cartridges for dual cartridge half-face respirators. Since combinations of filters and cartridges cannot be modified, a large variety of models must be maintained in stock to fit a variety of exposure conditions.

The normal protection factor provided by a half-face respirator which is accepted by OSHA is either 10 times the PEL or the cartridge limit, whichever is lower. In order to maintain the usefulness of cartridge respirators, they must be maintained and stored properly. Fit tests should be repeated periodically to insure that they still provide the required protection. Replacement parts for the exhalation valves should be maintained.

A problem alluded to in an earlier part of this section is the effort required to breathe through the filters. Although check valves can be designed so as to require little effort during the exhalation cycle, breathing air must pass through the cartridges on the intake cycle. Power-assisted breathing units are now available from a number of vendors which provide the flexibility of movement provided by the independent respirator, but remove much of the additional breathing effort incurred by wearing a respirator. In this system, air is fed into the face piece by a small battery-operated pump. The intake air is passed through cartridge filters located on the pump instead of on the face piece. The pump is usually attached to the belt and, including the weight of the battery, is still light enough not to represent any significant problem. The batteries are usually designed to provide 8 h of continuous operation if they are maintained properly. Most of the batteries are nickel-cadmium, which can lose capacity if they are not routinely taken through a complete discharge-charge cycle.

Since a power-assisted air-purifying (PAAP) respirator is a positive-pressure system, i.e., the air within the face piece is at a higher pressure than the outside air, this type of unit intrinsically provides more protection than the ordinary half-face respirator. If provision is made for an "escape" mode of operation, i.e., the wearer can continue to breathe through the filters should the pump fail and escape from the contaminated atmosphere, the ANSI Z88.2 standard would permit the use of this type of respirator in an IDLH atmosphere. Some of the early models of this type of unit had some problems with the seals on the pumps, but these problems have been corrected and they represent a desirable alternative if the wearer is to remain in a contaminated atmosphere for extended periods. There have been a few problems with the pumps overheating while being used at elevated temperatures.

Since a PAAP type of unit will permit entry into an IDLH atmosphere, the user should be aware of a number of essential safety practices associated with this use. At least one standby person with the proper equipment for entry and rescue must be present outside the affected area. Communications must be continuously maintained between the worker within the area and the standby person. The persons within the area must be equipped with devices such as harnesses and safety lines to facilitate rescue operations should they be necessary.

In the best of circumstances, speaking is difficult while wearing a respirator. Throat microphones are available which are connected to amplifiers or radios so that the wearer can communicate with others without having to use the hands to activate a microphone.

If the possibility of chemical splashing exists, the eyes can be protected by wearing chemical splash goggles and/or face masks in addition to the half-face respirator. Goggles which provide ports or other means of allowing air into the space behind the lens will not protect the eyes from vapors and gases in the air since the air behind the lens is room air. Some goggles and respirators are physically incompatible and, if both are needed, must be selected taking compatibility into account.

The use of any type of respirator in almost any type of interior laboratory should be the exception rather than the rule since these spaces should be engineered to be normally safe as far as atmospheric contaminants are concerned if workers follow safe laboratory practices. However, field workers often must depend upon the proper selection of personal protective equipment to protect them from the hazards of exposure to contaminants. In many cases, the half-face respirator is the minimum acceptable respiratory protection. It is often worn under adverse personal comfort conditions as well. The weather may be very hot and the respirator may be very uncomfortable, so the user may wish to forego wearing a unit and "take his chances." There is also a tendency for some workers to dismiss the need for the units and to make fun of persons asking for a respirator or wearing one. Management sometimes shares this attitude. It is extremely important that responsible respiratory protection programs be made available and the use of protective equipment required. This is especially true in academic institutions where future managers of commercial farm operations and agricultural research operations are being trained. They should be taught by formal instruction and example to follow good safety practices, including some that might appear inconvenient at the time, such as wearing the right respiratory protection in contaminated environments. The contaminated environments may not be immediately obvious. It is clear that when spraying operations are being conducted, respiratory protection is likely to be needed, but it is also likely to be needed when entering a treated field a day or two later, depending upon the rapidity of the biodegradation of the material used. In agricultural research, experimental chemicals are often used and the data to determine exposure problems may be incomplete or unavailable.

### 6.1.3. FULL-FACE RESPIRATORS

Full-face air-purifying respirators are similar in many respects to half-face respirators, with the obvious difference that the mask covers the upper part of the face, protecting the eyes. This has advantages and disadvantages. It is often easier to obtain a fit to the user than with a half-face unit. As a result, both the current ANSI Z88.2 standard and OSHA allow a higher protection factor (note that the present OSHA selection standards are based on the 1969 ANSI Z88.2 version instead of the current one), generally by a factor of five or less, depending upon the contaminant.

A major difficulty in wearing a full-face respirator exists for persons who require prescription lenses for seeing. The temple piece extending back over the ears will interfere with the seal at those points. Some units are built to accommodate eyeglasses. In other cases, the wearer may decide to temporarily remove the temple pieces from the glasses and tape them to the bridge of the nose. This solution is acceptable for occasional, sporadic use, but not for extended periods in which the face piece is removed frequently. One solution which is not acceptable is the wearing of contact lenses. These are not allowed under the OSHA standards.

Another problem with a full-face respirator is fogging due to the warm, moist air exhaled in breathing. Full-face units are designed so that the incoming air flows across the lens of the unit. This feature, plus antifogging coatings on the lens, will normally prevent fogging at normal temperatures. However, at temperatures below freezing, fogging becomes an increasingly serious problem as the temperature decreases. Some full-face respirators include nose cups, so that the warm air from the nose is directed through the exhalation valve and does not come into contact with the lens. These units should be able to go down to about −32°C (−25°F) and still allow adequate vision through the lens. At very low temperatures, the warm, moisture-laden air passing out through the exhalation valve may be a problem. The valve may stick open or closed because of ice or the moisture may freeze and block the free flow of air through the valve.

The cost of full-face respirators is substantially more than the half-face units, typically by a factor of four or five.

A variation of the full-face respirator is the powered air-purifying unit discussed in the previous section. In one version, the only difference is that it supplies a full-face mask instead of a half-face mask. However, a useful variation is for the powered air-purifying unit to supply air to the top of a hood which has a cape extending down to the shoulders. Some of these have a transparent section extending all the way around the upper part of the hood and provide an unusual degree of flexibility in vision and comfort with a minimal loss in protection.

For features of full-face respirators other than those covered here, the information on half-face units in Section 6.1.2 will apply.

### 6.1.4. AIR-SUPPLIED RESPIRATORS

An air-supplied respirator is intended to provide a source of breathing air to the user independent of the air in the surrounding space so that they can be used in oxygen-deficient atmospheres. However, depending upon the design, they may or may not be approved for IDLH atmospheres.

There are two basic designs, one in which the supply of air is from a source outside the contaminated area, while in the other, the wearer of the respirator unit carries his air supply with him. There are subdivisions within these two major types. A major subdivision common to both is whether the units operate as "demand" or "pressure-demand" units. For the former, the demand valve permits the flow of air only during inhalation, and a negative pressure exists at that time within the face piece, which may allow leakage from the contaminated atmosphere. The pressure-demand type maintains a positive pressure within the face piece at all times and is unlikely to allow leakage of outside air into the respirator. Pressure-demand units are much more desirable and should be used in most applications. Demand units are not approved for IDLH atmospheres.

Supplied-air units receive air through a hose from a source outside the contamination area, either from cylinders or air compressors, and must be of high purity. Cylinders may be used to supply oxygen instead of compressed air. Compressed air may contain low concentrations of oil. Since the contact of high-pressure oxygen with oil may result in a fire or explosion, it is not permissible to use oxygen with supplied-air units that have previously used compressed air. A compressor used to supply breathing air must be equipped with a high-temperature alarm, or carbon monoxide alarm, or both if the compressor is lubricated with oil. The air provided by the compressor must be passed through an absorbent and filter to insure that the air supplied is pure. A hose up to 300 ft long is permissible.

A major difficulty with the units supplied by air through a hose is the hose itself. The person wearing the unit is constrained to move only as permitted by the hose, the hose is subject to damage or kinking, and the wearer must retrace his steps when leaving the area. The major advantage is that the external source effectively provides an infinite supply of air if an ample number of cylinders are available or if the supply is from a compressor. An acceptable provision for escape from a contaminated area, should the pump fail or the hose fail or become constricted, would be an auxiliary tank of air carried by the user and connected to the respirator.

There are two different types of units in which the user carries his own air supply (in addition to the demand and pressure-demand versions). These self-contained breathing apparatus (SCBA) units have a basic limitation in that they provide only that amount of air that the user can carry with him. There are considerable differences in useful life among different types of units within this class. Most units incorporate a full-face piece, but other styles are available. Again, pressure-demand versions are the most desirable for most applications and, if equipped with escape provisions, are acceptable for IDLH atmospheres.

The basic type of SCBA unit is a tank of breathing air carried on the user's back and fed into the face piece through appropriate regulators and valves. If the tank holds ordinary air, a nominal 30-min tank may last only 15 min for a person engaged in strenuous activity. These units also suffer from being bulky and relatively heavy. Because of their limited life, there is often little time for productive labor while wearing one.

There are a number of commercial units which use pure compressed oxygen. After a reduction in pressure from that in the cylinder, the oxygen is used in a system in which the air exhaled by the wearer is passed through a chemical pack which removes the carbon dioxide from the exhaled air and returns the pure air to the system to supplement the oxygen from the portable cannister. The system is often much smaller and lighter than the basic system incorporating an air tank and, by appropriate sizing of the oxygen tank and air-purifying chemical, can be designed to last a fixed amount of time. A typical system will last 1 h although there are systems which are designed to last considerably longer. This usually is long enough to allow a significant amount of productive work. Some persons do not like to use this type when a fire is involved because of the pure oxygen.

Self-contained pressure-demand units which use full-face pieces are approved for IDLH atmospheres.

Exposure to contaminants can damage respirator components even after they have been cleaned and put in storage due to permeation of chemicals into the materials of which they are made. An examination of the units should be made each time they are worn and a careful check made on a definite schedule. This is important for all types of units, but especially for those which are intended to be used in unusually hazardous applications. Records should be maintained of all maintenance.

# 6.2. EYE PROTECTION

The primary concerns for laboratory workers are impact and chemical splash protection. There are numerous commercial products available which meet the OSHA standards for eye protection (29 CFR, Part 1910.133). This standard is based on ANSI standard Z87.1-1968. Later revisions of this standard have been issued, and as other revisions are adopted, eye protection programs should incorporate more protective portions of these standards.

Eye protection should be worn at all times while working with chemicals in the laboratory. A large percentage of the issues of *Chemical and Engineering News* contain letters reporting unexpected explosions in the laboratory, many of which could cause eye injuries for persons not protected by goggles. Visitors should be provided with temporary protective goggles or glasses if they are allowed in any area where the occupational use of eye protection is required.

### 6.2.1. CONTACT LENSES

Contact lenses provide no physical or chemical protection for the eye. However, by making it possible to trap contaminants under the lenses, they are likely to substantially reduce (or eliminate) the effectiveness of flushing with water from an eyewash station in removing the contaminant. In addition, they may, by capillary action, increase the amount of chemical trapped on the surface of the eye which otherwise might normally be removed by tearing. There are some eye conditions (other than solely for correction of vision) which are improved or controlled by wearing contact lenses. Except for individuals who need to wear contact lenses, it is recommended that contact lenses not be worn in the laboratory. Where it is necessary for medical reasons to wear contact lenses, the wearer should, as a minimum, wear chemical splash goggles at all times while in the laboratory and, under conditions of risk, supplement this practice by also wearing a face mask.

### 6.2.2. CHEMICAL SPLASH GOGGLES

There are dozens of brands of chemical splash goggles available, almost all of which meet the basic standards in ANSI Z87.1 for this type of eye protection. However, there are wide variations in the degree of acceptance of these goggles by users. Chemical splash goggles should fit snugly and comfortably around the eyes. The goggles should "breathe", i.e., the wearer should not overheat under them and perspire. They should not fog. They should provide good peripheral vision. Preferably, they should be compatible with the wearing of respirators. It would be desirable (if not essential) if prescription glasses could be worn under the goggles, although some styles of glasses are too big for most chemical splash goggles. The goggles should be easy to clean.

Just because chemical splash goggles are provided with "ports" to allow air into the space behind the lenses and an antifogging coating does not necessarily mean that all models with these features will be comparably effective. In many cases, ports on the side of the goggles appear to be less successful in eliminating fogging than are openings around the edge of the lens, where head motion moves air directly across the lens. The latter configuration also appears to work well in removing heat. Not all antifogging coatings appear equally effective, nor in some cases does product control quality appear to be uniform, even within a single brand. In at least one case involving over 1000 pairs of goggles, the coating worked very well on about half of the goggles, while for the remainder, the lens fogged up within 15 min of donning them.

Because it is so important to ensure that laboratory personnel wear eye protection under circumstances where it is needed, it is desirable to test goggles under actual use conditions. Not only must chemical splash goggles meet the required physical specifications, but they also must be sufficiently comfortable to be accepted by users as well. A pair of goggles pushed up on the forehead or lying on the work bench does not afford eye protection. Price is not necessarily a guarantee of quality. A mid-priced unit may perform as well as or better than a higher-priced unit. Products change over time and newer products are continually coming on the market. Prior to selecting a specific chemical splash goggle, it is recommended that it be tested under actual use conditions in comparison with a selection of other units which meet both the OSHA and ANSI standards and the criteria mentioned in the introductory paragraph of this section. A vendor's claim that a given product will perform as well as a tested unit needs to be verified. Making sure of the quality and efficacy of a goggle is especially important when buying large quantities for use in instructional laboratories or in a process involving a large number of employees.

If the probability of a vigorous reaction appears to be substantial or the material involved in the work is very corrosive to tissue, a face mask should be used to supplement the splash goggles and provide additional protection to the face and throat. In the event that there is a risk of a minor explosion, an explosion shield should be placed between the worker and the reaction vessel. A wrap-around shield will provide protection to the sides as well as directly in front of the shield.

The most commonly used lens material in safety goggles and safety glasses is polycarbonate. Typically, the material used in the goggles is 0.060 in. (1.52 mm) thick. It is also used in face masks and, in somewhat greater thicknesses, in explosion shields. This material is lightweight, tough, and resists impacts and scratches. Models coated with silicone are resistant to a number of chemicals.

The close fit to the face provided by chemical splash goggles and the strap around the head provides good stability against lateral impacts which might knock ordinary safety spectacles off.

### 6.2.3. SAFETY SPECTACLES

Safety spectacles, which resemble ordinary prescription glasses, that meet the ANSI Z87.1 standard for impact protection offer very limited protection against chemicals. They do not fit tightly against the face and would not prevent chemicals from running down the forehead and into the eyes. They would, however, provide protection against flying glass in the event a reaction vessel exploded. Side shields are used to protect the eyes from flying objects from the side. However, as noted in the previous section, chemical splash goggles are form fitting and are held on tightly, so they resist lateral impacts better than most safety spectacles.

Because they do not fit snugly against the face, safety spectacles do not have any more problems with heat and fogging than do ordinary glasses. In laboratory facilities which do not use chemicals, but do offer opportunities for mechanical injuries, safety glasses are recommended. Many companies offer safety spectacles as prescription glasses and in attractive choices of frames. Individuals who would resist wearing safety spectacles because they think that they are unattractive can be provided a choice of glasses which should prove satisfactory to almost any taste.

## 6.3. MATERIALS FOR PROTECTIVE APPAREL

As Keith[7] stated, "How many times have we seen the phrase 'Use appropriate protective materials'?" This is true in some other areas as well for specific types of protective gear. However, in the field of chemically protective materials where personnel must depend on protective gear, it is essential that they select effective protective clothing and articles, such as gloves, which will provide protection to the wearer against contact with the hazardous substances they use in their laboratories. An examination of the catalog description of many of the items of protective apparel shows that most of the advertisements at best provide only qualitative descriptions of the efficacy of the materials used in the products.* However, some firms will provide technical support data on request.

### 6.3.1. RECOMMENDED INFORMATION SOURCES

Two recent publications (one of which is a two-component computerized book and expert selection system) provide a substantial amount of detailed information, not only by types of materials, but also by brand names since not all versions of a given material have identical properties. These two publications are briefly described below. Both are recommended.

The first component of the system is a computerized reference book compiled by Forsberg,[8] *Chemical Permeation and Degradation Database and Selection Guide for Resistant Protective Materials.* According to the vendor, the database contains over 4200 permeation tests on more than 540 compounds and mixtures. Included are more than 6000 breakthrough times or permeation rates and over 20,000 pieces of associated data, which include information on the test material, manufacturer, model number, thickness, comments, a safety guide number, and references. The stress is

---

* The advertisements of NORTH HAND PROTECTION (Siebe North, Inc., Charleston, SC) provide detailed breakthrough times and permeation rates for 49 chemicals for their Silver Shield™ gloves and give comparison data for several other materials. They claim "It resists permeation and breakthrough by more toxic/hazardous chemicals than any other type of glove on the market today."

on gloves. The program is an outgrowth of "Guidelines for the Selection of Chemical Protective Clothing", which appeared in *Performance of Protective Clothing*, edited by R. L. Barker and G. C. Coletta, ASTM, Philadelphia, 1986.

The second component of the computerized system is an "expert" system called GlovEs which, given a set of initial conditions or parameters, screens the information in the database and makes recommendations based on the needs, as defined by the user of the system. The system allows considerable flexibility in seeking information and quickly provides the data in a useful form.

The program runs on a standard IBM or compatible PC with 640 k of memory and either two disk drives or a single disk drive and a hard disk. It requires DOS 2.0 or higher. An annual updating service is available.

The second source of data is the third edition of *Guidelines for the Selection of Chemical Protective Clothing*. The work was sponsored by the U.S. Environmental Protection Agency and the U.S. Coast Guard.

The publication is organized in two volumes, the first of which contains an overview of the general topic of chemical protective clothing (CPC) and tables which can be used to select and use CPC properly. Twelve major clothing materials are evaluated in the context of about 500 different chemicals and permeation data, including 25 multiple-component organic solutions. Of particular interest are Appendices I and J, which define equipment needed to provide different levels of protection and procedures for using protective gear in decontamination efforts. The latter information is of value to organizations which may have to respond occasionally to emergencies involving chemical spills, but, fortunately, not frequently enough to find it easy to maintain their skills.

The second volume is a technical support base for Volume 1. It provides the data on which the recommendations in Volume 1 are based and some of the theoretical material used in arriving at the recommendations.

### 6.3.2. OVERVIEW OF CHEMICAL PROTECTIVE CLOTHING*

The purpose of chemical protective clothing (CPC) is to prevent chemicals from reaching the skin. The chemicals can do this in two ways, permeate the material of which the clothing is made or enter through penetrations in the clothing. The two sources described in the previous section concentrate on the problem of permeation and breakthrough, although the second of the two, in the introductory portions, discusses the penetration issue.

Much of the data in the two resources is based on manufacturer's subjective evaluations, i.e., the material provided "excellent", "good", "fair", or "poor" protection and is based on visible degradation of the product. However, as noted in the *Guidelines*,[9] "it has been found that chemicals can permeate a material without there being any visible sign of problems". In addition, in the reported evaluations, the temperatures at which the tests were performed often are not given and the permeation rate has been found to depend significantly on the temperature. The thicknesses of the materials tested also are not given in many cases. More scientific means of comparative evaluations need to be adopted and there does appear to be interest in moving in this direction.

---

* Many of the concepts in this section are developed more fully in Reference 9.

In some cases, the chemical to which the barrier material is exposed will simply diffuse through the barrier. In others, the chemical will react with the barrier material and degrade the performance of the barrier, for example, by changing the chemical properties of the material or by leaching out some of the components in the material. As far as penetration is concerned, it is desirable for the material to be a poor absorber of the challenging chemical and for the rate of permeation to be slow.

Once they have begun to permeate a material, chemicals will continue to do so, even after the challenge has been removed from the surface, because of the chemical absorbed within the material. If the amount of material absorbed is large and the rate of permeation is relatively high, the chemical may eventually penetrate through the barrier material after the protective clothing has been removed and stored. The contaminant may cause exposure the next time it is used if it has not had an opportunity to diffuse away. If the clothing is carefully folded and placed in a container, the possibility of contaminant being trapped in the clothing is substantial. Another possibility is that the breakthrough will result in pinhole leaks and the item, which still afforded protection when removed, will no longer do so the next time it is used. If the amount absorbed is small and the permeation rate is slow, the diffusion back through the entering surface may result in a negligible amount penetrating to the interior surface.

Some articles of protective clothing, such as gloves, have no openings which should come into contact with chemicals. However, other items, such as coats, jackets, and trousers, usually have one or more edges which are intended to be opened and closed, and all of these garments and others have seams where sections of the materials are joined together.

Protective clothing is normally fabricated of sections of materials which are often welded together chemically or by heat. The resulting garments should have no penetrations for chemicals to penetrate due to the fabrication process, but the seams do have a tendency to split under strain. Some have sewn seams for added strength. The sewn seams, unless sealed afterwards, will leave pinholes through which chemicals can penetrate.

Normal openings such as the fronts of coats or overalls are fastened by zippers, pressure-locking lips, or buttons. All of these should be supplemented by inner and outer flaps if they are intended to eliminate the possibility of chemicals penetrating through these openings. For total encapsulating clothing, boots, hoods, and gloves may be integral parts of the suits, with a minimal number of openings for which secure seals have to be provided. Where separate items are used, arrangements for seals at the neck, feet, and hands must be provided. Many of these totally encapsulating suits will be used at a slight positive pressure to keep vapors from penetrating through any small gaps in the protective clothing which might occur.

The visor is a key component of any chemical splash suit. Not only must the visor material withstand the effects of the chemicals and maintain good visibilty, but the seal to the fabric portion of the suit should be checked frequently. Normally, visors are made of materials such as polycarbonates, acrylics, fluorinated ethylene propylene (FEP), and clear PVC. The first two are subject to "crazing" upon contact with some chemicals. Covering these two types of material with a thin layer of FEP is sometimes done to protect them. A good seal between the visor and the suit is essential. The seals should be checked frequently.

A major problem with a totally enclosing chemical protective suit is elimination of body heat. If the air supply is carried with the user (see the section on respiratory protection), a common practice is to wear a cooling vest which has pockets for ice cubes and cooling liquid. If the source of air is an external supply, a vortex tube cooling unit is often incorporated in the system. In order to protect the tubes, valves, etc. associated with the air supply system from the effects of chemicals, the air supply components are often worn inside the protective suit. High body temperature is an especially troublesome problem in extended field uses of protective clothing. Workers should take at least a 10-min break each hour during normal summer use, open the clothes in a safe location, and replenish the salt in the body with cool liquids containing a modest level of sodium.

Reference 9 provides a wealth of information on the properties of protective materials for individual chemicals. Table 6.1, which is adapted from this source, shows the resistance of materials for various classes of chemicals. RR, R, rr, and r represent positive degrees of resistance, while NN, N, nn, and n represent degrees of poor resistance. Double characters indicate that the rating is based on test data and single characters, on qualitative data. Upper-case letters indicate a large body of consistent data, while lower-case letters indicate either a small quantity of data or inconsistent information. Asterisks (**) mean that the material varied considerably in its resistance to chemicals within a given class and data for specific chemicals should be used if available or an alternative should be selected.

The column headings in Table 6.1 stand for the following materials:

Butyl rubber — Butyl
Chlorinated polyethylene —CPE
Viton/Neoprene — layered material, first material on surface
Natural rubber — same
Neoprene — same
Nitrile rubber + polyvinyl chloride — nitrile + PVC
Nitrile rubber — nitrile
Polyethylene — PE
Polyvinyl alcohol — PVA
Polyvinyl chloride — PVC
Viton — same
Butyl/Neoprene — layered material, first material on surface

In addition to the chemical properties of materials, a number of physical properties are also important in selecting an appropriate material. Table 6.2 is also adapted from Reference 9. The qualities listed in the table may also be affected by factors such as thickness, formulation, and whether there is a fabric backing to the material.

Much of the information used in the preparation of both resource books came from the use of materials tested in the fabrication of gloves, and there are some differences in how gloves and items of protective clothing are made. However, the basic information should be similar, if not identical.

The severity of the application will govern the choice of protective clothing in most cases. If the clothing is to be exposed to severe abuse and the environment is unusually hazardous, the choice should be the most durable and protective material

## TABLE 6.1
### Resistant Properties of Selected Materials by Chemical Class

| Chemical | Butyl | CPE | Viton/ neoprene | Natural rubber | Neoprene | Nitrile + PVC | Nitrile | PE | PVA | PVC | Viton | Butyl/ neoprene |
|---|---|---|---|---|---|---|---|---|---|---|---|---|
| **Acids, carboxylic and aliphatic** | | | | | | | | | | | | |
| Unsubstituted | R | r | r | ** | rr | ** | rr | NN | ** | ** | ** | r |
| Polybasic | RR | | | | rr | rr | rr | rr | n | rr | | |
| **Aldehydes** | | | | | | | | | | | | |
| Aliphatic and alicyclic | rr | NN | r | ** | NN | nn | NN | ** | NN | NN | ** | r |
| Aromatic and heterocyclic | rr | | n | nm | nm | n | nm | NN | rr | N | | r |
| **Amides** | | | | ** | nm | | nm | nm | | | nm | |
| **Amines, aliphatic and alicyclic** | | | | | | | | | | | | |
| Primary | ** | ** | n | NN | ** | | rr | | nn | ** | ** | |
| Secondary | ** | ** | n | NN | nn | | ** | | ** | NN | nm | n |
| Tertiary | ** | ** | | ** | ** | ** | ** | | ** | ** | rr | |
| Polyamine | ** | | | NN | ** | nm | | | | NN | rr | |
| **Cyanides** | | | | | r | | | | | | | |
| **Esters, carboxylic** | | | | | | | | | | | | |
| Formate | ** | | n | | | | | | | n | | n |
| Acetates | ** | ** | ** | NN | nn | | NN | NN | ** | NN | n | ** |
| Higher monobasic | nn | nn | r | NN | nn | nm | nm | NN | rr | NN | | ** |
| Polybasic | | | r | r | r | | ** | | | rr | | r |
| Aromatic phthalate | rr | | | ** | ** | | ** | | | nn | rr | r |
| **Ethers** | | | | | | | | | | | | |
| Aliphatic | ** | rr | ** | NN | ** | ** | ** | | ** | ** | | ** |
| **Halogen compounds** | | | | | | | | | | | | |
| Aliphatic, unsubstituted | nn | nn | r | NN | NN | NN | NN | NN | ** | NN | ** | n |
| Aliphatic, substituted | ** | nn | r | NN | rr | | nm | NN | ** | NN | rr | |
| Aromatic, unsubstituted | nn | nn | | N | | n | nm | | | N | rr | n |

| | | | | | | | | | | | | | |
|---|---|---|---|---|---|---|---|---|---|---|---|---|---|
| Polynuclear | ** | | | | NN | | | | | | n | rr | |
| Vinyl halides | nn | | | | | mn | | | | | n | rr | |
| Heterocyclic compounds | | | nn | | | | | | | | | | |
| Epoxy compounds | ** | | | ** | ** | mn | | | mm | | NN | NN | n |
| Furan derivatives | ** | mn | n | | ** | ** | | | ** | nn | ** | mn | n |
| Hydrazines | | | | | | | | | | | | | |
| Hydrocarbons | | | | | | | | | | | | | |
| Aliphatic and alicyclic | N | r | r | ** | NN | ** | | ** | ** | ** | NN | RR | r |
| Aromatic | ** | rr | r | NN | NN | NN | | NN | ** | ** | NN | RR | n |
| Hydroxyl compounds | | | | | | | | | | | | | |
| Aliphatic and alicyclic | | | | | | | | | | | | | |
| Primary | RR | rr | rr | nn | ** | nn | ** | ** | ** | ** | ** | rr | ** |
| Secondary | rr | rr | r | ** | rr | ** | rr | rr | ** | rr | *** | rr | r |
| Tertiary | r | | r | ** | rr | ** | rr | rr | *** | | *** | rr | *** |
| Polyois | r | | rr | rr | rr | *** | rr | *** | *** | mn | *** | *** | r |
| Aromatic | ** | ** | r | ** | ** | *** | *** | ** | *** | n | *** | *** | *** |
| Inorganic acids | ** | ** | | RR | RR | ** | RR | RR | RR | n | *** | rr | r |
| Inorganic bases | r | r | r | n | r | *** | r | r | r | n | r | r | *** |
| Inorganic gases | ** | r | n | ** | r | | r | r | R | | *** | rr | *** |
| Inorganic salts | r | | n | NN | n | | n | n | | rr | | | |
| Isocyanates | | | | NN | NN | | N | | ** | | | | |
| Ketones, aliphatic | ** | NN | n | NN | ** | ** | | ** | ** | NN | NN | NN | ** |
| Nitriles, aliphatic | rr | | | NN | | | | | NN | rr | NN | rr | |
| Nitro compounds | | | | NN | | | | | | | | | |
| Unsubstituted | rr | r | | NN | ** | ** | mn | ** | ** | ** | ** | ** | |
| Organo-phosphorous compounds | | | | r | | | | mn | | | | | r |
| Peroxides | | | | | | | | | | | | | |
| Sulfur compounds | | | ** | | | | | | | | | | |
| Thiois | | | | | | | | | | | | | n |

*Note:* See text for explanation of abbreviations.

**TABLE 6.2**
**Physical Characteristics of Chemically Resistant Materials**

| Material | Abrasion resistance | Cut resistance | Flexibility | Heat resistance | Ozone resistance | Puncture resistance | Tear resistance | Relative cost |
|---|---|---|---|---|---|---|---|---|
| Butyl rubber | F | G | G | E | E | G | G | High |
| Chlorinated polyethylene (CPE) | E | G | G | G | E | G | G | Low |
| Natural rubber | E | E | E | F | P | E | E | Medium |
| Nitrile-butadiene rubber (NBR) | E | E | E | G | F | E | G | Medium |
| Neoprene | E | E | G | G | E | G | G | Medium |
| Nitrile rubber (nitrile) | E | E | E | G | F | E | G | Medium |
| Nitrile rubber + polyvinyl chloride (nitrile + PVC) | G | G | G | F | E | G | G | Medium |
| Polyethylene | F | F | G | F | F | P | F | Low |
| Polyurethane | E | G | E | G | G | G | G | High |
| Polyvinyl alcohol (PVA) | F | F | P | G | E | F | G | Very high |
| Polyvinyl chloride (PVC) | G | P | F | P | E | G | G | Low |
| Styrene-butadiene rubber (SBR) | E | G | G | G | F | F | F | Low |
| Viton | G | G | G | G | E | G | G | Very high |

*Note:* E = excellent, G = good, F = fair, and P = poor.

available. In other cases, less protective and/or less durable units or even disposable items might prove entirely acceptable. Experience should assist in selecting brands. There have been instances where well-known brands have experienced high levels of problems with seam failures in their moderate-priced lines, while other brands of comparable price have not demonstrated these difficulties. Such problems should be documented and, unless the difficulties are resolved, used in the selection process. User confidence in the protection offered by the protective items is too important to allow competitive cost to be the only, or even primary, factor in selecting CPC items.

An important area not yet touched upon with regard to protective clothes is an alternative to the asbestos gloves used in laboratories to handle hot objects. Asbestos gloves observed in the laboratory often are in poor condition, e.g., the material of which they are made has become very friable. Any asbestos items in poor condition, especially ones employed as asbestos gloves are, should be discarded as hazardous material through the organization's hazardous waste program. It would be desirable to eliminate the asbestos gloves entirely. Alternatives that has been found acceptable for many high-temperature laboratory applications are gloves made of materials such as Kevlar™, Nomex™, Zetex™, and fiberglass or combinations of these materials. Zetex™, for example, is specifically advertised as a replacement for asbestos for high-temperature applications.

# 6.4. HEARING PROTECTION

The noise levels in most laboratories are usually not excessive, but there are

**TABLE 6.3**
**Permissible Noise Exposure**

| Duration (h/d) | Sound level (dBA slow response) |
|---|---|
| 8 | 90 |
| 6 | 92 |
| 4 | 95 |
| 3 | 97 |
| 2 | 100 |
| 1.5 | 102 |
| 1 | 105 |
| 0.5 | 110 |
| 0.25 or less | 115 |

laboratory facilities in which noise can reach levels where hearing protection should be provided or the employees required to be involved in a hearing conservation program. It would be preferable, of course, if the noise levels could be lowered, rather than depending upon personal protective devices.

OSHA adopted a comprehensive hearing conservation program in 29 CFR, Part 1910.95. Under this standard, any employee exposed to an 8-h, time-weighted average of 85 dB, as measured on a properly calibrated sound-level instrument on the slow response (A) scale, must be placed in an employer-run hearing conservation program. Among other requirements, an annual audiometric test is required. Loss of hearing as one grows older is normal, and there are diseases resulting in hearing losses which are not occupationally related. A properly administered audiometric test should be able to distinguish these two causes from a loss of hearing due to external factors. One of the difficulties in determining if the loss of hearing is occupationally related is that the individual's lifestyle during nonworking hours can also affect his hearing. Prolonged listening to loud music can cause problems, can frequent shooting of firearms for recreation. Workman's compensation claims are sometimes disallowed because the evidence is not clear that the hearing loss is occupationally related. Sound level measurements and sound dosimetry measurements in the workplace, as well as periodic audiometric tests, are both needed. At this time, the record of enforcement of the OSHA standard is not impressive, and the number of workman's compensation cases processed has been small. Hearing is vital, especially to a laboratory employee, since loss of hearing may make it impossible to perform many laboratory operations.

According to Table 6.3, at average noise levels in excess of 90 dB over an 8-h day, an unprotected worker must have his working hours reduced. The table provides that the work interval should be cut in half if the sound level goes up by 5 dB, implying that the sound level goes up by 2 when the measured level increases by 5 dB. Actually, the sound level increases by a factor of two for a measured increase of 3 dB, so the sound level at 105 dB is 32 times that at 90 dB instead of 8, as could be inferred from the table.

There are many different types of hearing protection on the market. The simplest type is ear plugs which are placed within the ear. Typically, these are soft foam which

## TABLE 6.4
### Laboratory First Aid Kit

| | |
|---|---|
| Adhesive bandages, various sizes | Antiseptic wipes |
| Sterile pads, various sizes | Cold packs |
| Sterile sponges | Burn cream |
| Bulk gauze | Antiseptic cream |
| Eye pads | Absorbent cotton |
| Adhesive tape | Scissors |
| First aid booklet | Tweezers |

conform to the ear canal and are effective in reducing noise levels. Many can be washed and reused, if desired. Some persons do not like to use this type of hearing protection because they do not like to keep the plugs in their ears and to be continually taking them in and out as they go from noisy areas to quieter ones. There are many inexpensive earmuffs on the market which attenuate sounds by 20 to 30 dB. Note that the attenuation of the hearing protection devices varies with frequency. The ear muffs can be taken on and off and kept available for reuse. Over a relatively short interval, they will be more economical than ear plugs and often have a higher degree of acceptance by users.

## 6.5. FIRST AID KITS

A first aid kit for a laboratory or for most areas should be intended to provide immediate treatment for most minor injuries or burns, not serve as a substitute for the family medical cabinet. It also need not be a large unit. Most injuries within a laboratory are to individuals. If a major accident were to occur, it would be necessary to call in emergency medical personnel rather than to attempt local treatment. The contents of a first-aid kit should include the supplies listed in Table 6.4.

Note that there are no tourniquets, aspirin (or other medicines to be taken internally), iodine, or merthiolate. Possible additions to this list could be Ipecac, used to induce vomiting, and activated charcoal to help absorb poisons internally, but if these are present, persons should receive specific training in how to use them properly. As noted in Chapter 1, it would be highly desirable if a number of persons in a laboratory facility would receive formal training in first aid and CPR. A specific person should be designated as responsible for maintaining the supplies in the first aid kit.

## REFERENCES

1. General Industry Standards, 29 CFR, § 1910.134, Respiratory Protection, OSHA, Washington, D.C., 1988.
2. *Practices for Respiratory Protection*, ANSI Z88.2, American National Standards Institute, New York, 1980.
3. *Practices for Respiratory Protection*, ANSI Z88.2, American National Standards Institute, New York, 1969.
4. *Practice for Occupational and Educational Eye and Face Protection*, ANSI Z87.1, American National Standards Institute, New York, 1979.
5. *Practice for Occupational and Educational Eye and Face Protection*, ANSI Z87.1, American National Standards Institute, New York, 1968.

6. *Recommendations for Prescription Ophthalmic Lenses,* ANSI Z80.1, American National Standards Institute, New York, 1979.
7. **Keith, V. H.,** Technology blends computers, books, for comparing protective materials, *Occup. Health Saf.,* 56(11), 74, 1987.
8. **Forsberg, K.,** *Chemical Permeation and Degradation Database and Selection Guide for Resistant Protective Materials,* Instant Reference Sources, Austin, TX, 1987.
9. *Guidelines for the Selection of Chemical Protective Clothing,* 3rd ed., Schwope, A. D., Costas, P. P., Jackson, J. O., Stull, J. O., and Weitzman, D. J., Eds., Arthur D. Little, Inc., U.S. Environmental Protection Agency, and U.S. Coastguard, American Conference of Governmental Industrial Hygienists, Cincinnati, OH, 1987.

# APPENDIX
# LABORATORY CHECKLIST

A.   General
1. Housekeeping satisfactory
2. Aisles not cluttered, paths of egress maintained free of obstructions
3. Hazard warning signs at entrance(s)
4. Laboratory authority list at entrances
5. Work area separated from study/social areas
6. Laboratory safety policies posted
7. Food not stored in laboratory refrigerators
8. Equipment maintained in good condition, preventative maintenance program in place

B.   Fire Safety
1. Flammables stored in flammable material storage cabinets
2. Class ABC fire extinguisher in laboratory, available on path of egress
3. Fire blankets available
4. Flammable material stocks maintained at minimal levels, within limits
5. Refrigerators used for flammables are flammable material storage units or explosive proof
6. Two well-separated exits, doors swing outward for hazard class labs
7. Flammables not stored along path of egress

C.   Chemical Handling
1. Chemicals stored according to compatibility
2. Ethers identified by date of receipt and latest date for disposal
3. Containers labeled with contents
4. Quantities of chemicals not excessive
5. Chemical stored at safe levels, in cabinets or on stable shelving
6. Chemical waste labeled and segregated prior to disposal, removed frequently
7. Gas cylinders strapped firmly in place, cylinders not in use capped, oxidizing and reducing agents properly segregated
8. Perchloric acid quantities maintained at minimal levels, used only in proper hoods
9. Apparatus marked with warning signs or protected by barriers if susceptible to damage
10. All work generating toxic and hazardous fumes done in hoods
11. Work capable of causing an explosion behind protective barriers, vacuum vessels taped, etc.

D.  Ventilation
    1.  Ventilation 100% fresh air, six air changes per hour or better
    2.  Laboratory at negative pressure with respect to corridors
    3.  Hoods located in low traffic, draft-free areas
    4.  Hoods maintain 100 fpm face velocity with sash wide open
    5.  Low-velocity warning alarm on hoods
    6.  Fume generating apparatus at least 20 cm from face of hood
    7.  Local exhaust units used where hoods not suitable
E.  Electrical
    1.  All electrical circuits are three wire
    2.  No circuits overloaded with extension cords or multiple connections
    3.  Circuits and circuit breakers labeled
    4.  Apparatus equipped with three-prong connectors or double insulated
    5.  Motors, nonsparking
    6.  Heating apparatus equipped with redundant temperature controls
    7.  Adequate lighting, lights in hoods protected from vapors
F.  Safety Devices
    1.  Eye wash station available
    2.  Deluge shower available
    3.  First aid kit available and supplied
    4.  Protective equipment — goggles, face masks, gloves, aprons, respirator available and used as needed
    5.  Evacuation route marked

# INDEX

# I

## M

# R